Dual Tableaux: Foundations, Methodology, Case Studies

TRENDS IN LOGIC
Studia Logica Library

VOLUME 36

SCOPE OF THE SERIES

Trends in Logic is a bookseries covering essentially the same area as the journal *Studia Logica* – that is, contemporary formal logic and its applications and relations to other disciplines. These include artificial intelligence, informatics, cognitive science, philosophy of science, and the philosophy of language. However, this list is not exhaustive, moreover, the range of applications, comparisons and sources of inspiration is open and evolves over time.

Volume Editor
Ryszard Wójcicki

For further volumes:
http://www.springer.com/series/6645

Ewa Orłowska • Joanna Golińska-Pilarek

Dual Tableaux: Foundations, Methodology, Case Studies

Prof. Ewa Orłowska
National Institute of
Telecommunications
Szachowa 1
04-894, Warszawa, Poland
orlowska@itl.waw.pl

Joanna Golińska-Pilarek
University of Warsaw
and National Institute
of Telecommunications
Krakowskie Przedmieście 3
00-927, Warszawa, Poland
j.golinska@uw.edu.pl

ISBN 978-94-007-0004-8 e-ISBN 978-94-007-0005-5
DOI 10.1007/978-94-007-0005-5
Springer Dordrecht Heidelberg London New York

Cover design: eStudio Calamar S.L.

Printed on acid-free paper

Springer is part of Springer Science+Business Media (www.springer.com)

To the Memory of Helena Rasiowa

Preface

The origin of dual tableaux goes back to the paper by Helena Rasiowa and Roman Sikorski 'On the Gentzen theorem' published in Fundamenta Mathematicae in 1960. The authors presented a cut free deduction system for the classical first-order logic without identity. Since then the deduction systems in the Rasiowa–Sikorski style have been constructed for a great variety of theories, ranging from well established non-classical logics such as intuitionistic, modal, relevant, and multiple-valued logics, to important applied theories such as, among others, temporal, in particular interval temporal logics, various logics of programs, fuzzy logics, logics of rough sets, theories of spatial reasoning including region connection calculus, theories of order of magnitude reasoning, and formal concept analysis.

Specific methodological principles of construction of dual tableaux which make possible such a broad applicability of these systems are:

- First, given a theory, a truth preserving translation is defined of the language of the theory into an appropriate language of relations (most often binary);
- Second, a dual tableau is constructed for this relational language so that it provides a deduction system for the original theory.

This methodology, reflecting the paradigm 'Formulas are Relations', enables us to represent within a uniform formalism the three basic components of formal systems: syntax, semantics, and deduction apparatus. The essential observation, leading to a relational formalization of theories, is that a standard relational structure (i.e., a Boolean algebra together with a monoid) constitutes a common core of a great variety of theories. Exhibiting this common core on all the three levels of syntax, semantics and deduction, enables us to create a general framework for representation, investigation and implementation of theories.

The relational approach enables us to build dual tableaux in a systematic, modular way. First, deduction rules are defined for the common relational core of the theories. These rules constitute a basis of all the relational dual tableau proof systems. Next, for any particular theory specific rules are added to the basic set of rules. They reflect the semantic constraints assumed in the models of the theory. As a consequence, we need not implement each deduction system from scratch, we should only extend the basic system with a module corresponding to the specific part of a theory under consideration.

Relational dual tableaux are powerful tools which perform not only verification of validity (i.e., verification of truth of the statements in all the models of a theory) but often they can also be used for proving entailment (i.e., verification that truth of a finite number of statements implies truth of some other statement), model checking (i.e., verification of truth of a statement in a particular fixed model), and satisfaction (i.e., verification that a statement is satisfied by some fixed objects of a model).

Part I of the book is concerned with the two systems which provide a foundation for all of the dual tableau systems presented in this book. In Chap. 1 we recall the original Rasiowa–Sikorski system and we extend it to the system for first-order logic with identity. We discuss relationships of dual tableaux with other deduction systems, namely, tableau systems, Hilbert-style systems, Gentzen-style systems, and resolution. In Chaps. 2 and 3 classical theories of binary relations and their dual tableaux are presented. It is shown how dual tableaux of these theories perform the above mentioned tasks of verification of validity, entailment, model checking, and verification of satisfaction. Some decidable classes of relational formulas are presented in this part together with dual tableau decision procedures.

Part II is concerned with some non-classical theories of relations. In Chap. 4 we present a theory of Peirce algebras and its dual tableau. Peirce algebras provide a means for representation of interactions between binary relations and sets. In Chap. 5 a theory of fork algebras and its dual tableau are presented. Fork algebras are the algebras of binary relations which, together with all the classical relational operations, have a special operation, referred to as fork of relations. While the relational theories of Chap. 2 serve as means of representation for propositional languages, the fork operation enables us a translation of first-order languages into a language of binary relations. In Chap. 6 we present a theory of typed relations and its dual tableau. The theory enables us to represent relations as they are understood in relational databases. The theory deals with relations of various finite arities and, moreover, each relation has its type which is meant to be a representation of a subset of attributes on which the relation is defined.

In Parts III–V relational formalizations of various theories are presented. In Part III relational dual tableaux are constructed for modal (Chap. 7), intuitionistic (Chap. 8), relevant (Chap. 9), and finitely many-valued (Chap. 10) logics.

Part IV is concerned with the major theories of reasoning with incomplete information. In Chaps. 11 and 12 we deal with logics of rough sets and their relational dual tableaux. Chapter 13 presents a relational treatment of formal concept analysis. In Chap. 14 a monoidal t-norm fuzzy logic is considered and a relational dual tableau for this logic is constructed. In this system ternary relations are needed for representation of the monoid product operation. Next, in Chap. 15 theories of order of magnitude reasoning are considered and their dual tableaux are presented.

Part V is concerned with dual tableaux for temporal reasoning, spatial reasoning, and for logics of programs. The first two chapters of that part refer to temporal logics. In Chap. 16 some classical temporal logics are dealt with and in Chap. 17 relational dual tableaux for a class of interval temporal logics are presented. In Chap. 18 dual tableaux for theories of spatial reasoning are constructed, including a system for the region connection calculus. Chapter 19 includes dual tableaux for various versions of propositional dynamic logic and for an event structure logic.

In Part VI we consider some theories for which dual tableau systems are constructed directly within the theory, without translation into any relational theory. In Chap. 20 we present a class of threshold logics where both weights of formulas and thresholds are elements of a commutative group. In Chap. 21 we present a construction of a signed dual tableau which is a decision procedure for a well known intermediate logic. Chapter 22 includes dual tableaux for a class of first-order Post logics. The reduct of this dual tableau for the propositional part of the logic is a decision procedure. Chapter 23 presents a propositional logic endowed with identity treated as a propositional operation and some theories based on this logic. Dual tableaux for all of these theories are presented. In Chap. 24 logics and algebras of conditional decisions are considered together with their dual tableau decision procedures.

The book concludes with Part VII. In the single Chap. 25 of this part we make a synthesis of what we learned in the process of developing dual tableaux in the preceding chapters. We collect observations on how the dual tableaux rules should be designed once the constraints on the models of the theories or definitions of some specific constants are given. We also discuss some useful strategies for construction of dual tableaux proofs.

All the dual tableau systems considered in the book are proved to be sound and complete. We present a general method of proving completeness of dual tableaux which is shown to be broadly applicable to many theories.

Researchers working in any of the theories mentioned in the titles of the chapters will receive in the book a formal tool of specification and verification of those problems in their theories which involve checking validity, satisfaction, or entailment. Every theory whose dual tableau is presented in a chapter of the book is briefly introduced at the beginning of the chapter and a bibliography is indicated where an interested reader could trace developments, major results, and applications of the theory.

To get an idea of what dual tableaux are and how they are related to the other major types of deduction systems, reading Chap. 1 is recommended. After reading the introductory material from Sects. 1.1, ..., 1.4, and Sects. 2.1, ..., 2.8, each chapter in Parts III, IV, and V may be read independently. The material of Chap. 7 may be helpful in reading Chapters 11, 12, 16, 17, and 19, since they are concerned with modal-style logics.

Readers interested in the formal methods of deduction and their application to specification and verification will find in the book an exhaustive exposition and discussion of dual tableaux and their methodology illustrated with several case studies.

Acknowledgments

Special thanks are due to Wendy MacCaull who suggested writing this book and discussed its scope with Ewa Orłowska during her stay as the F. W. James chair professor at St. Francis Xavier University in Antigonish, Canada. Ewa Orłowska is grateful to the colleagues from the RelMiCS (Relational Methods in Computer Science) community and the participants of the COST Action 274 TARSKI (Theory and Applications of Relational Structures as Knowledge Instruments) for cooperation, inspiration, and stimulating discussions on the subject of the book. The authors thank the colleagues who read and commented on some chapters of this book.

Partial supports from the Polish Ministry of Science and Higher Education grant 126 N N206 399134 and from the Spanish Ministry of Science Research Project TIN09-14562-C05-01 are gratefully acknowledged.

Warszawa, 2010 *Ewa Orłowska, Joanna Golińska-Pilarek*

Contents

Part I
Foundations

Chapter 1
Dual Tableau for Classical First-Order Logic

1.1 Introduction

In [RS60] Rasiowa and Sikorski developed a deduction system for classical first-order logic without identity. Their aim was to present a system which is a realization of the Beth idea of the analytic tableau [Bet59] and, in contrast with the Gentzen system [Gen34] which required the cut rule in the proof of completeness, was cut free. In this chapter we present an extension of the dual tableau of Rasiowa and Sikorski to first-order logic with the identity predicate. This deduction system is an implicit foundation of all the dual tableaux presented in this book.

In this chapter the notions and terminology which will be used throughout the book for presentation of dual tableaux is established. In particular, we discuss various types of dual tableaux rules, the notion of correctness of a rule in a proof system, and a form of dual tableaux proofs. We present a detailed proof of completeness of the dual tableau for first-order logic with identity. The main steps of this proof determine a paradigm which will be relevant to all the dual tableaux completeness proofs in the subsequent chapters of the book.

Next, we recall the tableau system for first-order logic introduced in [Smu68] and we discuss how it is related to the Rasiowa and Sikorski system. Following [GPO07b] and some ideas from [SOH04] we show that the two systems are dual to each other. We present a principle of this duality and we show how proofs in one of those systems can be transformed into proofs in the other system. We also discuss a relationship between dual tableaux and Hilbert-style systems, Gentzen-style systems, and resolution. Following [Kon02], we show that the dual tableau may be seen as Gentzen system with the rules where sequents have the empty precedents. We also compare dual tableaux proofs with resolution proofs in a similar way as tableaux and resolution are compared in [OdS93, Sch06]. A section of this chapter is devoted to a discussion of various ways the identity predicate may be treated in dual tableaux. We compare the dual tableaux rules for identity with the corresponding rules from some other deduction systems.

E. Orłowska and J. Golińska-Pilarek, *Dual Tableaux: Foundations, Methodology, Case Studies*, Trends in Logic 36, DOI 10.1007/978-94-007-0005-5_1,

1.2 Classical First-Order Logic with Identity

In this section we recall the language and the semantics of the classical first-order logic with identity. We consider the first-order logic without function symbols. It is known that these symbols are definable in terms of predicate symbols, therefore this is not a severe limitation. Throughout the book, this logic will be denoted by F.

The vocabulary of the logic F consists of the following pairwise disjoint sets of symbols:

- $\mathbb{OV}_{\mathsf{F}}$ – a countable infinite set of individual variables (also referred to as object variables);
- $\mathbb{P}_{\mathsf{F}}$ – a countable set of predicate symbols; we assume that the identity predicate '$=$' belongs to $\mathbb{P}_{\mathsf{F}}$;
- $\{\neg, \wedge, \vee\}$ – the set of propositional operations of negation, conjunction and disjunction, respectively;
- $\{\forall, \exists\}$ – the set of the universal and existential quantifier, respectively.

The set of *atomic formulas* of the logic F is the smallest set such that:

- $x = y$ is an atomic formula for all $x, y \in \mathbb{OV}_{\mathsf{F}}$;
- $P(x_1, \dots, x_k)$ is an atomic formula, for every k-ary predicate $P \in \mathbb{P}_{\mathsf{F}}$, $k \geq 1$, and for all $x_1, \dots, x_k \in \mathbb{OV}_{\mathsf{F}}$.

The set of F-formulas is the smallest set including the set of atomic formulas and closed on propositional operations and quantifiers. Throughout the book, a formula of the form $\neg(x = y)$ will be denoted by $x \neq y$. A *literal* is an atomic formula or a negated atomic formula.

As usual, propositional operations of implication, $\rightarrow$, and equivalence, $\leftrightarrow$, are definable:

For all F-formulas φ and ψ,

$$\varphi \rightarrow \psi \stackrel{\text{df}}{=} \neg\varphi \vee \psi,$$

$$\varphi \leftrightarrow \psi \stackrel{\text{df}}{=} (\varphi \rightarrow \psi) \wedge (\psi \rightarrow \varphi).$$

Let φ be an F-formula and let x be an individual variable occurring in φ. A variable x is said to be *free* in φ whenever at least one of its occurrences in φ is not in the scope of any quantifier, and it is said to be *bound* if it is not free. We write $\varphi(x)$ to say that a variable x is free in φ.

An F-*model* is a pair $\mathcal{M} = (U, m)$ satisfying the following conditions:

- U is a non-empty set;
- m is a meaning function assigning relations on U to predicates, i.e., for every k-ary predicate P, $m(P) \subseteq U^k$;
- $m(=)$ is an equivalence relation on U;

- The extensionality property (also referred to as a congruence property) is satisfied: for all $a_i, b_i \in U, i = 1, \ldots, k$, and for every k-ary predicate symbol P, if $(a_1, b_1) \in m(=), \ldots, (a_k, b_k) \in m(=)$, and $(a_1, \ldots, a_k) \in m(P)$, then $(b_1, \ldots, b_k) \in m(P)$.

An F-model is *standard* whenever the meaning of the predicate $=$ is the identity, i.e., $m(=) = \{(a, a) : a \in U\}$.

Let $\mathcal{M}$ be an F-model. A *valuation in* $\mathcal{M}$ is a mapping $v: \mathbb{OV}_{\mathsf{F}} \to U$. We write $\mathcal{M}, v \models \varphi$ to denote that φ *is satisfied in* $\mathcal{M}$ *by* v. The relation $\models$ is defined inductively as follows:

- $\mathcal{M}, v \models (x = y)$ iff $(v(x), v(y)) \in m(=)$;
- $\mathcal{M}, v \models P(x_1, \ldots, x_k)$ iff $(v(x_1), \ldots, v(x_k)) \in m(P)$;
- $\mathcal{M}, v \models \neg\varphi$ iff not $\mathcal{M}, v \models \varphi$;
- $\mathcal{M}, v \models \varphi \wedge \psi$ iff $\mathcal{M}, v \models \varphi$ and $\mathcal{M}, v \models \psi$;
- $\mathcal{M}, v \models \varphi \vee \psi$ iff $\mathcal{M}, v \models \varphi$ or $\mathcal{M}, v \models \psi$;
- $\mathcal{M}, v \models \forall x\varphi$ iff for every valuation v' in $\mathcal{M}$ such that v and v' coincide on $\mathbb{OV}_{\mathsf{F}} \setminus \{x\}$, $\mathcal{M}, v' \models \varphi$;
- $\mathcal{M}, v \models \exists x\varphi$ iff for some valuation v' in $\mathcal{M}$ such that v and v' coincide on $\mathbb{OV}_{\mathsf{F}} \setminus \{x\}$, $\mathcal{M}, v' \models \varphi$.

A formula φ is *true in* $\mathcal{M}$ if and only if $\mathcal{M}, v \models \varphi$ for every valuation v in $\mathcal{M}$. An F-formula is F-valid whenever it is true in all F-models. Throughout the book, 'not $\mathcal{M}, v \models \varphi$' will be written as $\mathcal{M}, v \not\models \varphi$.

Clearly, F-validity of a formula implies its truth in all standard F-models. The following fact is well known.

Proposition 1.2.1. *For every* F*-model* $\mathcal{M}$ *and for every valuation* v *in* $\mathcal{M}$*, there exist a standard* F*-model* $\mathcal{M}'$ *and a valuation* v' *in* $\mathcal{M}'$ *such that for every* F*-formula* φ*,* $\mathcal{M}, v \models \varphi$ *iff* $\mathcal{M}', v' \models \varphi$*.*

1.3 Rasiowa–Sikorski Proof System for Classical First-Order Logic with Identity

In this section we present the Rasiowa–Sikorski system (RS for short) for the logic F as presented in [RS63] and we expand it with a rule for identity. The rules of RS-system preserve and reflect validity of the sets of formulas, which are their conclusions and premises. Validity of a finite set of formulas is defined as validity of the disjunction of its elements.

The rules of dual tableau for logic F are of the forms:

$$(\text{rule}_1) \quad \frac{\Phi(\overline{x})}{\Phi_0(\overline{x}_0, z)} \qquad\qquad (\text{rule}_2) \quad \frac{\Phi(\overline{x})}{\Phi_0(\overline{x}_0, z) \mid \Phi_1(\overline{x}_1, z)}$$

where $\Phi(\overline{x})$ is a finite set of formulas whose individual variables are among the elements of set$(\overline{x})$, where $\overline{x}$ is a finite sequence of individual variables and set$(\overline{x})$ is the set of elements of sequence $\overline{x}$; every $\Phi_j(\overline{x}_j, z)$, $j = 0, 1$, is a finite non-empty set of formulas, whose individual variables are among the elements of set$(\overline{x}_j) \cup \{z\}$, where z is either instantiated to arbitrary individual variable (usually to the individual variable that appears in the set of formulas to which the rule is being applied) or z must be instantiated to a new variable (not appearing as a free variable in the formulas of the set to which the rule is being applied). A rule of the form (rule$_2$) is a branching rule. In a rule, the set above the line is referred to as its *premise* and the set(s) below the line is (are) its *conclusion(s)*. A rule of the form (rule$_1$) (resp. (rule$_2$)) is said to be *applicable* to a finite set X of formulas whenever $\Phi(\overline{x}) \subseteq X$. As a result of an application of a rule of the form (rule$_1$) (resp. (rule$_2$)) to a set X, we obtain the set $(X \setminus \Phi(\overline{x})) \cup \Phi_0(\overline{x}_0, z)$ (resp. the sets $(X \setminus \Phi(\overline{x})) \cup \Phi_i(\overline{x}_i, z)$, $i \in \{0, 1\}$). As usual, we will write premises and conclusions of the rules as sequences of formulas rather than sets.

Let φ and ψ be F-formulas. RS-dual tableau consists of *decomposition rules* of the following forms:

(RS$\vee$) $\dfrac{\varphi \vee \psi}{\varphi, \psi}$ (RS$\neg\vee$) $\dfrac{\neg(\varphi \vee \psi)}{\neg\varphi \mid \neg\psi}$

(RS$\wedge$) $\dfrac{\varphi \wedge \psi}{\varphi \mid \psi}$ (RS$\neg\wedge$) $\dfrac{\neg(\varphi \wedge \psi)}{\neg\varphi, \neg\psi}$

(RS$\neg$) $\dfrac{\neg\neg\varphi}{\varphi}$

(RS$\forall$) $\dfrac{\forall x\varphi(x)}{\varphi(z)}$ (RS$\neg\forall$) $\dfrac{\neg\forall x\varphi(x)}{\neg\varphi(z), \neg\forall x\varphi(x)}$

z is a new variable z is any variable

(RS$\exists$) $\dfrac{\exists x\varphi(x)}{\varphi(z), \exists x\varphi(x)}$ (RS$\neg\exists$) $\dfrac{\neg\exists x\varphi(x)}{\neg\varphi(z)}$

z is any variable z is a new variable

and the *specific rule* of the following form:

$$(\text{RS} =) \quad \frac{\varphi(x)}{x = z, \varphi(x) \mid \varphi(z), \varphi(x)}$$

where z is any variable, $\varphi(x)$ is an atomic formula, and $\varphi(z)$ is obtained from $\varphi(x)$ by replacing all the occurrences of x in $\varphi(x)$ with z.

A finite set of formulas is *RS-axiomatic* whenever it includes a subset of the form (RSAx1) or (RSAx2):

(RSAx1) $\{x = x\}$, where x is any variable;
(RSAx2) $\{\varphi, \neg\varphi\}$, where φ is any formula.

A finite set of formulas $\{\varphi_1, \varphi_2, \ldots, \varphi_n\}$, $n \geq 1$, is said to be an *RS-set* whenever the disjunction of its elements is F-valid. It follows that comma (,) in the rules is interpreted as disjunction.

A rule of the form (rule$_1$) (resp. (rule$_2$)) is *RS-correct* whenever for every finite set X of F-formulas, $X \cup \Phi(\overline{x})$ is an RS-set iff $X \cup \Phi_0(\overline{x}_0, z)$ is an RS-set(resp. $X \cup \Phi_0(\overline{x}_0, z)$ and $X \cup \Phi_1(\overline{x}_1, z)$ are RS-sets). It follows that branching (|) in the rules is interpreted as conjunction. Note that, as mentioned earlier, the definition of correctness establishes preservation and reflection of validity by the rules. It is a characteristic feature of all Rasiowa–Sikorski style deduction systems (see [RS63, GPO07b]). A transfer of validity from the conclusion of a rule to the premise is used for proving soundness of the system and the other direction for proving completeness.

According to the semantics of propositional operations and quantifiers we obtain:

Proposition 1.3.1.

1. *The RS-rules are RS-correct;*
2. *The RS-axiomatic sets are RS-sets.*

Proof. By way of example, we prove the proposition for rules (RS$\forall$), (RS$\exists$), and (RS=). Let X be a finite set of F-formulas and let $\varphi(x)$ be an F-formula with a free variable x.

(RS$\forall$) Let z be a variable that does not occur as a free variable in the formulas of the set $X \cup \{\forall x \varphi(x)\}$. Then $X \cup \{\varphi(z)\}$ is an RS-set if and only if for every F-model $\mathcal{M}$ and for every valuation v in $\mathcal{M}$, either there exists $\psi \in X$ such that $\mathcal{M}, v \models \psi$ or for every valuation v' in $\mathcal{M}$ such that v and v' coincide on $\mathbb{OV}_{\mathsf{F}} \setminus \{z\}$, $\mathcal{M}, v' \models \varphi(z)$. The latter is equivalent to F-validity of disjunction of formulas of the set $X \cup \{\forall x \varphi(x)\}$, from which RS-correctness of the rule (RS$\forall$) follows.

(RS$\exists$) Let z be any variable. Clearly, if the premise of the rule is an RS-set, then also the conclusion of the rule is an RS-set. Now, assume $X \cup \{\varphi(z), \exists x \varphi(x)\}$ is an RS-set and suppose $X \cup \{\exists x \varphi(x)\}$ is not an RS-set. Then there exist an F-model $\mathcal{M}$ and a valuation v in $\mathcal{M}$ such that $\mathcal{M}, v \not\models \exists x \varphi(x)$. However, by the assumption, $\mathcal{M}, v \models \varphi(z)$. Let v' be a valuation in $\mathcal{M}$ such that $v'(x) = v(z)$ and for every $y \in \mathbb{OV}_{\mathsf{F}} \setminus \{x\}$, $v'(y) = v(y)$. Thus, $\mathcal{M}, v \models \exists x \varphi(x)$, a contradiction.

(RS=) Let $\varphi(x)$ be an atomic formula. Clearly, if $X \cup \{\varphi(x)\}$ is an RS-set, then so are $X \cup \{x = z, \varphi(x)\}$ and $X \cup \{\varphi(z), \varphi(x)\}$. Assume that $X \cup \{x = z, \varphi(x)\}$ and $X \cup \{\varphi(z), \varphi(x)\}$ are RS-sets. Suppose $X \cup \{\varphi(x)\}$ is not an RS-set. Then there exist an F-model $\mathcal{M}$ and a valuation v in $\mathcal{M}$ such that for every formula $\vartheta \in X \cup \{\varphi(x)\}$, $\mathcal{M}, v \not\models \vartheta$. By the assumption, $\mathcal{M}, v \models x = z$ and $\mathcal{M}, v \models \varphi(z)$. Then by the extensionality property $\mathcal{M}, v \models \varphi(x)$, a contradiction. □

Given a formula, successive applications of the rules result in a tree whose nodes consist of finite sets of formulas.

Let φ be an F-formula. An *RS-proof tree for* φ is a tree with the following properties:

- The formula φ is at the root of this tree;
- Each node except the root is obtained by the application of an RS rule to its predecessor node;
- A node does not have successors whenever its set of formulas is an RS-axiomatic set or none of the rules is applicable to its set of formulas.

A branch of an RS-proof tree is said to be *closed* whenever it contains a node with an RS-axiomatic set of formulas. An RS-proof tree is closed whenever all of its branches are closed. Note that every closed branch is finite. A formula φ is *RS-provable* whenever there is a closed RS-proof tree for φ which is then referred to as its *RS-proof.*

From Proposition 1.3.1 we get soundness of RS-system.

Proposition 1.3.2. *If an* F*-formula* φ *is RS-provable, then* φ *is* F*-valid.*

Corollary 1.3.1. *If an* F*-formula* φ *is RS-provable, then* φ *is true in all standard* F*-models.*

As usual in proof theory a concept of completeness of a proof tree is needed. Intuitively, completeness of a tree means that all the rules that can be applied have been applied. By abusing the notation, for a branch b and a formula φ, we write $\varphi \in b$ if φ belongs to the set of formulas of a node of branch b.

A branch b of an RS-proof tree is said to be *complete* whenever it is closed or it satisfies the following completion conditions:

Cpl(RS$\vee$) (resp. Cpl(RS$\neg\wedge$)) If $(\varphi \vee \psi) \in b$ (resp. $\neg(\varphi \wedge \psi) \in b$), then both $\varphi \in b$ (resp. $\neg\varphi \in b$) and $\psi \in b$ (resp. $\neg\psi \in b$), obtained by an application of the rule (RS$\vee$) (resp. (RS$\neg\wedge$));

Cpl(RS$\wedge$) (resp. Cpl(RS$\neg\vee$)) If $(\varphi \wedge \psi) \in b$ (resp. $\neg(\varphi \vee \psi) \in b$), then either $\varphi \in b$ (resp. $\neg\varphi \in b$) or $\psi \in b$ (resp. $\neg\psi \in b$), obtained by an application of the rule (RS$\wedge$) (resp. (RS$\neg\vee$));

Cpl(RS$\neg$) If $(\neg\neg\varphi) \in b$, then $\varphi \in b$, obtained by an application of the rule (RS$\neg$);

Cpl(RS$\forall$) (resp. Cpl(RS$\neg\exists$)) If $\forall x\varphi(x) \in b$ (resp. $\neg\exists x\varphi(x) \in b$), then for some individual variable z, $\varphi(z) \in b$ (resp. $\neg\varphi(z) \in b$), obtained by an application of the rule (RS$\forall$) (resp. (RS$\neg\exists$));

Cpl(RS$\exists$) (resp. Cpl(RS$\neg\forall$)) If $\exists x\varphi(x) \in b$ (resp. $\neg\forall x\varphi(x) \in b$), then for every individual variable z, $\varphi(z) \in b$ (resp. $\neg\varphi(z) \in b$), obtained by an application of the rule (RS$\exists$) (resp. (RS$\neg\forall$));

Cpl(RS$=$) If $\varphi(x) \in b$ and $\varphi(x)$ is an atomic formula, then for every individual variable z, either $(x = z) \in b$ or $\varphi(z) \in b$, obtained by an application of the rule (RS$=$).

An RS-proof tree is said to be *complete* if and only if all of its branches are complete. A complete non-closed branch is said to be *open*. Note that the rules guarantee that every RS-proof tree can be extended to a complete RS-proof tree. A procedure for constructing a complete proof tree can be found in [DO96]. Observe also that every open branch of an F-proof tree that contains an atomic formula is infinite, since the specific rule (RS=) can be applied infinitely many times to any atomic formula.

Observe that the rules of RS-dual tableau preserve the literals, that is any application of a rule transfers the literals from the premises to the conclusions. Hence, we have:

Fact 1.3.1 (Preservation of literals). *If a node of an RS-proof tree contains a literal, then all of its successors contain this literal as well.*

Proposition 1.3.3. *For any branch of an RS-proof tree, if the literals φ and $\neg\varphi$ belong to the branch, then the branch is closed.*

Proof. Let b be a branch of an RS-proof tree. Fact 1.3.1 implies that if $\varphi \in b$ and $\neg\varphi \in b$, for an atomic formula φ, then eventually both of these formulas appear in a node of branch b. Since the set containing a subset $\{\varphi, \neg\varphi\}$ is F-axiomatic, b is closed. □

Let b be an open branch of an RS-proof tree. We define a *branch structure* $\mathcal{M}^b = (U^b, m^b)$ as follows:

- $U^b = \mathbb{OV}_\mathsf{F}$;
- $m^b(P) = \{(x_1, \ldots, x_k) \in (U^b)^k : P(x_1, \ldots, x_k) \notin b\}$, for every k-ary predicate symbol $P \in \mathbb{P}_\mathsf{F}, k \geq 1$.

Proposition 1.3.4. *For every open branch b of an RS-proof tree, $\mathcal{M}^b$ is an* F*-model.*

Proof. First, we show that $m^b(=)$ is an equivalence relation on the set U^b. If for some $x \in \mathbb{OV}_\mathsf{F}$, $(x, x) \notin m^b(=)$, then $(x = x) \in b$, which means that b is closed, a contradiction. Let $(x, y) \in m^b(=)$ and suppose $(y, x) \notin m^b(=)$. Then $(x = y) \notin b$ and $(y = x) \in b$. By completion condition Cpl(RS=), either $(x = y) \in b$ or $(y = y) \in b$. In the first case we have a contradiction, in the second case the branch b is closed, which contradicts the assumption. Let $(x, y) \in m^b(=)$ and $(y, z) \in m^b(=)$, which means that $(x = y), (y = z) \notin b$. Suppose $(x, z) \notin m^b(=)$, that is $(x = z) \in b$. By the completion condition Cpl(RS=), either $(x = y) \in b$ or $(y = z) \in b$, a contradiction.

Now, we show that $\mathcal{M}^b$ satisfies the extensionality property. We prove it for $k = 1$. In the general case the proof is similar. Let $(x, y) \in m^b(=)$ and let $x \in m^b(P)$, for some $x, y \in U^b$ and some unary predicate symbol P. Suppose $y \notin m^b(P)$. By the definition of $\mathcal{M}^b$, we obtain $(x = y) \notin b$, $P(x) \notin b$, and $P(y) \in b$. By the completion condition Cpl(RS=), either $(y = x) \in b$ or $P(x) \in b$. Applying once again the completion condition Cpl(RS=) with $\varphi(x)$ being $(y = x)$, we get either $(x = y) \in b$ or $P(x) \in b$, a contradiction. □

Any such model $\mathcal{M}^b$ is referred to as a *branch model*. It is constructed from the syntactic resources of the tree built during the proof search process.

Let $v^b : \mathbb{OV}_\mathsf{F} \to U^b$ be a valuation in $\mathcal{M}^b$ such that $v^b(x) = x$, for every $x \in \mathbb{OV}_\mathsf{F}$.

Proposition 1.3.5. *For every open branch b of an RS-proof tree and for every* F*-formula φ, if $\mathcal{M}^b, v^b \models \varphi$, then $\varphi \notin b$.*

Proof. The proof is by induction on the complexity of formulas. For atomic formulas the proposition holds by the definitions of $\mathcal{M}^b$ and v^b. If φ is a negated atomic formula, then the proposition follows from the definition of $\mathcal{M}^b$ and Proposition 1.3.3. Assume that the proposition holds for ψ, ϑ, and their negations.

Let $\varphi = \neg\neg\psi$. Assume $\mathcal{M}^b, v^b \models \neg\neg\psi$. Then $\mathcal{M}^b, v^b \models \psi$, hence by the induction hypothesis $\psi \notin b$. Suppose $\neg\neg\psi \in b$. By the completion condition Cpl(RS$\neg$), $\psi \in b$, a contradiction.

Let $\varphi = \forall x\psi(x)$. Assume that $\mathcal{M}^b, v^b \models \forall x\psi(x)$. Then for every $z \in U^b$, $\mathcal{M}^b, v^b \models \psi(z)$, thus by the induction hypothesis, $\psi(z) \notin b$. Suppose $\forall x\psi(x) \in b$. By the completion condition Cpl(RS$\forall$), for some $z \in U^b$, $\psi(z) \in b$, a contradiction.

Let $\varphi = \neg\forall x\psi(x)$. Assume $\mathcal{M}^b, v^b \models \neg\forall x\psi(x)$. Then for some $z \in U^b$, $\mathcal{M}^b, v^b \not\models \psi(z)$. Suppose that $\neg\forall x\psi(x) \in b$. By the completion condition Cpl(RS$\neg\forall$), for every $z \in U^b$, $\neg\psi(z) \in b$. Thus, by the induction hypothesis, $\mathcal{M}^b, v^b \models \psi(z)$, a contradiction.

In the remaining cases the proofs are similar. □

Given a branch model $\mathcal{M}^b$, we define the quotient model $\mathcal{M}^b_q = (U^b_q, m^b_q)$ as follows:

- $U^b_q = \{\|x\| : x \in U^b\}$, where $\|x\|$ is the equivalence class of $m^b(=)$ determined by x;
- $m^b_q(P) = \{(\|x_1\|, \ldots, \|x_k\|) \in (U^b_q)^k : (x_1, \ldots, x_k) \in m^b(P)\}$, for every k-ary predicate symbol P, $k \geq 1$.

Since the branch model satisfies the extensionality property, the definition of $m^b_q(P)$ is correct, i.e., if $(x_1, \ldots, x_k) \in m^b(P)$ and $(x_1, y_1), \ldots, (x_k, y_k) \in m^b(=)$, then $(y_1, \ldots, y_k) \in m^b(P)$.

Let v^b_q be a valuation in $\mathcal{M}^b_q$ such that $v^b_q(x) = \|x\|$, for every $x \in \mathbb{OV}_{\mathsf{F}}$.

Proposition 1.3.6.

1. *The model $\mathcal{M}^b_q$ is a standard* F*-model;*
2. *For every* F*-formula φ, $\mathcal{M}^b, v^b \models \varphi$ iff $\mathcal{M}^b_q, v^b_q \models \varphi$.*

Proof.

1. We have to show that $m^b_q(=)$ is the identity on U^b_q. Indeed, we have:

$$(\|x\|, \|y\|) \in m^b_q(=) \text{ iff } (x, y) \in m^b(=) \text{ iff } \|x\| = \|y\|.$$

2. The proof is by an easy induction on the complexity of formulas. For example, for the formulas of the form $x = y$ we have: $\mathcal{M}^b, v^b \models (x = y)$ iff $(x, y) \in m^b(=)$ iff $(\|x\|, \|y\|) \in m^b_q(=)$ $\mathcal{M}^b_q, v^b_q \models (x = y)$. □

Proposition 1.3.7. *If a formula φ is true in all standard* F*-models, then φ is RS-provable.*

Proof. Suppose there is no any closed RS-proof tree of φ. Consider a complete RS-proof tree with φ at its root. Let b be an open branch in this tree. Since $\varphi \in b$, by Proposition 1.3.5, $\mathcal{M}^b, v^b \not\models \varphi$. Therefore, by Proposition 1.3.6(2.), we have $\mathcal{M}^b_q, v^b_q \not\models \varphi$. Since $\mathcal{M}^b_q$ is a standard F-model, we get a contradiction. □

In this proof the branch model is constructed from a failed proof search.

Corollary 1.3.2. *If a formula φ is* F*-valid, then φ is RS-provable.*

Summarizing, RS-system provides a deduction tool for the logic F which has the same power as the Hilbert-style axiomatization, namely we have the following theorem which results from Corollaries 1.3.1 and 1.3.2, Propositions 1.3.2 and 1.3.7.

Theorem 1.3.1 (Soundness and Completeness of the RS-system). *Let φ be an* F*-formula. The following conditions are equivalent:*

1. *φ is* F*-valid;*
2. *φ is true in all standard* F*-models;*
3. *φ is RS-provable.*

Example. Consider the following F-formula:

$$\forall x(\varphi \vee \psi(x)) \rightarrow (\varphi \vee \forall x \psi(x)).$$

It can be equivalently presented in the form:

$$\neg\forall x(\varphi \vee \psi(x)) \vee (\varphi \vee \forall x \psi(x)).$$

This formula is F-valid. In Fig. 1.1 its RS-proof is presented.

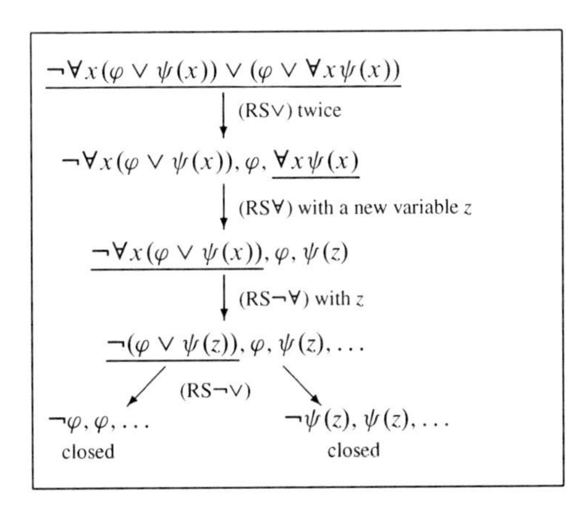

Fig. 1.1 An RS-proof of the formula $\forall x(\varphi \vee \psi(x)) \rightarrow (\varphi \vee \forall x \psi(x))$

Throughout the book, in each node of proof trees presented in the examples we underline the formulas which determine the rule that has been applied during the construction of the tree and we indicate which rule has been applied. If a rule introduces a variable, then we write how the variable has been instantiated. This concerns both the rules which introduce a new or an arbitrary variable. Furthermore, in each node we write only those formulas which are essential for the application of a rule and the succession of these formulas in the node is usually motivated by the reasons of formatting.

1.4 Tableau System for Classical First-Order Logic with Identity

In this section we present a tableau system for the logic F formulated in a way analogous to the formulation of the RS-system. In particular, we indicate explicitly in the rules the repetition of a decomposed formula if needed, in order to make the rules semantically correct. In the original presentation of Smullyan [Smu68] the repetition is shifted to a strategy of building a proof tree. Therefore in our case the Smullyan notation for the rules ($\alpha, \beta, \gamma, \delta$-rules) cannot be applied directly.

The rules of the tableau system preserve and reflect unsatisfiability of the sets of formulas which are their conclusions and premises. There are many versions of tableau systems. They were studied for example in [Fit90]. The specific rule for identity presented here differs from that known in the literature. Such a choice of the rules enables us to see an analogy between tableau and dual tableau treatment of identity (see Sect. 1.8).

Let φ and ψ be any F-formulas. The tableau system for the logic F consists of *decomposition rules* of the following forms:

$$(\mathrm{T}\vee)\quad \frac{\varphi \vee \psi}{\varphi \mid \psi} \qquad\qquad (\mathrm{T}\neg\vee)\quad \frac{\neg(\varphi \vee \psi)}{\neg\varphi, \neg\psi}$$

$$(\mathrm{T}\wedge)\quad \frac{\varphi \wedge \psi}{\varphi, \psi} \qquad\qquad (\mathrm{T}\neg\wedge)\quad \frac{\neg(\varphi \wedge \psi)}{\neg\varphi \mid \neg\psi}$$

$$(\mathrm{T}\neg)\quad \frac{\neg\neg\varphi}{\varphi}$$

$$(\mathrm{T}\forall)\quad \frac{\forall x\varphi(x)}{\varphi(z), \forall x\varphi(x)} \qquad\qquad (\mathrm{T}\neg\forall)\quad \frac{\neg\forall x\varphi(x)}{\neg\varphi(z)}$$

z is any variable $\qquad\qquad$ z is a new variable

$$(\mathrm{T}\exists)\quad \frac{\exists x\varphi(x)}{\varphi(z)} \qquad\qquad (\mathrm{T}\neg\exists)\quad \frac{\neg\exists x\varphi(x)}{\neg\varphi(z), \neg\exists x\varphi(x)}$$

z is a new variable $\qquad\qquad$ z is any variable

and the *specific rule* of the following form:

$$(\mathrm{T}{=})\ \frac{\neg\varphi(x)}{x \neq z, \neg\varphi(x) \quad | \quad \neg\varphi(z), \neg\varphi(x)}$$

where z is any variable, $\varphi(x)$ is an atomic formula, and $\varphi(z)$ is obtained from $\varphi(x)$ by replacing all the occurrences of x in $\varphi(x)$ with z.

A finite set of formulas is *T-axiomatic* whenever it includes a subset of the form (TAx1) or (TAx2):

(TAx1) $\{x \neq x\}$, where x is any variable;
(TAx2) $\{\varphi, \neg\varphi\}$, where φ is any formula.

A finite set of formulas $\{\varphi_1, \varphi_2, \ldots, \varphi_n\}$ is said to be a *T-set* whenever the conjunction of its elements is unsatisfiable, that is for every F-model $\mathcal{M}$ and for every valuation v in $\mathcal{M}$ there exists $i \in \{1, \ldots, n\}$ such that $\mathcal{M}, v \not\models \varphi_i$. It follows that in this case comma in the rules is interpreted as conjunction.

A rule of the form $\frac{\Phi(\overline{x})}{\Phi_0(\overline{x}_0, z)}$ (resp. $\frac{\Phi(\overline{x})}{\Phi_0(\overline{x}_0, z) \mid \Phi_1(\overline{x}_1, z)}$) is *T-correct* whenever for every finite set X of F-formulas, $X \cup \Phi(\overline{x})$ is a T-set if and only if $X \cup \Phi_0(\overline{x}_0, z)$ is a T-set (resp. $X \cup \Phi_0(\overline{x}_0, z)$ and $X \cup \Phi_1(\overline{x}_1, z)$ are T-sets). That is branching in the rules is interpreted as disjunction. Thus T-rules preserve and reflect unsatisfiability of the sets of formulas. The classical tableau system for first-order logic presented in [Smu68] has also the property of preserving and reflecting unsatisfiability. Although this fact is not provable directly from the definition of the classical tableau rules, it can be proved under the additional assumptions on repetition of some formulas in the process of application of the rules. In the classical tableau system this assumption is hidden, it is shifted to a strategy of building the proof trees. In our T-system the required repetitions are explicitly indicated in the rules.

It is easy to show that all the rules of T-system for the logic F are T-correct, and all its axiomatic sets are T-sets. These facts follow from the semantics of the propositional operations and quantifiers as in the case of the RS-system.

A proof in the T-system has the form of a finitely branching tree whose nodes are finite sets of formulas. Let φ be an F-formula. A *T-proof tree for* φ is a tree with the following properties:

- The formula $\neg\varphi$ is at the root of this tree;
- Each node except the root is obtained by the application of a T-rule to its predecessor node;
- A node does not have successors whenever its set of formulas is a T-axiomatic set or none of the rules is applicable to its set of formulas.

A branch of a T-proof tree is said to be *closed* whenever it contains a node with a T-axiomatic set of formulas. A T-proof tree is closed whenever all of its branches are closed. A formula φ is *T-provable* whenever there is a T-closed proof tree for φ which is then referred to as its *T-proof*.

Completion conditions and the branch model are defined in a similar way as in the RS-proof system. For instance, the completion conditions determined by the rules (T$\vee$), (T$\neg\vee$), (T$\forall$), and (T$\neg\forall$) are:

Cpl(T$\vee$) If $\varphi \vee \psi \in b$, then either $\varphi \in b$ or $\psi \in b$;
Cpl(T$\neg\vee$) If $\neg(\varphi \vee \psi) \in b$, then both $\neg\varphi \in b$ and $\neg\psi \in b$;
Cpl(T$\forall$) If $\forall x\varphi(x) \in b$, then for every individual variable z, $\varphi(z) \in b$;
Cpl(T$\neg\forall$) If $\neg\forall x\varphi(x) \in b$, then for some individual variable z, $\neg\varphi(z) \in b$.

Given an open branch b of a T-proof tree, we define a branch structure $\mathcal{M}^b = (U^b, m^b)$ as follows:

- $U^b = \mathbb{OV}_{\mathsf{F}}$;
- $m^b(P) = \{(x_1, \ldots, x_k) \in (U^b)^k : \neg P(x_1, \ldots, x_k) \in b\}$, for every k-ary predicate symbol $P \in \mathbb{P}_{\mathsf{F}}, k \geq 1$.

In a similar way as in RS-dual tableau, the following can be proved:

Proposition 1.4.1. *For every open branch b of a T-proof tree, $\mathcal{M}^b$ is an* F*-model.*

Proposition 1.4.2. *For every open branch b of a T-proof tree and for every* F*-formula φ, if $\mathcal{M}^b, v^b \models \varphi$, then $\neg\varphi \notin b$.*

The proof of soundness and completeness of the tableau proof system is based on the same idea as in the RS-proof system. Then, we have:

Theorem 1.4.1 (Soundness and Completeness of the T-system). *Let φ be an* F*-formula. Then the following conditions are equivalent:*

1. *φ is* F*-valid;*
2. *φ is true in all standard* F*-models;*
3. *φ is T-provable.*

1.5 Quasi Proof Trees

Let P $\in$ {RS, T} be one of the proof systems. Our aim is to define a transformation of a proof tree in one of the systems into a proof tree in the other system. For that purpose it is useful to modify the concept of a proof tree by defining a quasi proof tree. A quasi proof tree is in fact a proof tree modulo the double negation rule.

An F-formula is said to be positive whenever negation is not its principal operation. Let $n \geq 0$ and let φ be a positive F-formula. We define:

$$\neg^0\varphi \stackrel{\text{df}}{=} \varphi;$$
$$\neg^{n+1}\varphi \stackrel{\text{df}}{=} \neg(\neg^n\varphi).$$

We define the rules (P¬)*:

$$(\text{P}\neg)^* \quad \frac{\neg^n \varphi}{\neg^{n \bmod 2} \varphi}$$

where $n \geq 0$ and φ is a positive formula.

As usual, this rule is applicable to a set X of formulas whenever $\neg^n \varphi \in X$ for some $n \geq 0$ and for a positive formula φ. Its application to a set X may be seen as the iteration of applications of rule (P¬).

Let $\# \in \{\vee, \neg\vee, \wedge, \neg\wedge, \forall, \neg\forall, \exists, \neg\exists, =\}$. Let (P#¬*) be a rule defined as a composition of the rules (P#) and (P¬)* treated as maps on the family of finite subsets of formulas and returning a finite subset of formulas (or a pair of subsets in case (P#) is a branching rule).

$$(\text{P}\#\neg^*) \stackrel{\text{df}}{=} (\text{P}\neg)^* \circ (\text{P}\#)$$

This rule is applicable to a set X of formulas whenever the rule (P#) is applicable to X. Let X_0 (resp. X_0 and X_1 if (P#) is a branching rule) be the set(s) obtained from X by an application of rule (P#). Given a finite set Z of formulas, by $Z^{\bmod 2}$ we mean the set of formulas obtained from Z by replacing every formula of the form $\neg^l \varphi$, where $l \geq 0$ and φ is a positive formula, by the formula $\neg^{l \bmod 2}\varphi$. Then the result of application of rule (P#¬*) to X is the set ${X_0}^{\bmod 2}$ (resp. ${X_0}^{\bmod 2}$ and ${X_1}^{\bmod 2}$ if (P#) is a branching rule), where X_0 (resp. X_0 and X_1) is (are) the result(s) of application of rule (P#) to X.

Let $\neg^n \varphi$ be an F-formula, where $n \geq 0$ and φ is a positive formula. A *P-quasi proof tree for* $\neg^n \varphi$ is a tree with the following properties:

- Its root consists of the formula ψ, where:

$$\psi = \begin{cases} \neg^{n \text{mod} 2}\varphi, & \text{if P=RS,} \\ \neg^{(n+1)\text{mod} 2}\varphi, & \text{if P=T;} \end{cases}$$

- Each node except the root is obtained by the application of a rule (P#¬*) to its predecessor node;
- A node does not have successors if its set of formulas is a P-axiomatic set or none of the rules is applicable to its set of formulas.

An example of an RS-quasi proof tree is presented in Fig. 1.2, while Fig. 1.3 presents a T-quasi proof tree for the same formula. Observe that in a diagram of Fig. 1.2, after applying the rule (RS¬∃) to the set $Z_1 = \{\neg\exists x \exists y \neg(x \neq y \vee y = x)\}$ we obtain the set $\{\neg\exists y \neg(x_1 \neq y \vee y = x_1)\}$ to which the rule (RS¬)* is applied with $n = 1$. Thus, the application of the rule (RS¬∃¬*) to Z_1 results in Z_2. Then, we apply the rule (RS¬∃) to Z_2, so that we obtain the set $\{\neg\neg(x_1 \neq x_2 \vee x_2 = x_1)\}$ to which we apply the rule (RS¬)*. Since $\{\neg\neg(x_1 \neq x_2 \vee x_2 = x_1)\}^{\bmod 2} = Z_3$, the application of the rule (RS¬∃¬*) to Z_2 results in Z_3. The application of rule (RS∨) to Z_3 results in Z_4 such that $Z_4^{\bmod 2} = Z_4$. Therefore, Z_4 is the result of application of the rule (RS∨¬*) to Z_3. Similarly, the application of rule (RS=) to Z_4 results in

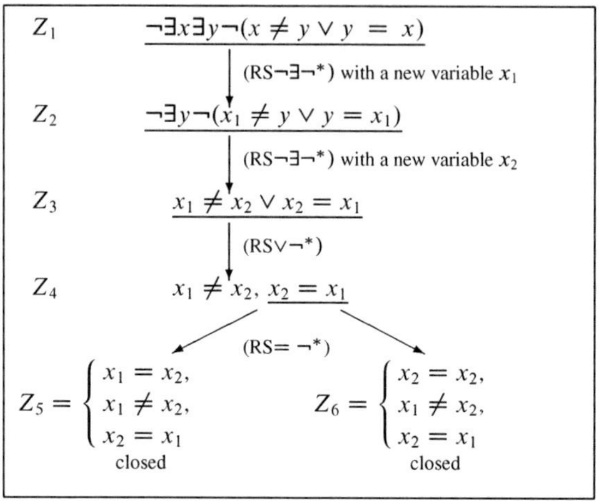

Z_1 $\neg\exists x \exists y \neg(x \neq y \vee y = x)$
(RS¬∃¬*) with a new variable x_1
Z_2 $\neg\exists y \neg(x_1 \neq y \vee y = x_1)$
(RS¬∃¬*) with a new variable x_2
Z_3 $x_1 \neq x_2 \vee x_2 = x_1$
(RS∨¬*)
Z_4 $x_1 \neq x_2, x_2 = x_1$
(RS= ¬*)
$Z_5 = \{x_1 = x_2, x_1 \neq x_2, x_2 = x_1\}$
closed
$Z_6 = \{x_2 = x_2, x_1 \neq x_2, x_2 = x_1\}$
closed

Fig. 1.2 An RS-quasi proof tree of the formula $\neg\neg\neg\exists x \exists y \neg(x \neq y \vee y = x)$

$\exists x \exists y \neg(x \neq y \vee y = x)$
(T∃¬*) with a new variable x_1
$\exists y \neg(x_1 \neq y \vee y = x_1)$
(T∃¬*) with a new variable x_2
$\neg(x_1 \neq x_2 \vee x_2 = x_1)$
(T¬ ∨ ¬*)
$x_1 = x_2, x_2 \neq x_1$
(T= ¬*)
$x_1 \neq x_2, x_1 = x_2, x_2 \neq x_1$
closed
$x_2 \neq x_2, x_1 = x_2, x_2 \neq x_1$
closed

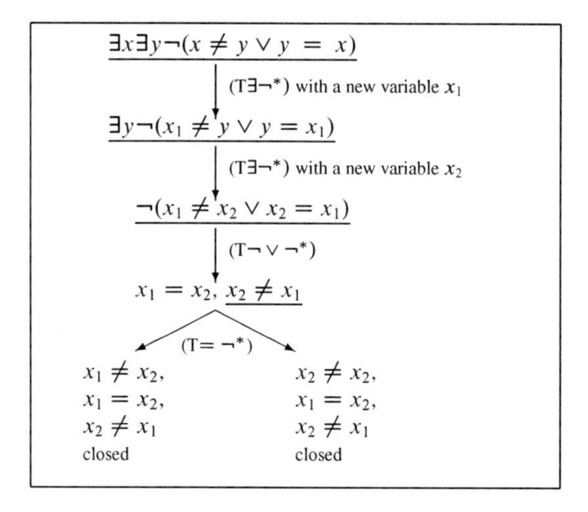

Fig. 1.3 A T-quasi proof tree of the formula $\neg\neg\neg\exists x \exists y \neg(x \neq y \vee y = x)$

Z_5 and Z_6. Since $Z_5^{\mathrm{mod}\,2} = Z_5$ and $Z_6^{\mathrm{mod}\,2} = Z_6$, the result of application of the rule (RS $= \neg^*$) to Z_4 are sets Z_5 and Z_6.

A branch of a P-quasi proof tree is said to be *closed* whenever it contains a node with a P-axiomatic set of formulas. A P-quasi proof tree is *closed* whenever all of its branches are closed. Quasi proof trees in Figs. 1.2 and 1.3 are closed.

It is easy to see that the following proposition holds:

Proposition 1.5.1. *Let φ be an* F*-formula and let $P \in \{RS, T\}$. Then there is a closed P-proof tree for a formula φ iff there is a closed P-quasi proof tree for φ.*

The proof of this proposition is by induction on the complexity of proof trees.

Theorem 1.5.1. *Let φ be an* F*-formula. Then the following conditions are equivalent:*

1. *φ is* F*-valid;*
2. *φ is true in all standard* F*-models;*
3. *There is a closed P-quasi proof tree for φ.*

1.6 Duality

A duality of the tableau and the dual tableau for logic F is presented in [GPO07b]. This duality can be observed at the syntactic level. It is manifested through a duality of formulas, a duality of rules, a duality of completion conditions, and a duality of quasi proof trees. We also indicate a duality of some constructions employed in the completeness proofs for the two systems under consideration. All these dualities will enable us, given a proof system, to construct a dual proof system and to prove its completeness.

Let φ be an F-formula and let $i = 0, 1$. We define a function of duality of formulas as follows:

$$dual(\neg^i \varphi) \stackrel{\text{df}}{=} \neg^{1-i} \varphi.$$

Given a finite set X of F-formulas, we set:

$$dual(X) \stackrel{\text{df}}{=} \{dual(\varphi) : \varphi \in X\}.$$

Now, we show the construction of a dual rule and a dual completion condition.

Let (R) $\frac{\Phi}{\Phi_0 \mid \Phi_1}$ (resp. $\frac{\Phi}{\Phi_0}$) be a rule of RS-system or T-system excluding (RS$\neg$) and (T$\neg$). Then *dual*(R) is the rule of the following form:

$$dual(R) = \frac{dual(\Phi)}{dual(\Phi_0) \mid dual(\Phi_1)} \left(\text{resp. } \frac{dual(\Phi)}{dual(\Phi_0)}\right).$$

For instance, the rule (T$\neg\vee$) is dual to the rule (RS$\vee$), since (T$\neg\vee$) can be obtained from (RS$\vee$) in the following way: the formula of the premise, $\varphi \vee \psi$, is replaced with a dual formula $\neg(\varphi \vee \psi)$, and formulas φ and ψ of conclusions are replaced with $\neg\varphi$ and $\neg\psi$, respectively. Rules (R) and *dual*(R) are said to be *dual*. Table 1.1 presents the dual pairs of rules.

The double negation rules are the only rules which do not appear in Table 1.1, since these rules are exactly the same in both systems.

Table 1.1 Dual rules

(RS$\vee$)	(RS$\wedge$)	(RS$\neg\vee$)	(RS$\neg\wedge$)	(RS$\forall$)	(RS$\exists$)	(RS$\neg\forall$)	(RS$\neg\exists$)	(RS=)
(T$\neg\vee$)	(T$\neg\wedge$)	(T$\vee$)	(T$\wedge$)	(T$\neg\forall$)	(T$\neg\exists$)	(T$\forall$)	(T$\exists$)	(T=)

Table 1.2 Dual operations

$\#$	$\vee$	$\wedge$	$\forall$	$\exists$
$dual(\#)$	$\wedge$	$\vee$	$\exists$	$\forall$

Table 1.3 Dual completion conditions

Cpl(RS$\vee$)	Cpl(RS$\wedge$)	Cpl(RS$\neg\vee$)	Cpl(RS$\neg\wedge$)
Cpl(T$\wedge$)	Cpl(T$\vee$)	Cpl(T$\neg\wedge$)	Cpl(T$\neg\vee$)

Cpl(RS$\forall$)	Cpl(RS$\exists$)	Cpl(RS$\neg\forall$)	Cpl(RS$\neg\exists$)	Cpl(RS$=$)
Cpl(T$\exists$)	Cpl(T$\forall$)	Cpl(T$\neg\exists$)	Cpl(T$\neg\forall$)	Cpl(T$=$)

Observe that if an RS-rule (resp. T-rule) introduces a variable (which may be arbitrary or new), then its dual T-rule (resp. RS-rule) introduces a variable of the same type.

For $\# \in \{\vee, \wedge, \forall, \exists\}$, let *dual*(#) be defined as in Table 1.2. Observe that *dual*(*dual*(#)) = #.

Let $i \in \{0, 1\}$ and let Cpl(P$\neg^i$#) be the completion condition of P-system for the rule (P$\neg^i$#), where P $\in$ {RS, T}, and $\# \in \{\vee, \wedge, \forall, \exists, =\}$. Then:

- If $\neg^i\#$ is not $=$, then *dual*(Cpl(P$\neg^i$#)) is obtained from Cpl(P$\neg^i$#) by replacing the symbol # occurring in the formula of the premise of the condition Cpl(P$\neg^i$#) with a dual operation, i.e., with *dual*(#);
- Otherwise, *dual*(Cpl(P=)) is obtained from Cpl(P=) by replacing all the formulas occurring in Cpl(P=) with their dual formulas.

Clearly, the duality defined above is symmetric. The completion conditions Cpl(P$\neg^i$#) and *dual*(Cpl(P$\neg^i$#)) are said to be *dual*. Table 1.3 shows pairs of dual completion conditions of the considered systems. Completion conditions for the double negation rule are the only that do not appear in Table 1.3, since they are exactly the same in both systems.

The following proposition summarizes the dualities defined above:

Proposition 1.6.1. *Let $i, j \in \{0, 1\}$, $P_i \in \{RS, T\}$, $P_0 \neq P_1$, and $\# \in \{\vee, \wedge, \forall, \exists\}$. Then:*

1. *The function of duality of rules satisfies:*
   - $dual(P_i\neg^j\#) = (P_{1-i}\neg^{1-j}\#)$, *if # is not the identity,*
   - $dual(P_i =) = (P_{1-i} =)$;
2. *The function of duality of completion conditions satisfies:*
   - $dual(Cpl(P_i\neg^j\#)) = (Cpl(P_{1-i}\neg^j dual(\#))$, *if* $\# \in \{\vee, \wedge, \forall, \exists\}$,
   - $dual(Cpl(P_i\#)) = (Cpl(P_{1-i}\#)$, *if* $\# \in \{\neg, =\}$.

Let $(P_i\#)$ be a rule of a P_i-system different from $(P_i\neg)$. Then a rule dual to $(P_i\#\neg^*)$ is defined by:

$$dual(P_i\#\neg^*) = dual(P_i\#) \circ (P_{1-i}\neg)^*.$$

1.7 Transformation of Proofs

Proof trees of RS-system and T-system are dual to each other. Consider an RS-proof tree and a T-proof tree for a formula φ. In the former, we start with the formula φ, in the latter with its negation, $\neg\varphi$. In the subsequent steps, every application of a rule in the RS-proof tree for φ can be mimicked in a T-proof tree starting with $\neg\varphi$ by the applications of the dual rule and/or the rule $(T\neg)$. As the result we obtain a node in the T-proof tree with formulas which are equivalent to the negations of the formulas from the corresponding node in the RS-proof tree. However, it can happen that in the RS-proof tree the rule $(RS\neg)$ was applied to some node while in the T-proof tree the rule $(T\neg)$ is not applicable to the corresponding node. It means that these proof trees may differ in the number of applications of the double negation rule, i.e. they may be seen as dual modulo applications of the double negation rule. The concept of a quasi proof tree enables us to define a transformation of a proof tree in the system RS (resp. T) into a proof tree in the system T (resp. RS).

The transformation Γ on the family of P-quasi proof trees is defined as follows. For a P-quasi proof tree $\mathcal{D}$, $\Gamma(\mathcal{D})$ is a quasi proof tree obtained from $\mathcal{D}$ by replacing all the formulas of $\mathcal{D}$ with their dual formulas and by replacing each rule $(P\#\neg^*)$ with $dual(P\#\neg^*)$. Observe that an RS-quasi proof tree from Fig. 1.2 and a T-quasi proof tree from Fig. 1.3 are transformations of each other.

Proposition 1.7.1. *Let $\mathcal{D}$ be an RS (resp. T) quasi proof tree. Then $\Gamma(\mathcal{D})$ is a T (resp. RS)-quasi proof tree.*

The above proposition follows from the definition of quasi proof trees. Now, it is easy to see that the following theorem holds:

Theorem 1.7.1. *Let $\mathcal{D}$ be an RS or T quasi proof tree. Then $\Gamma(\Gamma(\mathcal{D})) = \mathcal{D}$.*

It follows that the operator Γ expresses duality of quasi proof trees.

1.8 Discussion of Various Rules for Identity

Usually, a tableau system contains the following specific rules for identity (see [Fit90]):

$$(T^1_=)\ \frac{}{x = x} \qquad (T^2_=)\ \frac{x = y, \varphi(x)}{\varphi(y), x = y, \varphi(x)}$$

where $\varphi(x)$ is an atomic formula and $\varphi(y)$ is obtained from $\varphi(x)$ by replacing some occurrences of x in $\varphi(x)$ with y.

Moreover, there is only one type of axiomatic set: a finite set of formulas is axiomatic whenever it contains φ and $\neg\varphi$, for some formula φ. The completeness of the tableau system with these rules for identity can be proved in a similar way as the completeness of T-system (see Sect. 1.4). Observe that the tableau system in our presentation contains only one specific rule for identity while the above version of tableau includes two rules, and both of them are necessary for proving completeness of that version.

Similarly, we can admit in the RS-system the following two rules for identity:

$$(\mathrm{RS}^1_=)\ \frac{}{x \neq x} \qquad (\mathrm{RS}^2_=)\ \frac{x \neq y, \neg\varphi(x)}{\neg\varphi(y), x \neq y, \neg\varphi(x)}$$

where $\varphi(x)$ is an atomic formula and $\varphi(y)$ is obtained from $\varphi(x)$ by replacing some occurrences of x in $\varphi(x)$ with y.

It is easy to show that RS and T systems with these specific rules are dual to each other. Therefore, from a logical point of view it is immaterial which specific rules are taken: (RS=) or $(\mathrm{RS}^1_=)$ and $(\mathrm{RS}^2_=)$ (resp. (T=) or $(\mathrm{T}^1_=)$ and $(\mathrm{T}^2_=)$). As far as an implementation is concerned, the choice of the specific rules may be significant. Tableau system with rules $(\mathrm{T}^1_=)$ and $(\mathrm{T}^2_=)$ seems to be more 'deterministic' : the rule $(\mathrm{T}^2_=)$ is applicable to two formulas, while the rule (T=) is applicable only to one. Moreover, when the rule (T=) is applied to some formula we have to choose a variable, which is very nondeterministic. In other words, the rule $(\mathrm{T}^2_=)$ seems to be applicable in fewer cases than (T=). However, a tableau system with the rule $(\mathrm{T}^2_=)$ must contain also the rule $(\mathrm{T}^1_=)$. But the rule $(\mathrm{T}^1_=)$ is as nondeterministic as the rule (T=), in both of them we have to choose a variable. Note that $(\mathrm{T}^2_=)$ is not so deterministic as it seems to be: we must choose a variable x in a formula φ which is replaced with y. Even if some restrictions on a choice of a variable in the application of the rule $(\mathrm{T}^1_=)$ are added to a strategy of building a proof tree, similar restrictions may be made in the case of a tableau system with the rule (T=).

Fitting proposed a set of specific rules that does not include the rule $(\mathrm{T}^1_=)$ [Fit90]. Following this idea we may admit the following rules for identity in RS-system:

$$(\mathrm{RS}^{1'}_=)\ \frac{y \neq x, \varphi(x)}{\varphi(y), y \neq x, \varphi(x)} \qquad (\mathrm{RS}^{2'}_=)\ \frac{x \neq y, \varphi(x)}{\varphi(y), x \neq y, \varphi(x)}$$

where $\varphi(x)$ is an atomic formula or negation of an atomic formula and $\varphi(y)$ is obtained from $\varphi(x)$ by replacing some occurrences of x in $\varphi(x)$ with y.

However, from the implementation point of view also this set of specific rules seems to be comparable with the original one, since a variable x in a formula φ which is to be replaced by y must be chosen. It is difficult to compare the effectiveness of specific rules without circumscribing the set of formulas to which those rules are applied. Let us consider possible proof trees for formula $\forall x \forall y (x \neq y \vee y = x)$ that expresses symmetry of $=$. In the RS-proof tree for the above formula, after applications of the decomposition rules we obtain the subtree of Fig. 1.4. In this proof

$x \neq y, y = x$
(RS=)
$x = y, x \neq y, y = x$
closed
$y = y, x \neq y, y = x$
closed

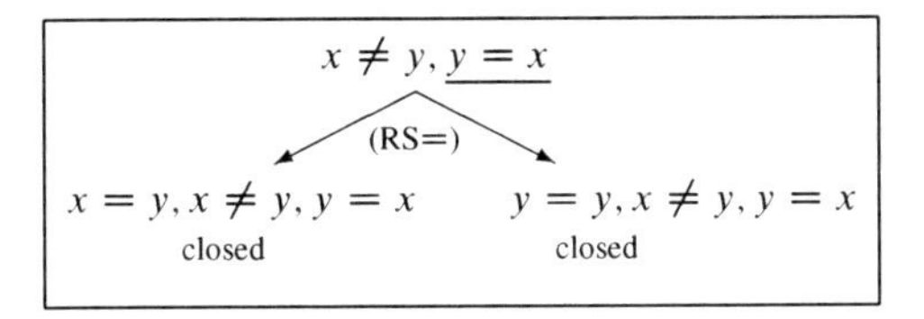

Fig. 1.4 A subtree of an RS-proof tree of the formula $\forall x \forall y(x \neq y \vee y = x)$

$x \neq y, y = x$
$(\mathrm{RS}^{1}_{=})$
$x \neq x, x \neq y, y = x$
$(\mathrm{RS}^{2}_{=})$, the first occurrence of x in $\varphi(x) = (x = x)$ is replaced by y
$y \neq x, y = x, \ldots$
closed

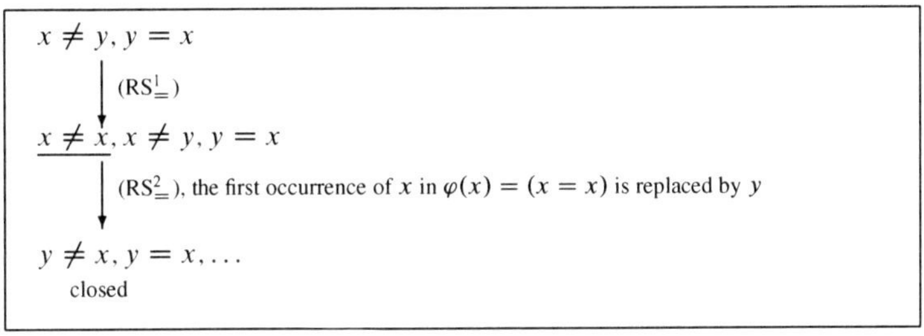

Fig. 1.5 A subtree of a tree of the formula $\forall x \forall y(x \neq y \vee y = x)$ built with the rules $(\mathrm{RS}^{1}_{=})$ and $(\mathrm{RS}^{2}_{=})$

$x \neq y, y = x$
$(\mathrm{RS}^{2'}_{=})$ applied to $\varphi(x) = (y = x)$
$y = y$
closed

Fig. 1.6 A subtree of a tree of the formula $\forall x \forall y(x \neq y \vee y = x)$ built with the rule $(\mathrm{RS}^{2'}_{=})$

tree the rule (RS=) is applied only once. If we admit the rules $(\mathrm{RS}^{1}_{=})$ and $(\mathrm{RS}^{2}_{=})$, then both rules have to be applied, so that we obtain the subtree of Fig. 1.5.

The first subtree is very short and involves branching, while the second one is longer and without branching. Therefore effectiveness of these specific rules seems to be comparable.

The system with rules $(\mathrm{RS}^{1'}_{=})$ and $(\mathrm{RS}^{2'}_{=})$ is more effective than the others. In this system the subtree of a proof tree for symmetry of $=$ is the shortest one and does not contain branching as it is presented in Fig. 1.6.

However, when the number of atomic formulas is less than the number of negated atomic formulas, the system with the rule (RS=) seems to be more effective than that with rules $(\mathrm{RS}^{1'}_{=})$ and $(\mathrm{RS}^{2'}_{=})$. In general, effectiveness of the systems depends both on a strategy of applications of the rules and on the formulas to which the specific rules are applied. Therefore we conclude that in order to construct a system for all the F-formulas the choice of the specific rules for identity is largely a matter of taste. Our choice was made for the reason of uniformity.

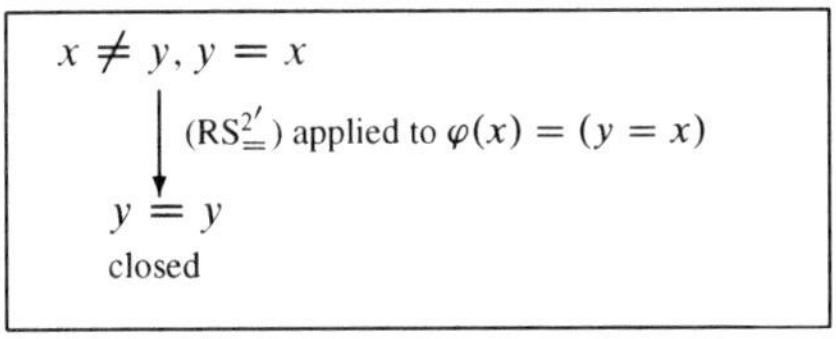

1.9 Dual Tableaux and Hilbert-Style Systems

In this section we discuss the relationship between dual tableaux and Hilbert-style systems. We show how the RS-proofs can be transformed into the proofs in a Hilbert system H. For simplicity of presentation, we focus on systems for first-order logic without identity and without function symbols, which is obtained from logic F, presented in Sect. 1.2, by deleting the identity predicate from its language. A dual tableau for this logic consists of the decomposition rules of the RS-system presented in Sect. 1.3 and the axiomatic sets being the supersets of $\{\varphi, \neg\varphi\}$, for any formula φ.

All Hilbert systems have the following features in common. Certain formulas are designated as *axioms* and some *rules of inference* are specified. A formula φ is said to be *provable* whenever there exists a finite sequence $\varphi_1, \ldots, \varphi_n$ of formulas, $n \geq 1$, such that $\varphi_n = \varphi$ and each φ_i, $i \in \{1, \ldots, n\}$, is an axiom or follows from earlier formulas in the sequence by one of the rules of inference. This sequence is then referred to as an *H-proof of* φ.

The axioms of the Hilbert system H are formulas which have the form of tautologies of the classical propositional logic, PC, together with:

(H1) $\forall x\varphi(x) \rightarrow \varphi(z)$;
(H2) $\forall x(\varphi \rightarrow \psi(x)) \rightarrow (\varphi \rightarrow \forall x\psi(x))$, provided that x is not free in φ.

The rules of inference are *modus ponens* and *generalization*:

$$\frac{\varphi, \varphi \rightarrow \psi}{\psi} \qquad\qquad \frac{\psi \rightarrow \varphi(z)}{\psi \rightarrow \forall x\varphi(x)}$$

provided that z does not occur in ψ

Note that the axioms and rules of system H involve formulas with the universal quantifier. The existential quantifier is introduced by defining $\exists x\varphi(x)$ as $\neg\forall x\neg\varphi(x)$.

Although, the rules of modus ponens and generalization are most often postulated in the H-system, some other rules may be allowed; these are the *derived rules*. A rule is said to be derived in system H whenever any of its applications can be replaced with a sequence of applications of the modus ponens rule and the generalization rule. It follows that the set of all provable formulas is closed on applications of the derived rules. In order to show that $\frac{\varphi_1 \quad \ldots \quad \varphi_n}{\psi}$ is a derived H-rule, it suffices to prove that $\varphi_1 \rightarrow (\ldots \rightarrow (\varphi_n \rightarrow \psi))$ has an H-proof.

Now, our aim is to define an effective procedure for transforming RS-proofs into H-proofs. The transformation consists of three steps. First, we replace the sets of formulas occurring in the nodes of the RS-proof tree by the disjunctions of its members. Second, we form a sequence of disjunctions of formulas obtained from the axiomatic sets of the RS-proof. Next, going along the levels of the RS-proof tree from the bottom to the top, we append to that sequence all the remaining disjunctions of formulas from the nodes of the tree. Clearly, the last formula in the sequence is φ. Third, we prove that: (1) disjunctions of formulas of the RS-axiomatic sets are instances of the axioms in H-system, and (2) each application of an RS-rule to a set X of formulas leading to set X_0 (resp. sets X_0 and X_1 in the case of a branching rule) can be converted into an application of a derived H-rule to the disjunction of formulas of X leading to the disjunction of formulas of X_0 (resp. disjunctions of formulas of X_0 and X_1).

Given a finite set $X = \{\varphi_1, \ldots, \varphi_n\}$ of formulas, we define:

$$\delta_X \stackrel{\text{df}}{=} \varphi_1 \vee \ldots \vee \varphi_n;$$

$$\text{Then } \neg\delta_X = \neg\varphi_1 \wedge \ldots \wedge \neg\varphi_n.$$

Consider an application of a rule $\frac{\Phi}{\Phi_0}$ (resp. $\frac{\Phi}{\Phi_1 \mid \Phi_2}$) to a finite set X of formulas. Then the corresponding derived H-rule is of the form:

$$\frac{\delta_X \vee \delta_{\Phi_0}}{\delta_X \vee \delta_{\Phi}} \qquad \left(\text{resp. } \frac{\delta_X \vee \delta_{\Phi_1} \qquad \delta_X \vee \delta_{\Phi_2}}{\delta_X \vee \delta_{\Phi}}\right).$$

Below we list derived H-rules induced by RS-rules applied to a set X. Due to disjunctive interpretation of ',' in the RS-system, the rule (H$\vee$) is a trivial rule which does not change its premise.

(RS$\neg\vee$) $\dfrac{\neg(\varphi \vee \psi)}{\neg\varphi \mid \neg\psi}$ (H$\neg\vee$) $\dfrac{\delta_X \vee \neg\varphi \qquad \delta_X \vee \neg\psi}{\delta_X \vee \neg(\varphi \vee \psi)}$

(RS$\wedge$) $\dfrac{\varphi \wedge \psi}{\varphi \mid \psi}$ (H$\wedge$) $\dfrac{\delta_X \vee \varphi \qquad \delta_X \vee \psi}{\delta_X \vee (\varphi \wedge \psi)}$

(RS$\neg\wedge$) $\dfrac{\neg(\varphi \wedge \psi)}{\neg\varphi, \neg\psi}$ (H$\neg\wedge$) $\dfrac{\delta_X \vee \neg\varphi \vee \neg\psi}{\delta_X \vee \neg(\varphi \wedge \psi)}$

(RS$\neg$) $\dfrac{\neg\neg\varphi}{\varphi}$ (H$\neg$) $\dfrac{\delta_X \vee \varphi}{\delta_X \vee \neg\neg\varphi}$

(RS$\forall$) $\dfrac{\forall x\varphi(x)}{\varphi(z)}$ (H$\forall$) $\dfrac{\delta_X \vee \varphi(z)}{\delta_X \vee \forall x\varphi(x)}$

z is a new variable z does not occur in X

(RS$\neg\forall$) $\dfrac{\neg\forall x\varphi(x)}{\neg\varphi(z), \neg\forall x\varphi(x)}$ (H$\neg\forall$) $\dfrac{\delta_X \vee \neg\varphi(z) \vee \neg\forall x\varphi(x)}{\delta_X \vee \neg\forall x\varphi(x)}$

z is any variable z may occur in X

(RS$\exists$) $\dfrac{\exists x\varphi(x)}{\varphi(z), \exists x\varphi(x)}$ (H$\exists$) $\dfrac{\delta_X \vee \varphi(z) \vee \exists x\varphi(x)}{\delta_X \vee \exists x\varphi(x)}$

z is any variable z may occur in X

(RS$\neg\exists$) $\dfrac{\neg\exists x\varphi(x)}{\neg\varphi(z)}$ (H$\neg\exists$) $\dfrac{\delta_X \vee \neg\varphi(z)}{\delta_X \vee \neg\exists x\varphi(x)}$

z is a new variable z does not occur in X

Proposition 1.9.1. *For every* $\# \in \{\vee, \wedge, \neg\vee, \neg\wedge, \neg, \forall, \exists, \neg\forall, \neg\exists\}$, *the rule (H#) is a derived H-rule.*

Proof. In order to prove that (H$\wedge$) is a derived rule, it suffices to show that $(\delta_X \vee \varphi) \rightarrow ((\delta_X \vee \psi) \rightarrow \delta_X \vee (\varphi \wedge \psi))$ has an H-proof. Indeed, since this formula is a tautology of the classical propositional logic, its H-proof consists of a single formula. The same holds for all the other rules for propositional operations. Now, we show that the rules for quantifiers are derived rules. By way of example, we prove it for (H$\forall$) and (H$\neg\forall$).

For (H$\forall$), we show that if $\delta_X \vee \varphi(z)$ has an H-proof, then $\delta_X \vee \forall x\varphi(x)$ has an H-proof provided that z does not occur in X.

1. $\delta_X \vee \varphi(z)$ — assumption
2. $\neg\delta_X \rightarrow \varphi(z)$ — modus ponens: 1, a PC axiom
3. $\forall x(\neg\delta_X \rightarrow \varphi(x))$ — generalization: 2
4. $\forall x(\neg\delta_X \rightarrow \varphi(x)) \rightarrow (\neg\delta_X \rightarrow \forall x\varphi(x))$ — axiom (H2)
5. $\neg\delta_X \rightarrow \forall x\varphi(x)$ — modus ponens: 3, 4
6. $\delta_X \vee \forall x\varphi(x)$ — modus ponens: 5, a PC axiom

For (H$\neg\forall$), we show that if $\delta_X \vee \neg\varphi(z) \vee \neg\forall x\varphi(x)$ has an H-proof, then $\delta_X \vee \forall x\varphi(x)$ has an H-proof for any variable z.

1. $\forall x\varphi(x) \rightarrow \varphi(z)$ — axiom (H1)
2. $\varphi(z) \vee \neg\forall x\varphi(x)$ — modus ponens: 1, a PC axiom
3. $\delta_X \vee \varphi(z) \vee \neg\forall x\varphi(x)$ — modus ponens: 2, a PC axiom
4. $\delta_X \vee \neg\varphi(z) \vee \neg\forall x\varphi(x)$ — assumption
5. $\delta_X \vee \neg\forall x\varphi(x)$ — modus ponens: 3, 4, a PC axiom

□

The method outlined above is applicable to the other dual tableaux presented in the book.

1.10 Dual Tableaux and Gentzen-Style Systems

In this section we present a transformation of the dual tableau for first-order logic without identity and without function symbols into a Gentzen sequent calculus G. This method is based on a close connection between validity of sequents and validity of sets. The method applies with minor adaptations to many dual tableaux presented in the book. We start with introducing basic notions of the Gentzen calculus.

A *sequent* is a pair (Γ, Δ) of finite sets Γ and Δ of formulas, written in the form $\Gamma \vdash \Delta$. A sequent $\Gamma \vdash \Delta$ is said to be valid whenever for any model $\mathcal{M}$, if all the formulas of Γ are true in $\mathcal{M}$, then at least one formula of Δ is true in $\mathcal{M}$.

Gentzen sequent calculus G consists of *axioms*, each of which has the form of a valid sequent, and *inference rules*, leading from valid sequents to valid sequents. Inference rules are of the form:

$$\frac{\Gamma_1 \vdash \Delta_1 \; \Gamma_2 \vdash \Delta_2 \ldots \Gamma_n \vdash \Delta_n}{\Gamma \vdash \Delta}$$

The sequents $\Gamma_i \vdash \Delta_i$, $i \in \{1, \ldots, n\}$ are premises of the rule and the sequent $\Gamma \vdash \Delta$ is its conclusion. A rule is said to be *G-correct* if and only if its conclusion is valid whenever all its premises are valid. A *G-proof of a sequent* S is a finite sequence of sequents $S_1, \ldots, S_n$ such that $S_n = S$ and every S_i is either an axiom or is deduced from previous sequents by the rules of Gentzen calculus.

The two major differences between decomposition rules of dual tableaux and inference rules of the sequent calculus are:

- The rules of dual tableaux preserve and reflect validity, i.e., the conclusion of a rule is valid iff all the premises are valid, whereas sequent calculus rules, in general, only preserve validity, that is the conclusion of a rule is valid whenever all the premises are valid;
- The rules of RS-system are used 'top down' to construct a proof tree of a formula, while the sequent calculus rules are used 'bottom up' to deduce a sequent from the axioms.

Directly from the definition of validity of a sequent and the definition of an RS-set we obtain:

Proposition 1.10.1. *A sequent* $\Gamma \vdash \Delta$ *is valid whenever the set* $\{\neg\varphi : \varphi \in \Gamma\} \cup \Delta$ *is an RS-set.*

We recall that for a finite set X of formulas, δ_X denotes disjunction of its members. As a consequence of Proposition 1.10.1 we have:

Proposition 1.10.2. *A sequent* $\Gamma \vdash \Delta$ *is valid whenever the formula* $\delta_{\{\neg\varphi:\varphi\in\Gamma\}} \vee \delta_{\{\varphi:\varphi\in\Delta\}}$ *is RS-provable.*

The above proposition suggests a simple method of developing a sequent calculus from the system RS. It suffices to derive the axioms and the inference rules of the sequent calculus from the axiomatic sets and the rules of RS-system.

Below we present the sequent calculus rules induced by the decomposition rules of the RS-system for classical predicate calculus presented in Sect. 1.2.

(RS∨) $\dfrac{\varphi \vee \psi}{\varphi, \psi}$ (G∨) $\dfrac{\Gamma \vdash \Delta \cup \{\varphi, \psi\}}{\Gamma \vdash \Delta \cup \{\varphi \vee \psi\}}$

(RS¬∨) $\dfrac{\neg(\varphi \vee \psi)}{\neg\varphi \mid \neg\psi}$ (G¬∨) $\dfrac{\Gamma \cup \{\varphi\} \vdash \Delta \quad \Gamma \cup \{\psi\} \vdash \Delta}{\Gamma \cup \{\varphi \vee \psi\} \vdash \Delta}$

(RS∧) $\dfrac{\varphi \wedge \psi}{\varphi \mid \psi}$ (G∧) $\dfrac{\Gamma \vdash \Delta \cup \{\varphi\} \quad \Gamma \vdash \Delta \cup \{\psi\}}{\Gamma \vdash \Delta \cup \{\varphi \wedge \psi\}}$

(RS¬∧) $\dfrac{\neg(\varphi \wedge \psi)}{\neg\varphi, \neg\psi}$ (G¬∧) $\dfrac{\Gamma \cup \{\varphi, \psi\} \vdash \Delta}{\Gamma \cup \{\varphi \wedge \psi\} \vdash \Delta}$

(RS¬) $\dfrac{\neg\neg\varphi}{\varphi}$ (G¬) $\dfrac{\Gamma \vdash \Delta \cup \{\varphi\}}{\Gamma \vdash \Delta \cup \{\neg\neg\varphi\}}$

(RS∀) $\dfrac{\forall x \varphi(x)}{\varphi(z)}$ (G∀) $\dfrac{\Gamma \vdash \Delta \cup \{\varphi(z)\}}{\Gamma \vdash \Delta \cup \{\forall x \varphi(x)\}}$

z is a new variable z occurs neither in Γ nor Δ

(RS¬∀) $\dfrac{\neg\forall x \varphi(x)}{\neg\varphi(z), \neg\forall x \varphi(x)}$ (G¬∀) $\dfrac{\Gamma \cup \{\varphi(z)\} \vdash \Delta}{\Gamma \cup \{\forall x \varphi(x)\} \vdash \Delta}$

z is any variable z may occur in Γ or Δ

(RS∃) $\dfrac{\exists x \varphi(x)}{\varphi(z), \exists x \varphi(x)}$ (G∃) $\dfrac{\Gamma \vdash \Delta \cup \{\varphi(z)\}}{\Gamma \vdash \Delta \cup \{\exists x \varphi(x)\}}$

z is any variable z may occur in Γ or Δ

(RS¬∃) $\dfrac{\neg\exists x \varphi(x)}{\neg\varphi(z)}$ (G¬∃) $\dfrac{\Gamma \cup \{\varphi(z)\} \vdash \Delta}{\Gamma \cup \{\exists x \varphi(x)\} \vdash \Delta}$

z is a new variable z occurs neither in Γ nor Δ

(RSAx) $\{\varphi, \neg\varphi\}$ (GAx) $\Gamma \cup \{\varphi\} \vdash \Delta \cup \{\varphi\}$

Proposition 1.10.3.

1. *The rule (RS#) is RS-correct iff the rule (G#) is G-correct, for every* $\# \in \{\vee, \neg\vee, \wedge, \neg\wedge, \neg, \forall, \neg\forall, \exists, \neg\exists\}$;
2. *For every formula* φ, *the sequent* $\Gamma \cup \{\varphi\} \vdash \Delta \cup \{\varphi\}$ *is valid.*

Now, we prove that the sequent calculus G presented above is sound and complete.

Theorem 1.10.1 (Soundness and Completeness of the Sequent Calculus G). *For every sequent* $\Gamma \vdash \Delta$, *the following conditions are equivalent:*

1. $\Gamma \vdash \Delta$ *is G-valid;*
2. $\Gamma \vdash \Delta$ *is G-provable.*

Proof. Soundness follows from Proposition 1.10.3. To prove completeness, assume that a sequent $\Gamma \vdash \Delta$ is G-valid. Then, by Proposition 1.10.2, the formula $\delta_{\{\neg\varphi:\varphi\in\Gamma\}} \vee \delta_{\{\varphi:\varphi\in\Delta\}}$ is RS-provable, that is it has a finite RS-proof tree with axiomatic sets at all its leaves. From that tree, we can easily construct a proof of the original sequent in the calculus G. Namely, we prove by induction that any sequent $\Gamma_i \vdash \Delta_i$ corresponding to the set $\{\neg\varphi : \varphi \in \Gamma_i\} \cup \Delta_i$ of formulas labelling a node of the tree is RS-provable. Indeed, the leaves are labelled with axiomatic sets, so the corresponding sequents are G-axioms. Going upwards in the tree, we can replace each downward application of an RS-decomposition rule by an upward application of the corresponding sequent calculus rule. We finish the induction at the root which is labelled by the formula $\delta_{\{\neg\varphi:\varphi\in\Gamma\}} \vee \delta_{\{\varphi:\varphi\in\Delta\}}$ corresponding to the original sequent. □

1.11 Dual Tableaux and Dual Resolution

In this section we compare resolution systems and dual tableaux. We focus on first-order logic with function symbols and without the identity predicate, which is obtained from logic F, presented in Sect. 1.2, by inclusion of function symbols and individual constants and by deleting the identity predicate. Let $\mathbb{C}$ (resp. $\mathbb{F}$) be a set of individual constants (resp. function symbols). The set of terms, $\mathbb{T}$, is the smallest set that includes individual variables and constants and such that if $f \in \mathbb{F}$ is an n-ary function symbol, $n \geq 1$, and $t_1, \ldots, t_n \in \mathbb{T}$, then $f(t_1, \ldots, t_n) \in \mathbb{T}$. Atomic formulas are of the form $P(t_1, \ldots, t_k)$, where P is a k-ary predicate symbol, $k \geq 1$, and $t_1, \ldots, t_k$ are terms.

The classical resolution is a refutation procedure; it is used to show that a formula is unsatisfiable. The rules operate on sets of clauses, where a clause is a disjunction of literals. They preserve satisfiability: if a set of clauses is satisfiable, then so is the set of clauses obtained by the application of a rule. A proof in the resolution system is obtained through a sequence of applications of the resolution rule or the factoring rule to sets of clauses leading to the empty clause at the end. If the unsatisfiable empty clause is ever obtained, the original set of clauses must have been unsatisfiable, hence the negation of the formula corresponding to the set of clauses in question is valid.

The connections between resolution systems and tableaux are extensively studied. It is known that tableaux for some logics (e.g., full first-order logic, some modal logics, description logics) can be translated into resolution systems, in the sense that the admissible resolution steps of deduction simulate their tableaux counterparts. Discussion of relationship between tableaux and the resolution can be found in [OdS93, Wol94, HS99, Sch06], among others. As it is shown in Sects. 1.6 and 1.7, the tableau system T and the Rasiowa–Sikorski system RS are dual, hence one can reasonably expect that the relationships of tableaux and resolution systems transfer to dual tableaux. However, in case of dual tableaux it is more natural to consider resolution in a dual form as suggested in [Sch06]. Accordingly, we present the dual resolution system for first-order logic with function symbols and without the identity predicate.

A *dual clause* is a conjunction of literals. Given the commutativity and idempotence of conjunction, a dual clause can be regarded as a set of literals and written simply as a list of its elements. A formula is in *prenex disjunctive normal form* whenever it is of the form $Q_1x_1 \ldots Q_nx_n\varphi$, where all Q_i, $i = 1, \ldots, n$, are quantifiers, φ is a quantifier free disjunction of dual clauses, and $x_1, \ldots, x_n$ are all the variables of φ. A sequence $Q_1x_1 \ldots Q_nx_n$ is called a *prefix* of φ. It is known that for every formula φ there exists a formula φ' in the prenex disjunctive normal form such that $\varphi \leftrightarrow \varphi'$ is valid. A formula is in *dual clause form* whenever it is in prenex disjunctive normal form and its prefix consists of the existential quantifiers. Usually, a formula in dual clause form is presented as the set of dual clauses appearing in it. As an example, consider the following formula in the prenex disjunctive normal form:

$$\exists x \exists y \exists z[(\neg P(x) \wedge P(z)) \vee (P(y) \wedge \neg P(z)) \vee (P(x) \wedge P(y))].$$

The above formula can be presented as:

$$\{\neg P(x), P(z)\}, \{P(y), \neg P(z)\}, \{P(x), P(y)\}.$$

Theorem 1.11.1 (Dual Skolem Normal Form). *For every formula φ there exists a formula φ' in dual clause form such that φ is valid iff φ' is valid.*

The proof of the above theorem is based on the method dual to the classical method of skolemization (see e.g., [EFT94]). A formula φ' postulated in the theorem is called a *dual Skolem form* of φ. It is obtained from prenex disjunctive normal form of φ by replacing each universally quantified variable y with a term $f(x_1, \ldots, x_k)$, $k \geq 1$, whose k-ary function symbol f is new (does not occur anywhere else in the formula) and $x_1, \ldots, x_k$ are the variables that are existentially quantified and such that a universally quantified variable y is in the scope of their quantifiers. The function f introduced in this process is called a *dual Skolem function* and $f(x_1, \ldots, x_k)$ is called a *dual Skolem term*. If a universally quantified variable y is not in the scope of any existential quantifier, then y is replaced with an individual constant, referred to as a *dual Skolem constant*. Therefore, apart from the symbols from φ, the formula φ' may contain new function symbols or constants used for elimination of universal quantifiers. Hence, the language of deduction in the dual resolution system must contain function and constant symbols, even if the language of the logic for which the system is constructed is function free.

Consider the formula $\varphi = \exists x_1 \forall y_1 \exists x_2 \forall y_2 (P(x_1, y_1) \wedge P(y_1, x_2) \wedge P(x_2, y_2))$ in the prenex disjunctive normal form. To obtain a dual Skolem form of φ we replace universally quantified variables y_1 and y_2 by terms $f(x_1)$ and $g(x_1, x_2)$, respectively, where f and g are new function symbols. Thus, a dual Skolem form of φ is:

$$\exists x_1 \exists x_2 (P(x_1, f(x_1)) \wedge P(f(x_1), x_2) \wedge P(x_2, g(x_1, x_2))).$$

The substitution is an assignment σ of terms to variables extended to the set of terms and to the set of formulas:

- $\sigma(a) = a$, for every individual constant a;
- $\sigma(f(t_1, \ldots, t_k)) = f(\sigma(t_1), \ldots, \sigma(t_k))$, for all terms $t_1, \ldots, t_k$ and for every k-ary function symbol f, $k \geq 1$;

- $\sigma(P(t_1,\dots,t_n)) = P(\sigma(t_1),\dots,\sigma(t_n))$, for all terms $t_1,\dots,t_n$ and for every n-ary predicate symbol P, $n \geq 1$;
- $\sigma(\varphi\#\psi) = \sigma(\varphi)\#\sigma(\psi)$, for $\# \in \{\vee,\wedge,\rightarrow\}$;
- $\sigma(\neg\varphi) = \neg\sigma(\varphi)$;
- $\sigma(Qx\varphi) = Qx\sigma(\varphi)$, for $Q \in \{\forall,\exists\}$.

A substitution σ is a *unifier* of the set $\{e_1,e_2\}$ of expressions, where e_1 and e_2 are terms or e_1 and e_2 are formulas, whenever $\sigma(e_1) = \sigma(e_2)$. A substitution σ is *the most general unifier* of $\{e_1,e_2\}$ whenever for every unifier σ' of $\{e_1,e_2\}$ there exists a substitution σ'' such that $\sigma' = \sigma'';\sigma$.

The dual resolution system is a validity procedure; it is used to show that a formula is valid. Its rules operate on sets of dual clauses. Let P be an n-ary predicate, let $t_1,\dots,t_n,t'_1,\dots,t'_n$, $n \geq 1$, be terms, and let C and D be dual clauses. The dual resolution rule has the form:

$$(dres) \qquad \frac{\{P(t_1,\dots,t_n)\}\cup C \qquad \{\neg P(t'_1,\dots,t'_n)\}\cup D}{\sigma(C\cup D)}$$

where σ is the most general unifier of $\{P(t_1,\dots,t_n), P(t'_1,\dots,t'_n)\}$.

The dual factorization rule has the form:

$$(dfac) \qquad \frac{\{P(t_1,\dots,t_n), P(t'_1,\dots,t'_n)\}\cup C}{\sigma(\{P(t_1,\dots,t_n)\}\cup C)}$$

where σ is the most general unifier of $\{P(t_1,\dots,t_n), P(t'_1,\dots,t'_n)\}$.

An application of a rule may lead to a multiset of literals, i.e., some literals may be identical. Therefore, usually we view the dual clauses as multisets of literals rather than sets.

Let φ be a formula and let $\{C_1,\dots,C_n\}$ be the set of dual clauses of the dual Skolem form of φ. An *R-proof* of φ is a sequence of sets of dual clauses starting with $\{C_1,\dots,C_n\}$ and such that each set in the sequence is obtained from the predecessor set by an application of the rule (*dres*) or (*dfac*) to some of its dual clauses, and the last set in the sequence is the empty clause.

Theorem 1.11.2 (Soundness and Completeness of Dual Resolution). *For every formula φ, the following conditions are equivalent:*

1. *φ is valid;*
2. *φ has an R-proof.*

It follows that in order to check validity of φ we should apply R-system to its dual clause form.

Example. Let $\varphi = \forall x\exists y(P(x) \rightarrow Q(y)) \rightarrow \exists y\forall x(P(x) \rightarrow Q(y))$.

The prenex disjunctive normal form of φ is:

$$\exists x_1\forall y_1\exists x_2\forall y_2[(P(x_1)\wedge\neg Q(y_1))\vee\neg P(y_2)\vee Q(x_2)].$$

After dual skolemization we obtain:

$$\exists x_1 \exists x_2[(P(x_1) \wedge \neg Q(f(x_1))) \vee \neg P(g(x_1, x_2)) \vee Q(x_2)]$$

or equivalently:

$$\exists x_1(P(x_1) \wedge \neg Q(f(x_1))) \vee \exists x_1' \exists x_2' \neg P(g(x_1', x_2')) \vee \exists x_2 Q(x_2).$$

Thus, φ can be presented in dual clause form as:

$$\{\{P(x_1), \neg Q(f(x_1))\}, \{\neg P(g(x_1', x_2'))\}, \{Q(x_2)\}\}.$$

An R-proof of φ is presented in Fig. 1.7.

$\underline{\{P(x_1), \neg Q(f(x_1))\}, \{\neg P(g(x_1', x_2'))\}}, \{Q(x_2)\}$

by (*dres*) with $\sigma(x_1) = g(x_1', x_2')$

$\underline{\{\neg Q(f(g(x_1', x_2')))\}, \{Q(x_2)\}}$

by (*dres*) with $\sigma(x_2) = f(g(x_1', x_2'))$

$\emptyset$

Fig. 1.7 An R-proof of $\forall x \exists y(P(x) \rightarrow Q(y)) \rightarrow \exists y \forall x(P(x) \rightarrow Q(y))$

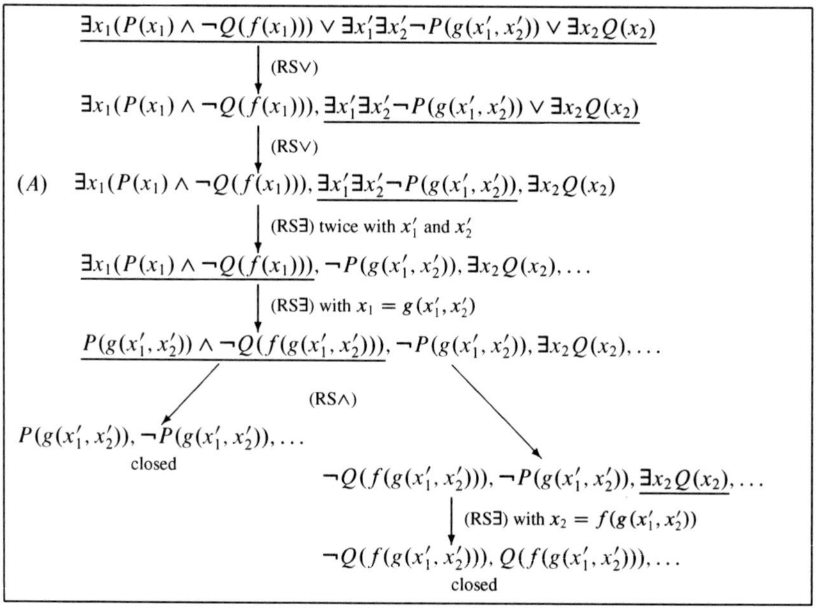

$\underline{\exists x_1(P(x_1) \wedge \neg Q(f(x_1))) \vee \exists x_1' \exists x_2' \neg P(g(x_1', x_2')) \vee \exists x_2 Q(x_2)}$

(RS∨)

$\exists x_1(P(x_1) \wedge \neg Q(f(x_1))), \underline{\exists x_1' \exists x_2' \neg P(g(x_1', x_2')) \vee \exists x_2 Q(x_2)}$

(RS∨)

(A) $\exists x_1(P(x_1) \wedge \neg Q(f(x_1))), \underline{\exists x_1' \exists x_2' \neg P(g(x_1', x_2'))}, \exists x_2 Q(x_2)$

(RS∃) twice with x_1' and x_2'

$\underline{\exists x_1(P(x_1) \wedge \neg Q(f(x_1)))}, \neg P(g(x_1', x_2')), \exists x_2 Q(x_2), \ldots$

(RS∃) with $x_1 = g(x_1', x_2')$

$\underline{P(g(x_1', x_2')) \wedge \neg Q(f(g(x_1', x_2')))}, \neg P(g(x_1', x_2')), \exists x_2 Q(x_2), \ldots$

(RS∧)

$P(g(x_1', x_2')), \neg P(g(x_1', x_2')), \ldots$

closed

$\neg Q(f(g(x_1', x_2'))), \neg P(g(x_1', x_2')), \underline{\exists x_2 Q(x_2)}, \ldots$

(RS∃) with $x_2 = f(g(x_1', x_2'))$

$\neg Q(f(g(x_1', x_2'))), Q(f(g(x_1', x_2'))), \ldots$

closed

Fig. 1.8 An RS-proof of $\forall x \exists y(P(x) \rightarrow Q(y)) \rightarrow \exists y \forall x(P(x) \rightarrow Q(y))$

One of the advantages of the resolution systems is that the computational complexity of the clause form of a given formula φ is lower than that of φ. This feature is one of the reasons for a success of resolution-based provers. The same concerns the dual resolution.

To see relationships between the dual resolution and the dual tableau consider the formula $\varphi = \forall x \exists y (P(x) \rightarrow Q(y)) \rightarrow \exists y \forall x (P(x) \rightarrow Q(y))$ from the previous example. An RS-proof starting from the dual Skolem form of φ is presented in Fig. 1.8. Applications of rule (RS$\vee$) to the dual Skolem form of φ result in the node (A) which consists of existentially quantified dual clauses. This is an equivalent representation of the dual clause form of φ from which its R-proof starts. Then the rule (RS$\exists$) is applied four times with such a choice of the terms which leads to axiomatic sets in the leaves of the proof tree. Between the last two applications of the rule (RS$\exists$), the rule (RS$\wedge$) is applied which completes the formation of the dual clause form.

From the dual resolution perspective, construction of an RS-proof of a formula consists of two interleaving processes: first, the process of forming the dual clause form of the formula and second, the process of finding the unifying substitutions. The first process is performed through the applications of decomposition rules for propositional operations. The second process is realized through the choice of the terms when the rule (RS$\exists$) is applied to a node. This choice is made so that eventually the branches including this node will close. To obtain an axiomatic set in a node one should apply a substitution to two of its formulas so that one of them will become the negation of the other.

Chapter 2
Dual Tableaux for Logics of Classical Algebras of Binary Relations

2.1 Introduction

The first steps in developing a mathematical theory of binary relations were taken by Augustus De Morgan in 1864 [Mor64]. His work was based on an earlier development by George Boole [Boo47] of a calculus of sets understood as an algebra of logic. Next, the theory was extensively developed by Charles Sanders Peirce [Pei83] and Ernst Schröder [Sch91]. This early theory studied binary relations between elements of a set and their properties. The research was mainly concerned with the arithmetic of binary relations and the relational notions that are expressible as equations in the calculus of relations. Alfred Tarski [Tar41] proposed a modern reformulation of the calculus of binary relations as a theory of abstract algebras called relation algebras. A relation algebra is a kind of a join of a Boolean algebra and an involuted monoid. The monoid operation of product is an abstract counterpart of the composition of two relations, the involution models the operation of forming the converse of a relation, and the neutral element of the product corresponds to the identity relation. An extensive study of the theory of relation algebras and the recent developments can be found in [TG87, HH02, Mad06], among others.

Many theories from a variety of fields, in particular from computer science and logic, can be interpreted as theories of algebras of relations. The fundamental relational structure consisting of a Boolean algebra together with a monoid constitutes a common core of a great variety of these theories. In Parts II, III, IV, and V of this book the relational interpretability of theories is a basis for the construction of dual tableau systems for them in a systematic modular way.

In this chapter we present a basic logic of binary relations which provides a framework for the relational interpretation of the theories considered in the book. The logic can be viewed as a generic logic for the development of dual tableaux for these theories. The rules of the dual tableau of the basic relational logic are included in all the proof systems presented in the chapters of the parts of the book listed above. For each particular theory we need, first, to expand the basic relational logic with specific relational constants and/or operations satisfying the appropriate axioms which enable us to prove relational interpretability of the given theory. Next, a dual tableau of the basic relational logic is extended with the rules characterizing

E. Orłowska and J. Golińska-Pilarek, *Dual Tableaux: Foundations, Methodology, Case Studies*, Trends in Logic 36, DOI 10.1007/978-94-007-0005-5_2,

those specific constants and operations. The method of proving correctness and completeness of dual tableaux presented in this book follows the major steps of the corresponding proofs for the dual tableau of the basic relational logic. In this way, the developments of this chapter establish the tools for relational interpretability of theories and for construction of their relational proof systems.

Some other proof systems for relation algebras are presented in [Wad75, Hen80, Sch82, Mad83, Gor95, Gor97, Gor01].

2.2 Algebras of Binary Relations

The *full algebra of binary relations on a set* U is a structure of the form:

$$\mathfrak{R}(U) = (\mathcal{P}(U \times U), \cup, \cap, -, \emptyset, U \times U, ;, ^{-1}, Id),$$

where $(\mathcal{P}(U \times U), \cup, \cap, -, \emptyset, U \times U)$ is the Boolean algebra of all the subsets of $U \times U$, ';' is the relative product of relations defined for all relations R, S on U as $R ; S = \{(x, y) \in U \times U : \exists z \in U\ (xRz \wedge zSy)\}$, '$^{-1}$' is the converse of relations defined as $R^{-1} = \{(x, y) \in U \times U : yRx\}$, and Id is the identity on U, i.e., $Id = \{(x, x) : x \in U\}$.

A *proper algebra of binary relations* is an algebra $(W, \cup, \cap, -, \emptyset, 1, ;, ^{-1}, 1')$ whose elements are binary relations, $\cup, \cap, ;, ^{-1}, \emptyset$ are as above, 1 is an equivalence relation and the largest element of W, $1'$ is the identity on the field of 1, and the complement operation '$-$' is defined with respect to 1.

A generalization of algebras of binary relations is provided by the notion of relation algebra [Tar41].

A relation algebra is a structure of the form:

$$\mathfrak{A} = (W, +, \cdot, -, 0, 1, ;, \breve{}, 1'), \text{ where :}$$

- $(W, +, \cdot, -, 0, 1)$ is a Boolean algebra,
- ; is an associative binary operation that distributes over $+$, i.e., $(x + y) ; z = x ; z + y ; z$ and $x ; (y + z) = x ; y + x ; z$,
- $\breve{}$ is an unary operation that distributes over $+$, i.e., $(x + y)\breve{} = x\breve{} + y\breve{}$, and satisfies $(x\breve{})\breve{} = x$ and $(x ; y)\breve{} = y\breve{} ; x\breve{}$, for all $x, y \in W$,
- $1' ; x = x ; 1' = x$, for every $x \in W$,
- $x\breve{}; (-(x; y)) \leq -y$, for all $x, y, z \in W$.

The class of relation algebras is denoted by RA. Clearly, every algebra of binary relations is a relation algebra. The similarity type of relation algebras usually contains the three constants 0, 1, and $1'$. However, 0 and 1 can be defined as $0 \stackrel{\mathrm{df}}{=} 1' \cdot (-1')$ and $1 \stackrel{\mathrm{df}}{=} 1' + (-1')$. We may also define $0' \stackrel{\mathrm{df}}{=} -1'$. The relation $0'$ is called the *diversity* relation.

A relation algebra is said to be *representable* whenever it is isomorphic to a proper algebra of binary relations. The class of all representable relation algebras is denoted by RRA. There exist relation algebras $\mathfrak{A}$ (both finite and infinite) such that $\mathfrak{A} \in$ RA and $\mathfrak{A} \notin$ RRA. Such algebras can be found in [Lyn50]. Some representation theorems, where each algebra from the class RA is embeddable into an algebra which is not necessarily an algebra of binary relations, can be found in [Dun01a]. In that paper each element of a relation algebra is mapped into a set of binary relations, not necessarily to a single relation. It is known that RRA is not finitely axiomatizable. Moreover, an infinite axiomatization of RRA requires infinitely many relation variables [Mon64]. RRA is a variety with a recursively enumerable but undecidable equational theory.

A relation R on U is said to be a *right* (resp. *left*) *ideal* relation whenever $R\,;1 = R$ (resp. $1\,;R = R$). Right ideal relations play an important role in relational formalization of logics discussed in Parts III, IV, and V of the book. Various properties of right ideal relations are presented in the relevant chapters, see e.g., Propositions 7.4.1 and 16.4.1.

Proposition 2.2.1. *For all relations $R, S \in \mathfrak{R}(U)$, the following conditions are satisfied:*

1. $R \subseteq S$ *iff* $-R \cup S = 1$,
2. $R = S$ *iff* $(-R \cup S) \cap (-S \cup R) = 1$,
3. $R \neq \emptyset$ *iff* $1\,;R\,;1 = 1$,
4. $R \neq 1$ *iff* $1\,;(-R)\,;1 = 1$,
5. $R = 1$ *and* $S = 1$ *iff* $R \cap S = 1$,
6. $R = 1$ *or* $S = 1$ *iff* $1\,;-(1\,;(-R)\,;1)\,;1 \cup S = 1$,
7. $R = 1$ *implies* $S = 1$ *iff* $(1\,;(-R)\,;1) \cup S = 1$.

Some arithmetic laws of relation algebras are collected in the following theorem:

Theorem 2.2.1. *The following are true in all relation algebras:*

1. $r\,;0 = 0\,;r = 0$,
2. $1^{\smile} = 1$,
3. $1\,;1 = 1$,
4. *If* $r \leq 1'$, *then* $r^{\smile} = r$,
5. *If* $r, s \leq 1'$, *then* $r\,;s = r \cdot s$,
6. *If* $r \leq 1'$, *then* $(r\,;1) \cdot s = r\,;s$ *and* $(1\,;r) \cdot s = s\,;r$,
7. $(-r)^{\smile} = -(r^{\smile})$,
8. $(r \cdot s)\,;t \leq (r\,;t) \cdot (s\,;t)$ *and* $r\,;(s \cdot t) \leq (r\,;s) \cdot (r\,;t)$,
9. *If* $t \leq 1'$, *then* $(t\,;r) \cdot s = t\,;(r \cdot s)$ *and* $(r\,;t) \cdot s = (r \cdot s)\,;t$,
10. *If* s *is right ideal, then* $r \cdot s = (s \cdot 1')\,;r$,
11. *If* s *is left ideal, then* $r \cdot s = r\,;(s \cdot 1')$.

Example. We present two examples of relation algebras quoted from [Mad91]. The full algebra of binary relations on a three element set $\{0, 1, 2\}$, $\mathfrak{R}(\{0, 1, 2\})$, has 512 elements and 9 atoms. Its atoms are $\{(0, 0)\}$, $\{(0, 1)\}$, $\{(0, 2)\}$, $\{(1, 0)\}$, $\{(1, 1)\}$, $\{(1, 2)\}$, $\{(2, 0)\}$, $\{(2, 1)\}$, and $\{(2, 2)\}$.

Table 2.1 Compositions of atoms of the pentagonal algebra

	$1'$	a	b
$1'$	$1'$	a	b
a	a	$1' + b$	$a + b$
b	b	$a + b$	$1' + a$

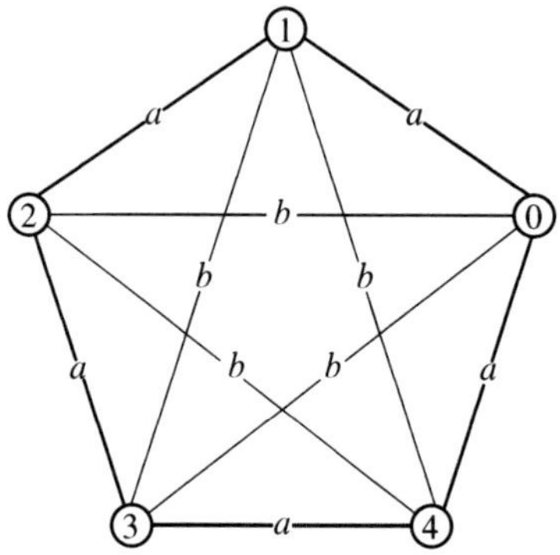

Fig. 2.1 Representation of the pentagonal algebra

The pentagonal relation algebra $\mathfrak{B}$ has three atoms $1', a$, and b. The elements of $\mathfrak{B}$ are $0, 1', a, b, 1' + b = -a, 1' + a = -b, a + b = 0'$, and $1 = 1' + a + b$. The compositions of atoms are given in Table 2.1. This table determines all the compositions of elements of $\mathfrak{B}$, since every element is a join of atoms and the composition ; distributees over $+$.

The algebra $\mathfrak{B}$ is representable, that is it can be embedded in some $\mathfrak{R}(U)$. It turns out that this is possible only if $card(U) = 5$. A copy of $\mathfrak{B}$ inside $\mathfrak{R}(\{0, \dots, 4\})$ can be defined as follows:

$$1' = \{(0,0), (1,1), (2,2), (3,3), (4,4)\};$$

$$a = \{(0,1), (1,0), (1,2), (2,1), (2,3), (3,2), (3,4), (4,3), (4,0), (0,4)\};$$

$$b = \{(0,2), (2,0), (1,3), (3,1), (2,4), (4,2), (3,0), (0,3), (4,1), (1,4)\}.$$

A picture of this representation in Fig. 2.1 shows a as a pentagon and b as a pentagram.

2.3 Logics of Binary Relations

In this chapter we present a general form of logics of binary relations considered in the book. There are two kinds of expressions of relational languages: terms and formulas. Terms represent relations and formulas express the facts that a pair of objects belongs to a relation.

Most of the relational logics of binary relations considered in the book are defined according to the following scheme. The vocabulary of the language of a relational logic L consists of the symbols from the following pairwise disjoint sets:

- $\mathbb{OV}_{\mathsf{L}}$ – a countable infinite set of object variables;
- $\mathbb{OC}_{\mathsf{L}}$ – a countable (possibly empty) set of object constants;
- $\mathbb{RV}_{\mathsf{L}}$ – a countable infinite set of binary relational variables;
- $\mathbb{RC}_{\mathsf{L}}$ – a countable (possibly empty) set of binary relational constants;
- A set of relational operations including the standard operations $-, \cup, \cap, ;$, and $^{-1}$.

Object constants are needed for relational representation of theories in Chaps. 3, 15, and Sect. 16.5. Relational representations of all the theories of Parts III, IV, and V require specific relational constants.

The set $\mathbb{RA}_{\mathsf{L}} = \mathbb{RV}_{\mathsf{L}} \cup \mathbb{RC}_{\mathsf{L}}$ is called the *set of atomic relational terms*. The set $\mathbb{OS}_{\mathsf{L}} = \mathbb{OV}_{\mathsf{L}} \cup \mathbb{OC}_{\mathsf{L}}$ is called the *set of object symbols*. The set $\mathbb{RT}_{\mathsf{L}}$ of *relational terms* is the smallest (with respect to inclusion) set of expressions that includes all the atomic relational terms and is closed with respect to all the relational operations. L-*formulas* are of the form xRy, where $x, y \in \mathbb{OS}_{\mathsf{L}}$ and $R \in \mathbb{RT}_{\mathsf{L}}$. An L-formula xRy is said to be *atomic* whenever $R \in \mathbb{RA}_{\mathsf{L}}$.

With an L-language a class of L-structures is associated. An L-*structure* has the form $\mathcal{M} = (U, m)$, where U is a non-empty set and m is a meaning function which assigns:

(m1) Elements of U to object constants, that is $m(c) \in U$, for every $c \in \mathbb{OC}_{\mathsf{L}}$;
(m2) Binary relations on U to atomic relational terms, that is $m(R) \subseteq U \times U$, for every $R \in \mathbb{RA}_{\mathsf{L}}$.

m extends to all the compound relational terms, in particular:

(m3) $m(-R) = (U \times U) \setminus m(R)$;
(m4) $m(R \cup S) = m(R) \cup m(S)$;
(m5) $m(R \cap S) = m(R) \cap m(S)$;
(m6) $m(R^{-1}) = (m(R))^{-1} = \{(x, y) \in U \times U : (y, x) \in m(R)\}$;
(m7) $m(R ; S) = m(R) ; m(S) = \{(x, y) \in U \times U : \exists z \in U, (x, z) \in m(R)$ and $(z, y) \in m(S)\}$.

So, for L-structures we only require that object constants are interpreted as elements of their universes, relational constants are interpreted as binary relations, and standard relational operations receive their usual meaning.

L-*models* are L-structures, where meaning function m satisfies the conditions specific for logic L. The conditions concern interpretation of relational operations specific for the given logic and/or interpretation of specific relational and/or object constants of the logic. Relational logics considered in the book are always defined in terms of their languages and the classes of models.

In order to avoid a confusion between syntactic objects in the languages of theories and semantic constraints on relations in the models of the theories, we write

xRy for the former, where x, y are object symbols and R is a relational term of a language, and $(x, y) \in R$ for the latter, where x and y are objects from the universe of a model and R is a relation in the model.

Let $\mathcal{M} = (U, m)$ be an L-structure. A *valuation* in $\mathcal{M}$ is any function $v: \mathbb{OS}_{\mathsf{L}} \to U$ such that $v(c) = m(c)$, for every $c \in \mathbb{OC}_{\mathsf{L}}$. *Satisfaction* of an L-formula xRy by a valuation v in an L-structure $\mathcal{M}$ is defined as:

- $\mathcal{M}, v \models xRy$ iff $(v(x), v(y)) \in m(R)$.

An L-formula xRy is *true* in $\mathcal{M}$ whenever it is satisfied by all the valuations in $\mathcal{M}$. If $\mathcal{K}$ is a class of L-structures, then an L-formula xRy is said to be $\mathcal{K}$-*valid* whenever it is true in every structure of $\mathcal{K}$, and it is L-*valid* whenever it is true in all L-models.

Fact 2.3.1. *Let L and L' be relational logics such that every L-structure is an L'-structure. Then for any relational formula xRy, if xRy is L'-valid, then it is L-valid.*

2.4 Relational Dual Tableaux

Relational dual tableaux are founded on the Rasiowa–Sikorski system for the first-order logic presented in Chap. 1. They are powerful tools for performing the four major reasoning tasks: verification of validity, verification of entailment, model checking, and verification of satisfaction. Relational proof systems are determined by their axiomatic sets of formulas and rules which most often apply to finite sets of relational formulas. Some relational proof systems with infinitary rules are known in the literature (see Sect. 19.2), but in the present chapter we confine ourselves to finitary rules only. The axiomatic sets take the place of axioms. The rules are intended to reflect properties of relational operations and constants. There are two groups of rules: decomposition rules and specific rules. Given a formula, the decomposition rules of the system enable us to transform it into simpler formulas, or the specific rules enable us to replace a formula by some other formulas. The rules have the following general form:

$$\text{(rule)} \qquad \frac{\Phi(\overline{x})}{\Phi_1(\overline{x}_1, \overline{u}_1, \overline{w}_1) \mid \ldots \mid \Phi_n(\overline{x}_n, \overline{u}_n, \overline{w}_n)}$$

where $\Phi(\overline{x})$ is a finite (possibly empty) set of formulas whose object symbols are among the elements of set$(\overline{x})$, where $\overline{x}$ is a finite sequence of object symbols and set$(\overline{x})$ is a set of elements of sequence $\overline{x}$; every $\Phi_j(\overline{x}_j, \overline{u}_j, \overline{w}_j)$, $1 \leq j \leq n$, is a finite non-empty set of formulas, whose object symbols are among the elements of set$(\overline{x}_j) \cup$ set$(\overline{u}_j) \cup$ set$(\overline{w}_j)$, where $\overline{x}_j, \overline{u}_j, \overline{w}_j$ are finite sequences of object symbols such that set$(\overline{x}_j) \subseteq$ set$(\overline{x})$, set$(\overline{u}_j)$ consists of the variables that may be instantiated to arbitrary object symbols when the rule is applied (usually to the object symbols that appear in the set to which the rule is being applied), set($\overline{w}_j$) consists of the variables that must be instantiated to pairwise distinct new variables (not appearing in the set to which the rule is being applied) and distinct from any variable of sequence $\overline{u}_j$. A rule of the form (rule) is the n-fold branching rule, where the jth successor of $\Phi(\overline{x})$ is the set $\Phi_j(\overline{x}_j, \overline{u}_j, \overline{w}_j)$. A rule of the form (rule) is *applicable* to a finite set X of formulas whenever $\Phi(\overline{x}) \subseteq X$. As a result of an application of a rule of the form (rule) to set X, we obtain the sets $(X \setminus \Phi(\overline{x})) \cup \Phi_j(\overline{x}_j, \overline{u}_j, \overline{w}_j)$, $j \in \{1, \ldots, n\}$. A set to which a rule has been applied is called the *premise* of the rule, and the sets obtained by an application of the rule are called its *conclusions*.

Let L be a relational logic. A finite set $\{\varphi_1, \ldots, \varphi_n\}$ of L-formulas is said to be an L-*set* whenever for every L-model $\mathcal{M}$ and for every valuation v in $\mathcal{M}$ there exists $i \in \{1, \ldots, n\}$ such that φ_i is satisfied by v in $\mathcal{M}$. It follows that the first-order disjunction of all the formulas from an L-set is valid in first-order logic. A rule of the form (rule) is L-*correct* whenever for every finite set X of L-formulas, $X \cup \Phi(\overline{x})$ is an L-set if and only if $X \cup \Phi_j(\overline{x}_j, \overline{u}_j, \overline{w}_j)$ is an L-set, for every $j \in \{1, \ldots, n\}$, i.e., the rule preserves and reflects validity. If $\mathcal{K}$ is a class of L-structures, then we define the notion of a $\mathcal{K}$-set and the notion of $\mathcal{K}$-correctness in a similar way. A finite set of L-formulas $\{\varphi_1, \ldots, \varphi_n\}$ is said to be a $\mathcal{K}$-*set* whenever for every $\mathcal{K}$-structure $\mathcal{M}$ and for every valuation v in $\mathcal{M}$ there exists $i \in \{1, \ldots, n\}$ such that φ_i is satisfied by v in $\mathcal{M}$. A rule of the form (rule) is $\mathcal{K}$-*correct* whenever for every finite set X of L-formulas, $X \cup \Phi(\overline{x})$ is a $\mathcal{K}$-set if and only if $X \cup \Phi_j(\overline{x}_j, \overline{u}_j, \overline{w}_j)$ is a $\mathcal{K}$-set, for every $j \in \{1, \ldots, n\}$.

Let xRy be an L-formula. The notion of an L-proof tree is defined in a similar way as in logic F in Sect. 1.3, that is an L-*proof tree for* xRy is a tree with the following properties:

- The formula xRy is at the root of the tree;
- Each node except the root is obtained by an application of an L-rule to its predecessor node;
- A node does not have successors whenever its set of formulas is an L-axiomatic set or none of the rules is applicable to its set of formulas.

Similarly as in the dual tableau for logic F, a branch of an L-proof tree is said to be *closed* whenever it contains a node with an L-axiomatic set of formulas. A tree is *closed* whenever all of its branches are closed. An L-formula xRy is L-*provable* whenever there is a closed L-proof tree for it which is then refereed to as its L-*proof*.

2.5 A Basic Relational Logic

The logic RL presented in this section is a common core of all the logics of binary relations presented in this book. RL represents, in fact, a family of logics which possibly differ in the object constants admitted in their languages. We do not assume here any specific properties of these constants, so all the developments of this section are relevant for any such logic. Throughout the book, by RL-logic we shall mean any logic from this family. The language of RL-logic is defined as in Sect. 2.3 with:

- $\mathbb{RC}_{\mathsf{RL}} = \emptyset$;
- The relational operations are $-, \cup, \cap, ;, ^{-1}$.

RL-*models* coincide with the RL-structures defined in Sect. 2.3 adjusted to the RL-language.

An RL-dual tableau consists of decomposition rules and axiomatic sets. Decomposition rules have the following forms:

For any object symbols x and y and for any relational terms R and S,

$$(\cup)\ \frac{x(R \cup S)y}{xRy, xSy} \qquad (-\cup)\ \frac{x-(R \cup S)y}{x-Ry \mid x-Sy}$$

$$(\cap)\ \frac{x(R \cap S)y}{xRy \mid xSy} \qquad (-\cap)\ \frac{x-(R \cap S)y}{x-Ry, x-Sy}$$

$$(-)\ \frac{x--Ry}{xRy}$$

$$(^{-1})\ \frac{xR^{-1}y}{yRx} \qquad (-^{-1})\ \frac{x-R^{-1}y}{y-Rx}$$

$$(;)\ \frac{x(R\,;S)y}{xRz, x(R\,;S)y \mid zSy, x(R\,;S)y} \quad z \text{ is any object symbol}$$

$$(-;)\ \frac{x-(R\,;S)y}{x-Rz, z-Sy} \quad z \text{ is a new object variable}$$

A set of RL-formulas is said to be an RL-*axiomatic set* whenever it includes a subset of the following form:

(Ax) $\{xRy, x-Ry\}$, where x, y are object symbols and R is a relational term.

Most of relational logics of binary relations studied in the book are obtained from RL by postulating some constraints on their object and/or relational constants. Dual tableaux for these logics include the decomposition rules of RL-dual tableau, where in each particular logic L the terms and object symbols range over the corresponding entities of the language of L. Specific rules reflect the properties of constants assumed in the L-language in question. In all relational dual tableaux considered in the book the sets including a subset of the form (Ax) are assumed to be L-*axiomatic sets*.

Proposition 2.5.1.

1. *The* RL*-rules are* RL*-correct;*
2. *The* RL*-axiomatic sets are* RL*-sets.*

Proof. By way of example, we prove correctness of the rules $(;)$ and $(-;)$. Let X be a finite set of RL-formulas.

($;$) Clearly, if $X \cup \{x(R\,;S)y\}$ is an RL-set, then so are $X \cup \{xRz, x(R\,;S)y\}$, $X \cup \{zSy, x(R\,;S)y\}$. Assume $X \cup \{xRz, x(R\,;S)y\}$ and $X \cup \{zSy, x(R\,;S)y\}$ are RL-sets, and suppose $X \cup \{x(R\,;S)y\}$ is not an RL-set. Then there exist an RL-model $\mathcal{M} = (U, m)$ and a valuation v in $\mathcal{M}$ such that for every $\varphi \in X \cup \{x(R\,;S)y\}$, $\mathcal{M}, v \not\models \varphi$. It follows that for every $a \in U$, $(v(x), a) \notin m(R)$ or $(a, v(y)) \notin m(S)$. However, by the assumption, model $\mathcal{M}$ and valuation v satisfy $(v(x), v(z)) \in m(R)$ and $(v(z), v(y)) \in m(S)$, a contradiction.

($-;$) Assume $X \cup \{x{-}(R\,;S)y\}$ is an RL-set. Suppose $X \cup \{x{-}Rz, z{-}Sy\}$ is not an RL-set, where z does not occur in X and $z \neq x, y$. Then there exist an RL-model $\mathcal{M} = (U, m)$ and a valuation v in $\mathcal{M}$ such that for every $\varphi \in X \cup \{x{-}Rz, z{-}Sy\}$, $\mathcal{M}, v \not\models \varphi$. Thus, $(v(x), v(z)) \in m(R)$ and $(v(z), v(y)) \in m(S)$. However, by the assumption, for every $a \in U$, $(v(x), a) \notin m(R)$ or $(a, v(y)) \notin m(S)$, a contradiction. Now, assume $X \cup \{x{-}Rz, z{-}Sy\}$ is an RL-set. Then, by the assumption on variable z, for every RL-model $\mathcal{M} = (U, m)$ and for every valuation v in $\mathcal{M}$, either there exists $\varphi \in X$ such that $\mathcal{M}, v \models \varphi$ or for every $a \in U$, either $(v(x), a) \notin m(R)$ or $(a, v(y)) \notin m(S)$. That is for every RL-model $\mathcal{M} = (U, m)$ and for every valuation v in $\mathcal{M}$ either there exists $\varphi \in X$ such that $\mathcal{M}, v \models \varphi$ or $\mathcal{M}, v \models x{-}(R\,;S)y$. Hence, $X \cup \{x{-}(R\,;S)y\}$ is an RL-set. □

Due to the above proposition, we obtain:

Proposition 2.5.2. *Let φ be an* RL*-formula. If φ is* RL*-provable, then it is* RL*-valid.*

A branch b of an RL-proof tree is said to be *complete* whenever it is closed or it satisfies the following RL-completion conditions:

For all object symbols x and y and for all relational terms R and S,

Cpl($\cup$) (resp. Cpl($-\cap$)) If $x(R \cup S)y \in b$ (resp. $x{-}(R \cap S)y \in b$), then both $xRy \in b$ (resp. $x{-}Ry \in b$) and $xSy \in b$ (resp. $x{-}Sy \in b$), obtained by an application of the rule ($\cup$) (resp. ($-\cap$));

Cpl($\cap$) (resp. Cpl($-\cup$)) If $x(R \cap S)y \in b$ (resp. $x{-}(R \cup S)y \in b$), then either $xRy \in b$ (resp. $x{-}Ry \in b$) or $xSy \in b$ (resp. $x{-}Sy \in b$), obtained by an application of the rule ($\cap$) (resp. ($-\cup$));

Cpl($-$) If $x(--R)y \in b$, then $xRy \in b$, obtained by an application of the rule ($-$);

Cpl($^{-1}$) If $xR^{-1}y \in b$, then $yRx \in b$, obtained by an application of the rule ($^{-1}$);

Cpl($-^{-1}$) If $x{-}R^{-1}y \in b$, then $y{-}Rx \in b$, obtained by an application of the rule ($-^{-1}$);

Cpl($;$) If $x(R\,;S)y \in b$, then for every object symbol z, either $xRz \in b$ or $zSy \in b$, obtained by an application of the rule ($;$);

Cpl($-;$) If $x{-}(R\,;S)y \in b$, then for some object variable z, both $x{-}Rz \in b$ and $z{-}Sy \in b$, obtained by an application of the rule ($-;$).

An RL-proof tree is said to be *complete* if and only if all of its branches are complete. A complete non-closed branch of an RL-proof tree is said to be *open*. Note that every RL-proof tree can be extended to a complete RL-proof tree, i.e., for every RL-formula φ there exists a complete RL-proof tree for φ.

Due to the forms of RL-decomposition rules, Fact 1.3.1 and Proposition 1.3.3 transfer to logic RL.

Fact 2.5.1. *If a node of an* RL*-proof tree contains an* RL*-formula* xRy *or* $x{-}Ry$*, for an atomic* R*, then all of its successors contain this formula as well.*

This property will be referred to as preservation of formulas built with atomic terms or their complements.

Proposition 2.5.3. *For every branch* b *of an* RL*-proof tree and for every atomic term* R*, if* $xRy \in b$ *and* $x{-}Ry \in b$*, then* b *is closed.*

We can prove a stronger form of Proposition 2.5.3:

Proposition 2.5.4. *Let* b *be a complete branch of an* RL*-proof tree. If there is a relational term* R *such that* $xRy \in b$ *and* $x{-}Ry \in b$*, then* b *is closed.*

Proof. Clearly, if branch b is closed, then the proposition holds. Assume that branch b is open. The proof of the proposition is by induction on the complexity of relational terms. For atomic relational terms the statement holds due to Proposition 2.5.3. By way of example, we prove the statement for terms of the form $R\,;S$. Suppose $x(R\,;S)y$ and $x{-}(R\,;S)y$ belong to b, for some relational terms R and S, and some object symbols x and y. Since $x{-}(R\,;S)y \in b$, by the completion condition Cpl(−;), both $x{-}Rz \in b$ and $z{-}Sy \in b$, for some object variable z. Furthermore, since $x(R\,;S)y \in b$, by the completion condition Cpl(;), for every object variable z, either $xRz \in b$ or $zSy \in b$. Thus either both xRz and $x{-}Rz$ belong to b or both zSy and $z{-}Sy$ belong to b. Hence, by the induction hypothesis, b is closed. □

In order to prove completeness of RL-dual tableau, first, we construct a branch structure $\mathcal{M}^b = (U^b, m^b)$ determined by an open branch b of a complete RL-proof tree as follows:

- $U^b = \mathbb{OS}_{\mathsf{RL}}$;
- $m^b(c) = c$, for every $c \in \mathbb{OC}_{\mathsf{RL}}$;
- $m^b(R) = \{(x, y) \in U^b \times U^b : xRy \notin b\}$, for every relational variable R;
- m^b extends to all the compound relational terms as in the RL-models.

Directly from this definition, we obtain:

Fact 2.5.2. *For every open branch* b *of an* RL*-proof tree,* $\mathcal{M}^b$ *is an* RL*-model.*

Any structure $\mathcal{M}^b$ is referred to as an RL-*branch model.* Let $v^b\colon \mathbb{OS}_{\mathsf{RL}} \to U^b$ be a valuation in $\mathcal{M}^b$ such that $v^b(x) = x$, for every $x \in \mathbb{OS}_{\mathsf{RL}}$.

Proposition 2.5.5. *For every open branch* b *of an* RL*-proof tree and for every* RL*-formula* φ*, if* $\mathcal{M}^b, v^b \models \varphi$*, then* $\varphi \notin b$*.*

Proof. The proof is by induction on the complexity of formulas.

Let $\varphi = xRy$ be an atomic RL-formula. Assume that $\mathcal{M}^b, v^b \models xRy$, that is $(x, y) \in m^b(R)$. By the definition of the branch model $xRy \notin b$. Let $R \in \mathbb{RV}_{\mathsf{RL}}$ and $\mathcal{M}^b, v^b \models x{-}Ry$, that is $(x, y) \notin m^b(R)$. Therefore $xRy \in b$. Then, by Proposition 2.5.3, $x{-}Ry \notin b$, for otherwise b would be closed.

Let $\mathcal{M}^b, v^b \models x(S\,;T)y$. Then $(x, y) \in m^b(S\,;T)$, that is there exists an object symbol $z \in U^b$ such that $(x, z) \in m^b(S)$ and $(z, y) \in m^b(T)$. By the induction hypothesis, $xSz \notin b$ and $zTy \notin b$. Suppose $x(S\,;T)y \in b$. By the completion condition Cpl(;), for every object symbol $z \in U^b$, either $xSz \in b$ or $zTy \in b$, a contradiction.

Let $\mathcal{M}^b, v^b \models x{-}(S\,;T)y$. Then $(x, y) \notin m^b(S\,;T)$, that is for every object symbol $z \in U^b$, either $(x, z) \notin m^b(S)$ or $(z, y) \notin m^b(T)$. Suppose that $x{-}(S\,;T)y \in b$. By the completion condition Cpl(−;), for some object variable $z \in U^b$, both $x{-}Sz \in b$ and $z{-}Ty \in b$. By the induction hypothesis, $(x, z) \in m^b(S)$ and $(z, y) \in m^b(T)$, a contradiction.

The proofs of the remaining cases are similar. □

Fact 2.5.2 and Proposition 2.5.5 enable us to prove:

Proposition 2.5.6. *Let φ be an* RL*-formula. If φ is* RL*-valid, then φ is* RL*-provable.*

Proof. Assume φ is RL-valid. Suppose there is no any closed RL-proof tree for φ. Then there exists a complete RL-proof tree for φ with an open branch, say b. Since $\varphi \in b$, by Proposition 2.5.5, φ is not satisfied by valuation v^b in the branch model $\mathcal{M}^b$. Hence φ is not RL-valid, a contradiction. □

By Propositions 2.5.2 and 2.5.6, we have:

Theorem 2.5.1 (Soundness and Completeness of RL). *For every* RL*-formula φ, the following conditions are equivalent:*

1. *φ is* RL*-valid;*
2. *φ is* RL*-provable.*

Example. We show that $(R \cap S)\,;T \subseteq (R\,;T) \cap (S\,;T)$ by proving the formula:

$$x(-((R \cap S)\,;T) \cup ((R\,;T) \cap (S\,;T)))y.$$

Figure 2.2 presents its RL-proof.

2.6 A Method of Proving Soundness and Completeness of Relational Dual Tableaux

The method applied to proving soundness and completeness of RL-dual tableau determines a paradigm for all the soundness and completeness proofs presented in this book.

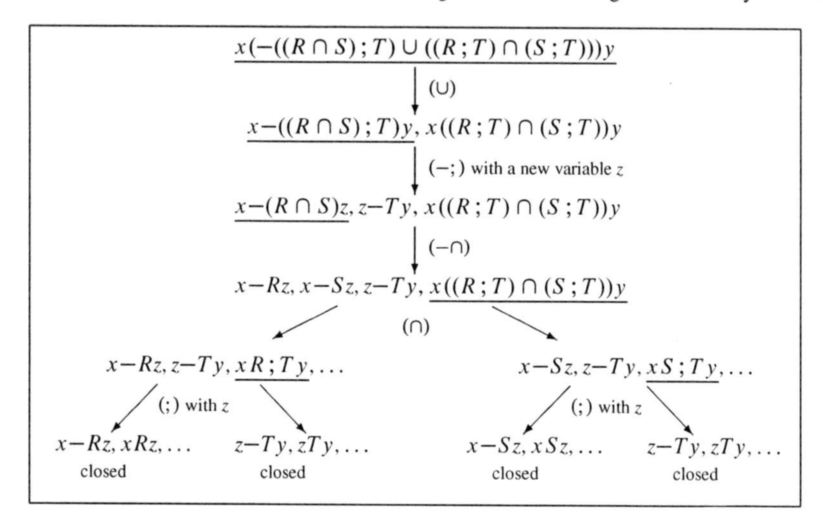

Fig. 2.2 An RL-proof of $(R \cap S) ; T \subseteq (R ; T) \cap (S ; T)$

Let L be a relational logic. In order to prove that an L-provable formula is L-valid it suffices to show that all the rules of an L-dual tableau are L-correct and all the axiomatic sets are L-sets. In some cases we also show correspondences between the semantic constraints posed on relational constants and/or relational operations and the rules and/or axiomatic sets reflecting those constraints. More precisely, we say that a rule (resp. an axiomatic set) of an L-dual tableau *reflects a constraint* assumed in L-models whenever for every class $\mathcal{K}$ of L-structures, $\mathcal{K}$-correctness of the rule (resp. $\mathcal{K}$-validity of the axiomatic set) implies that every structure from $\mathcal{K}$ satisfies the constraint.

In order to prove that an L-valid formula has an L-proof, we suppose that the formula does not have any L-proof. It follows that there exists a complete L-proof tree for this formula with an open branch b. Then we construct a branch structure $\mathcal{M}^b = (U^b, m^b)$ determined by b, where U^b consists of all the object symbols of L-language and m^b is the meaning function which assigns relations to atomic relational terms and extends homomorphically to all the terms as in RL-models. Most often the meaning function m^b of $\mathcal{M}^b$ is defined by:

$$m^b(R) \stackrel{\text{df}}{=} \{(x, y) \in U^b \times U^b : xRy \notin b\}, \text{ for every atomic relational term } R.$$

In that case we will say that $\mathcal{M}^b$ *is defined in a standard way*.We define the valuation v^b in the model $\mathcal{M}^b$ as the identity function, i.e., $v^b(x) = x$ for every object symbol x.

Then the major steps of the proof are the propositions showing the following three properties:

(1) **Closed Branch Property:** For any branch of an L-proof tree, if xRy and $x{-}Ry$, for an atomic term R, belong to the branch, then the branch can be closed.

(2) **Branch Model Property:** Let $\mathcal{M}^b$ be a branch structure determined by an open branch b of an L-proof tree. Then $\mathcal{M}^b$ is an L-model.
(3) **Satisfaction in Branch Model Property:** For every L-formula φ and for every open branch b of an L-proof tree for φ, the branch model $\mathcal{M}^b$ and valuation v^b in $\mathcal{M}^b$ satisfy:

$$\text{If } \mathcal{M}^b, v^b \models \varphi, \text{ then } \varphi \notin b.$$

Then we reason as in the proof of Proposition 2.5.6.

Roughly speaking, the method can be successfully applied provided that the following conditions will be satisfied: a sufficient condition for (1) is the preservation of formulas built with atomic terms or their complements by the applications of the rules; for (2), the L-dual tableau must have enough rules in order to reflect all the semantic properties of relational constants and relational operations of the L-language; for (3), the branch structure must provide a model that falsifies the non-provable formulas.

2.7 Relational Logic with Relations 1 and 1′

The logic considered in this section is obtained from RL-logic by expanding its language with constants 1 and 1′. The vocabulary of the language of $\mathsf{RL}(1,1')$-logic is defined as in Sect. 2.3 with:

- $\mathbb{RC}_{\mathsf{RL}(1,1')} = \{1, 1'\}$.

An $\mathsf{RL}(1,1')$-*structure* is an RL-model $\mathcal{M} = (U, m)$ such that $m(1)$ and $m(1')$ are binary relations on U. An $\mathsf{RL}(1,1')$-*model* is an $\mathsf{RL}(1,1')$-structure $\mathcal{M} = (U, m)$ such that the following conditions are satisfied:

- $m(1) = U \times U$;
- $m(1')$ is an equivalence relation on U;
- $m(1') ; m(R) = m(R) ; m(1') = m(R)$, for every atomic relational term R (extensionality);
- m extends to all the compound relational terms as in RL-models.

Proposition 2.7.1. *Let $\mathcal{M} = (U, m)$ be an $\mathsf{RL}(1,1')$-model. Then, for every relational term R of $\mathsf{RL}(1,1')$-language, the extensionality property holds:*

$$m(1') ; m(R) = m(R) ; m(1') = m(R).$$

Proof. The proof is by induction on the complexity of relational terms. By way of example, we show the extensionality property for $R = -S$. Since $m(1')$ is reflexive, $m(-S) \subseteq m(1') ; m(-S)$. For $\supseteq$, assume that there exists $z \in U$ such that $(x, z) \in m(1')$ and $(z, y) \notin m(S)$. By the induction hypothesis, for every $u \in U$, $(z, u) \notin m(1')$ or $(u, y) \notin m(S)$. In particular, $(z, x) \notin m(1')$ or $(x, y) \notin m(S)$. Since $m(1')$ is symmetric, $(x, y) \in m(-S)$.

The equality $m(-S) = m(-S)\,;m(1')$ can be proved in a similar way. The extensionality condition for the terms built with $\cup$, $\cap$, ;, and $^{-1}$ follows easily from the definition of the meaning function, properties of relational operations, and the induction hypothesis. □

Proposition 2.7.2. *Let $\mathcal{M} = (U, m)$ be an* $\mathsf{RL}(1, 1')$*-structure satisfying the following conditions:*

1. $m(1) = U \times U$;
2. $m(1')$ *is reflexive;*
3. $m(1')\,;m(R) = m(R)\,;m(1') = m(R)$, *for every relational term R.*

Then $\mathcal{M}$ is an $\mathsf{RL}(1, 1')$*-model.*

Proof. Since $m(1')$ is reflexive, so is $m(1'^{-1})$. Thus $m(1') \subseteq m(1'^{-1})\ ;\ m(1')$. By 3., $m(1'^{-1})\ ;\ m(1') = m(1'^{-1})$. Therefore $m(1')$ is symmetric. Transitivity follows directly from 3. □

It follows that an equivalent set of conditions characterizing the $\mathsf{RL}(1, 1')$-models could be reflexivity of $m(1')$ and the extensionality property for all the relational terms.

An $\mathsf{RL}(1, 1')$-model $\mathcal{M} = (U, m)$ is said to be *standard* whenever $m(1')$ is the identity on U, that is $m(1') = \{(x, x) : x \in U\}$.

Fact 2.7.1. *If a formula is* $\mathsf{RL}(1, 1')$*-valid, then it is true in all standard* $\mathsf{RL}(1, 1')$*-models.*

$\mathsf{RL}(1, 1')$-dual tableau includes the decomposition rules of RL-dual tableau (see Section 2.5) adjusted to $\mathsf{RL}(1, 1')$-language and the specific rules of the following forms:

For all object symbols x and y and for every atomic relational term R,

$$(1'1)\qquad \frac{xRy}{xRz, xRy \mid y1'z, xRy}\qquad z \text{ is any object symbol}$$

$$(1'2)\qquad \frac{xRy}{x1'z, xRy \mid zRy, xRy}\qquad z \text{ is any object symbol}$$

The rule $(1'1)$ reflects symmetry, transitivity, and the part $m(R)\,;m(1') \subseteq m(R)$ of the extensionality property of $1'$. The rule $(1'2)$ reflects the part $m(1')\,;m(R) \subseteq m(R)$ of extensionality property of $1'$. The formal presentation of these facts is given in Theorem 2.7.1.

A finite set of formulas is said to be $\mathsf{RL}(1, 1')$-axiomatic whenever it is an RL-axiomatic set adjusted to $\mathsf{R}(1, 1')$-language or it includes either of the sets of the following forms:

For all object symbols x and y,

(Ax1) $\{x1'x\}$;
(Ax2) $\{x1y\}$.

The axiomatic set $\{x1'x\}$ reflects reflexivity of $1'$.

Theorem 2.7.1 (Correspondence). *Let $\mathcal{K}$ be a class of* $\mathsf{RL}(1, 1')$*-structures. Then the following conditions are equivalent:*

1. *$\mathcal{K}$ is a class of* $\mathsf{RL}(1, 1')$*-models;*
2. *The* $\mathsf{RL}(1, 1')$*-axiomatic sets (Ax1) and (Ax2) are $\mathcal{K}$-sets and the rules* $(1'1)$ *and* $(1'2)$ *are $\mathcal{K}$-correct.*

Proof. (1. $\rightarrow$ 2.) Let $\mathcal{K}$ be a class of $\mathsf{RL}(1, 1')$-models. Since $m(1) = U \times U$, every superset of $\{x1y\}$ is an $\mathsf{RL}(1, 1')$-set. Furthermore, since $m(1')$ is reflexive, $X \cup \{x1'x\}$ is a $\mathcal{K}$-set, for every finite set X of $\mathsf{RL}(1, 1')$-formulas.

Now, we prove that the rule $(1'1)$ is $\mathcal{K}$-correct. Let X be a finite set of $\mathsf{RL}(1, 1')$-formulas. It is easy to see that if $X \cup \{xRy\}$ is a $\mathcal{K}$-set, then so are $X \cup \{xRz, xRy\}$ and $X \cup \{y1'z, xRy\}$. Now, assume $X \cup \{xRz, xRy\}$ and $X \cup \{y1'z, xRy\}$ are $\mathcal{K}$-sets. Suppose $X \cup \{xRy\}$ is not a $\mathcal{K}$-set, that is there exist an $\mathsf{RL}(1, 1')$-model $\mathcal{M}$ in $\mathcal{K}$ and a valuation v in $\mathcal{M}$ such that for every formula $\varphi \in X \cup \{xRy\}$, $\mathcal{M}, v \not\models \varphi$, in particular $\mathcal{M}, v \not\models xRy$. Then, by the assumption and since $m(1')$ is symmetric, $\mathcal{M}, v \models xRz$ and $\mathcal{M}, v \models z1'y$, that is $(v(x), v(z)) \in m(R)$ and $(v(z), v(y)) \in m(1')$. Since $m(R) ; m(1') \subseteq m(R)$, $(v(x), v(y)) \in m(R)$. Hence, $\mathcal{M}, v \models xRy$, a contradiction. The proof for the rule $(1'2)$ is similar.

(2. $\rightarrow$ 1.) Let $\mathcal{K}$ be a class of $\mathsf{RL}(1, 1')$-structures. We need to show that every $\mathcal{K}$-structure $\mathcal{M} = (U, m)$ is an $\mathsf{RL}(1, 1')$-model, i.e., $m(1) = U \times U$, $m(1')$ is an equivalence relation on U, and the extensionality property is satisfied. Since every finite superset of $\{x1y\}$ is a $\mathcal{K}$-set, the formula $x1y$ is valid in every structure from $\mathcal{K}$. Hence, the first condition holds.

For reflexivity, observe that $\{x1'x\}$ is a $\mathcal{K}$-set. Thus, in every $\mathcal{K}$-structure $\mathcal{M}$, $\mathcal{M} \models x1'x$.

For symmetry, let $X \stackrel{\text{df}}{=} \{y{-}1'x\}$. Since the rule $(1'1)$ is $\mathcal{K}$-correct and the sets $X \cup \{x1'x, x1'y\}$ and $X \cup \{y1'x, x1'y\}$ are $\mathcal{K}$-sets, $X \cup \{x1'y\}$ is also a $\mathcal{K}$-set. Therefore, for every $\mathcal{K}$-structure $\mathcal{M} = (U, m)$ and for every valuation v in $\mathcal{M}$, if $(v(y), v(x)) \in m(1')$, then $(v(x), v(y)) \in m(1')$.

For transitivity, let $X \stackrel{\text{df}}{=} \{x{-}1'z, z{-}1'y\}$. Then, $X \cup \{x1'z, x1'y\}$ and $X \cup \{z1'y, x1'y\}$ are $\mathcal{K}$-sets. Thus, by $\mathcal{K}$-correctness of the rule $(1'2)$, $X \cup \{x1'y\}$ is also a $\mathcal{K}$-set. Therefore, for every $\mathcal{K}$-structure $\mathcal{M} = (U, m)$ and for every valuation v in $\mathcal{M}$, if $(v(x), v(z)) \in m(1')$ and $(v(z), v(y)) \in m(1')$, then $(v(x), v(y)) \in m(1')$.

Now, we prove the extensionality property, i.e., we show that for every atomic relational term R, $m(1') ; m(R) = m(R) ; m(1') = m(R)$.

Since $m(1')$ is reflexive, $m(R) \subseteq m(1') ; m(R)$ and $m(R) \subseteq m(R) ; m(1')$.

To show $m(1') ; m(R) \subseteq m(R)$, consider $X \stackrel{\text{df}}{=} \{x{-}1'z, z{-}Ry\}$. By $\mathcal{K}$-correctness of the rule $(1'2)$, $X \cup \{xRy\}$ is a $\mathcal{K}$-set. Therefore, for every $\mathcal{K}$-structure $\mathcal{M} = (U, m)$ and for every valuation v in $\mathcal{M}$, if $(v(x), v(z)) \in m(1')$ and $(v(z), v(y)) \in m(R)$, then $(v(x), v(y)) \in m(R)$, which implies that if $(v(x), v(y)) \in m(1') ; m(R)$, then $(v(x), v(y)) \in m(R)$.

For $m(R) ; m(1') \subseteq m(R)$, let $X \stackrel{\text{df}}{=} \{x{-}Rz, y{-}1'z\}$. Then, by $\mathcal{K}$-correctness of the rule $(1'1)$, $X \cup \{xRy\}$ is a $\mathcal{K}$-set. Therefore, for every $\mathcal{K}$-structure $\mathcal{M} = (U, m)$

and for every valuation v in $\mathcal{M}$, if $(v(x), v(z)) \in m(R)$ and $(v(y), v(z)) \in m(1')$, then $(v(x), v(y)) \in m(R)$. $\mathcal{K}$-correctness of the rule $(1'1)$ implies symmetry of $m(1')$, so if $(v(x), v(y)) \in m(R)\,;m(1')$, then $(v(x), v(y)) \in m(R)$.

Hence, the extensionality property follows. □

$\mathsf{RL}(1, 1')$-correctness of the decomposition rules can be proved as in Proposition 2.5.1. Thus, by Theorem 2.7.1, we have:

Proposition 2.7.3.

1. *The* $\mathsf{RL}(1, 1')$*-rules are* $\mathsf{RL}(1, 1')$*-correct;*
2. *The* $\mathsf{RL}(1, 1')$*-axiomatic sets are* $\mathsf{RL}(1, 1')$*-sets.*

Following the general method of proving soundness described in Sect. 2.6, the above proposition implies:

Proposition 2.7.4. *Let* φ *be an* $\mathsf{RL}(1, 1')$*-formula. If* φ *is* $\mathsf{RL}(1, 1')$*-provable, then it is* $\mathsf{RL}(1, 1')$*-valid.*

Corollary 2.7.1. *Let* φ *be an* $\mathsf{RL}(1, 1')$*-formula. If* φ *is* $\mathsf{RL}(1, 1')$*-provable, then it is true in all standard* $\mathsf{RL}(1, 1')$*-models.*

The notions of an $\mathsf{RL}(1, 1')$-proof tree, a closed branch, a closed $\mathsf{RL}(1, 1')$-proof tree, and an $\mathsf{RL}(1, 1')$-proof of an $\mathsf{RL}(1, 1')$-formula are defined as in Sect. 2.4. Observe that any application of the rules of $\mathsf{RL}(1, 1')$-dual tableau, in particular an application of the specific rules $(1'1)$ and $(1'2)$, preserves the formulas built with atomic terms or their complements (see Fact 2.5.1). Therefore, whenever an atomic formula xRy and the formula $x{-}Ry$ appear in a branch, then the branch is closed. Thus, the closed branch property holds.

A branch b of an $\mathsf{RL}(1, 1')$-proof tree is said to be $\mathsf{RL}(1, 1')$-complete whenever it is closed or it satisfies $\mathsf{RL}(1')$-completion conditions which consist of the completion conditions of RL-dual tableau adjusted to $\mathsf{RL}(1, 1')$-language and the following completion conditions determined by the specific rules of $\mathsf{RL}(1, 1')$-dual tableau:

For all object symbols x and y and for every atomic relational term R,

Cpl$(1'1)$ If $xRy \in b$, then for every object symbol z, either $xRz \in b$ or $y1'z \in b$, obtained by an application of the rule $(1'1)$;

Cpl$(1'2)$ If $xRy \in b$, then for every object symbol z, either $x1'z \in b$ or $zRy \in b$, obtained by an application of the rule $(1'1)$.

Let b be an open branch of an $\mathsf{RL}(1, 1')$-proof tree. We define a branch structure $\mathcal{M}^b = (U^b, m^b)$ as in RL-logic adapting it to the $\mathsf{RL}(1, 1')$-language. In particular, $m^b(R) \stackrel{\text{df}}{=} \{(x, y) \in U^b \times U^b : xRy \notin b\}$, for $R \in \{1, 1'\}$.

Proposition 2.7.5 (Branch Model Property). *For every open branch* b *of an* $\mathsf{RL}(1, 1')$*-proof tree, a branch structure* $\mathcal{M}^b$ *is an* $\mathsf{RL}(1, 1')$*-model.*

Proof. We need to prove: (1) $m^b(1) = U^b \times U^b$, (2) $m^b(1')$ is an equivalence relation on U^b, and (3) $m^b(1');m^b(R) = m^b(R);m^b(1') = m^b(R)$, for every atomic relational term R.

Proof of (1). Clearly, for all object symbols x and y it must be $x1y \notin b$, for otherwise b would be closed. Thus $m^b(1) = U^b \times U^b$.

Proof of (2). For every $x \in U^b$, $x1'x \notin b$, for otherwise b would be closed. Therefore $(x, x) \in m^b(1')$, hence $m^b(1')$ is reflexive. Assume $(x, y) \in m^b(1')$, that is $x1'y \notin b$. Suppose $(y, x) \notin m^b(1')$. Then $y1'x \in b$. By the completion condition Cpl(1′1), either $y1'y \in b$ or $x1'y \in b$, a contradiction. Therefore $m^b(1')$ is symmetric. To prove transitivity, assume that $(x, y) \in m^b(1')$ and $(y, z) \in m^b(1')$, that is $x1'y \notin b$ and $y1'z \notin b$. Suppose $(x, z) \notin m^b(1')$. Then $x1'z \in b$. By the completion condition Cpl(1′1), either $x1'y \in b$ or $z1'y \in b$. In the first case we get a contradiction, so $z1'y \in b$. By the completion condition Cpl(1′1) applied to $z1'y$, either $z1'z \in b$ or $y1'z \in b$. In both cases we get a contradiction.

Proof of (3). Since $m^b(1')$ is reflexive, we have $m^b(R) \subseteq m^b(1');m^b(R)$ and $m^b(R) \subseteq m^b(R);m^b(1')$.

Now, assume $(x, y) \in m^b(1');m^b(R)$, that is there exists $z \in U^b$ such that $x1'z \notin b$ and $zRy \notin b$. Suppose $(x, y) \notin m^b(R)$. Then $xRy \in b$. By the completion condition Cpl(1′2), for every $z \in U^b$, either $x1'z \in b$ or $zRy \in b$, a contradiction.

Assume $(x, y) \in m^b(R);m^b(1')$, that is, by symmetry of $m^b(1')$, there exists $z \in U^b$ such that $xRz \notin b$ and $y1'z \notin b$. Suppose $(x, y) \notin m^b(R)$. Then $xRy \in b$. By the completion condition Cpl(1′1), for every $z \in U^b$, either $xRz \in b$ or $y1'z \in b$, a contradiction. □

Any structure $\mathcal{M}^b$ defined as above is referred to as an RL(1, 1′)-branch model. Let $v^b\colon \mathbb{OS}_{\mathsf{RL}(1,1')} \rightarrow U^b$ be a valuation in $\mathcal{M}^b$ such that $v^b(x) = x$, for every object symbol x.

Proposition 2.7.6 (Satisfaction in Branch Model Property). *For every open branch b of an* RL(1, 1′)*-proof tree and for every* RL(1, 1′)*-formula φ, if $\mathcal{M}^b, v^b \models \varphi$, then $\varphi \notin b$.*

The proof of the above proposition is similar to the proof of Proposition 2.5.5.

Given an RL(1, 1′)-branch model $\mathcal{M}^b$, since $m^b(1')$ is an equivalence relation on U^b, we may define the quotient model $\mathcal{M}^b_q = (U^b_q, m^b_q)$ as follows:

- $U^b_q = \{\|x\| : x \in U^b\}$, where $\|x\|$ is the equivalence class of $m^b(1')$ generated by x;
- $m^q_b(c) = \|c\|$, for every object constant c;
- $m^b_q(R) = \{(\|x\|, \|y\|)) \in U^b_q \times U^b_q : (x, y) \in m^b(R)\}$, for every atomic relational term R;
- m^b_q extends to all the compound relational terms as in the RL(1, 1′)-models.

Since a branch model satisfies the extensionality property, the definition of $m_q^b(R)$ is correct, that is:

$$\text{If } (x,y) \in m^b(R) \text{ and } (x,z),(y,t) \in m^b(1'), \text{ then } (z,t) \in m^b(R).$$

Let v_q^b be a valuation in $\mathcal{M}_q^b$ such that $v_q^b(x) = \|x\|$, for every object symbol x.

Proposition 2.7.7.

1. *The model $\mathcal{M}_q^b$ is a standard* $\mathsf{RL}(1,1')$*-model;*
2. *For every* $\mathsf{RL}(1,1')$*-formula φ, $\mathcal{M}^b, v^b \models \varphi$ iff $\mathcal{M}_q^b, v_q^b \models \varphi$.*

Proof.

1. We have to show that $m_q^b(1')$ is the identity on U_q^b . Indeed, we have:

$$(\|x\|, \|y\|) \in m_q^b(1') \text{ iff } (x,y) \in m^b(1') \text{ iff } \|x\| = \|y\|.$$

2. The proof is by an easy induction on the complexity of relational terms. □

Proposition 2.7.8. *Let φ be an* $\mathsf{RL}(1,1')$*-formula. If φ is true in all standard* $\mathsf{RL}(1,1')$*-models, then φ is* $\mathsf{RL}(1,1')$*-provable.*

Proof. Assume φ is true in all standard $\mathsf{RL}(1,1')$-models. Suppose there is no any closed $\mathsf{RL}(1,1')$-proof tree for φ. Then there exists a complete $\mathsf{RL}(1,1')$-proof tree for φ with an open branch, say b. Since $\varphi \in b$, by Proposition 2.7.6, φ is not satisfied by v^b in the branch model $\mathcal{M}^b$. By Proposition 2.7.7(2.), φ is not satisfied by v_q^b in the quotient model $\mathcal{M}_q^b$. Since $\mathcal{M}_q^b$ is a standard $\mathsf{RL}(1,1')$-model, φ is not true in all standard $\mathsf{RL}(1,1')$-models, a contradiction. □

From Fact 2.7.1, Propositions 2.7.4 and 2.7.8 we get:

Theorem 2.7.2 (Soundness and Completeness of $\mathsf{RL}(1,1')$). *Let φ be an* $\mathsf{RL}(1,1')$*-formula. Then the following conditions are equivalent:*

1. *φ is* $\mathsf{RL}(1,1')$*-valid;*
2. *φ is true in all standard* $\mathsf{RL}(1,1')$*-models;*
3. *φ is* $\mathsf{RL}(1,1')$*-provable.*

The class of $\mathsf{RL}(1,1')$-models corresponds to the class FRA of full relation algebras, as it will be proved in Sect. 2.9.

2.8 Discussion of Various Rules for Relation 1′

As in the case of F-proof system, in $\mathsf{RL}(1,1')$-dual tableau we can admit some other rules for relation $1'$. There are at least three possibilities of choosing rules for $1'$. In this section, rules $(1'1)$ and $(1'2)$ (see p. 46) will be called *classical*.

In many relational proof systems in the style of Rasiowa–Sikorski the following specific rules are used:

Standard

$(1'1)^1 \quad \dfrac{xRy}{xRz, xRy \mid z1'y, xRy} \qquad (1'2)^1 \quad \dfrac{xRy}{x1'z, xRy \mid zRy, xRy}$

$(\mathrm{sym})^1 \quad \dfrac{x1'y}{y1'x}$

where R is any relational variable or relational constant, x, y, z are any object symbols.

Then, the axiomatic sets are those of $\mathsf{RL}(1, 1')$-dual tableau, i.e., (Ax) from Sect. 2.5 and (Ax1) and (Ax2) from Sect. 2.7.

Completion conditions for rules $(1'1)^1$ and $(1'2)^2$ are analogous to those for the classical rules, and for the rule $(\mathrm{sym})^1$ we have:

Cpl$(\mathrm{sym})^1$ If $x1'y \in b$, then $y1'x \in b$.

Note that these standard specific rules differ from the classical ones in the rules $(1'2)^1$ and $(\mathrm{sym})^1$. The rule $(1'2)^1$ seems to be more natural than the rule $(1'2)$. On the metalogical level the rule $(1'2)^1$ says: xRy is valid iff the conjunction of $x1'z \vee xRy$ and $zRy \vee xRy$ is valid. In the rule $(1'2)$ positions of variables x and z in a formula $x1'z$ are interchanged and this, in some sense, expresses symmetry of $1'$. From a logical point of view it is immaterial which of these sets of specific rules is used. The standard rules are correct and give the complete proof system. The proof of completeness is similar to the proof for the system with the classical rules. The only difference is that symmetry of $m^b(1')$ follows from the completion condition corresponding to the rule $(\mathrm{sym})^1$.

The following rules are essentially different from the classical ones.

Negative standard

$(1'1)^2 \quad \dfrac{x-1'y, y-Rz}{x-Rz, x-1'y, y-Rz} \qquad (1'2)^2 \quad \dfrac{x-1'y, z-Rx}{z-Ry, x-1'y, z-Rx}$

$(\mathrm{sym})^2 \quad \dfrac{x-1'y}{y-1'x} \qquad (\mathrm{ref})^2 \quad \dfrac{}{x-1'x}$

$(1)^2 \quad \dfrac{}{x-1y}$

where R is any relational variable or relational constant, and x, y, z are any object symbols.

The axiomatic sets are those of RL-dual tableau, i.e., (Ax) from Sect. 2.5.

The negative forms of specific rules are dual to the rules of a tableau system for the relational logic. Note that the rules $(1'1)^2$ and $(1'2)^2$ can be applied only to negated atomic formulas. Moreover, contrary to the classical rules, they do not branch a proof tree and do not involve introduction of a variable which makes them more suitable for implementation. Our conclusions from the discussion of different forms of the specific rule for the first-order logic in Sect. 1.8 are also relevant to this case.

Below we prove completeness of the proof system with the negative form of the rules for $1'$. Moreover, we show that all these rules are needed to prove it.

The completion conditions corresponding to the negative rules are:

For all object symbols x, y, and z and for every atomic relational term R,

Cpl$(1'1)^2$ If $x{-}1'y \in b$ and $y{-}Rz \in b$, then $x{-}Rz \in b$, obtained by an application of the rule $(1'1)^2$;

Cpl$(1'2)^2$ If $x{-}1'y \in b$ and $z{-}Rx \in b$, then $z{-}Ry \in b$, obtained by an application of the rule $(1'2)^2$;

Cpl(sym)^2 If $x{-}1'y \in b$, then $y{-}1'x \in b$, obtained by an application of the rule (sym)^2;

Cpl(ref)^2 For every object symbol x, $x{-}1'x \in b$, obtained by an application of the rule (ref)^2;

Cpl$(1)^2$ For all object symbols x and y, $x{-}1y \in b$, obtained by an application of the rule $(1)^2$.

Proposition 2.8.1 (Closed Branch Property). *For every branch of a proof tree in the relational dual tableau with negative standard rules, if xRy and $x{-}Ry$, for an atomic R, belong to the branch, then the branch can be closed.*

Proof. Although the rule (sym)^2 does not preserve the formula $x{-}1'y$, if $x{-}1'y$ appears in a node of a branch, say b, then all the successors of this node in b include either $x{-}1'y$, if the rule (sym)^2 has not been applied yet, or $y{-}1'x$ if the rule (sym)^2 has been applied. If $x1'y$ is in a node, say n, of branch b, then either $x{-}1'y \in n$ and $x1'y \in n$ or $y{-}1'x \in n$ and $x1'y \in n$. In the former case, the branch is closed. In the latter, we can apply rule (sym)^2 to $y{-}1'x$ in node n. Then we obtain a node which includes $x{-}1'y$ and $x1'y$ and we close the branch. All the remaining rules preserve the formulas built with atomic terms or their complements. Thus, the closed branch property holds. □

Since the rules are applied to formulas built with complements of atomic terms, we need to change the definition of a branch structure. Let b be an open branch of a proof tree. We define a branch structure $\mathcal{M}^b = (U^b, m^b)$ as follows:

- $U^b = \mathbb{OS}_{\mathsf{RL}(1,1')}$;
- $m^b(R) = \{(x, y) \in U^b \times U^b : x{-}Ry \in b\}$, for every atomic relational term R;
- m^b extends to all the compound terms as in RL-models.

Let $v^b\colon \mathbb{OS}_{\mathsf{RL}(1,1')} \to U^b$ be a valuation in $\mathcal{M}^b$ such that $v^b(x) = x$, for every object symbol x. First, we prove that $m^b(1')$ is an equivalence relation.

Proposition 2.8.2. *$m^b(1')$ is an equivalence relation.*

Proof. By the completion condition Cpl(ref)^2, for every $x \in U^b$, $x{-}1'x \in b$, which means that $(x,x) \in m^b(1')$. Hence, $m^b(1')$ is reflexive. Assume $(x,y) \in m^b(1')$, that is $x{-}1'y \in b$. By the completion condition Cpl(sym)^2, $y{-}1'x \in b$, hence $(y,x) \in m^b(1')$. Thus, $m^b(1')$ is symmetric. Assume $(x,y) \in m^b(1')$, $(y,z) \in m^b(1')$, that is $x{-}1'y \in b$ and $y{-}1'z \in b$. By the completion condition Cpl$(1'1)^2$, $x{-}1'z \in b$, hence $(x,z) \in m^b(1')$, which proves transitivity of $m^b(1')$. □

Note that the rules (ref)^2 and (sym)^2 are needed to prove reflexivity and symmetry of $m^b(1')$, respectively. The rule $(1'1)^2$ is needed to prove transitivity of $m^b(1')$.

Proposition 2.8.3. *For every atomic relational term R:*

1. $m^b(1') ; m^b(R) \subseteq m^b(R)$*;*
2. $m^b(R) ; m^b(1') \subseteq m^b(R)$.

Proof. For 1., assume $(x,y) \in m^b(1') ; m^b(R)$. Then there exists $z \in U^b$ such that $(x,z) \in m^b(1')$ and $(z,y) \in m^b(R)$, that is $x{-}1'z \in b$ and $z{-}Ry \in b$. By the completion condition Cpl$(1'1)^2$, $x{-}Ry \in b$, hence $(x,y) \in m^b(R)$.

The proof of 2. is similar. □

By the above proposition, the branch structure satisfies the extensionality property assumed in RL$(1,1')$-models. Moreover, due to the completion condition Cpl$(1)^2$, it satisfies the constraint $m^b(1) = U^b \times U^b$. We conclude that the branch structure $\mathcal{M}^b$ is an RL$(1,1')$-model. Therefore, the quotient model can be defined in a similar way as in the proof of the completeness of RL$(1,1')$ (see p. 49). The satisfaction in branch model property can be proved by induction on the complexity of relational terms, that is we have:

Proposition 2.8.4 (Satisfaction in Branch Model Property). *For every open branch b and for every* RL$(1,1')$*-formula* φ*, if* $\mathcal{M}^b, v^b \models \varphi$*, then* $\varphi \notin b$.

Moreover, the following can be proved in a similar way as in RL$(1,1')$-dual tableau:

Proposition 2.8.5.

1. $\mathcal{M}^b_q$ *is a standard* RL$(1,1')$*-model;*
2. $\mathcal{M}^b, v^b \models \varphi$ *iff* $\mathcal{M}^b_q, v^b_q \models \varphi$.

Then completeness of RL$(1,1')$-dual tableau with negative specific rules follows from the above propositions.

We may refine the set of negative rules in analogy with the classical set of specific rules by deleting the symmetry rule and interchanging the positions of x and y in the rule $(1'2)^2$. Then we have less specific rules than in the negative standard set.

$$(1'1)^3 \quad \frac{x{-}1'y,\ y{-}Rz}{x{-}Rz,\ x{-}1'y,\ y{-}Rz} \qquad\qquad (1'2)^3 \quad \frac{y{-}1'x,\ z{-}Rx}{z{-}Ry,\ y{-}1'x,\ z{-}Rx}$$

$$(\mathrm{ref})^3 \quad \frac{}{x{-}1'x} \qquad\qquad (1)^3 \quad \frac{}{x{-}1y}$$

where R is any relational variable or relational constant, and x, y, z are any object symbols.

Then, the axiomatic sets are those of RL-dual tableau, i.e., (Ax) from Sect. 2.5.

This set of specific rules together with the decomposition rules provides a complete proof system for $\mathsf{RL}(1, 1')$. The proof of completeness is similar to the previous one except for the parts where rules $(1'2)^2$ and $(\mathrm{sym})^2$ are used.

2.9 Full Relation Algebras and Relational Logics

We recall that the class of full relation algebras, FRA, is the class of algebras of the form $(\mathcal{P}(U \times U), \cup, \cap, -, \emptyset, U \times U, ;, ^{-1}, 1')$, where U is a non-empty set, $1'$ is the identity on U, operations $-$, $\cup$, and $\cap$ are Boolean operations, $^{-1}$ and ; are converse and composition of binary relations, respectively.

Let $\mathfrak{C}$ be a class of relation algebras. An equation $R_1 = R_2$, where R_1 and R_2 are relation algebra terms, is said to be $\mathfrak{C}$-valid whenever it is true in all algebras of $\mathfrak{C}$. The following theorem follows directly from the definition of validity in relational logics.

Theorem 2.9.1. *Let R be any relational term and let x and y be any two different object variables. Then xRy is true in all standard* $\mathsf{RL}(1, 1')$*-models iff $R = 1$ is* FRA*-valid.*

Proof. $(\rightarrow)$ Assume xRy is true in all standard $\mathsf{RL}(1, 1')$-models and suppose that $R = 1$ is not FRA-valid. Then there exists a FRA-algebra $\mathfrak{A}$ of binary relations on a set U such that $R^{\mathfrak{A}} \neq 1^{\mathfrak{A}}$, where $1^{\mathfrak{A}}$ is the universal relation on U. Hence, for some $u, u' \in U$, $(u, u') \notin R^{\mathfrak{A}}$. Consider an $\mathsf{RL}(1, 1')$-model $\mathcal{M}_{\mathfrak{A}} = (U, m_{\mathfrak{A}})$ such that:

- $m_{\mathfrak{A}}(P) = P^{\mathfrak{A}}$, for every relational variable P and for $P \in \{1, 1'\}$;
- $m_{\mathfrak{A}}$ extends to all the compound relational terms as in $\mathsf{RL}(1, 1')$-models (see Sect. 2.3).

It follows that for every relational term Q, $m_{\mathfrak{A}}(Q) = Q^{\mathfrak{A}}$. Consider a valuation v in model $\mathcal{M}_{\mathfrak{A}}$ such that $v(x) = u$ and $v(y) = u'$. Such a valuation v exists, since variables x and y are different. Then $(v(x), v(y)) \notin m_{\mathfrak{A}}(R)$, a contradiction.

$(\leftarrow)$ Assume $R = 1$ is FRA-valid and suppose xRy is not true in a standard $\mathsf{RL}(1, 1')$-model $\mathcal{M} = (U, m)$. Then there exists a valuation v in $\mathcal{M}$ such that $(v(x), v(y)) \notin m(R)$. Consider a FRA-algebra $\mathfrak{A}_{\mathcal{M}} = (\mathcal{P}(U \times U), \cup, \cap, -, \emptyset, U \times U, ;, ^{-1}, m(1'))$ such that $P^{\mathfrak{A}_{\mathcal{M}}} = m(P)$, for every relational variable P. It follows that for every relational term Q, $Q^{\mathfrak{A}_{\mathcal{M}}} = m(Q)$. Since $R = 1$ is FRA-valid, it must be true in $\mathfrak{A}_{\mathcal{M}}$. Hence, $R^{\mathfrak{A}_{\mathcal{M}}} = 1^{\mathfrak{A}_{\mathcal{M}}} = U \times U$. This yields $m(R) = U \times U$, a contradiction. □

Due to the above theorem and Theorem 2.7.2, we obtain:

Theorem 2.9.2. *Let xRy be an* $\mathsf{RL}(1, 1')$*-formula. Then xRy is* $\mathsf{RL}(1, 1')$*-provable iff $R = 1$ is* FRA*-valid.*

An example of $\mathsf{RL}(1, 1')$-provable formula which represents a FRA-valid equation which is not RA-valid is presented in Sect. 2.10.

As a consequence of Proposition 2.2.1(1.), we have:

Proposition 2.9.1. *Let R and S be* $\mathsf{RL}(1, 1')$*-terms. Then $R \subseteq S$ holds in every* FRA*-algebra iff $x(-R \cup S)y$ is* $\mathsf{RL}(1, 1')$*-valid.*

Since the class RRA is a variety generated by FRA, it is know that:

Theorem 2.9.3. *The set of equations valid in* RRA *coincides with the set of equations valid in* FRA.

Hence, $\mathsf{RL}(1, 1')$-logic is a tool for verification of validity of equations in the class RRA of representable relation algebras.

2.10 An Example of a Relational Dual Tableau Proof

As an example of an $\mathsf{RL}(1, 1')$-provable equation let us consider the equation $\tau = 1$, for $\tau \stackrel{\text{df}}{=} (1 ; \rho ; 1)$ and $\rho \stackrel{\text{df}}{=} (A \cup B \cup C \cup D \cup E)$, where A, B, C, D, and E are defined as follows, for relational variables R and N:

- $A = -(1 ; R ; 1)$,
- $B = [R \cap -[(N ; N) \cap (R ; N)]]$,
- $C = (N ; N \cup R ; R) \cap N$,
- $D = [(R \cup R^{-1} \cup 1') \cap N]$,
- $E = -(R \cup R^{-1} \cup 1' \cup N)$.

This equation is RRA-valid, while it is not RA-valid.

Proposition 2.10.1. *The equation τ is true in all representable relation algebras.*

Proof. Suppose τ is not true in some $\mathfrak{A} \in \mathsf{FRA}$, that is $\tau \neq 1$. Then, by Proposition 2.2.1(3.), $\rho = \emptyset$ in $\mathfrak{A}$. It means that the following are true in $\mathfrak{A}$:

(1) $-(1 ; R ; 1) = \emptyset$,
(2) $(R \cap -[(N ; N) \cap (R ; N)]) = \emptyset$,
(3) $[(R ; R \cup N ; N) \cap N] = \emptyset$,
(4) $[(R \cup R^{-1} \cup 1') \cap N] = \emptyset$,
(5) $-(R \cup R^{-1} \cup 1' \cup N) = \emptyset$.

The first condition implies that $R \neq \emptyset$, since the following holds:

$$-(1\,;R\,;1) = \emptyset \text{ iff } (1\,;R\,;1) = 1 \text{ iff } R \neq \emptyset.$$

The condition (2) is equivalent to $R \subseteq (N\,;N) \cap (R\,;N)$, which means that $R \subseteq N\,;N$ and $R \subseteq R\,;N$. From condition (3) it follows that $(R\,;R) \cap N = \emptyset$ and $(N\,;N) \cap N = \emptyset$. Condition (4) implies that relations $R \cap N$, $R^{-1} \cap N$, $1' \cap N$ are empty. The condition (5) means that $R \cup R^{-1} \cup 1' \cup N$ is the universal relation 1.

Summarizing, if τ is not true in $\mathfrak{A}$, then the following must hold:

(a) $R \neq \emptyset$,
(b1) $R \subseteq N\,;N$,
(b2) $R \subseteq R\,;N$,
(c1) $(R\,;R) \cap N = \emptyset$,
(c2) $(N\,;N) \cap N = \emptyset$,
(d1) $R \cap N = \emptyset$,
(d2) $R^{-1} \cap N = \emptyset$,
(d3) $1' \cap N = \emptyset$,
(e) $(R \cup R^{-1} \cup 1' \cup N) = 1$.

By (a), (b2), and (d1) there are different x, y in the universe of $\mathfrak{A}$ such that $(x, y) \in R$. By (b1) an element z must exist such that $(x, z), (z, y) \in N$. Since $R \cap N = \emptyset$, we have $z \neq x, y$. Figure 2.3 presents this part of $\mathfrak{A}$. Thick lines and thin lines correspond to relations R and N, respectively.

By (b2) there is v such that $(x, v) \in R$ and $(v, y) \in N$. Since by (d3) $1' \cap N = \emptyset$, $v \neq y$. If $v = x$ or $v = z$, then $R \cap N \neq \emptyset$. Hence $v \neq x, y, z$, $(x, v) \in R$, and $(v, y) \in N$ as presented in Fig. 2.4a. By condition (e), $(z, v) \in R$ or $(v, z) \in R$ or $(z, v) \in N$. Suppose $(v, z) \in R$. Then $(x, v), (v, z) \in R$ and $(x, z) \in N$ which implies that $(R\,;R) \cap N$ is not empty, a contradiction with condition (c1). Suppose $(z, v) \in N$. Then $(z, v), (v, y), (z, y) \in N$, which contradicts condition (c2). So $(z, v) \in R$, and hence the algebra $\mathfrak{A}$ must look as shown in Fig. 2.4b. Since $(x, v) \in R$ and by (b1) $R \subseteq (N\,;N)$, there must exist an element s such that $(x, s), (s, v) \in N$. By condition (d1), $s \neq y, v$, for otherwise $R \cap N \neq \emptyset$. By condition (d3), $s \neq x$, for otherwise $1' \cap N \neq \emptyset$. Suppose $s = z$. Then $(z, y), (v, y), (z, v) \in N$, which means that $(N\,;N) \cap N \neq \emptyset$, a contradiction with condition (c2). Therefore, s is distinct from x, y, z, and v (see Fig. 2.5a). Hence, $(x, y), (x, v) \in R$ and $(z, y), (x, z), (v, y), (x, s), (s, v) \in N$.

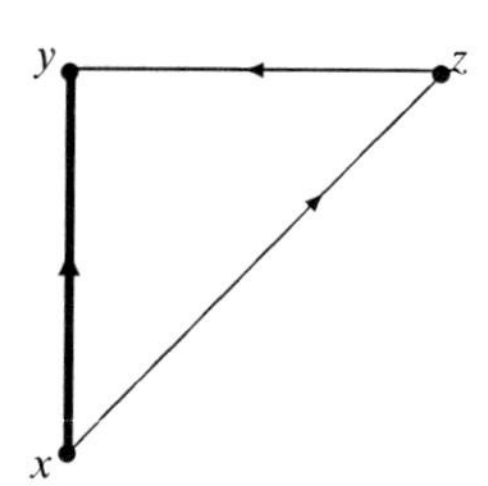

Fig. 2.3 Algbra $\mathfrak{A}$, step 1

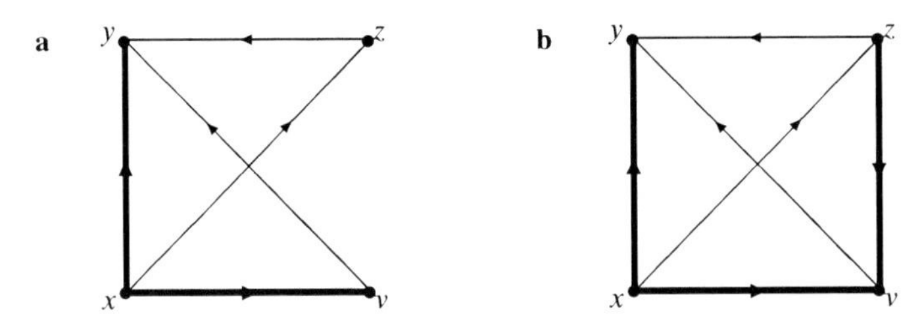

Fig. 2.4 Algebra $\mathfrak{A}$, step 2

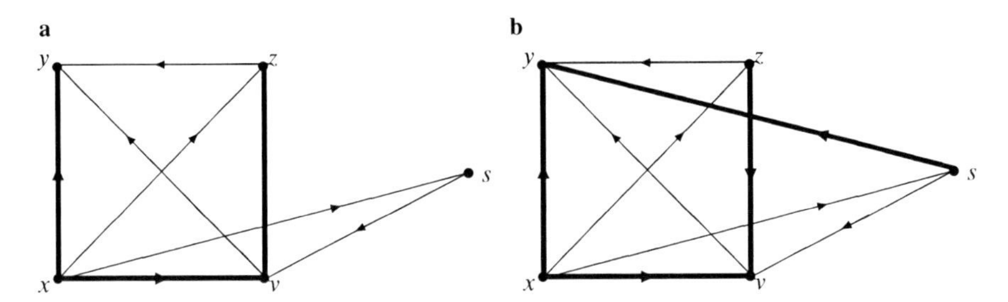

Fig. 2.5 Algebra $\mathfrak{A}$, step 3

Again by condition (e), $(y, s) \in R$ or $(s, y) \in R$ or $(s, y) \in N$. Assumptions $(s, y) \in N$ and $(y, s) \in R$ lead to a contradiction with conditions (c2) and (c1), respectively. Therefore $(s, y) \in R$ as shown in Fig. 2.5b. By condition (e), s and z must be in one of the relations R, R^{-1} or N. If $(s, z) \in R$, then we have $(s, z), (z, v) \in R$ and $(s, v) \in N$, a contradiction with condition (c1). Similarly, we can prove that $(z, s) \notin R$. Therefore $(s, z) \in N$, but then $(N \,;\, N) \cap N$ is non-empty, a contradiction. □

Proposition 2.10.2. *There exists a relation algebra $\mathfrak{A}$ in which the equation τ is not true.*

Proof. Let $\mathfrak{A} = (W, +, \cdot, -, 0, 1, ;, \breve{}, 1')$ be a relation algebra such that $(W, +, \cdot, -, 0, 1)$ is a Boolean algebra generated by the four atoms a, b, c, and d and the operations ; and $\breve{}$ are defined as follows:

	$\breve{}$
a	a
b	c
c	b
d	d

;	a	b	c	d
a	a	b	c	d
b	b	b	1	$b+d$
c	c	1	c	$c+d$
d	d	$b+d$	$c+d$	$a+b+c$

Let v be a valuation such that $v(1') = a, v(R) = b, v(N) = d$. We show that the equation τ is not satisfied in $\mathfrak{A}$ by v.

From Proposition 2.10.1 it is known that unsatisfiability of τ in a relation algebra implies that all the equations (1)–(5) are true in such an algebra. Therefore it suffices to show that these equations are true in $\mathfrak{A}$.

Since $(W, +, \cdot, -, 0, 1)$ is a Boolean algebra generated by the atoms a, b, c, d and $1 = a + b + c + d$, the following hold:

(1) $-(1 ; b ; 1) = -[(a + b + c + d) ; b ; (a + b + c + d)] = -(1 ; 1) = 0$, since ; is an associative operation that distributes over $+$,
(2) $(b \cdot -[(d ; d) \cdot (b ; d)]) = b \cdot -[(a + b + c) \cdot (b + d)] = b \cdot -b = 0$,
(3) $[(b ; b + d ; d) \cdot d] = [b + (a + b + c)] \cdot d = 0$,
(4) $[(b + b^{\smile} + 1') \cdot d] = (a + b + c) \cdot d = 0$,
(5) $-(b \cdot b^{\smile} + 1' + d) = -(a + b + c + d) = -1 = 0$.

Therefore we may conclude that τ is not true in $\mathfrak{A}$. □

Below we present a construction of an RL(1, 1′)-proof of the formula $u\tau w$ which provides a proof of RRA-validity of the equation $\tau = 1$.

It is easy to show that in a proof tree of $u\tau w$, if a formula $u\tau w$ occurs in a node of this tree, then it is possible to build a subtree of this proof tree with this formula at the root such that it ends with exactly one non-axiomatic node containing at least one of the following formulas: zAv, zBv, zCv, zDv, and zEv, for some variables z and v. Therefore, in such cases instead of building long subtrees we will use the following derived rules:

For all object symbols u, w, z, and v,

$$(Azv)\ \frac{u\tau w}{zAv, u\tau w} \qquad (Bzv)\ \frac{u\tau w}{zBv, u\tau w} \qquad (Czv)\ \frac{u\tau w}{zCv, u\tau w}$$

$$(Dzv)\ \frac{u\tau w}{zDv, u\tau w} \qquad (Ezv)\ \frac{u\tau w}{zEv, u\tau w}$$

By way of example, in Fig. 2.6 we present a derivation of rule (Azv).

Similarly, we can admit the following derived rules:

$$(1'^{*})\ \frac{x1'y}{y1'x} \qquad (Rxyz)\ \frac{x{-}Ry, u\tau w}{x{-}Nz, z{-}Ny, x{-}(R ; N)y, x{-}Ry, u\tau w}$$

z is a new variable

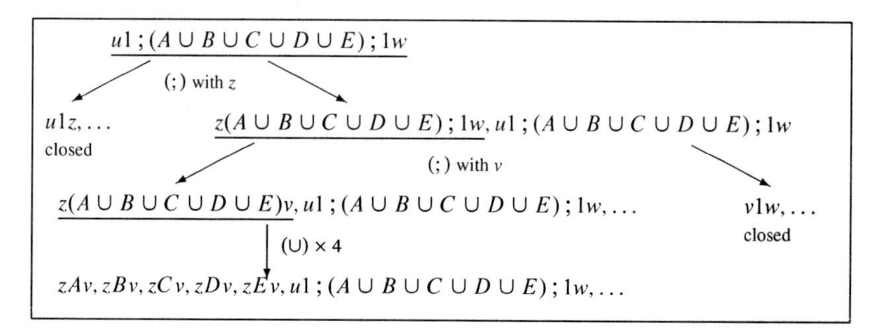

Fig. 2.6 A derivation of rule (Azv)

$$(RN1'xyz)_1 \quad \frac{z-Rx, z-Ny, u\tau w}{x1'y, z-Rx, z-Ny, u\tau w \quad y1'x, z-Rx, z-Ny, u\tau w}$$

$$(RN1'xyz)_2 \quad \frac{x-Rz, y-Nz, u\tau w}{x1'y, x-Rz, y-Nz, u\tau w \quad y1'x, x-Rz, y-Nz, u\tau w}$$

$$(RRNxyz) \quad \frac{x-Ry, y-Rz, x-Nz, u\tau w}{closed}$$

$$(NNNxyz) \quad \frac{x-Ny, y-Nz, x-Nz, u\tau w}{closed}$$

By way of example, in Figs. 2.10 and 2.11 we show how to obtain the derived rules $(Rxyz)$ and $(RN1'xyz)_1$, respectively. The remaining derived rules are obtained in a similar way. It is easy to check that the derived rule (Cxy) is needed to get $(RRNxyz)$ and $(NNNxyz)$, while (Dxy) is needed in the proofs of $(RN1'xyz)_1$ and $(RN1'xyz)_2$. Figure 2.7 presents an $\mathsf{RL}(1, 1')$-proof of $u\tau w$.

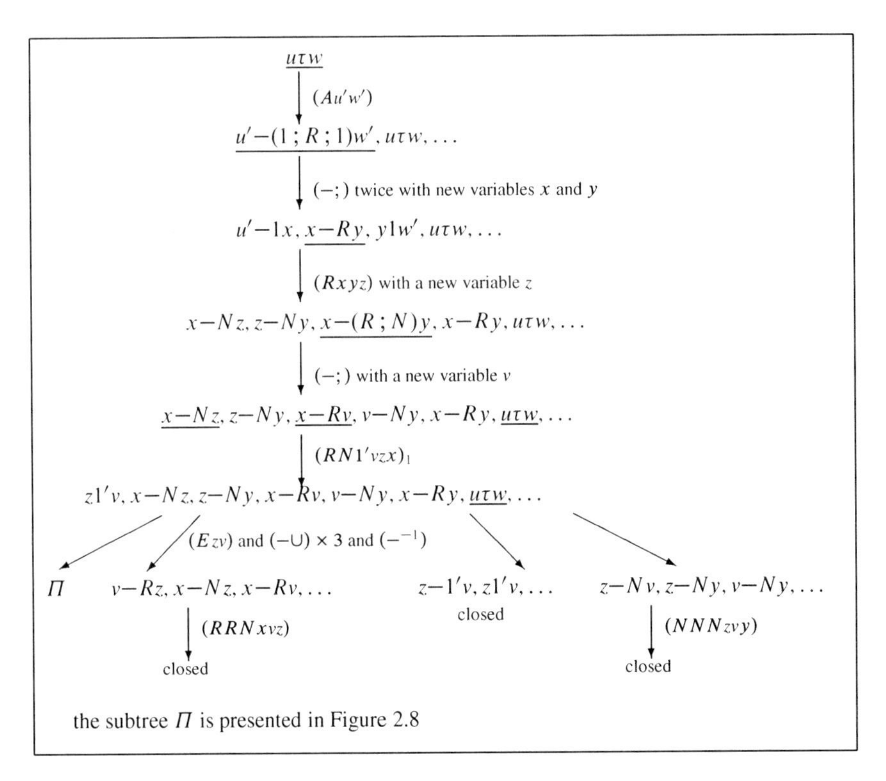

Fig. 2.7 An $\mathsf{RL}(1, 1')$-proof of $u\tau w$

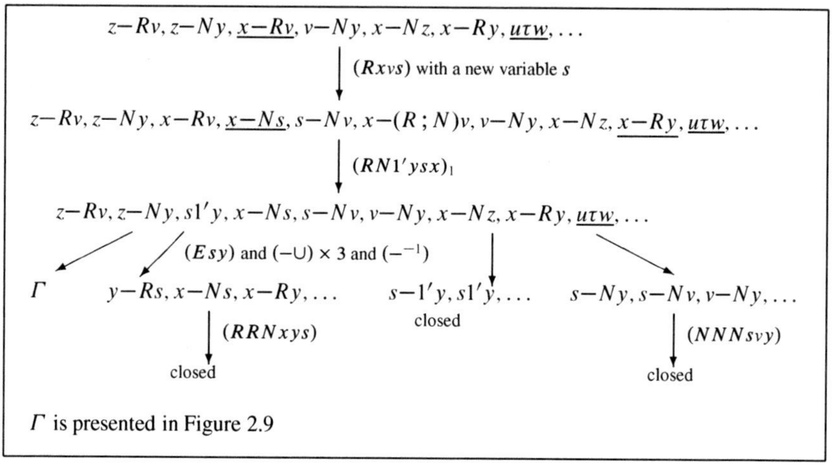

z−Rv, z−Ny, x−Rv, v−Ny, x−Nz, x−Ry, uτw, ...
(Rxvs) with a new variable s
z−Rv, z−Ny, x−Rv, x−Ns, s−Nv, x−(R ; N)v, v−Ny, x−Nz, x−Ry, uτw, ...
(RN1′ysx)₁
z−Rv, z−Ny, s1′y, x−Ns, s−Nv, v−Ny, x−Nz, x−Ry, uτw, ...
(Esy) and (−∪) × 3 and (−⁻¹)
Γ
y−Rs, x−Ns, x−Ry, ...
s−1′y, s1′y, ...
s−Ny, s−Nv, v−Ny, ...
(RRNxys)
closed
(NNNsvy)
closed
closed
Γ is presented in Figure 2.9

Fig. 2.8 The subtree Π

s−Ry, z−Rv, z−Ny, x−Ns, s−Nv, v−Ny, x−Nz, uτw, ...
(RN1′zsv)₂
s1′z, s−Ry, z−Rv, z−Ny, x−Ns, s−Nv, v−Ny, x−Nz, uτw, ...
(Esz) and (−∪) × 3 and (−⁻¹)
s−Rz, z−Rv, s−Nv ...
z−Rs, s−Ry z−Ny ...
s−1′z, s1′z, ...
closed
s−Nz, x−Ns, x−Nz, ...
(RRNszv)
closed
(RRNzsy)
closed
(NNNxsz)
closed

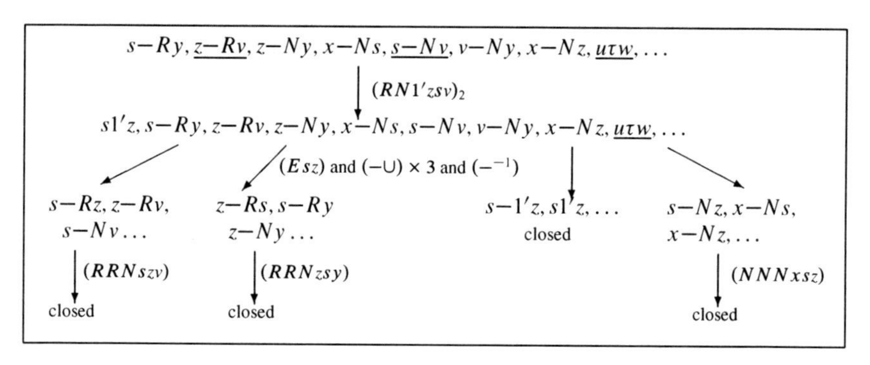

Fig. 2.9 The subtree Γ

x−Ry, uτw
(Bxy)
x(R ∩ −((N ; N) ∩ (R ; N)))y, x−Ry, uτw, ...
(∩)
xRy, x−Ry, ...
closed
x−((N ; N) ∩ (R ; N))y, x−Ry, uτw, ...
(−∩) and (−;) with a new variable z
x−Nz, z−Ny, x−(R ; N)y, x−Ry, uτw, ...

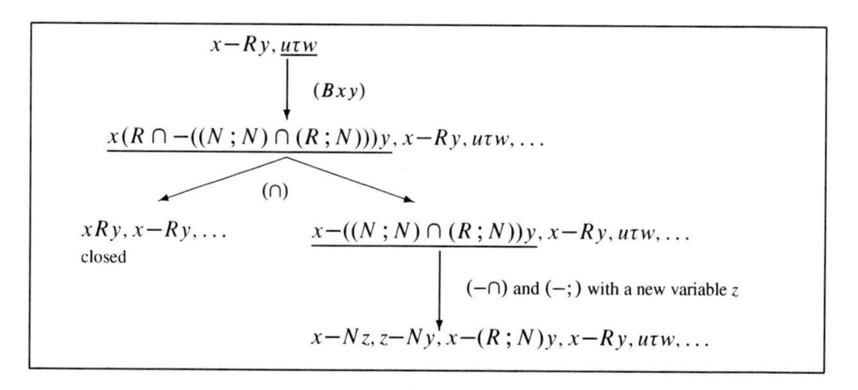

Fig. 2.10 A derivation of the rule $(Rxyz)$

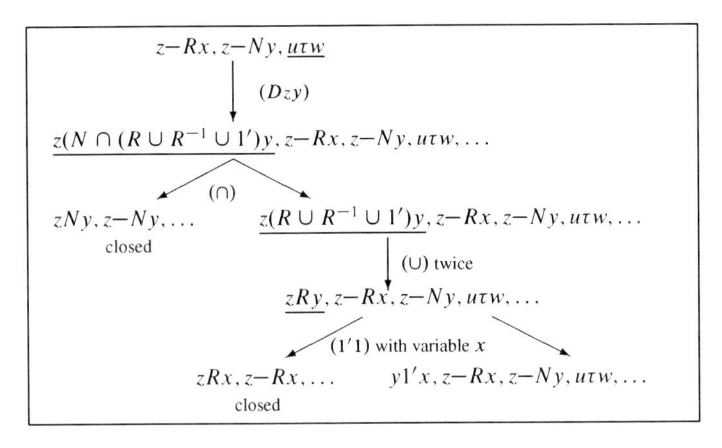

Fig. 2.11 A derivation of the rule $(RN1'xyz)_1$

2.11 Relational Entailment

The logic $\mathsf{RL}(1, 1')$ can be used to verify entailment of formulas from a finite set of formulas. The method is based on the following fact. Let $n \geq 1$ and let $R_1, \ldots, R_n, R$ be binary relations on a set U and let $1 = U \times U$. It is known ([Tar41], see also Proposition 2.2.1(7.)), that $R_1 = 1, \ldots, R_n = 1$ imply $R = 1$ iff $(1 ; -(R_1 \cap \ldots \cap R_n) ; 1) \cup R = 1$. It follows that for every $\mathsf{RL}(1, 1')$-model $\mathcal{M}$, $\mathcal{M} \models xR_1y, \ldots, \mathcal{M} \models xR_ny$ imply $\mathcal{M} \models xRy$ iff $\mathcal{M} \models x(1 ; -(R_1 \cap \ldots \cap R_n) ; 1) \cup R)y$ which means that entailment in $\mathsf{RL}(1, 1')$ can be expressed in its language.

Example. We prove that $x-(R ; -P)y$ and $x-(R ; -(-P \cup Q))y$ entail $x-(R ; -Q)y$, by showing that the formula:

$$x[(1 ; -(-(R ; -P) \cap -(R ; -(-P \cup Q))) ; 1) \cup -(R ; -Q)]y$$

is $\mathsf{RL}(1, 1')$-provable. Figure 2.12 presents an $\mathsf{RL}(1, 1')$-proof of this formula.

Now, we show that for every relation R, $R ; 1 = 1$ or $1 ; -R = 1$. Applying Proposition 2.2.1(6.), we need to show that the formula

$$x1 ; (-(1 ; (-(R ; 1) ; 1)) ; 1) \cup (1 ; -R)y$$

is $\mathsf{RL}(1, 1')$-provable. Figure 2.13 presents an $\mathsf{RL}(1, 1')$-proof of this formula.

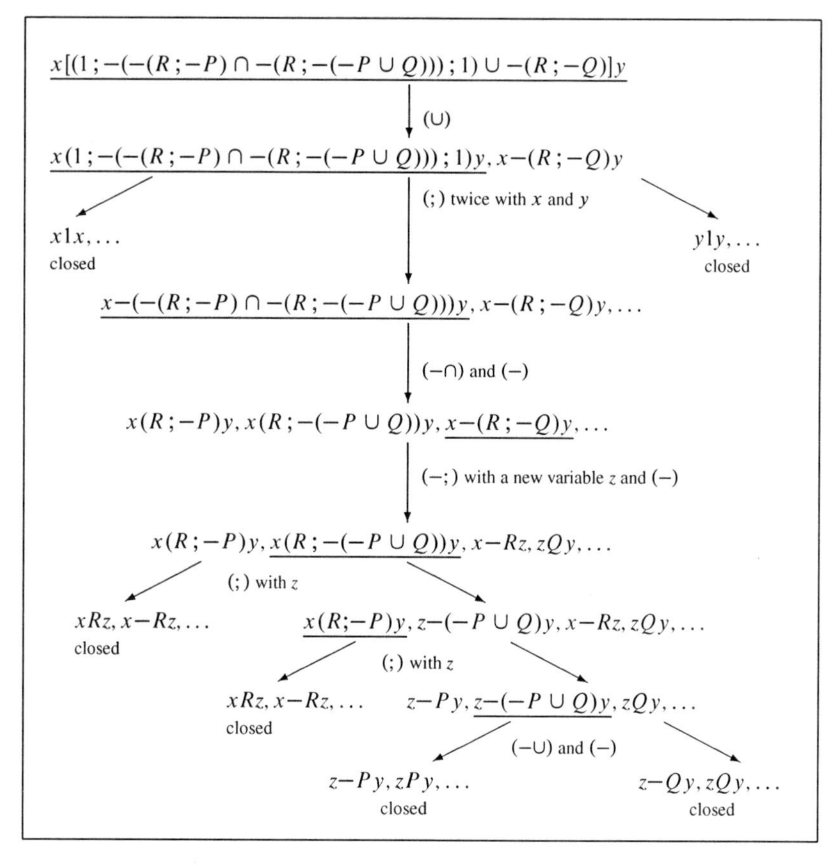

Fig. 2.12 An $\mathsf{RL}(1, 1')$-proof showing that $x{-}(R\,;-P)y$ and $x{-}(R\,;-(-P \cup Q))y$ entail $x{-}(R\,;-Q)y$

2.12 Decision Procedures for Some Relational Logics

In this section we present some decidable subclasses of formulas of the basic relational logic RL chosen from those presented in [BO97, Dob96a, Dob96b]. The classes are described in the syntactic terms. Usually, a distinguishing feature of a class is formulated in terms of a condition on the occurrences of the composition symbol in the formulas of the class. For the sake of simplicity, we will often say composition when referring to its occurrences.

An occurrence of the symbol of composition (;) is said to be *positive* (resp. *negative*) in a formula iff it is in the scope of an even (resp. odd) number of the complement symbols in the formula. Similarly, an occurrence of the symbol of composition is *negative-positive* (resp. *positive-negative*) iff it is positive (resp. negative) and it falls into the scope of a negative (resp. positive) composition. An occurrence c of the composition is between the compositions c_1 and c_2 iff c falls into the scope of c_1 and c_2 falls into the scope of c.

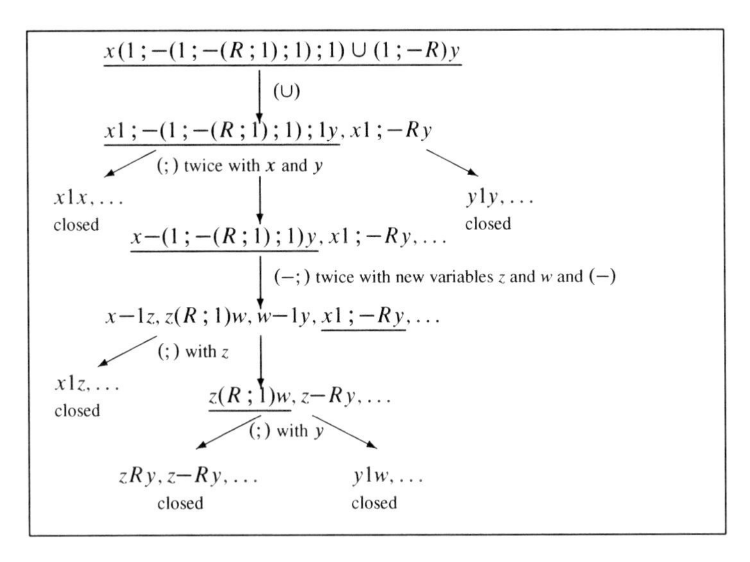

Fig. 2.13 An $\mathsf{RL}(1, 1')$-proof showing that $R ; 1 = 1$ or $1 ; -R = 1$

A formula φ of the logic RL is said to be:

- *Positive* (resp. *negative*) iff all the compositions in φ are positive (resp. negative);
- *Negative-positive* if there is no any positive-negative composition in φ;
- 1-*positive* iff there is exactly one positive composition in φ (and possibly some negative ones);
- 2-*positive* iff there are exactly two positive compositions in φ and there is no any negative composition in φ between the positive compositions.

The classes of formulas considered above will be referred to as:

- $\mathsf{RL}_{\mathsf{POS}}$ – the class of all positive formulas;
- $\mathsf{RL}_{\mathsf{NEG}}$ – the class of all negative formulas;
- $\mathsf{RL}_{\mathsf{NP}}$ – the class of all negative-positive formulas;
- $\mathsf{RL}_{\mathsf{POS(1)}}$ – the class of all 1-positive formulas;
- $\mathsf{RL}_{\mathsf{POS(2)}}$ – the class of all 2-positive formulas.

Example. Let P, Q be relational variables. Consider the following RL-terms:

$$T_1 = -(-(P ; (Q \cap P)) \cup -(Q ; P)),$$
$$T_2 = (-(P ; Q) \cap -(Q \cup (Q ; P))),$$
$$T_3 = -((-(P ; Q) ; -(Q ; P)) ; (-(Q ; -P) ; -(P ; -Q))),$$
$$T_4 = -((P \cup Q) ; -(-(Q ; P) ; Q)),$$
$$T_5 = -(-P ; -((Q ; -(Q ; P)) ; Q)).$$

The formula xT_1y is positive, since all of its compositions are in the scope of an even number of complements, and similarly, xT_2y is negative, since all of its compositions are negative. The formula xT_3y is negative–positive, and its positive compositions occur in the following terms that do not contain any negative occurrence of composition: $(P ; Q)$, $(Q ; P)$, $(Q ; -P)$, and $(P ; -Q)$. The formula xT_4y is 1-positive, since the only positive composition of this formula is the principal composition of $(-(Q ; P) ; Q)$ and all the other compositions are negative. The formula xT_5y is 2-positive. In order to see this, observe that xT_5y contains only two positive compositions, namely those in the term $((Q ; -(Q ; P)) ; Q)$, and there is no any negative composition between these two positive compositions.

Decidability of these classes of formulas is proved by showing that the first-order translations of their members form decidable subclasses of the classical first-order predicate logic F. A natural translation τ of relational formulas into first-order formulas is defined as follows. Let a one-to-one assignment τ' be given of object variables and relational variables of RL to individual variables and predicate symbols from the language of F, respectively. Then we define:

- $\tau(xPy) = \tau'(P)(\tau'(x), \tau'(y))$, for every atomic relational variable P;

For all relational terms R and S:

- $\tau(x{-}Ry) = \neg\tau(xRy)$;
- $\tau(x(R \cup S)y) = \tau(xRy) \vee \tau(xSy)$;
- $\tau(x(R \cap S)y) = \tau(xRy) \wedge \tau(xSy)$;
- $\tau(xR^{-1}y) = \tau(yRx)$;
- $\tau(x(R ; S)y) = \exists\tau'(z)(\tau(xRz) \wedge \tau(zSy))$.

The translations of relational formulas from the classes defined above belong to the following decidable classes of first-order formulas. Let $(\forall)$ and $(\exists)$ denote finite strings of universal and existential quantifiers, respectively. The classes are denoted by indicating a form of the string of quantifiers in the prenex normal form of the formulas in the class:

- $(\exists)$ – the class of existential formulas;
- $(\forall)$ – the class of universal formulas;
- $(\forall)(\exists)$ – the class of formulas with universal quantifiers followed by existential quantifiers;
- $(\forall)\exists(\forall)$ – the class of formulas with exactly one existential quantifier between the universal quantifiers;
- $(\forall)\exists_1\exists_2(\forall)$ – the class of formulas with exactly two existential quantifiers between the universal quantifiers.

The following is well known (see [Ack54]):

Proposition 2.12.1. *The classes* $(\exists)$, $(\forall)$, $(\forall)(\exists)$, $(\forall)\exists(\forall)$, *and* $(\forall)\exists_1\exists_2(\forall)$ *have a decidable validity problem.*

The translations of relational formulas from the classes listed above fall into the following classes of first-order logic:

Proposition 2.12.2. *For every relational formula φ, the prenex form of $\tau(\varphi)$, $pf(\tau(\varphi))$, satisfies:*

1. *If φ is positive, then $pf(\tau(\varphi)) \in \forall_1\forall_2(\exists)$;*
2. *If φ is negative, then $pf(\tau(\varphi)) \in (\forall)$;*
3. *If φ is negative-positive, then $pf(\tau(\varphi)) \in (\forall)(\exists)$;*
4. *If φ is 1-positive, then $pf(\tau(\varphi)) \in (\forall)\exists(\forall)$;*
5. *If φ is 2-positive, then $pf(\tau(\varphi)) \in (\forall)\exists_1\exists_2(\forall)$.*

By Propositions 2.12.1 and 2.12.2, we get:

Theorem 2.12.1 (Decidable Classes of Relational Formulas). *The classes* $\mathsf{RL}_{\mathsf{POS}}$, $\mathsf{RL}_{\mathsf{NEG}}$, $\mathsf{RL}_{\mathsf{NP}}$, $\mathsf{RL}_{\mathsf{POS(1)}}$, *and* $\mathsf{RL}_{\mathsf{POS(2)}}$ *of relational formulas have a decidable validity problem.*

Below we present sound and complete dual tableaux that are decision procedures for the classes $\mathsf{RL}_{\mathsf{POS}}$, $\mathsf{RL}_{\mathsf{NEG}}$, $\mathsf{RL}_{\mathsf{NP}}$. Note that RL-dual tableau is not a decision procedure for these classes, since if the rule $(;)$ is applicable, then it can be applied infinitely many times. Observe that the rule $(;)$ is the only rule of RL-dual tableau with that property. Therefore, in order to obtain decision procedures for the classes defined in terms of positive formulas we need to modify the rule for the composition or to restrict its applicability. According to the definition of the classes $\mathsf{RL}_{\mathcal{K}}$, for $\mathcal{K} \in \{\mathsf{POS}, \mathsf{NEG}, \mathsf{NP}\}$ given above, we observe that:

- If φ is positive, then none of its proof trees involves any application of the rule $(-;)$,
- If φ is negative, then none of its proof trees involves any application of the rule $(;)$,
- If φ is negative-positive, then in each of its proof trees the applications of the rule $(-;)$ precede the applications of the rule $(;)$.

These observations suggest that dual tableaux for these classes should be constructed from RL-dual tableau as follows:

- $\mathsf{RL}_{\mathsf{POS}}$-dual tableau consists of the axiomatic sets and decomposition rules of RL-dual tableau except the rule $(-;)$,
- $\mathsf{RL}_{\mathsf{NEG}}$-dual tableau consists of the axiomatic sets and decomposition rules of RL-dual tableau except the rule $(;)$,
- $\mathsf{RL}_{\mathsf{NP}}$-dual tableau consists of the axiomatic sets and decomposition rules of RL-dual tableau.

In order to obtain decision procedures for the classes listed above we modify the applicability of the rule $(;)$ as follows. The rule $(;)$ of the form:

$$\frac{x(R\,;S)y}{xRz, x(R\,;S)y \mid zSy, x(R\,;S)y}$$

may be applied to a finite set X of relational formulas if and only if the following conditions are satisfied:

- $x(R;S)y \in X$;
- z occurs in X;
- The rule (;) introduces a new formula, i.e., $xRz \notin X$ or $zSy \notin X$;
- No other rule is applicable to X.

The second condition implies that there are finitely many possibilities of choosing a variable z. Hence, together with the third condition, they prevent needless expansion of sets of formulas ad infinitum by repeated applications of the rule. The last condition guarantees that the rule (;) can be applied only after the applications of the rule $(-;)$, since all the classes in question do not contain positive–negative formulas.

As in RL-logic, the rules of all dual tableaux considered above guarantee that whenever xRy and $x{-}Ry$, for an atomic R, belong to a branch of a proof tree, then the branch is closed. Thus the closed branch property holds. Since all the rules are RL-correct, we have:

Proposition 2.12.3. *For every $\mathcal{K} \in \{\mathsf{POS}, \mathsf{NEG}, \mathsf{NP}\}$, all $\mathsf{RL}_{\mathcal{K}}$-rules are $\mathsf{RL}_{\mathcal{K}}$-correct.*

Moreover, since formulas of the class $\mathsf{RL}_{\mathsf{NEG}}$ do not contain positive compositions, the completeness of the reduct of RL-dual tableau for $\mathsf{RL}_{\mathsf{NEG}}$ can be proved as for the appropriate part of RL-dual tableau.

Theorem 2.12.2 (Soundness and Completeness of $\mathsf{RL}_{\mathsf{NEG}}$). *For every $\mathsf{RL}_{\mathsf{NEG}}$-formula φ, the following conditions are equivalent:*

1. *φ is $\mathsf{RL}_{\mathsf{NEG}}$-valid;*
2. *φ is $\mathsf{RL}_{\mathsf{NEG}}$-provable.*

By the above theorem and due to the forms of $\mathsf{RL}_{\mathsf{NEG}}$-rules, $\mathsf{RL}_{\mathsf{NEG}}$-dual tableau is a decision procedure for the class $\mathsf{RL}_{\mathsf{NEG}}$.

Now, we aim at proving completeness of $\mathsf{RL}_{\mathsf{POS}}$-dual tableau. Let b be a branch of an $\mathsf{RL}_{\mathsf{POS}}$-proof tree. Let $\mathbb{OV}(b)$ be the set of all variables that occur in formulas appearing in the nodes of b. The branch b is said to be complete if it is closed or it satisfies all the completion conditions of RL-dual tableau except Cpl(;) and, in addition, the following completion condition:

Cpl'(;) If $x(R;S)y \in b$, then for every $z \in \mathbb{OV}(b)$, either $xRz \in b$ or $zSy \in b$.

We define the branch structure $\mathcal{M}^b = (U^b, m^b)$ as follows:

- $U^b = \mathbb{OV}(b)$;
- $m^b(P) = \{(x, y) \in U^b \times U^b : xPy \notin b\}$, for every atomic term P;
- m^b extends to all the compound relational terms as in RL-models.

Clearly, $\mathcal{M}^b$ is an RL-model. Hence, the branch model property holds. Let v^b be a valuation in $\mathcal{M}^b$ such that $v^b(x) = x$ for every object variable x.

Proposition 2.12.4 (Satisfaction in Branch Model Property). *Let b be an open branch of an* $\mathsf{RL_{POS}}$*-proof tree. Then, for every* $\mathsf{RL_{POS}}$*-formula φ, if $\varphi \in b$, then $\mathcal{M}^b, v^b \not\models \varphi$.*

Proof. The proof is by induction on the complexity of terms. If P is a relational variable and $xPy \in b$, then the proposition holds by the definition of $\mathcal{M}^b$. If $x{-}Py \in b$, then $xPy \notin b$, since otherwise, due to the closed branch property, b would be closed. Thus, $(x, y) \in m^b(P)$. Suppose $\mathcal{M}^b, v^b \models x{-}Py$. Then $(x, y) \notin m^b(P)$, a contradiction.

Now, assume $x(R\,;S)y \in b$. Then, by the completion condition Cpl'(;), for every $z \in U^b$, either $xRz \in b$ or $zSy \in b$. By the induction hypothesis, for every $z \in U^b$, either $(x, z) \notin m^b(R)$ or $(z, y) \notin m^b(S)$. Suppose $\mathcal{M}^b, v^b \models x(R\,;S)y$. Then there exists $z \in U^b$ such that $(x, z) \in m^b(R)$ and $(z, y) \in m^b(S)$, a contradiction.

The proofs for terms of the form $R \cup S$, $R \cap S$, and their complements are similar as in the completeness proof of RL-dual tableau. □

We conclude:

Theorem 2.12.3 (Soundness and Completeness of $\mathsf{RL_{POS}}$). *For every* $\mathsf{RL_{POS}}$*-formula φ, the following conditions are equivalent:*

1. *φ is* $\mathsf{RL_{POS}}$*-valid;*
2. *φ is* $\mathsf{RL_{POS}}$*-provable.*

Note that none of the $\mathsf{RL_{POS}}$-rules introduces a new variable. Therefore, the rule (;) can be applied to a given set of formulas only finitely many times. Hence, $\mathsf{RL_{POS}}$-dual tableau is a decision procedure for the class $\mathsf{RL_{POS}}$.

Completeness of $\mathsf{RL_{NP}}$-dual tableau can be proved in a similar way as completeness of $\mathsf{RL_{POS}}$-dual tableau. Namely, $\mathsf{RL_{NP}}$-completion conditions are $\mathsf{RL_{POS}}$-completion condition and the following:

Cpl'(−;) If $x{-}(R\,;S)y \in b$, then for some $z \in \mathbb{OV}(b)$, both $xRz \in b$ and $zSy \in b$.

We have:

Theorem 2.12.4 (Soundness and Completeness of $\mathsf{RL_{NP}}$). *For every* $\mathsf{RL_{NP}}$*-formula φ, the following conditions are equivalent:*

1. *φ is* $\mathsf{RL_{NP}}$*-valid;*
2. *φ is* $\mathsf{RL_{NP}}$*-provable.*

Recall that $\mathsf{RL_{NP}}$-formulas do not contain positive-negative compositions, which means that each positive composition is in the scope of a negative composition. Although, the rule (−;) introduces a new variable, it is never applied after the application of the rule (;). Consequently, there are finitely many possibilities of choosing a variable in the application of the rule (;), and hence the rule (;) can be applied to a given set of formulas only finitely many times. Therefore, we conclude that $\mathsf{RL_{NP}}$-dual tableau is a decision procedure for the class $\mathsf{RL_{NP}}$.

Chapter 3
Theories of Point Relations and Relational Model Checking

3.1 Introduction

In this chapter we consider logics providing a means of relational reasoning in the theories which refer to individual objects of their domains. There are two relational formalisms for coping with the objects. A logic $\mathsf{RL}_{ax}(\mathbb{C})$ presented in Sect. 3.2 is a purely relational formalism, where objects are introduced through point relations which, in turn, are presented axiomatically with a well known set of axioms. The axioms say that a binary relation is a point relation whenever it is a non-empty right ideal relation with one-element domain. We recall that a binary relation R on a set U is right ideal whenever $R\,;1 = R$, where $1 = U \times U$. In other words such an R is of the form $X \times U$, for some $X \subseteq U$. We may think of right ideal relations as representing sets, they are sometimes referred to as vectors (see [SS93]). If the domain of a right ideal relation is a singleton set, the relation may be seen as a representation of an individual object. A logic $\mathsf{RL}_{df}(\mathbb{C})$ presented in Sect. 3.3 includes object constants in its language interpreted as singletons. Moreover, associated with each object constant c is a binary relation C, such that its meaning in every model is defined as a right ideal relation with the domain consisting of the single element being the meaning of c. The logic $\mathsf{RL}_{ax}(\mathbb{C})$ will be applied in Sect. 16.5 to the relational representation of some temporal logics. The logic $\mathsf{RL}_{df}(\mathbb{C})$ will be applied in Chap. 15 to the relational representation of the logic for order of magnitude reasoning. In Sects. 3.4 and 3.5 we present the methods of model checking and verification of satisfaction of a formula by some given objects in a finite model, respectively. The methods are based on the development of a relational logic which enables us to replace the problems of model checking and verification of satisfaction by the problems of verification of validity of some formulas of this logic. The logic is obtained from $\mathsf{RL}(1, 1')$-logic by an appropriate choice of object constants and relational constants in its language and by some specific postulates concerning its models. Then, a dual tableau for the logic is obtained from the $\mathsf{RL}(1, 1')$-dual tableau by adapting it to this language and by adding the rules which reflect these specific semantic postulates.

E. Orłowska and J. Golińska-Pilarek, *Dual Tableaux: Foundations, Methodology, Case Studies*, Trends in Logic 36, DOI 10.1007/978-94-007-0005-5_3,

3.2 Relational Logics with Point Relations Introduced with Axioms

The language of the logics considered in this section includes, apart from the relational constants 1 and $1'$, a family $\mathbb{C}$ of relational constants interpreted as point relations. The language of $\mathsf{RL}_{ax}(\mathbb{C})$-logic is a relational language as presented in Sect. 2.3 such that:

- $\mathbb{RC}_{\mathsf{RL}_{ax}(\mathbb{C})} = \{1', 1\} \cup \mathbb{C}$;
- The relational operations are $-$, $\cup$, $\cap$, $^{-1}$, and ;.

As in the case of RL-logic (see Sect. 2.5), $\mathsf{RL}_{ax}(\mathbb{C})$ represents, in fact, a whole family of logics which possibly differ in the object constants admitted in their languages.

An $\mathsf{RL}_{ax}(\mathbb{C})$-*model* is an $\mathsf{RL}(1, 1')$-model $\mathcal{M} = (U, m)$ such that for every $C \in \mathbb{C}$:

(1) $m(C) \neq \emptyset$,
(2) $m(C) = m(C) ; m(1)$,
(3) $m(C) ; m(C)^{-1} \subseteq m(1')$.

An $\mathsf{RL}_{ax}(\mathbb{C})$-model $\mathcal{M} = (U, m)$ is said to be *standard* whenever $m(1')$ is the identity on U.

The conditions (1), (2), and (3) say that relations from $\mathbb{C}$ are point relations. Condition (2) guarantees that every $C \in \mathbb{C}$ is a right ideal relation, and condition (3) together with the remaining axioms says that in the standard models the domains of relations from $\mathbb{C}$ are singleton sets.

$\mathsf{RL}_{ax}(\mathbb{C})$-dual tableau consists of decomposition rules and specific rules of $\mathsf{RL}(1, 1')$-system adjusted to the $\mathsf{RL}_{ax}(\mathbb{C})$-language and the specific rules that characterize relational constants from the set $\mathbb{C}$:

For all object symbols x and y and for every $C \in \mathbb{C}$,

$$(C1) \quad \frac{}{z{-}Ct} \qquad z, t \text{ are new object variables and } z \neq t$$

$$(C2) \quad \frac{xCy}{xCz, xCy} \qquad z \text{ is any object symbol}$$

$$(C3) \quad \frac{x1'y}{xCz, x1'y \mid yCz, x1'y} \qquad z \text{ is any object symbol}$$

$\mathsf{RL}_{ax}(\mathbb{C})$-axiomatic sets are those of $\mathsf{RL}(1, 1')$ adapted to the $\mathsf{RL}_{ax}(\mathbb{C})$-language.

It is easy to see that the specific rules for relational constants from $\mathbb{C}$ have the property of preservation of formulas built with atomic terms or their complements. Hence, the closed branch property holds.

As usual, an $\mathsf{RL}_{ax}(\mathbb{C})$-set is a finite set of $\mathsf{RL}_{ax}(\mathbb{C})$-formulas such that the first-order disjunction of its members is true in all $\mathsf{RL}_{ax}(\mathbb{C})$-models. Correctness of a rule is defined as in Sect. 2.4.

Proposition 3.2.1.

1. *The* $\mathsf{RL}_{ax}(\mathbb{C})$*-rules are* $\mathsf{RL}_{ax}(\mathbb{C})$*-correct;*
2. *The* $\mathsf{RL}_{ax}(\mathbb{C})$*-axiomatic sets are* $\mathsf{RL}_{ax}(\mathbb{C})$*-sets.*

Proof. Correctness of the rules $(C1)$, $(C2)$, and $(C3)$ follows directly from the semantic conditions (1), (2), and (3), respectively. By way of example, we prove correctness of the rule $(C1)$. Let X be a finite set of $\mathsf{RL}_{ax}(\mathbb{C})$-formulas and let z and t be object variables that do not occur in X and such that $z \neq t$. If X is an $\mathsf{RL}_{ax}(\mathbb{C})$-set, then so is $X \cup \{z{-}Ct\}$. Assume that $X \cup \{z{-}Ct\}$ is an $\mathsf{RL}_{ax}(\mathbb{C})$-set, that is for every $\mathsf{RL}_{ax}(\mathbb{C})$-model $\mathcal{M}$ and for every valuation v in $\mathcal{M}$, $(z,t) \notin m(C)$ or $\mathcal{M}, v \models \varphi$ for some formula $\varphi \in X$. By the assumption on variables z and t, for every $\mathsf{RL}_{ax}(\mathbb{C})$-model $\mathcal{M} = (U, m)$ and for every valuation v in $\mathcal{M}$, either $(a,b) \notin m(C)$ for all $a, b \in U$ or $\mathcal{M}, v \models \varphi$ for some formula $\varphi \in X$. Since in every $\mathsf{RL}_{ax}(\mathbb{C})$-model $m(C_i) \neq \emptyset$, for every $\mathsf{RL}_{ax}(\mathbb{C})$-model $\mathcal{M}$ and for every valuation v in $\mathcal{M}$ there exists $\varphi \in X$ such that $\mathcal{M}, v \models \varphi$. Therefore, X is an $\mathsf{RL}_{ax}(\mathbb{C})$-set. □

The notions of an $\mathsf{RL}_{ax}(\mathbb{C})$-proof tree, a closed branch of such a tree, a closed $\mathsf{RL}_{ax}(\mathbb{C})$-proof tree, and an $\mathsf{RL}_{ax}(\mathbb{C})$-proof of an $\mathsf{RL}_{ax}(\mathbb{C})$-formula are defined as in Sect. 2.4.

Following the general method of proving soundness presented in Sect. 2.6, Proposition 3.2.1 implies:

Proposition 3.2.2. *Let* φ *be an* $\mathsf{RL}_{ax}(\mathbb{C})$*-formula. If* φ *is* $\mathsf{RL}_{ax}(\mathbb{C})$*-provable, then it is* $\mathsf{RL}_{ax}(\mathbb{C})$*-valid.*

Corollary 3.2.1. *Let* φ *be an* $\mathsf{RL}_{ax}(\mathbb{C})$*-formula. If* φ *is* $\mathsf{RL}_{ax}(\mathbb{C})$*-provable, then it is true in all standard* $\mathsf{RL}_{ax}(\mathbb{C})$*-models.*

A branch b of an $\mathsf{RL}_{ax}(\mathbb{C})$-proof tree is said to be $\mathsf{RL}_{ax}(\mathbb{C})$-complete whenever it is closed or it satisfies $\mathsf{RL}_{ax}(\mathbb{C})$-completion conditions which consist of the completion conditions of $\mathsf{RL}(1, 1')$-dual tableau and the following:

For every $C \in \mathbb{C}$ and for all object symbols x and y,

Cpl($C1$) There exist object variables z and t such that $z{-}Ct \in b$, obtained by an application of the rule $(C1)$;

Cpl($C2$) If $xCy \in b$, then for every object symbol z, $xCz \in b$, obtained by an application of the rule $(C2)$;

Cpl($C3$) If $x1'y \in b$, then for every object symbol z, either $xCz \in b$ or $yCz \in b$, obtained by an application of the rule $(C3)$.

The notions of a complete $\mathsf{RL}_{ax}(\mathbb{C})$-proof tree and an open branch of an $\mathsf{RL}_{ax}(\mathbb{C})$-proof tree are defined as in RL-logic (see Sect. 2.5).

Let b be an open branch of an $\mathsf{RL}_{ax}(\mathbb{C})$-proof tree. We define a branch structure $\mathcal{M}^b = (U^b, m^b)$ with $U^b = \mathbb{OS}_{\mathsf{RL}_{ax}(\mathbb{C})}$ in a similar way as in $\mathsf{RL}(1, 1')$-logic, that is $m^b(R) = \{(x, y) \in U^b \times U^b : xRy \notin b\}$, for every $R \in \mathbb{RA}_{\mathsf{RL}_{ax}(\mathbb{C})}$.

Proposition 3.2.3 (Branch Model Property). *For every open branch b of an* $\mathsf{RL}_{ax}(\mathbb{C})$*-proof tree, the branch structure $\mathcal{M}^b$ is an* $\mathsf{RL}_{ax}(\mathbb{C})$*-model.*

Proof. We have to show that meaning function m^b satisfies conditions (1), (2), and (3) of $\mathsf{RL}_{ax}(\mathbb{C})$-models.

For (1), by the completion condition Cpl($C1$), there are $z, t \in U^b$ such that $z{-}Ct \in b$. Thus, $zCt \notin b$, since otherwise b would be closed. Hence, $(z,t) \in m^b(C)$.

For (2), note that $m^b(1) = U^b \times U^b$ implies $m^b(C) \subseteq m^b(C) ; m^b(1)$. Assume there exists $z \in U^b$ such that $(x,z) \in m^b(C)$ and $(z,y) \in m^b(1)$, that is $xCz \notin b$ and $z1y \notin b$. Suppose $(x,y) \notin m^b(C)$. Then $xCy \in b$. By the completion condition Cpl($C2$), for every $z \in U^b$, $xCz \in b$, a contradiction.

The proof of (3) is similar. □

Since the branch model $\mathcal{M}^b$ is defined in a standard way (see Sect. 2.6, p. 44), the satisfaction in branch model property can be proved as in $\mathsf{RL}(1, 1')$-logic (see Sects. 2.5 and 2.7). Hence, completeness of $\mathsf{RL}_{ax}(\mathbb{C})$-dual tableau follows.

Proposition 3.2.4. *Let φ be an* $\mathsf{RL}_{ax}(\mathbb{C})$*-formula. If φ is true in all standard* $\mathsf{RL}_{ax}(\mathbb{C})$*-models, then it is* $\mathsf{RL}_{ax}(\mathbb{C})$*-provable.*

The proof of this theorem can be obtained applying the general method described in Sect. 2.6 (p. 44), see also Propositions 2.5.6 and 2.7.8.

Corollary 3.2.2. *Let φ be an* $\mathsf{RL}_{ax}(\mathbb{C})$*-formula. If φ is* $\mathsf{RL}_{ax}(\mathbb{C})$*-valid, then it is* $\mathsf{RL}_{ax}(\mathbb{C})$*-provable.*

Due to Proposition 3.2.2 and 3.2.4, we get:

Theorem 3.2.1 (Soundness and Completeness of $\mathsf{RL}_{ax}(\mathbb{C})$). *Let φ be an* RL_{ax} $(\mathbb{C})$*-formula. Then the following conditions are equivalent:*

1. *φ is* $\mathsf{RL}_{ax}(\mathbb{C})$*-valid;*
2. *φ is true in all standard* $\mathsf{RL}_{ax}(\mathbb{C})$*-models;*
3. *φ is* $\mathsf{RL}_{ax}(\mathbb{C})$*-provable.*

3.3 Relational Logics with Point Relations Introduced with Definitions

The language of $\mathsf{RL}_{df}(\mathbb{C})$-logic is a relational language with relational constants which are explicitly defined in such a way that in the standard models they are right ideal relations with singleton domains. For that purpose we include in the language the object constants interpreted as elements of the models. Thus, the vocabulary of $\mathsf{RL}_{df}(\mathbb{C})$-language is a relational language as defined in Sect. 2.3 such that:

- $\mathbb{RC}_{\mathsf{RL}_{df}(\mathbb{C})} = \{1', 1\} \cup \mathbb{C}$;
- $\mathbb{OC}_{\mathsf{RL}_{df}(\mathbb{C})}$ is a set of object constants, which includes the set $\{c_C : C \in \mathbb{C}\}$ of object constants needed for definitions of point relations.

An $\mathsf{RL}_{df}(\mathbb{C})$-*model* is an $\mathsf{RL}(1, 1')$-model $\mathcal{M} = (U, m)$ such that for every $C \in \mathbb{C}$ the following hold:

- $m(c_C) \in U$;
- $m(C) = \{(x, y) \in U \times U : (x, m(c_C)) \in m(1')\}$.

The following proposition shows that for the formulas of $\mathsf{RL}_{ax}(\mathbb{C})$-language the notions of validity in the logics $\mathsf{RL}_{ax}(\mathbb{C})$ and $\mathsf{RL}_{df}(\mathbb{C})$ coincide.

Proposition 3.3.1. *For every* $\mathsf{RL}_{ax}(\mathbb{C})$*-formula* φ*, the following conditions are equivalent:*

1. φ *is* $\mathsf{RL}_{ax}(\mathbb{C})$*-valid;*
2. φ *is* $\mathsf{RL}_{df}(\mathbb{C})$*-valid.*

Proof. Let φ be an $\mathsf{RL}_{ax}(\mathbb{C})$-formula.

(1. $\rightarrow$ 2.) Assume that φ is $\mathsf{RL}_{ax}(\mathbb{C})$-valid. Suppose φ is not $\mathsf{RL}_{df}(\mathbb{C})$-valid, that is there exists an $\mathsf{RL}_{df}(\mathbb{C})$-model $\mathcal{M} = (U, m)$ such that $\mathcal{M} \not\models \varphi$. Consider a model $\mathcal{M}' = (U, m')$ with the same universe as $\mathcal{M}$ and such that $m'(R) = m(R)$, for every $R \in \mathbb{RA}_{\mathsf{RL}_{ax}(\mathbb{C})}$. Model $\mathcal{M}'$ is an $\mathsf{RL}_{ax}(\mathbb{C})$-model. Indeed, $m'(C)$ is a non-empty right ideal binary relation on U, hence conditions (1) and (2) from definition of $\mathsf{RL}_{ax}(\mathbb{C})$-models in Sect. 3.2 are satisfied. Moreover, by the definition of $m(C)$, if $(x, z) \in m'(C)$ and $(y, z) \in m'(C)$, then by symmetry and transitivity of 1', $(x, y) \in m'(1')$. Therefore, the condition (3) is satisfied. Clearly, models $\mathcal{M}$ and $\mathcal{M}'$ satisfy the same $\mathsf{RL}_{ax}(\mathbb{C})$-formulas. Thus, by the assumption, $\mathcal{M}' \not\models \varphi$, a contradiction.

(2. $\rightarrow$ 1.) Now, assume that φ is $\mathsf{RL}_{df}(\mathbb{C})$-valid. Suppose that φ is not $\mathsf{RL}_{ax}(\mathbb{C})$-valid, that is there exists an $\mathsf{RL}_{ax}(\mathbb{C})$-model $\mathcal{M} = (U, m)$ such that $\mathcal{M} \not\models \varphi$. Note that by condition (3) from definition of $\mathsf{RL}_{ax}(\mathbb{C})$-models, for every relational constant $C \in \mathbb{C}$, if x and y belong to the domain of $m(C)$, then $(x, y) \in m(1')$. We construct a model $\mathcal{M}' = (U, m')$ with the same universe as in $\mathcal{M}$ as follows: $m'(R) = m(R)$, for every $R \in \mathbb{RA}_{\mathsf{RL}_{ax}(\mathbb{C})}$, and $m'(c_C)$ is defined as an arbitrary element from the domain of $m(C)$. Now, by the above definition and condition (3), it follows that $m'(C) = \{x \in U : (x, m'(c_C)) \in m'(1')\} \times U$, hence model $\mathcal{M}'$ is an $\mathsf{RL}_{df}(\mathbb{C})$-model satisfying the same $\mathsf{RL}_{ax}(\mathbb{C})$-formulas as $\mathcal{M}$. Therefore, by the assumption, $\mathcal{M}' \not\models \varphi$, a contradiction. □

$\mathsf{RL}_{df}(\mathbb{C})$-dual tableau consists of the rules and axiomatic sets of $\mathsf{RL}(1, 1')$-dual tableau adjusted to $\mathsf{RL}_{df}(\mathbb{C})$-language and the specific rules that characterize relational constants from the set $\mathbb{C}$:

For every $C \in \mathbb{C}$ and for all object symbols x and y,

$$(CD1) \quad \frac{xCy}{x1'c_C, xCy}$$

$$(CD2) \quad \frac{x{-}Cy}{x{-}1'c_C, x{-}Cy}$$

The notions of $\mathsf{RL}_{df}(\mathbb{C})$-set, correctness of a rule, an $\mathsf{RL}_{df}(\mathbb{C})$-proof tree, a closed branch of an $\mathsf{RL}_{df}(\mathbb{C})$-proof tree, a closed $\mathsf{RL}_{df}(\mathbb{C})$-proof tree, and $\mathsf{RL}_{df}(\mathbb{C})$-provability are defined in a standard way as in Sect. 2.4.

Proposition 3.3.2.

1. *The* $\mathsf{RL}_{df}(\mathbb{C})$*-rules are* $\mathsf{RL}_{df}(\mathbb{C})$*-correct;*
2. *The* $\mathsf{RL}_{df}(\mathbb{C})$*-axiomatic sets are* $\mathsf{RL}_{df}(\mathbb{C})$*-sets.*

Proof. By way of example, we show correctness of the specific rules for relational constants from the set $\mathbb{C}$. It is easy to see that correctness of the rule $(CD1)$ follows from the property: for every $\mathsf{RL}_{df}(\mathbb{C})$-model $\mathcal{M} = (U, m)$, if $(x, m(c_C)) \in m(1')$, then for every $y \in U$, $(x, y) \in m(C)$. Correctness of the rule $(CD2)$ follows from the property: if $(x, m(c_C)) \notin m(1')$, then for every $y \in U$, $(x, y) \notin m(C)$. □

Due to the above proposition, we have:

Proposition 3.3.3. *Let* φ *be an* $\mathsf{RL}_{df}(\mathbb{C})$*-formula. If* φ *is* $\mathsf{RL}_{df}(\mathbb{C})$*-provable, then it is* $\mathsf{RL}_{df}(\mathbb{C})$*-valid.*

Corollary 3.3.1. *Let* φ *be an* $\mathsf{RL}_{df}(\mathbb{C})$*-formula. If* φ *is* $\mathsf{RL}_{df}(\mathbb{C})$*-provable, then it is true in all standard* $\mathsf{RL}_{df}(\mathbb{C})$*-models.*

To prove completeness of $\mathsf{RL}_{df}(\mathbb{C})$-dual tableau we define, as usual, the branch structure and we prove that branch model property and satisfaction in branch model property are satisfied. Note that in view of Fact 2.5.1 and since any application of the rules $(CD1)$ and $(CD2)$ to a set X of formulas preserves the formulas of X built with atomic terms or their complements, for every branch b of an $\mathsf{RL}_{df}(\mathbb{C})$-proof tree whenever an atomic formula xRy and the formula $x{-}Ry$ appear in b, then the branch is closed. As stated in Sects. 2.5 and 2.7, the same holds for the remaining rules.

A branch b of an $\mathsf{RL}_{df}(\mathbb{C})$-proof tree is said to be $\mathsf{RL}_{df}(\mathbb{C})$-complete whenever it is closed or it satisfies $\mathsf{RL}_{df}(\mathbb{C})$-completion conditions which consist of the completion conditions of $\mathsf{RL}(1, 1')$-system adjusted to $\mathsf{RL}_{df}(\mathbb{C})$-language and the following completion conditions determined by the specific rules for relational constants from $\mathbb{C}$.

For every $C \in \mathbb{C}$ and for all object symbols x and y,

Cpl($CD1$) If $xCy \in b$, then $x1'c_C \in b$, obtained by an application of the rule $(CD1)$;

Cpl($CD2$) If $x{-}Cy \in b$, then $x{-}1'c_C \in b$, obtained by an application of the rule $(CD2)$.

The notions of a complete $\mathsf{RL}_{df}(\mathbb{C})$-proof tree and an open branch of an $\mathsf{RL}_{df}(\mathbb{C})$-proof tree are defined as usual (see Sect. 2.5).

Let b be an open branch of an $\mathsf{RL}_{df}(\mathbb{C})$-proof tree. We define a branch structure $\mathcal{M}^b = (U^b, m^b)$ as follows:

- $U^b = \mathbb{OS}_{\mathsf{RL}_{df}(\mathbb{C})}$;
- $m^b(c) = c$, for every $c \in \mathbb{OC}_{\mathsf{RL}_{df}(\mathbb{C})}$;

- $m^b(R) = \{(x, y) \in U^b \times U^b : xRy \notin b\}$, for every $R \in \mathbb{RV}_{\mathsf{RL}_{df}(\mathbb{C})} \cup \{1, 1'\}$;
- $m^b(C) = \{x \in U^b : (x, c_C) \in m^b(1')\} \times U^b$, for every $C \in \mathbb{C}$;
- m extends to all the compound relational terms as in $\mathsf{RL}_{df}(\mathbb{C})$-models.

Directly from the above definition we get:

Proposition 3.3.4 (Branch Model Property). *Let b be an open branch of an $\mathsf{RL}_{df}(\mathbb{C})$-proof tree. Then $\mathcal{M}^b$ is an $\mathsf{RL}_{df}(\mathbb{C})$-model.*

Proposition 3.3.5 (Satisfaction in Branch Model Property). *For every open branch b of an $\mathsf{RL}_{df}(\mathbb{C})$-proof tree and for every $\mathsf{RL}_{df}(\mathbb{C})$-formula φ, if $\mathcal{M}^b, v^b \models \varphi$, then $\varphi \notin b$.*

Proof. By way of example, we prove that the proposition holds for formulas of the form xCy and $x{-}Cy$.

Assume $(x, y) \in m^b(C)$, that is $(x, c_C) \in m^b(1')$. Then $x1'c_C \notin b$. Suppose $xCy \in b$. By the completion condition Cpl($CD1$), $x1'c_C \in b$, a contradiction.

Now, assume $(x, y) \in m^b(-C)$, that is $(x, c_C) \notin m^b(1')$. Suppose $x{-}Cy \in b$. By the completion condition Cpl($CD2$), $x{-}1'c_C \in b$. Thus $x1'c_C \notin b$, hence $(x, c_C) \in m^b(1')$, contradiction. □

Finally, we get:

Proposition 3.3.6. *Let φ be an $\mathsf{RL}_{df}(\mathbb{C})$-formula. If φ is true in all standard $\mathsf{RL}_{df}(\mathbb{C})$-models, then it is $\mathsf{RL}_{df}(\mathbb{C})$-provable.*

The proof of the above proposition follows the general method described in Sect. 2.6 (p. 44), see also Propositions 2.5.6 and 2.7.8.

Corollary 3.3.2. *Let φ be an $\mathsf{RL}_{df}(\mathbb{C})$-formula. If φ is $\mathsf{RL}_{df}(\mathbb{C})$-valid, then it is $\mathsf{RL}_{df}(\mathbb{C})$-provable.*

Due to Proposition 3.3.3 and 3.3.6, we have:

Theorem 3.3.1 (Soundness and Completeness of $\mathsf{RL}_{df}(\mathbb{C})$). *Let φ be an $\mathsf{RL}_{df}(\mathbb{C})$-formula. Then the following conditions are equivalent:*

1. *φ is $\mathsf{RL}_{df}(\mathbb{C})$-valid;*
2. *φ is true in all standard $\mathsf{RL}_{df}(\mathbb{C})$-models;*
3. *φ is $\mathsf{RL}_{df}(\mathbb{C})$-provable.*

3.4 Model Checking in Relational Logics

The $\mathsf{RL}(1, 1')$-dual tableau can also be used for model checking in relational logics, besides verification of validity and entailment. Let $\mathcal{M} = (U, m)$ be a fixed standard $\mathsf{RL}(1, 1')$-model with a finite universe U and let $\varphi = xRy$ be an $\mathsf{RL}(1, 1')$-formula, where R is a relational term and x, y are any object symbols. For the simplicity of the presentation we assume that R does not contain 1 or 1′, although the presented method applies to all $\mathsf{RL}(1, 1')$-terms. In order to obtain a relational formalism appropriate for representing and solving the problem '$\mathcal{M} \models \varphi$?', we consider an instance $\mathsf{RL}_{\mathcal{M},\varphi}$ of the logic $\mathsf{RL}(1, 1')$. Its language provides a code of model $\mathcal{M}$ and formula φ, and in its models the syntactic elements of φ are interpreted as in the model $\mathcal{M}$.

The vocabulary of language of the logic $\mathsf{RL}_{\mathcal{M},\varphi}$ consists of symbols from the following pairwise disjoint sets:

- $\mathbb{OV}_{\mathsf{RL}_{\mathcal{M},\varphi}}$ – a countable infinite set of object variables;
- $\mathbb{OC}_{\mathsf{RL}_{\mathcal{M},\varphi}} = \mathbb{OC}^0_{\mathsf{RL}_{\mathcal{M},\varphi}} \cup \mathbb{OC}^1_{\mathsf{RL}_{\mathcal{M},\varphi}}$ – a finite set of object constants, where $\mathbb{OC}^0_{\mathsf{RL}_{\mathcal{M},\varphi}} = \{c_{\mathsf{a}} : \mathsf{a} \in U\}$ is such that if $\mathsf{a} \neq \mathsf{b}$, then $c_{\mathsf{a}} \neq c_{\mathsf{b}}$, $\mathbb{OC}^1_{\mathsf{RL}_{\mathcal{M},\varphi}} = \{c \in \mathbb{OC}_{\mathsf{RL}(1,1')} : c \text{ occurs in } \varphi\}$ and $\mathbb{OC}^1_{\mathsf{RL}_{\mathcal{M},\varphi}} \cap \mathbb{OC}^0_{\mathsf{RL}_{\mathcal{M},\varphi}} = \emptyset$;
- $\mathbb{RC}_{\mathsf{RL}_{\mathcal{M},\varphi}} = \mathbb{S} \cup \{1, 1'\}$ – the set of relational constants, where $\mathbb{S}$ is the set of all the atomic subterms of R;
- $\{-, \cup, \cap, ;, ^{-1}\}$ – the set of relational operations.

An $\mathsf{RL}_{\mathcal{M},\varphi}$-model is a pair $\mathcal{N} = (W, n)$, where:

- $W = U$;
- $n(c) = m(c)$, for every $c \in \mathbb{OC}^1_{\mathsf{RL}_{\mathcal{M},\varphi}}$;
- $n(c_{\mathsf{a}}) = \mathsf{a}$, for every $c_{\mathsf{a}} \in \mathbb{OC}^0_{\mathsf{RL}_{\mathcal{M},\varphi}}$;
- $n(S) = m(S)$, for every $S \in \mathbb{S}$;
- $n(1), n(1')$ are defined as in standard $\mathsf{RL}(1, 1')$-models;
- n extends to all the compound terms as in $\mathsf{RL}(1, 1')$-models.

A valuation in $\mathcal{N}$ is a function v: $\mathbb{OS}_{\mathsf{RL}_{\mathcal{M},\varphi}} \to W$ such that $v(c) = n(c)$, for every $c \in \mathbb{OC}_{\mathsf{RL}_{\mathcal{M},\varphi}}$. Observe that any valuation v in model $\mathcal{N}$ restricted to $\mathbb{OS}_{\mathsf{RL}_{\mathcal{M},\varphi}} \setminus \mathbb{OC}^0_{\mathsf{RL}_{\mathcal{M},\varphi}}$ is a valuation in model $\mathcal{M}$. Moreover, the above definition implies that for every $S \in \mathbb{S}$ and for all $x, y \in \mathbb{OS}_{\mathsf{RL}_{\mathcal{M},\varphi}}$, $\mathcal{N}, v \models xSy$ iff $\mathcal{M}, v \models xSy$. Therefore, it is easy to prove that $n(R) = m(R)$. Note also that the class of $\mathsf{RL}_{\mathcal{M},\varphi}$-models has exactly one element up to isomorphism. Therefore, $\mathsf{RL}_{\mathcal{M},\varphi}$-validity of xRy is equivalent to its truth in a single $\mathsf{RL}_{\mathcal{M},\varphi}$-model $\mathcal{N}$, that is the following holds:

Proposition 3.4.1. *The following statements are equivalent:*

1. $\mathcal{M} \models \varphi$;
2. φ *is* $\mathsf{RL}_{\mathcal{M},\varphi}$*-valid.*

Dual tableau for the logic $\mathsf{RL}_{\mathcal{M},\varphi}$ includes all the rules and axiomatic sets of $\mathsf{RL}(1, 1')$-system adapted to the language of $\mathsf{RL}_{\mathcal{M},\varphi}$ and, in addition, the rules and axiomatic sets listed below.

Specific Rules for Model Checking

For every $S \in \mathbb{S}$, for any object symbols x and y, and for any $\mathtt{a}, \mathtt{b} \in U$,

$$(-S\mathtt{ab}) \quad \frac{x{-}Sy}{x1'c_{\mathtt{a}}, x{-}Sy \mid y1'c_{\mathtt{b}}, x{-}Sy} \quad \text{for } (\mathtt{a},\mathtt{b}) \notin m(S)$$

$$(1') \quad \frac{}{x{-}1'c_1 \mid \ldots \mid x{-}1'c_n} \quad \text{whenever } \mathbb{OC}^0_{\mathsf{RL}_{\mathcal{M},\varphi}} = \{c_1, \ldots, c_n\}, \mathtt{n} \geq 1$$

$$(\mathtt{a} \neq \mathtt{b}) \quad \frac{}{c_{\mathtt{a}} 1' c_{\mathtt{b}}} \quad \text{for } \mathtt{a} \neq \mathtt{b}$$

Observe that these rules preserve the formulas built with atomic terms and their complements.

Specific Axiomatic Sets for Model Checking

$\{c_{\mathtt{a}} S c_{\mathtt{b}}\}$, for any $\mathtt{a}, \mathtt{b} \in U$ such that $(\mathtt{a},\mathtt{b}) \in m(S)$;
$\{c_{\mathtt{a}} {-} S c_{\mathtt{b}}\}$, for any $\mathtt{a}, \mathtt{b} \in U$ such that $(\mathtt{a},\mathtt{b}) \notin m(S)$.

The rules $(-S\mathtt{ab})$ and the axiomatic sets reflect the meaning of atomic subterms of R, while the rules $(1')$ and $(\mathtt{a} \neq \mathtt{b})$ guarantee that the universe of an $\mathsf{RL}_{\mathcal{M},\varphi}$-model is of the same cardinality as that of $\mathcal{M}$.

The notions of $\mathsf{RL}_{\mathcal{M},\varphi}$-set and $\mathsf{RL}_{\mathcal{M},\varphi}$-correctness of a rule are defined as in Sect. 2.4.

Proposition 3.4.2. *The rules* $(-S\mathtt{ab})$, $(1')$, *and* $(\mathtt{a} \neq \mathtt{b})$ *are* $\mathsf{RL}_{\mathcal{M},\varphi}$*-correct.*

Proof. For the rule $(-S\mathtt{ab})$, let $\mathtt{a}, \mathtt{b} \in U$ be such that $(\mathtt{a},\mathtt{b}) \notin m(S)$ and let X be any finite set of $\mathsf{RL}_{\mathcal{M},\varphi}$-formulas. Assume $X \cup \{x1'c_{\mathtt{a}}, x{-}Sy\}$ and $X \cup \{y1'c_{\mathtt{b}}, x{-}Sy\}$ are $\mathsf{RL}_{\mathcal{M},\varphi}$-sets. Suppose $X \cup \{x{-}Sy\}$ is not $\mathsf{RL}_{\mathcal{M},\varphi}$-set, that is for some valuation v in $\mathcal{N}$, $(v(x), v(y)) \in m(S)$. It follows from the assumption that the valuation v satisfies $v(x) = \mathtt{a}$ and $v(y) = \mathtt{b}$. Since $(\mathtt{a},\mathtt{b}) \notin m(S)$, $(v(x), v(y)) \notin m(S)$, a contradiction. On the other hand, if $X \cup \{x{-}Sy\}$ is an $\mathsf{RL}_{\mathcal{M},\varphi}$-set, then so are $X \cup \{x1'c_{\mathtt{a}}, x{-}Sy\}$ and $X \cup \{y1'c_{\mathtt{b}}, x{-}Sy\}$.

For the rule $(1')$, note that for every $x \in \mathbb{OS}_{\mathsf{RL}_{\mathcal{M},\varphi}}$ and for every valuation v in $\mathcal{N}$, there exists $c \in \mathbb{OC}^0_{\mathsf{RL}_{\mathcal{M},\varphi}}$ such that $v(x) = n(c)$, hence the rule $(1')$ is $\mathsf{RL}_{\mathcal{M},\varphi}$-correct.

The rule $(\mathtt{a} \neq \mathtt{b})$ is correct, since for all $\mathtt{a}, \mathtt{b} \in U$, if $\mathtt{a} \neq \mathtt{b}$, then $n(c_{\mathtt{a}}) \neq n(c_{\mathtt{b}})$. □

Validity of specific axiomatic sets follows directly from the definition of semantics of $\mathsf{RL}_{\mathcal{M},\varphi}$.

The notions of an $\mathsf{RL}_{\mathcal{M},\varphi}$-proof tree, a closed branch of such a tree, a closed $\mathsf{RL}_{\mathcal{M},\varphi}$-proof tree, and an $\mathsf{RL}_{\mathcal{M},\varphi}$-proof of an $\mathsf{RL}_{\mathcal{M},\varphi}$-formula are defined as in Sect. 2.4.

The notions of a complete branch of an $\mathsf{RL}_{\mathcal{M},\varphi}$-proof tree and a complete $\mathsf{RL}_{\mathcal{M},\varphi}$-proof tree are defined as in Sect. 2.5. The completion conditions are those of $\mathsf{RL}(1,1')$-dual tableau adapted to the language of $\mathsf{RL}_{\mathcal{M},\varphi}$ and the conditions listed below:

For all object symbols x and y, for every $S \in \mathbb{S}$, and for all $\mathtt{a},\mathtt{b} \in U$ such that $(\mathtt{a},\mathtt{b}) \notin m(S)$,

Cpl($-S\mathtt{ab}$) If $x{-}Sy \in b$, then either $x1'c_{\mathtt{a}} \in b$ or $y1'c_{\mathtt{b}} \in b$, obtained by an application of the rule ($-S\mathtt{ab}$);

Cpl($1'$) There exists $c \in \mathbb{OC}^0_{\mathsf{RL}_{\mathcal{M},\varphi}}$ such that $x{-}1'c \in b$, obtained by an application of the rule ($1'$);

Cpl($\mathtt{a} \neq \mathtt{b}$) $c_{\mathtt{a}}1'c_{\mathtt{b}} \in b$, for all $\mathtt{a},\mathtt{b} \in U$ such that $\mathtt{a} \neq \mathtt{b}$, obtained by an application of the rule ($\mathtt{a} \neq \mathtt{b}$).

An open branch of an $\mathsf{RL}_{\mathcal{M},\varphi}$-proof tree is defined as in Sect. 2.5. A branch structure $\mathcal{N}^b = (W^b, n^b)$ is defined as follows:

- $W^b = \mathbb{OC}^0_{\mathsf{RL}_{\mathcal{M},\varphi}}$;
- $n^b(c) = c_{\mathtt{a}}$, where $\mathtt{a} = n(c)$, for every $c \in \mathbb{OC}^1_{\mathsf{RL}_{\mathcal{M},\varphi}}$;
- $n^b(c) = c$, for every $c \in \mathbb{OC}^0_{\mathsf{RL}_{\mathcal{M},\varphi}}$;
- $n^b(S) = \begin{cases} \{(c_{\mathtt{a}}, c_{\mathtt{b}}) \in W^b \times W^b : c_{\mathtt{a}}Sc_{\mathtt{b}} \notin b\}, & \text{if } S \in \{1,1'\} \\ \{(c_{\mathtt{a}}, c_{\mathtt{b}}) \in W^b \times W^b : (\mathtt{a},\mathtt{b}) \in m(S)\}, & \text{if } S \in \mathbb{S}; \end{cases}$
- n^b extends to all the compound terms as in $\mathsf{RL}(1,1')$-models.

Similarly as in $\mathsf{RL}(1,1')$-logic it is easy to prove that $n^b(1')$ and $n^b(1)$ are an equivalence relation and a universal relation, respectively. Therefore, $\mathcal{N}^b$ is an $\mathsf{RL}(1,1')$-model. Note that $\mathcal{N}^b$ is not necessarily an $\mathsf{RL}_{\mathcal{M},\varphi}$-model, since $n^b(1')$ may not be the identity.

Let $v^b \colon \mathbb{OS}_{\mathsf{RL}_{\mathcal{M},\varphi}} \to W^b$ be a valuation in $\mathcal{N}^b$ such that:

$v^b(c) = n^b(c)$, for $c \in \mathbb{OC}_{\mathsf{RL}_{\mathcal{M},\varphi}}$;
$v^b(x) = c_{\mathtt{a}}$, where $c_{\mathtt{a}} \in \mathbb{OC}_{\mathsf{RL}_{\mathcal{M},\varphi}}$ is such that $x1'c_{\mathtt{a}} \notin b$, for $x \in \mathbb{OV}_{\mathsf{RL}_{\mathcal{M},\varphi}}$.

The valuation v^b is well defined, that is for every $x \in \mathbb{OV}_{\mathsf{RL}_{\mathcal{M},\varphi}}$, there exists exactly one $c \in \mathbb{OC}^0_{\mathsf{RL}_{\mathcal{M},\varphi}}$ such that $x1'c \notin b$. Indeed, by the completion condition Cpl($1'$), there exists $c \in \mathbb{OC}^0_{\mathsf{RL}_{\mathcal{M},\varphi}}$ such that $x{-}1'c \in b$. So $x1'c \notin b$. Suppose there exist two different $c_{\mathtt{a}}, c_{\mathtt{b}} \in \mathbb{OC}^0_{\mathsf{RL}_{\mathcal{M},\varphi}}$ such that $x1'c_{\mathtt{a}} \notin b$ and $x1'c_{\mathtt{b}} \notin b$. By the completion condition Cpl($\mathtt{a} \neq \mathtt{b}$), $c_{\mathtt{a}}1'c_{\mathtt{b}} \in b$. Then, by the completion conditions Cpl($1'1$) and Cpl($1'2$), $x1'c_{\mathtt{a}} \in b$ or $x1'c_{\mathtt{b}} \in b$, a contradiction.

Proposition 3.4.3 (Satisfaction in Branch Model Property). *For every open branch b of an $\mathsf{RL}_{\mathcal{M},\varphi}$-proof tree and for every $\mathsf{RL}_{\mathcal{M},\varphi}$-formula ψ, if $\mathcal{N}^b, v^b \models \psi$, then $\psi \notin b$.*

Proof. First, we need to show that the proposition holds for formulas xSy and $x{-}Sy$, where $S \in \mathbb{S}$.

Let $\psi = xSy$, for some $S \in \mathbb{S}$. Assume $\mathcal{N}^b, v^b \models xSy$. Let $\mathsf{a}, \mathsf{b} \in U$ be such that $v^b(x) = c_{\mathsf{a}}$ and $v^b(y) = c_{\mathsf{b}}$, that is $x1'c_{\mathsf{a}} \notin b$ and $y1'c_{\mathsf{b}} \notin b$. Since $\mathcal{N}^b, v^b \models xSy$, $(c_{\mathsf{a}}, c_{\mathsf{b}}) \in n^b(S)$, hence $(\mathsf{a}, \mathsf{b}) \in m(S)$. Thus $c_{\mathsf{a}}Sc_{c_{\mathsf{b}}} \notin b$, otherwise b would be closed. Suppose $xSy \in b$. By the completion conditions for the rules $(1'1)$ and $(1'2)$ presented in Sect. 2.7, at least one of the following holds: $x1'c_{\mathsf{a}} \in b$ or $y1'c_{\mathsf{b}} \in b$ or $c_{\mathsf{a}}Sc_{c_{\mathsf{b}}} \in b$, a contradiction.

Let $\psi = x{-}Sy$, for some $S \in \mathbb{S}$. Assume $\mathcal{N}^b, v^b \models x{-}Sy$. Let $\mathsf{a}, \mathsf{b} \in U$ be such that $v^b(x) = c_{\mathsf{a}}$ and $v^b(y) = c_{\mathsf{b}}$, that is $x1'c_{\mathsf{a}} \notin b$ and $y1'c_{\mathsf{b}} \notin b$. Since $\mathcal{N}^b, v^b \models x{-}Sy$, $(c_{\mathsf{a}}, c_{\mathsf{b}}) \notin n^b(S)$. Therefore $(\mathsf{a}, \mathsf{b}) \notin m(S)$. Suppose $x{-}Sy \in b$. Then by the completion condition Cpl($-S\mathsf{ab}$), either $x1'c_{\mathsf{a}} \in b$ or $y1'c_{\mathsf{b}} \in b$, a contradiction.

The rest of the proof is similar to the proofs of the analogous propositions for logics RL and $\mathsf{RL}(1, 1')$ (see Proposition 2.5.5 and 2.7.6). □

Since $n^b(1')$ is an equivalence relation on W^b, we may define the quotient structure $\mathcal{N}^b_q = (W^b_q, n^b_q)$:

- $W^b_q = \{\|c\| : c \in W^b\}$, where $\|c\|$ is the equivalence class of $n^b(1')$ determined by c;
- $n^b_q(c) = \|n^b(c)\|$, for every $c \in \mathbb{OC}_{\mathsf{RL}_{\mathcal{M},\varphi}}$;
- $n^b_q(S) = \{(\|c_{\mathsf{a}}\|, \|c_{\mathsf{b}}\|) \in W^b_q \times W^b_q : (c_{\mathsf{a}}, c_{\mathsf{b}}) \in n^b(S)\}$, for every $S \in \mathbb{RC}_{\mathsf{RL}_{\mathcal{M},\varphi}}$;
- n^b_q extends to all the compound terms as in $\mathsf{RL}(1, 1')$-models.

Proposition 3.4.4 (Branch Model Property). *Let b be an open branch of an $\mathsf{RL}_{\mathcal{M},\varphi}$-proof tree. Then models $\mathcal{N}^b_q$ and $\mathcal{N}$ are isomorphic.*

Proof. Since constants $c \in \mathbb{OC}^0_{\mathsf{RL}_{\mathcal{M},\varphi}}$ are uniquely assigned to the elements of model $\mathcal{M}$, $card(W^b_q) = card(W)$. Let $f: W \to W^b$ be a function defined as $f(\mathsf{a}) \stackrel{\text{df}}{=} \|c_{\mathsf{a}}\|$, for every $\mathsf{a} \in W$. By the definition of model $\mathcal{N}^b_q$, the function f is an isomorphism between $\mathcal{N}^b_q$ and $\mathcal{N}$. □

Let v^b_q be a valuation in $\mathcal{N}^b_q$ defined as $v^b_q(x) = \|v^b(x)\|$, for every $x \in \mathbb{OS}_{\mathsf{RL}_{\mathcal{M},\varphi}}$. As in RL-logic, it is easy to show that the sets of formulas satisfied in $\mathcal{N}^b$ and $\mathcal{N}^b_q$ by valuations v^b and v^b_q, respectively, coincide. Moreover, since $\mathcal{N}^b_q$ and $\mathcal{N}$ are isomorphic, they satisfy exactly the same formulas. Now the completeness of $\mathsf{RL}_{\mathcal{M},\varphi}$ can be proved following Theorems 2.5.1 and 2.7.2.

Theorem 3.4.1 (Soundness and Completeness of $\mathsf{RL}_{\mathcal{M},\varphi}$). *For every $\mathsf{RL}_{\mathcal{M},\varphi}$-formula ψ, the following conditions are equivalent:*

1. *ψ is $\mathsf{RL}_{\mathcal{M},\varphi}$-valid;*
2. *ψ is $\mathsf{RL}_{\mathcal{M},\varphi}$-provable.*

Due to the above theorem and Proposition 3.4.1, we have:

Theorem 3.4.2 (Model Checking in $\mathsf{RL}(1,1')$). *For every* $\mathsf{RL}(1,1')$*-formula* φ *and for every finite standard* $\mathsf{RL}(1,1')$*-model* $\mathcal{M}$*, the following statements are equivalent:*

1. $\mathcal{M} \models \varphi$;
2. φ *is* $\mathsf{RL}_{\mathcal{M},\varphi}$*-provable.*

Example. Consider $\mathsf{RL}(1,1')$-formula $\varphi = xR\,;Py$ and the standard $\mathsf{RL}(1,1')$-model $\mathcal{M} = (U,m)$ defined as follows:

- $U = \{\mathtt{a},\mathtt{b}\}$;
- $m(1) = U \times U$;
- $m(P) = \{(\mathtt{a},\mathtt{a}),(\mathtt{a},\mathtt{b})\}$;
- $m(R) = \{(\mathtt{a},\mathtt{a}),(\mathtt{b},\mathtt{a})\}$;
- $m(1') = \{(\mathtt{a},\mathtt{a}),(\mathtt{b},\mathtt{b})\}$;
- m extends to all the compound terms as in $\mathsf{RL}(1,1')$-models.

We apply the method presented above to checking whether φ is true in $\mathcal{M}$. The vocabulary of $\mathsf{RL}_{\mathcal{M},\varphi}$-language adequate for expressing this problem consists of the following sets of symbols:

- $\mathbb{OV}_{\mathsf{RL}_{\mathcal{M},\varphi}}$ – a countable infinite set of object variables;
- $\mathbb{OC}_{\mathsf{RL}_{\mathcal{M},\varphi}} = \mathbb{OC}^0_{\mathsf{RL}_{\mathcal{M},\varphi}} = \{c_{\mathtt{a}}, c_{\mathtt{b}}\}$ – the set of object constants;
- $\mathbb{RC}_{\mathsf{RL}_{\mathcal{M},\varphi}} = \{R,P,1,1'\}$ – the set of relational constants;
- $\{-,\cup,\cap,;,^{-1}\}$ – the set of relational operations.

An $\mathsf{RL}_{\mathcal{M},\varphi}$-model is the structure $\mathcal{N} = (U,n)$ defined as model $\mathcal{M}$ with the following additional conditions: $n(c_{\mathtt{a}}) = \mathtt{a}$, $n(c_{\mathtt{b}}) = \mathtt{b}$.

The specific rules of $\mathsf{RL}_{\mathcal{M},\varphi}$-dual tableau are: $(-R\mathtt{ab})$, $(-R\mathtt{bb})$, $(-P\mathtt{ba})$, $(-P\mathtt{bb})$, $(\mathtt{a} \neq \mathtt{b})$, $(\mathtt{b} \neq \mathtt{a})$, and the rule $(1')$ of the following form:

$$(1') \quad \frac{}{x{-}1'c_{\mathtt{a}} \mid x{-}1'c_{\mathtt{b}}}$$

Specific $\mathsf{RL}_{\mathcal{M},\varphi}$-axiomatic sets are those including either of the following sets: $\{c_{\mathtt{a}}Rc_{\mathtt{a}}\}$, $\{c_{\mathtt{b}}Rc_{\mathtt{a}}\}$, $\{c_{\mathtt{b}}{-}Rc_{\mathtt{b}}\}$, $\{c_{\mathtt{a}}{-}Rc_{\mathtt{b}}\}$, $\{c_{\mathtt{a}}Pc_{\mathtt{a}}\}$, $\{c_{\mathtt{a}}Pc_{\mathtt{b}}\}$, $\{c_{\mathtt{b}}{-}Pc_{\mathtt{b}}\}$ or $\{c_{\mathtt{b}}{-}Pc_{\mathtt{a}}\}$.

By Theorem 3.4.2, truth of φ in $\mathcal{M}$ is equivalent to $\mathsf{RL}_{\mathcal{M},\varphi}$-provability of φ. Figure 3.1 presents an $\mathsf{RL}_{\mathcal{M},\varphi}$-proof of φ.

3.5 Verification of Satisfaction in Relational Logics

The logic $\mathsf{RL}(1,1')$ can also be used for verification of satisfaction of a formula in a fixed finite model. Let $\varphi = xRy$ be an $\mathsf{RL}(1,1')$-formula, where R is a relational term and x,y are any object symbols, let $\mathcal{M} = (U,m)$ be a fixed standard $\mathsf{RL}(1,1')$-model with a finite universe U, and let v be a valuation in $\mathcal{M}$ such that

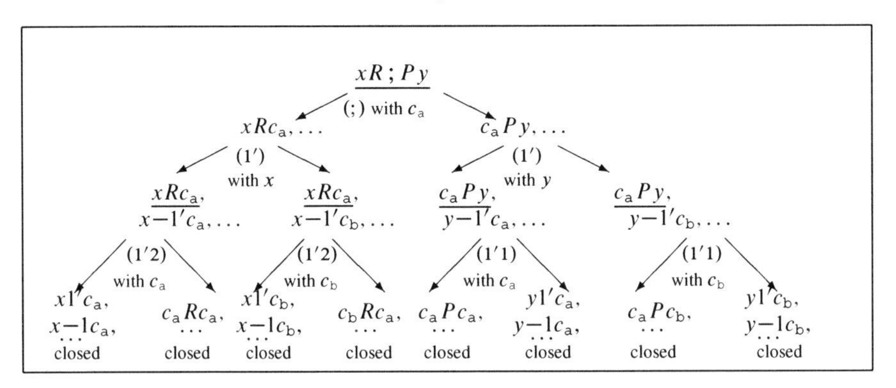

Fig. 3.1 An $\mathsf{RL}_{\mathcal{M},\varphi}$-proof showing that $xR\,;Py$ is true in the model $\mathcal{M}$

$v(x) = \mathtt{a}$ and $v(y) = \mathtt{b}$, for some elements $\mathtt{a}, \mathtt{b} \in U$. Recall that for every object constant c, for every relational model $\mathcal{M} = (U, m)$, and for every valuation v in $\mathcal{M}$, $v(c) = m(c)$ by definition. In particular, if both x and y are object constants, then there is exactly one pair $(\mathtt{a}, \mathtt{b})$ of elements of U such that $v(x) = \mathtt{a}$ and $v(y) = \mathtt{b}$.

The relational formalism appropriate for solving the problem '$(\mathtt{a}, \mathtt{b}) \in m(R)$?' is the logic $\mathsf{RL}_{\mathcal{M},\varphi}$ defined in the previous section with $\varphi = xRy$. Since for every $\mathsf{RL}_{\mathcal{M},\varphi}$-model $\mathcal{N} = (U, n)$, for every valuation v in $\mathcal{N}$, and for every $\mathtt{a} \in U$, $v(c_{\mathtt{a}}) = \mathtt{a}$, the following holds:

Proposition 3.5.1. *The following statements are equivalent:*

1. $(\mathtt{a}, \mathtt{b}) \in m(R)$;
2. $c_{\mathtt{a}} R c_{\mathtt{b}}$ *is* $\mathsf{RL}_{\mathcal{M},\varphi}$*-valid.*

Due to the above proposition and Theorem 3.4.1, we get:

Theorem 3.5.1 (Satisfaction in $\mathsf{RL}(1, 1')$-models). *For every relational term R of* $\mathsf{RL}(1, 1')$*-language, for every finite standard* $\mathsf{RL}(1, 1')$*-model $\mathcal{M} = (U, m)$, and for all $\mathtt{a}, \mathtt{b} \in U$, the following statements are equivalent:*

1. $(\mathtt{a}, \mathtt{b}) \in m(R)$;
2. $c_{\mathtt{a}} R c_{\mathtt{b}}$ *is* $\mathsf{RL}_{\mathcal{M},\varphi}$*-provable.*

Example. Consider $\mathsf{RL}(1, 1')$-formula $\varphi = x(P\,;-(R\,;P))y$ and the standard $\mathsf{RL}(1, 1')$-model $\mathcal{M} = (U, m)$ such that:

- $U = \{\mathtt{a}, \mathtt{b}, \mathtt{c}\}$;
- $m(1) = U \times U$;
- $m(P) = \{(\mathtt{a}, \mathtt{a}), (\mathtt{a}, \mathtt{b}), (\mathtt{a}, \mathtt{c})\}$;
- $m(R) = \{(\mathtt{b}, \mathtt{a}), (\mathtt{c}, \mathtt{c})\}$;
- $m(1') = \{(\mathtt{a}, \mathtt{a}), (\mathtt{b}, \mathtt{b}), (\mathtt{c}, \mathtt{c})\}$;
- m extends to all the compound terms as in $\mathsf{RL}(1, 1')$-models.

Let v be a valuation such that $v(x) = \mathtt{a}$ and $v(y) = \mathtt{b}$. By Theorem 3.5.1 the satisfaction problem 'is the formula $\varphi = x(P\,;-(R\,;P))y$ satisfied in $\mathcal{M}$ by v?' is equivalent to $\mathsf{RL}_{\mathcal{M},\varphi}$-provability of $c_{\mathtt{a}}(P\,;-(R\,;P))c_{\mathtt{b}}$.

$\mathsf{RL}_{\mathcal{M},\varphi}$-dual tableau contains the rules and axiomatic sets of $\mathsf{RL}(1,1')$-proof system adjusted to $\mathsf{RL}_{\mathcal{M},\varphi}$-language, the rules $(-R\mathtt{aa})$, $(-R\mathtt{ab})$, $(-R\mathtt{ac})$, $(-R\mathtt{bb})$, $(-R\mathtt{bc})$, $(-R\mathtt{ca})$, $(-R\mathtt{cb})$, $(-P\mathtt{ba})$, $(-P\mathtt{bb})$, $(-P\mathtt{bc})$, $(-P\mathtt{ca})$, $(-P\mathtt{cb})$, $(-P\mathtt{cc})$, $(\mathtt{a} \neq \mathtt{b})$, $(\mathtt{a} \neq \mathtt{c})$, $(\mathtt{b} \neq \mathtt{c})$, and the rule $(1')$ of the following form:

$$(1') \quad \frac{}{x-1'c_{\mathtt{a}}\,|\,x-1'c_{\mathtt{b}}\,|\,x-1'c_{\mathtt{c}}}$$

The axiomatic sets specific for $\mathsf{RL}_{\mathcal{M},\varphi}$ are those including one of the following sets: $\{c_{\mathtt{b}}Rc_{\mathtt{a}}\}$, $\{c_{\mathtt{c}}Rc_{\mathtt{c}}\}$, $\{c_{\mathtt{a}}Pc_{\mathtt{a}}\}$, $\{c_{\mathtt{a}}Pc_{\mathtt{b}}\}$, $\{c_{\mathtt{a}}Pc_{\mathtt{c}}\}$, $\{c_{\mathtt{a}}-Rc_{\mathtt{a}}\}$, $\{c_{\mathtt{a}}-Rc_{\mathtt{b}}\}$, $\{c_{\mathtt{a}}-Rc_{\mathtt{c}}\}$, $\{c_{\mathtt{b}}-Rc_{\mathtt{b}}\}$, $\{c_{\mathtt{b}}-Rc_{\mathtt{c}}\}$, $\{c_{\mathtt{c}}-Rc_{\mathtt{a}}\}$, $\{c_{\mathtt{c}}-Rc_{\mathtt{b}}\}$, $\{c_{\mathtt{b}}-Pc_{\mathtt{a}}\}$, $\{c_{\mathtt{b}}-Pc_{\mathtt{b}}\}$, $\{c_{\mathtt{b}}-Pc_{\mathtt{c}}\}$, $\{c_{\mathtt{c}}-Pc_{\mathtt{a}}\}$, $\{c_{\mathtt{c}}-Pc_{\mathtt{b}}\}$ or $\{c_{\mathtt{c}}-Pc_{\mathtt{c}}\}$.

Figure 3.2 presents an $\mathsf{RL}_{\mathcal{M},\varphi}$-proof of $c_{\mathtt{a}}(P\,;-(R\,;P))c_{\mathtt{b}}$ that shows satisfaction of $x(P\,;-(R\,;P))y$ in the model $\mathcal{M}$ by the valuation v defined above.

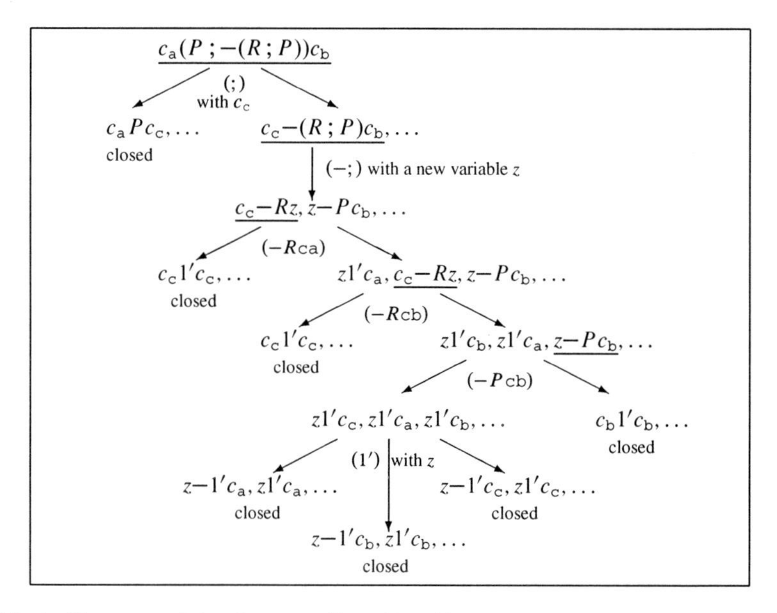

Fig. 3.2 An $\mathsf{RL}_{\mathcal{M},\varphi}$-proof showing that $x(P\,;-(R\,;P))y$ is satisfied in the model $\mathcal{M}$ by the pair $(\mathtt{a},\mathtt{b})$

Part II
Reasoning in Logics of Non-classical Algebras of Relations

Chapter 4
Dual Tableaux for Peirce Algebras

4.1 Introduction

Peirce algebras are two-sorted algebras of relations and sets. A Peirce algebra consists of a Boolean algebra where sets interact with each other, a relation algebra where relations interact with each other, and two operators that relate these two structures: a set-forming operator acting on a relation and a set, and a relation-forming operator acting on a set. Peirce algebras were first introduced in a modern form in [Bri88] and studied in [BBS94, dR99, SOH04, Hir07]. However, the history of these algebras can be traced back to the work of Charles Sanders Peirce who gave the first algebraic treatment of the algebra of relations interacting with sets. Whereas De Morgan was primarily interested in the formalization of statements within the paradigm of binary relations, Peirce considered the expressions obtained as a product of a relation and a set. Brink [Bri81] named this operation the Peirce product and axiomatized it within the framework of Boolean modules. The relation forming operator of Peirce algebras may be viewed as a cylindrification. Peirce algebras provide tools for modelling program constructors in programming languages, for natural language analysis [Böt92a, Böt92b] and for knowledge representation. In particular, they provide semantics for terminological languages [Sch91, WS92].

In this chapter we present a dual tableau for Peirce logic based on Peirce algebras and we prove its completeness. The logic has the expressions of the two sorts representing Boolean and relational elements, respectively, of Peirce algebras. It follows that in the dual tableau there are three types of decomposition rules: those applicable to purely relational expressions, those to purely Boolean expressions, and those to mixed expressions built with the Peirce product or the cylindrification. Next, we discuss a representation of terminological languages with Peirce logic. Some other logic of Peirce algebras can be found in [dR95].

E. Orłowska and J. Golińska-Pilarek, *Dual Tableaux: Foundations, Methodology, Case Studies*, Trends in Logic 36, DOI 10.1007/978-94-007-0005-5_4,

4.2 Peirce Algebras

A *full Peirce algebra* on a non-empty set U is a two-sorted structure of the form $(\mathfrak{B}(U), \mathfrak{R}(U), :, ^c)$, where:

- $\mathfrak{B}(U) = (\mathcal{P}(U), \cup, \cap, -, \emptyset, U)$ is a Boolean algebra of all subsets of U;
- $\mathfrak{R}(U) = (\mathcal{P}(U \times U), \cup, \cap, -, \emptyset, U \times U, ;, ^{\smile}, Id)$ is a full algebra of binary relations on U;
- $:$ is a mapping $\mathcal{P}(U \times U) \times \mathcal{P}(U) \rightarrow \mathcal{P}(U)$ referred to as a Peirce product and defined for any $R \subseteq U \times U$ and for any $A \subseteq U$ as $R : A \stackrel{\text{df}}{=} \{x \in U : \exists y \in U, xRy \text{ and } y \in A\}$;
- $^c : \mathcal{P}(U) \rightarrow \mathcal{P}(U \times U)$ is a right cylindrification operation defined for any $A \subseteq U$ as $A^c \stackrel{\text{df}}{=} \{(x, y) \in U \times U : x \in A\}$.

In terms of the operations of Peirce algebras some other useful operations can be defined:

- Left cylindrification ${}^cA \stackrel{\text{df}}{=} (A^c)^{-1}$,
- Product of sets $A \times B \stackrel{\text{df}}{=} (A^c) \cap ({}^cB) = \{(x, y) \in U \times U : x \in A \text{ and } y \in B\}$,
- Domain of a relation $Dom(R) \stackrel{\text{df}}{=} R : 1$,
- Range of a relation $Rng(R) \stackrel{\text{df}}{=} R^{-1} : 1$.

In an abstract setting, a Peirce algebra is a two-sorted structure $(\mathfrak{B}, \mathfrak{R}, :, ^c)$ such that:

- $\mathfrak{B} = (B, -, +, \cdot, 1, 0)$ is a Boolean algebra;
- $\mathfrak{R} = (W, -, +, \cdot, ;, ^{\smile}, 1, 0, 1')$ is a relation algebra;
- $:$ is a function $W \times B \rightarrow B$, ^{c} is a function $B \rightarrow W$, and for all $a, b \in B$ and for all $r, s \in W$ the following conditions are satisfied:

(P1) $r : (a + b) = r : a + r : b$,
(P2) $(r + s) : a = r : a + s : a$,
(P3) $r : (s : a) = (r ; s) : a$,
(P4) $1' : a = a$,
(P5) $0 : a = 0$,
(P6) $r^{\smile} : -(r : a) \leq -a$,
(P7) $a^c : 1 = a$,
(P8) $(r : 1)^c = r ; 1$.

Every full Peirce algebra is a Peirce algebra. The following theorem presents some arithmetic laws of Peirce algebras.

Theorem 4.2.1. *Let $(\mathfrak{B}, \mathfrak{R}, :, ^c)$ be a Peirce algebra. Then for all $a, b \in B$ and for all $r, s \in W$:*

1. *a^c is a right ideal element,*
2. $0^c = 0$, $1^c = 1$,
3. *$a \leq b$ iff $a^c \leq b^c$,*

4. $(-a)^c = -(a^c)$,
5. $(a+b)^c = a^c + b^c$,
6. $(a \cdot b)^c = a^c \cdot b^c$,
7. $(r : a)^c = r ; a^c$,
8. $(r ; a)^c : 1 = r : a$,
9. $r : a = b$ *iff* $r ; a^c = b^c$,
10. $r : 1 = 0$ *iff* $r = 0$,
11. $(r \cdot (a^c)^{\smile}) : b = r : (a \cdot b)$.

A verification of these laws can be found in [BBS94].

Peirce algebras are expressively equivalent to relation algebras. More precisely, it is shown in [BBS94] that the Boolean elements of a Peirce algebra are up to isomorphism the right ideal elements of its relation algebra part.

Theorem 4.2.2. *Let* $(\mathfrak{B}, \mathfrak{R}, :, ^c)$ *be a Peirce algebra. Then* $\mathfrak{B}$ *is isomorphic to* $\{r \in \mathfrak{R} : r = r ; 1\}$.

A Peirce algebra is representable whenever its relation algebra part is representable.

4.3 Peirce Logic

The language of the Peirce logic, PL, is built with the symbols from the following pairwise disjoint sets:

- $\mathbb{OV}_{\mathsf{PL}}$ – a countable infinite set of object variables;
- $\mathbb{BV}_{\mathsf{PL}}$ – a countable infinite set of Boolean variables;
- $\mathbb{RV}_{\mathsf{PL}}$ – a countable set of relational variables;
- $\mathbb{BC}_{\mathsf{PL}} = \{1_B\}$ – the set consisting of the Boolean unit element;
- $\mathbb{RC}_{\mathsf{PL}} = \{1_R, 1'\}$ – the set of relational constants;
- $\{-, \cup, \cap, ;, ^{-1}, :, ^c\}$ – the set of operations.

The sets $\mathbb{BA}_{\mathsf{PL}} = \mathbb{BV}_{\mathsf{PL}} \cup \{1_B\}$ and $\mathbb{RA}_{\mathsf{PL}} = \mathbb{RV}_{\mathsf{PL}} \cup \{1_R, 1'\}$ are called the sets of atomic Boolean terms and atomic relational terms, respectively. The set $\mathbb{BT}_{\mathsf{PL}}$ of Boolean terms and the set $\mathbb{RT}_{\mathsf{PL}}$ of relational terms of the logic PL are defined inductively as the smallest sets satisfying the following conditions:

- $\mathbb{BA}_{\mathsf{PL}} \subseteq \mathbb{BT}_{\mathsf{PL}}$;
- $\mathbb{RA}_{\mathsf{PL}} \subseteq \mathbb{RT}_{\mathsf{PL}}$;
- If $A, B \in \mathbb{BT}_{\mathsf{PL}}$ and $R \in \mathbb{RT}_{\mathsf{PL}}$, then $-A$, $A \cup B$, $A \cap B$, $R : A \in \mathbb{BT}_{\mathsf{PL}}$;
- If $A \in \mathbb{BT}_{\mathsf{PL}}$ and $R, S \in \mathbb{RT}_{\mathsf{PL}}$, then $-R$, $R \cup S$, $R \cap S$, $R ; S$, R^{-1}, $A^c \in \mathbb{RT}_{\mathsf{PL}}$.

Formulas of the Peirce logic are of the form xA or xRy, where $x, y \in \mathbb{OV}_{\mathsf{PL}}$, $A \in \mathbb{BT}_{\mathsf{PL}}$, and $R \in \mathbb{RT}_{\mathsf{PL}}$.

A PL-*model* is a structure $\mathcal{M} = (U, m)$ where U is a non-empty set and $m : \mathbb{BA}_{\mathsf{PL}} \cup \mathbb{RA}_{\mathsf{PL}} \to \mathcal{P}(U) \cup \mathcal{P}(U \times U)$ is a meaning function satisfying the following conditions:

- $m(A) \subseteq U$, for every $A \in \mathbb{BV}_{\mathsf{PL}}$;
- $m(1_B) = U$;

- m extends to all the Boolean terms:
  - $m(A \cup B) = m(A) \cup m(B)$,
  - $m(A \cap B) = m(A) \cap m(B)$,
  - $m(-A) = -m(A)$;
- $m(R) \subseteq U \times U$, for every $R \in \mathbb{RV}_{\mathsf{PL}}$;
- $m(1_R) = U \times U$;
- $m(1')$ is an equivalence relation on U and for every $R \in \mathbb{RA}_{\mathsf{PL}}$ the following extensionality property is satisfied: $m(1' ; R) = m(R ; 1') = m(R)$;
- m extends to all the relational terms:
  - $m(R \cup S) = m(R) \cup m(S)$,
  - $m(R \cap S) = m(R) \cap m(S)$,
  - $m(-R) = -m(S)$,
  - $m(R^{-1}) = m(R)^{-1}$,
  - $m(R ; S) = m(R) ; m(S)$;
- $m(A^c) = \{(a,b) \in U \times U : a \in m(A)\}$, for every $A \in \mathbb{BT}_{\mathsf{PL}}$;
- $m(R : A) = \{a \in U : \exists b \in U, (a,b) \in m(R) \text{ and } b \in m(A)\}$, for every $A \in \mathbb{BT}_{\mathsf{PL}}$ and for every $R \in \mathbb{RT}_{\mathsf{PL}}$.

A PL-model $\mathcal{M}$ is said to be *standard* whenever $m(1')$ is the identity on U.

A valuation in a PL-model $\mathcal{M} = (U, m)$ is any mapping $v \colon \mathbb{OV}_{\mathsf{PL}} \to U$. Let $A \in \mathbb{BT}_{\mathsf{PL}}$ and $R \in \mathbb{RT}_{\mathsf{PL}}$. Satisfaction of a formula by a valuation v in a PL-model $\mathcal{M} = (U, m)$ is defined as:

$$\mathcal{M}, v \models xA \text{ iff } v(x) \in m(A);$$

$$\mathcal{M}, v \models xRy \text{ iff } (v(x), v(y)) \in m(R).$$

A formula is *true* in a model $\mathcal{M}$ provided that it is satisfied by all the valuations in $\mathcal{M}$. A formula is PL-*valid* whenever it is true in all PL-models.

4.4 Dual Tableau for Peirce Logic

The PL-dual tableau consists of decomposition rules for relational formulas, decomposition rules for Boolean formulas, and two mixed rules, where decomposition of a relational formula built with the cylindrification operation results in a Boolean formula, and decomposition of a Boolean formula built with the Peirce product results in a relational formula and a Boolean formula. Decomposition rules for relational formulas are those of $\mathsf{RL}(1, 1')$-dual tableau, i.e., ($\cup$), ($-\cup$), ($\cap$), ($-\cap$), ($-$), ($^{-1}$), ($-^{-1}$), (;), and ($-$;) adjusted to PL-language (see Sect. 2.4).

Both a dual tableau and a tableau for Peirce logic are presented in [SOH04]. Decomposition rules of PL-dual tableau for Boolean formulas have the following forms:

For all object variables x and y, for all Boolean terms A and B, and for every relational term R,

$$(\cup\text{Bool})\ \frac{x(A \cup B)}{xA, xB} \qquad (-\cup\text{Bool})\ \frac{x{-}(A \cup B)}{x{-}A \mid x{-}B}$$

$$(\cap\text{Bool})\ \frac{x(A \cap B)}{xA \mid xB} \qquad (-\cap\text{Bool})\ \frac{x{-}(A \cap B)}{x{-}A, x{-}B}$$

$$(-\text{Bool})\ \frac{x{-}{-}A}{xA}$$

$$({}^{c})\ \frac{xA^{c}y}{xA} \qquad (-{}^{c})\ \frac{x{-}A^{c}y}{x{-}A}$$

$$(:)\ \frac{x(R:A)}{xRz, x(R:A) \mid zA, x(R:A)} \qquad (-:)\ \frac{x{-}(R:A)}{x{-}Rz, z{-}A}$$

z is any object variable (for $(:)$); z is a new object variable (for $(-:)$)

The specific rules of PL-dual tableau are the rules $(1'1)$ and $(1'2)$ of $\mathsf{RL}(1, 1')$-dual tableau (see Sect. 2.7) adjusted to PL-language and the rules of the following forms:

For every object variable x and for every atomic Boolean term A,

$$(1'3)\ \frac{xA}{zA, xA \mid x1'z, xA} \quad z \text{ is any object variable}$$

The rule $(1'3)$ reflects the law of replacement of equivalent objects in Boolean formulas.

A finite set of formulas is said to be PL-axiomatic whenever it includes either of the subsets (Ax1), ..., (Ax5):

For all object variables x and y, for every Boolean term A, and for every relational term R,

(Ax1) $\{x1'x\}$;
(Ax2) $\{x1_R y\}$;
(Ax3) $\{x1_B\}$;
(Ax4) $\{xA, x{-}A\}$;
(Ax5) $\{xRy, x{-}Ry\}$.

As in RL-logic, a PL-set is a finite set of PL-formulas such that the first-order disjunction of its members is true in all PL-models. Correctness of a rule is defined in a similar way as in the logic RL (see Sect. 2.4).

Proposition 4.4.1.

1. *The* PL-*rules are* PL-*correct;*
2. *The* PL-*axiomatic sets are* PL-*sets.*

Proof. By way of example, we show correctness of the rule $(- :)$.

Let X be a finite set of PL-formulas. Assume $X \cup \{x-(R : A)\}$ is a PL-set. Suppose $X \cup \{x-Rz, z-A\}$, where z does not occur in X and $z \neq x$, is not a PL-set. Then there exist a PL-model $\mathcal{M} = (U, m)$ and a valuation v in $\mathcal{M}$ such that for every $\varphi \in X \cup \{x-Rz, z-A\}$, $\mathcal{M}, v \not\models \varphi$. Thus, $(v(x), v(z)) \in m(R)$ and $v(z) \in m(A)$. By the definition of Peirce product, $v(x) \in m(R : A)$. On the other hand, since $X \cup \{x-(R : A)\}$ is a PL-set, $\mathcal{M}, v \models x-(R : A)$, and hence $v(x) \notin m(R : A)$, a contradiction. Now, assume that $X \cup \{x-Rz, z-A\}$ is a PL-set. Then, for every PL-model $\mathcal{M} = (U, m)$ and for every valuation v in $\mathcal{M}$, either there exists $\varphi \in X$ such that $\mathcal{M}, v \models \varphi$ or $\mathcal{M}, v \models x-Rz$ or $\mathcal{M}, v \models z-A$. By the assumption on variable z, either there exists $\varphi \in X$ such that $\mathcal{M}, v \models \varphi$ or for every $t \in U$, $(v(x), t) \notin m(R)$ or $t \notin m(A)$, so $v(x) \notin m(R : A)$. Therefore, if $X \cup \{x-Rz, z-A\}$ is a PL-set, then for every PL-model $\mathcal{M} = (U, m)$ and for every valuation v in $\mathcal{M}$, either there exists $\varphi \in X$ such that $\mathcal{M}, v \models \varphi$ or for every $t \in U$, $v(x) \in m(-R : A)$. Hence, $X \cup \{x-(R : A)\}$ is a PL-set. □

The notions of a PL-proof tree, a closed branch of such a tree, a closed PL-proof tree, and a PL-proof of a PL-formula are defined as in Sect. 2.4.

The rules of PL-dual tableau, in particular the specific rules, guarantee that for any branch of a PL-proof tree, if xRy and $x-Ry$ (resp. xA and $x-A$) for an atomic relational term R (resp. for an atomic Boolean term A) belong to the branch, then the branch is closed, i.e., both of these formulas appear in a node of that branch. Thus, the closed branch property holds.

Completion conditions corresponding to the rules that are specific for PL-logic are as follows:

For all object variables x and y, for all Boolean terms A and B, and for every relational term R,

Cpl($\cup$Bool) (resp. Cpl($-\cap$Bool)) If $x(A \cup B) \in b$ (resp. $x-(A \cap B) \in b$), then both $xA \in b$ (resp. $x-A \in b$) and $xB \in b$ (resp. $x-B \in b$), obtained by an application of the rule ($\cup$Bool) (resp. ($-\cap$Bool));

Cpl($\cap$Bool) (resp. Cpl($-\cup$Bool)) If $x(A \cap B) \in b$ (resp. $x-(A \cup B) \in b$), then either $xA \in b$ (resp. $x-A \in b$) or $xB \in b$ (resp. $x-B \in b$), obtained by an application of the rule ($\cap$Bool) (resp. ($-\cup$Bool));

Cpl($-$Bool) If $x--A \in b$, then $xA \in b$, obtained by an application of the rule ($-$Bool);

Cpl(^{c}) If $xA^c y \in b$, then $xA \in b$, obtained by an application of the rule (^{c});

Cpl($-^c$) If $x-A^c y \in b$, then $x-A \in b$, obtained by an application of the rule ($-^c$);

Cpl(:) If $x(R : A) \in b$, then for every object variable z, either $xRz \in b$ or $zA \in b$, obtained by an application of the rule (:);

Cpl($-$:) If $x-(R : A) \in b$, then for some object variable z, both $x-Rz \in b$ and $z-A \in b$, obtained by an application of the rule ($-$:);

Cpl($1'3$) If $xA \in b$ for some atomic Boolean term A, then for every object variable z, either $zA \in b$ or $x1'z \in b$, obtained by an application of the rule ($1'3$).

The remaining completion conditions determined by the rules for relational operations are the same as those presented in Sects. 2.5 and 2.7 for the relational logics of classical algebras of binary relations. The notions of a complete PL-proof tree and an open branch of a PL-proof tree are defined as in RL-logic (see Sect. 2.5).

Let b be an open branch of a PL-proof tree. We define a branch structure $\mathcal{M}^b = (U^b, m^b)$:

- $U^b = \mathbb{OV}_{\mathsf{PL}}$;
- $m^b(R) = \{(x, y) \in U^b \times U^b : xRy \notin b\}$, for every $R \in \mathbb{RA}_{\mathsf{PL}}$;
- $m^b(A) = \{x \in U^b : xA \notin b\}$, for every $A \in \mathbb{BA}_{\mathsf{PL}}$;
- m^b extends to all the Boolean and relational terms as in PL-models.

Let $v^b : \mathbb{OV}_{\mathsf{PL}} \to U^b$ be a valuation in $\mathcal{M}^b$ such that $v^b(x) = x$, for every $x \in \mathbb{OV}_{\mathsf{PL}}$.

According to the method of proving completeness described in Sect. 2.6 we need to prove the branch model property and the satisfaction in branch model property. The branch model property can be proved as in RL(1, 1′)-logic.

Proposition 4.4.2 (Branch Model Property). *For every open branch b of a PL-proof tree, $\mathcal{M}^b$ is a PL-model.*

Proposition 4.4.3 (Satisfaction in Branch Model Property). *For every open branch b of a PL-proof tree and for every PL-formula φ, if $\mathcal{M}^b, v^b \models \varphi$, then $\varphi \notin b$.*

Proof. The proof is by induction on the complexity of formulas. The atomic case can be proved as in Sect. 2.5 and it uses the closed branch property. Let $R \in \mathbb{RT}_{\mathsf{PL}}$ and $A \in \mathbb{BT}_{\mathsf{PL}}$. By way of example, we show the proposition for the formulas of the form $xA^c y$ and $x(R : A)$.

Assume $\mathcal{M}^b, v^b \models xA^c y$. Then $x \in m^b(A)$. By the induction hypothesis, $xA \notin b$. Suppose $xA^c y \in b$. By the completion condition Cpl(^{c}), $xA \in b$, a contradiction.

Assume $\mathcal{M}^b, v^b \models x(R : A)$. Then there exists $y \in U^b$ such that $(x, y) \in m^b(R)$ and $y \in m^b(A)$. By the induction hypothesis, $xRy \notin b$ and $yA \notin b$. Suppose $x(R : A) \in b$. By the completion condition Cpl(:), either $xRy \in b$ or $yA \in b$, a contradiction.

The proofs of the remaining cases are similar as in the relational logics of classical algebras of binary relations (see Sects. 2.5 and 2.7). □

The quotient model $\mathcal{M}_q^b$ is defined as in relational logics in Sect. 2.7. Then the following proposition can be proved:

Proposition 4.4.4. *For every open branch b of a PL-proof tree and for every PL-formula φ:*

1. *$\mathcal{M}_q^b$ is a standard PL-model;*
2. *$\mathcal{M}^b, v^b \models \varphi$ iff $\mathcal{M}_q^b, v_q^b \models \varphi$.*

Due to Propositions 4.4.1–4.4.4, we have:

Theorem 4.4.1 (Soundness and Completeness of PL). *For every PL-formula φ, the following conditions are equivalent:*

1. *φ is PL-valid;*
2. *φ is true in all standard PL-models;*
3. *φ is PL-provable.*

Example. We consider inclusions from right to left of the set instances of the Eq. 11. from Theorem 4.2.1 and the axiom (P8) of Peirce algebras. Applying Proposition 2.2.1(1.) of Sect. 2.2 we conclude that their validity in Peirce algebras is equivalent to PL-validity of the following PL-formulas:

$$x[-((R \cap (A^c)^{-1}) : B) \cup (R : (A \cap B))],$$

$$x[-(R : 1)^c \cup (R ; 1)]y.$$

Figures 4.1 and 4.2 present PL-proofs of these formulas.

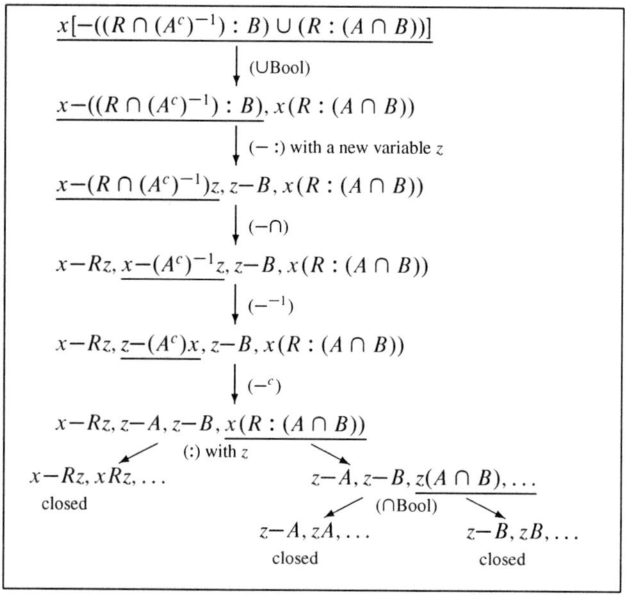

$x[-((R \cap (A^c)^{-1}) : B) \cup (R : (A \cap B))]$

↓ ($\cup$Bool)

$x-((R \cap (A^c)^{-1}) : B), x(R : (A \cap B))$

↓ ($-$:) with a new variable z

$x-(R \cap (A^c)^{-1})z, z-B, x(R : (A \cap B))$

↓ ($-\cap$)

$x-Rz, x-(A^c)^{-1}z, z-B, x(R : (A \cap B))$

↓ ($-^{-1}$)

$x-Rz, z-(A^c)x, z-B, x(R : (A \cap B))$

↓ ($-^c$)

$x-Rz, z-A, z-B, x(R : (A \cap B))$

↙ (:) with z ↘

$x-Rz, xRz, \ldots$ closed | $z-A, z-B, z(A \cap B), \ldots$

↙ ($\cap$Bool) ↘

$z-A, zA, \ldots$ closed | $z-B, zB, \ldots$ closed

Fig. 4.1 A PL-proof of $x[-((R \cap (A^c)^{-1}) : B) \cup (R : (A \cap B))]$

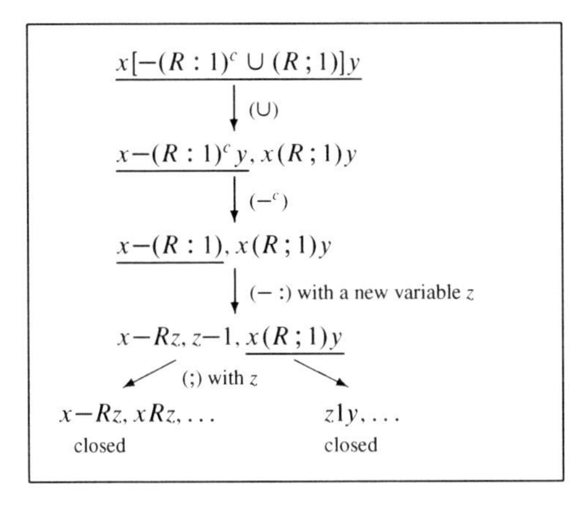

Fig. 4.2 A PL-proof of $x[-(R:1)^c \cup (R;1)]y$

4.5 Entailment, Model Checking, and Satisfaction in Peirce Logic

The method for verification of entailment presented in Sect. 2.11 can be applied to Peirce logic provided that the Boolean terms of the logic will be replaced with relational terms representing right ideal relations. To obtain a relational formalism appropriate for verification of entailment of PL-formulas from a finite set of PL-formulas, we consider an instance PL_{ent} of the logic $\mathsf{RL}(1, 1')$. The vocabulary of the language of PL_{ent}-logic is defined as in Sect. 2.7 with:

- $\mathbb{RV}_{\mathsf{PL}_{ent}} = \mathbb{RV}^0_{\mathsf{PL}_{ent}} \cup \mathbb{RV}^1_{\mathsf{PL}_{ent}}$, where $\mathbb{RV}^0_{\mathsf{PL}_{ent}}$ and $\mathbb{RV}^1_{\mathsf{PL}_{ent}}$ are infinite sets of relational variables such that $\mathbb{RV}^0_{\mathsf{PL}_{ent}} \cap \mathbb{RV}^1_{\mathsf{PL}_{ent}} = \emptyset$; relational variables from the set $\mathbb{RV}^0_{\mathsf{PL}_{ent}}$ are intended to represent Boolean variables of PL-logic as right ideal relations.

A PL_{ent}-*model* is an $\mathsf{RL}(1, 1')$-model $\mathcal{M} = (U, m)$ such that $m(R)$ is a right ideal relation for every $R \in \mathbb{RV}^0_{\mathsf{PL}_{ent}}$.

The translation of PL-terms into PL_{ent}-terms starts with a one-to-one assignment of relational variables from $\mathbb{RV}^0_{\mathsf{PL}_{ent}}$ and relational variables from $\mathbb{RV}^1_{\mathsf{PL}_{ent}}$ to Boolean variables and relational variables of PL-logic, respectively. Let τ' be such an assignment. Then the translation τ is defined inductively as follows:

- $\tau(A) = \tau'(A)$, for every $A \in \mathbb{BV}_{\mathsf{PL}}$;
- $\tau(R) = \tau'(R)$, for every $R \in \mathbb{RV}_{\mathsf{PL}}$;
- $\tau(1_B) = \tau(1_R) = 1$ and $\tau(1') = 1'$;

For PL-terms T, T_1, and T_2 and for $\# \in \{\cup, \cap\}$ such that $T_1 \# T_2$ is a PL-term,

- $\tau(-T) = -\tau(T)$;
- $\tau(T_1 \# T_2) = \tau(T_1) \# \tau(T_2)$;

For any relational PL-terms T, T_1, and T_2,

- $\tau(T^{-1}) = \tau(T)^{-1}$;
- $\tau(T_1; T_2) = \tau(T_1); \tau(T_2)$;

For any relational term R and for any Boolean term A,

- $\tau(A^c) = \tau(A)$;
- $\tau(R : A) = \tau(R); \tau(A)$.

Now, we extend translation τ to PL-formulas as follows.

For any Boolean PL-term A, a relational PL-term R, and for any object variables x and y,

- $\tau(xA) = x\tau(A)y$;
- $\tau(xRy) = x\tau(R)y$.

Observe that since being a right ideal relation is preserved by making Boolean operations and by composition of an arbitrary relation with a right ideal relation, every Boolean term of PL-logic is translated into a relational PL_{ent}-term representing a right ideal relation.

Proposition 4.5.1. *Let φ be a PL-formula. Then:*

1. *For every PL-model $\mathcal{M}$ there exists a PL_{ent}-model $\mathcal{M}'$ such that $\mathcal{M} \models \varphi$ iff $\mathcal{M}' \models \tau(\varphi)$;*
2. *For every PL_{ent}-model $\mathcal{M}'$ there exists a PL-model $\mathcal{M}$ such that $\mathcal{M} \models \varphi$ iff $\mathcal{M}' \models \tau(\varphi)$.*

The above proposition implies:

Theorem 4.5.1 (Entailment in PL). *Let $\varphi_1, \ldots, \varphi_n$, $n \geq 1$, and φ be PL-formulas. Then the following conditions are equivalent:*

1. *$\varphi_1, \ldots, \varphi_n$ entail φ;*
2. *$\tau(\varphi_1), \ldots, \tau(\varphi_n)$ entail $\tau(\varphi)$.*

Therefore, entailment in PL-logic can be represented as entailment in PL_{ent}-logic. As in $\mathsf{RL}(1, 1')$-logic, entailment in PL_{ent}-logic is expressible in its language and we apply the method presented in Sect. 2.11 to verify it.

PL_{ent}-dual tableau includes the rules and axiomatic sets of $\mathsf{RL}(1, 1')$-dual tableau (see Sects. 2.5 and 2.7) adjusted to PL_{ent}-language and the specific rule of the following form:

For all object variables x and y and for every relational variable $R \in \mathbb{RV}^0_{\mathsf{PL}_{ent}}$,

$$\text{(ideal)} \quad \frac{xRy}{xRz, xRy} \qquad z \text{ is any object variable}$$

The rule (ideal) reflects the fact that relational variables representing Boolean PL-variables are interpreted as right ideal relations. PL_{ent}-correctness of the rule (ideal) follows directly from the definition of PL_{ent}-models.

The completion condition determined by the rule (ideal) is:

For all object variables x and y and for every relational variable $R \in \mathbb{RV}^0_{\mathsf{PL}_{ent}}$,

Cpl(ideal) If $xRy \in b$, then for every object variable z, $xRz \in b$.

Now, completeness of PL_{ent}-dual tableau can be proved in a similar way as for $\mathsf{RL}(1, 1')$-logic.

Theorem 4.5.2 (Soundness and Completeness of PL_{ent}). *Let φ be a PL_{ent}-formula. Then the following conditions are equivalent:*

1. *φ is PL_{ent}-valid;*
2. *φ is true in all standard PL_{ent}-model;*
3. *φ is PL_{ent}-provable.*

Example. Let A and B be Boolean variables of PL-logic. Consider PL-formulas:

$$\varphi = x(1 ; B^c ; 1)y,$$
$$\psi = x(-A \cup (A^c : B)).$$

The translations of these formulas into PL_{ent}-formulas are:

$$\tau(\varphi) = x(1 ; R_B ; 1)y,$$
$$\tau(\psi) = x(-R_A \cup (R_A ; R_B))y,$$

where for simplicity of notation $\tau'(A) = R_A$ and $\tau'(B) = R_B$, for $R_A, R_B \in \mathbb{RV}^0_{\mathsf{PL}_{ent}}$. By Theorem 4.5.1, the problem of verification whether φ entails ψ is equivalent to the problem 'Does $\tau(\varphi)$ entails $\tau(\psi)$'. According to the method described in Sect. 2.11, the latter can be verified by showing that the formula:

$$x((1 ; -(1 ; R_B ; 1) ; 1) \cup -R_A \cup (R_A ; R_B))y$$

is PL_{ent}-provable. Figure 4.3 presents its PL_{ent}-proof.

The method of model checking and verification of satisfaction presented in Sects. 3.4 and 3.5 can also be applied to PL-formulas. Let $\mathcal{M} = (U, m)$ be a fixed standard PL-model with a finite universe U and let φ be a PL-formula. As in $\mathsf{RL}(1, 1')$-logic, in order to obtain a relational formalism appropriate for representing and solving the problem '$\mathcal{M} \models \varphi$?', we consider an instance $\mathsf{PL}_{\mathcal{M},\varphi}$ of the logic PL. The vocabulary of the logic $\mathsf{PL}_{\mathcal{M},\varphi}$ consists of symbols from the following pairwise disjoint sets:

- $\mathbb{OV}_{\mathsf{PL}_{\mathcal{M},\varphi}}$ – a countable infinite set of object variables;
- $\mathbb{OC}_{\mathsf{PL}_{\mathcal{M},\varphi}} = \{c_{\mathtt{a}} : \mathtt{a} \in U\}$ is the set of object constants such that if $\mathtt{a} \neq \mathtt{b}$, then $c_{\mathtt{a}} \neq c_{\mathtt{b}}$;

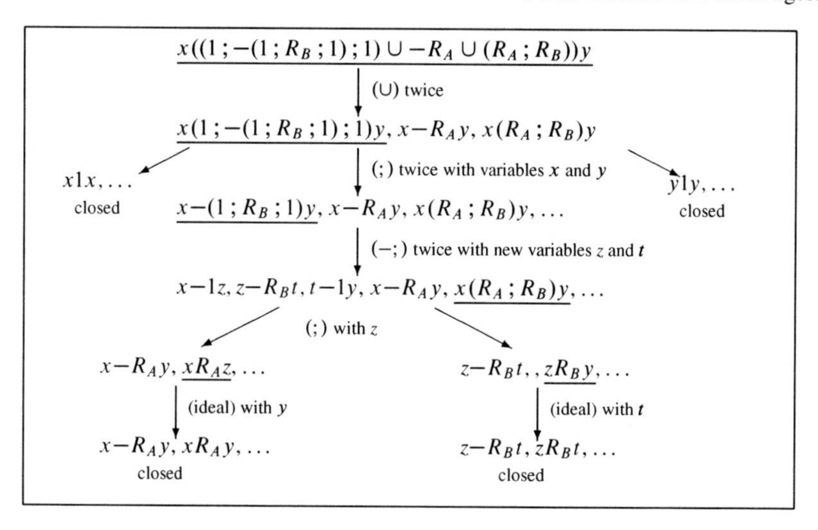

Fig. 4.3 A PL_{ent}-proof showing that $x(1 ; B^c ; 1)y$ entails $x(-A \cup (A^c : B))$

- $\mathbb{BC}_{\mathsf{PL}_{\mathcal{M},\varphi}} = \mathbb{B} \cup \{1_B\}$ – the set of Boolean constants, where $\mathbb{B}$ is the set of all the atomic Boolean terms occurring in φ;
- $\mathbb{RC}_{\mathsf{PL}_{\mathcal{M},\varphi}} = \mathbb{S} \cup \{1_R, 1'\}$ – the set of relational constants, where $\mathbb{S}$ is the set of all the atomic relational terms occurring in φ;
- $\{-, \cup, \cap, ;, ^{-1}, :, ^c\}$ – the set of operations.

A $\mathsf{PL}_{\mathcal{M},\varphi}$-model is a pair $\mathcal{N} = (W, n)$, where $W = U$ and n is a meaning function satisfying:

- $n(c_{\mathsf{a}}) = \mathsf{a}$, for every $c_{\mathsf{a}} \in \mathbb{OC}_{\mathsf{PL}_{\mathcal{M},\varphi}}$;
- $n(B) = m(B)$, for every $B \in \mathbb{B}$;
- $n(S) = m(S)$, for every $S \in \mathbb{S}$;
- $n(1_B), n(1_R), n(1')$ are defined as in standard PL-models;
- n extends to all the compound terms as in PL-models.

A valuation in $\mathcal{N}$ is a function $v\colon \mathbb{OS}_{\mathsf{PL}_{\mathcal{M},\varphi}} \to W$ such that $v(c) = n(c)$, for every $c \in \mathbb{OC}_{\mathsf{PL}_{\mathcal{M},\varphi}}$. As in the logic $\mathsf{RL}(1, 1')$, it is easy to prove that for every Boolean $\mathsf{PL}_{\mathcal{M},\varphi}$-term A and for every relational $\mathsf{PL}_{\mathcal{M},\varphi}$-term Q the following hold: $n(A) = m(A)$ and $n(Q) = m(Q)$. Therefore, $\mathsf{PL}_{\mathcal{M},\varphi}$-validity of φ is equivalent to its truth in a single $\mathsf{PL}_{\mathcal{M},\varphi}$-model $\mathcal{N}$. Hence, the following holds:

Proposition 4.5.2. *The following statements are equivalent:*

1. $\mathcal{M} \models \varphi$;
2. φ *is* $\mathsf{PL}_{\mathcal{M},\varphi}$*-valid.*

Dual tableau for the logic $\mathsf{PL}_{\mathcal{M},\varphi}$ is constructed as follows. We extend PL-dual tableau with the rules and the axiomatic sets of $\mathsf{RL}_{\mathcal{M},\varphi}$-dual tableau presented in Sect. 3.4 adapted to the language of $\mathsf{PL}_{\mathcal{M},\varphi}$. In particular, in the rules $(;)$, $(:)$, and

(1′3), z is an object variable or an object constant. Furthermore, we add the rules and axiomatic sets of the following forms:

- For every $B \in \mathbb{B}$, for any object symbol x, and for any $\mathsf{a} \in U$,

$$(-B\mathsf{a}) \quad \frac{x{-}B}{x1'c_{\mathsf{a}}, x{-}B} \quad \text{for every } \mathsf{a} \notin m(B);$$

- For every $B \in \mathbb{B}$ and for every $\mathsf{a} \in U$ such that $\mathsf{a} \in m(B)$, any set including $\{c_{\mathsf{a}}B\}$ is assumed to be a $\mathsf{PL}_{\mathcal{M},\varphi}$-axiomatic set;
- For every $B \in \mathbb{B}$ and for every $\mathsf{a} \in U$ such that $\mathsf{a} \notin m(B)$, any set including $\{c_{\mathsf{a}}{-}B\}$ is assumed to be a $\mathsf{PL}_{\mathcal{M},\varphi}$-axiomatic set.

These rules and axiomatic sets reflect the meaning of atomic Boolean terms occurring in φ.

A $\mathsf{PL}_{\mathcal{M},\varphi}$-set of formulas, $\mathsf{PL}_{\mathcal{M},\varphi}$-correctness of a rule, a $\mathsf{PL}_{\mathcal{M},\varphi}$-proof tree, a closed branch and an open branch of a $\mathsf{PL}_{\mathcal{M},\varphi}$-proof tree are defined in a standard way as in Sect. 2.4.

Proposition 4.5.3. *The rule* $(-B\mathsf{a})$ *is* $\mathsf{PL}_{\mathcal{M},\varphi}$*-correct.*

Proof. Let $\mathsf{a} \in U$ be such that $\mathsf{a} \notin m(B)$ and let X be a finite set of $\mathsf{PL}_{\mathcal{M},\varphi}$-formulas. Assume $X \cup \{x1'c_{\mathsf{a}}, x{-}B\}$ is a $\mathsf{PL}_{\mathcal{M},\varphi}$-set. Suppose $X \cup \{x{-}B\}$ is not a $\mathsf{PL}_{\mathcal{M},\varphi}$-set, that is for some valuation v in $\mathcal{N}$, $v(x) \in m(B)$. By the assumption, $v(x) = \mathsf{a}$. Since $\mathsf{a} \notin m(B)$, $v(x) \notin m(B)$, a contradiction. On the other hand, if $X \cup \{x{-}B\}$ is a $\mathsf{PL}_{\mathcal{M},\varphi}$-set, then so is $X \cup \{x1'c_{\mathsf{a}}, x{-}B\}$. Therefore, the rule $(-B\mathsf{a})$ is $\mathsf{PL}_{\mathcal{M},\varphi}$-correct. □

Correctness of the remaining rules follows from the corresponding propositions for the relational logics of classical algebras of binary relations and for PL-logic (see Sects. 2.5, 2.7, 3.4, and 4.4). The validity of all the added axiomatic sets follows directly from the definition of semantics of $\mathsf{PL}_{\mathcal{M},\varphi}$.

The completion conditions are those of $\mathsf{RL}_{\mathcal{M},\varphi}$-dual tableau adapted to the language of $\mathsf{PL}_{\mathcal{M},\varphi}$ and the condition determined by the rule $(-B\mathsf{a})$. For every object symbol x, for every $B \in \mathbb{B}$, and for every $\mathsf{a} \in U$ such that $\mathsf{a} \notin m(B)$, we postulate:

Cpl$(-B\mathsf{a})$ If $x{-}B \in b$, then $x1'c_{\mathsf{a}} \in b$, obtained by an application of the rule $(-B\mathsf{a})$.

A branch structure $\mathcal{N}^b = (W^b, n^b)$ is defined as follows:

- $W^b = \mathbb{OC}_{\mathsf{PL}_{\mathcal{M},\varphi}}$;
- $n^b(c) = c$, for every $c \in \mathbb{OC}_{\mathsf{PL}_{\mathcal{M},\varphi}}$;
- $n^b(1_B) = W^b$;
- $n^b(B) = \{c_{\mathsf{a}} \in W^b : \mathsf{a} \in m(B)\}$, for every $B \in \mathbb{B}$;

- $n^b(S) = \begin{cases} \{(c_{\mathsf{a}}, c_{\mathsf{b}}) \in W^b \times W^b : c_{\mathsf{a}} S c_{\mathsf{b}} \notin b\}, & \text{if } S \in \{1_R, 1'\} \\ \{(c_{\mathsf{a}}, c_{\mathsf{b}}) \in W^b \times W^b : (\mathsf{a}, \mathsf{b}) \in m(S)\}, & \text{if } S \in \mathbb{S}; \end{cases}$
- n^b extends to all the compound terms as in PL-models.

In a similar way as in $\mathsf{RL}_{\mathcal{M},\varphi}$-dual tableau, it is easy to prove that $\mathcal{N}^b$ is a PL-model.

Let $v^b \colon \mathbb{OS}_{\mathsf{PL}_{\mathcal{M},\varphi}} \to W^b$ be a valuation in $\mathcal{N}^b$ such that:

$v^b(c) = n^b(c)$, for $c \in \mathbb{OC}_{\mathsf{PL}_{\mathcal{M},\varphi}}$;
$v^b(x) = c_{\mathsf{a}}$, where $c_{\mathsf{a}} \in \mathbb{OC}_{\mathsf{PL}_{\mathcal{M},\varphi}}$ is such that $x1'c_{\mathsf{a}} \notin b$, for any object variable x.

As in the completeness proof of $\mathsf{RL}_{\mathcal{M},\varphi}$-dual tableau, it can be proved that the valuation v^b is well defined, that is for every $x \in \mathbb{OV}_{\mathsf{PL}_{\mathcal{M},\varphi}}$, there exists exactly one $c \in \mathbb{OC}_{\mathsf{PL}_{\mathcal{M},\varphi}}$ such that $x1'c \notin b$.

Proposition 4.5.4 (Satisfaction in Branch Model Property). *For every open branch b of a $\mathsf{PL}_{\mathcal{M},\varphi}$-proof tree and for every $\mathsf{PL}_{\mathcal{M},\varphi}$-formula ψ, if $\mathcal{N}^b, v^b \models \psi$, then $\psi \notin b$.*

Proof. It suffices to show the proposition for formulas of the form xB and $x{-}B$, where $B \in \mathbb{B}$.

Let $\psi = xB$ for some $B \in \mathbb{B}$. Assume $\mathcal{N}^b, v^b \models xB$. Let $\mathsf{a} \in U$ be such that $v^b(x) = c_{\mathsf{a}}$, that is $x1'c_{\mathsf{a}} \notin b$. Since $\mathcal{N}^b, v^b \models xB$, $c_{\mathsf{a}} \in n^b(B)$, hence $\mathsf{a} \in m(B)$. Thus $c_{\mathsf{a}}B \notin b$, otherwise b would be closed. Suppose $xB \in b$. By the completion condition Cpl(1′3), at least one of the following holds: $c_{\mathsf{a}}B \in b$ or $x1'c_{\mathsf{a}} \in b$, a contradiction.

Let $\psi = x{-}B$, for some $B \in \mathbb{B}$. Assume $\mathcal{N}^b, v^b \models x{-}B$. Let $\mathsf{a} \in U$ be such that $v^b(x) = c_{\mathsf{a}}$, that is $x1'c_{\mathsf{a}} \notin b$. Since $\mathcal{N}^b, v^b \models x{-}B$, $c_{\mathsf{a}} \notin n^b(B)$. Therefore $\mathsf{a} \notin m(B)$. Suppose $x{-}B \in b$. Then by the completion condition Cpl($-B\mathsf{a}$), $x1'c_{\mathsf{a}} \in b$, a contradiction.

The rest of the proof is similar to the proofs of the analogous propositions for logics RL, RL(1, 1′), $\mathsf{RL}_{\mathcal{M},\varphi}$, and PL (see Sects. 2.5, 2.7, 3.4, and 4.4). □

Since $n^b(1')$ is an equivalence relation on W^b, we may define the quotient structure $\mathcal{N}_q^b = (W_q^b, n_q^b)$ as follows:

- $W_q^b = \{\|c\| : c \in W^b\}$, where $\|c\|$ is the equivalence class of $n^b(1')$ determined by c;
- $n_q^b(c) = \|n^b(c)\|$, for every $c \in \mathbb{OC}_{\mathsf{PL}_{\mathcal{M},\varphi}}$;
- $n_q^b(B) = \{\|c_{\mathsf{a}}\| \in W_q^b : c_{\mathsf{a}} \in n^b(B)\}$, for every $B \in \mathbb{BC}_{\mathsf{PL}_{\mathcal{M},\varphi}}$;
- $n_q^b(S) = \{(\|c_{\mathsf{a}}\|, \|c_{\mathsf{b}}\|) \in W_q^b \times W_q^b : (c_{\mathsf{a}}, c_{\mathsf{b}}) \in n^b(S)\}$, for every $S \in \mathbb{RC}_{\mathsf{PL}_{\mathcal{M},\varphi}}$;
- n_q^b extends to all the compound terms as in PL-models.

Proposition 4.5.5 (Branch Model Property). *Let b be an open branch of a $\mathsf{PL}_{\mathcal{M},\varphi}$-proof tree. Then models $\mathcal{N}_q^b$ and $\mathcal{N}$ are isomorphic.*

The above proposition can be proved following the proof of Proposition 3.4.4.

Let v_q^b be a valuation in $\mathcal{N}_q^b$ defined as $v_q^b(x) = \|v^b(x)\|$, for every $x \in \mathbb{OS}_{\mathsf{PL}_{\mathcal{M},\varphi}}$. As in RL-logic, it is easy to show that the sets of $\mathsf{PL}_{\mathcal{M},\varphi}$-formulas satisfied in $\mathcal{N}^b$ and $\mathcal{N}_q^b$ by valuations v^b and v_q^b, respectively, coincide. Moreover, since $\mathcal{N}_q^b$ and $\mathcal{N}$ are isomorphic, they satisfy exactly the same formulas. Now, the completeness of $\mathsf{PL}_{\mathcal{M},\varphi}$ can be proved in a similar way as in $\mathsf{RL}(1, 1')$-logic.

Theorem 4.5.3 (Soundness and Completeness of $\mathsf{PL}_{\mathcal{M},\varphi}$). *For every $\mathsf{PL}_{\mathcal{M},\varphi}$-formula ψ, the following conditions are equivalent:*

1. *ψ is $\mathsf{PL}_{\mathcal{M},\varphi}$-valid;*
2. *ψ is $\mathsf{PL}_{\mathcal{M},\varphi}$-provable.*

Therefore, we have:

Theorem 4.5.4 (Model Checking in PL). *For every PL-formula φ and for every finite standard PL-model $\mathcal{M}$, the following statements are equivalent:*

1. *$\mathcal{M} \models \varphi$;*
2. *φ is $\mathsf{PL}_{\mathcal{M},\varphi}$-provable.*

In a similar way, we can use the method for verification of satisfaction presented in Sect. 3.5. Let $\varphi = xRy$ (resp. $\varphi = xB$) be a PL-formula, where R is a relational term (resp. B is a Boolean term) and let x, y be any object variables. Let $\mathcal{M} = (U, m)$ be a fixed finite standard PL-model, and let v be a valuation in $\mathcal{M}$ such that $v(x) = \mathsf{a}$ and $v(y) = \mathsf{b}$, for some elements a, b of U. To solve the problem 'Is φ satisfied in $\mathcal{M}$ by valuation v?' we consider the logic $\mathsf{PL}_{\mathcal{M},\varphi}$ defined in this section. For every $\mathsf{PL}_{\mathcal{M},\varphi}$-model $\mathcal{N} = (U, n)$, for every valuation v in $\mathcal{N}$, and for all $\mathsf{a}, \mathsf{b} \in U$, $(\mathsf{a}, \mathsf{b}) \in m(R)$ (resp. $\mathsf{a} \in m(B)$) if and only if $c_{\mathsf{a}} R c_{\mathsf{b}}$ (resp. $c_{\mathsf{a}} B$) is $\mathsf{PL}_{\mathcal{M},\varphi}$-valid. Thus, we have:

Theorem 4.5.5 (Satisfaction in PL-models). *For every relational term R of PL-language, for every Boolean term B of PL-language, for every finite standard PL-model $\mathcal{M} = (U, m)$, and for all $\mathsf{a}, \mathsf{b} \in U$, the following statements are equivalent:*

1. *$(\mathsf{a}, \mathsf{b}) \in m(R)$ (resp. $\mathsf{a} \in m(B)$);*
2. *$c_{\mathsf{a}} R c_{\mathsf{b}}$ (resp. $c_{\mathsf{a}} B$) is $\mathsf{PL}_{\mathcal{M},\varphi}$-provable.*

Following the methods described above, in the next section we present examples of applications of PL-dual tableau to model checking and verification of satisfaction of PL-formulas.

4.6 Peirce Algebras and Terminological Languages

Peirce algebras provide an algebraic semantics for the terminological languages from the family originated with KL–ONE [BS85].

Terminological representation languages are extensively studied in description logics. The languages have two kinds of syntactic primitives, called *concepts* and *roles*. Concepts are interpreted as sets and roles as binary relations. Concepts consist of *atomic* concepts, *constant* concepts $\top$ and $\bot$, and *compound* concepts of the following forms, where C, C_1, C_2 are concepts and R is a role:

- $(and\, C_1 C_2)$;
- $(not\, C)$;
- $(some\, RC)$;
- $(all\, RC)$.

Roles consist of *atomic* roles, *constant* roles 0, 1, and $self$, and *compound* roles of the following forms, where C is a concept and R, R_1, R_2 are roles:

- $(and\, R_1 R_2)$;
- $(not\, R)$;
- $(inverse\, R)$;
- $(compose\, R_1 R_2)$;
- $(restrict\, RC)$.

A model for a terminological language, T-language for short, is a pair $\mathcal{M} = (U, m)$, where U is a non-empty set and m is a meaning function satisfying the following conditions:

- $m(C) \subseteq U$, for every atomic concept C;
- $m(R) \subseteq U \times U$, for every atomic role R;
- $m(\top) = U, m(\bot) = \emptyset$;
- $m(and\, C_1 C_2) = m(C_1) \cap m(C_2)$;
- $m(not\, C) = U \setminus m(C)$;
- $m(some\, RC) = \{x \in U : \exists y \in U, (x, y) \in m(R)$ and $y \in m(C)\}$;
- $m(all\, RC) = \{x \in U : \forall y \in U, (x, y) \in m(R)$ implies $y \in m(C)\}$;
- $m(1) = U \times U, m(0) = \emptyset, m(self) = \{(x, y) \in U \times U : x = y\}$;
- $m(and\, R_1 R_2) = m(R_1) \cap m(R_2)$;
- $m(not\, R) = U \times U - m(R)$;
- $m(inverse\, R) = \{(x, y) \in U \times U : (y, x) \in m(R)\}$;
- $m(compose\, R_1 R_2)$
  $= \{(x, y) \in U \times U : \exists z \in U, (x, z) \in m(R_1)$ and $(z, y) \in m(R_2)\}$;
- $m(restrict\, RC) = \{(x, y) \in U \times U : (x, y) \in m(R)$ and $y \in m(C)\}$.

Given a T-language, the important questions in terminological reasoning are:

(T_1) The satisfiability problem: given a concept C (resp. a role R) is there a model $\mathcal{M} = (U, m)$ of the T-language such that $m(C) \neq \emptyset$ (resp. $m(R) \neq \emptyset$);

(T_2) The subsumption problem: given concepts C and D (resp. roles R and Q) is it true that in every model $\mathcal{M} = (U, m)$ of the T-language C subsumes D, i.e., $m(D) \subseteq m(C)$ (resp. R subsumes Q, i.e., $m(Q) \subseteq m(R)$).

There is a direct relationship between Peirce algebras and terminological languages. Concepts can be viewed as Boolean terms of Peirce algebras and roles as their relational terms. The following translation function transforms concepts and roles into Peirce algebra terms.

Let τ' be a one-to-one assignment of Boolean variables to atomic concepts and relation variables to atomic roles. Then a translation function τ is defined as follows:

- $\tau(\bot) = -1_B$, $\tau(\top) = 1_B$;
- $\tau(0) = -1_R$, $\tau(1) = 1_R$, $\tau(self) = 1'$;
- $\tau(and\, C_1 C_2) = \tau(C_1) \cap \tau(C_2)$;
- $\tau(not\, C) = -\tau(C)$;
- $\tau(some\; RC) = \tau(R) : \tau(C)$;
- $\tau(all\; RC) = -(\tau(R) : -\tau(C))$;
- $\tau(and\; R_1 R_2) = \tau(R_1) \cap \tau(R_2)$;
- $\tau(not\; R) = -\tau(R)$;
- $\tau(inverse\; R) = \tau(R)^{-1}$;
- $\tau(compose\; R_1 R_2) = \tau(R_1) \,;\, \tau(R_2)$;
- $\tau(restrict\; RC) = \tau(R) \cap (\tau(C)^c)^{-1}$.

The problems (T_1) and (T_2) can be represented in Peirce logic:

($\mathsf{T}_1^{\mathsf{PL}}$) Let C be a concept (resp. let R be a role) of a T-language and let $\mathcal{M} = (U, m)$ be a model of this language. The problem 'Does $m(C) \neq \emptyset$ (resp. $m(R) \neq \emptyset$) hold in $\mathcal{M}$?' is equivalent to the problem of verifying whether $x(1_R : \tau(C))$ (resp. $x(1_R \,;\, \tau(R) \,;\, 1_R)y$) is true in model $\mathcal{M}$; so we may apply the method of model checking presented in Sect. 4.5;

($\mathsf{T}_2^{\mathsf{PL}}$) Let C and D be concepts (resp. let R and Q be roles) of a T-language. The problem 'Is it true that for every model $\mathcal{M} = (U, m)$ of the T-language $m(D) \subseteq m(C)$ (resp. $m(Q) \subseteq m(R)$)?' is equivalent to PL-validity of the formula $x(-D \cup C)$ (resp. $x(-Q \cup R)y$); we may apply the dual tableau presented in Section 4.4 for verification of validity of these formulas.

Example. Consider a semantic network given in Fig. 4.4, which is a reformulation of a network presented in [Sch91].

The universe consists of three people: Anne (a), Charles (c), and William (w). The nodes represent concepts: 'Vegetarians' (Vg) and 'Mathematicians' (Ma). Thick lines indicate their members. The directed edges marked with thin lines represent roles: 'sister-of' (S) and 'parent-of' (P). Thus, the above semantic network represents some explicit facts like 'William is a mathematician', 'Charles is a parent of William', 'Anne is a sister of Charles', etc. The semantic network also contains some *implicit* facts, such as: 'Charles is a parent of some mathematician', 'Anne is an aunt of some mathematician', 'Some vegetarian is a parent of William'.

A terminological language for representation of the information given above is a T-language such that its atomic concepts are Vg and Ma and its atomic roles are S and P. A model that represents the information given in Fig. 4.4 is a T-model

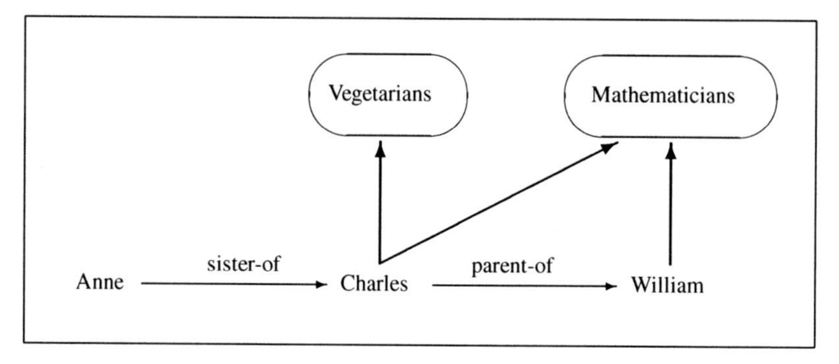

Fig. 4.4 A semantic network sample

$\mathcal{M} = (U, m)$ such that:

- $U = \{\mathtt{a}, \mathtt{c}, \mathtt{w}\}$;
- $m(Vg) = \{\mathtt{c}\}$;
- $m(Ma) = \{\mathtt{c}, \mathtt{w}\}$;
- $m(S) = \{(\mathtt{a}, \mathtt{c})\}$;
- $m(P) = \{(\mathtt{c}, \mathtt{w})\}$.

Consider the following problem which is an instance of (T_1): Is it true that some vegetarian is a parent of some mathematician? This problem can be represented in a terminological language as:

$$\text{Does } \mathit{and}\ Vg\ (\mathit{some}\ P\ Ma) \neq \emptyset?$$

Let τ be a translation from the T-language in question to a corresponding Peirce logic. In what follows, we will identify symbols of concepts and roles of the T-language with their translations into Peirce logic. Let $\mathcal{M}$ be a model defined above that represents the information of Fig. 4.4. Then, the problem '$and\ Vg$ ($some$ $P\ Ma) \neq \emptyset$?' can be expressed in Peirce logic:

$$\text{Does } x(1_R : Vg \cap (P : Ma)) \text{ is true in } \mathcal{M}?$$

By Theorem 4.5.4, the above problem is equivalent to $\mathsf{PL}_{\mathcal{M},\varphi}$-provability of $\varphi = x(1_R : Vg \cap (P : Ma))$, where $\mathsf{PL}_{\mathcal{M},\varphi}$ is constructed as described in Sect. 4.5.

The specific symbols of the language of $\mathsf{PL}_{\mathcal{M},\varphi}$-logic are:

- $\mathbb{OC}_{\mathsf{PL}_{\mathcal{M},\varphi}} = \{c_{\mathtt{a}}, c_{\mathtt{c}}, c_{\mathtt{w}}\}$;
- $\mathbb{BC}_{\mathsf{PL}_{\mathcal{M},\varphi}} = \{Vg, Ma, 1_B\}$ – the set of Boolean constants;
- $\mathbb{RC}_{\mathsf{PL}_{\mathcal{M},\varphi}} = \{1_R, 1', P\}$ – the set of relational constants.

A $\mathsf{PL}_{\mathcal{M},\varphi}$-model is a PL-model $\mathcal{N} = (W, n)$ such that:

- $W = \{\mathtt{a}, \mathtt{c}, \mathtt{w}\}$;
- $n(c_{\mathtt{i}}) = \mathtt{i}$, for every $c_{\mathtt{i}} \in \mathbb{OC}_{\mathsf{PL}_{\mathcal{M},\varphi}}$;

- $n(B) = m(B)$, for every $B \in \{1_B, Vg, Ma\}$;
- $n(S) = m(S)$, for every $S \in \{1_R, 1', P\}$.

The rules that are specific for $\mathsf{PL}_{\mathcal{M},\varphi}$-dual tableau are:

$$(-Vg\ \mathtt{i}) \qquad \frac{x-Vg}{x1'c_{\mathtt{i}},\ x-Vg} \qquad \text{for every } \mathtt{i} \in \{\mathtt{a}, \mathtt{w}\}$$

$$(-Ma\ \mathtt{a}) \qquad \frac{x-Ma}{x1'c_{\mathtt{a}},\ x-Ma}$$

$$(-P\mathtt{ij}) \qquad \frac{x-Py}{x1'c_{\mathtt{i}},\ x-Py \mid y1'c_{\mathtt{j}},\ x-Py} \qquad \text{for every } (\mathtt{i}, \mathtt{j}) \notin m(P)$$

$$(1') \qquad \frac{}{x-1'c_{\mathtt{a}} \mid x-1'c_{\mathtt{c}} \mid x-1'c_{\mathtt{w}}} \qquad \text{for every object variable } x$$

$$(\mathtt{i} \neq \mathtt{j}) \qquad \frac{}{c_{\mathtt{i}}1'c_{\mathtt{j}}} \qquad \text{for all } \mathtt{i}, \mathtt{j} \text{ such that } \mathtt{i} \neq \mathtt{j}$$

The axiomatic sets that are specific for $\mathsf{PL}_{\mathcal{M},\varphi}$-dual tableau are:

- $\{c_{\mathtt{c}}Vg\}$, $\{c_{\mathtt{a}}-Vg\}$, and $\{c_{\mathtt{w}}-Vg\}$;
- $\{c_{\mathtt{c}}Ma\}$, $\{c_{\mathtt{w}}Ma\}$, and $\{c_{\mathtt{a}}-Ma\}$;
- $\{c_{\mathtt{c}}Pc_{\mathtt{w}}\}$, $\{c_{\mathtt{a}}-Pc_{\mathtt{a}}\}$, $\{c_{\mathtt{a}}-Pc_{\mathtt{c}}\}$, $\{c_{\mathtt{a}}-Pc_{\mathtt{w}}\}$, $\{c_{\mathtt{c}}-Pc_{\mathtt{a}}\}$, $\{c_{\mathtt{c}}-Pc_{\mathtt{c}}\}$, $\{c_{\mathtt{w}}-Pc_{\mathtt{a}}\}$, $\{c_{\mathtt{w}}-Pc_{\mathtt{c}}\}$, $\{c_{\mathtt{w}}-Pc_{\mathtt{w}}\}$.

Figure 4.5 presents a proof of $x(1_R : Vg \cap (P : Ma))$. Thus, it follows that the problem in question has the positive solution.

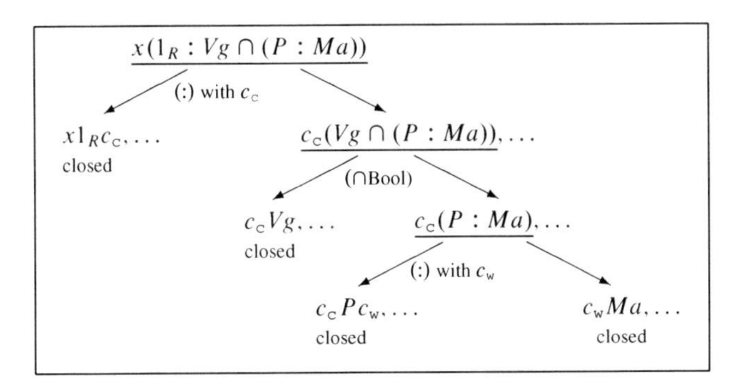

Fig. 4.5 A $\mathsf{PL}_{\mathcal{M},\varphi}$-proof solving the problem '*and* Vg (*some* P Ma) $\neq \emptyset$?'

Chapter 5
Dual Tableaux for Fork Algebras

5.1 Introduction

The origin of fork algebras can be traced back to the paper [HV91a]. Fork algebras are the algebras of binary relations which together with all the classical relational operations have a specific binary operation referred to as fork of relations. In proper fork algebras the operation fork is induced by an injective operation which performs a coding of a pair of objects into a single object. Abstract fork algebras are defined with a finite set of axioms [FHV97]. In contrast to abstract relation algebras, abstract fork algebras are representable in such a way that the representation algebras are proper fork algebras [VHF95, Gyu95]. As a consequence, the first-order theory of proper fork algebras and the first-order theory of abstract fork algebras are equivalent, and hence any first-order property of binary relations which is true in all proper fork algebras can be proved from the axioms of abstract fork algebras. This has a direct application in the theory of programming. Since programs are often viewed as input-output relations, their specifications can be conveniently expressed in the language of fork algebras. The first ideas of application of fork algebras to specification, derivation, and verification of programs can be found in [HV91b]. Since then this issue has been extensively explored, see e.g., [HBS93, BFHL96, FBH98, FBH01, Fri02, FGSB06].

Another very important application of fork algebras is that classical first-order logic is interpretable in a logic of fork algebras [VHF95]. In this chapter we present, first, a generic logic of fork algebras, FL, and a dual tableau for this logic. We prove its soundness and completeness. Next, we show how this logic can be specialized to logics appropriate for the translation of first-order languages each of which is determined by fixed families of object constants, function symbols and predicate symbols. The dual tableau for logic FL could be extended to provide deduction mechanisms for first-order theories.

Yet another stream of applications of fork algebras and their logics is concerned with interpretability of first-order non-classical logics see e.g., [FO95, FO98, FBM02, FP06].

E. Orłowska and J. Golińska-Pilarek, *Dual Tableaux: Foundations, Methodology, Case Studies*, Trends in Logic 36, DOI 10.1007/978-94-007-0005-5_5,

5.2 Fork Algebras

A *full fork algebra* is an algebra of the form:

$$(\mathcal{P}(U \times U), \cup, \cap, -, \emptyset, U \times U, ;, ^{-1}, Id, *, \nabla), \text{ where:}$$

- $(\mathcal{P}(U \times U), \cup, \cap, -, \emptyset, U \times U, ;, ^{-1}, Id)$ is a full algebra of binary relations on U;
- $*: U \times U \to U$ is a binary function on U which is injective, i.e., for all $x, y, u, v \in U$, if $*(x, y) = *(u, v)$, then $x = u$ and $y = v$;
- ∇ is a binary relational operation defined for all relations R and S on U as:

$$(\nabla) \quad R\nabla S \stackrel{\text{df}}{=} \{(x, *(y, z)) : (x, y) \in R \wedge (x, z) \in S\}.$$

It is referred to as *fork of relations R and S*. A graphical representation of fork is given in Fig. 5.1. Set U is referred to as the underlying domain of the algebra.

The function $*$ plays the role of pairing, encoding a pair of objects into a single object. Clearly, there are $*$ functions which are distinct from set-theoretical pair formation, that is $*(x, y)$ may differ from $\{x, \{x, y\}\}$.

A *proper fork algebra* is an algebra of the form:

$$(W, \cup, \cap, -, \emptyset, 1, ;, ^{-1}, 1', *, \nabla), \text{ where:}$$

- $(W, \cup, \cap, -, \emptyset, 1, ;, ^{-1}, 1')$ is a proper algebra of binary relations (see Sect. 2.2);
- $*: U \times U \to U$ is a binary function which is injective on $(U \times U) \cap 1$;
- The operation ∇ is defined with condition (∇).

The class of proper fork algebras is denoted by PFA. Clearly, any full fork algebra is a proper fork algebra.

Given a pair of binary relations, the operation called *cross* performs a kind of parallel product. Its definition is:

$$R \otimes S \stackrel{\text{df}}{=} \{(*(x, y), *(w, z)) : (x, w) \in R \wedge (y, z) \in S\}.$$

It is easy to show that in full fork algebras the cross operation is definable from the other relational operations with the use of the fork operation, namely:

$$R \otimes S = ((Id\nabla U \times U)^{-1} ; R)\nabla((U \times U\nabla Id)^{-1} ; S).$$

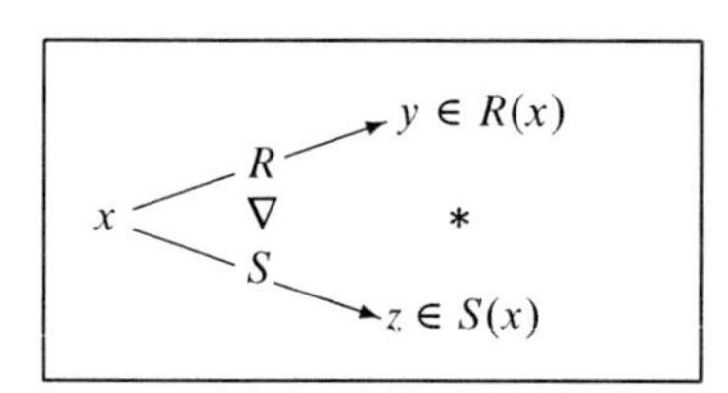

Fig. 5.1 The operation ∇

An abstract *fork algebra* is a structure of the form:

$$(W, +, \cdot, -, 0, 1, ;, \breve{}, 1', \nabla),$$

where $(W, +, \cdot, -, 0, 1, ;, \breve{}, 1')$ is a relation algebra and for all $r, s, t, q \in W$, the following conditions are satisfied:

- $r\nabla s = (r ; (1'\nabla 1)) \cdot (s ; (1\nabla 1'))$;
- $(r\nabla s) ; (t\nabla q)\breve{} = (r ; t\breve{}) \cdot (s ; q\breve{})$;
- $(1'\nabla 1)\breve{}\ \nabla(1\nabla 1')\breve{} \leq 1'$.

The cross operation is defined by the equation:

$$r \otimes s \stackrel{\text{df}}{=} ((1'\nabla 1)\breve{} ; r)\nabla((1\nabla 1')\breve{} ; s).$$

The class of all fork algebras is denoted by FA. Next theorem states a relationship between proper and abstract fork algebras. In contrast with relation algebras, fork algebras are representable.

Theorem 5.2.1. *Every fork algebra is isomorphic to a proper fork algebra.*

A first proof of this representation theorem for complete and atomistic fork algebras can be found in [FBHV93] and [FBHV95]. Later on Gyuris [Gyu95] and Frias et al. [FHV97] proved a representation theorem for the whole class of fork algebras.

In proper fork algebras the relations $\pi \stackrel{\text{df}}{=} (1'\nabla 1)\breve{}$ and $\rho \stackrel{\text{df}}{=} (1\nabla 1')\breve{}$ appearing in the third specific axiom of fork algebras behave as projections, projecting components of pairs constructed with the function $*$. They are presented in Fig. 5.2. An element f of a fork algebra is said to be *functional* whenever $f\breve{} ; f \leq 1'$. The domain of a relation r is defined as $Dom(r) \stackrel{\text{df}}{=} (r ; r\breve{}) \cdot 1'$. Similarly, the range of r is defined as $Rng(r) \stackrel{\text{df}}{=} (r\breve{} ; r) \cdot 1'$.

We define a constant $_{\mathsf{Ur}}1$ by setting:

$$_{\mathsf{Ur}}1 \stackrel{\text{df}}{=} -((1\nabla 1)\breve{}).$$

This constant is a right ideal relation whose domain consists of what is called *urelements*, that is, atomic objects which do not involve the pairing function $*$.

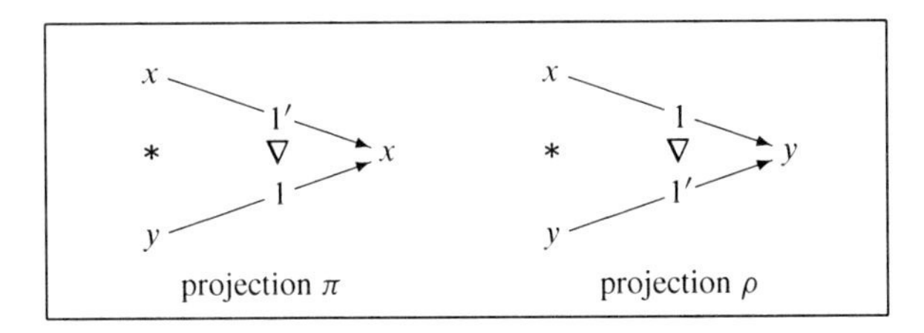

Fig. 5.2 The projections π and ρ

Furthermore, we consider the constants 1_{Ur} and $1'_{\mathsf{Ur}}$ defined as:

$$1_{\mathsf{Ur}} \stackrel{\text{df}}{=} -(1\nabla 1);$$

$$1'_{\mathsf{Ur}} \stackrel{\text{df}}{=} ({}_{\mathsf{Ur}}1\,;1_{\mathsf{Ur}})\cdot 1'.$$

1_{Ur} has urelements in the range, and $1'_{\mathsf{Ur}}$ is a partial identity ranging over urelements. Let (Ur) be the following equation:

$$(\mathsf{Ur})\quad 1\,;1'_{\mathsf{Ur}}\,;1 = 1.$$

It is true in a proper fork algebra whenever its set of urelements is non-empty.

The class of fork algebras with urelements, FAU, is a subclass of FA satisfying the equation (Ur). The class of proper fork algebras with urelements, PFAU, is the subclass of PFA of those algebras $\mathfrak{A}$ whose underlying domain $U_{\mathfrak{A}}$ contains a subset $Ur_{\mathfrak{A}}$ of the elements that are not pairs.

Next proposition lists some arithmetic properties of fork algebras.

Proposition 5.2.1. *The following properties hold in all fork algebras for all relations f, r, s, t, u:*

1. $r\,;(s\nabla t) \leq (r\,;s)\nabla(r\,;t)$;
2. *Let f be functional, then* $f\,;(r\nabla s) = (f\,;r)\nabla(f\,;s)$;
3. *If $f \leq 1'$, then* $(f\,;r)\nabla s = f\,;(r\nabla s)$;
4. $(r\nabla s)\cdot(t\nabla u) = (r\cdot t)\nabla(s\cdot u)$;
5. $(r+s)\otimes t = (r\otimes t)+(s\otimes t)$ *and* $r\otimes(s+t) = (r\otimes s)+(r\otimes t)$;
6. *The relations π and ρ are functional;*
7. $\pi^{\smile}\,;\rho = 1$;
8. $Dom(\pi) = Dom(\rho) = 1'\otimes 1'$;
9. $(r\nabla s)\,;\pi = Dom(s)\,;r$ *and* $(r\nabla s)\,;\rho = Dom(r)\,;s$.

In the next section we present a logic corresponding to fork algebras.

5.3 Fork Logic

The vocabulary of the language of fork logic, FL, consists of the following pairwise disjoint sets:

- $\mathbb{OV}_{\mathsf{FL}}$ – a countable infinite set of object variables;
- $\mathbb{OU}r_{\mathsf{FL}}$ – a non-empty set of object variables representing urelements;
- $\mathbb{RV}_{\mathsf{FL}}$ – a countable set of relational variables;
- $\mathbb{RC}_{\mathsf{FL}} = \{1, 1'\}$ – the set of the relational constants;
- $\{-, \cup, \cap, ;, ^{-1}, \nabla\}$ – the set of relational operations;
- $\{*\}$ – the set consisting of a binary object operation.

The set $\mathbb{OT}_{\mathsf{FL}}$ of object terms is the smallest set satisfying:

- $\mathbb{OV}_{\mathsf{FL}} \cup \mathbb{OU}r_{\mathsf{FL}} \subseteq \mathbb{OT}_{\mathsf{FL}}$;
- If $t_1, t_2 \in \mathbb{OT}_{\mathsf{FL}}$, then $t_1 * t_2 \in \mathbb{OT}_{\mathsf{FL}}$.

The set $\mathbb{RA}_{\mathsf{FL}} = \mathbb{RV}_{\mathsf{FL}} \cup \mathbb{RC}_{\mathsf{FL}}$ is called the set of *atomic relational terms*. The set $\mathbb{RT}_{\mathsf{FL}}$ of relational terms is the smallest set such that:

- $\mathbb{RA}_{\mathsf{FL}} \subseteq \mathbb{RT}_{\mathsf{FL}}$;
- If $R, S \in \mathbb{RT}_{\mathsf{FL}}$, then $R \cup S, R \cap S, R ; S, -R, R^{-1}, R \nabla S \in \mathbb{RT}_{\mathsf{FL}}$.

Formulas of the logic FL have the form xRy, where $x, y \in \mathbb{OT}_{\mathsf{FL}}$, $R \in \mathbb{RT}_{\mathsf{FL}}$. In semantics of FL-language we will slightly abuse the notation and use the symbol $*$ for the function which provides an interpretation of the object operation $*$.

A structure $\mathcal{M} = (U, Ur, *, m)$ is an FL-*model* whenever the following conditions are satisfied:

- U is a non-empty set;
- $m(R) \subseteq U \times U$, for every $R \in \mathbb{RA}$;
- $m(1)$ and $m(1')$ are defined as in $\mathsf{RL}(1, 1')$-models, i.e., $m(1) = U \times U$ and $m(1')$ is an equivalence relation on U such that the extensionality condition $m(1') ; m(R) = m(R) ; m(1') = m(R)$ is satisfied for every atomic relational term R;
- $*: U \times U \to U$ is a binary operation on U such that the following conditions are satisfied:
  - $(*1)$ $Ur \stackrel{\text{df}}{=} \{x \in U : \forall y, z \in U\ (x, y * z) \notin m(1')\}$ is a non-empty proper subset of U,
  - $(*2)$ $((x_1 * x_2), (y_1 * y_2)) \in m(1')$ iff $(x_1, y_1) \in m(1')$ and $(x_2, y_2) \in m(1')$, for all $x_1, x_2, y_1, y_2 \in U$;
- m extends to all the compound relational terms as follows:
  - $m(-R), m(R^{-1}), m(R \cup S), m(R \cap S), m(R ; S)$ are defined as in RL-models;
  - $m(R \nabla S) = \{(x, y) \in U \times U : \exists u, t \in U[(y, (u * t)) \in m(1') \wedge (x, u) \in m(R) \wedge (x, t) \in m(S)]\}$.

An FL-model $\mathcal{M} = (U, Ur, *, m)$ is said to be *standard* whenever $m(1')$ is the identity on U. Note the condition $(*1)$ implies that for every FL-model $\mathcal{M} = (U, Ur, *, m)$ and for all $x, y \in U$, $x * y \in U \setminus Ur$. Moreover, in standard FL-models, $*$ is an injective function from $U \times U$ to $U \setminus Ur$.

Let $\mathcal{M}$ be an FL-model. A valuation in $\mathcal{M}$ is a mapping $v: \mathbb{OT}_{\mathsf{FL}} \to U$ defined inductively as follows:

- $v(x) \in U$, for any $x \in \mathbb{OV}_{\mathsf{FL}}$;
- $v(x) \in Ur$, for any $x \in \mathbb{OU}r_{\mathsf{FL}}$;
- $v(x * y) = v(x) * v(y)$, for any $x, y \in \mathbb{OT}_{\mathsf{FL}}$.

An FL-formula xRy is said to be *satisfied* in an FL-model $\mathcal{M}$ by a valuation v whenever $(v(x), v(y)) \in m(R)$. An FL-formula xRy is *true* in $\mathcal{M}$ if and only if for every valuation v in $\mathcal{M}$, $\mathcal{M}, v \models xRy$. An FL-formula is FL-*valid* whenever it is true in all FL-models.

Observe that every standard FL-model determines a full fork algebra with urelements and a theorem analogous to Theorem 2.9.1 holds:

Theorem 5.3.1. *For every* FL-*term* R *and for any two different variables* $x, y \in \mathbb{OV}_{\mathsf{FL}}$, *the following conditions are equivalent:*

1. $R = 1$ *is true in all full fork algebras with urelements;*
2. xRy *is true in all standard* FL-*models.*

Theorem 5.3.2. *For every* FL-*term* R *and for any two different variables* $x, y \in \mathbb{OV}_{\mathsf{FL}}$, *if* $R = 1$ *is true in all full fork algebras, then* xRy *is true in all standard* FL-*models.*

5.4 Dual Tableau for Fork Logic

As usual, a dual tableau for the logic FL consists of decomposition rules and specific rules. Specific rules reflect properties of the function $*$ and the relational constant $1'$.

FL-dual tableau includes the *decomposition rules* $(\cup)$, $(\cap)$, $(-\cup)$, $(-\cap)$, $(-)$, $(^{-1})$, $(-^{-1})$, $(;)$, and $(-;)$ from Sect. 2.5 adapted to FL-language. In all the rules x, y denote arbitrary FL-object terms, in the rule $(;)$ z is any FL-object term, and in the rule $(-;)$ z is a new object variable. Furthermore, we admit the decomposition rules of the following forms:

For all $x, y \in \mathbb{OT}_{\mathsf{FL}}$ and $R, S \in \mathbb{RT}_{\mathsf{FL}}$,

$$(\nabla)\quad \frac{x(R\nabla S)y}{y1'(u*t), x(R\nabla S)y \mid xRu, x(R\nabla S)y \mid xSt, x(R\nabla S)y}$$

u, t are any object terms

$$(-\nabla)\quad \frac{x-(R\nabla S)y}{y-1'(u*t), x-Ru, x-St}\quad u, t \text{ are new object variables and } u \neq t$$

The specific rules of FL-dual tableau have the following forms:

For every atomic relational term R and for all object terms x, y, x_1, x_2, y_1, y_2,

$$(1')\quad \frac{(x_1 * x_2)1'(y_1 * y_2)}{x_1 1' y_1 \mid x_2 1' y_2} \qquad (-1')\quad \frac{(x_1 * x_2)-1'(y_1 * y_2)}{x_1-1'y_1, x_2-1'y_2}$$

$$(1'1)\quad \frac{xRy}{xRz, xRy \mid y1'z, xRy}\quad z \text{ is any object term}$$

$$(1'2)\quad \frac{xRy}{x1'z, xRy \mid zRy, xRy}\quad z \text{ is any object term}$$

$$(1'\text{Cut}) \quad \frac{x1'y}{(x*u)1'(y*t), x1'y \mid (x*u){-}1'(y*t), x1'y}$$

u, t are any object terms

$$(r\mathsf{Ur}) \quad \frac{}{u1'(t_1 * t_2)} \quad u \in \mathbb{OU}r_{\mathsf{FL}} \text{ and } t_1, t_2 \text{ are any object terms}$$

The rules ($1'$) and ($-1'$) reflect condition ($*2$) assumed in FL-models. ($1'1$) and ($1'2$) are the usual rules for relation $1'$, the same as in RL($1, 1'$)-dual tableau (see Sect. 2.7). The rule (rUr) reflects the condition ($*1$). The relationships between these rules and the corresponding semantic constraints will be shown in detail in the proof of Proposition 5.4.1.

A finite set of formulas is FL-*axiomatic* whenever it includes either of the following subsets:

For $x, y \in \mathbb{OT}_{\mathsf{FL}}$ and $R \in \mathbb{RT}_{\mathsf{FL}}$,

(Ax1) $\{x1y\}$;
(Ax2) $\{x1'x\}$;
(Ax3) $\{xRy, x{-}Ry\}$.

The notions of FL-set, FL-correctness of a rule, an FL-proof tree, a closed FL-proof tree for an FL-formula, and FL-provability of an FL-formula are defined as in relational logics in Sect. 2.4.

Proposition 5.4.1.

1. *The* FL-*decomposition rules are* FL-*correct;*
2. *The* FL-*specific rules are* FL-*correct;*
3. *The* FL-*axiomatic sets are* FL-*sets.*

Proof. We prove some parts of 1. and 2.

1. By way of example, we prove correctness of the rule ($-\nabla$). Let X be a finite set of FL-formulas.

Let $u, t \in \mathbb{OV}_{\mathsf{FL}}$ be variables that do not occur in $X \cup \{x{-}(R\nabla S)y\}$ and such that $u \neq t$. Assume $X \cup \{x{-}(R\nabla S)y\}$ is an FL-set. Suppose that $X \cup \{y{-}1'(u*t), x{-}Ru, x{-}St\}$ is not an FL-set. Then there exist an FL-model $\mathcal{M} = (U, Ur, *, m)$ and a valuation v in $\mathcal{M}$ such that for every $\varphi \in X \cup \{y{-}1'(u*t), x{-}Ru, x{-}St\}$, $\mathcal{M}, v \not\models \varphi$, in particular $(v(y), (v(u) * v(t))) \in m(1')$, $(v(x), v(u)) \in m(R)$, and $(v(x), v(t)) \in m(S)$. Hence, $(v(x), v(y)) \in m(R\nabla S)$. However, since $X \cup \{x{-}(R\nabla S)y\}$ is an FL-set and for every $\varphi \in X$ $\mathcal{M}, v \not\models \varphi$, $\mathcal{M}, v \models x{-}(R\nabla S)y$, that is $(v(x), v(y)) \notin m(R\nabla S)$, a contradiction.

Now, assume that $X \cup \{y{-}1'(u*t), x{-}Ru, x{-}St\}$ is an FL-set. Then for every FL-model $\mathcal{M}$ and for every valuation v in $\mathcal{M}$ either there exists $\varphi \in X$ such that $\mathcal{M}, v \models \varphi$ or the following hold: $(v(y), (v(u) * v(t))) \notin m(1')$ or $(v(x), v(u)) \notin m(R)$ or $(v(x), v(t)) \notin m(S)$. By the assumption on variables u and t, it follows that for every FL-model $\mathcal{M}$ and for every valuation v in $\mathcal{M}$ either there exists $\varphi \in X$ such that $\mathcal{M}, v \models \varphi$, or for all $a, b \in U$ the following hold: either $(v(y), (a * b)) \notin$

$m(1')$ or $(v(x), a) \notin m(R)$, or $(v(x), b) \notin m(S)$. Thus, either there exists $\varphi \in X$ such that $\mathcal{M}, v \models \varphi$ or $(v(x), v(y)) \notin m(R\nabla S)$. Therefore, $X \cup \{x-(R\nabla S)y\}$ is an FL-set, and hence the rule $(-\nabla)$ is FL-correct.

Correctness of the rule (∇) can be proved in a similar way. Correctness of the remaining decomposition rules can be proved as in RL-logic (see Sect. 2.5).

2. By way of example, we prove correctness of the rules $(1')$ and $(r\mathsf{Ur})$. For $(1')$, let X be a finite set of FL-formulas. Assume $X \cup \{(x_1 * x_2)1'(y_1 * y_2)\}$ is an FL-set. Suppose $X \cup \{x_1 1' y_1\}$ is not an FL-set. Then there exist an FL-model $\mathcal{M}$ and a valuation v in $\mathcal{M}$ such that for every $\varphi \in X \cup \{x_1 1' y_1\}$, $\mathcal{M}, v \not\models \varphi$. Hence, $(v(x_1), v(y_1)) \notin m(1')$. It follows that $\mathcal{M}, v \models (x_1 * x_2)1'(y_1 * y_2)$. Then $((v(x_1) * v(x_2)), (v(y_1) * v(y_2))) \in m(1')$. By the condition $(*2)$, $(v(x_1), v(y_1)) \in m(1')$, a contradiction. If $X \cup \{x_2 1' y_2\}$ is not an FL-set, then the proof is similar.

For $(r\mathsf{Ur})$, let $u \in \mathbb{OU}r_{\mathsf{FL}}$ and let $t_1, t_2 \in \mathbb{OT}_{\mathsf{FL}}$. Clearly, if X is an FL-set, then $X \cup \{u1'(t_1 * t_2)\}$ is also an FL-set. Assume $X \cup \{u1'(t_1 * t_2)\}$ is an FL-set and suppose that X is not an FL-set. Then there exist an FL-model $\mathcal{M}$ and a valuation v in $\mathcal{M}$ such that $\mathcal{M}, v \models u1'(t_1 * t_2)$. Thus, there exist $v(u) \in Ur$ and $v(t_1 * t_2) \in U \setminus Ur$ such that $(v(u), v(t_1 * t_2)) \in m(1')$ a contradiction with condition $(*1)$. □

A branch b of an FL-proof tree is complete whenever it is closed or it satisfies the completion conditions of RL$(1, 1')$-dual tableau adjusted to FL-language (see Sects. 2.5 and 2.7) and the following conditions which are specific for the FL-dual tableau:

For all $x, y \in \mathbb{OT}_{\mathsf{FL}}$ and $R, S \in \mathbb{RT}_{\mathsf{FL}}$,

Cpl(∇) If $x(R\nabla S)y \in b$, then for all $u, t \in \mathbb{OT}_{\mathsf{FL}}$, either $y1'u * t \in b$ or $xRu \in b$ or $xSt \in b$, obtained by an application of the rule (∇);

Cpl$(-\nabla)$ If $x-(R\nabla S)y \in b$, then for some $u, t \in \mathbb{OV}_{\mathsf{FL}}$ all the formulas $y-1'(u * t)$, $x-Ru$, and $x-St$ are in b, obtained by an application of the rule $(-\nabla)$;

For every atomic relational term R and for all object terms x, y, x_1, x_2, y_1, y_2,

Cpl$(1')$ If $(x_1 * x_2)1'(y_1 * y_2) \in b$, then either $x_1 1' y_1 \in b$ or $x_2 1' y_2 \in b$, obtained by an application of the rule $(1')$;

Cpl$(-1')$ If $(x_1 * x_2)-1'(y_1 * y_2) \in b$, then both $x_1-1'y_1 \in b$ and $x_2-1'y_2 \in b$, obtained by an application of the rule $(-1')$;

Cpl$(1'\mathrm{Cut})$ If $x1'y \in b$, then for all object terms u and t, either $(x * u)1'(y * t) \in b$ or $(x * u)-1'(y * t) \in b$, obtained by an application of the rule $(1'\mathrm{Cut})$;

Cpl$(r\mathsf{Ur})$ For every $u \in \mathbb{OU}r_{\mathsf{FL}}$ and for all object terms t_1 and t_2, $u1'(t_1 * t_2) \in b$, obtained by an application of the rule $(r\mathsf{Ur})$.

The notions of a complete FL-proof tree and an open branch of an FL-proof tree are defined as in RL-logic (see Sect. 2.5).

Although, the rules $(1')$ and $(-1')$ do not preserve the formulas of the form $(x_1 * x_2)1'(y_1 * y_2)$ and $(x_1 * x_2)-1'(y_1 * y_2)$, respectively, we can show that the closed branch property holds. The proof is similar to the proof of Proposition 2.8.1.

Let b be an open branch of an FL-proof tree. We define a branch structure $\mathcal{M}^b = (U^b, Ur^b, *^b, m^b)$ as follows:

- $U^b = \mathbb{OT}_{\mathsf{FL}}$;
- $x *^b y = x * y$, for all $x, y \in U^b$;
- $m^b(R) = \{(x, y) \in U^b \times U^b : xRy \notin b\}$, for every atomic relational FL-term R;
- $Ur^b = \{x \in U^b : \forall u, t \in U^b\ (x, u *^b t) \notin m^b(1')\}$;
- m^b extends to all the compound relational terms as in FL-models.

Proposition 5.4.2 (Branch Model Property). *For every open branch b of an* FL*-proof tree, $\mathcal{M}^b$ is an* FL*-model.*

Proof. Clearly, U^b is a non-empty set. Moreover, as in RL$(1, 1')$-logic, $m^b(1)$ and $m^b(1')$ satisfy all the required conditions. Now we show that $\mathcal{M}^b$ satisfies all the conditions that are specific for FL-logic, i.e., the conditions concerning operation $*$.

To show condition $(*1)$ observe that by the completion condition Cpl(rUr), for every $x \in \mathbb{OU}r_{\mathsf{FL}}$ and for all $y_1, y_2 \in \mathbb{OT}_{\mathsf{FL}}$, $x1'(y_1 * y_2) \in b$. Thus $(x, y_1 * y_2) \notin m^b(1')$, and hence Ur^b is a non-empty set such that $\mathbb{OU}r_{\mathsf{FL}} \subseteq Ur^b$. Since $((y_1 * y_2), (y_1 * y_2)) \in m^b(1')$ for all $y_1, y_2 \in U^b$, we have $y_1 * y_2 \in U^b \setminus Ur^b$. Therefore, Ur^b is a proper subset of U^b. Hence, the condition $(*1)$ is satisfied.

Now, we show condition $(*2)$. Assume $((x_1 *^b x_2), (y_1 *^b y_2)) \in m^b(1')$, that is $(x_1 * x_2)1'(y_1 * y_2) \notin b$. Suppose $(x_1, y_1) \notin m^b(1')$ or $(x_2, y_2) \notin m^b(1')$, so $x_1 1' y_1 \in b$ or $x_2 1' y_2 \in b$. If $x_1 1' y_1 \in b$, then by Cpl($1'$Cut), either $(x_1 * x_2)1'(y_1 * y_2) \in b$ or $(x_1 * x_2){-}1'(y_1 * y_2) \in b$. The first case is not possible, therefore $(x_1 * x_2){-}1'(y_1 * y_2) \in b$. Thus, by the completion condition Cpl($-1'$), $x_1{-}1'y_1 \in b$, a contradiction. In case $x_2 1' y_2 \in b$, the proof is similar. Now, assume that $(x_1, y_1) \in m^b(1')$ and $(x_2, y_2) \in m^b(1)$, that is $x_1 1' y_1 \notin b$ and $x_2 1' y_2 \notin b$. Suppose that $((x_1 *^b x_2), (y_1 *^b y_2)) \notin m^b(1')$, that is $(x_1 * x_2)1'(y_1 * y_2) \in b$. Thus by the completion condition Cpl($1'$), either $x_1 1' y_1 \in b$ or $x_2 1' y_2 \in b$, a contradiction. Therefore, the condition $(*2)$ is satisfied. Hence, $\mathcal{M}^b$ is an FL-model. □

Let v^b be a valuation in $\mathcal{M}^b$ such that:

- $v^b(x) = x$, for every $x \in \mathbb{OV}_{\mathsf{FL}} \cup \mathbb{OU}r_{\mathsf{FL}}$;
- $v^b(x * y) = v^b(x) *^b v^b(y)$, for all $x, y \in \mathbb{OT}_{\mathsf{FL}}$.

It is easy to see that $v^b(x * y) = x * y$, for all $x, y \in \mathbb{OT}_{\mathsf{FL}}$. Note also that v^b is well defined. Indeed, since $\mathbb{OU}r_{\mathsf{FL}} \subseteq Ur^b$, for every $x \in \mathbb{OU}r_{\mathsf{FL}}$, $v^b(x) \in Ur^b$.

Proposition 5.4.3 (Satisfaction in Branch Model Property). *Let b be an open branch of an* FL*-proof tree. Then for every* FL*-formula φ, if $\mathcal{M}^b, v^b \models \varphi$, then $\varphi \notin b$.*

Proof. The proof is by induction on the complexity of formulas. The atomic case can be proved as in Sect. 2.5 and it uses the closed branch property. By way of example, we show the proposition for specific compound fork formulas.

Assume that $(x, y) \in m^b(R\nabla S)$, that is there are $u, t \in \mathbb{OT}_{\mathsf{FL}}$ such that $(y, (u *^b t)) \in m^b(1')$, $(x, u) \in m^b(R)$, and $(x, t) \in m^b(S)$. Then, by the induction hypothesis, $y1'u * t \notin b$, $xRu \notin b$, and $xSt \notin b$. Suppose that $x(R\nabla S)y \in b$. By the completion condition Cpl(∇), either $y1'u * t \in b$ or $xRu \in b$ or $xSt \in b$, a contradiction.

Now, assume that $(x, y) \in m^b(-(R\nabla S))$. Then, for all $u, t \in \mathbb{OT}_{\mathsf{FL}}$ either $(y, (u *^b t)) \notin m^b(1')$ or $(x, u) \notin m^b(R)$ or $(x, t) \notin m^b(S)$. Suppose $x-(R\nabla S)y \in b$. By the completion condition Cpl($-\nabla$), for some $u, t \in \mathbb{OV}_{\mathsf{FL}}$, $y-1'(u * t) \in b$, $x-Ru \in b$, and $x-St \in b$. Then, by the induction hypothesis, $(y, (u *^b t)) \in m^b(1')$, $(x, u) \in m^b(R)$, and $(x, t) \in m^b(S)$, a contradiction. □

The quotient model $\mathcal{M}_q^b = (U_q^b, Ur_q^b, *_q^b, m_q^b)$ is defined in an analogous way as in Sect. 2.7:

- $U_q^b = \{\|x\| : x \in U^b\}$ and $Ur_q^b = \{\|x\| : x \in Ur^b\}$, where $\|x\|$ is the equivalence class of $m^b(1')$ generated by x;
- $m_q^b(R) = \{(\|x\|, \|y\|) \in U_q^b \times U_q^b : (x, y) \in m^b(R)\}$, for every atomic relational term R;
- $\|x\| *_q^b \|y\| = \|x * y\|$, for all $\|x\|, \|y\| \in U_q^b$;
- m_q^b extends to all the compound relational terms as in the FL-models.

The valuation v_q^b is defined as usual, i.e., $v_q^b(x) \stackrel{\text{df}}{=} \|x\|$, for all $x \in \mathbb{OT}_{\mathsf{FL}}$. It is easy to see that v_q^b is a valuation in an FL-model $\mathcal{M}_q^b$. As in Sect. 2.7, it can be proved that $\mathcal{M}_q^b$ is a standard FL-model satisfying the same formulas as branch model $\mathcal{M}^b$. Therefore, the following theorem holds:

Theorem 5.4.1 (Soundness and Completeness of FL). *For every FL-formula φ, the following conditions are equivalent:*

1. *φ is FL-valid;*
2. *φ is true in all standard FL-models;*
3. *φ is FL-provable.*

By the above theorem and Theorem 5.3.1, we have:

Theorem 5.4.2. *For every FL-term R and for any two different variables $x, y \in \mathbb{OV}_{\mathsf{FL}}$, the following conditions are equivalent:*

1. *$R = 1$ is true in all full fork algebras with urelements;*
2. *xRy is FL-provable.*

As a consequence, by Theorem 5.3.2, the following holds:

Theorem 5.4.3. *For every FL-term R and for any two different variables $x, y \in \mathbb{OV}_{\mathsf{FL}}$, if $R = 1$ is true in all full fork algebras, then xRy is FL-provable.*

Example. Consider the following equations:

$$(1\,;[(-(1\nabla 1)^{-1}\,;-(1\nabla 1))\cap 1']\,;1 = 1,$$

$$((1'\nabla 1)^{-1})^{-1}\,;(1\nabla 1')^{-1} = 1.$$

The first equation is true in the fork algebras with urelements, while the second equation, $\pi^{\smile}\,;\rho = 1$, is true in all fork algebras (see Proposition 5.2.1(7.)). Let x and y be any different object variables. Due to Theorems 5.3.1 and 5.4.1 the truth of the above equations is equivalent to FL-provability of the following formulas, respectively:

$$x1\,;[(-(1\nabla 1)^{-1}\,;-(1\nabla 1))\cap 1']\,;1y,$$

$$x((1'\nabla 1)^{-1})^{-1}\,;(1\nabla 1')^{-1}y.$$

Figures 5.3 and 5.4 present FL-proofs of these formulas, respectively.

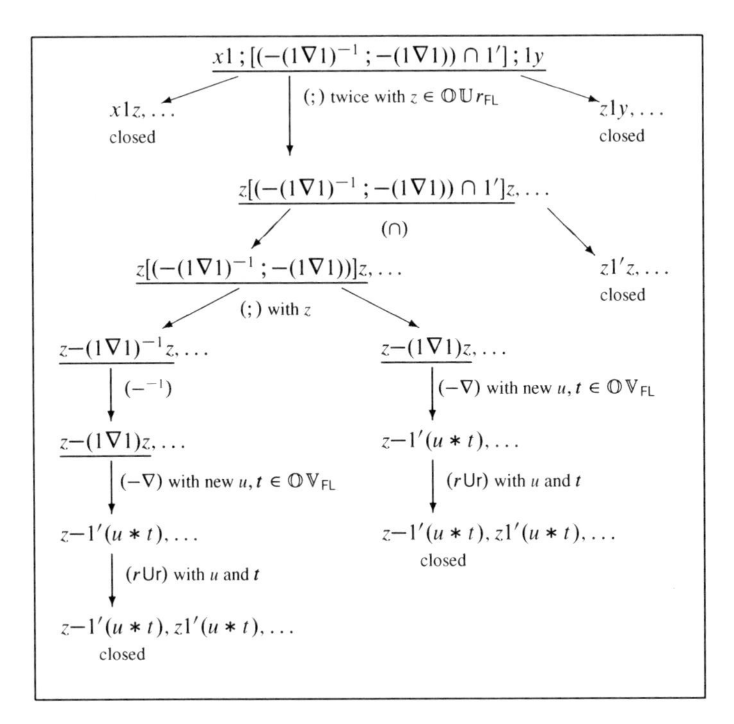

Fig. 5.3 An FL-proof of $x1\,;[(-(1\nabla 1)^{-1}\,;-(1\nabla 1))\cap 1']\,;1y$

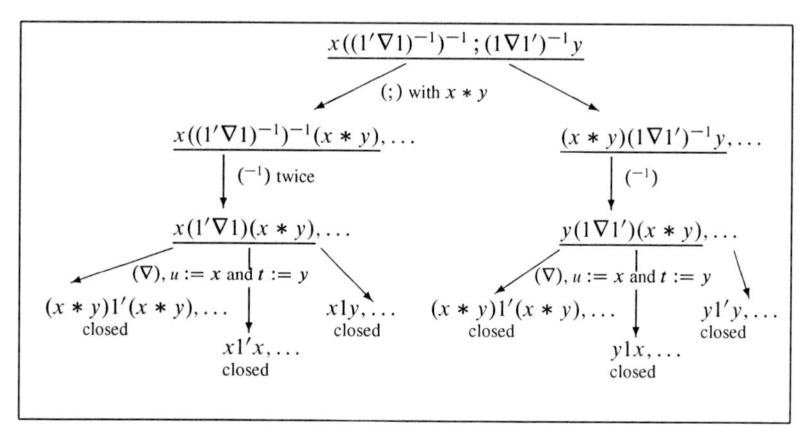

Fig. 5.4 An FL-proof of $x((1'\nabla 1)^{-1})^{-1} ; (1\nabla 1')^{-1}y$

5.5 Relational Interpretation of First-Order Theories

Much the same as relational logic $\mathsf{RL}(1, 1')$ serves as a tool for presentation of languages of propositional logics and for relational reasoning in these logics, fork logic is expressive enough for relational interpretation of first-order languages. The computer science applications of fork logic are concerned with program construction based on binary relations. Within the FL-framework any program is thought of as a relation which it establishes between input and output data. Since program specifications are often presented in a first-order language, fork logic and its deductive tools enable us to derive programs from their specifications, see e.g., [FBH01, Fri02, FGSB06].

In this section we briefly present the main steps of the translation of a first-order language with function symbols and without the identity predicate into terms of the logic FL. A detailed description of the translation procedure and the proofs of the theorems of this section can be found in [Fri02].

Now, we recall the definition of the first-order language dealt with in this section. Let FOL be a first-order language built from the symbols of the following pairwise disjoint sets:

- $\mathbb{OV}_{\mathsf{FOL}}$ – a set of individual variables;
- $\mathbb{OC}_{\mathsf{FOL}}$ – a set of individual constants;
- $\mathbb{F}_{\mathsf{FOL}}$ – a set of function symbols;
- $\mathbb{P}_{\mathsf{FOL}}$ – a set of predicate symbols;
- $\{\neg, \vee, \wedge, \exists, \forall\}$ – the set of propositional operations and quantifiers.

In order to reflect the intuition of relations as programs, we assume that with every n-ary predicate symbol P, $n \geq 1$, there are associated two natural numbers r and s such that $r + s = n$ with intended meaning that P acts on r input objects and s output objects.

The set $\mathbb{T}_{\mathsf{FOL}}$ of FOL-terms is the smallest set satisfying:

- $\mathbb{OV}_{\mathsf{FOL}} \cup \mathbb{OC}_{\mathsf{FOL}} \subseteq \mathbb{T}_{\mathsf{FOL}}$;
- If $f \in \mathbb{F}_{\mathsf{FOL}}$ is an n-ary function symbol, $n \geq 1$, and $t_1, \ldots, t_n \in \mathbb{T}_{\mathsf{FOL}}$, then $f(t_1, \ldots, t_n) \in \mathbb{T}_{\mathsf{FOL}}$.

Atomic FOL-formulas are of the form $P(t_1, \ldots, t_n)$, where $P \in \mathbb{P}_{\mathsf{FOL}}$ is an n-ary predicate symbol, $n \geq 1$, and $t_1, \ldots, t_n \in \mathbb{T}_{\mathsf{FOL}}$. The set of FOL-formulas is the smallest set including all atomic FOL-formulas and closed with respect to the propositional operations and quantifiers.

A FOL-model is a structure $\mathcal{M} = (U, m)$ such that U is a non-empty set and m is the meaning function satisfying:

- $m(c) \in U$, for any $c \in \mathbb{OC}_{\mathsf{FOL}}$;
- $m(f){:}\, U^n \rightarrow U$ is an n-ary function on U, for every n-ary function symbol $f \in \mathbb{F}_{\mathsf{FOL}}$;
- $m(P) \subseteq U^n$, for any n-ary predicate symbol $P \in \mathbb{P}_{\mathsf{FOL}}$;
- m extends to all terms so that $m(t){:}\, U^{\mathbb{OV}_{\mathsf{FOL}}} \rightarrow U$ is a function from the set of valuations in $\mathcal{M}$ into U such that for a valuation $v{:}\, \mathbb{OV}_{\mathsf{FOL}} \rightarrow U$ its value is defined as:

  if $t = x$, $x \in \mathbb{OV}_{\mathsf{FOL}}$, then $m(x)(v) = v(x)$,
  if $t = c$, $c \in \mathbb{OC}_{\mathsf{FOL}}$, then $m(c)(v) = m(c)$,
  $m(f(t_1, \ldots, t_n))(v) = m(f)(m(t_1)(v), \ldots, m(t_n)(v))$.

Let $\mathcal{M}$ be a FOL-model. Satisfaction of a FOL-formula φ in $\mathcal{M}$ by a valuation v, $\mathcal{M}, v \models \varphi$, is defined inductively as follows:

- $\mathcal{M}, v \models P(t_1, \ldots, t_n)$ iff $(m(t_1)(v), \ldots, m(t_n)(v)) \in m(P)$;
- $\mathcal{M}, v \models \neg\varphi$ iff $\mathcal{M}, v \not\models \varphi$;
- $\mathcal{M}, v \models \varphi \vee \psi$ iff $\mathcal{M}, v \models \varphi$ or $\mathcal{M}, v \models \psi$;
- $\mathcal{M}, v \models \varphi \wedge \psi$ iff $\mathcal{M}, v \models \varphi$ and $\mathcal{M}, v \models \psi$;
- $\mathcal{M}, v \models \exists x\varphi$ iff there exists v' in $\mathcal{M}$ such that for all $z \neq x$, $v(z) = v'(z)$ and $\mathcal{M}, v' \models \varphi$;
- $\mathcal{M}, v \models \forall x\varphi$ iff for all v' in $\mathcal{M}$, if $v(z) = v'(z)$ for all $z \neq x$, then $\mathcal{M}, v' \models \varphi$.

The language of $\mathsf{FL_{FOL}}$ is an FL-language endowed with the symbols of binary relational constants corresponding to the object constants, function symbols, and predicate symbols of FOL. More precisely, the language of $\mathsf{FL_{FOL}}$ is an FL-language with the following set of relational constants:

$$\mathbb{RC}_{\mathsf{FL_{FOL}}} = \{1, 1'\} \cup \{R_c : c \in \mathbb{OC}_{\mathsf{FOL}}\} \cup \{R_f : f \in \mathbb{F}_{\mathsf{FOL}}\} \cup \{R_P : P \in \mathbb{P}_{\mathsf{FOL}}\}.$$

The semantics of the $\mathsf{FL_{FOL}}$-language is defined so that in the models of $\mathsf{FL_{FOL}}$ the sets of urelements represent the universes of FOL-models. The function $*$ in an $\mathsf{FL_{FOL}}$-model enables us to represent each n-tuple of elements of the model as a single element. The specific relational constants of $\mathsf{FL_{FOL}}$ are interpreted as binary relations satisfying the conditions which reflect their role in the language of FOL.

An $\mathsf{FL}_{\mathsf{FOL}}$-model is a standard FL-model $\mathcal{M} = (U, Ur, *, m)$ where:

(R_c) If c is an object constant of FOL-language, then $m(R_c)$ is a binary relation C which is a point relation, i.e., it satisfies the following conditions:

(C1) $C^{-1} ; C \subseteq 1'_{\mathsf{Ur}}$,
(C2) $1 ; C = C$,
(C3) $C ; 1 = 1$,

(R_f) If f is an n-ary function symbol of FOL-language, then $m(R_f)$ is a binary functional relation F whose domain consists of elements of the form $a_1 * \ldots * a_n$ with $a_1, \ldots, a_n \in Ur$. This is guaranteed by the following conditions:

(F1) $F^{-1} ; F \subseteq 1'_{\mathsf{Ur}}$,
(F2) $(1'_{\mathsf{Ur}})^n ; F = F$,

(R_P) If P is an n-ary predicate of FOL-language such that $n = r + s$, then $m(R_P)$ is a binary relation P' such that its domain consists of elements of the form $a_1 * \ldots * a_r$, for $a_1, \ldots, a_r \in Ur$, which represent strings of r urelements, and its range consists of objects of the form $a'_1 * \ldots * a'_s$, for $a'_1, \ldots, a'_s \in Ur$, which represent strings of s urelements. This is guaranteed by the following condition:

(P) $(1'_{\mathsf{Ur}})^r ; P' ; (1'_{\mathsf{Ur}})^s = P'$.

The elements of set Ur in $\mathsf{FL}_{\mathsf{FOL}}$-models are intended to represent objects of the universe of a FOL-model. Condition (C1) says that C is a functional relation on a set of urelements, (C2) says that C is left ideal, and (C3) says that C is non-empty. It is intended that the elements from the range of C represent objects of a FOL-model. Condition (F1) says that F is a function on the set of urelements. Condition (F2) says that domain of relation F represents n-tuples of arguments of function f. Condition (P) guarantees that relation P' represents n-tuples of arguments of predicate P with an indication of input and output arguments. In Propositions 5.5.1 and 5.5.2 we show how these postulates can be realized.

The first step towards a translation of the formulas of FOL into terms of $\mathsf{FL}_{\mathsf{FOL}}$ is to assign binary relation symbols to all the symbols of $\mathsf{FL}_{\mathsf{FOL}}$. We define the mapping τ' from the symbols of FOL to relation variables or relation constants of $\mathsf{FL}_{\mathsf{FOL}}$ as follows:

- $\tau'(c) = R_c$, for any $c \in \mathbb{OC}_{\mathsf{FOL}}$;
- $\tau'(f) = R_f$, for any n-ary function symbol $f \in \mathbb{F}_{\mathsf{FOL}}$;
- $\tau'(P) = R_P$, for any n-ary predicate symbol $P \in \mathbb{P}_{\mathsf{FOL}}$.

The translation of FOL-terms into FL-terms is defined relative to a string of free variables in a FOL-formula, where the terms appear. Let φ be a FOL-formula with the free variables $x_1, \ldots, x_m$, $m \geq 1$. Then we define the translation τ_φ of FOL-terms as follows:

- If $x_i \in \mathbb{OV}_{\mathsf{FOL}}$ appears among $x_1, \ldots, x_m$, then:

$$\tau_\varphi(x_i) \stackrel{\text{df}}{=} \begin{cases} \underbrace{\rho\,;\ldots;\rho}_{i-1 \text{ times}}\,;\pi & \text{if } i \neq m \\ \underbrace{\rho\,;\ldots;\rho}_{m-1 \text{ times}} & \text{if } i = m \end{cases}$$

where ρ and π are projection relations defined in Sect. 5.2 and for $i = 1$:

$$\underbrace{\rho\,;\ldots;\rho}_{i-1 \text{ times}} \stackrel{\text{df}}{=} 1',$$

- If $c \in \mathbb{OC}_{\mathsf{FOL}}$ and c appears in φ, then $\tau_\varphi(c) \stackrel{\text{df}}{=} \tau'(c)$,
- If $f(t_1, \ldots, t_n) \in \mathbb{T}_{\mathsf{FOL}}$ and $f(t_1, \ldots, t_n)$ appears in φ, then:

$$\tau_\varphi(f(t_1, \ldots t_n)) \stackrel{\text{df}}{=} (\tau_\varphi(t_1)\nabla \ldots \nabla\tau_\varphi(t_n))\,;\tau'(f).$$

It follows that the translation of a term $f(t_1, \ldots, t_n)$ is obtained by making a contraction of the string $t_1, \ldots, t_n$ of terms to an element $t = t_1 * (t_2 * \ldots (t_{n-1} * t_n) \ldots)$ and by replacing f by a functional relation $\tau'(f)$ which takes t as its argument.

Now, we define a translation τ of FOL-formulas into $\mathsf{FL}_{\mathsf{FOL}}$-terms:

- If φ is an atomic formula of the form $P(t_1, \ldots, t_r, t'_1, \ldots, t'_s)$, then:

$$\tau(\varphi) \stackrel{\text{df}}{=} [(\tau_\varphi(t_1)\nabla \ldots \nabla\tau_\varphi(t_r))\nabla(\tau_\varphi(t'_1)\nabla \ldots \nabla\tau_\varphi(t'_s))\,;\tau'(P)^{-1}]\,;2^{-1}\,;1,$$

  and $2 \stackrel{\text{df}}{=} 1'\nabla 1'$, i.e., 2 consists of the elements of the form $(a, a * a)$, where $a \in U$ for some set U;
- $\tau(\neg\varphi) \stackrel{\text{df}}{=} -\tau(\varphi)$;
- $\tau(\varphi \vee \psi) \stackrel{\text{df}}{=} \tau(\varphi) \cup \tau(\psi)$;
- $\tau(\varphi \wedge \psi) \stackrel{\text{df}}{=} \tau(\varphi) \cap \tau(\psi)$;
- $\tau(\exists x\varphi) \stackrel{\text{df}}{=} (\tau_\varphi(x_1)\nabla \ldots \nabla\tau_\varphi(x_{i-1})\nabla 1_{\mathsf{Ur}}\nabla\tau_\varphi(x_{i+1})\nabla \ldots \nabla\tau_\varphi(x_m))\,;\tau(\varphi)$, where $x_1, \ldots, x_m$ is a string of all the free variables of φ and x appears in it on the ith position;
- $\tau(\forall x\varphi) \stackrel{\text{df}}{=} \tau(\neg\exists x\neg\varphi)$.

Note that every FOL-formula $\tau(\varphi)$ represents a right ideal relation.

Proposition 5.5.1. *For every* FOL*-model* $\mathcal{M} = (U, m)$ *there exists an* $\mathsf{FL}_{\mathsf{FOL}}$*-model* $\mathcal{M}' = (U', Ur, *, m')$ *such that* $Ur = U \subset U'$ *and for every* FOL*-formula* φ *with k free variables,* $\mathcal{M} \models \varphi$ *iff* $m'(\tau(\varphi)) = (1'_{\mathsf{Ur}})^k\,;1$.

Proof. The model $\mathcal{M}'$ is defined to be a standard FL-model, i.e., relational variables are interpreted as binary relations on U' and constants 1 and $1'$ are interpreted as $m'(1) = U \times U$ and $m'(1') = Id_{U'}$, respectively. Furthermore, we define:

- For all $a, b, c \in U'$, $a * b = c$ iff $(c, a) \in m'(\pi)$ and $(c, b) \in m'(\rho)$;
- $m'(R_c) \stackrel{\text{df}}{=} \{(a, m(c)) : a \in U'\}$, for any object constant $c \in \mathbb{OC}_{\mathsf{FOL}}$;
- $m'(R_f) \stackrel{\text{df}}{=} \{(a_1 * \ldots * a_n, b) : m(f)(a_1, \ldots, a_n) = b\}$, for any n-ary function symbol $f \in \mathbb{F}_{\mathsf{FOL}}$;
- $m'(R_P) \stackrel{\text{df}}{=} \{((a_1 * \ldots * a_r,) * (a'_1 * \ldots * a'_s), b) : (a_1, \ldots, a_r, a'_1, \ldots, a'_s) \in m(P), b \in U'\}$, for any n-ary predicate symbol $P \in \mathbb{P}_{\mathsf{FOL}}$ such that $n = r + s$.

The model defined above satisfies all the required conditions. □

Proposition 5.5.2. *For every* $\mathsf{FL}_{\mathsf{FOL}}$*-model* $\mathcal{M}' = (U', Ur, *, m')$ *there exists a* FOL*-model* $\mathcal{M} = (U, m)$ *such that* $U = Ur$ *and for every* FOL*-formula* φ *with free variables* $x_1, \ldots, x_m$, $m \geq 1$, *(or in case of formulas* $\exists x\varphi$ *or* $\forall x\varphi$, $x_1, \ldots, x_m$ *are free variables of* φ*), and for every* $s = a_1 * \ldots * a_m \in U'$ *with* $a_1, \ldots, a_m \in Ur$, *the following holds:*

$s \in Dom(m'(\tau(\varphi)))$ *iff* $\mathcal{M}, v \models \varphi$ *for all* v *in* $\mathcal{M}$ *such that* $v(x_i) = a_i$, $i = 1, \ldots, m$.

Proof. The model $\mathcal{M}$ is constructed as follows:

- $m(c) \stackrel{\text{df}}{=} Rng(m'(R_c))$;
- If $(a_1 * \ldots * a_n, b) \in m'(R_f)$, then we put $m(f)(a_1, \ldots, a_n) = b$;
- If $((a_1 * \ldots * a_r) * (a'_1 * \ldots * a'_s), b) \in m'(R_P)$, then we put $(a_1, \ldots, a_r, a'_1, \ldots, a'_s) \in m(P)$.

The conditions (R_c), (R_f), and (R_P) assumed in $\mathsf{FL}_{\mathsf{FOL}}$-models guarantee that the meanings of the relational constants R_c, R_f, and R_P in model $\mathcal{M}'$ are binary relations on set Ur and satisfy the required conditions. □

Theorem 5.5.1. *For every* FOL*-formula* φ *and for all object variables* $x, y \in \mathbb{OV}_{\mathsf{FL}_{\mathsf{FOL}}}$, *the following conditions are equivalent:*

1. φ *is* FOL*-valid;*
2. *In every* $\mathsf{FL}_{\mathsf{FOL}}$*-model* $\mathcal{M} = (U, Ur, *, m)$, $m(\tau(\varphi)) = 1$*;*
3. $x\tau(\varphi)y$ *is* $\mathsf{FL}_{\mathsf{FOL}}$*-valid.*

Chapter 6
Dual Tableaux for Relational Databases

6.1 Introduction

We present a calculus TRL of typed relations introduced in [MO04] which is intended to be a formal tool both for representing relational databases [Cod70] and also for reasoning with them. Typed relations are heterogeneous relations, i.e., the objects related with a relation may range over different domains. Three features of this calculus distinguish it from the calculus of ordinary relations in the Tarski-style.

First, associated with each relation is its type, which is a finite subset of a set whose members are interpreted as attributes. In this way we cope with the fact that database relations are determined by (finite) subsets of a set of attributes. Therefore, the relations of the calculus are relative in the sense suggested in [Orł88] and [DO02].

Second, as with ordinary relations, each typed relation has an arity, which is the cardinality of its type. However, for any $n \geq 1$, the order of the elements in the n-tuples belonging to a relation does not matter. This reflects the well-known property of database relations that the order of the attributes in the data table is immaterial. Tuples are treated as mappings that assign to an attribute an element of its domain.

Third, the calculus is comprised of relations of various arities and some operations may act on relations of not necessarily the same arity.

The basic operations on typed relations are intersection, complement, product, and projection. The other typical operations used in relational databases, namely, union, complement of a relation with respect to another relation, natural join, division, and selection are definable in terms of these basic operations. Analogous to the logics of binary relations, formulas of the logic of typed relations are built with typed relational terms and a string of object variables and/or constants of the appropriate length. We present a dual tableau for this logic and prove its completeness.

Next, following [BO97], we discuss a relational representation of database dependencies. All the typical dependencies can be expressed in terms of indiscernibility relations in the set of tuples of a relation in a relational database. Along these lines, any relation generated from indiscernibility relations with the operations from algebras of binary relations may be regarded as a generalized database dependency. This allows us to apply a rule extension of the dual tableau of logic $\mathsf{RL}(1, 1')$ to

E. Orłowska and J. Golińska-Pilarek, *Dual Tableaux: Foundations, Methodology, Case Studies*, Trends in Logic 36, DOI 10.1007/978-94-007-0005-5_6,

the verification of these dependencies and their implications. Since indiscernibility relations are equivalence relations we have to add the rules reflecting reflexivity, symmetry, and transitivity to the rules of the system for $\mathsf{RL}(1, 1')$. The dual tableau obtained in this way is sound and complete.

6.2 The Calculus of Typed Relations

Let Ω be an infinite set whose elements are referred to as *attributes*. To each $a \in \Omega$ there is associated a non-empty set D_a called the *domain of attribute a*. Types of relations, usually denoted by capital letters $A, B, \ldots$, etc., are finite subsets of Ω, including the empty set; clearly, if A and B are types, then $A \cup B$, $A \cap B$, and $A-B$ are types. $A \uplus B$ denotes the union of disjoint sets A' and B' obtained from A and B, respectively, by renaming their elements, if necessary. Consequently, card(A) = card(A'), card(B) = card(B'), and $A' \cap B' = \emptyset$. In particular, $A \uplus A$ is the union of two disjoint sets each of which has the same cardinality as A. This understanding of $\uplus$ allows us to assume that $\uplus$ is commutative and associative and that $A \uplus \emptyset = A$. It is a common practice in database systems to rename attributes as needed. To enable renaming, we assume that for every attribute $a \in \Omega$, there are infinitely many attributes a_i such that $D_{a_i} = D_a$. When forming $A \uplus B$, if $a' \in A'$ and $b' \in B'$ correspond to $a \in A$ and $b \in B$, respectively, it is necessary that $D_{a'} = D_a$ and $D_{b'} = D_b$. The set of all types will be denoted by $\mathbb{T}_\Omega$. Our definition of the disjoint union involves renaming implicitly.

Let $D_A = \bigcup\{D_a : a \in A\}$; then in particular, $D_\emptyset = \emptyset$. A tuple of type A is a map $u : A \to D_A$ such that for every $a \in A$, $u(a) \in D_a$. The collection of all tuples of type A is called the relation 1^A; for each $a \in \Omega$ the collection of tuples of type $\{a\}$ is denoted by 1^a. Let $1^\emptyset = \{e\}$, where e is the empty tuple. For each $a \in \Omega$, $D_a \neq \emptyset$; therefore $1^a \neq \emptyset$. Consequently, $1^A \neq \emptyset$ for all $A \in \mathbb{T}_\Omega$.

The above definitions imply that $1^{A \uplus B} = \{t : \exists u \in 1^A, \exists v \in 1^B$ such that if $a \in A$ then $t(a) = u(a)$ and if $b \in B$ then $t(b) = v(b)\}$. Observe that we have defined $\uplus$ so that $1^{A \uplus B} = 1^{B \uplus A}$, $1^{A \uplus (B \uplus C)} = 1^{(A \uplus B) \uplus C}$, and if $B \subseteq A$, then $1^A = 1^{B \uplus (A-B)}$. We often denote tuples $t \in 1^{A \uplus B}$ by uv, and say $t = uv$, where $u \in 1^A$ and $v \in 1^B$. Clearly, uv is a mapping, $uv : A \uplus B \to D_{A \uplus B}$, where $D_{A \uplus B} = D_A \cup D_B$. Our notation uv is analogous to the relational database notation for unions of sets of attributes: uv is the union of two mappings, where a mapping is understood as a set of pairs satisfying the well known functionality requirements. Thus $uv = vu$; similarly, $(uv)w = u(vw)$. Finally, for any $A \in \mathbb{T}_\Omega$, and for any $u \in 1^A$, $ue = eu = u$.

A relation R is said to be *of type A* whenever it is a subset of 1^A.

The basic operations on typed relations are as follows. Let $A, B \in \mathbb{T}_\Omega$:

- Intersection ($\cap$)
  Let $R, S \subseteq 1^A$; then $R \cap^A S \stackrel{\text{df}}{=} \{u \in 1^A : u \in R$ and $u \in S\}$.

- Projection (Π)
  Let $B \subseteq A$ and let $R \subseteq 1^A$; then $\Pi^A_B R \stackrel{\text{df}}{=} \{u \in 1^B : \exists v \in 1^{A-B}$ such that $uv \in R\}$.
- Product ($\times$)
  Let $R \subseteq 1^A$ and $S \subseteq 1^B$; then $R \times^{A \uplus B} S \stackrel{\text{df}}{=} \{uv \in 1^{A \uplus B} : u \in 1^A, v \in 1^B$, $u \in R$, and $v \in S\}$.
- Complement ($-$)
  Let $R \subseteq 1^A$; then $-^A R \stackrel{\text{df}}{=} (1^A - R) = \{u \in 1^A : u \notin R\}$.

We define the constant 0^A as follows: $0^A \stackrel{\text{df}}{=} -^A 1^A$. Clearly, $0^A = \emptyset$ for all $A \in \mathbb{T}_\Omega$. Union, $R \cup^A S$, and complement of S with respect to R, $R -^A S$, can easily be defined in terms of the above operations. Other operations are typically used in databases; we give their set theoretic definition and the corresponding expression in terms of the above four basic operations:

- Natural join ($\rhd\lhd$)
  Let $R \subseteq 1^A$ and $S \subseteq 1^B$; then $R \rhd\lhd^{A \cup B} S \stackrel{\text{df}}{=} \{uvw \in 1^{A \cup B} : u \in 1^{A-(A \cap B)}$, $v \in 1^{A \cap B}$, $w \in 1^{B-(A \cap B)}$, $uv \in R$, and $vw \in S\}$.
  The corresponding term of the calculus of typed relations is:

$$(R \times^{A \uplus (B-(A \cap B))} 1^{B-(A \cap B)}) \cap^{A \uplus (B-(A \cap B))} (S \times^{B \uplus (A-(B \cap A))} 1^{A-(B \cap A)}).$$

- Division ($\div$)
  Let $B \subseteq A$, let $R \subseteq 1^A$ and $S \subseteq 1^B$, $S \neq 0^B$; then $R \div^A_B S \stackrel{\text{df}}{=} \{t \in 1^{A-B} : \forall s \in S, ts \in R\}$.
  The representation in the calculus of typed relations is:

$$\Pi^A_{A-B} R -^{A-B} (\Pi^A_{A-B}((\Pi^A_{A-B} R \times^{(A-B) \uplus B} S) -^A R)).$$

A general notion of selection operation, namely *select S in R* is defined for any $B \subseteq A$, $S \subseteq 1^B$ and $R \subseteq 1^A$ as follows. Its application to such S and R yields the tuples $ut \in R$ such that $u \in S$. We give its set theoretic definition and an equivalent formulation in terms of the four basic operations:

- Selection (σ)
  Let $B \subseteq A$, let $R \subseteq 1^A$ and let $S \subseteq 1^B$; then $\sigma^A_B(S, R) \stackrel{\text{df}}{=} \{ut \in 1^A : u \in 1^B$, $t \in 1^{A-B}$, $u \in S$, and $ut \in R\}$.
  The representation in the calculus of typed relations is:

$$(S \times^{B \uplus (A-B)} 1^{A-B}) \cap^A R.$$

Now, we define a binary operation, $\odot_{a,b}$, which is useful to represent entailment. Let $R, S \subseteq 1^A$ and let $a, b \in A, a \neq b$. Then:

$$R \odot_{a,b} S \stackrel{\text{df}}{=} (1^{A-\{a,b\}} \times \Pi^A_a R) \times \Pi^A_b S.$$

In [IL84] it is shown that the operations of Codd's relational algebra [Cod70] may be defined in terms of intersection, complement, and cylindrification. It follows that Codd's relational algebra can be treated as a disguised version of cylindric set algebra. The approach to relational databases via cylindrification operation has the advantage that all operations are total, since all relations are of the same type. The disadvantage of this approach is that all relations are forced to be of the same arity, and in real-life databases query checking is computationally more efficient if we use relations with varying arities. Moreover, there is no completeness theorem for the cylindrical version of relational database theory. For this reason, we choose to develop a typed calculus, with the four basic operations defined above. Other operations definable in terms of intersection, projection, product, and complement include the update operations (see [Ngu91]) and other joins.

We can easily see that with typed relations we can express all of the fundamental notions of relational database theory: *schema* – a set of attributes, *relation over a schema* – a typed relation, *tuple* and *database* – a set of typed relations. We often write explicitly the types of both the operations and their argument relations, although some typing information may be redundant.

Some arithmetics laws of the calculus of typed relations are listed in the following propositions. The proofs can be found in [MO06].

Proposition 6.2.1 (Properties of Π^A_B). *For all $A, B \in \mathbb{T}_\Omega$ and for all $R, S \subseteq 1^A$, if $B \subseteq A$, then:*

1. $\Pi^A_B(1^A) = 1^B$, $\Pi^A_B(0^A) = 0^B$;
2. $\Pi^A_B(R \cup^A S) = \Pi^A_B R \cup^B \Pi^A_B S$;
3. $\Pi^A_B(R \cap^A S) \subseteq \Pi^A_B R \cap^B \Pi^A_B S$;
4. $-^B \Pi^A_B R \subseteq \Pi^A_B(-^A R)$;
5. $\Pi^A_A R = R$;
6. $\Pi^A_\emptyset R \subseteq 1^\emptyset$ *and if* $R \neq 0^A$, *then* $\Pi^A_\emptyset R = 1^\emptyset$.

Proposition 6.2.2 (Properties of σ^A_B). *For all $A, B \in \mathbb{T}_\Omega$, for all $R \subseteq 1^A$, and for all $S, T \subseteq 1^B$, if $B \subseteq A$, then:*

1. $\sigma^A_B(S \cup^B T, R) = \sigma^A_B(S, R) \cup^A \sigma^A_B(T, R)$;
2. $\sigma^A_B(S, R \cup^A T) = \sigma^A_B(S, R) \cup^A \sigma^A_B(S, T)$;
3. $\sigma^A_B(S \cap^B T, R) = \sigma^A_B(S, R) \cap^A \sigma^A_B(T, R)$;
4. $\sigma^A_B(S, R \cap^A T) = \sigma^A_B(S, R) \cap^A \sigma^A_B(S, T)$;
5. $\sigma^A_B(-^B S, R) \subseteq -^A \sigma^A_B(S, R)$;
6. $\sigma^A_B(0^B, R) = 0^A$;
7. $\sigma^A_B(S, 0^A) = 0^A$.

Proposition 6.2.3 (Properties of $\times$). *For all $A, B, C \in \mathbb{T}_\Omega$, for all $R \in 1^A, S \in 1^B, T \in 1^C$:*

1. $R \times^{A \uplus B} S = S \times^{B \uplus A} R$;
2. $(R \times^{A \uplus B} S) \times^{(A \uplus B) \uplus C} T = R \times^{A \uplus (B \uplus C)} (S \times^{B \uplus C} T)$.

Proposition 6.2.4 (Properties of $\triangleright\triangleleft$). *For all $A, B, C \in \mathbb{T}_\Omega$, for all $R \in 1^A$, $S \in 1^B, T \in 1^C$:*

1. $R \triangleright\triangleleft^{A\cup A} R = R$;
2. $R \triangleright\triangleleft^{A\cup B} S = S \triangleright\triangleleft^{B\cup A} R$;
3. $(R \triangleright\triangleleft^{A\cup B} S) \triangleright\triangleleft^{(A\cup B)\cup C} T = R \triangleright\triangleleft^{A\cup(B\cup C)} (S \triangleright\triangleleft^{B\cup C} T)$.

Proposition 6.2.5 (Properties of $\odot_{a,b}$). *For all $A \in \mathbb{T}_\Omega$, for all $a, b \in A$ such that $a \neq b$, and for all $R, S, P, Q \subseteq 1^A$, if $Q \neq \emptyset$, then:*

1. $1^A \odot_{a,b} -^A 1^A = 0^A$;
2. $(1^A \odot_{a,b} -^A 1^A) \odot_{a,b} 1^A = 0^A$;
3. $1^A \odot_{a,b} 1^A = 1^A$;
4. $R \odot_{a,b} 0^A = 0^A = 0^A \odot_{a,b} R$;
5. $R \odot_{a,b} (S \odot_{a,b} P) = (R \odot_{a,b} S) \odot_{a,b} P$.

Let $R, S \subseteq 1^A$ and let $a, b \in A, a \neq b$. Then:

$$R \supset^A_{a,b} S \stackrel{\text{df}}{=} ((1^A \odot_{a,b} -^A R) \odot_{a,b} 1^A) \cup^A S.$$

Proposition 6.2.6. *Let $R, S \subseteq 1^A$. Then for all $a, b \in A$ such that $a \neq b$ the following hold:*

1. $1^A \supset^A_{a,b} S = S$;
2. *If $R \neq 1^A$, then $R \supset^A_{a,b} S = 1^A$.*

The following theorem shows that entailment in the calculus of typed relations can be expressed in its language.

Theorem 6.2.1. *Let $A, B \subset \Omega$, let $R \subseteq 1^A, S \subseteq 1^B$, $a, b \in A, a \neq b$, and let $C = A \cup B$. Then the following are equivalent:*

1. *$R = 1^A$ implies $S = 1^B$;*
2. $R \times 1^{C-A} \supset^C_{a,b} S \times 1^{C-B} = 1^C$.

6.3 A Logic of Typed Relations

In this section we present a language of typed relations introduced in [MO06] whose intended models are databases. Let Ω be a set of attributes as defined in Sect. 6.2. Let $\mathbb{T}_\Omega$ be a set of types based on Ω. Then the expressions of the language of the logic TRL of typed relations over Ω are built from the following pairwise disjoint sets of symbols:

- $\{e\}$ – the empty tuple;
- $\mathbb{OV}^a_{\mathsf{TRL}}$ – an infinite set of object variables of type a, for each $a \in \Omega$; by $\mathbb{OV}^A_{\mathsf{TRL}}$ we denote the set $\bigcup_{a\in A} \mathbb{OV}^a_{\mathsf{TRL}}$, where $A \in \mathbb{T}_\Omega$;

- $\mathbb{OC}^a_{\mathsf{TRL}}$ – a set of object constants of type a, for each $a \in \Omega$; by $\mathbb{OC}^A_{\mathsf{TRL}}$ we denote the set $\bigcup_{a\in A} \mathbb{OC}^A_{\mathsf{TRL}}$, where $A \in \mathbb{T}_\Omega$;
- $\mathbb{OS}^a_{\mathsf{TRL}} = \mathbb{OV}^a_{\mathsf{TRL}} \cup \mathbb{OC}^a_{\mathsf{TRL}}$; by $\mathbb{OS}^A_{\mathsf{TRL}}$ we denote the set $\bigcup_{a\in A} \mathbb{OS}^a_{\mathsf{TRL}}$, where $A \in \mathbb{T}_\Omega$; we presume $\mathbb{OC}^\emptyset_{\mathsf{TRL}} = \{e\}$;
- $\mathbb{RV}^A_{\mathsf{TRL}}$ – a set of relational variables of type A, for each $A \in \mathbb{T}_\Omega$;
- $\mathbb{RC}^A_{\mathsf{TRL}}$ – a set of relational constants of type A, for each $A \in \mathbb{T}_\Omega$; $0^A, 1^A \in \mathbb{RC}^A_{\mathsf{TRL}}$;
- $\mathbb{OP}_{\mathsf{TRL}}$ – a set of operations of varying arities such that: for every k-ary operation $\# \in \mathbb{OP}_{\mathsf{TRL}}$, $k \geq 1$, there is associated a sequence $\tau(\#) = (A_1, \ldots, A_k, A)$ of $k+1$ elements of $\mathbb{T}_\Omega$, where A_i is the type of the ith argument of #, $i = 1, \ldots, k$, A is the type of the expression obtained by performing the operation #.

We presume: $\mathbb{OP}_{\mathsf{TRL}} \supseteq \{\Pi^A_B, \cup^A, \cap^A, -^A, \times^{A\uplus C} : A, B, C \in \mathbb{T}_\Omega, B \subseteq A\}$, where $\tau(\Pi^A_B) = (A, B)$, $\tau(\cup^A) = \tau(\cap^A) = (A, A, A)$, $\tau(-^A) = (A, A)$, and $\tau(\times^{A\uplus C}) = (A, C, A \uplus C)$. It follows that $\tau(\div^A_B) = (A, B, A - B)$, $\tau(\sigma^A_B) = (B, A, A)$, and for any $D \in \mathbb{T}_\Omega$, $\tau(\rhd\lhd^{A\cup D}) = (A, D, A \cup D)$.

Assumptions concerning the elements of $\mathbb{OS}^{A\uplus B}_{\mathsf{TRL}}$ and $\mathbb{OS}^\emptyset_{\mathsf{TRL}}$, analogous to the corresponding assumptions on the set of tuples, are assumed to hold. It follows from the definitions that $\mathbb{OS}^{A\uplus B}_{\mathsf{TRL}} = \mathbb{OS}^{B\uplus A}_{\mathsf{TRL}}$ and $\mathbb{OS}^{(A\uplus B)\uplus C}_{\mathsf{TRL}} = \mathbb{OS}^{A\uplus(B\uplus C)}_{\mathsf{TRL}}$. As with the notation defined for the tuples, if u denotes a variable from $\mathbb{OV}^{A\uplus B}_{\mathsf{TRL}}$, it may be replaced by an expression vw, where $v \in \mathbb{OV}^A_{\mathsf{TRL}}$ and $w \in \mathbb{OV}^B_{\mathsf{TRL}}$. The set of atomic terms $\mathbb{RA}^A_{\mathsf{TRL}}$ is defined as the set $\mathbb{RV}^A_{\mathsf{TRL}} \cup \mathbb{RC}^A_{\mathsf{TRL}}$. We assume that for all A, $\mathbb{OS}^A_{\mathsf{TRL}} \neq \emptyset$ and $\mathbb{OS}^\emptyset_{\mathsf{TRL}} = \{e\}$.

For each $A \in \mathbb{T}_\Omega$, the set of terms of type A, $\mathbb{RT}^A_{\mathsf{TRL}}$, is the smallest set such that:

- $\mathbb{RA}^A_{\mathsf{TRL}} \subseteq \mathbb{RT}^A_{\mathsf{TRL}}$;
- If $\# \in \mathbb{OP}_{\mathsf{TRL}}$ is such that $\tau(\#) = (A_1, \ldots, A_m, A)$ and $F_i \in \mathbb{RT}^{A_i}_{\mathsf{TRL}}$, $i = 1, \ldots, m$, then $\#(F_1, \ldots, F_m) \in \mathbb{RT}^A_{\mathsf{TRL}}$.

A formula in the language of the logic TRL of typed relations over Ω is an expression of the form $F(u)$, where $F \in \mathbb{RT}^A_{\mathsf{TRL}}$ and $u \in \mathbb{OS}^A_{\mathsf{TRL}}$, for some $A \in \mathbb{T}_\Omega$.

A TRL-*model* for the language of the logic TRL of typed relations over Ω is a system $\mathcal{M} = \{\{A : A \in \mathbb{T}_\Omega\}, \{U^A : A \in \mathbb{T}_\Omega\}, e, m\}$, where U^A is a non-empty set of tuples of type A, $U^\emptyset = \{e\}$, and m is a meaning function subject to the following conditions:

- If $u \in \mathbb{OC}^A_{\mathsf{TRL}}$, then $m(u) \in U^A$ and if $u = vw$, then $m(u) = m(v)m(w)$; in addition, $m(e) = e$;
- If $R \in \mathbb{RA}^A_{\mathsf{TRL}}$, then $m(R) \subseteq U^A$; in particular, $m(0^A) = \emptyset$ and $m(1^A) = U^A$;
- If $\# \in \mathbb{OP}_{\mathsf{TRL}}$ and $\tau(\#) = (A_1, \ldots, A_k, A)$, then $m(\#)$ is a k-ary operation acting on relations of types $A_1, \ldots, A_k$ and returning a relation of type A: $m(\#) : 2^{U^{A_1}} \times \cdots \times 2^{U^{A_k}} \to 2^{U^A}$;
- m extends to all the compound relational terms as follows: if $F = \#(F_1, \ldots, F_k)$, then $m(F) = m(\#)(m(F_1), \ldots, m(F_k))$.

It is easy to see that each database over Ω determines a TRL-model.

A valuation in a TRL-model $\mathcal{M} = \{\{A : A \in \mathbb{T}_\Omega\}, \{U^A : A \in \mathbb{T}_\Omega\}, e, m\}$ is a function $v: \bigcup\{\mathbb{OS}^A_{\mathsf{TRL}} : A \in \mathbb{T}_\Omega\} \to \bigcup\{U^A : A \in \mathbb{T}_\Omega\}$ such that:

- If $u \in \mathbb{OV}^A_{\mathsf{TRL}}$, then $v(u) \in U^A$;
- If $u \in \mathbb{OC}^A_{\mathsf{TRL}}$, then $v(u) = m(u)$;
- If $u \in \mathbb{OS}^A_{\mathsf{TRL}}$ and $w \in \mathbb{OS}^B_{\mathsf{TRL}}$, then $v(uw) = v(u)v(w)$.

It follows from the definition of tuple that $v(ut) = v(tu)$, $v((ut)w) = v(u(tw))$, and $v(ue) = v(eu) = v(u)$.

Let $F \in \mathbb{RT}^A_{\mathsf{TRL}}$ and let $u \in \mathbb{OS}^A_{\mathsf{TRL}}$. The formula $F(u)$ is said to be satisfied in a TRL-model $\mathcal{M}$ by a valuation v whenever $v(u) \in m(F)$. We say that the formula $F(u)$ is true in the model $\mathcal{M}$ if and only if it is satisfied by all valuations in $\mathcal{M}$. Therefore if $u \in \mathbb{OV}^A_{\mathsf{TRL}}$, then $F(u)$ is true in the model $\mathcal{M}$ if and only if $m(F) = U^A$. We say that $F(u)$ is TRL-valid if it is true in all TRL-models.

6.4 Dual Tableau for the Logic of Typed Relations

We present a dual tableau for the language of typed relations whose semantics is determined by the class of TRL-models.

Decomposition rules have the following forms:

For all $F, G \in \mathbb{RT}^A_{\mathsf{TRL}}$, $B \subseteq A$, $u \in \mathbb{OS}^A_{\mathsf{TRL}}$, and $w \in \mathbb{OS}^B_{\mathsf{TRL}}$,

$$(\cup)\quad \frac{(F \cup^A G)(u)}{F(u), G(u)} \qquad (-\cup)\quad \frac{-^A(F \cup^A G)(u)}{-^A F(u) \mid -^A G(u)}$$

$$(\cap)\quad \frac{(F \cap^A G)(u)}{F(u) \mid G(u)} \qquad (-\cap)\quad \frac{-^A(F \cap^A G)(u)}{-^A F(u), -^A G(u)}$$

$$(-)\quad \frac{-^A -^A F(u)}{F(u)}$$

$$(\Pi)\quad \frac{(\Pi^A_B F)(w)}{F(wt), (\Pi^A_B F)(w)} \qquad t \in \mathbb{OS}^{A-B}_{\mathsf{TRL}} \text{ is any object symbol}$$

$$(-\Pi)\quad \frac{-^B(\Pi^A_B F)(w)}{-^A F(wt)} \qquad t \in \mathbb{OV}^{A-B}_{\mathsf{TRL}} \text{ is a new object variable}$$

For all $F \in \mathbb{RT}^A_{\mathsf{TRL}}$, $G \in \mathbb{RT}^B_{\mathsf{TRL}}$, $v \in \mathbb{OS}^A_{\mathsf{TRL}}$, $w \in \mathbb{OS}^B_{\mathsf{TRL}}$, and $z \in \mathbb{OS}^{A-B}_{\mathsf{TRL}}$,

$$(\times)\quad \frac{(F \times^{A \uplus B} G)(vw)}{F(v) \mid G(w)} \qquad (-\times)\quad \frac{-^{A\uplus B}(F \times^{A \uplus B} G)(vw)}{-^A F(v), -^B G(w)}$$

Specific rules have the following forms:

For all $F \in \mathbb{RT}^A_{\mathsf{TRL}}$, $u_i \in \mathbb{OS}^{A_i}_{\mathsf{TRL}}$, $A = A_1 \uplus \ldots \uplus A_n$, $1 \leq i \leq n$,

$$(e)\quad \frac{F(u_1 \dots u_{i-1}(eu_i) \dots u_n)}{F(u_1 \dots u_{i-1}u_i \dots u_n),\ F(u_1 \dots u_{i-1}(eu_i) \dots u_n)}$$

$$(e')\quad \frac{F(u_1 \dots u_{i-1}u_i \dots u_n)}{F(u_1 \dots u_{i-1}(eu_i) \dots u_n),\ F(u_1 \dots u_{i-1}u_i \dots u_n)}$$

$$(\pi)\quad \frac{F(u_1 \dots u_n)}{F(u_{\pi(1)} \dots u_{\pi(n)})}$$

where π is a permutation on $\{1, \dots, n\}$

$$(\uplus)\quad \frac{F(u_1 \dots u_n)}{F(u_1 \dots u_{i-1}v_1v_2u_{i+1} \dots u_n),\ F(u_1 \dots u_n)}$$

$A_i = B_1 \uplus B_2$,
$v_1 \in \mathbb{OS}^{B_1}_{\mathsf{TRL}}$, $v_2 \in \mathbb{OS}^{B_2}_{\mathsf{TRL}}$ are any object symbols

The rules (e) and (e') reflect the interpretation of e as the empty tuple, the rule (π) reflects the fact that objects can be permuted without changing the meaning of a formula, and the rule $(\uplus)$ reflects the fact that any variable can be split into components in such a way that its type is preserved. The language includes variables of the empty type so that the rule $\uplus$ holds for all types.

A set of TRL-formulas is said to be TRL-axiomatic whenever it includes either of the sets of the following forms:

For any $u \in \mathbb{OS}^{A}_{\mathsf{TRL}}$, $R \in \mathbb{RA}^{A}_{\mathsf{TRL}}$, and $F \in \mathbb{RT}^{A}_{\mathsf{TRL}}$, $F \neq 0^A$,

(Ax1) $\{R(u), (-^A R)(u)\}$;
(Ax2) $\{1^A(u)\}$;
(Ax3) $\{-^A 0^A(u)\}$;
(Ax4) $\{(\Pi^A_\emptyset F)(e)\}$.

Rules for the defined operations may be given explicitly using their set theoretic formulation. We present some examples below.

Let $F \in \mathbb{RT}^{A}_{\mathsf{TRL}}$, $G \in \mathbb{RT}^{B}_{\mathsf{TRL}}$, $B \subseteq A$, $w \in \mathbb{OS}^{B}_{\mathsf{TRL}}$, $t \in \mathbb{OS}^{A-B}_{\mathsf{TRL}}$, then:

$$(\sigma)\quad \frac{(\sigma^A_B(G, F))(wt)}{G(w) \mid F(wt)} \qquad (-\sigma)\quad \frac{-^A(\sigma^A_B(G, F))(wt)}{-^B G(w),\ -^A F(wt)}$$

if, in addition, $B \neq \emptyset$, then:

$$(\div)\quad \frac{(F \div^A_B G)(t)}{-^B G(w),\ F(tw)}$$ $w \in \mathbb{OV}^{B}_{\mathsf{TRL}}$ is a new object variable

$$(-\div)\quad \frac{-^{A-B}(F \div^A_B G)(t)}{G(w),\ -^{A-B}(F \div^A_B G)(t) \mid -^A F(tw),\ -^{A-B}(F \div^A_B G)(t)}$$

$w \in \mathbb{OS}^{B}_{\mathsf{TRL}}$ is any object symbol

Let $F \in \mathbb{RT}^A_{\mathsf{TRL}}$, $G \in \mathbb{RT}^B_{\mathsf{TRL}}$, $u \in \mathbb{OS}^{A-(A\cap B)}_{\mathsf{TRL}}$, $v \in \mathbb{OS}^{A\cap B}_{\mathsf{TRL}}$, $w \in \mathbb{OS}^{B-(A\cap B)}_{\mathsf{TRL}}$, then:

$$(\triangleright\triangleleft)\ \frac{(F \triangleright\triangleleft^{A\cup B} G)(uvw)}{F(uv) \mid G(vw)} \qquad (-\triangleright\triangleleft)\ \frac{-^{A\cup B}(F \triangleright\triangleleft^{A\cup B} G)(uvw)}{-^{A}F(uv), -^{B}G(vw)}$$

As usual, a TRL-set is a finite set of TRL-formulas such that the first-order disjunction of its members is true in all TRL-models. Correctness of a rule is defined in a similar way as in the relational logics of classical algebras of binary relations (see Sect. 2.4).

Proposition 6.4.1.

1. *The* TRL-*rules are* TRL-*correct;*
2. *The* TRL-*axiomatic sets are* TRL-*sets.*

Proof. By way of example, we show correctness of the rules $(-\Pi)$ and (e).

$(-\Pi)$ Let X be a finite set of TRL-formulas and let t be such that $t \neq w$ and it does not occur in X. Assume $X \cup \{-^B(\Pi^A_B F)(w)\}$ is a TRL-set. Suppose $X \cup \{-^A F(wt)\}$ is not a TRL-set. Then, there exist a TRL-model $\mathcal{M}$ and a valuation v in $\mathcal{M}$ such that for every $\varphi \in X \cup \{-^A F(wt)\}$, $\mathcal{M}, v \not\models \varphi$, which means that $v(w)v(t) \in m(F)$, where $v(w) \in 1^B$ and $v(t) \in 1^{A-B}$. Since $X \cup \{-^B(\Pi^A_B F)(w)\}$ is a TRL-set, $\mathcal{M}, v \models -^B(\Pi^A_B F)(w)$. Thus, for every $u \in 1^{A-B}$, $v(w)u \notin m(F)$, a contradiction. Now, assume that $X \cup \{-^A F(wt)\}$ is a TRL-set. Then for every TRL-model $\mathcal{M}$ and for every valuation v in $\mathcal{M}$, either there exists $\varphi \in X$ such that $\mathcal{M}, v \models \varphi$ or $\mathcal{M}, v \models -^A F(wt)$. By the assumption on variable t, it follows that either there exists $\varphi \in X$ such that $\mathcal{M}, v \models \varphi$ or for every $u \in 1^{A-B}$, $v(w)u \notin m(F)$. Hence, $X \cup \{-^B(\Pi^A_B F)(w)\}$ is a TRL-set.

(e) Let X be a finite set of TRL-formulas. If $X \cup \{F(u_1 \dots u_{i-1}(eu_i) \dots u_n)\}$ is a TRL-set, then so is $X \cup \{F(u_1 \dots u_{i-1}(eu_i) \dots u_n), F(u_1 \dots u_{i-1}u_i \dots u_n)\}$. Assume $X \cup \{F(u_1 \dots u_{i-1}(eu_i) \dots u_n), F(u_1 \dots u_{i-1}u_i \dots u_n)\}$ is a TRL-set. Suppose $X \cup \{F(u_1 \dots u_{i-1}(eu_i) \dots u_n)\}$ is not a TRL-set. Then, there exist a TRL-model $\mathcal{M}$ and a valuation v in $\mathcal{M}$ such that for every $\varphi \in X \cup \{F(u_1 \dots u_{i-1}(eu_i) \dots u_n)\}$, $\mathcal{M}, v \not\models \varphi$, so $v(u_1) \dots v(u_{i-1})v(eu_i) \dots v(u_n) \notin m(F)$. Since $v(eu_i) = v(u_i)$, we obtain $v(u_1) \dots v(u_{i-1})v(u_i) \dots v(u_n) \notin m(F)$. However, by the assumption $\mathcal{M}, v \models F(u_1 \dots u_{i-1}u_i \dots u_n)$. Hence, $v(u_1) \dots v(u_{i-1})v(u_i) \dots v(u_n) \in m(F)$, a contradiction. □

The notions of a TRL-proof tree, a closed branch of such a tree, a closed TRL-proof tree, and TRL-provability are defined as in Sect. 2.4.

Due to Proposition 6.4.1, we obtain:

Proposition 6.4.2. *Let φ be a* TRL-*formula. If φ is* TRL-*provable, then it is* TRL-*valid.*

Below we list the completion conditions that are specific for the logic TRL. The completion conditions for the rules $(\cap)$, $(-\cap)$, and $(-)$ are as in Sect. 2.5.

For all $F, G \in \mathbb{RT}^A_{\mathsf{TRL}}$, $B \subseteq A$, $u \in \mathbb{OS}^A_{\mathsf{TRL}}$, and $w \in \mathbb{OS}^B_{\mathsf{TRL}}$,

Cpl(Π) (resp. Cpl($-\Pi$)) If $(\Pi^A_B F)(u) \in b$ (resp. $(-^B \Pi^A_B F)(u) \in b$), then for every $t \in \mathbb{OS}^{A-B}_{\mathsf{TRL}}$, $F(ut) \in b$ (resp. for some $t \in \mathbb{OV}^{A-B}_{\mathsf{TRL}}$, $-^A F(ut) \in b$), obtained by an application of the rule (Π) (resp. ($-\Pi$));

For all $F \in \mathbb{RT}^A_{\mathsf{TRL}}$, $G \in \mathbb{RT}^B_{\mathsf{TRL}}$, $v \in \mathbb{OS}^A_{\mathsf{TRL}}$, $w \in \mathbb{OS}^B_{\mathsf{TRL}}$, and $z \in \mathbb{OS}^{A-B}_{\mathsf{TRL}}$,

Cpl($\times$) (resp. Cpl($-\times$)) If $(F \times^{A \uplus B} G)(vw) \in b$ (resp. $-^{A \uplus B}(F \times^{A \uplus B} G)(vw) \in b$), for some $v \in \mathbb{OS}^A_{\mathsf{TRL}}$ and $w \in \mathbb{OS}^B_{\mathsf{TRL}}$, then either $F(v) \in b$ or $G(w) \in b$ (resp. both $-^A F(v) \in b$ and $-^B G(w) \in b$), obtained by an application of the rule ($\times$) (resp. ($-\times$));

For all $F \in \mathbb{RT}^A_{\mathsf{TRL}}$, $u_i \in \mathbb{OS}^{A_i}_{\mathsf{TRL}}$, $A = A_1 \uplus \ldots \uplus A_n$, $1 \leq i \leq n$,

Cpl(e) If $F(u_1 \ldots u_{i-1}(eu_i) \ldots u_n) \in b$, for $n \geq 1$ and $1 \leq i \leq n$, then $F(u_1 \ldots u_{i-1} u_i \ldots u_n) \in b$, obtained by an application of the rule (e);

Cpl(e') If $F(u_1 \ldots u_{i-1}(u_i) \ldots u_n) \in b$, for $n \geq 1$ and $1 \leq i \leq n$, then $F(u_1 \ldots u_{i-1}(eu_i) \ldots u_n) \in b$, obtained by an application of the rule (e');

Cpl(π) If $F(u_1 \ldots u_n) \in b$, then for any permutation π of the indices $1, \ldots, n$, $F(u_{\pi(1)} \ldots u_{\pi(n)}) \in b$, obtained by an application of the rule (π);

Cpl($\uplus$) If $F(u_1 \ldots u_n) \in b$, for $u_i \in \mathbb{OS}^{A_i}_{\mathsf{TRL}}$, $1 \leq i \leq n$, and $A_i = B_1 \uplus B_2$, then for all $v_1 \in \mathbb{OS}^{B_1}_{\mathsf{TRL}}$, $v_2 \in \mathbb{OS}^{B_2}_{\mathsf{TRL}}$, $F(u_1 \ldots u_{i-1} v_1 v_2 u_{i+1} \ldots u_n) \in b$, obtained by an application of the rule ($\uplus$).

The notions of a complete TRL-proof tree and an open branch of a TRL-proof tree are defined as in RL-logic (see Sect. 2.5). Due to the forms of the rules of TRL-dual tableau, we can prove that whenever a branch of a TRL-proof tree contains both of the formulas $R(u)$ and $(-^A R)(u)$, for $u \in \mathbb{OS}^A_{\mathsf{TRL}}$ and for an atomic $R \in \mathbb{RA}^A_{\mathsf{TRL}}$, then the branch can be closed. Thus, the closed branch property can be proved.

Let b be an open branch of a TRL-proof tree. A branch structure

$$\mathcal{M}^b = \{\{A : A \in \mathbb{T}_\Omega\}, \{U^A : A \in \mathbb{T}_\Omega\}, e^b, m^b\}$$

is defined as follows:

- $U^A = \mathbb{OS}^A_{\mathsf{TRL}}$, for any $\emptyset \neq A \in \mathbb{T}_\Omega$;
- $U^\emptyset = \{e\}$ and $e^b = m^b(e) = e$;
- $m^b(c) = c$, for every $c \in \mathbb{OC}^A_{\mathsf{TRL}}$;
- $m^b(R) = \{u \in \mathbb{OS}^A_{\mathsf{TRL}} : R(u) \notin b\}$, for every $R \in \mathbb{RA}^A_{\mathsf{TRL}}$;
- For every $\# \in \mathbb{OP}_{\mathsf{TRL}}$, $m^b(\#)$ is defined as in TRL-models;
- m^b extends to all the compound relational terms as in TRL-models.

It is easy to see that the branch structure defined above is a TRL-model. Hence, the branch model property holds. Let v^b be a valuation in $\mathcal{M}^b$ such that $v^b(u) = u$, for every $u \in \mathbb{OS}_{\mathsf{TRL}}$.

Proposition 6.4.3 (Satisfaction in Branch Model Property). *For every open branch b of a* TRL*-proof tree and for every* TRL*-formula φ, if $\mathcal{M}^b, v^b \models \varphi$, then $\varphi \notin b$.*

Proof. The proof is by induction on the complexity of formulas. The atomic case can be proved as in Sect. 2.5, it uses the closed branch property. By way of example, we show the proposition for some compound formulas that are specific for TRL-logic.

Let $\varphi = (\Pi_B^A F)(w)$. Assume $\mathcal{M}^b, v^b \models (\Pi_B^A F)(w)$. Then there exists $u \in \mathbb{OS}_{\mathsf{TRL}}^{A-B}$ such that $wu \in m^b(F)$, and by the induction hypothesis, $F(wu) \notin b$. Suppose $(\Pi_B^A F)(w) \in b$. Then by the completion condition Cpl(Π), for every $u \in \mathbb{OS}_{\mathsf{TRL}}^{A-B}$, $F(wu) \in b$, a contradiction.

Let $\varphi = -^{A \uplus B}(F \times^{A \uplus B} G)(uw)$. Assume $\mathcal{M}^b, v^b \models -^{A \uplus B}(F \times^{A \uplus B} G)(uw)$. Then $u \notin m^b(F)$ or $w \notin m^b(G)$. Suppose $-^{A \uplus B}(F \times^{A \uplus B} G)(uw) \in b$. Then, by the completion condition Cpl($-\times$), $-^A F(u) \in b$ and $-^B G(w) \in b$. Thus, by the induction hypothesis, $u \in m^b(F)$ and $w \in m^b(G)$, a contradiction. □

As usual, Proposition 6.4.3 enables us to prove:

Proposition 6.4.4. *Let φ be a TRL-formula. If φ is TRL-valid, then it is TRL-provable.*

Finally, by Propositions 6.4.2 and 6.4.4, we have:

Theorem 6.4.1 (Soundness and Completeness of TRL). *For every TRL-formula φ, the following conditions are equivalent:*

1. *φ is TRL-valid;*
2. *φ is TRL-provable.*

Some other relational approaches to databases can be found in [DM01, Mac00, Mac01].

Theorem 6.4.2.

1. *Each of the Boolean algebra identities is TRL-provable;*
2. *Each of the expressions in Propositions 6.2.1, ..., 6.2.5 is TRL-provable.*

Example. Consider the following property: if R is a relation of type A and $B \subseteq A$, then $R \subseteq (\Pi_B^A R) \times^{B \uplus (A-B)} 1^{A-B}$. To prove that this property holds in all TRL-models it suffices to demonstrate that the formula $(R \rightarrow^A (\Pi_B^A R) \times^{B \uplus (A-B)} 1^{A-B})(u)$ is TRL-provable, where $F \rightarrow^A G$ is defined as $-^A F \cup^A G$. If $u \in \mathbb{OV}_{\mathsf{TRL}}^A$, then $(F \rightarrow^A G)(u)$ is TRL-valid iff for all TRL-models, $m(F) \subseteq m(G)$. Figure 6.1 presents a TRL-proof of the formula

$$(-^A R \cup^A (\Pi_B^A R) \times^{B \uplus (A-B)} 1^{A-B})(u).$$

Figure 6.2 shows TRL-provability of the formula:

$$(T \times^{A_1 \uplus A_2} \sigma_B^{A_2}(S, R) \rightarrow^{A_1 \uplus A_2} \sigma_B^{A_1 \uplus A_2}(S, T \times R))(u).$$

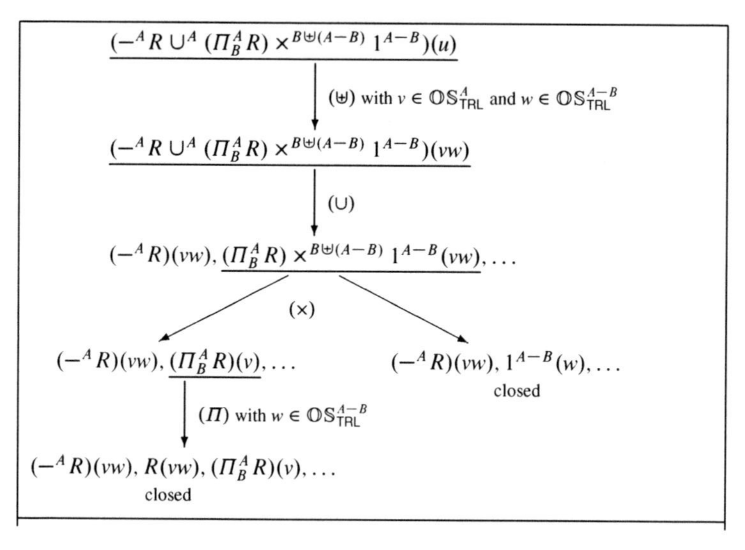

Fig. 6.1 A TRL-proof of the formula $(-^A R \cup^A (\Pi^A_B R) \times^{B \uplus (A-B)} 1^{A-B})(u)$

Its equivalent form is:

$$(-^{A_1 \uplus A_2}(T \times^{A_1 \uplus A_2} \sigma^{A_2}_B(S, R)) \cup^{A_1 \uplus A_2} \sigma^{A_1 \uplus A_2}_B(S, T \times R))(u).$$

6.5 Relational Representation of Database Dependencies

Let a universe Ω of attributes be given and let A be a finite subset of Ω. We recall that a database relation over A, R^A, is a set of tuples of type A. We shall often omit the index A in the name of a database relation. Given a database relation R, by AT_R we mean the underlying set of attributes. Following the usual database notation, for a subset B of A and for a tuple $t \in R^A$ we write $t[B]$ for a restriction of t (as a mapping) to set B.

Given a database relation R over A and a subset B of A, we define an *indiscernibility relation in* R as follows. For any $t, u \in R$, $(t, u) \in ind(B)$ iff $t[B] = u[B]$.

Example. Consider relation R defined in Table 6.1. Indiscernibility relation $ind(a)$ determined by attribute a consists of the following pairs of tuples:

$$ind(a) = \{(t_1, t_1), (t_2, t_2), (t_3, t_3), (t_4, t_4), (t_1, t_4), (t_2, t_3), (t_3, t_2)\}.$$

Indiscernibility relation $ind(bc)$ determined by attributes b and c consists of the following tuples:

$$ind(bc) = \{(t_1, t_1), (t_2, t_2), (t_3, t_3), (t_4, t_4), (t_1, t_2), (t_2, t_1)\}.$$

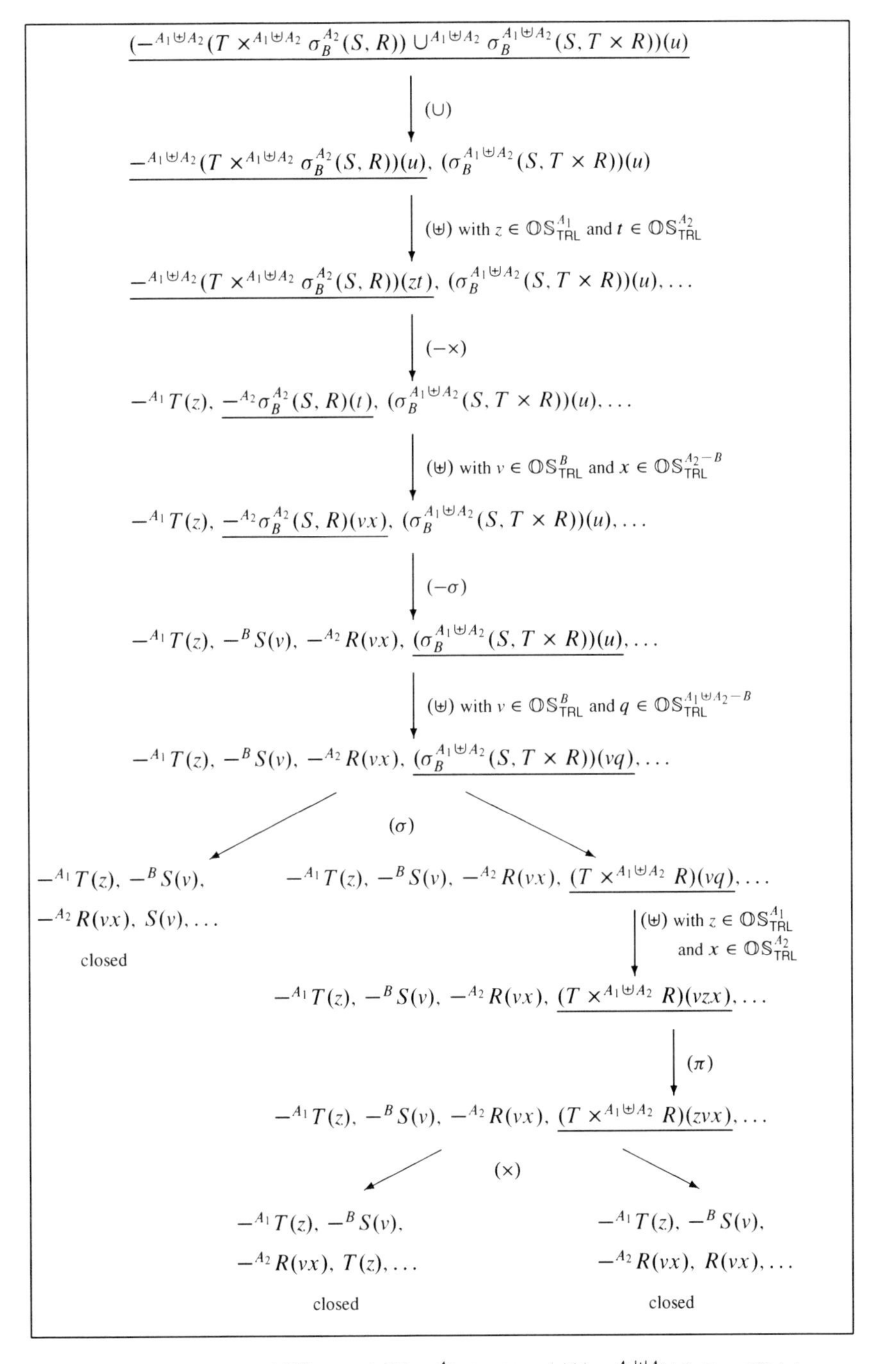

Fig. 6.2 A TRL-proof of $(-^{A_1 \uplus A_2}(T \times^{A_1 \uplus A_2} \sigma_B^{A_2}(S, R)) \cup^{A_1 \uplus A_2} \sigma_B^{A_1 \uplus A_2}(S, T \times R))(u)$

Table 6.1 A database relation

Tuples	Attributes		
	a	b	c
t_1	1	2	3
t_2	4	2	3
t_3	4	5	3
t_4	1	5	6

In the following proposition we recall some basic properties of the indiscernibility relations.

Proposition 6.5.1. *For any database relation R and for all $A, B \subseteq AT_R$, the following conditions are satisfied:*

1. $ind(AT_R) = \{(t,t) : t \in R\}$, $ind(\emptyset) = R \times R$;
2. $ind(A \cup B) = ind(A) \cap ind(B)$;
3. $ind(A) \cup ind(B) \subseteq ind(A) ; ind(B)$;
4. *If* $A \subseteq B$, *then* $ind(B) \subseteq ind(A)$;
5. $ind(A) = \bigcap\{ind(a) : a \in A\}$.

Let a database relation R be given. Various attribute dependencies in R can be defined in terms of indiscernibility relations. We recall the standard definitions of those dependencies and their relational representation developed in [Orł87]. Let $A, B, C \subseteq AT_R$.

Functional Dependency

$A \rightarrow B$ holds in R iff for all tuples $t, u \in R$, if $t[A] = u[A]$, then $t[B] = u[B]$.

Proposition 6.5.2. *The following conditions are equivalent:*

1. *Functional dependency* $A \rightarrow B$ *holds in* R;
2. $ind(A) \subseteq ind(B)$.

Multivalued Dependency

$A \rightarrow\rightarrow B$ holds in R iff for all tuples $t, u \in R$, if $t[A] = u[A]$, then there exists $t' \in R$ such that the following conditions are satisfied:

- $t'[A \cup B] = t[A \cup B]$;
- $t'[AT_R - (A \cup B)] = u[AT_R - (A \cup B)]$.

Proposition 6.5.3. *The following conditions are equivalent:*

1. *Multivalued dependency* $A \rightarrow\rightarrow B$ *holds in* R;
2. $ind(A) \subseteq ind(A \cup B) ; ind(AT_R - (A \cup B))$.

Embedded Multivalued Dependency

$A \rightarrow\rightarrow B|C$ holds in R iff $A \rightarrow\rightarrow B$ holds in the set $\{t[A \cup B \cup C] : t \in R\}$.

Proposition 6.5.4. *The following conditions are equivalent:*

1. *Embedded multivalued dependency* $A \rightarrow\!\rightarrow B|C$ *holds in* R*;*
2. $ind(A) \subseteq ind(A \cup B) ; ind((C-(A \cup B)))$.

Decomposition

(A, B) holds in R iff $A \cup B = AT_R$ and for all $t, u \in R$, if $t[A \cap B] = u[A \cap B]$, then there exists $t' \in R$ such that $t'[A] = t[A]$ and $t'[B] = u[B]$.

Proposition 6.5.5. *The following conditions are equivalent:*

1. *Decomposition* (A, B) *holds in* R*;*
2. $A \cup B = AT_R$ *and* $ind(A \cap B) \subseteq ind(A) ; ind(B)$.

Join Dependency

$^*(A_1, \ldots, A_n)$ holds in R iff $A_1, \ldots, A_n = AT_R$ and R is the join of relations $\{t[A_i] : t \in R\}$, for $i = 1, \ldots, n$.

Proposition 6.5.6. *Join dependency* $^*(A_1, \ldots, A_n)$ *holds in* R *implies* $ind(A_1 \cap \ldots \cap A_n) \subseteq ind(A_1) ; \ldots ; ind(A_n)$.

A dependency defined by the condition $ind(A_1 \cap \ldots \cap A_n) \subseteq ind(A_1) ; \ldots ; ind(A_n)$, for $A_1, \ldots, A_n$ such that $A_1 \cup \ldots A_n = AT_R$, is called a *generalized join dependency*.

We conclude that given a database relation R, any relation generated from relations $ind(a)$, for $a \in AT_R$, with the operations of algebras of binary relations may be viewed as a kind of database dependency.

6.6 Dual Tableau for Database Dependencies

Due to the relational representation of database dependencies we can verify dependencies and implications of dependencies within a relational logic. The logic adequate for this purpose, $\mathsf{RL}_{\mathsf{EQ}}$, is obtained from $\mathsf{RL}(1, 1')$ by restricting its class of models to the models where the meanings of relation variables are assumed to be equivalence relations. This assumption is in agreement with the fact that dependencies are represented with relations generated by indiscernibility relations which in turn are equivalence relations.

A relational logic obtained from $\mathsf{RL}_{\mathsf{EQ}}$ by restricting its class of models to the models whose universes are sets of tuples of a database relation, is referred to as a *relational logic with database models*, $\mathsf{RL}_{\mathsf{DEP}}$.

Given a database relation R, an $\mathsf{RL}_{\mathsf{DEP}}$-*model* is a structure $\mathcal{M} = (R, m)$ such that:

- R is the set of tuples;
- $m(P) \in \{ind_R(a) : a \in AT_R\}$, for every relational variable P;
- $m(1) = R \times R$;
- $m(1') = Id_R$.

Since AT_R is finite for any database relation R, the image under m of the set of relational variables is finite in any database model. Clearly, $\mathsf{RL_{DEP}}$-model is an $\mathsf{RL_{EQ}}$-model. On the other hand, for every $\mathsf{RL_{EQ}}$-model we can construct an $\mathsf{RL_{DEP}}$-model such that the models verify the same formulas, that is we have the following proposition.

Proposition 6.6.1.

1. *For every* $\mathsf{RL_{DEP}}$*-model there exists an* $\mathsf{RL_{EQ}}$*-model* $\mathcal{M} = (U, m)$ *such that* $m(\{P : P \text{ is a relational variable}\})$ *is finite and the models verify the same formulas;*
2. *For every* $\mathsf{RL_{EQ}}$*-model* $\mathcal{M} = (U, m)$*, if* $m(\{P : P \text{ is a relational variable}\})$ *is finite, then there is an* $\mathsf{RL_{DEP}}$*-model such that the models verify the same formulas.*

Proof. The part 1. of the proposition is obvious in view of the corresponding definitions. To prove 2. we construct an $\mathsf{RL_{DEP}}$-model determined by the given $\mathsf{RL_{EQ}}$-model $\mathcal{M} = (U, m)$. We define a database relation R so that $AT_R = m(\{P : P \text{ is a relational variable}\})$, that is the attributes of R are the equivalence relations determined by the meanings of relational variables. It follows that for every relational variable P, $\{m(P)\}$ is an attribute; it will be denoted by a_P. For every $x \in U$ we define a tuple t_x as $t_x(a_P) = \|x\|_{m(P)}$ which assigns an equivalence class of x with respect to the relation $m(P)$ to the attribute a_P. Then we have $(t_x, t_y) \in ind_R(a_P)$ iff $(x, y) \in m(P)$. It is easy to check that the sets of formulas true in these models coincide. □

A dual tableau for the logic $\mathsf{RL_{EQ}}$ consists of the rules and axiomatic sets of $\mathsf{RL}(1, 1')$-dual tableau endowed with the rules which reflect interpretation of relational variables as equivalence relations:

For all object symbols x and y and for any relational variable P,

$$(\text{ref } P) \quad \frac{xPy}{x1'y, xPy} \qquad\qquad (\text{sym } P) \quad \frac{xPy}{yPx}$$

$$(\text{tran } P) \quad \frac{xPy}{xPz, xPy \mid zPy, xPy} \qquad z \text{ is any object symbol}$$

The notions of an $\mathsf{RL_{EQ}}$-set and $\mathsf{RL_{EQ}}$-correctness of a rule are defined as in Sect. 2.4.

Theorem 6.6.1 (Correspondence). *Let* $\mathcal{K}$ *be a class of* $\mathsf{RL}(1, 1')$*-models. Then* $\mathcal{K}$ *is a class of* $\mathsf{RL_{EQ}}$*-models iff the rules (ref* P*), (sym* P*), and (tran* P*) are* $\mathcal{K}$*-correct.*

Proof. Let $\mathcal{K}$ be a class of $\mathsf{RL}(1, 1')$-models.

($\rightarrow$) Assume that $\mathcal{K}$ is the class of $\mathsf{RL_{EQ}}$-models. By way of example, we show correctness of the rule (tran P). Let X be any finite set of formulas. Clearly, if $X \cup \{xPy\}$ is an $\mathsf{RL_{EQ}}$-set, then so are $X \cup \{xPz, xPy\}$ and $X \cup \{zPy, xPy\}$.

Now, assume that $X \cup \{xPz, xPy\}$ and $X \cup \{zPy, xPy\}$ are $\mathsf{RL_{EQ}}$-sets, and suppose $X \cup \{xPy\}$ is not an $\mathsf{RL_{EQ}}$-set. Then, there exist an $\mathsf{RL_{EQ}}$-model $\mathcal{M} = (U, m)$ and a valuation v in $\mathcal{M}$ such that for every $\varphi \in X \cup \{xPy\}$, $\mathcal{M}, v \not\models \varphi$. Thus, $(v(x), v(y)) \notin m(P)$. By the assumption, $\mathcal{M}, v \models xPz$ and $\mathcal{M}, v \models zPy$. Hence, $(v(x), v(z)) \in m(P)$ and $(v(z), v(y)) \in m(P)$, and by transitivity of $m(P)$, $(v(x), v(y)) \in m(P)$, a contradiction. Therefore, the rule (tran P) is $\mathsf{RL_{EQ}}$-correct.

($\leftarrow$) Assume that for all relational variables P the rules (ref P), (sym P), and (tran P) are $\mathcal{K}$-correct. We show that for every relational variable P, and for every $\mathcal{K}$-model $\mathcal{M} = (U, m)$, $m(P)$ is an equivalence relation. By way of example, we show transitivity of the relation $m(P)$. Observe that due to correctness of rule (tran P), for every finite set X of relational formulas the following holds: $X \cup \{xPy\}$ is a $\mathcal{K}$-set iff $X \cup \{xPz, xPy\}$ and $X \cup \{zPy, xPy\}$ are $\mathcal{K}$-sets. Let $X \stackrel{\text{df}}{=} \{x{-}Pz, z{-}Py\}$. Since $\{xPz, xPy, x{-}Pz, z{-}Py\}$ and $\{zPy, xPy, x{-}Pz, z{-}Py\}$ are $\mathcal{K}$-sets, then so is the set $\{xPy, x{-}Pz, z{-}Py\}$. Thus, for every $\mathcal{K}$-model $\mathcal{M} = (U, m)$ and for every valuation v in $\mathcal{M}$, if $(v(x), v(z)) \in m(P)$ and $(v(z), v(y)) \in m(P)$, then $(v(x), v(y)) \in m(P)$. Hence, $m(P)$ is transitive in every $\mathcal{K}$-model. □

Proposition 6.6.2.

1. *The* $\mathsf{RL_{EQ}}$*-rules are* $\mathsf{RL_{EQ}}$*-correct;*
2. *The* $\mathsf{RL_{EQ}}$*-axiomatic sets are* $\mathsf{RL_{EQ}}$*-sets.*

Correctness of the rules (ref P), (sym P), and (tran P) follows from Theorem 6.6.1. Correctness of the remaining rules can be proved as in $\mathsf{RL}(1, 1')$-logic (see Sects. 2.5 and 2.7).

The notions of an $\mathsf{RL_{EQ}}$-proof tree, a closed branch of such a tree, a closed $\mathsf{RL_{EQ}}$-proof tree, and $\mathsf{RL_{EQ}}$-provability are defined as in Sect. 2.4.

The completion conditions that are specific for $\mathsf{RL_{EQ}}$-dual tableau are:

For all object symbols x and y and for any relational variable P,

Cpl(ref P) If $xPy \in b$, then $x1'y \in b$, obtained by an application of the rule (ref P);

Cpl(sym P) If $xPy \in b$, then $yPx \in b$, obtained by an application of the rule (sym P);

Cpl(tran P) If $xPy \in b$, then for every object symbol z, either $xPz \in b$ or $zPy \in b$, obtained by an application of the rule (tran P).

The notions of a complete $\mathsf{RL_{EQ}}$-proof tree and an open branch of an $\mathsf{RL_{EQ}}$-proof tree are defined as in RL-logic (see Sect. 2.5).

Although, the rule (sym P) does not preserve the formulas of the form xPy, for a relational variable P, we can show that the closed branch property holds. The proof is similar to the proof of Proposition 2.8.1.

The branch structure $\mathcal{M}^b = (U^b, m^b)$ determined by an open branch b of an $\mathsf{RL_{EQ}}$-proof tree is defined as in the completeness proof of RL-dual tableau, in particular $m^b(P) = \{(x, y) \in U^b \times U^b : xPy \notin b\}$, for every relational variable P (see Sect. 2.6, p. 44).

Proposition 6.6.3 (Branch Model Property). *For every open branch b of an* $\mathsf{RL_{EQ}}$*-proof tree,* $\mathcal{M}^b$ *is an* $\mathsf{RL_{EQ}}$*-model.*

Proof. Following the method of proving the branch model property in the completeness proof of $\mathsf{RL}(1,1')$-dual tableau (see Sects. 2.5 and 2.7), we show that $\mathcal{M}^b$ satisfies specific properties of $\mathsf{RL_{EQ}}$-models, namely, we need to prove that every relation $m^b(P)$ is an equivalence relation. For reflexivity, assume that $(x,y) \in m^b(1')$ and suppose that $(x,y) \notin m^b(P)$. Then $x1'y \notin b$ and $xPy \in b$. By the completion condition Cpl(ref P), $x1'y \in b$, a contradiction. For symmetry, assume that $(x,y) \in m^b(P)$ and suppose that $(y,x) \notin m^b(P)$. Then $xPy \notin b$ and $yPx \in b$. By the completion condition Cpl(sym R), $xPy \in b$, a contradiction. For transitivity, assume that $(x,y) \in m^b(P)$ and $(y,z) \in m^b(P)$, that is $xPy \notin b$ and $yPz \notin b$. Suppose that $(x,z) \notin m^b(P)$. Then $xPz \in b$, and by the completion condition Cpl(tran P), either $xPy \in b$ or $yPz \in b$, a contradiction. □

Now, the completeness of the $\mathsf{RL_{EQ}}$-dual tableau can be proved as in $\mathsf{RL}(1,1')$-logic.

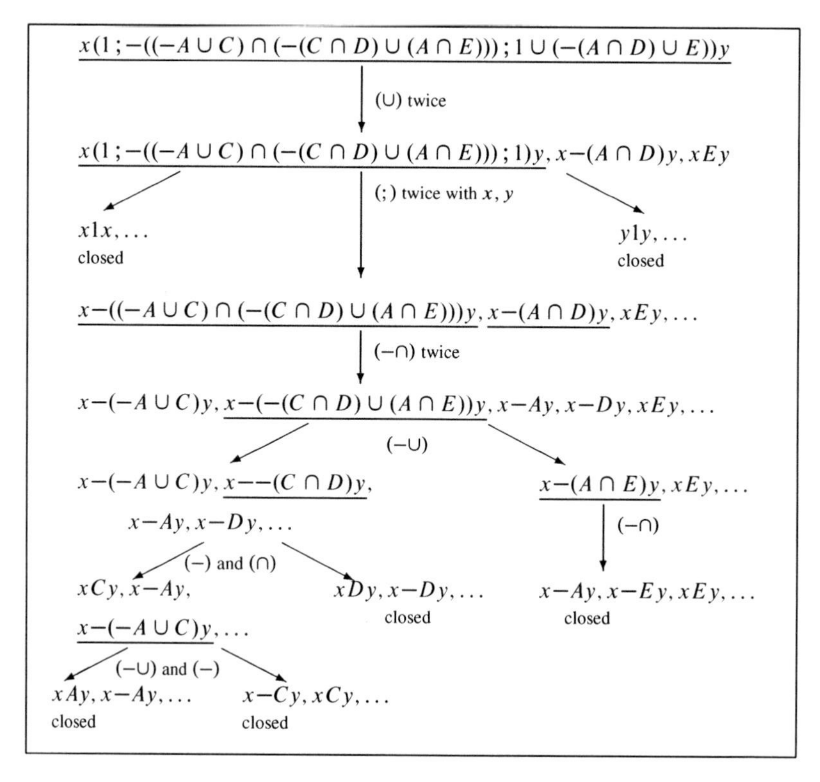

Fig. 6.3 An $\mathsf{RL_{EQ}}$-proof showing that $\{A \to C, CD \to AE\}$ implies $AD \to E$

Theorem 6.6.2 (Soundness and Completeness of $\mathsf{RL_{EQ}}$). *For every $\mathsf{RL_{EQ}}$-formula φ, the following conditions are equivalent:*

1. *φ is $\mathsf{RL_{EQ}}$-valid;*
2. *φ is $\mathsf{RL_{EQ}}$-provable.*

Theorems on entailment, model checking, and verification of satisfaction presented in Sects. 2.11, 3.4, and 3.5, respectively, apply to $\mathsf{RL_{EQ}}$-logic.

Example. We show that the set $\{A \to C, CD \to AE\}$ of functional dependencies implies dependency $AD \to E$. By Propositions 6.5.1(2.) and 6.5.2, it suffices to show that $ind(A) \subseteq ind(C)$ and $ind(C) \cap ind(D) \subseteq ind(A) \cap ind(E)$ entail $ind(A) \cap ind(D) \subseteq ind(E)$. In what follows, for the sake of simplicity, we shall write Z instead of $ind(Z)$ for $Z = A, C, D, E$. We apply the method of verification of entailment presented in Sect. 2.11, thus we verify $\mathsf{RL_{EQ}}$-validity of the following relational formula:

$$x[1 ; -[(-A \cup C) \cap (-(C \cap D) \cup (A \cap E))] ; 1 \cup (-(A \cap D) \cup E)]y.$$

Figure 6.3 presents its $\mathsf{RL_{EQ}}$-proof.

Part III
Relational Reasoning in Traditional Non-classical Logics

Chapter 7
Dual Tableaux for Classical Modal Logics

7.1 Introduction

In a narrow sense, modal logic is a logic obtained from the classical logic by endowing it with unary propositional operations intuitively corresponding to 'it is necessary that' and 'it is possible that'. These operations are intensional, i.e., the truth of a formula built with the operation does not depend only on the truth of the subformula to which the operation is applied but also on a relevant state or a situation in which the truth is considered. A development of the semantics of modal logics in terms of a relational structure of states is due to Stig Kanger [Kan57] and Saul Kripke [Kri63]. Algebraic semantics of these standard modal logics is provided by Boolean algebras with normal and additive operations [JT52]. Since the origin of Kripke semantics, intensional logics have been introduced to computer science as an important tool for its formal methods.

In a broad sense modal logic is a field of studies of logics with intensional operations. The operations may have various intuitive interpretations and are relevant in a variety of fields. In logical theories intensional operations enable us to express qualitative degrees of truth, belief, knowledge, obligation, permission, etc. Elements of the relevant relational structures are then interpreted as possible worlds, situations, instants of time, etc. In computer science intensional operations serve as formal tools for expressing dynamic aspects of physical or cognitive processes. In these cases elements of the relational structures are interpreted as the states of a computer program, the tuples of a relational database, the objects of an information system with incomplete information, the agents of multiagents systems, etc.

The basic systems of modal logic in a modern form are due to Clarence Irving Lewis (see [Lew20, LL59, Zem73]). They evolved from his treatment of implication in search for elimination of the paradoxes of the classical implication of Frege–Russell. A development and broad range of research in modal logic and its applications can be traced through an extensive literature of the subject, see e.g., [Fey65, HC68, Seg71, Gal75, Gab76, Boo79, Che80, HC84, vB85, Boo93, Gol93, vB96, CZ97, HKT00, BdRV01, BvBW06].

E. Orłowska and J. Golińska-Pilarek, *Dual Tableaux: Foundations, Methodology, Case Studies*, Trends in Logic 36, DOI 10.1007/978-94-007-0005-5_7,

In this chapter we present a relational formalization of modal logics which originated in [Orł91]. Given a modal logic L, we show how one can construct a relational logic, $\mathsf{RL_L}$, and a dual tableau for $\mathsf{RL_L}$ so that it provides a validity checker for the logic L. We show that in fact the $\mathsf{RL_L}$-dual tableau does more: it can be used for proving entailment of an L-formula from a finite set of L-formulas, model checking of L-formulas in finite L-models, and verification of satisfaction of L-formulas in finite L-models. The relational formalization of modal logics presented in this chapter provides a paradigm for all the relational formalisms and dual tableaux considered in Parts III, IV, and V of the book.

7.2 Classical Propositional Logic

The vocabulary of the language of the classical propositional logic, PC, consists of the following pairwise disjoint sets:

- $\mathbb{V}$ – a countable set of propositional variables;
- $\{\neg, \vee, \wedge\}$ – the set of propositional operations of negation $\neg$, disjunction $\vee$, and conjunction $\wedge$.

The set of PC-formulas is the smallest set including $\mathbb{V}$ and closed with respect to the propositional operations. We admit the operations of implication, $\rightarrow$, and equivalence, $\leftrightarrow$, as the standard abbreviations, that is for all PC-formulas φ and ψ:

- $\varphi \rightarrow \psi \stackrel{\text{df}}{=} \neg\varphi \vee \psi$;
- $\varphi \leftrightarrow \psi \stackrel{\text{df}}{=} (\neg\varphi \vee \psi) \wedge (\neg\psi \vee \varphi)$.

Let $\{0, 1\}$ be a two-element Boolean algebra whose elements represent truth-values 'false' and 'true', respectively. A PC-model is a structure $\mathcal{M} = (\{0, 1\}, v)$, where $v: \mathbb{V} \rightarrow \{0, 1\}$ is a valuation of propositional variables in $\{0, 1\}$. A PC-formula φ is said to be true in a PC-model $\mathcal{M} = (\{0, 1\}, v)$ whenever the following conditions hold:

- $\mathcal{M} \models p$ iff $v(p) = 1$, for every propositional variable p;
- $\mathcal{M} \models \neg\varphi$ iff $\mathcal{M} \not\models \varphi$;
- $\mathcal{M} \models \varphi \vee \psi$ iff $\mathcal{M} \models \varphi$ or $\mathcal{M} \models \psi$;
- $\mathcal{M} \models \varphi \wedge \psi$ iff $\mathcal{M} \models \varphi$ and $\mathcal{M} \models \psi$.

A PC-formula φ is PC-*valid* whenever it is true in all PC-models.

7.3 Propositional Modal Logics

The vocabulary of a *modal language* consists of the following pairwise disjoint sets:

- $\mathbb{V}$ – a countable set of propositional variables and/or constants;
- A set of relational constants;

- $\{\neg, \vee, \wedge\}$ – a set of classical propositional operations;
- A set of modal propositional operations is included in the set:

$$\{[R], \langle R\rangle, [[R]], \langle\langle R\rangle\rangle : R \text{ is a relational constant}\}.$$

The modal operations $\langle R\rangle$ and $\langle\langle R\rangle\rangle$ are definable in terms of $[R]$ and $[[R]]$, respectively, as follows:

$$\langle R\rangle\varphi \stackrel{\text{df}}{=} \neg[R]\neg\varphi, \quad \langle\langle R\rangle\rangle \stackrel{\text{df}}{=} \neg[[R]]\neg\varphi.$$

The propositional operations $[R]$, $\langle R\rangle$, $[[R]]$, and $\langle\langle R\rangle\rangle$ are referred to as *necessity, possibility, sufficiency*, and *dual sufficiency*, respectively.

The set of modal formulas is the smallest set including the set $\mathbb{V}$ and closed with respect to the propositional operations.

If the set of relational constants is a singleton set, say $\{R\}$, then the modal propositional operations $[R]$, $\langle R\rangle$, $[[R]]$, and $\langle\langle R\rangle\rangle$ are often written as $\Box$, $\Diamond$, $[[\,]]$, and $\langle\langle\rangle\rangle$, respectively. Some modal languages include propositional constants which are interpreted as singletons; they are referred to as *nominals*. If the cardinality of the set of relational constants is at least 2 or the language includes at least two different and not mutually definable modal operations, then the logic is referred to as *multimodal*.

Let a modal language be given. A *model* for the modal language is a structure $\mathcal{M} = (U, m)$ such that U is a non-empty set, whose elements are referred to as *states*, and m is a meaning function such that the following conditions are satisfied:

- $m(p) \subseteq U$, for every $p \in \mathbb{V}$;
- $m(R) \subseteq U \times U$, for every relational constant R.

The relations $m(R)$ are referred to as the *accessibility relations*.

A *frame* is a structure $\mathcal{F} = (U, m)$ such that U is a non-empty set and m is a map which assigns binary relations on U to relational constants. If there are finitely many relational constants in a modal language, then in the frames we often list explicitly all the corresponding relations and usually we denote them with the same symbols as the corresponding constants in the language. A model $\mathcal{M} = (U, m')$ is said to be *based on* a frame $\mathcal{F} = (U, m)$ whenever m is the restriction of m' to the set of relational constants.

A formula φ is said to be satisfied in a model $\mathcal{M}$ by a state $s \in U$, $\mathcal{M}, s \models \varphi$, whenever the following conditions are satisfied:

- $\mathcal{M}, s \models p$ iff $s \in m(p)$ for $p \in \mathbb{V}$;
- $\mathcal{M}, s \models \varphi \vee \psi$ iff $\mathcal{M}, s \models \varphi$ or $\mathcal{M}, s \models \psi$;
- $\mathcal{M}, s \models \varphi \wedge \psi$ iff $\mathcal{M}, s \models \varphi$ and $\mathcal{M}, s \models \psi$;
- $\mathcal{M}, s \models \neg\varphi$ iff $\mathcal{M}, s \not\models \varphi$;
- $\mathcal{M}, s \models [R]\varphi$ iff for every $s' \in U$, if $(s, s') \in m(R)$, then $\mathcal{M}, s' \models \varphi$;
- $\mathcal{M}, s \models \langle R\rangle\varphi$ iff there is $s' \in U$ such that $(s, s') \in m(R)$ and $\mathcal{M}, s' \models \varphi$;

- $\mathcal{M}, s \models [[R]]\varphi$ iff for every $s' \in U$, if $\mathcal{M}, s' \models \varphi$, then $(s, s') \in m(R)$;
- $\mathcal{M}, s \models \langle\langle R \rangle\rangle\varphi$ iff there is $s' \in U$ such that $(s, s') \notin m(R)$ and $\mathcal{M}, s' \not\models \varphi$.

As usual, given a model $\mathcal{M}$ and a state s, '$\mathcal{M}, s \not\models \varphi$' is an abbreviation of 'not $\mathcal{M}, s \models \varphi$'. Throughout the book, by a modal logic we mean the pair $\mathsf{L} =$ (a modal language, a class of models of the language). Given a modal logic L, we write L-language and L-model for the relevant components of L. Formulas of the L-language are referred to as L-formulas.

An L-formula φ is said to be *true* in an L-model $\mathcal{M} = (U, m)$, $\mathcal{M} \models \varphi$, whenever for every $s \in U$, $\mathcal{M}, s \models \varphi$, and it is L-valid whenever it is true in all L-models. An L-formula φ is said to be *true* in the L-frame $\mathcal{F}$, $\mathcal{F} \models \varphi$ for short, whenever φ is true in all the L-models based on $\mathcal{F}$. Note that in every L-model the propositional operations $\neg$, $\vee$, and $\wedge$ receive their standard meaning as classical operations of negation, disjunction, and conjunction of PC, respectively.

Standard Modal Logics

The standard modal logics are K, T, B, $\mathsf{S4}$, and $\mathsf{S5}$. Their common language is a modal language with a single relational constant R and with the modal operations $[R]$ and $\langle R \rangle$.

The models of these logics are of the form $\mathcal{M} = (U, R, m)$ where:

- $\mathcal{M}$ is a K-model iff R is a binary relation on U;
- $\mathcal{M}$ is a T-model iff R is a reflexive relation on U (i.e., $1' \subseteq R$);
- $\mathcal{M}$ is a B-model iff it is a T-model such that R is a symmetric relation on U (i.e., $R^{-1} \subseteq R$);
- $\mathcal{M}$ is an $\mathsf{S4}$-model iff it is a T-model such that R is a transitive relation on U (i.e., $R ; R \subseteq R$);
- $\mathcal{M}$ is a $\mathsf{S5}$-model iff it is a B-model and $\mathsf{S4}$-model (i.e., R is an equivalence relation on U).

All the standard modal logics are decidable. For details of the proof see e.g., [BvBW06].

7.4 Relational Formalization of Modal Logics

The logic $\mathsf{RL}(1, 1')$ serves as a basis for the relational formalisms for modal logics whose Kripke-style semantics is determined by frames with binary accessibility relations (see [Orł97b]). Let L be a modal logic. The relational logic RL_{L} appropriate for expressing L-formulas is obtained from $\mathsf{RL}(1, 1')$ by endowing its language with relational constants representing the accessibility relations from the models of L-language and with propositional constants of L (if there are any) which will be interpreted appropriately as relations.

The vocabulary of the relational logic $\mathsf{RL_L}$ consists of the symbols from the following pairwise disjoint sets:

- $\mathbb{OV}_{\mathsf{RL_L}}$ – a countable infinite set of object variables;
- $\mathbb{OC}_{\mathsf{RL_L}}$ – a countable (possibly empty) set of object constants;
- $\mathbb{RV}_{\mathsf{RL_L}}$ – a countable infinite set of relational variables;
- $\mathbb{RC}_{\mathsf{RL_L}} = \{1, 1'\} \cup \{R : R$ is a relational constant of $\mathsf{L}\} \cup \{C_c : c$ is a propositional constant of $\mathsf{L}\}$ – a set of relational constants;
- $\{-, \cup, \cap, ;, ^{-1}\}$ – the set of relational operations.

Object symbols, relational terms, and $\mathsf{RL_L}$-formulas are defined as in $\mathsf{RL}(1, 1')$-logic (see Sect. 2.3).

An $\mathsf{RL_L}$-structure is an $\mathsf{RL}(1, 1')$-model $\mathcal{M} = (U, m)$ (see Sect. 2.7) such that:

- $m(R) \subseteq U \times U$, for every relational constant R of L;
- $m(C_c) \subseteq U \times U$, for every propositional constant c of L.

An $\mathsf{RL_L}$-model is an $\mathsf{RL}(1, 1')$-model $\mathcal{M} = (U, m)$ such that:

- $m(C_c) = X \times U$, where $X \subseteq U$, for any propositional constant c of L; it follows that propositional constants of L are represented in $\mathsf{RL_L}$ as right ideal relations;
- The domains of relations $m(C_c)$ satisfy the constraints posed on propositional constants c in L-models; examples of such constants can be found in Sects. 11.3, 15.2, and 16.5;
- For all relational constants representing the accessibility relations of L, all the properties of these relations from L-models are assumed in $\mathsf{RL_L}$-models; many examples of such properties can be found in Sects. 7.5, 11.4, and 16.3.

If in a modal logic L there are finitely many accessibility relations, then in the $\mathsf{RL_L}$-models we list explicitly these relations and we denote them with the same symbols as the corresponding constants in the language.

As established in Sect. 2.7, the models of $\mathsf{RL_L}$ with $1'$ interpreted as the identity are referred to as standard $\mathsf{RL_L}$-models.

Translation

The translation of modal formulas into relational terms starts with a one-to-one assignment of relational variables to the propositional variables. Let τ' be such an assignment. Then the translation τ of formulas is defined inductively:

- $\tau(p) = \tau'(p) ; 1$, for any propositional variable $p \in \mathbb{V}$;
- $\tau(c) = C_c ; 1$, for any propositional constant $c \in \mathbb{V}$;
- $\tau(\neg\varphi) = -\tau(\varphi)$;
- $\tau(\varphi \vee \psi) = \tau(\varphi) \cup \tau(\psi)$;
- $\tau(\varphi \wedge \psi) = \tau(\varphi) \cap \tau(\psi)$;

and for every relational constant R of L:

- $\tau(\langle R \rangle \varphi) = R ; \tau(\varphi)$;
- $\tau([R]\varphi) = -(R ; -\tau(\varphi))$;

- $\tau(\langle\langle R\rangle\rangle\varphi) = -R\,;-\tau(\varphi)$;
- $\tau([[R]]\varphi) = -(-R\,;\tau(\varphi))$.

It follows that translation of defined propositional operations $\rightarrow$ and $\leftrightarrow$ is:

- $\tau(\varphi \rightarrow \psi) = -\tau(\varphi) \cup \tau(\psi)$;
- $\tau(\varphi \leftrightarrow \psi) = (-\tau(\varphi) \cup \tau(\psi)) \cap (-\tau(\psi) \cup \tau(\varphi)))$.

Hence, when passing from modal formulas to relational terms we replace propositional variables and constants by relational variables and relational constants, respectively, and propositional operations by relational operations. The crucial point here is that the accessibility relation is 'taken out' of the modal operation and it becomes an argument of an appropriate relational operation. In particular, possibility operation is replaced by the relational composition of two relations: the relation representing an accessibility relation and the relation resulting from the translation of the formula which is in the scope of the possibility operation. In this way to any formula φ of a modal logic there is associated a relational term $\tau(\varphi)$. The above translation assigns to any modal formula a right ideal relation i.e., a relation Q that satisfies $Q = Q\,;1$. It follows from the following proposition:

Proposition 7.4.1. *For every set U, the following conditions are satisfied:*

1. *The family of right ideal relations on U is closed on $-$, $\cup$, and $\cap$;*
2. *For any relation R on U and any right ideal relation P on U, $R\,;P$ is a right ideal relation;*
3. *If P is a right ideal relation on U, then for all $s, s' \in U$: $(s, s') \in P$ iff for every $t \in U$, $(s, t) \in P$.*

Proof. 1. and 3. follow directly from the definition of right ideal relations. For 2., note that since $P = P\,;1$, $R\,;P = R\,;(P\,;1) = (R\,;P)\,;1$. □

In some examples of dual tableaux proofs presented in the book a simpler translation of the formula to be proved is sufficient such that $\tau(p) = \tau'(p)$ for every propositional variable p appearing in the formula.

The translation τ is defined so that it preserves validity of formulas.

Proposition 7.4.2. *Let L be a modal logic and let φ be an L-formula. Then, for every L-model $\mathcal{M} = (U, m)$ there exists an $\mathsf{RL_L}$-model $\mathcal{M}' = (U, m')$ with the same universe as that of $\mathcal{M}$ such that for all $s, s' \in U$, $\mathcal{M}, s \models \varphi$ iff $(s, s') \in m'(\tau(\varphi))$.*

Proof. Let φ be an L-formula and let $\mathcal{M} = (U, m)$ be an L-model. We define an $\mathsf{RL_L}$-model $\mathcal{M}' = (U, m')$ as follows:

- $m'(1) = U \times U$;
- $m'(1')$ is the identity on U;
- $m'(\tau(p)) = m(p) \times U$, for every propositional variable p;
- $m'(\tau(c)) = m(c) \times U$, for any propositional constant $c \in \mathbb{V}$;

- $m'(R) = m(R)$;
- m' extends to all the compound terms as in $\mathsf{RL}(1, 1')$-models.

Now, we prove the proposition by induction on the complexity of formulas. Let $s, s' \in U$.

Let $\varphi = p$, $p \in \mathbb{V}$. Then $\mathcal{M}, s \models p$ iff $s \in m(p)$ iff $(s, s') \in m'(\tau(p))$, since $m'(\tau(p))$ is a right ideal relation.

Let $\varphi = \psi \vee \vartheta$. Then $\mathcal{M}, s \models \psi \vee \vartheta$ iff $\mathcal{M}, s \models \psi$ or $\mathcal{M}, s \models \vartheta$ iff, by the induction hypothesis, $(s, s') \in m'(\tau(\psi))$ or $(s, s') \in m'(\tau(\vartheta))$ iff $(s, s') \in m'(\tau(\psi)) \cup m'(\tau(\vartheta))$ iff $(s, s') \in m'(\tau(\psi \vee \vartheta))$.

Let $\varphi = \langle R \rangle \psi$. By the induction hypothesis, for all $t, s' \in U$, we have $\mathcal{M}, t \models \psi$ iff $(t, s') \in m'(\tau(\psi))$. Therefore, $\mathcal{M}, s \models \varphi$ iff there exists $t \in U$ such that $(s, t) \in m(R)$ and $\mathcal{M}, t \models \psi$ iff, by the induction hypothesis, there exists $t \in U$ such that $(s, t) \in m'(R)$ and $(t, s') \in m'(\tau(\psi))$ iff $(s, s') \in m'(R \,;\, \tau(\psi))$ iff $(s, s') \in m'(\tau(\varphi))$.

Let $\varphi = [[R]]\psi$. Then $\mathcal{M}, s \models \varphi$ iff for every $t \in U$, if $\mathcal{M}, t \models \psi$, then $(s, t) \in m(R)$ iff, by the induction hypothesis, for every $t \in U$, if $(t, s') \in m'(\tau(\psi))$, then $(s, t) \in m'(R)$ iff $(s, s') \notin m'(-R \,;\, \tau(\psi))$ iff $(s, s') \in m'(\tau(\varphi))$.

In the remaining cases the proofs are similar. □

Proposition 7.4.3. *Let* L *be a modal logic and let* φ *be an* L*-formula. Then, for every standard* $\mathsf{RL_L}$*-model* $\mathcal{M}' = (U, m')$ *there exists an* L*-model* $\mathcal{M} = (U, m)$ *with the same universe as that of* $\mathcal{M}'$ *such that for all* $s, s' \in U$*, the condition of Proposition 7.4.2 holds.*

Proof. Let φ be an L-formula and let $\mathcal{M}' = (U, m')$ be a standard $\mathsf{RL_L}$-model. We define an L-model $\mathcal{M} = (U, m)$ as follows:

- $m(p) = \{x \in U :$ there exists $y \in U$, $(x, y) \in m'(\tau(p))\}$, for every propositional variable p;
- For every propositional constant c, $s \in m(c)$ iff there is $s' \in U$ such that $(s, s') \in m'(\tau(c))$;
- $m(R) = m'(R)$.

We can prove that $\mathcal{M}, s \models \varphi$ iff $(s, s') \in m'(\tau(\varphi))$ in a similar way as in Proposition 7.4.2. □

Proposition 7.4.4. *Let* L *be a modal logic and let* φ *be an* L*-formula. Then, for every* L*-model* $\mathcal{M}$ *there exists an* $\mathsf{RL_L}$*-model* $\mathcal{M}'$ *such that for all object variables* x *and* y*,* $\mathcal{M} \models \varphi$ *iff* $\mathcal{M}' \models x\tau(\varphi)y$*.*

Proof. Let φ be an L-formula, let $\mathcal{M} = (U, m)$ be an L-model. We construct a standard $\mathsf{RL_L}$-model $\mathcal{M}' = (U, m')$ as in the proof of Proposition 7.4.2. Let x and y be any object variables. Assume $\mathcal{M} \models \varphi$. Suppose there exists a valuation v in $\mathcal{M}'$ such that $\mathcal{M}', v \not\models x\tau(\varphi)y$. Then $(v(x), v(y)) \notin m'(\tau(\varphi))$. However, by Proposition 7.4.2, models $\mathcal{M}$ and $\mathcal{M}'$ satisfy $\mathcal{M}, v(x) \models \varphi$ iff $(v(x), v(y)) \in m'(\tau(\varphi))$. Therefore, $\mathcal{M}, v(x) \not\models \varphi$, and hence $\mathcal{M} \not\models \varphi$, a contradiction. Now, assume $\mathcal{M}' \models x\tau(\varphi)y$. Suppose there is $s \in U$ such that $\mathcal{M}, s \not\models \varphi$. Let s' be any element of U. By Proposition 7.4.2, $\mathcal{M}, s \models \varphi$ if and only if $(s, s') \in m'(\tau(\varphi))$. Let v be a valuation in $\mathcal{M}'$ such that $v(x) = s$ and $v(y) = s'$. Since $\mathcal{M}, s \not\models \varphi$, $(v(x), v(y)) \notin m'(\tau(\varphi))$, so $\mathcal{M}', v \not\models x\tau(\varphi)y$, and hence $\mathcal{M}' \not\models x\tau(\varphi)y$, a contradiction. □

Due to Proposition 7.4.3, the following can be proved:

Proposition 7.4.5. *Let* L *be a modal logic and let* φ *be an* L*-formula. Then, for every standard* $\mathsf{RL_L}$*-model* $\mathcal{M}'$ *there exists an* L*-model* $\mathcal{M}$ *such that for all object variables* x *and* y*, the condition of Proposition 7.4.4 holds.*

From Theorem 2.7.2, Propositions 7.4.4, and 7.4.5, we get:

Theorem 7.4.1. *Let* L *be a modal logic. Then, for every* L*-formula* φ *and for all object variables* x *and* y*, the following conditions are equivalent:*

1. φ *is* L*-valid;*
2. $x\tau(\varphi)y$ *is* $\mathsf{RL_L}$*-valid.*

Proof. (1. → 2.) Let φ be L-valid. Suppose $x\tau(\varphi)y$ is not $\mathsf{RL_L}$-valid. Then, there exists a standard $\mathsf{RL_L}$-model $\mathcal{M}$ such that $\mathcal{M} \not\models x\tau(\varphi)y$. By Proposition 7.4.5, there is an L-model $\mathcal{M}'$ such that $\mathcal{M}' \not\models \varphi$, which contradicts the assumption of L-validity of φ.

(2. → 1.) Let φ be an L-formula such that $x\tau(\varphi)y$ is $\mathsf{RL_L}$-valid. Suppose φ is not L-valid. Then there exists an L-model $\mathcal{M}$ such that $\mathcal{M} \not\models \varphi$. By Proposition 7.4.4, there exists an $\mathsf{RL_L}$-model $\mathcal{M}'$ such that $\mathcal{M}' \not\models x\tau(\varphi)y$, a contradiction. □

Relational dual tableaux for modal logics are extensions of $\mathsf{RL}(1, 1')$-dual tableau. We add to the $\mathsf{RL}(1,' 1)$-system the specific rules and/or axiomatic sets that reflect properties of the specific relational constants corresponding to accessibility relations and propositional constants (if there are any) of the logic in question. Given a logic L, a relational dual tableau for L, $\mathsf{RL_L}$-dual tableau, enables us to prove facts about the relations from the models of L expressed in the language of L or in the language of $\mathsf{RL_L}$.

Given a modal logic L, an $\mathsf{RL_L}$-set is a finite set of $\mathsf{RL_L}$-formulas such that the first-order disjunction of its members is true in all $\mathsf{RL_L}$-models. Correctness of a rule is defined in a similar way as in the relational logics of classical algebras of binary relations (see Sect. 2.4).

Following the method of proving soundness and completeness of RL-dual tableau described in Sect. 2.6, we can prove soundness and completeness of the $\mathsf{RL_L}$-dual tableau in a similar way. To prove soundness, it suffices to show that all the rules are $\mathsf{RL_L}$-correct and all the axiomatic sets are $\mathsf{RL_L}$-sets (see Proposition 7.5.1). In order to prove completeness we need to prove the closed branch property, the branch model property, and the satisfaction in branch model property. Then the completeness proof is the same as the completeness proof of $\mathsf{RL}(1, 1')$-dual tableau (see Sects. 2.5 and 2.7). In Table 7.1 we recall the main facts that have to be proved.

Table 7.1 The key steps in the completeness proof of relational dual tableaux

(1) **Closed Branch Property**: For any branch of an $\mathsf{RL_L}$-proof tree, if $xRy \in b$ and $x{-}Ry \in b$ for an atomic term R, then the branch can be closed;
(2) **Branch Model Property**: Define the branch model $\mathcal{M}^b$ determined by an open branch b of an $\mathsf{RL_L}$-proof tree and prove that it is an $\mathsf{RL_L}$-model;
(3) **Satisfaction in Branch Model Property**: For every branch b of an $\mathsf{RL_L}$-proof tree and for every $\mathsf{RL_L}$-formula φ, the branch model $\mathcal{M}^b$ and the identity valuation v^b in $\mathcal{M}^b$ satisfy: If $\mathcal{M}^b, v^b \models \varphi$, then $\varphi \notin b$.

7.5 Dual Tableaux for Standard Modal Logics

Let L be a standard modal logic as presented in Sect. 7.3 (see p. 146). The relational logic appropriate for expressing L-formulas, $\mathsf{RL_L}$, is obtained from logic $\mathsf{RL}(1, 1')$ by expanding its language with a relational constant R representing the accessibility relation from the models of L-language. If a relation R in the models of logic L is assumed to satisfy some conditions, e.g., reflexivity (logic T), symmetry (logic B), transitivity (logic $\mathsf{S4}$) etc., then in the corresponding logic $\mathsf{RL_L}$ we add the respective conditions as the axioms of its models. The translation of a modal formula of L into a relational term of $\mathsf{RL_L}$ is defined as in Sect. 7.4.

By Theorem 7.4.1 the following holds:

Theorem 7.5.1. *For every formula φ of a standard modal logic L and for all object variables x and y, the following conditions are equivalent:*

1. *φ is L-valid;*
2. *$xt(\varphi)y$ is $\mathsf{RL_L}$-valid.*

Dual tableaux for standard modal logics in their relational formalizations are constructed as follows. We add to the $\mathsf{RL}(1, 1')$-dual tableau the following rules:

- Logic T: (ref R),
- Logic B: (ref R) and (sym R),
- Logic $\mathsf{S4}$: (ref R) and (tran R),
- Logic $\mathsf{S5}$: (ref R), (sym R), and (tran R).

We recall that these rules are of the form (see Sect. 6.6):

For all object symbols x and y,

$$(\text{ref } R)\ \frac{xRy}{x1'y,\ xRy} \qquad (\text{sym } R)\ \frac{xRy}{yRx}$$

$$(\text{tran } R)\ \frac{xRy}{xRz,\ xRy \mid zRy,\ xRy} \qquad z \text{ is any object symbol}$$

As defined in Sect. 7.4, given a standard modal logic L, an $\mathsf{RL_L}$-structure is of the form $\mathcal{M} = (U, R, m)$, where (U, m) is an $\mathsf{RL}(1, 1')$-model and R is a binary relation on U. An $\mathsf{RL_L}$-model is an $\mathsf{RL_L}$-structure such that relation R satisfies the constraints posed in L-models.

Theorem 7.5.2 (Correspondence). *Let* L *be a standard modal logic and let* $\mathcal{K}$ *be a class of* $\mathsf{RL_L}$*-structures. Relation* R *is reflexive (resp. symmetric, transitive) in all structures of* $\mathcal{K}$ *iff the rule (ref* R*), (resp. (sym* R*), (tran* R*)) is* $\mathcal{K}$*-correct.*

For the proof see Theorem 6.6.1. Theorem 7.5.2 leads to:

Proposition 7.5.1. *Let* L *be a standard modal logic. Then:*

1. *The* $\mathsf{RL_L}$*-rules are* $\mathsf{RL_L}$*-correct;*
2. *The* $\mathsf{RL_L}$*-axiomatic sets are* RL_L*-sets.*

The notions of an $\mathsf{RL_L}$-proof tree, a closed branch of such a tree, a closed $\mathsf{RL_L}$-proof tree, and $\mathsf{RL_L}$-provability are defined as in Sect. 2.4.

We recall that the completion conditions determined by the rules (ref R), (sym R), and (tran R) are:

For all object symbols x and y,

Cpl(ref R) If $xRy \in b$, then $x1'y \in b$;
Cpl(sym R) If $xRy \in b$, then $yRx \in b$;
Cpl(tran R) If $xRy \in b$, then for every object symbol z, either $xRz \in b$ or $zRy \in b$.

The notions of a complete branch of an $\mathsf{RL_L}$-proof tree, a complete $\mathsf{RL_L}$-proof tree, and an open branch of an $\mathsf{RL_L}$-proof tree are defined as in RL-logic (see Sect. 2.5). In order to prove completeness, we need to define a branch model and to show the three theorems of Table 7.1.

The branch model is defined as in the completeness proof of $\mathsf{RL}(1, 1')$-dual tableau, that is $R^b = m^b(R) = \{(x, y) \in U^b \times U^b : xRy \notin b\}$. Using the completion conditions, it is easy to show that R^b satisfies the conditions assumed in the corresponding L-models (see the proof of Proposition 6.6.3). Hence, the branch model property is satisfied. In similar way as in $\mathsf{RL}(1, 1')$-dual tableau (see also $\mathsf{RL_{EQ}}$-dual tableau in Sect. 6.6), we can prove the closed branch property and the satisfaction in branch model property:

Proposition 7.5.2 (Satisfaction in Branch Model Property). *For every open branch* b *of an* $\mathsf{RL_L}$*-proof tree and for every* $\mathsf{RL_L}$*-formula* φ*, if* $\mathcal{M}^b, v^b \models \varphi$*, then* $\varphi \notin b$*.*

Then, completeness can be proved as for $\mathsf{RL}(1, 1')$-dual tableau.

Theorem 7.5.3 (Soundness and Completeness of Relational Logics for Standard Modal Logics). *Let* L *be a standard modal logic and let* φ *be an* $\mathsf{RL_L}$*-formula. Then, the following conditions are equivalent:*

1. φ *is* $\mathsf{RL_L}$*-valid;*
2. φ *is true in all standard* $\mathsf{RL_L}$*-models;*
3. φ *is* $\mathsf{RL_L}$*-provable.*

Finally, by Theorem 7.5.1 and Theorem 7.5.3, we obtain:

Theorem 7.5.4 (Relational Soundness and Completeness of Standard Modal Logics). *Let* L *be a standard modal logic and let* φ *be an* L*-formula. Then for all object variables x and y, the following conditions are equivalent:*

1. φ *is* L*-valid;*
2. $x\tau(\varphi)y$ *is* $\mathsf{RL_L}$*-provable.*

Example. We present a translation and a relational proof of a formula of logic K. Note that $\mathsf{RL_K}$-proof system is exactly the same as $\mathsf{RL}(1,1')$-system, because in K-models the accessibility relation is an arbitrary binary relation. Consider the following K-formula:

$$\varphi = ([R]p \wedge [R]q) \rightarrow [R](p \wedge q).$$

For reasons of simplicity, let $\tau(p) = P$ and $\tau(q) = Q$. The translation $\tau(\varphi)$ of the formula φ into a relational term of logic $\mathsf{RL_K}$ is:

$$-[-(R\,;-P) \cap -(R\,;-Q)] \cup -(R\,;-(P \cap Q)).$$

We show that the formula φ is K-valid, by showing that $x\tau(\varphi)y$ is $\mathsf{RL_K}$-valid. Figure 7.1 presents its $\mathsf{RL_K}$-proof.

7.6 Entailment in Modal Logics

The logic $\mathsf{RL}(1,1')$ can be used to verify entailment of formulas of non-classical logics, provided that they can be translated into binary relations. Let L be a modal logic. In order to verify the entailment we apply the method presented in Sect. 2.11. We translate L-formulas in question into relational terms of the logic $\mathsf{RL_L}$ and then we use the method of verification of entailment for $\mathsf{RL_L}$-logic as shown in Sect. 2.11.

For example, in every model $\mathcal{M}$ of K-logic the truth of the formula $p \rightarrow q$ in $\mathcal{M}$ implies the truth of $[R]p \rightarrow [R]q$ in $\mathcal{M}$. The translation of these formulas to $\mathsf{RL_K}$-terms is:

$$\tau(p \rightarrow q) = -P \cup Q,$$

$$\tau([R]p \rightarrow [R]q) = --(R\,;-P) \cup -(R\,;-Q),$$

where for simplicity $\tau(p) = P$ and $\tau(q) = Q$. To verify the entailment we need to show that $-P \cup Q = 1$ implies $--(R\,;-P) \cup -(R\,;-Q) = 1$. According to Proposition 2.2.1(7.), we need to show that the formula:

$$x[(1\,;-(-P \cup Q)\,;1) \cup --(R\,;-P) \cup -(R\,;-Q)]y$$

is $\mathsf{RL_K}$-provable. Figure 7.2 presents an $\mathsf{RL_K}$-proof of this formula.

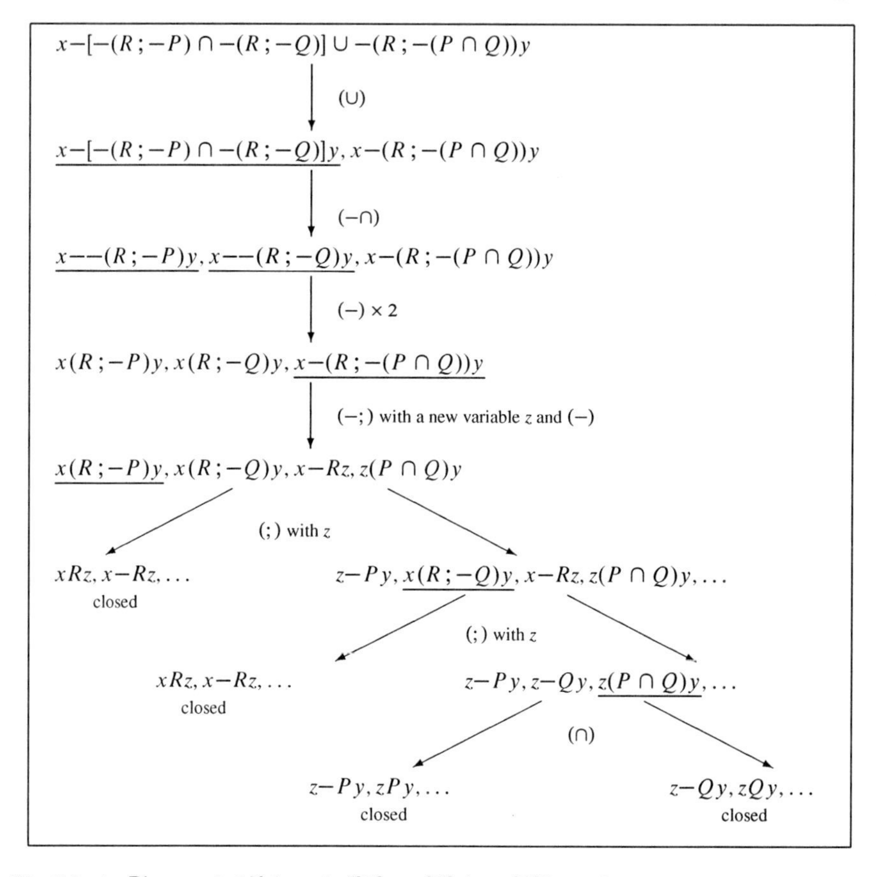

Fig. 7.1 An $\mathsf{RL_K}$-proof of K-formula $([R]p \wedge [R]q) \rightarrow [R](p \wedge q)$

Furthermore, observe that in every model $\mathcal{M}$ of K-logic the truth of $R ; R \subseteq R$ in $\mathcal{M}$ implies the truth of $[R]p \rightarrow [R][R]p$ in $\mathcal{M}$. The translation of the latter formula to $\mathsf{RL_K}$-term is:

$$--(R;-P) \cup -(R;--(R;-P)),$$

where for simplicity $\tau(p) = P$. To verify the entailment we need to show that $-(R;R) \cup R = 1$ implies $--(R;-P) \cup -(R;--(R;-P)) = 1$. According to Proposition 2.2.1(7.), we need to show that the formula:

$$x[(1;-(-(R;R) \cup R);1) \cup --(R;-P) \cup -(R;--(R;-P))]y$$

is $\mathsf{RL_K}$-provable. Figure 7.3 presents its $\mathsf{RL_K}$-proof.

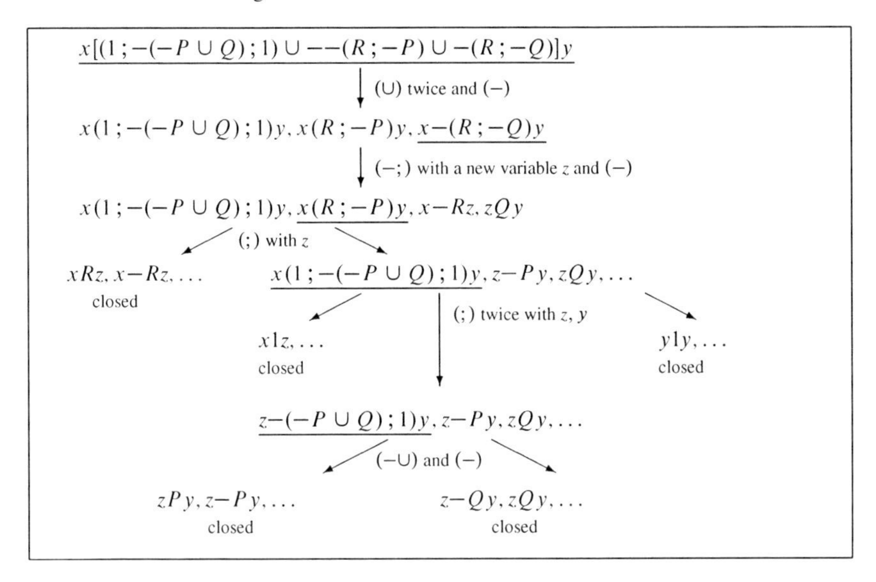

Fig. 7.2 An $\mathsf{RL_K}$-proof showing that $p \to q$ entails $[R]p \to [R]q$

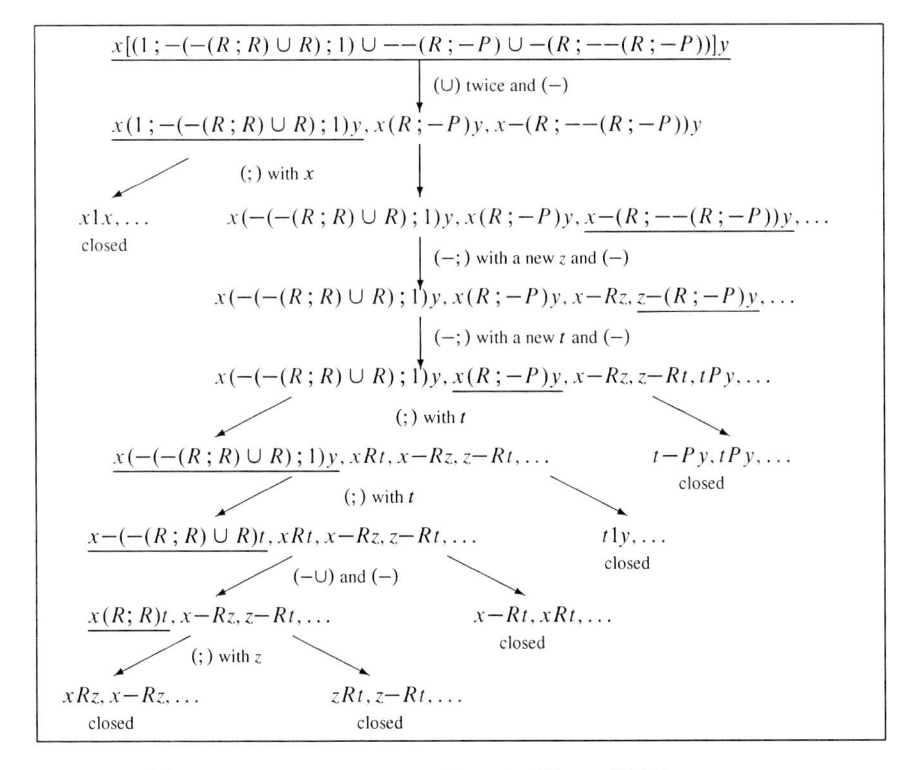

Fig. 7.3 An $\mathsf{RL_K}$-proof showing that $R ; R \subseteq R$ entails $[R]p \to [R][R]p$

7.7 Model Checking in Modal Logics

The method presented in Sect. 3.4 can be used for model checking in finite models of modal logics. The general idea is as follows. Let $\mathcal{M}$ be a finite model of a modal logic L and let φ be an L-formula. In order to verify whether φ is true in $\mathcal{M}$, we construct a relational logic $\mathsf{RL_L}$ and an $\mathsf{RL_L}$-model $\mathcal{M}'$ such that for all object variables x and y, the problem '$\mathcal{M} \models \varphi$?' is equivalent to the problem '$\mathcal{M}' \models x\tau(\varphi)y$?'. Then, we apply the method of model checking for the relational logic $\mathsf{RL_L}$, the model $\mathcal{M}'$, and the formula $x\tau(\varphi)y$ as presented in Sect. 3.4. For that purpose, we consider an instance $\mathsf{RL}_{\mathcal{M}',x\tau(\varphi)y}$ of the logic $\mathsf{RL_L}$. Then, we obtain:

Theorem 7.7.1 (Relational Model Checking in Modal Logics). *For every* L*-formula* φ*, for every finite* L*-model* $\mathcal{M}$*, and for all object variables* x *and* y*, the following statements are equivalent:*

1. $\mathcal{M} \models \varphi$;
2. $x\tau(\varphi)y$ *is* $\mathsf{RL}_{\mathcal{M}',x\tau(\varphi)y}$*-provable.*

By way of example, consider modal logic K. Let $\mathcal{M} = (U, R, m)$ be a K-model such that $U = \{\mathtt{a}, \mathtt{b}, \mathtt{c}\}$, $m(p) = \{\mathtt{a}\}$, and the accessibility relation is $R = \{(\mathtt{a}, \mathtt{b}), (\mathtt{b}, \mathtt{c}), (\mathtt{a}, \mathtt{c})\}$. Let φ be the formula of the form $\neg\langle R\rangle\langle R\rangle p$. Let us consider the problem: 'is φ true in $\mathcal{M}$?'. The translation of the formula φ is:

$$\tau(\varphi) = -(R\,;(R\,;(P\,;1))),$$

where $\tau'(p) = P$. Using the construction from the proof of Proposition 7.4.2 it is easy to prove that there exists an $\mathsf{RL_K}$-model $\mathcal{M}'$ such that for all object variables x and y, $\mathcal{M} \models \varphi$ iff $\mathcal{M}' \models x\tau(\varphi)y$.

The model $\mathcal{M}' = (U', R', m')$ is an $\mathsf{RL_K}$-model such that $U' = U$, $R' = R$, and m' is the meaning function satisfying:

$$m'(P) = \{(\mathtt{a}, \mathtt{a}), (\mathtt{a}, \mathtt{b}), (\mathtt{a}, \mathtt{c})\}.$$

Let x and y be any object variables. The model checking problem 'is φ true in $\mathcal{M}$?' is equivalent to the problem 'is the formula $x\tau(\varphi)y$ true in $\mathcal{M}'$?'. For the latter we apply the method presented in Sect. 3.4. The vocabulary of the language adequate for testing whether $\mathcal{M}' \models x\tau(\varphi)y$ consists of the following pairwise disjoint sets of symbols:

- $\mathbb{OV}_{\mathsf{RL}_{\mathcal{M}',x\tau(\varphi)y}}$ - a countable infinite set of object variables;
- $\mathbb{OC}_{\mathsf{RL}_{\mathcal{M}',x\tau(\varphi)y}} = \{c_\mathtt{a}, c_\mathtt{b}, c_\mathtt{c}\}$ – the set of object constants;
- $\mathbb{RC}_{\mathsf{RL}_{\mathcal{M}',x\tau(\varphi)y}} = \{R, P, 1, 1'\}$ – the set of relational constants;
- $\{-, \cup, \cap, ;, ^{-1}\}$ – the set of relational operations.

An $\mathsf{RL}_{\mathcal{M}',x\tau(\varphi)y}$-model is the structure $\mathcal{N} = (W, R, n)$, where:

- $W = \{\mathtt{a}, \mathtt{b}, \mathtt{c}\}$;
- $n(c_\mathtt{a}) = \mathtt{a}$, $n(c_\mathtt{b}) = \mathtt{b}$, and $n(c_\mathtt{c}) = \mathtt{c}$;

- $n(1) = W \times W$;
- $n(1') = \{(\mathtt{a}, \mathtt{a}), (\mathtt{b}, \mathtt{b}), (\mathtt{c}, \mathtt{c})\}$;
- $n(P) = \{(\mathtt{a}, \mathtt{a}), (\mathtt{a}, \mathtt{b}), (\mathtt{a}, \mathtt{c})\}$;
- $R = n(R) = \{(\mathtt{a}, \mathtt{b}), (\mathtt{b}, \mathtt{c}), (\mathtt{a}, \mathtt{c})\}$;
- n extends to all the compound terms as in RL-models.

The rules of $\mathsf{RL}_{\mathcal{M}',x\tau(\varphi)y}$-dual tableau which are specific for the model checking problem in question have the following forms:

$$(-R\mathtt{ij}) \quad \frac{x{-}Ry}{x1'c_{\mathtt{i}}, x{-}Ry \mid y1'c_{\mathtt{j}}, x{-}Ry}$$

for every $(\mathtt{i}, \mathtt{j}) \in \{(\mathtt{a}, \mathtt{a}), (\mathtt{b}, \mathtt{a}), (\mathtt{b}, \mathtt{b}), (\mathtt{c}, \mathtt{a}), (\mathtt{c}, \mathtt{b}), (\mathtt{c}, \mathtt{c})\}$;

$$(-P\mathtt{ij}) \quad \frac{x{-}Py}{x1'c_{\mathtt{i}}, x{-}Py \mid y1'c_{\mathtt{j}}, x{-}Py}$$

for every $(\mathtt{i}, \mathtt{j}) \in \{(\mathtt{b}, \mathtt{a}), (\mathtt{b}, \mathtt{b}), (\mathtt{b}, \mathtt{c}), (\mathtt{c}, \mathtt{a}), (\mathtt{c}, \mathtt{b}), (\mathtt{c}, \mathtt{c})\}$;

$$(1') \quad \frac{}{x{-}1'c_{\mathtt{a}} \mid x{-}1'c_{\mathtt{b}} \mid x{-}1'c_{\mathtt{c}}}$$

$$(\mathtt{i} \neq \mathtt{j}) \quad \frac{}{c_{\mathtt{i}}1'c_{\mathtt{j}}} \quad \text{for any } \mathtt{i}, \mathtt{j} \in \{\mathtt{a}, \mathtt{b}, \mathtt{c}\}, \mathtt{i} \neq \mathtt{j}$$

The $\mathsf{RL}_{\mathcal{M}',x\tau(\varphi)y}$-axiomatic sets are:

- $\{c_{\mathtt{i}}Rc_{\mathtt{j}}\}$, for every $(\mathtt{i}, \mathtt{j}) \in \{(\mathtt{a}, \mathtt{b}), (\mathtt{b}, \mathtt{c}), (\mathtt{a}, \mathtt{c})\}$;
- $\{c_{\mathtt{i}}{-}Rc_{\mathtt{j}}\}$, for every$(\mathtt{i}, \mathtt{j}) \in \{(\mathtt{a}, \mathtt{a}), (\mathtt{b}, \mathtt{a}), (\mathtt{b}, \mathtt{b}), (\mathtt{c}, \mathtt{a}), (\mathtt{c}, \mathtt{b}), (\mathtt{c}, \mathtt{c})\}$;
- $\{c_{\mathtt{i}}Pc_{\mathtt{j}}\}$, for every $(\mathtt{i}, \mathtt{j}) \in \{(\mathtt{a}, \mathtt{a}), (\mathtt{a}, \mathtt{b}), (\mathtt{a}, \mathtt{c})\}$;
- $\{c_{\mathtt{i}}{-}Pc_{\mathtt{j}}\}$, for every $(\mathtt{i}, \mathtt{j}) \in \{(\mathtt{b}, \mathtt{a}), (\mathtt{b}, \mathtt{b}), (\mathtt{b}, \mathtt{c}), (\mathtt{c}, \mathtt{a}), (\mathtt{c}, \mathtt{b}), (\mathtt{c}, \mathtt{c})\}$.

Let x and y be object variables. The truth of φ in model $\mathcal{M}$ is equivalent to $\mathsf{RL}_{\mathcal{M}',x\tau(\varphi)y}$-provability of $x\tau(\varphi)y$. Figure 7.4 presents an $\mathsf{RL}_{\mathcal{M}',x\tau(\varphi)y}$-proof of $x\tau(\varphi)y$. The subtree Π_1 is presented in Fig. 7.5. The subtrees Π_2 and Π_3 are presented in Figs. 7.6 and 7.7, respectively. Observe that in a diagram of Fig. 7.7 the applications of the rules $(-P\mathtt{ca})$, $(-P\mathtt{cb})$, and $(-P\mathtt{cc})$ result in the nodes with formulas $v1'c_{\mathtt{a}}$, $v1'c_{\mathtt{b}}$, and $v1'c_{\mathtt{c}}$. Therefore, in the picture we identify all these nodes.

7.8 Verification of Satisfaction in Modal Logics

The method of verification of satisfaction in finite models presented in Sect. 3.5 can be also used in the case of standard modal logics. Let $\mathcal{M} = (U, R, m)$ be a finite model of a modal logic L, let φ be an L-formula, and let $\mathtt{a} \in U$ be a state. In order to verify whether φ is satisfied in $\mathcal{M}$ by the state $\mathtt{a}$, we construct a relational logic

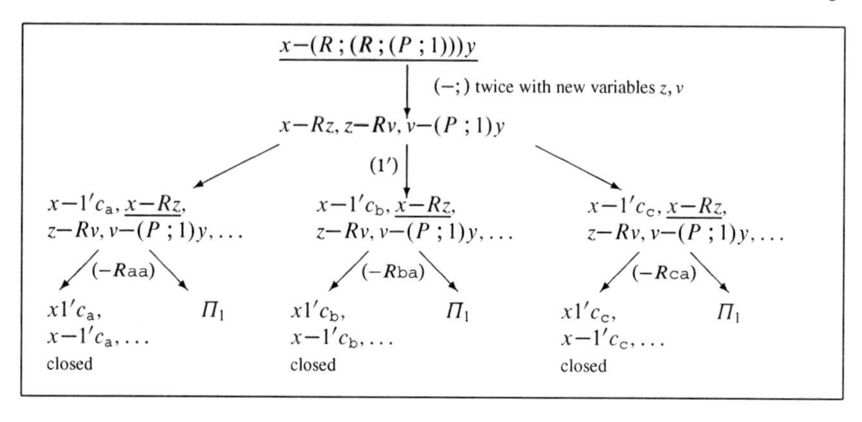

Fig. 7.4 An $\mathsf{RL}_{\mathcal{M}',x\tau(\varphi)y}$-proof of the truth of K-formula $\langle R\rangle\langle R\rangle p \rightarrow \langle R\rangle p$ in the model $\mathcal{M}$

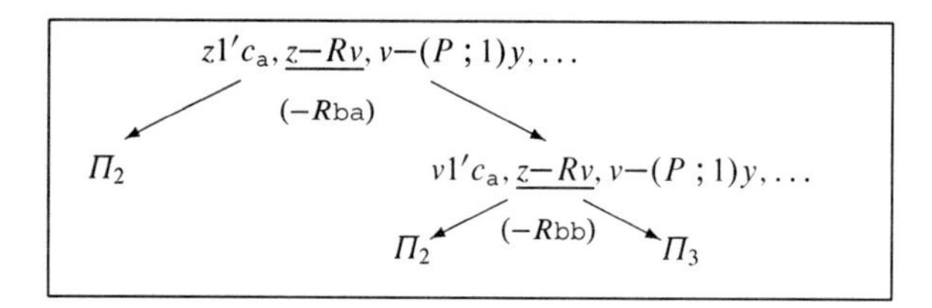

Fig. 7.5 The subtree Π_1

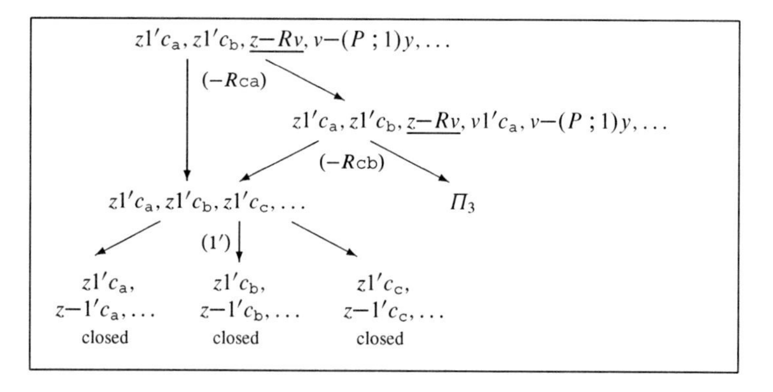

Fig. 7.6 The subtree Π_2

$\mathsf{RL_L}$, an $\mathsf{RL_L}$-model $\mathcal{M}' = (U', R', m')$, and a valuation v_a in $\mathcal{M}'$ such that for all object variables x and y, the problem '$\mathcal{M}, \mathsf{a} \models \varphi$?' is equivalent to the problem '$\mathcal{M}', v_\mathsf{a} \models x\tau(\varphi)y$?'. Then, we apply the method of verification of satisfaction as presented in Sect. 3.5 to the relational logic $\mathsf{RL_L}$, the model $\mathcal{M}'$, the formula $x\tau(\varphi)y$, and elements $v_\mathsf{a}(x)$ and $v_\mathsf{a}(y)$ of U'. We construct an instance $\mathsf{RL}_{\mathcal{M}',x\tau(\varphi)y}$ of the logic $\mathsf{RL_L}$ and we obtain:

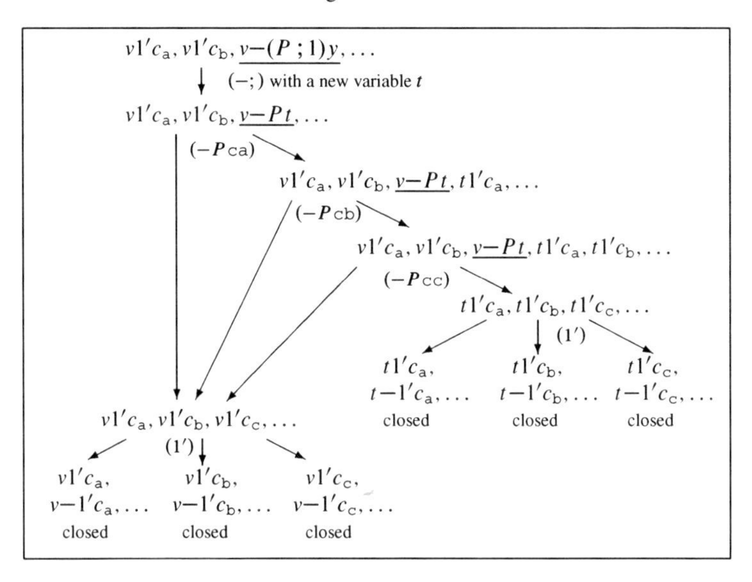

Fig. 7.7 The subtree Π_3

Theorem 7.8.1 (Relational Satisfaction in Standard Modal Logics). *For every* L*-formula* φ*, for every finite* L*-model* $\mathcal{M} = (U, R, m)$*, and for every state* $\mathtt{a} \in U$*, the following statements are equivalent:*

1. $\mathcal{M}, \mathtt{a} \models \varphi$;
2. $c_{v_\mathtt{a}(x)}\tau(\varphi)v_\mathtt{a}(y)$ *is* $\mathsf{RL}_{\mathcal{M}',x\tau(\varphi)y}$*-provable.*

As an example of an application of the method, consider the modal logic K. Let $\mathcal{M} = (U, R, m)$ be a K-model such that $U = \{\mathtt{a}, \mathtt{b}\}$, $m(p) = \{\mathtt{a}\}$ and the accessibility relation is $R = \{(\mathtt{b}, \mathtt{a})\}$. Let φ be the formula $\langle R\rangle p$. Let us consider the problem: 'is φ satisfied in $\mathcal{M}$ by state $\mathtt{b}$?' The translation of the formula φ is $\tau(\varphi) = (R;(P;1))$, where $\tau'(p) = P$. By Proposition 7.4.2, there exist a standard RL_K-model $\mathcal{M}'$ and a valuation $v_\mathtt{b}$ in $\mathcal{M}'$ such that for all object variables x and y, $\mathcal{M}, \mathtt{b} \models \varphi$ iff $\mathcal{M}', v_\mathtt{b} \models x\tau(\varphi)y$.

The RL_K-model $\mathcal{M}' = (U', R', m')$ is such that $U' = U$, $R' = R$, and m' is the meaning function satisfying:

$$m'(P) = \{(\mathtt{a}, \mathtt{a}), (\mathtt{a}, \mathtt{b})\}.$$

Let $v_\mathtt{b}$ be a valuation such that $v_\mathtt{b}(x) = b$ and $v_\mathtt{b}(y) = a$. Then $\mathcal{M}'$ and $v_\mathtt{b}$ satisfy the condition: $\mathcal{M}, \mathtt{b} \models \varphi$ iff $\mathcal{M}', v_\mathtt{b} \models x\tau(\varphi)y$.

Therefore the satisfaction problem 'is φ satisfied in $\mathcal{M}$ by state $\mathtt{b}$?' is equivalent to the problem '$(\mathtt{b}, \mathtt{a}) \in m'(\tau(\varphi))$?'. By Proposition 3.5.1 this is equivalent to $\mathsf{RL}_{\mathcal{M}',x\tau(\varphi)y}$-provability of $c_\mathtt{b}\tau(\varphi)c_\mathtt{a}$.

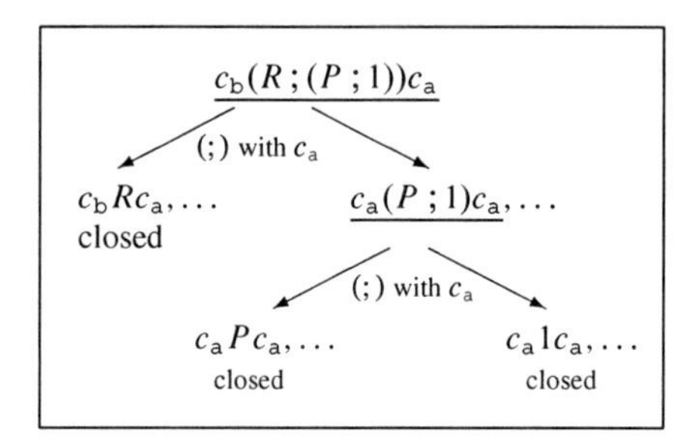

Fig. 7.8 An $\mathsf{RL}_{\mathcal{M}',x\tau(\varphi)y}$-proof showing that $\langle R\rangle p$ is satisfied in $\mathcal{M}$ by state b

$\mathsf{RL}_{\mathcal{M}',x\tau(\varphi)y}$-dual tableau contains the rules and axiomatic sets of $\mathsf{RL}(1,1')$-dual tableau adjusted to $\mathsf{RL}_{\mathcal{M}',x\tau(\varphi)y}$-language and the rules and axiomatic sets specific for the satisfaction problem as presented in Sect. 3.5:

- The rules are: $(-R\mathrm{aa})$, $(-R\mathrm{ab})$, $(-R\mathrm{bb})$, $(-P\mathrm{bb})$, $(-P\mathrm{ba})$, $(1')$, and $(\mathrm{a}\neq\mathrm{b})$;
- The axiomatic sets are those that include either of the following subsets: $\{c_{\mathrm{b}}Rc_{\mathrm{a}}\}$, $\{c_{\mathrm{a}}Pc_{\mathrm{a}}\}$, $\{c_{\mathrm{a}}Pc_{\mathrm{b}}\}$, $\{c_{\mathrm{a}}-Rc_{\mathrm{a}}\}$, $\{c_{\mathrm{b}}-Rc_{\mathrm{b}}\}$, $\{c_{\mathrm{a}}-Rc_{\mathrm{b}}\}$, $\{c_{\mathrm{b}}-Pc_{\mathrm{b}}\}$, and $\{c_{\mathrm{b}}-Pc_{\mathrm{a}}\}$.

It is easy to see that the formula $x\tau(\varphi)y$ is satisfied in $\mathcal{M}'$ by valuation v_{b}, though it is not true in $\mathcal{M}'$. Therefore, the formula $\langle R\rangle p$ is satisfied in $\mathcal{M}$ by state b, while it is not true in $\mathcal{M}$. Figure 7.8 presents an $\mathsf{RL}_{\mathcal{M}',x\tau(\varphi)y}$-proof of $c_{\mathrm{b}}\tau(\varphi)c_{\mathrm{a}}$, which shows that the formula $\langle R\rangle p$ is satisfied in $\mathcal{M}$ by state b.

Chapter 8
Dual Tableaux for Some Logics Based on Intuitionism

8.1 Introduction

Intuitionistic logic encompasses the principles of reasoning which were used by L. Brouwer in developing his intuitionistic mathematics, beginning in [Bro07]. Intuitionistic logic can be succinctly described as classical logic without the law $\varphi \vee \neg\varphi$ of excluded middle. Brouwer observed that this law was abstracted from finite situations and its application to statements about infinite collections is not justified. One of the consequences of the rejection of the law of excluded middle is that every intuitionistic proof of an existential sentence can be effectively transformed into an intuitionistic proof of an instance of that sentence. More precisely, if a formula of the form $\exists x\varphi(x)$ without free variables is provable in the intuitionistic predicate logic, then there is a term t without free variables such that $\varphi(t)$ is provable. In particular, if $\varphi \vee \psi$ is provable, then either φ or ψ is provable. In that sense intuitionistic logic may provide a logical basis for constructive reasoning. A formal system of intuitionistic logic was proposed in [Hey30]. A relationship between the classical propositional logic PC and intuitionistic propositional logic INT was proved by Glivenko in [Gli29], namely a PC-formula φ is PC-provable if and only if $\neg\neg\varphi$ is INT-provable. Kripke semantics for intuitionistic logic was developed in [Kri65].

In this chapter we present relational dual tableaux for the propositional intuitionistic logic INT, for the minimal intuitionistic logic of Johansson, and for some axiomatic extensions of INT, namely for the Scott's logic and the logic of weak excluded middle. All these logics are decidable. We also develop a dual tableau for the logic proposed in [FM95] for hardware verification. It is an intuitionistic logic endowed with a propositional operation which enables us to specify the propagation of signals along the gates of combinatorial circuits. Dual tableaux presented in this chapter are modifications of the deduction systems presented in [FO95] within the context of relational logics of Chap. 2.

E. Orłowska and J. Golińska-Pilarek, *Dual Tableaux: Foundations, Methodology, Case Studies*, Trends in Logic 36, DOI 10.1007/978-94-007-0005-5_8,

8.2 Relational Formalization of Intuitionistic Logic

In this section we present the relational formalization of an intuitionistic logic, INT. The vocabulary of INT-language consists of symbols from the following pairwise disjoint sets:

- $\mathbb{V}$ – a countable set of propositional variables;
- $\{\neg, \vee, \wedge, \rightarrow\}$ – the set of propositional operations.

The set of INT-formulas is the smallest set including the set $\mathbb{V}$ and closed with respect to all the propositional operations.

An INT-*model* is a structure $\mathcal{M} = (U, R, m)$ such that U is a non-empty set of states, R is a reflexive and transitive relation on U, and m is a meaning function such that the following conditions are satisfied:

- $m(p) \subseteq U$, for every propositional variable $p \in \mathbb{V}$;
- For all $s, s' \in U$ the following heredity condition holds:

$$\text{(her)} \quad \text{If } (s, s') \in R \text{ and } s \in m(p), \text{ then } s' \in m(p).$$

Let $\mathcal{M} = (U, R, m)$ be an INT-model and let $s \in U$. Satisfaction of INT-formula φ in $\mathcal{M}$ by state s, $\mathcal{M}, s \models \varphi$ for short, is defined inductively as follows:

- $\mathcal{M}, s \models p$ iff $s \in m(p)$, for any propositional variable p;
- $\mathcal{M}, s \models \neg\varphi$ iff for every $s' \in U$, if $(s, s') \in R$, then $\mathcal{M}, s' \not\models \varphi$;
- $\mathcal{M}, s \models \varphi \vee \psi$ iff $\mathcal{M}, s \models \varphi$ or $\mathcal{M}, s \models \psi$;
- $\mathcal{M}, s \models \varphi \wedge \psi$ iff $\mathcal{M}, s \models \varphi$ and $\mathcal{M}, s \models \psi$;
- $\mathcal{M}, s \models (\varphi \rightarrow \psi)$ iff for every $s' \in U$, if $(s, s') \in R$ and $\mathcal{M}, s' \models \varphi$, then $\mathcal{M}, s' \models \psi$.

It is known that the heredity condition holds for every formula φ, i.e., if $(s, s') \in R$ and $\mathcal{M}, s \models \varphi$, then $\mathcal{M}, s' \models \varphi$.

An INT-formula φ is said to be *true* in an INT-model $\mathcal{M} = (U, R, m)$, $\mathcal{M} \models \varphi$, whenever for every $s \in U$, $\mathcal{M}, s \models \varphi$, and it is said to be INT-*valid* whenever it is true in all INT-models.

In the relational formalization of logic INT we follow the main ideas of the method presented in Sect. 7.4. The vocabulary of the language of the relational logic $\mathsf{RL_{INT}}$ consists of the symbols from the following pairwise disjoint sets:

- $\mathbb{OV}_{\mathsf{RL_{INT}}}$ – a countable infinite set of object variables;
- $\mathbb{RV}_{\mathsf{RL_{INT}}}$ – a countable infinite set of relational variables;
- $\mathbb{RC}_{\mathsf{RL_{INT}}} = \{1, 1', R\}$ – the set of relational constants;
- $\{-, \cup, \cap, ;, ^{-1}\}$ – the set of relational operations.

As in case of modal logics of Sect. 7.4, the constant R represents the accessibility relation from INT-models. Relational terms and formulas of the logic $\mathsf{RL_{INT}}$ are defined as described in Sect. 2.3. The relational variables are intended to represent intuitionistic formulas.

An $\mathsf{RL}_{\mathsf{INT}}$-model is an $\mathsf{RL}(1,1')$-model $\mathcal{M} = (U, R, m)$ such that the following conditions are satisfied:

- $m(P)$ is a right ideal relation on U, for every $P \in \mathbb{RV}_{\mathsf{RL}_{\mathsf{INT}}}$;
- $R = m(R)$ is a reflexive and transitive relation on U such that for every $P \in \mathbb{RV}_{\mathsf{RL}_{\mathsf{INT}}}$ and for all $x, y, z \in U$, the following holds:

(her') If $(x, y) \in R$ and $(x, z) \in m(P)$, then $(y, z) \in m(P)$.

The translation of INT-formulas into relational terms starts with a one-to-one assignment of relational variables to the propositional variables. Let τ' be such an assignment. Then the translation τ of formulas is defined inductively:

- $\tau(p) = \tau'(p)$, for any propositional variable p;
- $\tau(\neg\varphi) = -(R\,;\tau(\varphi))$;
- $\tau(\varphi \vee \psi) = \tau(\varphi) \cup \tau(\psi)$;
- $\tau(\varphi \wedge \psi) = \tau(\varphi) \cap \tau(\psi)$;
- $\tau(\varphi \rightarrow \psi) = -(R\,;(\tau(\varphi) \cap -\tau(\psi)))$.

The translation of atomic INT-formulas cannot be defined as in modal logics (see Sect. 7.4) due to the assumption of heredity. Therefore, in the $\mathsf{RL}_{\mathsf{INT}}$-models relational variables are assumed to be right ideal relations and in the $\mathsf{RL}_{\mathsf{INT}}$-dual tableau the rule (ideal) reflecting this condition is included.

The translation τ is defined so that it preserves validity of formulas.

Proposition 8.2.1. *Let φ be an INT-formula. Then for every INT-model $\mathcal{M} = (U, R, m)$ there exists an $\mathsf{RL}_{\mathsf{INT}}$-model $\mathcal{M}' = (U, R, m')$ with the same universe and the same relation R as those in $\mathcal{M}$ such that for all $s, s' \in U$, $\mathcal{M}, s \models \varphi$ iff $(s, s') \in m'(\tau(\varphi))$.*

Proof. The proof is similar to the proof of Proposition 7.4.2. Let φ be an INT-formula and let $\mathcal{M} = (U, R, m)$ be an INT-model. We define an $\mathsf{RL}_{\mathsf{INT}}$-model $\mathcal{M}' = (U, R, m')$ as follows:

- $m'(1) = U \times U$;
- $m'(1')$ is the identity on U;
- $m'(\tau(p)) = \{(x, y) \in U \times U : x \in m(p)\}$, for every propositional variable p;
- $m'(R) = R$;
- m' extends to all the compound terms as in $\mathsf{RL}(1, 1')$-models.

The proof is by induction on the complexity of formulas. By way of example, we show the proposition for a compound formula built with the implication operation which has a non-classical semantics in INT-logic.

Let $\varphi = \psi \rightarrow \vartheta$ and let $s' \in U$. Then $\mathcal{M}, s \models \psi \rightarrow \vartheta$ iff for every $t \in U$, if $(s, t) \in R$ and $\mathcal{M}, t \models \psi$, then $\mathcal{M}, t \models \vartheta$ iff, by the induction hypothesis, for every $t \in U$, if $(s, t) \in R$ and $(t, s') \in m'(\tau(\psi))$, then $(t, s') \in m'(\tau(\vartheta))$ iff $(s, s') \in m'(-(R\,;(\tau(\psi) \cap -\tau(\vartheta))))$ iff $(s, s') \in m'(\tau(\varphi))$. $\square$

Proposition 8.2.2. *Let φ be an* INT*-formula. Then for every standard* $\mathsf{RL}_{\mathsf{INT}}$*-model* $\mathcal{M}' = (U, R, m')$ *there exists an* INT*-model* $\mathcal{M} = (U, R, m)$ *with the same universe and the same relation R as those in $\mathcal{M}'$ such that for all $s, s' \in U$, the condition of Proposition 8.2.1 holds.*

Proof. Let φ be an INT-formula, let $\mathcal{M}' = (U, R, m')$ be a standard $\mathsf{RL}_{\mathsf{INT}}$-model. We define an INT-model $\mathcal{M} = (U, R, m)$ as follows:

$m(p) = \{(x \in U :$ for some $y \in U$, $(x, y) \in m'(\tau(p))\}$, for every propositional variable p.

Now, the proposition can be proved as Proposition 8.2.1. □

Due to the above propositions we can prove the following (see also Propositions 7.4.4 and 7.4.5):

Proposition 8.2.3. *Let φ be an* INT*-formula. Then for every* INT*-model $\mathcal{M}$ there exists an* $\mathsf{RL}_{\mathsf{INT}}$*-model $\mathcal{M}'$ such that for any object variables x and y, $\mathcal{M} \models \varphi$ iff $\mathcal{M}' \models x\tau(\varphi)y$.*

Proposition 8.2.4. *Let φ be an* INT*-formula. Then for every standard* $\mathsf{RL}_{\mathsf{INT}}$*-model $\mathcal{M}'$ there exists an* INT*-model $\mathcal{M}$ such that for any object variables x and y, the condition of Proposition 8.2.3 holds.*

Finally, from Propositions 8.2.3 and 8.2.4, we get:

Theorem 8.2.1. *For every* INT*-formula φ and for all object variables x and y, the following conditions are equivalent:*

1. *φ is* INT*-valid;*
2. *$x\tau(\varphi)y$ is* $\mathsf{RL}_{\mathsf{INT}}$*-valid.*

A dual tableau for the logic $\mathsf{RL}_{\mathsf{INT}}$ consists of the axiomatic sets and the rules of dual tableau for $\mathsf{RL}(1, 1')$-logic adjusted to the language of $\mathsf{RL}_{\mathsf{INT}}$, the rules (ref R) and (tran R) presented in Sects. 6.6 (see also Sect. 7.4) that reflect reflexivity and transitivity of relation R from INT-models, respectively. Furthermore, we have the rules of the following forms:

For every relational variable P and for all object variables x and y,

$$\text{(rher')}\ \frac{xPy}{zRx, xPy \mid zPy, xPy} \qquad z \text{ is any object variable}$$

$$\text{(ideal)}\ \frac{xPy}{xPz, xPy} \qquad z \text{ is any object variable}$$

The rule (rher') reflects the heredity condition, while the rule (ideal) reflects the fact that every relational variable is interpreted in an $\mathsf{RL}_{\mathsf{INT}}$-model as a right ideal relation. Note that any application of the rules of $\mathsf{RL}_{\mathsf{INT}}$-dual tableau, in particular an application of the specific rules listed above, preserves the formulas built with atomic terms or their complements, and hence the closed branch property holds.

The notions of an $\mathsf{RL}_{\mathsf{INT}}$-set of formulas and correctness of a rule are defined as in Sect. 2.4.

Proposition 8.2.5. *The specific* $\mathsf{RL}_{\mathsf{INT}}$*-rules are* $\mathsf{RL}_{\mathsf{INT}}$*-correct.*

Proof. By way of example, we show correctness of the rule (rher'). Correctness of the remaining rules can be proved as in standard modal logics. Let X be a finite set of $\mathsf{RL}_{\mathsf{INT}}$-formulas. Clearly, if $X \cup \{xPy\}$ is an $\mathsf{RL}_{\mathsf{INT}}$-set, then so are $X \cup \{zRx, xPy\}$ and $X \cup \{zPy, xPy\}$. Now, assume that $X \cup \{zRx, xPy\}$ and $X \cup \{zPy, xPy\}$ are $\mathsf{RL}_{\mathsf{INT}}$-sets and suppose $X \cup \{xPy\}$ is not an $\mathsf{RL}_{\mathsf{INT}}$-set. Then there exist an $\mathsf{RL}_{\mathsf{INT}}$-model $\mathcal{M} = (U, R, m)$ and a valuation v in $\mathcal{M}$ such that for every $\varphi \in X \cup \{xPy\}$ $\mathcal{M}, v \not\models \varphi$, which implies $(v(x), v(y)) \notin m(P)$. By the assumption, $\mathcal{M}, v \models zRx$ and $\mathcal{M}, v \models zPy$. Thus, $(v(z), v(x)) \in R$ and $(v(z), v(y)) \in m(P)$. Then, by the condition (her'), $(v(x), v(y)) \in m(P)$, a contradiction. □

It is easy to see that all the remaining rules of $\mathsf{RL}_{\mathsf{INT}}$-dual tableau are $\mathsf{RL}_{\mathsf{INT}}$-correct and all the $\mathsf{RL}_{\mathsf{INT}}$-axiomatic sets are $\mathsf{RL}_{\mathsf{INT}}$-sets.

The notions of an $\mathsf{RL}_{\mathsf{INT}}$-proof tree, a closed branch of such a tree, a closed $\mathsf{RL}_{\mathsf{INT}}$-proof tree, and $\mathsf{RL}_{\mathsf{INT}}$-provability are defined as in Sect. 2.4.

A branch b of an $\mathsf{RL}_{\mathsf{INT}}$-proof tree is complete whenever it is either closed or it satisfies the completion conditions of $\mathsf{RL}(1, 1')$-dual tableau and the following completion conditions determined by the rules specific for the $\mathsf{RL}_{\mathsf{INT}}$-dual tableau:

For every relational variable P and for all object variables x and y,

Cpl(rher') If $xPy \in b$, then for every object variable z, either $zRx \in b$ or $zPy \in b$, obtained by an application of the rule (rher');

Cpl(ideal) If $xPy \in b$, then for every object variable z, $xPz \in b$, obtained by an application of the rule (ideal).

The notions of a complete $\mathsf{RL}_{\mathsf{INT}}$-proof tree and an open branch of an $\mathsf{RL}_{\mathsf{INT}}$-proof tree are defined as in RL-logic (see Sect. 2.5).

The branch structure $\mathcal{M}^b = (U^b, R^b, m^b)$ is defined in a standard way, that is $m^b(Q) = \{(x, y) \in U^b \times U^b : xQy \notin b\}$, for every $Q \in \mathbb{RV}_{\mathsf{RL}_{\mathsf{INT}}} \cup \{R\}$, and $R^b = m^b(R)$. The completion conditions enable us to show that R^b satisfies all the conditions assumed in $\mathsf{RL}_{\mathsf{INT}}$-models. Therefore, the following holds:

Proposition 8.2.6 (Branch Model Property). *For every open branch b of an* $\mathsf{RL}_{\mathsf{INT}}$*-proof tree, $\mathcal{M}^b$ is an* $\mathsf{RL}_{\mathsf{INT}}$*-model.*

Proof. We show that $\mathcal{M}^b$ satisfies the condition (her') and every relational variable P is interpreted in $\mathcal{M}^b$ as a right ideal relation. Assume that $(x, y) \in R^b$ and $(x, z) \in m^b(P)$, that is $xRy \notin b$ and $xPz \notin b$. Suppose $(y, z) \notin m^b(P)$, that is $yPz \in b$. Then, by the completion condition Cpl(rher'), either $xRy \in b$ or $xPz \in b$, a contradiction. Hence, $\mathcal{M}^b$ satisfies the condition (her'). Now, assume that $(x, y) \in m^b(P)$, that is $xPy \notin b$. Suppose that for some $z \in U^b$, $(x, z) \notin m^b(P)$. Then $xPz \in b$ and by the completion condition Cpl(ideal), $xPy \in b$, a contradiction. Therefore, for every relational variable P, $m^b(P)$ is a right ideal relation on U^b. □

Since branch model $\mathcal{M}^b$ is defined in a standard way, the satisfaction in branch model property can be proved as in $\mathsf{RL}(1, 1')$-logic (see Sects. 2.5 and 2.7).

Proposition 8.2.7 (Satisfaction in Branch Model Property). *For every open branch b of an $\mathsf{RL_{INT}}$-proof tree and for every $\mathsf{RL_{INT}}$-formula φ, if $\mathcal{M}^b, v^b \models \varphi$, then $\varphi \notin b$.*

Hence, the completeness of $\mathsf{RL_{INT}}$-dual tableau follows.

Theorem 8.2.2 (Soundness and Completeness of $\mathsf{RL_{INT}}$). *Let φ be an $\mathsf{RL_{INT}}$-formula. Then the following conditions are equivalent:*

1. *φ is $\mathsf{RL_{INT}}$-valid;*
2. *φ is true in all standard $\mathsf{RL_{INT}}$-models;*
3. *φ is $\mathsf{RL_{INT}}$-provable.*

By the above theorem and Theorem 8.2.1, we obtain:

Theorem 8.2.3 (Relational Soundness and Completeness of INT). *Let φ be an INT-formula. Then for all object variables x and y, the following conditions are equivalent:*

1. *φ is INT-valid;*
2. *$x(\tau(\varphi))y$ is $\mathsf{RL_{INT}}$-provable.*

Example. Consider the following INT-formula φ:

$$\varphi = \neg p \rightarrow (p \rightarrow t).$$

The translation of this formula into $\mathsf{RL_{INT}}$-term is:

$$\tau(\varphi) = -(R\,;(-(R\,;P) \cap --(R\,;(P \cap -T)))),$$

where $\tau(p) = P$ and $\tau(t) = T$. INT-validity of φ is equivalent to $\mathsf{RL_{INT}}$-provability of the formula $x\tau(\varphi)y$. Figure 8.1 presents its $\mathsf{RL_{INT}}$-proof.

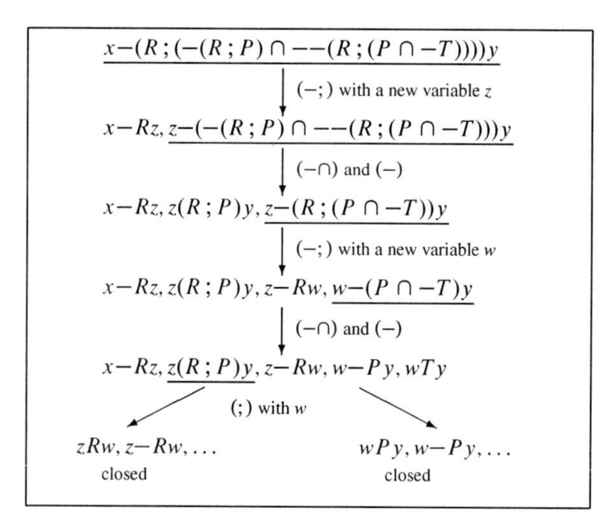

Fig. 8.1 An $\mathsf{RL_{INT}}$-proof of the formula $\neg p \rightarrow (p \rightarrow t)$

8.3 Relational Formalization of Minimal Intuitionistic Logic

Minimal intuitionistic logic J was introduced by Johansson in [Joh36]. It differs from the intuitionistic logic INT in that the formula $\neg\varphi \rightarrow (\varphi \rightarrow \psi)$ is not a theorem of the logic. The language of the logic J is the same as INT-language. However, these logics differ in semantics. A J-model is a structure $\mathcal{M} = (U, R, Q, m)$ such that the following conditions are satisfied:

- (U, R, m) is an INT-model;
- $Q \subseteq U$ is such that for every $s \in U$, if $s \in Q$ and $(s, s') \in R$, then $s' \in Q$.

The satisfaction of a J-formula φ in a J-model $\mathcal{M}$ by a state $s \in U$ is defined as in INT-logic except for the clause for negated formulas:

$$\mathcal{M}, s \models \neg\psi \text{ iff for all } s' \in U, \text{ if } (s, s') \in R, \text{ then } \mathcal{M}, s' \not\models \psi \text{ or } s' \in Q.$$

Intuitively, Q is thought of as the set of those states which are inconsistent.

The language of the relational logic $\mathsf{RL_J}$ is an $\mathsf{RL_{INT}}$-language with the following set of relational constants: $\mathbb{RC}_{\mathsf{RL_J}} = \{1, 1', R, Q\}$.

An $\mathsf{RL_J}$-model is an $\mathsf{RL_{INT}}$-model $\mathcal{M} = (U, R, Q, m)$ such that the following conditions are satisfied:

$(Q1)$ Q is a right ideal relation on U;
$(Q2)$ If $(x, z) \in Q$ and $(x, y) \in R$, then $(y, z) \in Q$, for all $x, y, z \in U$.

The translation τ from J-formulas into relational terms coincides with the translation of INT-formulas defined in Sect. 8.2 except for the translation of negated formulas:

$$\tau(\neg\varphi) \stackrel{\text{df}}{=} -(R\,;(\tau(\varphi) \cap -Q)).$$

Following the method of proving Theorem 8.2.1, we can prove:

Theorem 8.3.1. *For every* J*-formula* φ *and for all object variables* x *and* y*, the following conditions are equivalent:*

1. φ *is* J*-valid;*
2. $x\tau(\varphi)y$ *is* $\mathsf{RL_J}$*-valid.*

A dual tableau for the logic $\mathsf{RL_J}$ consists of the axiomatic sets and the rules of dual tableau for $\mathsf{RL_{INT}}$-logic presented in Sect. 8.2 adjusted to the language of $\mathsf{RL_J}$ and in addition the rules of the following forms:

For all object variables x and y,

$$(\mathrm{r}Q1)\ \frac{xQy}{xQz, xQy} \quad z \text{ is any object variable}$$

$$(\mathrm{r}Q2)\ \frac{xQy}{zQy, xQy \mid zRx, xQy} \quad z \text{ is any object variable}$$

The above rules reflect the properties of relation Q assumed in $\mathsf{RL_J}$-models. As in $\mathsf{RL_{INT}}$-dual tableau, any application of the rules listed above preserves the formulas built with atomic terms or their complements, and hence the closed branch property holds.

Proposition 8.3.1. *The* $\mathsf{RL_J}$*-rules are* $\mathsf{RL_J}$*-correct.*

Proof. By way of example, we prove correctness of the rule (rQ2). Correctness of the other rules can be proved as in standard modal logics. Let X be any finite set of $\mathsf{RL_J}$-formulas. Clearly, if $X \cup \{xQy\}$ is an $\mathsf{RL_J}$-set, then so are $X \cup \{zQy, xQy\}$ and $X \cup \{zRx, xQy\}$. Now, assume that $X \cup \{zQy, xQy\}$ and $X \cup \{zRx, xQy\}$ are $\mathsf{RL_J}$-sets. Suppose $X \cup \{xQy\}$ is not an $\mathsf{RL_J}$-set. Then there exist an $\mathsf{RL_J}$-model $\mathcal{M} = (U, R, Q, m)$ and a valuation v in $\mathcal{M}$ such that for every $\varphi \in X \cup \{xQy\}$ $\mathcal{M}, v \not\models \varphi$, which implies $(v(x), v(y)) \notin m(Q)$. By the assumption, $\mathcal{M}, v \models zQy$ and $\mathcal{M}, v \models zRx$. Thus, $(v(z), v(y)) \in Q$ and $(v(z), v(x)) \in R$. By property (Q2) assumed in $\mathsf{RL_J}$-models, $(v(x), v(y)) \in Q$, a contradiction. Therefore, the rule (rQ2) is $\mathsf{RL_J}$-correct. □

The completion conditions determined by the rules (rQ1) and (rQ2) are:
For all object variables x and y,

Cpl(rQ1) If $xQy \in b$, then for every object variable z, $xQz \in b$;
Cpl(rQ2) If $xQy \in b$, then for every object variable z, either $zQy \in b$ or $zRx \in b$.

The branch model is defined in a standard way, i.e., $m^b(T) = \{(x, y) \in U^b \times U^b : xTy \notin b\}$, for every $T \in \mathbb{RV}_{\mathsf{RL_J}} \cup \{R, Q\}$. The relation Q^b satisfies all the conditions assumed in $\mathsf{RL_J}$-models, therefore we have:

Proposition 8.3.2 (Branch Model Property). *For every open branch b of an* $\mathsf{RL_J}$*-proof tree, $\mathcal{M}^b$ is an* $\mathsf{RL_J}$*-model.*

Proof. We show that Q^b is well defined. It can be easily proved that Q^b is a right ideal relation on U^b. Now, assume that $(x, z) \in Q^b$ and $(x, y) \in R^b$, that is $xQz \notin b$ and $xRy \notin b$. Suppose $(y, z) \notin Q^b$, that is $yQz \in b$. Then, by the completion condition Cpl(rQ2), either $xQz \in b$ or $xRy \in b$, a contradiction. □

Satisfaction in branch model property can be proved as in $\mathsf{RL}(1, 1')$-logic, hence we get:

Theorem 8.3.2 (Soundness and Completeness of $\mathsf{RL_J}$). *Let φ be an* $\mathsf{RL_J}$*-formula. Then the following conditions are equivalent:*

1. *φ is* $\mathsf{RL_J}$*-valid;*
2. *φ is true in all standard* $\mathsf{RL_J}$*-models;*
3. *φ is* $\mathsf{RL_J}$*-provable.*

By the above theorem and Theorem 8.3.1, we have:

Theorem 8.3.3 (Relational Soundness and Completeness of J). *Let φ be a J-formula. Then for all object variables x and y, the following conditions are equivalent:*

1. *φ is J-valid;*
2. *$x(\tau(\varphi))y$ is $\mathsf{RL_J}$-provable.*

Example. Consider the following J-formula φ:

$$\varphi = p \rightarrow \neg\neg p.$$

The translation of φ into an $\mathsf{RL_J}$-term is:

$$\tau(\varphi) = -(R\,;(P \cap --(R\,;(-(R\,;(P \cap -Q)) \cap -Q)))),$$

where $\tau(p) = P$. J-validity of φ is equivalent to $\mathsf{RL_J}$-provability of the formula $x\tau(\varphi)y$. Figure 8.2 presents its $\mathsf{RL_J}$-proof.

In the previous section we constructed an $\mathsf{RL_{INT}}$-proof for the formula $\varphi = \neg p \rightarrow (p \rightarrow t)$ which showed its INT-validity. This formula is not J-valid. Its translation into $\mathsf{RL_J}$-term is:

$$\tau(\varphi) = -(R\,;(-(R\,;(P \cap -Q)) \cap --(R\,;(P \cap -T)))),$$

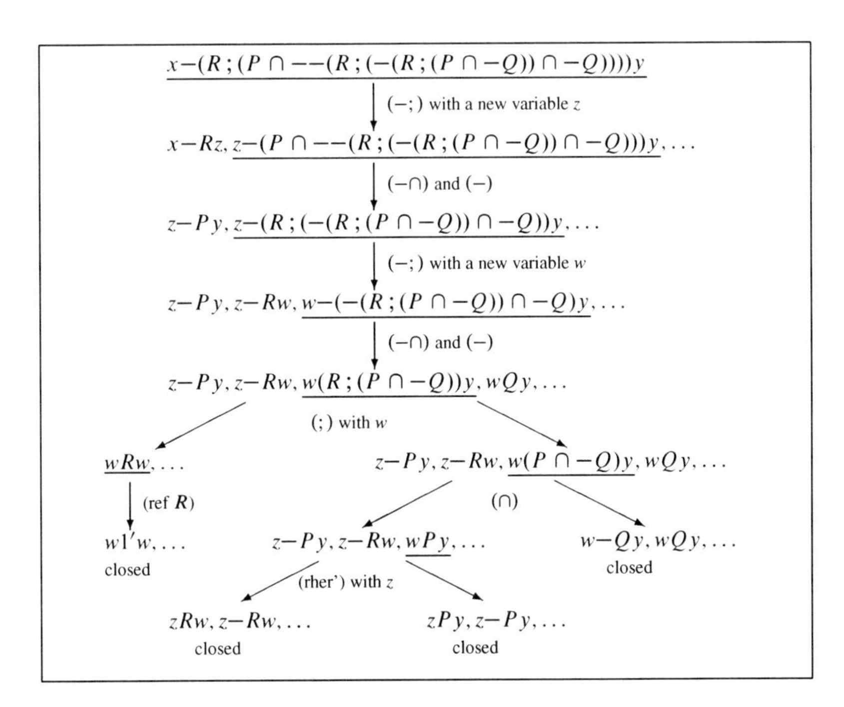

Fig. 8.2 An $\mathsf{RL_J}$-proof of the formula $p \rightarrow \neg\neg p$

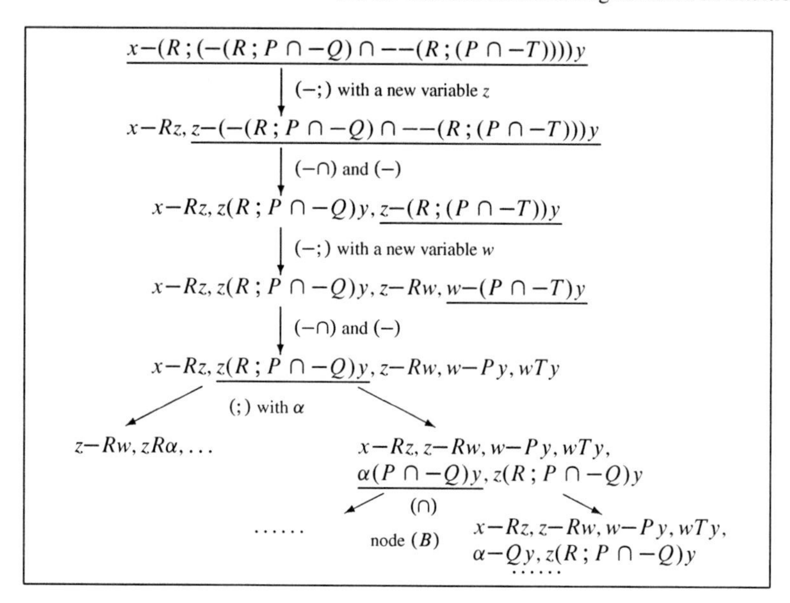

Fig. 8.3 A failed proof search of the formula $\neg p \rightarrow (p \rightarrow t)$

where $\tau(p) = P$ and $\tau(t) = T$. Since φ is not J-valid, any $\mathsf{RL_J}$-proof tree must not close. Figure 8.3 presents an $\mathsf{RL_J}$-proof tree of $x\tau(\varphi)y$ in which there is a branch that cannot be closed. Consider the node (B) in that tree which consists of the following formulas: $x{-}Rz$, $z{-}Rw$, $w{-}Py$, wTy, $\alpha{-}Qy$, $z(R\,;(P \cap -Q))y$, where a variable α is arbitrary. The only decomposition rule that can be applied to node (B) is the rule $(;)$, which generates, among others, a node consisting of the same formulas as (B) and a formula $\beta{-}Qy$, where β may be an arbitrary variable. We can observe that no application of the rule $(;)$ would close the branch. Also, no application of a specific rule would close the branch. The rules $(1'1)$ and $(1'2)$ can be applied only to the formula wTy. As the result we obtain a node that consists of the same formulas as node (B) and the formula $y1'\beta$ or $w1'\beta$. Since node (B) does not contain a formula built with the complement of $1'$, the branch can be closed neither by an application of the rule $(1'1)$ nor $(1'2)$. The rules (ref R) and (tran R) cannot be applied to the node (B), because it does not contain a formula built with the atomic term R. Similarly, the rules (rQ1) and (rQ2) cannot be applied. The rules (rher') and (ideal) can be applied only to the formula wTy. However, observe that any application of the rule (rher') generates a node consisting of the same formulas as (B) together with the formula βTy. It is easy to see that it cannot close the branch. In a similar way, one can show that the same holds for the rule (ideal). Thus, the branch cannot be closed.

8.4 Relational Formalization of Some Intermediate Logics

Intermediate logics are the logics whose valid formulas include all the formulas that are valid in intuitionistic logic but not necessarily all the tautologies of classical propositional logic. In that sense, intermediate logics are between intuitionistic and classical logic. The common language of intermediate logics is the INT-language.

In this section we consider two intermediate logics $\mathsf{INT}_{\mathsf{L}_1}$ and $\mathsf{INT}_{\mathsf{L}_2}$ whose models are the INT-models satisfying the following conditions, respectively:

(L_1) $\forall x \forall y \forall z \forall t[((x,y) \in R \wedge (x,z) \in R \wedge (y,t) \in R) \rightarrow ((y,z) \in R \vee (z,y) \in R \vee (z,t) \in R)]$;

(L_2) $\forall x \forall y \forall z[((x,y) \in R \wedge (x,z) \in R) \rightarrow \exists t((y,t) \in R \wedge (z,t) \in R)]$.

The specific axioms of the logics $\mathsf{INT}_{\mathsf{L}_1}$ and $\mathsf{INT}_{\mathsf{L}_2}$ in their Hilbert-style axiomatizations are:

(A1) $[(\neg\neg\varphi \rightarrow \varphi) \rightarrow (\neg\varphi \vee \varphi)] \rightarrow (\neg\varphi \vee \neg\neg\varphi)$;

(A2) $\neg\varphi \vee \neg\neg\varphi$.

$\mathsf{INT}_{\mathsf{L}_1}$ is referred to as Scott's logic and $\mathsf{INT}_{\mathsf{L}_2}$ is the logic of weak excluded middle.

The common language of a relational logic $\mathsf{RL}_{\mathsf{INT}_\mathsf{L}}$ for an intermediate logic $\mathsf{L} \in \{\mathsf{L}_1, \mathsf{L}_2\}$ is the $\mathsf{RL}_{\mathsf{INT}}$-language. Models of $\mathsf{RL}_{\mathsf{INT}_\mathsf{L}}$ satisfy all the conditions assumed in INT_L-models for the accessibility relation R and (L_1) or (L_2), respectively. The translation of INT_L-formulas into relational terms is defined as for INT-formulas in Sect. 8.2. To obtain dual tableaux for these logics, we add to $\mathsf{RL}_{\mathsf{INT}}$-dual tableau the rules reflecting the specific semantic conditions that are assumed in their models. The specific rules reflecting the conditions (L_1) and (L_2) have the following forms:

For all object variables x, y, z, and t,

$$(r\mathsf{L}_1) \quad \frac{yRz, zRy, zRt}{xRy, yRz, zRy, zRt \mid xRz, yRz, zRy, zRt \mid yRt, yRz, zRy, zRt}$$

$$(r\mathsf{L}_2) \quad \frac{}{xRy \mid xRz \mid y-(R ; R^{-1})z}$$

The above rules have the property of preserving formulas built with atomic terms or their complements. Hence, the closed branch property holds. An alternative form of rule ($r\mathsf{L}_2$) is discussed in Sect. 25.9.

Theorem 8.4.1 (Correspondence). *Let $\mathcal{K}$ be a class of $\mathsf{RL}_{\mathsf{INT}}$-models and let $\mathsf{i} \in \{1, 2\}$. Then, $\mathcal{K}$ is a class of $\mathsf{RL}_{\mathsf{INT}_{\mathsf{L}_\mathsf{i}}}$-models iff the rule ($r\mathsf{L}_\mathsf{i}$) is $\mathcal{K}$-correct.*

Proof.

($\rightarrow$) By way of example, we show it for $\mathsf{i} = 1$. Let $\mathcal{K}$ be a class of $\mathsf{RL}_{\mathsf{INT}_{\mathsf{L}_1}}$-models, that is every model from $\mathcal{K}$ satisfies the condition (L_1). Let X be a finite set of $\mathsf{RL}_{\mathsf{INT}}$-formulas. Clearly, if $X \cup \{yRz, zRy, zRt\}$ is a $\mathcal{K}$-set, then so are $X \cup \{xRy, yRz, zRy, zRt\}$, $X \cup \{xRz, yRz, zRy, zRt\}$, and $X \cup \{yRt, yRz, zRy, zRt\}$. Assume $X \cup \{xRy, yRz, zRy, zRt\}$, $X \cup \{xRz, yRz,$

$zRy, zRt\}$, and $X \cup \{yRt, yRz, zRy, zRt\}$ are $\mathcal{K}$-sets. Suppose $X \cup \{yRz, zRy, zRt\}$ is not a $\mathcal{K}$-set. Then there exist an $\mathsf{RL_{INT_{L_1}}}$-model $\mathcal{M} = (U, R, m)$ and a valuation v in $\mathcal{M}$ such that for every $\varphi \in X \cup \{yRz, zRy, zRt\}$, $\mathcal{M}, v \not\models \varphi$, in particular, $(v(y), v(z)) \notin R$, $(v(z), v(y)) \notin R$, and $(v(z), v(t)) \notin R$. However, by the assumption, model $\mathcal{M}$ and valuation v satisfy $(v(x), v(y)) \in R$, $(v(x), v(z)) \in R$, and $(v(y), v(t)) \in R$. By the condition ($\mathsf{L_1}$), it means that either $(v(y), v(z)) \in R$ or $(v(z), v(y)) \in R$ or $(v(z), v(t)) \in R$, a contradiction.

($\leftarrow$) By way of example, we show it for $\mathsf{i} = 2$. Let $\mathcal{K}$ be a class of $\mathsf{RL_{INT}}$-models. Assume the rule (r$\mathsf{L_2}$) is $\mathcal{K}$-correct. Let $X \stackrel{\text{df}}{=} \{x-Ry, x-Rz, y(R\,;R^{-1})z\}$. Clearly, the sets $X \cup \{xRy\}$, $X \cup \{xRz\}$, and $X \cup \{y-(R\,;R^{-1})z\}$ are $\mathcal{K}$-sets. Thus, by the assumption, X must be a $\mathcal{K}$-set. Therefore, for every model $\mathcal{M} = (U, R, m)$ from $\mathcal{K}$ and for all $x, y, z \in U$, if $(x, y) \in R$ and $(x, z) \in R$, then $(y, z) \in R\,;R^{-1}$. Since $(y, z) \in R\,;R^{-1}$ is equivalent to $\exists t((y, t) \in R$ and $(z, t) \in R)$, $\mathcal{K}$ is a class of $\mathsf{RL_{INT_{L_2}}}$-models. □

The completion conditions determined by the rules (r$\mathsf{L_1}$) and (r$\mathsf{L_2}$) are:

For all object variables x, y, z, and t,

Cpl(r$\mathsf{L_1}$) If the formulas yRz, zRy, and zRt are in b, then either $xRy \in b$ or $xRz \in b$ or $yRt \in b$;

Cpl(r$\mathsf{L_2}$) Either $xRy \in b$ or $xRz \in b$ or $y-(R\,;R^{-1})z \in b$.

In order to prove completeness, we need to show that the branch structure is an $\mathsf{RL_{INT_L}}$-model, for $\mathsf{L} \in \{\mathsf{L_1}, \mathsf{L_2}\}$.

Proposition 8.4.1 (Branch Model Property). *Let* $\mathsf{L} \in \{\mathsf{L_1}, \mathsf{L_2}\}$ *and let* b *be an open branch of an* $\mathsf{RL_{INT_L}}$*-proof tree. Then,* $\mathcal{M}^b$ *is an* $\mathsf{RL_{INT_L}}$*-model.*

Proof. It suffices to show that if $\mathsf{RL_{INT_L}}$-dual tableau includes a rule (r$\mathsf{L_i}$), then $\mathcal{M}^b$ satisfies the condition ($\mathsf{L_i}$), for $\mathsf{i} \in \{1, 2\}$. By way of example, we show it for $\mathsf{i} = 2$. Indeed, by the completion condition Cpl(r$\mathsf{L_2}$), for all $x, y, z \in U^b$, either $xRy \in b$ or $xRz \in b$ or $y-(R\,;R^{-1})z \in b$. Thus, by the completion conditions Cpl(;) and Cpl($-^{-1}$), for all $x, y, z \in U^b$, either $xRy \in b$ or $xRz \in b$ or there exists $t \in U^b$ such that $y-Rt \in b$ and $z-Rt \in b$. So $(x, y) \notin R^b$ or $(x, z) \notin R^b$ or there exists $t \in U^b$ such that $(y, t) \in R^b$ and $(z, t) \in R^b$. Therefore, for all $x, y, z \in U^b$, if $(x, y) \in R^b$ and $(x, z) \in R^b$, then there exists $t \in U^b$ such that $(y, t) \in R^b$ and $(z, t) \in R^b$, hence the condition ($\mathsf{L_2}$) is satisfied. □

Since the language of the logics $\mathsf{RL_{INT_{L_1}}}$ and $\mathsf{RL_{INT_{L_2}}}$ is the same as in $\mathsf{RL_{INT}}$-logic, the satisfaction in branch model property holds. Thus, the proof of completeness is similar as in $\mathsf{RL_{INT}}$-logic.

Theorem 8.4.2 (Soundness and Completeness of $\mathsf{RL_{INT_L}}$). *Let* $\mathsf{L} \in \{\mathsf{L_1}, \mathsf{L_2}\}$ *and let* φ *be an* $\mathsf{RL_{INT_L}}$*-formula. Then the following conditions are equivalent:*

1. φ *is* $\mathsf{RL}_{\mathsf{INT}_\mathsf{L}}$*-valid;*
2. φ *is true in all standard* $\mathsf{RL}_{\mathsf{INT}_\mathsf{L}}$*-models;*
3. φ *is* $\mathsf{RL}_{\mathsf{INT}_\mathsf{L}}$*-provable.*

Theorem 8.4.3 (Relational Soundness and Completeness of INT_L). *Let* $\mathsf{L} \in \{\mathsf{L}_1, \mathsf{L}_2\}$ *and let* φ *be an* INT_L*-formula. Then for all object variables x and y, the following conditions are equivalent:*

1. φ *is* INT_L*-valid;*
2. $x(\tau(\varphi))y$ *is* $\mathsf{RL}_{\mathsf{INT}_\mathsf{L}}$*-provable.*

Example. Consider the following $\mathsf{INT}_{\mathsf{L}_2}$-formula:

$$\varphi = \neg p \vee \neg\neg p.$$

The translation of φ into $\mathsf{RL}_{\mathsf{INT}_{\mathsf{L}_2}}$-term is:

$$\tau(\varphi) = -(R\,;P) \cup -(R\,;-(R\,;P)),$$

where $\tau(p) = P$. $\mathsf{INT}_{\mathsf{L}_2}$-validity of φ is equivalent to $\mathsf{RL}_{\mathsf{INT}_{\mathsf{L}_2}}$-provability of the formula $x\tau(\varphi)y$. Figure 8.4 presents its $\mathsf{RL}_{\mathsf{INT}_{\mathsf{L}_2}}$-proof.

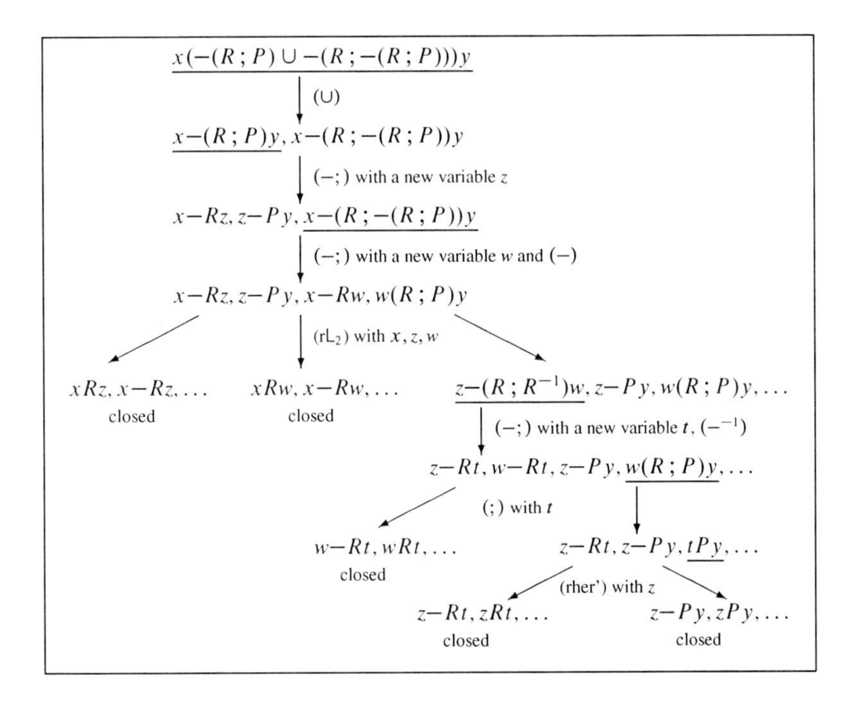

Fig. 8.4 An $\mathsf{RL}_{\mathsf{INT}_{\mathsf{L}_2}}$-proof of the formula $\neg p \vee \neg\neg p$

8.5 Relational Formalization of a Logic for Hardware Verification

In this section we consider a propositional logic PLL which has been proposed as a tool for a formal verification of computer hardware (see [FM95, Men90, Men93]).

The logic PLL is obtained from the logic INT by endowing its language with a unary propositional operation $\bigcirc$ that models a delay of propagation of signals. Signals are conceived as Boolean valued functions, where Boolean values are denoted by 1 and 0. To input and output signals of the gates we assign propositional variables. If p is such a variable, then truth of p in a model is interpreted as 'p is stable at 1'', truth of $\neg p$ means "p is stable at 0', truth of $\bigcirc p$ means 'p is going to stabilize to 1', and truth of $\bigcirc\neg p$ means 'p is going to stabilize to 0'.

A PLL-model is a structure $\mathcal{M} = (U, R, S, m)$ such that (U, R, m) is an INT-model and S is a reflexive and transitive relation on U such that $S \subseteq R$. Then, the satisfaction of formulas of the form $\bigcirc\varphi$ is defined as follows:

$\mathcal{M}, s \models \bigcirc\varphi$ iff for every $s' \in U$, if $(s, s') \in R$, then there exists $t \in \mathrm{U}$ such that $(s', t) \in S$ and $\mathcal{M}, t \models \varphi$.

The language of the relational logic $\mathsf{RL}_{\mathsf{PLL}}$ is the language of $\mathsf{RL}_{\mathsf{INT}}$-logic endowed with the relational constant S.

An $\mathsf{RL}_{\mathsf{PLL}}$-model is a structure $\mathcal{M} = (U, R, S, m)$ such that (U, R, m) is an $\mathsf{RL}_{\mathsf{INT}}$-model and S is a reflexive and transitive relation on U satisfying $S \subseteq R$.

The translation τ from PLL-formulas into relational terms is defined as the translation in Sect. 8.2 with the following clause for formulas of the form $\bigcirc\varphi$:

$$\tau(\bigcirc\varphi) \stackrel{\text{df}}{=} -(R\,;-(S\,;(\tau(\varphi)))).$$

Much the same as we proved Theorem 8.2.1, we can show:

Theorem 8.5.1. *For every* PLL*-formula φ and for all object variables x and y, the following conditions are equivalent:*

1. *φ is* PLL*-valid;*
2. *$x\tau(\varphi)y$ is* $\mathsf{RL}_{\mathsf{PLL}}$*-valid.*

A dual tableau for the logic $\mathsf{RL}_{\mathsf{PLL}}$ consists of the axiomatic sets and the rules of dual tableau for $\mathsf{RL}_{\mathsf{INT}}$-logic (see Sect. 8.2) adjusted to the language of $\mathsf{RL}_{\mathsf{PLL}}$, the rules (ref S) and (tran S) reflecting reflexivity and transitivity of S, respectively (see Sect. 6.6), and in addition the rule (SR) of the following form:

For all object variables x and y,

$$(SR) \quad \frac{xRy}{xSy, xRy}$$

The above rule reflects the condition $S \subseteq R$. The rule (SR) preserves formulas built with atomic terms. Hence, the closed branch property holds. It is easy to see that the rule (SR) is $\mathsf{RL}_{\mathsf{PLL}}$-correct.

Proposition 8.5.1. *The* $\mathsf{RL}_{\mathsf{PLL}}$*-rules are* $\mathsf{RL}_{\mathsf{PLL}}$*-correct.*

The completion condition determined by the rule (SR) is:
For all object variables x and y,

Cpl(SR) If $xRy \in b$, then $xSy \in b$.

The branch model is defined in a standard way, that is $m^b(T) = \{(x, y) \in U^b \times U^b : xTy \notin b\}$, for every $T \in \mathbb{RV}_{\mathsf{RL}_{\mathsf{PLL}}} \cup \{R, S\}$.

Proposition 8.5.2 (Branch Model Property). *For every open branch b of an* $\mathsf{RL}_{\mathsf{PLL}}$*-proof tree, $\mathcal{M}^b$ is an* $\mathsf{RL}_{\mathsf{PLL}}$*-model.*

Proof. It suffices to show that S^b satisfies all the conditions assumed in $\mathsf{RL}_{\mathsf{PLL}}$-models. Reflexivity and transitivity of S^b can be proved as in Theorem 6.6.1. Now, assume that $(x, y) \in S^b$, that is $xSy \notin b$. Suppose $(x, y) \notin R^b$, that is $xRy \in b$. Then, by the completion condition Cpl(SR), we get $xSy \in b$, a contradiction. Hence, $\mathcal{M}^b$ is an $\mathsf{RL}_{\mathsf{PLL}}$-model. □

Since the branch model $\mathcal{M}^b$ is defined in a standard way (see Sect. 2.6, p. 44), the satisfaction in branch model property can be proved as in $\mathsf{RL}(1, 1')$-logic (see Sects. 2.5 and 2.7). Hence, completeness of $\mathsf{RL}_{\mathsf{PLL}}$-dual tableau follows.

Theorem 8.5.2 (Soundness and Completeness of $\mathsf{RL}_{\mathsf{PLL}}$). *Let φ be an* $\mathsf{RL}_{\mathsf{PLL}}$*-formula. Then the following conditions are equivalent:*

1. *φ is* $\mathsf{RL}_{\mathsf{PLL}}$*-valid;*
2. *φ is true in all standard* $\mathsf{RL}_{\mathsf{PLL}}$*-models;*
3. *φ is* $\mathsf{RL}_{\mathsf{PLL}}$*-provable.*

By the above theorem and Theorem 8.5.1, we have:

Theorem 8.5.3 (Relational Soundness and Completeness of PLL). *Let φ be a PLL-formula. Then for all object variables x and y, the following conditions are equivalent:*

1. *φ is PLL-valid;*
2. *$x(\tau(\varphi))y$ is* $\mathsf{RL}_{\mathsf{PLL}}$*-provable.*

Example. Consider the following PLL-formula:

$$\varphi = p \rightarrow \bigcirc p.$$

The translation of φ into $\mathsf{RL}_{\mathsf{PLL}}$-term is:

$$\tau(\varphi) = -(R ; (P \cap --(R ; -(S ; P)))),$$

where $\tau(p) = P$. PLL-validity of φ is equivalent to $\mathsf{RL}_{\mathsf{PLL}}$-provability of the formula $x\tau(\varphi)y$. Figure 8.5 presents its $\mathsf{RL}_{\mathsf{PLL}}$-proof.

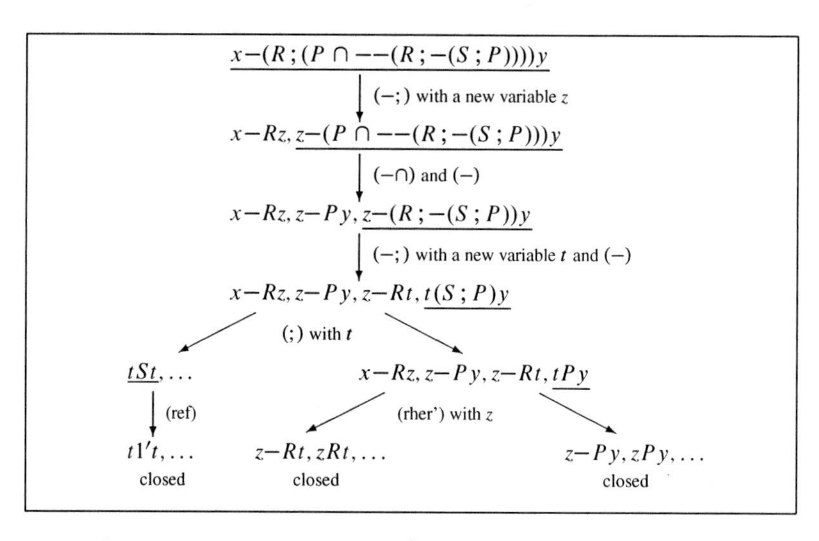

Fig. 8.5 An $\mathsf{RL}_{\mathsf{PLL}}$-proof of the formula $p \to \bigcirc p$

Chapter 9
Dual Tableaux for Relevant Logics

9.1 Introduction

Relevant logics developed as attempts to avoid the paradoxes of classical implication such as e.g., validity in the classical propositional logic PC of the formulas of the form $(\varphi \wedge \neg\varphi) \rightarrow \psi$ or $\varphi \rightarrow (\psi \rightarrow \varphi)$. The source of these paradoxes was identified as irrelevance of the antecedent of the implication to the consequent. The history and developments of relevant logics can be found e.g., in [AB75, ABD92, RMPB83, Bra03a].

Like in the semantics of modal logics, in a Kripke-style semantics for relevant logics truth of formulas depends on states. However, the accessibility relations in models of relevant logic must be ternary; binary relations are not sufficient for eliminating the paradoxical tautology $\varphi \rightarrow (\psi \rightarrow \varphi)$. The first contributions to semantics of relevant logics are [Urq72, RM73, Mak73].

Relevant logics have many applications in computer science where they provide an insight into the nature of information, see e.g., [Res96]. With this interpretation, states of the models of the logics are thought of as pieces of information. The languages of many relevant logics include a binary product operation which is interpreted as a fusion, in the sense that it combines the information of two states. In the linear logic [Gir87], which belongs to the family of relevant logics, formulas provide an information about computational resources and implication $\varphi \rightarrow \psi$ is read as saying that we can get ψ from a single resource of type φ. Since in the linear logic the contraction law $(\varphi \rightarrow (\varphi \rightarrow \psi)) \rightarrow (\varphi \rightarrow \psi)$ is not assumed, the truth of formula $\varphi \rightarrow (\varphi \rightarrow \psi)$ says that we can obtain a resource of type ψ from two resources of type φ, but it does not necessarily imply that we can get ψ from a single resource φ. A dual tableau for the linear logic is developed in [Mac97].

In this chapter we present relational dual tableaux for various relevant logics. The method applied here extends the method followed in the preceding chapters in that we interpret relevant logics in the logics of ternary relations. However, the main principle of relational proof systems for non-classical logics is the same: we interpret formulas of the logics as relations of the same arity as the relations from the models of the logics. Dual tableaux presented in this chapter originated in [Orł92]. They provide a uniform deduction tool for a variety of relevant logics.

E. Orłowska and J. Golińska-Pilarek, *Dual Tableaux: Foundations, Methodology, Case Studies*, Trends in Logic 36, DOI 10.1007/978-94-007-0005-5_9,

Relevant logics belong to a larger family of substructural logics. Dual tableaux for some substructural logics not treated in this chapter can be found in [Mac97, Mac98, Mac99]. Decidability and undecidability problems of relevant logics are discussed in [Urq84, Gia85, Bra90, Bra03b], among others.

9.2 Relevant Logics

The vocabulary of the language of propositional relevant logics consists of symbols from the following pairwise disjoint sets:

- $\mathbb{V}$ – a countable infinite set of propositional variables;
- $\{\neg, \vee, \wedge, \odot, \rightarrow\}$ – a set of propositional operations of negation, disjunction, conjunction, intensional conjunction (also referred to as fusion or product), and implication, respectively.

The set of relevant formulas is the smallest set including the set $\mathbb{V}$ and closed with respect to the propositional operations.

Models of relevant logics are the structures of the form $\mathcal{M} = (U, O, *, R, m)$ such that U is a non-empty set of states, $O \in U$ is a distinguished 'real' state, $*: U \rightarrow U$ is a function on the set of states, $R \subseteq U \times U \times U$ is a ternary relation on U, and $m: \mathbb{V} \rightarrow \mathcal{P}(U)$ is a meaning function which assigns sets of states to propositional variables. Function $*$ provides an interpretation of negation and relation R provides an interpretation of the intensional conjunction and implication.

A relevant formula φ is said to be satisfied in a model $\mathcal{M}$ by a state $s \in U$, $\mathcal{M}, s \models \varphi$, whenever the following conditions are satisfied:

- $\mathcal{M}, s \models p$ iff $s \in m(p)$, for every $p \in \mathbb{V}$;
- $\mathcal{M}, s \models \neg\varphi$ iff not $\mathcal{M}, s^* \models \varphi$;
- $\mathcal{M}, s \models \varphi \vee \psi$ iff $\mathcal{M}, s \models \varphi$ or $\mathcal{M}, s \models \psi$;
- $\mathcal{M}, s \models \varphi \wedge \psi$ iff $\mathcal{M}, s \models \varphi$ and $\mathcal{M}, s \models \psi$;
- $\mathcal{M}, s \models \varphi \odot \psi$ iff for some $x, y \in U$, $(x, y, s) \in R$ and $\mathcal{M}, x \models \varphi$ and $\mathcal{M}, y \models \psi$;
- $\mathcal{M}, s \models \varphi \rightarrow \psi$ iff for all $x, y \in U$, if $(s, x, y) \in R$ and $\mathcal{M}, x \models \varphi$, then $\mathcal{M}, y \models \psi$.

A formula φ is said to be *true* in a relevant model $\mathcal{M}$ if and only if $\mathcal{M}, O \models \varphi$. A formula φ is valid in a relevant logic whenever φ is true in all models for the logic.

In every relevant logic function $*$ and relation R satisfy some constraints specific for the logic. A minimal set of constraints which are assumed in many relevant logics is:

For all $x, y, z, w \in U$,

(M1) $(O, x, x) \in R$;
(M2)(i) If $(O, x, y) \in R$ and $(y, z, w) \in R$, then $(x, z, w) \in R$;
(M2)(ii) If $(O, x, y) \in R$ and $x \in m(p)$, then $y \in m(p)$, for every $p \in \mathbb{V}$;

(M3) If $(x, y, z) \in R$, then $(x, z^*, y^*) \in R$;
(M4) $(O, x^{**}, x) \in R$ and $(O, x, x^{**}) \in R$.

Algebras of relevant logics are distributive residuated lattices, where product $\odot$ has a left identity and $\rightarrow$ is the left residuum of $\odot$. Dualities for relevant algebras (Priestley duality of [Urq96] and discrete duality of [ORR10]) suggest that holding $R(x, y, z)$ in a frame may be viewed as holding $x \odot y \leq z$ in a relevant algebra and element O of a frame is a counterpart to the left identity of $\odot$. Then condition (M1) corresponds to reflexivity of $\leq$, (M2) (i) simulates left monotonicity of $\odot$, (M2) (ii) is the atomic heredity, (M3) says that negation is order reversing, and (M4) simulates the law of double negation.

In what follows, by RLV we mean the logic whose models satisfy the conditions (M1), ..., (M4).

For $x, y, z, t \in U$, we define:

$$(x, y, z, t) \in R_1 \overset{\text{df}}{\Leftrightarrow} \text{there is } a \in U \text{ such that } (x, y, a) \in R \text{ and } (a, z, t) \in R;$$

$$(x, y, z, t) \in R_2 \overset{\text{df}}{\Leftrightarrow} \text{there is } a \in U \text{ such that } (y, z, a) \in R \text{ and } (x, a, t) \in R.$$

In the relevant logic literature $R_1(x, y, z, t)$ is usually denoted by R^2xyzt and $R_2(x, y, z, t)$ is denoted by $R^2x(yz)t$. Our notation enables us presentation of these constants without using object variables which is useful in the definition of relational logics associated to relevant logics.

Various relevant logics can be defined by adding postulates to (M1), ..., (M4). Below we give the list of some such typical conditions:

(M5) $(x, y, z, t) \in R_1$ implies $(x, y, z, t) \in R_2$;
(M6) $(x, y, z, t) \in R_1$ implies $(y, x, z, t) \in R_2$;
(M7) $(x, y, z, t) \in R_1$ implies $(x, z, y, t) \in R_1$;
(M8) $(x, y, z) \in R$ implies $(x, y, y, z) \in R_1$;
(M9) $(x, x^*, x) \in R$;
(M10) If $(x, y, z) \in R$, then $(O, x, z) \in R$ or $(O, y, z) \in R$;
(M11) If $(x, y, z) \in R$, then $(O, x, z) \in R$;
(M12) $(x, y, z) \in R$ implies $(z, x, y, z) \in R_2$.

9.3 Translation of Relevant Logics into Relational Logics

In this section we define a translation of relevant logics into relational logics. Let L be a relevant logic. We define a relational logic $\mathsf{RL_L}$ corresponding to L. Formulas of $\mathsf{RL_L}$, interpreted as ternary or quaternary relations, will represent formulas of the relevant logic L, the accessibility relation from the models of L, and the constants R_1 and R_2. The vocabulary of the language of $\mathsf{RL_L}$ consists of the symbols from the following pairwise disjoint sets:

- $\mathbb{OV}_{\mathsf{RL_L}}$ – a countable infinite set of object variables;
- $\{O\}$ – the set consisting of an object constant;

- $\mathbb{RV}_{\mathsf{RL_L}}$ – a countable infinite set of relational variables corresponding to the propositional variables of L;
- $\{R, R_1, R_2\}$ – the set of the relational constants;
- $\{*\}$ – the set consisting of an unary object operation;
- $\{-, \neg, \cup, \cap, \odot, \rightarrow\}$ – the set of relational operations.

The operations $\neg, \odot, \rightarrow$ will be defined so that they will be the counterparts to the specific operations of logic L and $-$, $\cup$, and $\cap$ are the classical complement, union, and intersection of relations, respectively. We slightly abuse the notation here by using the same symbols $\neg$, $\odot$, $\rightarrow$ in relevant logics and in the corresponding relational logics.

The set $\mathbb{OT}_{\mathsf{RL_L}}$ of object terms is the smallest set including $\mathbb{OV}_{\mathsf{RL_L}} \cup \{O\}$ and closed on the operation $*$. The set of relational terms is defined as $\mathbb{RT}_{\mathsf{RL_L}} \stackrel{\text{df}}{=} \mathbb{RT}^0_{\mathsf{RL_L}} \cup \{R_1, R_2, -R_1, -R_2\}$, where $\mathbb{RT}^0_{\mathsf{RL_L}}$ is the smallest set including $\mathbb{RV}_{\mathsf{RL_L}} \cup \{R\}$ and closed on the relational operations. Formulas are of the form $S(x, y, z)$ or $Q(x, y, z, t)$, where $S \in \mathbb{RT}^0_{\mathsf{RL_L}}$, $Q \in \{R_1, R_2, -R_1, -R_2\}$, and $x, y, z, t \in \mathbb{OT}_{\mathsf{RL_L}}$.

Now, we define models of the logic $\mathsf{RL_L}$. As usual, we use the same symbols for operations and constants in the language and for the corresponding entities in the models. An $\mathsf{RL_{RLV}}$-model is a structure $\mathcal{M} = (U, O, *, R, R_1, R_2, m)$ such that U is a non-empty set, m is a meaning function that provides the interpretation of the symbols of the language, and the following conditions are satisfied:

- $O \in U$ is an object which provides interpretation of constant O;
- $*\colon U \rightarrow U$ is a function in U which provides interpretation of operation $*$;
- $R \subseteq U^3$ is a ternary relation which provides interpretation of relational constant R;
- R_1 and R_2 are quaternary relations on U providing the interpretation of relational constants R_1 and R_2, respectively, defined as in RLV-models in Sect. 9.2;
- $m(P) = X \times U \times U$, for some $X \subseteq U$, for every relational variable $P \in \mathbb{RV}_{\mathsf{RL_L}}$;
- The conditions (M1), (M2)(i), (M3), (M4) of L-models are satisfied together with the following modification of (M2) (ii):

  (M2)'(ii) If $(O, x, y) \in R$ and $(x, t, u) \in m(P)$, then $(y, t, u) \in m(P)$, for every relational variable P;

- m extends to all the compound relational terms as follows:

  For $Q \in \{R_1, R_2\}$,

  $$m(-Q) = U^4 - m(Q),$$

  For all ternary relational terms S and T,

  $$m(-S) = U^3 - m(S),$$
  $$m(\neg S) = \neg m(S) \stackrel{\text{df}}{=} \{(x, y, z) \in U^3 : (x^*, y, z) \in m(-S)\},$$

$$m(S \cup T) = m(S) \cup m(T)$$
$$\stackrel{\text{df}}{=} \{(x, y, z) \in U^3 : (x, y, z) \in m(S) \text{ or } (x, y, z) \in m(T)\},$$

$$m(S \cap T) = m(S) \cap m(T)$$
$$\stackrel{\text{df}}{=} \{(x, y, z) \in U^3 : (x, y, z) \in m(S) \text{ and } (x, y, z) \in m(T)\},$$

$$m(S \odot T) = m(S)m(T)$$
$$\stackrel{\text{df}}{=} \{(x, y, z) \in U^3 : \text{ there are } t, w \in U \text{ such that } (t, w, x) \in R \text{ and } (t, y, z) \in m(S) \text{ and, } (w, y, z) \in m(T)\},$$

$$m(S \to T) = m(S) \to m(T)$$
$$\stackrel{\text{df}}{=} \{(x, y, z) \in U^3 : \text{ for all } t, w \in U \text{ if } (x, t, w) \in R \text{ and } (t, y, z) \in m(S), \text{ then } (w, y, z) \in m(T))\}.$$

In analogy to binary right ideal relations, the ternary relations on U which are of the form $X \times U \times U$, for some $X \subseteq U$, will be referred to as ideal relations.

Proposition 9.3.1. *For every* $\mathsf{RL}_{\mathsf{RLV}}$*-model* $\mathcal{M} = (U, O, *, R, R_1, R_2, m)$*, the set of ideal relations in* U *is closed with respect to the relational operations* $-, \neg, \cup, \cap, \odot$*, and* $\to$*.*

Proof. Let S and T be ternary ideal relations on U. Consider relation $S \to T$. Assume $(x, y, z) \in S \to T$. Then, for all $t, w \in U$, if $(x, t, w) \in R$ and $(t, y, z) \in S$, then $(w, y, z) \in T$. Suppose there are $s, u \in U$ such that $(x, s, u) \notin S \to T$. Thus, there are $t', w' \in U$ such that $(x, t', w') \in R$ and $(t', s, u) \in S$ and $(w', s, u) \notin T$. Since S is an ideal relation, $(t', y, z) \in S$. Since $(x, t', w') \in R$ and $(t', y, z) \in S$, we have $(w', y, z) \in T$. Hence, since T is an ideal relation, $(w', s, u) \in T$, a contradiction.

In the remaining cases the proofs are similar. □

If L is obtained from RLV by assuming some of the conditions (M5), ..., (M12) in its models, then RL_{L}-models are $\mathsf{RL}_{\mathsf{RLV}}$-models which in addition satisfy the conditions corresponding to those specific conditions of L-models (see e.g., Sect. 9.5).

Let $\mathcal{M}$ be an RL_{L}-model. A valuation in $\mathcal{M}$ is a function $v: \mathbb{OT}_{\mathsf{RL}_{\mathsf{L}}} \to U$ such that $v(O) = O$ and $v(x^*) = v(x)^*$. We say that an RL_{L}-formula $S(x, y, z)$ (resp. $Q(x, y, z, t)$) is satisfied in $\mathcal{M}$ by a valuation v, $\mathcal{M}, v \models S(x, y, z)$ (resp. $\mathcal{M}, v \models Q(x, y, z, t)$), whenever $(v(x), v(y), v(z)) \in m(S)$ (resp. $(v(x), v(y), v(z), v(t)) \in m(Q)$). An RL_{L}-formula is true in $\mathcal{M}$ if and only if it is satisfied in $\mathcal{M}$ by all valuations, and it is RL_{L}-valid whenever it is true in all RL_{L}-models.

Now, we define a translation function τ from formulas of relevant logics into relational terms. Let $\tau': \mathbb{V} \to \mathbb{RV}_{\mathsf{RL}_{\mathsf{L}}}$ be a one-to-one mapping assigning relational variables to propositional variables. Then we define:

- $\tau(p) = \tau'(p)$, for every $p \in \mathbb{V}$;
- $\tau(\neg\varphi) = \neg\tau(\varphi)$;
- $\tau(\varphi \vee \psi) = \tau(\varphi) \cup \tau(\psi)$;

- $\tau(\varphi \wedge \psi) = \tau(\varphi) \cap \tau(\psi)$;
- $\tau(\varphi \odot \psi) = \tau(\varphi) \odot \tau(\psi)$;
- $\tau(\varphi \rightarrow \psi) = \tau(\varphi) \rightarrow \tau(\psi)$.

Clearly, on the right hand side of these equalities we have the relational operations of RL_L-logic. Observe also that every relevant formula is translated into ternary relational term.

Proposition 9.3.2. *For every* L*-model* $\mathcal{M} = (U, O, *, R, m)$ *there exists an* RL_L*-model* $\mathcal{M}' = (U, O, *, R, R_1, R_2, m')$ *such that its set of objects, element* O*, operation* $*$*, and relation* R *are the same as those in* $\mathcal{M}$*, and for every* L*-formula* φ *and for all* $s, t, u \in U$*,* $\mathcal{M}, s \models \varphi$ *iff* $(s, t, u) \in m'(\tau(\varphi))$*.*

Proof. Let $\mathcal{M} = (U, O, *, R, m)$ be an L-model. We construct an RL_L-model $\mathcal{M}' = (U, O, *, R, R_1, R_2, m')$ defining m' by:

- $m'(P) = m(p) \times U^2$, for every relational variable P, where p is a propositional variable such that $\tau(p) = P$;
- R_1 and R_2 are quaternary relations on U providing the interpretation of relational constants R_1 and R_2, respectively, defined as in RLV-models in Sect. 9.2;
- m' extends to all the compound relational terms as in RL_L-models.

It is easy to see that since R is the same in both models, conditions (M1), (M2)(i), (M2)'(ii), (M3), and (M4) are satisfied in $\mathcal{M}'$, and if $\mathcal{M}$ satisfies some of the conditions (Mj), for $j \in \{5, \ldots, 12\}$, then the corresponding conditions are also satisfied in $\mathcal{M}'$.

Now, we prove the proposition by induction on the complexity of formulas. Let $s, t, u \in U$. Let $p \in \mathbb{V}$. Then $\mathcal{M}, s \models p$ iff $s \in m(p)$ iff, by the definition of $m'(P)$, $(s, t, u) \in m'(P)$ iff $(s, t, u) \in m(\tau(p))$.

Assume that the proposition holds for formulas φ and ψ.

$\mathcal{M}, s \models \neg\varphi$ iff $\mathcal{M}, s^* \not\models \varphi$ iff, by the induction hypothesis, $(s^*, t, u) \notin m'(\tau(\varphi))$ iff $(s, t, u) \in m'(\tau(\neg\varphi))$

$\mathcal{M}, s \models \varphi \odot \psi$ iff there exist $x, y \in U$ such that $(x, y, s) \in R$ and $\mathcal{M}, x \models \varphi$ and $\mathcal{M}, y \models \psi$ iff, by the induction hypothesis, there exist $x, y \in U$ such that $(x, y, s) \in R$ and $(x, t, u) \in m'(\tau(\varphi))$ and $(y, t, u) \in m'(\tau(\psi))$ iff $(s, t, u) \in m'(\tau(\varphi\psi))$.

In the remaining cases the proofs are similar. □

Proposition 9.3.3. *For every* RL_L*-model* $\mathcal{M}' = (U, O, *, R, R_1, R_2, m')$ *there exists an* L*-model* $\mathcal{M} = (U, O, *, R, m)$ *such that its set of objects, element* O*, operation* $*$*, and relation* R *are the same as those in* $\mathcal{M}'$*, and for every* L*-formula* φ *and for all* $s, t, u \in U$*,* $\mathcal{M}, s \models \varphi$ *iff* $(s, t, u) \in m'(\tau(\varphi))$*.*

Proof. Let $\mathcal{M}' = (U, O, *, R, R_1, R_2, m')$ be an RL_L-model. We construct an L-model $\mathcal{M} = (U, O, *, R, m)$ such that for every $p \in \mathbb{V}$, $m(p)$ is defined as:

$m(p) = \{x \in U : \text{ for some } y, z \in U, (x, y, z) \in m'(P)\}$, where P is a relational variable such that $P = \tau'(p)$.

Then the proof is similar to the proof of Proposition 9.3.2. □

Propositions 9.3.2 and 9.3.3 lead to the following theorem:

Theorem 9.3.1. *For every* L*-formula* φ *and for all object variables* x *and* y*, the following conditions are equivalent:*

1. φ *is* L*-valid;*
2. $\tau(\varphi)(O, x, y)$ *is* $\mathsf{RL_L}$*-valid.*

Proof. Assume that φ is true in all L-models. Suppose $\tau(\varphi)(O, x, y)$ is not $\mathsf{RL_L}$-valid, that is there exist an $\mathsf{RL_L}$-model $\mathcal{M} = (U, O, *, R, R_1, R_2, m)$ and a valuation v in $\mathcal{M}$ such that $(O, v(x), v(y)) \notin m(\tau(\varphi))$. By Proposition 9.3.3, there exists an L-model $\mathcal{M}'$ such that $\mathcal{M}', O \not\models \varphi$, a contradiction. Now, assume that $\tau(\varphi)(O, x, y)$ is $\mathsf{RL_L}$-valid and suppose φ is not L-valid. Then there exists an L-model $\mathcal{M} = (U, O, *, R, m)$ in which φ is not true, that is $\mathcal{M}, O \not\models \varphi$. By Proposition 9.3.2, there exists $\mathsf{RL_L}$-model $\mathcal{M}'$ such that $\mathcal{M}' \not\models \tau(\varphi)(O, x, y)$, which means that $\tau(\varphi)(O, x, y)$ is not $\mathsf{RL_L}$-valid, a contradiction. □

9.4 Relational Dual Tableau for Logic RLV

In this section we present a dual tableau for the logic RLV defined in Sect. 9.2. $\mathsf{RL_{RLV}}$-dual tableau contains decomposition rules of the following forms:

For all object terms x, y, z and for all relational terms S and T,

$$(-)\quad \frac{--S(x,y,z)}{S(x,y,z)}$$

$$(\neg)\quad \frac{\neg S(x,y,z)}{-S(x^*,y,z)} \qquad (-\neg)\quad \frac{-\neg S(x,y,z)}{S(x^*,y,z)}$$

$$(\cup)\quad \frac{(S \cup T)(x,y,z)}{S(x,y,z),\, T(x,y,z)} \qquad (-\cup)\quad \frac{-(S \cup T)(x,y,z)}{-S(x,y,z) \mid -T(x,y,z)}$$

$$(\cap)\quad \frac{(S \cap T)(x,y,z)}{S(x,y,z) \mid T(x,y,z)} \qquad (-\cap)\quad \frac{-(S \cap T)(x,y,z)}{-S(x,y,z),\, -T(x,y,z)}$$

$$(\odot)\quad \frac{(S \odot T)(x,y,z)}{R(t,w,x), \Psi \mid S(t,y,z), \Psi \mid T(w,y,z), \Psi}$$

t, w are any object terms and $\Psi = (S \odot T)(x, y, z)$

$$(-\odot)\quad \frac{-(S \odot T)(x,y,z)}{-R(t,w,x), -S(t,y,z), -T(w,y,z)}$$

t, w are new object variables such that $t \neq w$

$$(\rightarrow) \quad \frac{(S \rightarrow T)(x, y, z)}{-R(x, t, w), -S(t, y, z), T(w, y, z)}$$

t, w are new object variables such that $t \neq w$

$$(-\rightarrow) \quad \frac{-(S \rightarrow T)(x, y, z)}{R(x, t, w), \Psi \mid S(t, y, z), \Psi \mid -T(w, y, z), \Psi}$$

t, w are any object terms and $\Psi = -(S \rightarrow T)(x, y, z)$

Specific rules of $\mathsf{RL}_{\mathsf{RLV}}$-dual tableau are of the following forms:

For all object terms x, y, z, t, and w and for any relational variable P,

$$(R_1) \quad \frac{R_1(x, y, z, t)}{R(x, y, u), R_1(x, y, z, t) \mid R(u, z, t), R_1(x, y, z, t)}$$

u is any object term

$$(-R_1) \quad \frac{-R_1(x, y, z, t)}{-R(x, y, u), -R(u, z, t)}$$

u is a new object variable

$$(R_2) \quad \frac{R_2(x, y, z, t)}{R(y, z, u), R_2(x, y, z, t) \mid R(x, u, t), R_2(x, y, z, t)}$$

u is any object term

$$(-R_2) \quad \frac{-R_2(x, y, z, t)}{-R(y, z, u), -R(x, u, t)} \qquad u \text{ is a new object variable}$$

$$(\text{ideal}) \quad \frac{P(x, y, z)}{P(x, t, u), P(x, y, z)} \qquad t, u \text{ are any object terms}$$

$$(\text{rM2i}) \quad \frac{R(x, z, w)}{R(O, x, y), R(x, z, w) \mid R(y, z, w), R(x, z, w)} \qquad y \text{ is any object term}$$

$$(\text{rM2'ii}) \quad \frac{P(y, t, u)}{R(O, x, y), P(y, t, u) \mid P(x, t, u), P(y, t, u)} \qquad x \text{ is any object term}$$

$$(\text{rM3}) \quad \frac{R(x, z^*, y^*)}{R(x, y, z), R(x, z^*, y^*)}$$

A set of RLV-formulas is said to be $\mathsf{RL}_{\mathsf{RLV}}$-axiomatic whenever it includes either of the following subsets:

For all object terms x, y, z and t, for any ternary relational term S, and for any quaternary relational term Q,

(Ax0) $\{S(x, y, z), -S(x, y, z)\}$;
(Ax1) $\{Q(x, y, z, t), -Q(x, y, z, t)\}$;
(Ax2) $\{R(O, x, x)\}$;
(Ax3) $\{R(O, x^{**}, x)\}$;
(Ax4) $\{R(O, x, x^{**})\}$.

As usual, an $\mathsf{RL_{RLV}}$-set is a finite set of $\mathsf{RL_{RLV}}$-formulas whose first-order disjunction is true in all $\mathsf{RL_{RLV}}$-models. Correctness of a rule is defined in a similar way as in the relational logics of classical algebras of binary relations, i.e., a rule is $\mathsf{RL_{RLV}}$-correct whenever is preserves and reflects $\mathsf{RL_{RLV}}$-validity of its premise and conclusion(s) (see Sect. 2.4).

Proposition 9.4.1.

1. *The* $\mathsf{RL_{RLV}}$*-axiomatic sets are* $\mathsf{RL_{RLV}}$*-sets;*
2. *The* $\mathsf{RL_{RLV}}$*-decomposition rules are* $\mathsf{RL_{RLV}}$*-correct;*
3. *The* $\mathsf{RL_{RLV}}$*-specific rules are* $\mathsf{RL_{RLV}}$*-correct.*

Proof. For 1., note that sets including subsets of the form (Ax0) or (Ax1) are $\mathsf{RL_{RLV}}$-sets by the definition of relational complement. Sets including subsets of the form (Ax2) or (Ax3) or (Ax4) are $\mathsf{RL_{RLV}}$-sets, because (Ax2) reflects condition (M1) assumed in $\mathsf{RL_{RLV}}$-models, while (Ax3) and (Ax4) reflect condition (M4).

By way of example, we prove 2. for the rule ($\odot$). Let X be a finite set of $\mathsf{RL_{RLV}}$-formulas. Clearly, if $X \cup \{(S \odot T)(x, y, z)\}$ is an $\mathsf{RL_{RLV}}$-set, then so are the sets $X \cup \{R(t, w, x), (S \odot T)(x, y, z)\}$, $X \cup \{S(t, y, z), (S \odot T)(x, y, z)\}$, and $X \cup \{T(w, y, z), (S \odot T)(x, y, z)\}$. Now, assume that the sets $X \cup \{R(t, w, x), (S \odot T)(x, y, z)\}$, $X \cup \{S(t, y, z), (S \odot T)(x, y, z)\}$, and $X \cup \{T(w, y, z), (S \odot T)(x, y, z)\}$ are $\mathsf{RL_{RLV}}$-sets and suppose that $X \cup \{(S \odot T)(x, y, z)\}$ is not an $\mathsf{RL_{RLV}}$-set. Then there exist an $\mathsf{RL_{RLV}}$-model $\mathcal{M} = (U, O, *, R, R_1, R_2, m)$ and a valuation v in $\mathcal{M}$ such that $\mathcal{M}, v \not\models \varphi$, for every $\varphi \in X$, and $\mathcal{M}, v \not\models (S \odot T)(x, y, z)$. Hence, $(v(x), v(y), v(z)) \notin m(S \odot T)$. By the assumption, we obtain:

$\mathcal{M}, v \models R(t, w, x)$;
$\mathcal{M}, v \models S(t, y, z)$;
$\mathcal{M}, v \models T(w, y, z)$.

Therefore, we obtain $(v(t), v(w), v(x)) \in R$, $(v(t), v(y), v(z)) \in m(S)$, and $(v(w), v(y), v(z)) \in m(T)$. Hence, by the definition of the operation $\odot$, $(v(x), v(y), v(z)) \in m(S \odot T)$, a contradiction.

The proofs of correctness of the remaining decomposition rules are similar.

By way of example, we show 3. for the rules (R_1) and $(-R_2)$. Let X be a finite set of $\mathsf{RL_{RLV}}$-formulas.

For (R_1), note that if $X \cup \{R_1(x, y, z, t)\}$ is an $\mathsf{RL_{RLV}}$-set, then so are $X \cup \{R(x, y, u), R_1(x, y, z, t)\}$ and $X \cup \{R(u, z, t), R_1(x, y, z, t)\}$. Assume $X \cup \{R(x, y, u), R_1(x, y, z, t)\}$ and $X \cup \{R(u, z, t), R_1(x, y, z, t)\}$ are $\mathsf{RL_{RLV}}$-sets and suppose $X \cup \{R_1(x, y, z, t)\}$ is not an $\mathsf{RL_{RLV}}$-set. Then, there is an $\mathsf{RL_{RLV}}$-model $\mathcal{M}$ and a valuation v in $\mathcal{M}$ such that $(v(x), v(y), v(z), v(t)) \notin R_1$. Since the assumption implies $(v(x), v(y), v(u)) \in R$ and $(v(u), v(z), v(t)) \in R$, we get a contradiction with the definition of R_1.

For $(-R_2)$, let u be an object variable that does not occur in the set X and $x, y, z, t \neq u$. Assume $X \cup \{-R(y, z, u), -R(x, u, t)\}$ is an $\mathsf{RL}_{\mathsf{RLV}}$-set and suppose that $X \cup \{-R_2(x, y, z, t)\}$ is not an $\mathsf{RL}_{\mathsf{RLV}}$-set. Then, there exist an $\mathsf{RL}_{\mathsf{RLV}}$-model $\mathcal{M}$ and a valuation v in $\mathcal{M}$ such that $(v(x), v(y), v(z), v(t)) \in R_2$ and, by the assumption on variable u, for every $a \in U$ either $(v(y), v(z), a) \notin R$ or $(v(x), a, v(t)) \notin R$, a contradiction. The remaining implication can be proved in a similar way. □

The notion of an $\mathsf{RL}_{\mathsf{RLV}}$-proof tree, a closed branch of such a tree, a closed $\mathsf{RL}_{\mathsf{RLV}}$-proof tree, and $\mathsf{RL}_{\mathsf{RLV}}$-provability are defined in a similar way as in the logic RL (see Sect. 2.4). As usual, to prove completeness we need the notion of a complete proof tree. A branch b of an $\mathsf{RL}_{\mathsf{RLV}}$-proof tree is said to be complete whenever it is closed or it satisfies the following completion conditions.

The completion conditions for the decomposition rules for set operations $-$, $\cup$, and $\cap$ are analogous to the corresponding conditions in Sect. 2.5.

For all object terms x, y, z and for all relational terms S and T,

Cpl($\neg$) If $\neg S(x, y, z) \in b$, then $-S(x^*, y, z) \in b$, obtained by an application of the rule ($\neg$);

Cpl($-\neg$) If $-\neg S(x, y, z) \in b$, then $S(x^*, y, z) \in b$, obtained by an application of the rule ($-\neg$);

Cpl($\odot$) If $(S \odot T)(x, y, z) \in b$, then for all object terms t and w, either $R(t, w, x) \in b$ or $S(t, y, z) \in b$ or $T(w, y, z) \in b$, obtained by an application of the rule ($\odot$);

Cpl($-\odot$) If $-(S \odot T)(x, y, z) \in b$, then for some object variables t and w, $-R(t, w, x) \in b$, $-S(t, y, z) \in b$, and $-T(w, y, z) \in b$, obtained by an application of the rule ($-\odot$);

Cpl($\rightarrow$) If $(S \rightarrow T)(x, y, z) \in b$, then for some object variables t and w, $-R(x, t, w) \in b$, $-S(t, y, z) \in b$, and $T(w, y, z) \in b$, obtained by an application of the rule ($\rightarrow$);

Cpl($-\rightarrow$) If $-(S \rightarrow T)(x, y, z) \in b$, then for all object terms t and w, either $R(x, t, w) \in b$ or $S(t, y, z) \in b$ or $-T(w, y, z) \in b$, obtained by an application of the rule ($-\rightarrow$);

For all object terms x, y, z, t, and w and for any relational variable P,

Cpl(R_1) If $R_1(x, y, z, t) \in b$, then for every object term u, either $R(x, y, u) \in b$ or $R(u, z, t) \in b$, obtained by an application of the rule (R_1);

Cpl($-R_1$) If $-R_1(x, y, z, t) \in b$, then for some object variable u, both $-R(x, y, u) \in b$ and $-R(u, z, t) \in b$, obtained by an application of the rule ($-R_1$);

Cpl(R_2) If $R_2(x, y, z, t) \in b$, then for every object term u, either $R(y, z, u) \in b$ or $R(x, u, t) \in b$, obtained by an application of the rule (R_2);

Cpl($-R_2$) If $-R_2(x, y, z, t) \in b$, then for some object variable u, both $-R(y, z, u) \in b$ and $-R(x, u, t) \in b$, obtained by an application of the rule ($-R_2$);

Cpl(ideal) If $P(x, y, z) \in b$ for some relational variable P, then for all object terms t and u, $P(x, t, u) \in b$, obtained by an application of the rule (ideal);
Cpl(rM2i) If $R(x, z, w) \in b$, then for every object term y, either $R(O, x, y) \in b$ or $R(y, z, w) \in b$, obtained by an application of the rule (rM2i);
Cpl(rM2'ii) If $P(y, t, u) \in b$ for some relational variable P, then for every object term x, either $R(O, x, y) \in b$ or $P(x, t, u) \in b$, obtained by an application of the rule (rM2'ii);
Cpl(rM3) If $R(x, z^*, y^*) \in b$, then $R(x, y, z) \in b$, obtained by an application of the rule (rM3).

The notions of a complete $\mathsf{RL}_{\mathsf{RLV}}$-proof tree and an open branch of an $\mathsf{RL}_{\mathsf{RLV}}$-proof tree are defined as in RL-logic (see Sect. 2.5).

The following form of the closed branch property holds:

Fact 9.4.1 (Closed Branch Property). *For every branch b of an $\mathsf{RL}_{\mathsf{RLV}}$-proof tree, if $S(x, y, z) \in b$ and $-S(x, y, z) \in b$ for some atomic ternary relational term S, then branch b is closed.*

Although, it does not concern the constants R_1 and R_2, it is sufficient for proving satisfaction in branch model property (Proposition 9.4.3). The reason being that the rules for these constants reflect their corresponding definitions in logic RLV.

Let b be an open branch of an $\mathsf{RL}_{\mathsf{RLV}}$-proof tree for a formula φ. We define a branch structure $\mathcal{M}^b = (U^b, O^b, *^b, R^b, R_1^b, R_2^b, m^b)$ as follows:

- $U^b = \mathbb{OT}_{\mathsf{RL}_{\mathsf{RLV}}}$, that is the set of objects coincides with the set of object terms;
- $O^b = O = m^b(O)$;
- The function $*^b$ is defined by the clause $*^b(x) = x^*$;
- $R^b = m^b(R) = \{(x, y, z) \in (U^b)^3 : R(x, y, z) \notin b\}$;
- $R_i^b = m^b(R_i)$, for $i \in \{1, 2\}$, are defined as:

$$R_1^b = \{(x, y, z, t) \in (U^b)^4 : \exists a \in U^b, (x, y, a) \in R^b \wedge (a, z, t) \in R^b\},$$
$$R_2^b = \{(x, y, z, t) \in (U^b)^4 : \exists a \in U^b, (y, z, a) \in R^b \wedge (x, a, t) \in R^b\};$$

- $m^b(P) = \{(x, y, z) \in (U^b)^3 : P(x, y, z) \notin b\}$, for $P \in \mathbb{RV}_{\mathsf{RL}_{\mathsf{RLV}}}$;
- m^b extends to all the compound relational terms as in $\mathsf{RL}_{\mathsf{RLV}}$-models.

Proposition 9.4.2 (Branch Model Property). *For every open branch b of an $\mathsf{RL}_{\mathsf{RLV}}$-proof tree, the branch structure $\mathcal{M}^b = (U^b, O^b, *^b, R^b, R_1^b, R_2^b, m^b)$ is an $\mathsf{RL}_{\mathsf{RLV}}$-model.*

Proof. Let b be an open branch of an $\mathsf{RL}_{\mathsf{RLV}}$-proof tree. In order to prove that $\mathcal{M}^b$ is an $\mathsf{RL}_{\mathsf{RLV}}$-model, it suffices to show that $\mathcal{M}^b$ satisfies the conditions (M1), (M2)(i), (M2)'(ii), (M3), and (M4). By way of example, we prove it for (M2)'(ii) and (M4).

For (M2)'(ii), assume $(O, x, y) \in R^b$ and $(x, t, u) \in m^b(P)$, that is $R(O, x, y) \notin b$ and $P(x, t, u) \notin b$. Suppose $(y, t, u) \notin m^b(P)$. Then $P(y, t, u) \in b$. By the completion condition Cpl(rM2'ii), for every object term x, either $R(O, x, y) \in b$ or $P(x, t, u) \in b$, a contradiction.

For (M4), since $\{R(O, x^{**}, x)\}$ and $\{R(O, x, x^{**})\}$ are $\mathsf{RL_{RLV}}$-axiomatic sets, for every object term x, $R(O, x^{**}, x) \notin b$ and $R(O, x, x^{**}) \notin b$. Thus, for every object term x, $(O, x^{**}, x) \in R^b$ and $(O, x, x^{**}) \in R^b$. □

Let v^b be the identity valuation, that is $v^b(x) = x$, for any object term x. Then, the following can be proved:

Proposition 9.4.3 (Satisfaction in Branch Model Property). *For every open branch b of an $\mathsf{RL_{RLV}}$-proof tree and for every $\mathsf{RL_{RLV}}$-formula φ, if $\mathcal{M}^b, v^b \models \varphi$, then $\varphi \notin b$.*

Proof. Let φ be an $\mathsf{RL_{RLV}}$-formula. The proof is by induction on the complexity of formulas.

If $\varphi = S(x, y, z)$ for an atomic ternary relational term, then the required condition holds by the definition of the meaning function m^b. Let $\varphi = -S(x, y, z)$, for an atomic ternary term S. If $\mathcal{M}^b, v^b \models -S(x, y, z)$, then $S(x, y, z) \in b$. By Fact 9.4.1, $-S(x, y, z) \notin b$.

Let $\varphi = R_2(x, y, z, t)$. Assume $\mathcal{M}^b, v^b \models R_2(x, y, z, t)$, that is there exists $u \in U^b$ such that $(y, z, u) \in R^b$ and $(x, u, t) \in R^b$. Suppose $R_2(x, y, z, t) \in b$. Then, by the completion condition Cpl(R_2), for every $u \in U^b$, either $(y, z, u) \notin R^b$ or $(x, u, t) \notin R^b$, a contradiction. Let $\varphi = -R_1(x, y, z, t)$. Assume $\mathcal{M}^b, v^b \models -R_1(x, y, z, t)$, that is for all $u \in U^b$, either $(x, y, u) \notin R^b$ or $(u, z, t) \notin R^b$. Suppose $-R_1(x, y, z, t) \in b$. Then, by the completion condition Cpl($-R_1$), for some $u \in U^b$, both $(x, y, u) \in R^b$ and $(u, z, t) \in R^b$, a contradiction.

Now, assume that the required condition holds for formulas built with relational terms S and T and their complements. By way of example, we show that it holds for $\neg S(x, y, z)$, and $(S \rightarrow T)(x, y, z)$.

Let $\varphi = \neg S(x, y, z)$. Assume that $\mathcal{M}^b, v^b \models \neg S(x, y, z)$. Then we have $(x^*, y, z) \notin m^b(S)$. Suppose $\neg S(x, y, z) \in b$. Then, by the completion condition Cpl($\neg$), $-S(x^*, y, z) \in b$ and then, by the induction hypothesis, $\mathcal{M}^b, v^b \not\models -S(x^*, y, z)$. Thus, $(x^*, y, z) \in m^b(S)$, a contradiction.

Let $\varphi = (S \rightarrow T)(x, y, z)$. Assume $\mathcal{M}^b, v^b \models (S \rightarrow T)(x, y, z)$, that is $(x, y, z) \in m^b(S \rightarrow T)$. Thus, for all object terms t and w, if $(x, t, w) \in R^b$ and $(t, y, z) \in m^b(S)$, then $(w, y, z) \in m^b(T)$. Then, by the induction hypothesis, if $R(x, t, w) \notin b$ and $S(t, y, z) \notin b$, then $T(w, y, z) \notin b$. Suppose $(S \rightarrow T)(x, y, z) \in b$. By the completion condition Cpl($\rightarrow$), there are object variables t and w such that $-R(x, t, w) \in b$ and $-S(t, y, z) \in b$, and $T(w, y, z) \in b$, a contradiction.

The proofs of the remaining cases are similar. □

Due to Propositions 9.4.1–9.4.3, we obtain:

Theorem 9.4.1 (Soundness and Completeness of $\mathsf{RL_{RLV}}$). *Let φ be an $\mathsf{RL_{RLV}}$–formula. Then the following conditions are equivalent:*

1. *φ is $\mathsf{RL_{RLV}}$-valid;*
2. *φ is $\mathsf{RL_{RLV}}$-provable.*

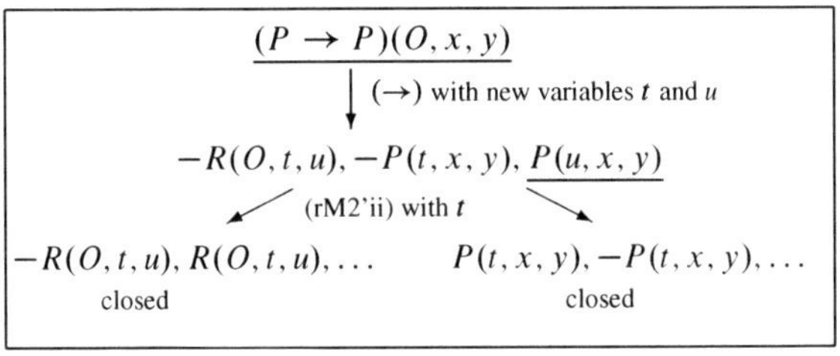

$$\underline{(P \to P)(O, x, y)}$$
$\downarrow$ $(\to)$ with new variables t and u
$$-R(O, t, u), -P(t, x, y), \underline{P(u, x, y)}$$
(rM2'ii) with t

$-R(O, t, u), R(O, t, u), \ldots$ closed | $P(t, x, y), -P(t, x, y), \ldots$ closed

Fig. 9.1 An $\mathsf{RL_{RLV}}$-proof of RLV-formula $p \to p$

Due to the above theorem and Theorem 9.3.1, we have:

Theorem 9.4.2 (Relational Soundness and Completeness of RLV-logic). *Let φ be an RLV-formula. Then for all object variables x and y, the following conditions are equivalent:*

1. *φ is RLV-valid;*
2. *$\tau(\varphi)(O, x, y)$ is $\mathsf{RL_{RLV}}$-provable.*

Example. Consider the following RLV-formulas:

$$\varphi = p \to p,$$
$$\psi = (p \to \neg q) \to (q \to \neg p).$$

Let $\tau'(p) = P$ and $\tau'(q) = Q$. By Theorem 9.4.2, for all object variables x and y, RLV-validity of formulas φ and ψ is equivalent to $\mathsf{RL_{RLV}}$-provability of the formulas $\tau(\varphi)(O, x, y)$ and $\tau(\psi)(O, x, y)$, respectively. Figures 9.1 and 9.2 present $\mathsf{RL_{RLV}}$-proofs of these formulas, respectively.

9.5 Relational Dual Tableaux for Axiomatic Extensions of Logic RLV

In this section we present dual tableaux for logics which are obtained from the logic RLV by assuming that its models satisfy some of the conditions (M5), ..., (M12). In order to construct dual tableaux for these axiomatic extensions we add to $\mathsf{RL_{RLV}}$-dual tableau the specific rules corresponding to the chosen constraints. In what follows, rule (rMi) corresponds to condition (Mi), for $i \in \{5, \ldots, 8, 10, \ldots, 12\}$.

For all object terms x, y, z, and t,

(rM5) $\dfrac{}{R_1(x, y, z, t) \,|\, -R_2(x, y, z, t)}$ (rM6) $\dfrac{}{R_1(x, y, z, t) \,|\, -R_2(y, x, z, t)}$

(rM7) $\dfrac{}{R_1(x, y, z, t) \,|\, -R_1(x, z, y, t)}$ (rM8) $\dfrac{}{R(x, y, z) \,|\, -R_1(x, y, y, z)}$

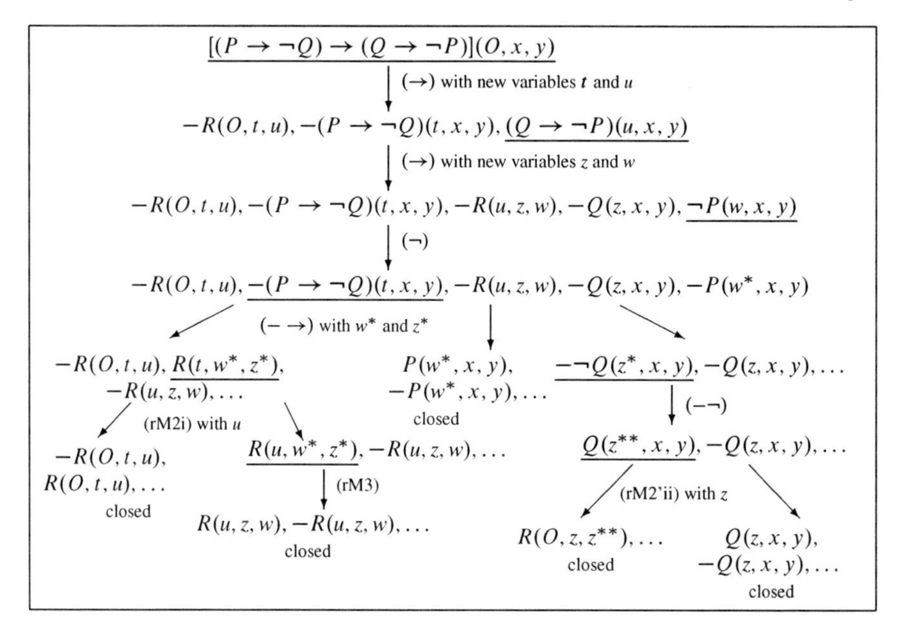

Fig. 9.2 An $\mathsf{RL}_{\mathsf{RLV}}$-proof of RLV-formula $(p \to \neg q) \to (q \to \neg p)$

Condition (M9) leads to an axiomatic set:

(AxM9) any set including a formula $R(x, x^*, x)$ is an axiomatic set.

(rM10) $$\frac{R(O, x, z), R(O, y, z)}{R(x, y, z), R(O, x, z), R(O, y, z)}$$

(rM11) $$\frac{R(O, x, z)}{R(x, y, z), R(O, x, z)} \qquad y \text{ is any object term}$$

(rM12) $$\frac{}{R(x, y, z) \mid -R_2(z, x, y, z)}$$

The rules (rM5), (rM6), (rM7), (rM8), and (rM12) are specialized cut rules. In the presence of the ordinary cut rule they could be replaced by the rules with a non-empty premise. This issue is discussed in detail in Sect. 25.9.

Let $\mathcal{K}$ be a class of $\mathsf{RL}_{\mathsf{RLV}}$-models. A finite set X of $\mathsf{RL}_{\mathsf{RLV}}$-formulas is said to be a $\mathcal{K}$-set whenever for every model $\mathcal{M}$ of $\mathcal{K}$ and for every valuation v in $\mathcal{M}$ there exists a formula $\varphi \in X$ such that $\mathcal{M}, v \models \varphi$. Then, $\mathcal{K}$-correctness of a rule is defined in a similar way as $\mathsf{RL}_{\mathsf{RLV}}$-correctness in Sect. 9.4.

Theorem 9.5.1 (Correspondence). *Let $\mathcal{K}$ be a class of* $\mathsf{RL}_{\mathsf{RLV}}$*-models and let $i \in \{5, \ldots, 8, 10, \ldots, 12\}$. Then, the condition (Mi) is true in all $\mathcal{K}$-models iff the rule (rMi) is $\mathcal{K}$-correct.*

Proof.

(→) By way of example, we prove the statement for $i = 5$.

(M5) Assume that (M5) is true in all $\mathcal{K}$-models. Let X be a finite set of formulas. Clearly, if X is a $\mathcal{K}$-set, then so are $X \cup \{R_1(x,y,z,t)\}$ and $X \cup \{-R_2(x,y,z,t)\}$. Let $X \cup \{R_1(x,y,z,t)\}$ and $X \cup \{-R_2(x,y,z,t)\}$ be $\mathcal{K}$-sets. Suppose X is not a $\mathcal{K}$-set. Then there exist a $\mathcal{K}$-model $\mathcal{M}$ and a valuation v in $\mathcal{M}$ such that $(v(x),v(y),v(z),v(t)) \in R_1$, but $(v(x),v(y),v(z),v(t)) \notin R_2$, which contradicts condition (M5).

(←) By way of example, we prove the statement for $i = 10$.

(M10) Assume that the rule (rM10) is $\mathcal{K}$-correct. Then for every finite set X, $X \cup \{R(O,x,z), R(O,y,z)\}$ is a $\mathcal{K}$-set iff $X \cup \{R(x,y,z), R(O,x,z), R(O,y,z)\}$ is a $\mathcal{K}$-set. Since $\{-R(x,y,z), R(x,y,z), R(O,x,z), R(O,y,z)\}$ is a $\mathcal{K}$-set, $\{-R(x,y,z), R(O,x,z), R(O,y,z)\}$ is also a $\mathcal{K}$-set. Therefore, for every $\mathcal{K}$-model $\mathcal{M}$ and for every valuation v in $\mathcal{M}$, if $(v(x),v(y),v(z)) \in R$, then $(O,v(x),v(z)) \in R$ or $(O,v(y),v(z)) \in R$, hence condition (M10) is satisfied. □

Completion conditions determined by the above rules are:

For all object terms x, y, z, and t,

Cpl(rM5) Either $R_1(x,y,z,t) \in b$ or $-R_2(x,y,z,t) \in b$, obtained by an application of the rule (rM5);

Cpl(rM6) Either $R_1(x,y,z,t) \in b$ or $-R_2(y,x,z,t) \in b$, obtained by an application of the rule (rM6);

Cpl(rM7) Either $R_1(x,y,z,t) \in b$ or $-R_1(x,z,y,t) \in b$, obtained by an application of the rule (rM7);

Cpl(rM8) Either $R(x,y,z) \in b$ or $-R_1(x,y,y,z) \in b$, obtained by an application of the rule (rM8);

Cpl(rM10) If $R(O,x,z) \in b$ and $R(O,y,z) \in b$, then $R(x,y,z) \in b$, obtained by an application of the rule (rM10);

Cpl(rM11) If $R(O,x,z) \in b$, then for every object term y, $R(x,y,z) \in b$, obtained by an application of the rule (rM11);

Cpl(rM12) Either $R(x,y,z) \in b$ or $-R_2(z,x,y,z) \in b$, obtained by an application of the rule (rM12).

Let L_i be an axiomatic extension of RLV such that all L_i-models satisfy condition (Mi), for some $i \in \{5,\dots,12\}$. Then the corresponding $\mathsf{RL}_{\mathsf{L}_i}$-models are $\mathsf{RL}_{\mathsf{RLV}}$-models satisfying all the specific conditions assumed in L_i-models and $\mathsf{RL}_{\mathsf{L}_i}$-dual tableau consists of the rules and axiomatic sets of $\mathsf{RL}_{\mathsf{RLV}}$-dual tableau and in addition:

- The rule (rMi), if $i \in \{5,\dots,8,10,\dots,12\}$;
- Axiomatic set (AxM9), if $i = 9$.

Consider an open branch of an $\mathsf{RL}_{\mathsf{L}_i}$-proof tree. We define the branch structure $\mathcal{M}^b = (U^b, O^b, *^b, R^b, R_1^b, R_2^b, m^b)$ as in the completeness proof of $\mathsf{RL}_{\mathsf{RLV}}$-dual tableau (see p. 187).

Proposition 9.5.1 (Branch Model Property). *Let b be an open branch of an $\mathsf{RL}_{\mathsf{L}_i}$-proof tree and let $i \in \{5, \dots, 12\}$. The branch structure $\mathcal{M}^b = (U^b, O^b, *^b, R^b, R_1^b, R_2^b, m^b)$ is an $\mathsf{RL}_{\mathsf{L}_i}$-model.*

Proof. Let b be an open branch of an $\mathsf{RL}_{\mathsf{L}_i}$-proof tree. In order to show that the branch structure $\mathcal{M}^b = (U^b, O^b, *^b, R^b, R_1^b, R_2^b, m^b)$ is an $\mathsf{RL}_{\mathsf{L}_i}$-model, it suffices to show that $\mathcal{M}^b$ satisfies the condition (Mi). By way of example, we prove the statement for $i \in \{6, 10\}$.

(M6) By the completion condition Cpl(rM6), for all object terms x, y, z, and t, either $R_1(x, y, z, t) \in b$ or $-R_2(y, x, z, t) \in b$. By the completion condition Cpl(R_1), for every $u \in U^b$ either $(x, y, u) \notin R^b$ or $(u, z, t) \notin R^b$, hence $(x, y, z, t) \notin R_1^b$. On the other hand, by the completion condition Cpl($-R_2$), for some $u \in U^b$ both $(x, z, u) \in R^b$ and $(y, u, t) \in R^b$, hence $(y, x, z, t) \in R_2^b$. Therefore, if $(x, y, z, t) \in R_1^b$, then $(y, x, z, t) \in R_2^b$, hence condition (M6) is satisfied.

(M10) Assume that $(x, y, z) \in R^b$, that is $R(x, y, z) \notin b$. Suppose that $(O, x, z) \notin R^b$ and $(O, y, z) \notin R^b$. Then $R(O, x, z) \in b$ and $R(O, y, z) \in b$, and by the completion condition Cpl(rM10), $R(x, y, z) \in b$, a contradiction. □

Now, the satisfaction in branch model property can be proved in exactly the same way as in $\mathsf{RL}_{\mathsf{RLV}}$-dual tableau (see Proposition 9.4.3). Thus, we have:

Proposition 9.5.2 (Satisfaction in Branch Model Property). *Let $i \in \{5, \dots, 12\}$. For every open branch b of an $\mathsf{RL}_{\mathsf{L}_i}$-proof tree and for every $\mathsf{RL}_{\mathsf{L}_i}$-formula φ, if $\mathcal{M}^b, v^b \models \varphi$, then $\varphi \notin b$.*

Finally, we obtain:

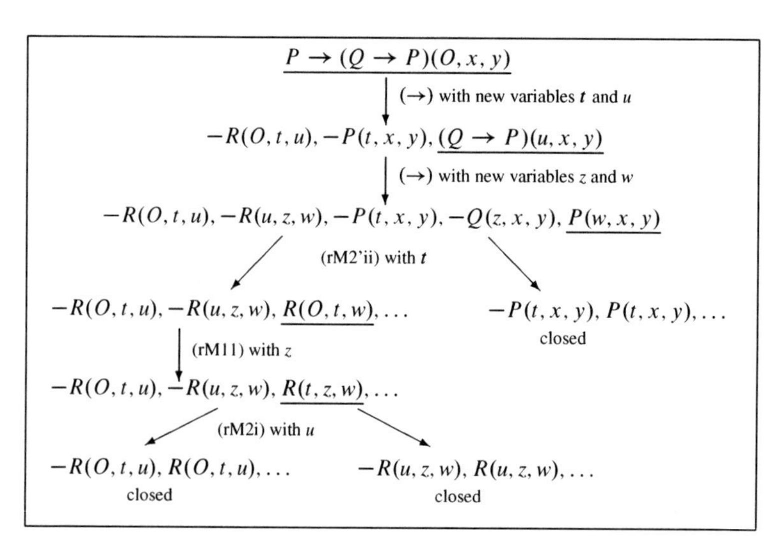

Fig. 9.3 An $\mathsf{RL}_{\mathsf{L}_{11}}$-proof of the formula $p \to (q \to p)$

Theorem 9.5.2 (Soundness and Completeness of $\mathsf{RL}_{\mathsf{L}_i}$). *Let $i \in \{5, \ldots, 12\}$. For every $\mathsf{RL}_{\mathsf{L}_i}$-formula φ, the following conditions are equivalent:*

1. *φ is $\mathsf{RL}_{\mathsf{L}_i}$-valid;*
2. *φ is $\mathsf{RL}_{\mathsf{L}_i}$-provable.*

By the above theorem and Theorem 9.3.1, we obtain:

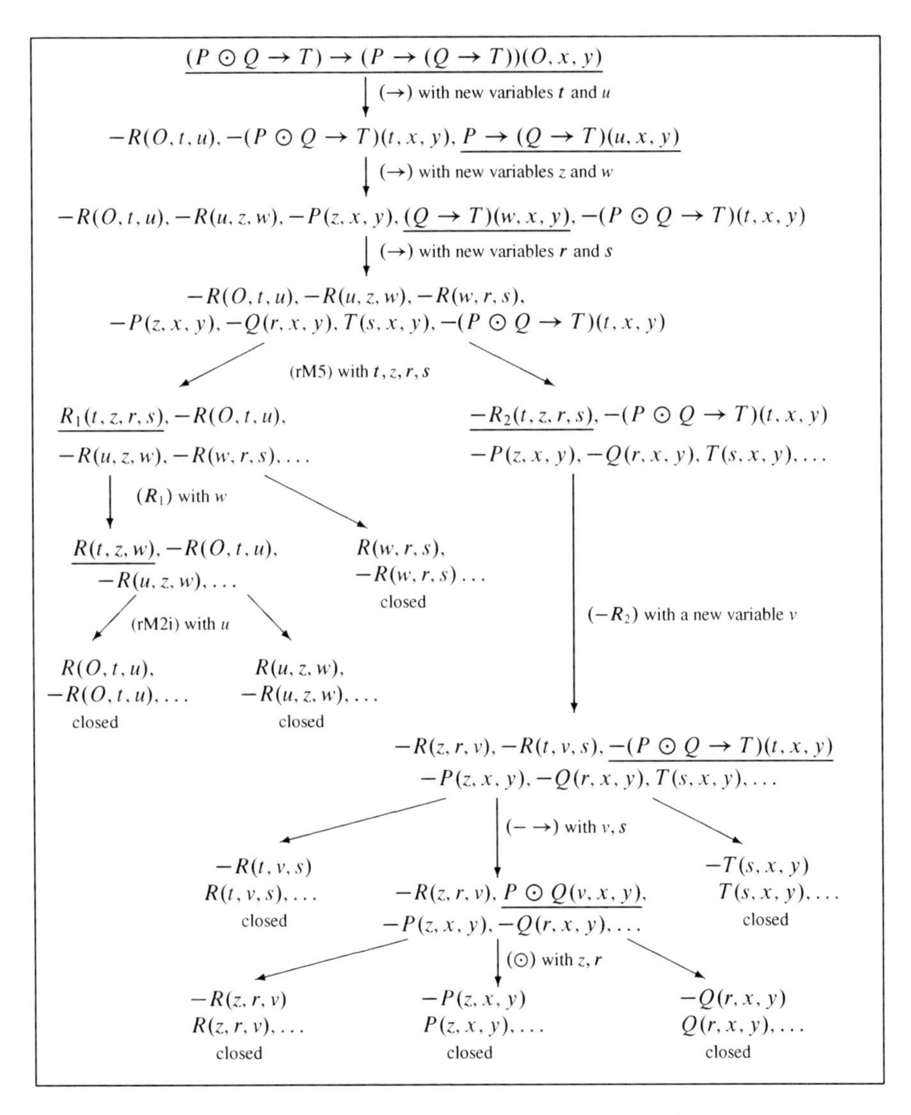

Fig. 9.4 An $\mathsf{RL}_{\mathsf{L}_5}$-proof of the formula $(p \odot q \to t) \to (p \to (q \to t))$

Theorem 9.5.3 (Relational Soundness and Completeness of L_i). *Let $i \in \{5, \ldots, 12\}$. For every L_i-formula φ and for object variables x and y, the following conditions are equivalent:*

1. *φ is L_i-valid;*
2. *$\tau(\varphi)(O, x, y)$ is $\mathsf{RL}_{\mathsf{L}_i}$-provable.*

Example. Consider the following RLV-formulas:

$$\varphi = p \rightarrow (q \rightarrow p),$$
$$\psi = (p \odot q \rightarrow t) \rightarrow (p \rightarrow (q \rightarrow t)).$$

The formula φ is $\mathsf{RL}_{\mathsf{L}_{11}}$-valid, while the formula ψ is $\mathsf{RL}_{\mathsf{L}_5}$-valid. Let $\tau'(p) = P$, $\tau'(q) = Q$, and $\tau'(t) = T$. By Theorem 9.5.3, for all object variables x and y, L_{11}-validity of ψ is equivalent to $\mathsf{RL}_{\mathsf{L}_{11}}$-provability of the formula $\tau(\psi)(O, x, y)$, and L_5-validity of φ is equivalent to $\mathsf{RL}_{\mathsf{L}_5}$-provability of the formula $\tau(\varphi)(O, x, y)$. Figures 9.3 and 9.4 present relational proofs of the formulas $\tau(\varphi)(O, x, y)$ and $\tau(\psi)(O, x, y)$, which show L_{11}-validity and L_5-validity of formulas φ and ψ, respectively.

Chapter 10
Dual Tableaux for Many-Valued Logics

10.1 Introduction

The development of multiple-valued logic in its modern form began with the work of Jan Łukasiewicz [Łuk20] and Emil Post [Pos20, Pos21]. Since the emergence of computer science as an independent discipline, there have been an extensive interplay and mutual inspiration between the two fields. Apart from its logical and philosophical motivation, multiple-valued logic has applications, among others, in hardware design and artificial intelligence. In the field of hardware design classical propositional logic is used as a tool for specification and analysis of electrical switching circuits with two stable voltage levels. A generalization to a finitely-valued logic allows the analogous applications with possibly many stable states. In artificial intelligence multiple-valued logic provides models of vagueness or uncertainty of information and contributes to the development of formal methods simulating commonsense reasoning. Some recent developments and applications of multiple-valued logics can be found e.g., in [Mal93, Got00, Häh01, FO03].

In this chapter we develop a method of designing a dual tableau for an arbitrary finite-valued propositional logic. We follow a relational approach, however, in this case the relational logic appropriate for the translation of formulas of an n-valued logic is not $\mathsf{RL}(1, 1')$, a logic of n-ary relations is employed, where $n \geq 2$. The relational operations of the logic include n specific unary relational operators such that kth operator, $k \leq n$, applied to a relation selects, in a sense, the kth components of the n-tuples belonging to that relation and represents them as an n-ary relation. We apply the method to three multiple-valued logics: Rosser and Turquette logic, RT [RT52], symmetric Heyting logic of order n, $n \geq 2$, SH_n [Itu83, IO06], and a finite poset-based generalization of Post logic, L_T [Ras91]. Decidability of the logic RT follows from the developments in [Got00, Häh03]. Decidability of the logic SH_n follows from the results presented in [Itu82]. Decidability of logic L_T is proved in [Nou99]. The present chapter is based on the developments in [KMO98]. Dual tableaux for many-valued modal logics can be found in [KO01].

E. Orłowska and J. Golińska-Pilarek, *Dual Tableaux: Foundations, Methodology, Case Studies*, Trends in Logic 36, DOI 10.1007/978-94-007-0005-5_10,

10.2 Finitely Many-Valued Logics

In this chapter we consider n_{L}-valued, $n_{\mathsf{L}} \geq 2$, propositional logics, L, whose languages are built from the following pairwise disjoint sets of symbols:

- $\mathbb{V}$ – a countable infinite set of propositional variables;
- $\{o_j : 1 \leq j \leq j_{\mathsf{L}}\}$ – a set of propositional operations where $j_{\mathsf{L}} \geq 1$ and the arity of o_j is $a(j) \geq 1$.

The set of L-formulas is generated from the propositional variables with the propositional operations. The standard many-valued semantics for L is based on a semantic range $\mathcal{SR}_{\mathsf{L}} = \{0, 1, \ldots, s, \ldots, n_{\mathsf{L}} - 1\}$ consisting of n_{L} logical values indexed by integers, where $0 < s \leq n_{\mathsf{L}} - 1$. We assume that the values $\{s, \ldots, n_{\mathsf{L}} - 1\}$ are designated, and all the other ones undesignated. Though in notation we identify the values with their natural number indices, in general we do not assume any kind of ordering, in particular, any linear ordering corresponding to that of the natural indices is not necessarily assumed in the set of the logical values.

With the family of propositional operations $\{o_1, \ldots, o_{i_{\mathsf{L}}}\}$, we associate a family of semantic functions $\{f_{o_j} : j = 1, \ldots, i_{\mathsf{L}}\}$, where f_{o_j} maps $\mathcal{SR}_{\mathsf{L}}^{a(j)}$ into $\mathcal{SR}_{\mathsf{L}}$.

An L-model is a structure of the form $\mathcal{M} = (U, m, \{m_k : k \in \mathcal{SR}_{\mathsf{L}}\})$, where U is a non-empty set of states, $m \colon \mathbb{V} \times U \to \mathcal{SR}_{\mathsf{L}}$, and $\{m_k : k \in \mathcal{SR}_{\mathsf{L}}\}$ is a family of meaning functions such that:

- $m_k(p) = \{w \in U : m(p, w) = k\}$, for $p \in \mathbb{V}$;
- $m_k(o_j(\varphi_1, \ldots, \varphi_{a(j)})) = \bigcup_{f_{o_j}(k_1, \ldots, k_{a(j)}) = k} (m_{k_1}(\varphi_1) \cap \ldots \cap m_{k_{a(j)}}(\varphi_{a(j)}))$.

Intuitively, $m_k(\varphi)$ is the set of states at which formula φ takes value k.

The standard notions of satisfaction of a formula at a state in a model, truth in a model, and validity are defined as follows. An L-formula φ is satisfied in a model $\mathcal{M}$ at a state w, written $\mathcal{M}, w \models \varphi$, whenever $w \in m_k(\varphi)$ for some $k \in \{s, \ldots, n_{\mathsf{L}} - 1\}$; φ is true in a model $\mathcal{M}$, written $\mathcal{M} \models \varphi$, if and only if $\mathcal{M}, w \models \varphi$, for all $w \in U$, and it is L-valid whenever it is true in all L-models.

Along these lines we present three examples of multiple-valued logics.

Rosser-Turquette Logic

Rosser-Turquette logic RT is an n_{RT}-valued logic with $\mathcal{SR}_{\mathsf{RT}} = \{0, \ldots, n - 1\}$, designated values are $s, \ldots, n - 1$, where $0 < s \leq n - 1$, and the logical values are linearly ordered consistently with the order of their natural indices. The propositional operations of RT include the family $\{J_k : k \in \mathcal{SR}_{\mathsf{RT}}\}$ of unary operations. The operations J_k play a special role. Namely, J_k is a unary operation 'selecting' the logical value k. Other propositional operations are $\vee$, $\wedge$, and $\neg$.

The respective semantic functions are defined as follows:

- $f_{\vee}(k, l) = \max(k, l)$;
- $f_{\wedge}(k, l) = \min(k, l)$;

- $f_{J_k}(k) = n-1$, and for $l \neq k$, $f_{J_k}(l) = 0$;
- $f_\neg(l) = \max(f_{J_0}(l), \ldots, f_{J_{s-1}}(l))$.

An RT-model is a structure of the form $\mathcal{M} = (U, m, \{m_k : k \in \mathcal{SR}_{\mathsf{RT}}\})$, where U is a non-empty set of states, $m\colon \mathbb{V} \times U \to \mathcal{SR}_{\mathsf{RT}}$, and $\{m_k : k \in \mathcal{SR}_{\mathsf{RT}}\}$ is a family of meaning functions such that:

- $m_k(p) = \{w \in U : m(p, w) = k\}$, for every $k \in \mathcal{SR}_{\mathsf{RT}}$;
- $m_k(\varphi \vee \psi) = \bigcup_{i=0}^{k}((m_k(\varphi) \cap m_i(\psi)) \cup (m_i(\varphi) \cap m_k(\psi)))$;
- $m_k(\varphi \wedge \psi) = \bigcup_{i=k}^{n-1}((m_k(\varphi) \cap m_i(\psi)) \cup (m_i(\varphi) \cap m_k(\psi)))$;
- $m_{n-1}(J_l(\varphi)) = m_l(\varphi)$, $m_0(J_l(\varphi)) = \bigcup_{i \neq l} m_i(\varphi)$, and $m_k(J_l(\varphi)) = \emptyset$, for $k \neq 0, n-1$;
- $m_{n-1}(\neg\varphi) = \bigcup_{i=0}^{s-1} m_i(\varphi)$, $m_0(\neg\varphi) = \bigcup_{i=s}^{n-1} m_i(\varphi)$, and $m_k(\neg\varphi) = \emptyset$, for $k \neq 0, n-1$.

Symmetric Heyting Logic of Order n, SH_n

The formulas of logic SH_n, $n \geq 2$, are constructed from propositional variables with the operations $\wedge, \vee, \to, \neg, \sim$, and with a family $\{\sigma_i\}_{i=1,\ldots,n-1}$ of unary operations. The semantic range for logic SH_n is:

$$\mathcal{SR}_{\mathsf{SH}_n} = \left\{(x, y) : x, y \in \left\{0, \frac{1}{n-1}, \frac{2}{n-1}, \ldots, 1\right\}\right\}.$$

We can treat the set $\left\{0, \frac{1}{n-1}, \frac{2}{n-1}, \ldots, 1\right\}$ as a symmetric Heyting algebra

$$((n), \vee, \wedge, \to, \neg, \sim, 0, 1),$$

where $(n) = \left\{0, \frac{1}{n-1}, \frac{2}{n-1}, \ldots, 1\right\}$, and

$$x \vee y = \max(x, y) \qquad x \wedge y = \min(x, y) \qquad \sim x = 1 - x$$

$$x \to y = \begin{cases} 1 & \text{if } x \leq y \\ y & \text{otherwise} \end{cases} \qquad \neg x = \begin{cases} 1 & \text{if } x = 0 \\ 0 & \text{if } x > 0 \end{cases}$$

We define the unary operations $\{\sigma_i\}_{i=1,\ldots,n-1}$ in this algebra by:

$$\sigma_i\left(\frac{x}{n-1}\right) = \begin{cases} 1 & \text{if } x \geq n-i \\ 0 & \text{otherwise} \end{cases}$$

Then the semantic functions for the propositional operations in $\mathcal{SR}_{\mathsf{SH}_n}$ are given by:

- $f_\vee((x, y), (x', y')) = (x \vee x', y \vee y')$;
- $f_\wedge((x, y), (x', y')) = (x \wedge x', y \wedge y')$;
- $f_\to((x, y), (x', y')) = (x \to x', y \to y')$;
- $f_\neg((x, y)) = (x \to 0, y \to 0)$;

- $f_{\sim}((x, y)) = (1 - y, 1 - x)$;
- $f_{\sigma_i}((x, y)) = (\sigma_i(x), \sigma_i(y))$.

An SH_n-model is a structure of the form $\mathcal{M} = (U, m, \{m_{(k,l)} : (k,l) \in \mathcal{SR}_{\mathsf{SH}_n}\})$, where U is a non-empty set of states, $m \colon \mathbb{V} \times U \to \mathcal{SR}_{\mathsf{SH}_n}$, and $\{m_{(k,l)} : (k,l) \in \mathcal{SR}_{\mathsf{SH}_n}\}$ is a family of meaning functions reflecting the intended interpretation of the operations. For example, the functions for the implication and negation $\sim$ are defined as follows:

- $m_{(1,1)}(\varphi \to \psi) = \bigcup_{x \leq x',\ y \leq y'}(m_{(x,y)}(\varphi) \cap m_{(x',y')}(\psi))$;
- $m_{(1,l)}(\varphi \to \psi) = \bigcup_{x \leq x',\ y > l}(m_{(x,y)}(\varphi) \cap m_{(x',l)}(\psi))$;
- $m_{(k,1)}(\varphi \to \psi) = \bigcup_{x > k,\ y \leq y'}(m_{(x,y)}(\varphi) \cap m_{(k,y')}(\psi))$;
- $m_{(k,l)}(\varphi \to \psi) = \bigcup_{x > k,\ y > l}(m_{(x,y)}(\varphi) \cap m_{(x',y')}(\psi))$, where $k, l \neq 1$;
- $m_{(k,l)}(\sim \varphi) = m_{(1-l,1-k)}(\varphi)$.

The Logic $\mathsf{L_T}$

The logic $\mathsf{L_T}$ is a certain version of the logics introduced in [CH73], see also [Ras91]. The logic is based on a generalization of Post algebras investigated in [CHR89] where a chain of Post constants is replaced by a finite partially ordered set. The elements of this set are the indices of the unary Post operations but, in contrast with the classical Post logic, they do not have syntactic counterparts in the language.

$\mathsf{L_T}$ is a propositional logic whose formulas are built from propositional variables with the operations $\vee, \wedge, \to, \neg$, and a family $\{d_t\}_{t \in \mathsf{T}}$ of unary operations, where T is a finite set partially ordered by a relation $\leq$. The semantic range for the logic $\mathsf{L_T}$ is the set of all the increasing subsets of T, that is:

$$\mathcal{SR}_{\mathsf{L_T}} = \{s \subseteq \mathsf{T} : \text{ for all } x, y,\ x \in s \text{ and } x \leq y \text{ imply } y \in s\} \cup \{\emptyset\}.$$

The semantic functions providing meaning of the propositional operations are defined as follows:

- $f_{\vee}(s, s') = s \cup s'$;
- $f_{\wedge}(s, s') = s \cap s'$;
- $f_{\to}(s, s') = \bigcup\{u \in \mathcal{SR}_{\mathsf{L_T}} : s \cap u \subseteq s'\}$;
- $f_{\neg}(s) = s \to \emptyset$;
- $f_{d_t}(s) = \begin{cases} \mathsf{T} & \text{if } t \in s, \\ \emptyset & \text{otherwise.} \end{cases}$

The only distinguished element of $\mathcal{SR}_{\mathsf{L_T}}$ is the set T.

In an $\mathsf{L_T}$-model $\mathcal{M} = (U, m, \{m_k : k \in \mathcal{SR}_{\mathsf{L_T}}\})$, the meaning functions for operations d_t are defined as follows, for $l \in \mathcal{SR}_{\mathsf{L_T}}$:

- $m_{\mathsf{T}}(d_t(\varphi)) = \bigcup_{t \in l} m_l(\varphi)$;
- $m_{\emptyset}(d_t(\varphi)) = \bigcup_{t \notin l} m_l(\varphi)$;
- $m_l(d_t(\varphi)) = \emptyset$, for $l \neq \emptyset, \mathsf{T}$.

10.3 Relational Formalization of Finitely Many-Valued Logics

In the relational formalization of many-valued logics we apply the standard method of interpreting formulas of an n_{L}-valued logic L as n_{L}-ary relations. The vocabulary of the language of relational logic RL_{L} adequate for logic L consists of symbols from the following pairwise disjoint sets:

- $\mathbb{RV}_{\mathsf{RL}_{\mathsf{L}}}$ – a countable infinite set of relational variables representing n_{L}-ary relations;
- $\{o_j : 1 \leq j \leq i_{\mathsf{L}}\} \cup \{J_t : 0 \leq t \leq n_{\mathsf{L}} - 1\}$ – the set of relational operations, where o_j is of the arity $a(j)$, for $j = 1, \ldots, i_{\mathsf{L}}$, and every J_t is unary.

We slightly abuse the notation here by denoting the relational operations of RL_{L} with the same symbols as the operations of L.

The set $\mathbb{RT}_{\mathsf{RL}_{\mathsf{L}}}$ of relational terms is the smallest set that includes $\mathbb{RV}_{\mathsf{RL}_{\mathsf{L}}}$ and is closed with respect to all the relational operations. The set of formulas is simply the set of relational terms $\mathbb{RT}_{\mathsf{RL}_{\mathsf{L}}}$. An RL_{L}-formula is said to be *indecomposable* whenever it is of the form $J_t(P)$, for some $t \in \{0, \ldots, n_{\mathsf{L}} - 1\}$ and $P \in \mathbb{RV}_{\mathsf{RL}_{\mathsf{L}}}$. A finite set of RL_{L}-formulas is indecomposable whenever all of its formulas are indecomposable.

An RL_{L}-model is a structure $\mathcal{M} = (U \cup \{\emptyset\}, m, \{m_k : k \in \mathcal{SR}_{\mathsf{L}}\})$ such that U is a non-empty set, $m: \mathbb{RT}_{\mathsf{RL}_{\mathsf{L}}} \to \mathcal{P}((U \cup \{\emptyset\})^{n_{\mathsf{L}}-1})$ and for every $k \in \mathcal{SR}_{\mathsf{L}}$, $m_k: \mathbb{RT}_{\mathsf{RL}_{\mathsf{L}}} \to \mathcal{P}(U \cup \{\emptyset\})$ are meaning functions such that the following conditions are satisfied:

- $m_k(P) \subseteq U \cup \{\emptyset\}$ and $m(P) = m_0(P) \times \cdots \times m_{n_{\mathsf{L}}-1}(P)$, for $P \in \mathbb{RV}_{\mathsf{RL}_{\mathsf{L}}}$;
- If $m(P) = P_0 \times \cdots \times P_{n_{\mathsf{L}}-1}$, then:

  (1) For any i, $0 \leq i \leq n_{\mathsf{L}} - 1$, $P_i \in \mathcal{P}(U \cup \{\emptyset\}) \setminus \{\emptyset\}$, i.e., P_i is a non-empty subset of $U \cup \{\emptyset\}$,
  (2) If $i \neq j$, then $P_i \cap P_j \in \{\emptyset, \{\emptyset\}\}$, i.e., $P_i \cap P_j$ is either empty or equals $\{\emptyset\}$,
  (3) $U \subseteq \bigcup_{k=0}^{n_{\mathsf{L}}-1} P_k$;

- For any j and t the operations o_j and J_t are interpreted as functions on relations on $(U \cup \{\emptyset\})^{n_{\mathsf{L}}-1}$ such that:

  (i) For all terms $P^1, \ldots, P^{a(j)}$ such that $m(P^l) = (m_0(P^l) \times \cdots \times m_{n_{\mathsf{L}}-1}(P^l))$, we have:

$$m\left(o_j\left(P^1, \ldots, P^{a(j)}\right)\right) = Q_0 \times \cdots \times Q_{n_{\mathsf{L}}-1},$$

  where

$$Q_k = \bigcup_{f_{o_j}(g_1, \ldots, g_{a(j)}) = k} m_{g_1}(P^1) \cap \cdots \cap m_{g_{a(j)}}\left(P^{a(j)}\right)$$

  if the above union is non-empty, and $Q_k = \{\emptyset\}$ otherwise;

(ii) For any relational term P such that $m(P) = m_0(P) \times \cdots \times m_{n_{\mathsf{L}}-1}(P)$ we have:

$$m(J_t(P)) = Q_0 \times \cdots \times Q_{n_{\mathsf{L}}-1},$$

where

$$Q_{n_{\mathsf{L}}-1} = m_t(P),\ Q_0 = \bigcup_{l \neq t} m_l(P),\ \text{ and } Q_i = \{\emptyset\} \text{ for } i \neq 0, n_{\mathsf{L}} - 1.$$

Note that $m(J_t(P))$ partitions the set U into the sets $m_t(P)$ and $U - m_t(P)$.

An $\mathsf{RL_L}$-formula P is said to be true in an $\mathsf{RL_L}$-model $\mathcal{M}$, written $\mathcal{M} \models P$, whenever $\bigcup_{i=s}^{n_{\mathsf{L}}-1}(m_i(P)) = U \cup \{\emptyset\}$. A formula P is $\mathsf{RL_L}$-valid whenever it is true in all $\mathsf{RL_L}$-models.

Proposition 10.3.1. $\mathsf{RL_L}$*-models are well-defined.*

Proof. We have to show that $\mathcal{P}((U \cup \{\emptyset\})^{n_{\mathsf{L}}-1})$ is closed under the interpretations of operations o_j and J_t, that is for all relational terms $P, P^1, \ldots, P^{a(j)}$, the relations $m\left(o_j\left(P^1, \ldots, P^{a(j)}\right)\right)$ and $m(J_t(P))$ satisfy the conditions (1), (2), and (3). The preservation of condition (1) is quite obvious, so we concentrate on conditions (2) and (3).

Assume $m(P^r) = m_0(P^r) \times \cdots \times m_{n_{\mathsf{L}}-1}(P^r)$ satisfy the conditions (1), (2), and (3), for $r = 1, \ldots, a(j)$. Let $Q = m\left(o_j\left(P^1, \ldots, P^{a(j)}\right)\right)$, where $Q = Q_0 \times \cdots \times Q_{n_{\mathsf{L}}-1}$. Let $0 \leq l, k \leq n_{\mathsf{L}} - 1$, and $l \neq k$. Then either at least one of Q_k, Q_l is $\{\emptyset\}$, and then obviously $Q_k \cap Q_l$ is either $\emptyset$ or $\{\emptyset\}$ or

$$Q_k = \bigcup_{f_{o_j}(g_1,\ldots,g_{a(j)})=k} m_{g_1}\left(P^1\right) \cap \ldots \cap m_{g_{a(j)}}\left(P^{a(j)}\right),$$

$$Q_l = \bigcup_{f_{o_j}(h_1,\ldots,h_{a(j)})=l} m_{h_1}\left(P^1\right) \cap \ldots \cap m_{h_{a(j)}}\left(P^{a(j)}\right),$$

hence

$$Q_k \cap Q_l = \bigcup_{f_{o_j}(g_1,\ldots,g_{a(j)})=k}\ \bigcup_{f_{o_j}(g_1,\ldots,g_{a(j)})=l} m_{g_1}\left(P^1\right) \cap \ldots \cap m_{g_{a(j)}}\left(P^{a(j)}\right) \cap$$

$$\cap\, m_{h_1}\left(P^1\right) \cap \ldots \cap m_{h_{a(j)}}\left(P^{a(j)}\right).$$

By our assumption about $m(P^r)$'s we have $m_s(P^r) \cap m_t(P^r) = \emptyset$ or $\{\emptyset\}$, for $s \neq t$. On the other hand, in each of the summands above $g_r = h_r$, for $r = 1, \ldots, a(j)$ would imply

$$k = f_{o_j}(g_1, \ldots, g_{a(j)}) = f_{o_j}(h_1, \ldots, h_{a(j)}) = l$$

which is a contradiction. Thus $g_r \neq h_r$, for some r, so $m_{g_r}(P^r) \cap m_{h_r}(P^r) \in \{\emptyset, \{\emptyset\}\}$. Obviously, this means that the same holds for each summand in $Q_k \cap Q_l$, and in consequence for the latter intersection, therefore Q satisfies the condition (2).

To prove that Q satisfies condition (3), consider any $u \in U$. Since all the P^r's satisfy condition (3), then for each r, $1 \leq r \leq a(j)$, there exists k_r, $0 \leq k_r \leq n_{\mathsf{L}}-1$ such that $u \in m_{k_r}(P^r)$. Obviously, $u \in m_{k_1}(P^1) \cap \ldots \cap m_{k_{a(j)}}(P^{a(j)})$. Therefore, for $k = f_{o_j}(k_1, \ldots, k_{a(j)})$, we obtain:

$$u \in \bigcup_{f_{o_j}(g_1,\ldots,g_{a(j)})=k} m_{g_1}(P^1) \cap \ldots \cap m_{g_{a(j)}}\left(P^{a(j)}\right).$$

Since u is in the above union, the union is non-empty. Thus, by the definition it equals Q_k. Therefore, $u \in \bigcup_{l=0}^{n_{\mathsf{L}}-1} Q_l$. Since u is an arbitrary element of U, Q satisfies condition (3).

Assume that $m(P) = m_0(P) \times \cdots \times m_{n_{\mathsf{L}}-1}(P)$ satisfies conditions (1), (2), and (3). Let $Q = m(J_t P)$, where $Q = Q_0 \times \cdots \times Q_{n_{\mathsf{L}}-1}$. Then obviously, Q satisfies condition (2), because $Q_1 = \{\emptyset\}$ for $l \neq 0, n_{\mathsf{L}} - 1$ and $Q_0 \cap Q_{n_{\mathsf{L}}} = (\bigcup_{l \neq t} m_l(P)) \cap m_t(P)$. Since for $l \neq t$ the intersection $m_l(P) \cap m_t(P)$ is either $\emptyset$ or $\{\emptyset\}$ by the assumption on $m(P)$, the same holds for the union representing $Q_0 \cap Q_{n_{\mathsf{L}}}$. Finally, Q satisfies also condition (3), since:

$$\bigcup_{i=0}^{n_{\mathsf{L}}} Q_i = \bigcup_{l \neq k} m_l(P) \cup \{\emptyset\} \cup m_t(P) \subseteq \bigcup_{i=0}^{n_{\mathsf{L}}-1} m_i(P) \supseteq U,$$

by the assumption on $m(P)$. □

The translation of L-formulas into relational terms starts with a one-to-one assignment $\tau' \colon \mathbb{V} \to \mathbb{RV}_{\mathsf{RL_L}}$ of relational variables to the propositional variables. Then, the translation τ of formulas is defined by:

$$\tau(o_j(\varphi_1, \ldots, \varphi_{a(j)})) = o_j(\tau(\varphi_1), \ldots, \tau(\varphi_{a(j)})).$$

Proposition 10.3.2. *For every* L*-formula* φ *and for every* L*-model* $\mathcal{M}$ *there exists an* $\mathsf{RL_L}$*-model* $\mathcal{M}'$ *such that* $\mathcal{M} \models \varphi$ *iff* $\mathcal{M}' \models \tau(\varphi)$.

Proof. Let φ be an L-formula and let $\mathcal{M} = (U, m, \{m_k : k \in \mathcal{SR}_{\mathsf{L}}\})$ be an L-model. We define the corresponding $\mathsf{RL_L}$-model $\mathcal{M}' = (U' \cup \{\emptyset\}, m', \{m'_k : k \in \mathcal{SR}_{\mathsf{L}}\})$ as follows:

- $U' = U$;
- For a relational variable P such that $\tau'(p) = P$ we define $m'_k(P) = m_k(p) \cup \{\emptyset\}$ and $m'(P) = m'_0(P) \times \cdots \times m'_{n_{\mathsf{L}}-1}(P)$;
- m' extends to all the compound relational terms as in $\mathsf{RL_L}$-models.

We show that for every L-formula φ, condition (1) $m'_k(\tau(\varphi)) = m_k(\varphi) \cup \{\emptyset\}$ holds. Then, models $\mathcal{M}$ and $\mathcal{M}'$ clearly satisfy the proposition.

We prove (1) by induction on the complexity of formulas. If $\varphi = p$, for $p \in \mathbb{V}$, then (1) holds by the definition of $m'_k(\tau(p))$. Let $\varphi = o_j(p^1, \ldots, p^{a(j)})$. Then $m'_k(\tau(\varphi)) = m'_k(o_j(\tau(p^1), \ldots, \tau(p^{a(j)})))$. By the definition of the model $\mathcal{M}'$, $m'_k(o_j(\tau(p^1), \ldots, \tau(p^{a(j)})))$ equals:

$$\bigcup_{f_{o_j}(k_1,\ldots,k_{a(j)})=k} (m'_{k_1}(\tau(p^1)) \cap \ldots \cap m'_{k_{a(j)}}(\tau(p^{a(j)})))$$

if the union is non-empty, otherwise it equals $\{\emptyset\}$. If the latter holds, then $m_k(\varphi) = \emptyset$, hence $m'_k(\tau(\varphi)) = \{\emptyset\} = m_k(\varphi) \cup \{\emptyset\}$. If the union is non-empty, then by the induction hypothesis, we obtain:

$$\begin{aligned}
&\bigcup_{f_{o_j}(k_1,\ldots,k_{a(j)})=k} (m'_{k_1}(\tau(p^1)) \cap \ldots \cap m'_{k_{a(j)}}(\tau(p^{a(j)}))) \\
&= \bigcup_{f_{o_j}(k_1,\ldots,k_{a(j)})=k} ((m_{k_1}(p^1) \cup \{\emptyset\}) \cap \ldots \cap (m_{k_{a(j)}}(p^{a(j)})) \cup \{\emptyset\}) \\
&= \left(\bigcup_{f_{o_j}(k_1,\ldots,k_{a(j)})=k} (m_{k_1}(p^1) \cap \ldots \cap m_{k_{a(j)}}(p^{a(j)})) \right) \cup \{\emptyset\} = m_k(\varphi) \cup \{\emptyset\}.
\end{aligned}$$

Therefore $m'_k(\tau(\varphi)) = m_k(\varphi) \cup \{\emptyset\}$. □

In a similar way we can prove:

Proposition 10.3.3. *For every* L*-formula* φ *and for every* $\mathsf{RL_L}$*-model* $\mathcal{M}$ *there exists an* L*-model* $\mathcal{M}'$ *such that* $\mathcal{M}' \models \varphi$ *iff* $\mathcal{M} \models \tau(\varphi)$.

The above two propositions lead to:

Theorem 10.3.1. *Let* φ *be an* L*-formula. Then the following conditions are equivalent:*

1. φ *is* L*-valid;*
2. $\tau(\varphi)$ *is* $\mathsf{RL_L}$*-valid.*

Relational Formalization of Rosser–Turquette Logic RT

The vocabulary of the language of the relational logic $\mathsf{RL_{RT}}$ consists of the set of relational variables, two binary operations $\vee$ and $\wedge$, and unary operations $\neg$ and J_t, for $t = 0, \ldots, n_{\mathsf{RT}} - 1$.

In $\mathsf{RL_{RT}}$-models we define the meaning functions in a standard way with the following clauses for the compound terms:

- $m_k(P \vee Q) = \bigcup_{l=0}^{k}(m_k(P) \cap m_l(Q) \cap m_l(P) \cap m_k(Q))$ if this union is non-empty, and $m_k(P \vee Q) = \{\emptyset\}$ otherwise;
- $m_k(P \wedge Q) = \bigcup_{l=k}^{n_{\mathsf{RT}}-1}(m_k(P) \cap m_l(Q) \cap m_l(P) \cap m_k(Q))$ if this union is non-empty, and $m_k(P \wedge Q) = \{\emptyset\}$ otherwise;

- $m_0(\neg P) = \bigcup_{l=s}^{n_{\mathsf{RT}}-1} m_l(P)$, $m_{n_{\mathsf{RT}}-1}(\neg P) = \bigcup_{l=0}^{s-1} m_l(P)$, and for $l \neq 0, n_{\mathsf{RT}}-1$, $m_l(\neg P) = \{\emptyset\}$;
- For any relational term P such that $m(P) = m_0(P) \times \cdots \times m_{n_{\mathsf{RT}}-1}(P)$ we have:

$$m(J_k(P)) = Q_0 \times \cdots \times Q_{n_{\mathsf{RT}}-1},$$

where

$$Q_{n_{\mathsf{RT}}-1} = m_k(P), \; Q_0 = \bigcup_{l \neq k} m_l(P), \; \text{ and } Q_i = \{\emptyset\} \text{ for } i \neq 0, n_{\mathsf{RT}}-1.$$

Relational Formalization of Symmetric Heyting Logics SH_n

The vocabulary of the relational logic $\mathsf{RL}_{\mathsf{SH}_n}$ consists of relational variables and the relational operations which are the direct counterparts to the propositional operations of the SH_n logics. $\mathsf{RL}_{\mathsf{SH}_n}$-models are defined so that the meaning of compound relational terms reflects properties of the corresponding propositional operations. For example, for any relational term P such that:

$$m(P) = \times_{(k,l) \in \mathcal{SR}_{\mathsf{SH}_n}} m_{(k,l)}(P)$$

we have

$$m(\sigma_i(P)) = \times_{(k,l) \in \mathcal{SR}_{\mathsf{SH}_n}} R_{(k,l)},$$

where $\times_i A_i$ denotes the direct product of sets A_i and:

$R_{(k,l)} = \{\emptyset\}$ if either $k \neq 0, 1$ or $l \neq 0, 1$,
$R_{(1,1)} = \bigcup_{x \geq n-i, \; y \geq n-i} m_{(\frac{x}{n-1}, \frac{y}{n-1})}(P)$
if the above union is non-empty, and $R_{(1,1)} = \{\emptyset\}$ otherwise,

$R_{(1,0)} = \bigcup_{x \geq n-i, \; y < n-i} m_{(\frac{x}{n-1}, \frac{y}{n-1})}(P)$
if the above union is non-empty, and $R_{(1,0)} = \{\emptyset\}$ otherwise,

$R_{(0,1)} = \bigcup_{x < n-i, \; y \geq n-i} m_{(\frac{x}{n-1}, \frac{y}{n-1})}(P)$
if the above union is non-empty, and $R_{(0,1)} = \{\emptyset\}$ otherwise,

$R_{(0,0)} = \bigcup_{x < n-i, \; y < n-i} m_{(\frac{x}{n-1}, \frac{y}{n-1})}(P)$
if the above union is non-empty, and $R_{(0,0)} = \{\emptyset\}$ otherwise.

Relational Formalization of the Logic L_T

The formulas of relational logic $\mathsf{RL}_{\mathsf{L}_\mathsf{T}}$ are built from relational variables with the relational operations which are the direct counterparts of the propositional operations of the logic $\mathsf{L_T}$. $\mathsf{RL_{L_T}}$-models are defined in a standard way, so that the meaning of compound relational terms reflects properties of the corresponding propositional operations. For example, if $R = P \to Q$, and $m(Z) = \times_{k \in \mathcal{SR}_{\mathsf{L_T}}} m_k(Z)$, for $Z = P, Q, R$, then we have:

$$m_k(R) = \bigcup_{\bigcup\{u \in \mathcal{SR}_{\mathsf{L_T}} : l \cap u \subseteq t\} = k} m_l(P) \cap m_t(Q),$$

if the above union is non-empty, and $m_k(R) = \{\emptyset\}$ otherwise.

10.4 Dual Tableaux for Finitely Many-Valued Logics

Let L be a finitely many-valued logic. A relational dual tableau for L consist of axiomatic sets of formulas and decomposition rules that apply to finite sets of formulas. The notion of a rule is defined as in Sect. 2.4.

Decomposition rules have the following forms:

$(J{-}in)$ $\quad \dfrac{R}{J_s(R), \ldots, J_{n_{\mathsf{L}}-1}(R)}$

where R is not of the form $J_l(Q)$

for any $l \in \mathcal{SR}_{\mathsf{L}}$ and for any $\mathsf{RL_L}$-formula Q

$(o_j)_1$ $\quad \dfrac{K, J_t(o_j(R_1, \ldots, R_i))}{K, H, J_{t_1}(R_1) \mid \ldots \mid K, H, J_{t_i}(R_i)}$

for all $t_1, \ldots, t_i$ such that $f_{o_j}(t_1, \ldots, t_i) = t$, where

$i = a(j)$, $H = J_t(o_j(R_1, \ldots, R_i))$, and

for some $l \in \{1, \ldots, i\}$ sequence K of formulas

does not contain $J_{t_l}(R_l)$

$(o_j)_2$ $\quad \dfrac{K, J_t(o_j(R_1, \ldots, R_i))}{K}$

where $K = J_{t_1}(R_1), \ldots, J_{t_i}(R_i)$ and $f_{o_j}(t_1, \ldots, t_i) = t$

$(J_0(J_l))$ $\quad \dfrac{J_0(J_l(R))}{J_0(R), \ldots, J_{l-1}(R), J_{l+1}(R), \ldots, J_{n_{\mathsf{L}}-1}(R)}$

for any $l \in \mathcal{SR}_{\mathsf{L}}$ such that $l \neq 0$

$(J_0(J_0))$ $\quad \dfrac{J_0(J_0(R))}{J_1(R), \ldots, J_{n_{\mathsf{L}}-1}(R)}$

$(J_{n_{\mathsf{L}}-1}(J_l))$ $\quad \dfrac{J_{n_{\mathsf{L}}-1}(J_l(R))}{J_l(R)}$, for any $l \in \mathcal{SR}_{\mathsf{L}}$

$(J_t(J_l))$ $\quad \dfrac{K, J_t(J_l(R))}{K}$

for any $l \in \mathcal{SR}_{\mathsf{L}}$ and for any $t \neq 0, n_{\mathsf{L}} - 1$,

and for any sequence K of $\mathsf{RL_L}$-formulas.

A set of $\mathsf{RL_L}$-formulas is said to be an $\mathsf{RL_L}$-axiomatic set whenever it includes either of the following sets:

For any relational term R,

- $\{J_0(R), \ldots, J_{n_{\mathsf{L}}-1}(R)\}$;
- $\{J_0(J_t(J_l(R)))\}$, for $t \neq 0, n_{\mathsf{L}} - 1$.

A finite set $X = \{P_1, \ldots, P_n\}$ of $\mathsf{RL_L}$-formulas is said to be $\mathsf{RL_L}$-set whenever for every $\mathsf{RL_L}$-model $\mathcal{M}$, $\bigcup_{j=1}^{n} \bigcup_{i=s}^{n_{\mathsf{L}}-1} m_i(P_j) = U \cup \{\emptyset\}$. Correctness of a rule is defined in a similar way as in the relational logics of classical algebras of binary relations (see Sect. 2.4).

Proposition 10.4.1.

1. *The* $\mathsf{RL_L}$*-rules are* $\mathsf{RL_L}$*-correct;*
2. *The* $\mathsf{RL_L}$*-axiomatic sets are* $\mathsf{RL_L}$*-sets.*

Proof. By way of example, we prove $\mathsf{RL_L}$-correctness of the rule $(J_{n_{\mathsf{L}}-1}(J_l))$. First, note that by the definition of $\mathsf{RL_L}$-models we have:

$$m_{n_{\mathsf{L}}-1}(J_{n_{\mathsf{L}}-1}(J_l(R))) = m_{n_{\mathsf{L}}-1}(J_l(R)) = m_l(R),$$

$$m_i(J_{n_{\mathsf{L}}-1}(J_l(R))) = m_i(J_l(R)) = \{\emptyset\}, \text{ for any } i \neq 0, n_{\mathsf{L}} - 1.$$

Therefore, $\bigcup_{k=s}^{n_{\mathsf{L}}-1} m_k(J_{n_{\mathsf{L}}-1}(J_l(R))) = \bigcup_{k=s}^{n_{\mathsf{L}}-1} m_k(J_l(R)) = \{\emptyset\} \cup m_l(R)$. Thus, for every $\mathsf{RL_L}$-model $\mathcal{M}$, $\mathcal{M} \models J_{n_{\mathsf{L}}-1}(J_l(R))$ iff $\mathcal{M} \models J_l(R)$. Hence, the correctness follows. □

The notions of an $\mathsf{RL_L}$-proof tree and $\mathsf{RL_L}$-provability of an $\mathsf{RL_L}$-formula are defined as in Sect. 2.4.

Theorem 10.4.1 (Soundness and Completeness of $\mathsf{RL_L}$). *Let φ be an* $\mathsf{RL_L}$*-formula. Then, the following conditions are equivalent:*

1. *φ is* $\mathsf{RL_L}$*-valid;*
2. *φ is* $\mathsf{RL_L}$*-provable.*

Proof. The implication 2. → 1. follows from Proposition 10.4.1, hence $\mathsf{RL_L}$-dual tableau is sound. Moreover, it can be easily proved that every $\mathsf{RL_L}$-proof tree is finite. Assume that φ is $\mathsf{RL_L}$-valid. Suppose φ does not have a closed $\mathsf{RL_L}$-proof tree. Let us consider a non-closed $\mathsf{RL_L}$-proof tree for φ. This tree has to contain a branch b which ends with a non-axiomatic set Δ of $\mathsf{RL_L}$-formulas. By the construction of the tree, each element of Δ is of the form $J_t(P)$, for some $P \in \mathbb{RV}_{\mathsf{RL_L}}$ and

$t \in \{0, \ldots, n_{\mathsf{L}} - 1\}$, since otherwise we could apply to Δ one of the rules. Define the branch structure $\mathcal{M}^b = (U^b \cup \{\emptyset\}, m^b, \{m_k^b : k \in \mathcal{SR}_{\mathsf{L}}\})$ as follows:

- $U^b = \{w\}$;
- For any relational variable P,

$$m_k^b(P) = \begin{cases} \{w\} & \text{if } k = \min\{l : J_l(P) \notin \Delta\}, \\ \{\emptyset\} & \text{otherwise;} \end{cases}$$

- m_k^b extends to all compound relational terms as in RL_{L}-models.

Since Δ is not an axiomatic set, $\{l : J_l(P) \notin \Delta\} \neq \emptyset$. Therefore, for every $P \in \mathbb{RV}_{\mathsf{RL}_{\mathsf{L}}}$, there exists exactly one $i \in \{0, \ldots, n_{\mathsf{L}} - 1\}$ such that $m_i(P) = \{w\}$. Hence, it can be easily proved that $\mathcal{M}^b$ is an RL_{L}-model. Moreover, if $J_l(P) \in \Delta$ for some $l \in \{0, \ldots, n_{\mathsf{L}} - 1\}$, then by the definition of the branch structure $m_l(P) = \{\emptyset\}$. Therefore, if $J_l(P) \in \Delta$, then $\bigcup_{k=s}^{n_{\mathsf{L}}-1} m_k(J_l(P)) = \{\emptyset\}$, so $\mathcal{M}^b \not\models J_l(P)$. Hence, Δ is not RL_{L}-valid. Since every node of the branch b is obtained from its predecessor node by means of some RL_{L}-rule, and the rule preserves and reflects validity, and since Δ is not RL_{L}-valid, the formula $\varphi \in b$ is not RL_{L}-valid, a contradiction. □

Finally, by the above theorem and Theorem 10.3.1, we obtain:

Theorem 10.4.2 (Relational Soundness and Completeness of L). *Let φ be an L-formula. Then, the following conditions are equivalent:*

1. *φ is L-valid;*
2. *$\tau(\varphi)$ is RL_{L}-provable.*

Dual Tableaux for Rosser–Turquette Logics

$\mathsf{RL}_{\mathsf{RT}}$-dual tableau consists of the rules (J-in), $(J_0(J_l))$, $(J_{n_{\mathsf{RT}}-1}(J_l))$, $(J_k(J_l))$, and the rules of introduction and elimination of disjunction, conjunction, and negation:

$$(\vee\text{-}in)_1 \quad \frac{J_t(P \vee Q)}{J_t(P \vee Q), J_t(P) \mid J_t(P \vee Q), J_l(Q)}$$
for $0 \leq l \leq t$

$$(\vee\text{-}in)_2 \quad \frac{J_t(P \vee Q)}{J_t(P \vee Q), J_l(P) \mid J_t(P \vee Q), J_t(Q)}$$
for $0 \leq l \leq t$

$$(\vee\text{-}el) \quad \frac{K, J_t(P \vee Q)}{K}$$
$K = J_0(P), J_0(Q), \ldots, J_t(P), J_t(Q)$

$$(\wedge\text{-}in)_1 \quad \frac{J_t(P \wedge Q)}{J_t(P \wedge Q), J_t(P) \mid J_t(P \wedge Q), J_l(Q)}$$
for $t \leq l \leq n_{\mathsf{RT}} - 1$

$(\wedge\text{--}in)_2$ $$\frac{J_t(P \wedge Q)}{J_t(P \wedge Q), J_l(P) \mid J_t(P \wedge Q), J_t(Q)}$$

for $t \leq l \leq n_{\mathsf{RT}} - 1$

$(\wedge\text{--}el)$ $$\frac{K, J_t(P \wedge Q)}{K}$$

$K = J_t(P), J_t(Q), \ldots, J_{n_{\mathsf{RT}}-1}(P), J_{n_{\mathsf{RT}}-1}(Q)$

$(\neg\text{--}in)$ $$\frac{J_t(\neg R)}{J_i(R), \ldots, J_{n_{\mathsf{RT}}-1-t}(R)}$$

for $t = 0, n_{\mathsf{RT}} - 1$, where $i = \begin{cases} s, & \text{if } t = 0 \\ s-1, & \text{if } t = n_{\mathsf{RT}} - 1 \end{cases}$

$(\neg\text{--}el)$ $$\frac{K, J_t(\neg R)}{K}$$

$$K = \begin{cases} \text{any set of } \mathsf{RL_{RT}}\text{-formulas}, & \text{if } t \neq 0, n_{\mathsf{RT}} - 1 \\ J_0(R), \ldots, J_{s-1}(R), & \text{if } t = n_{\mathsf{RT}} - 1 \\ J_s(R), \ldots, J_{n_{\mathsf{RT}}-1}(R), & \text{if } t = 0 \end{cases}$$

Dual Tableaux for Symmetric Heyting Logics SH_n

We present $\mathsf{RL}_{\mathsf{SH}_n}$-rules for the operations σ_i, $i = 1, \ldots, n-1$:

For any sequence K of $\mathsf{RL}_{\mathsf{SH}_n}$-formulas,

$(J_{(k,l)}\sigma_i\text{--}el)_1$ $$\frac{K, J_{(k,l)}(\sigma_i(R))}{K}, \quad \text{if either } k \neq 0, 1 \text{ or } l \neq 0, 1$$

$(J_{(k,l)}\sigma_i\text{--}in)_1$ $$\frac{K, J_{(1,1)}(\sigma_i(R))}{K, J_{(1,1)}(\sigma_i(R)), J_{(\frac{x}{n-1}, \frac{y}{n-1})}(R)}$$

for $x \geq n-i$, $y \geq n-i$, if $J_{(\frac{x}{n-1}, \frac{y}{n-1})}(R) \notin K$

$(J_{(k,l)}\sigma_i\text{--}in)_2$ $$\frac{K, J_{(1,0)}(\sigma_i(R))}{K, J_{(1,0)}(\sigma_i(R)), J_{(\frac{x}{n-1}, \frac{y}{n-1})}(R)}$$

for $x \geq n-i$, $y < n-i$, if $J_{(\frac{x}{n-1}, \frac{y}{n-1})}(R) \notin K$

$(J_{(k,l)}\sigma_i\text{--}in)_3$ $$\frac{K, J_{(0,1)}(\sigma_i(R))}{K, J_{(0,1)}(\sigma_i(R)), J_{(\frac{x}{n-1}, \frac{y}{n-1})}(R)}$$

for $x < n-i$, $y \geq n-i$, if $J_{(\frac{x}{n-1}, \frac{y}{n-1})}(R) \notin K$

$(J_{(k,l)}\sigma_i\text{–}in)_4 \quad \dfrac{K, J_{(0,0)}(\sigma_i(R))}{K, J_{(0,0)}(\sigma_i(R)), J_{(\frac{x}{n-1},\frac{y}{n-1})}(R)}$

for $x < n - i$, $y < n - i$, if $J_{(\frac{x}{n-1},\frac{y}{n-1})}(R) \notin K$

$(J_{(k,l)}\sigma_i\text{–}el)_2 \quad \dfrac{K, J_{(k,l)}(\sigma_i(R))}{K}$

for $(k, l) \in \{0, 1\}^2$, if for any x, y, $J_{(\frac{x}{n-1},\frac{y}{n-1})}(R) \in K$

Dual Tableaux for Logics $\mathsf{L_T}$

We present $\mathsf{RL}_{\mathsf{L_T}}$-rules for the operations d_t:

For any sequence K of $\mathsf{RL}_{\mathsf{L_T}}$-formulas,

$(J_l(d_t)\text{–}el)_1 \quad \dfrac{K, J_l(d_t(R))}{K}$, for $l \neq \emptyset, \mathsf{T}$

$(J_l(d_t)\text{–}in)_1 \quad \dfrac{K, J_{\mathsf{T}}(d_t(R))}{K, J_{\mathsf{T}}(d_t(R)), J_l(R)}$

for $l \in \mathcal{SR}_{\mathsf{L_T}}$ such that $t \in l$, if $J_l(R) \notin K$

$(J_l(d_t)\text{–}in)_2 \quad \dfrac{K, J_{\emptyset}(d_t(R))}{K, J_{\emptyset}(d_t(R)), J_l(R)}$

for $l \in \mathcal{SR}_{\mathsf{L_T}}$ such that $t \notin l$, if $J_l(R) \notin K$

$(J_l(d_t)\text{–}el)_2 \quad \dfrac{K, J_k(d_t(R))}{K}$

for $k \in \{\emptyset, \mathsf{T}\}$, if $J_l(R) \in K$ for every $l \in \mathcal{SR}_{\mathsf{L_T}}$

10.5 Three-Valued Logics

In this section we present examples of dual tableau proofs in three-valued instances of Rosser–Turquette logic, symmetric Heyting logic, and logic $\mathsf{L_T}$.

Consider a three-valued Rosser–Turquette logic $\mathsf{RT}_{(3,1)}$ with $\mathcal{SR}_{\mathsf{RT}_{(3,1)}} = \{0, 1, 2\}$ where 1 and 2 are the designated values. Let φ be the following $\mathsf{RT}_{(3,1)}$-formula:

$$\varphi = \neg J_1(\neg p).$$

Its translations into $\mathsf{RL}_{\mathsf{RT}_{(3,1)}}$-term is:

$$\tau(\varphi) = \neg J_1(\neg P),$$

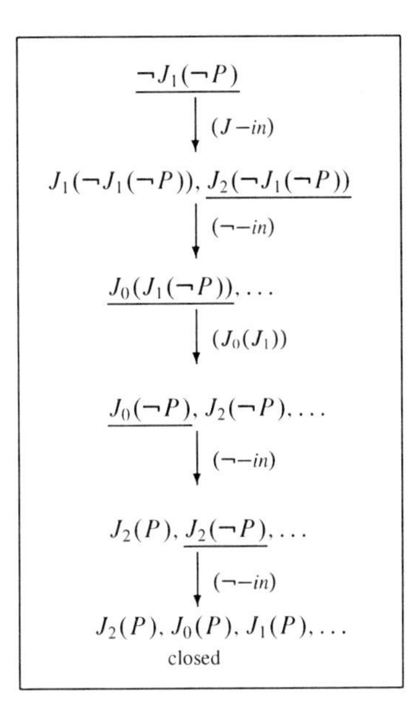

Fig. 10.1 An $\mathsf{RL}_{\mathsf{RT}_{(3,1)}}$-proof of $\mathsf{RT}_{(3,1)}$-formula $\neg J_1(\neg p)$

where $\tau'(p) = P$. $\mathsf{RT}_{(3,1)}$-validity of this formula is equivalent with $\mathsf{RL}_{\mathsf{RT}_{(3,1)}}$-provability of its translation. Figure 10.1 presents an $\mathsf{RL}_{\mathsf{RT}_{(3,1)}}$-proof of $\tau(\varphi)$.

Now, consider a three-valued Rosser–Turquette logic $\mathsf{RT}_{(3,2)}$ with $\mathcal{SR}_{\mathsf{RT}_{(3,2)}} = \{0, 1, 2\}$, where 2 is the only designated value. Let ψ be the following $\mathsf{RT}_{(3,2)}$-formula:

$$\psi = J_0(p) \vee J_1(p) \vee J_2(p).$$

Its translation into $\mathsf{RL}_{\mathsf{RT}_{(3,2)}}$-term is:

$$\tau(\psi) = J_0(P) \vee J_1(P) \vee J_2(P),$$

where $\tau'(p) = P$. Figure 10.2 presents an $\mathsf{RL}_{\mathsf{RT}_{(3,2)}}$-proof of $\tau(\psi)$ which shows $\mathsf{RT}_{(3,2)}$-validity of the formula ψ.

Now, consider symmetric Heyting logic of order 3, SH_3, with $(1, 1)$ as the only designated value. The semantic range for the logic SH_3 is $\mathcal{SR}_{\mathsf{SH}_3} = \{(x, y) : x, y \in \{0, \frac{1}{2}, 1\}\}$. Among the rules of $\mathsf{RL}_{\mathsf{SH}_3}$-dual tableau are the rules of the form:

$$(\rightarrow)_{\mathsf{SH}_3} \quad \frac{J_{(1,1)}(R \rightarrow Q)}{J_{(t_1,t_2)}(R), J_{(1,1)}(R \rightarrow Q) \mid J_{(t_1',t_2')}(Q), J_{(1,1)}(R \rightarrow Q)}$$

for any $t_1, t_2, t_1', t_2' \in \mathcal{SR}_{\mathsf{SH}_3}$ such that $t_1 \leq t_1'$ and $t_2 \leq t_2'$

$$(\sim)_{\mathsf{SH}_3} \quad \frac{J_{(k,l)}(\sim R)}{J_{(1-l,1-k)}(R)} \quad \text{for any } k, l \in \mathcal{SR}_{\mathsf{SH}_3}$$

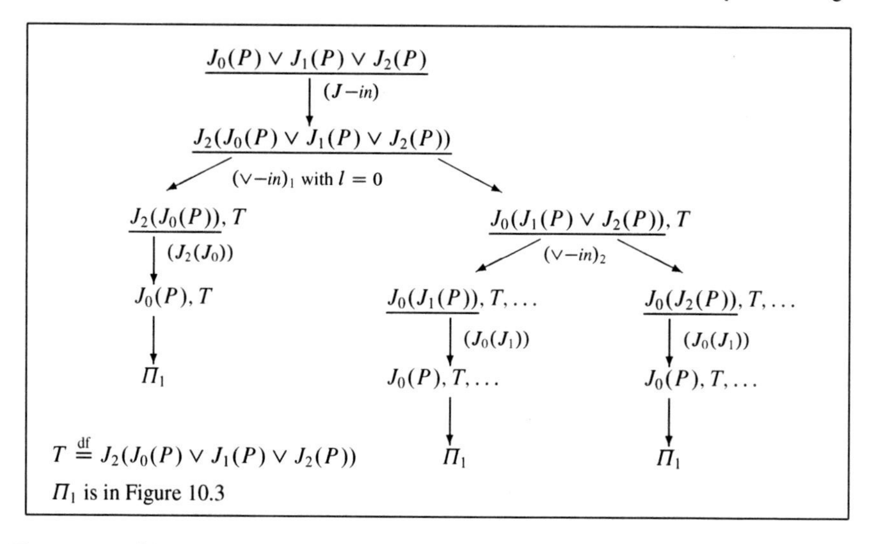

Fig. 10.2 An $\mathsf{RL}_{\mathsf{RT}_{(3,2)}}$-proof of $\mathsf{RT}_{(3,2)}$-formula $J_0(p) \vee J_1(p) \vee J_2(p)$.

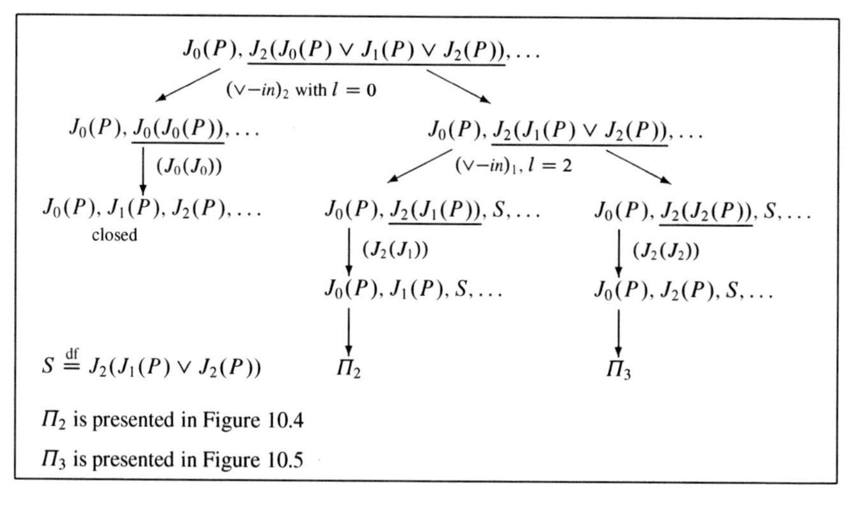

Fig. 10.3 A subtree Π_1

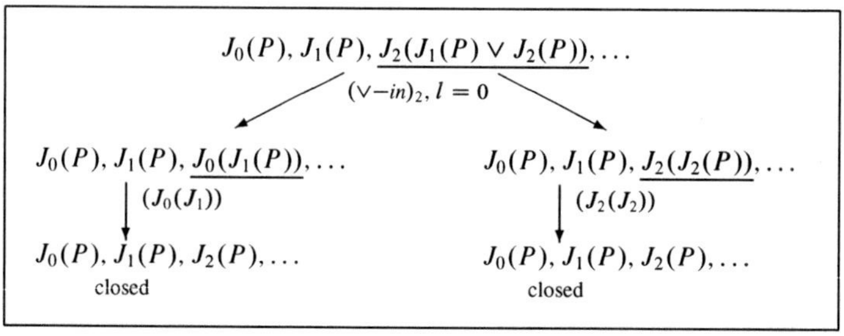

$J_0(P), J_1(P), J_2(J_1(P) \vee J_2(P)), \ldots$

$(\vee-in)_2, l = 0$

$J_0(P), J_1(P), J_0(J_1(P)), \ldots$ — $(J_0(J_1))$ — $J_0(P), J_1(P), J_2(P), \ldots$ closed

$J_0(P), J_1(P), J_2(J_2(P)), \ldots$ — $(J_2(J_2))$ — $J_0(P), J_1(P), J_2(P), \ldots$ closed

Fig. 10.4 A subtree Π_2

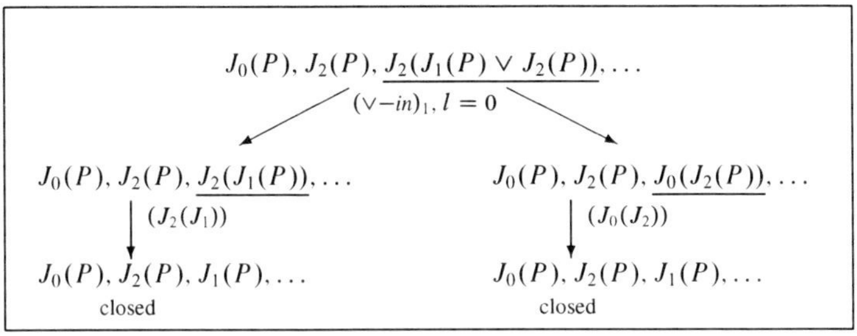

$J_0(P), J_2(P), \underline{J_2(J_1(P) \vee J_2(P))}, \ldots$

$(\vee{-}in)_1, l = 0$

$J_0(P), J_2(P), \underline{J_2(J_1(P))}, \ldots$ — $(J_2(J_1))$ — $J_0(P), J_2(P), J_1(P), \ldots$ closed

$J_0(P), J_2(P), \underline{J_0(J_2(P))}, \ldots$ — $(J_0(J_2))$ — $J_0(P), J_2(P), J_1(P), \ldots$ closed

Fig. 10.5 A subtree Π_3

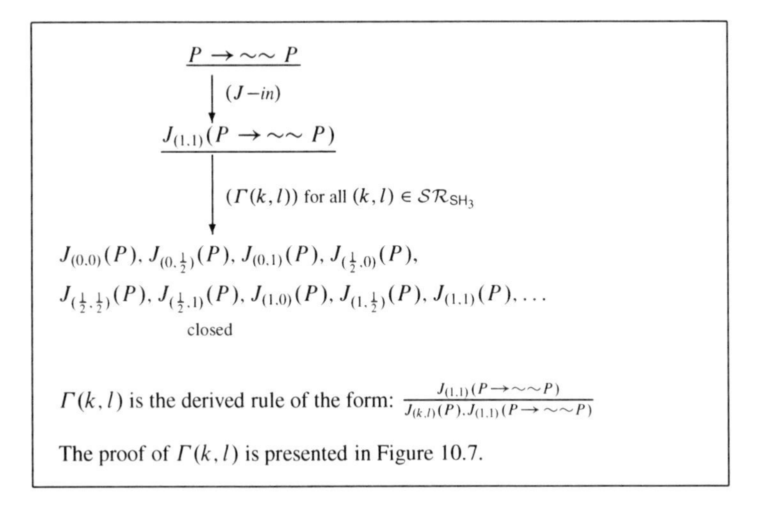

Fig. 10.6 An $\mathsf{RL}_{\mathsf{SH}_3}$-proof of SH_3-formula $p \rightarrow \sim\sim p$

These rules are the instances of the scheme $(o_j)_1$ presented in Section 10.4. Note that the result of an application of the rule $(\sim)_{\mathsf{SH}_3}$ to the formula $J_{(k,l)}(\sim\sim R)$ is the formula $J_{(1-l,1-k)}(\sim R)$, while the result of an application of the rule $(\sim)_{\mathsf{SH}_3}$ to $J_{(1-l,1-k)}(\sim R)$ is the formula $J_{(k,l)}(R)$. Thus, we can introduce the derived rule of the form:

$$(\sim\sim)_{\mathsf{SH}_3} \qquad \frac{J_{(k,l)}(\sim\sim R)}{J_{(k,l)}(R)} \qquad \text{for any } k, l \in \mathcal{SR}_{\mathsf{SH}_3}$$

Let φ be an SH_3-formula $p \rightarrow \sim\sim p$. The translation of φ into $\mathsf{RL}_{\mathsf{SH}_3}$-term is $\tau(\varphi) = P \rightarrow \sim\sim P$, where $\tau'(p) = P$. Figure 10.6 presents an $\mathsf{RL}_{\mathsf{SH}_3}$-proof of $\tau(\varphi)$ that shows SH_3-validity of φ.

Observe that in a diagram of Fig. 10.7 an application of a derived rule $(\sim\sim)_{\mathsf{SH}_3}$ to $J_{(k,l)}(\sim\sim P)$ results in the node which has the same formulas as those obtained by an application of rule $(\rightarrow)_{\mathsf{SH}_3}$ to formula $J_{(1,1)}(P \rightarrow \sim\sim P)$. Therefore, we identify the two nodes.

As the last example, consider logic L_{T} with $\mathsf{T} = \{0, 1\}$. The semantic range for this logic is $\mathcal{SR}_{\mathsf{L}_{\mathsf{T}}} = \{\emptyset, \{1\}, \{0, 1\}\}$.

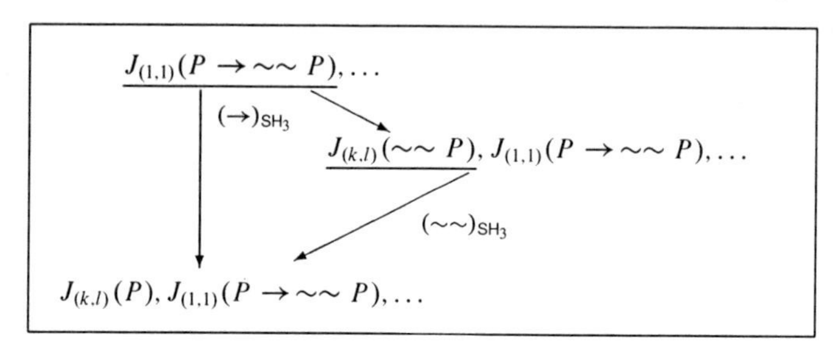

Fig. 10.7 An $\mathsf{RL}_{\mathsf{SH}_3}$-proof of a derived rule $\Gamma(k,l) = \frac{J_{(1,1)}(P \rightarrow \sim\sim P)}{J_{(k,l)}(P), J_{(1,1)}(P \rightarrow \sim\sim P)}$.

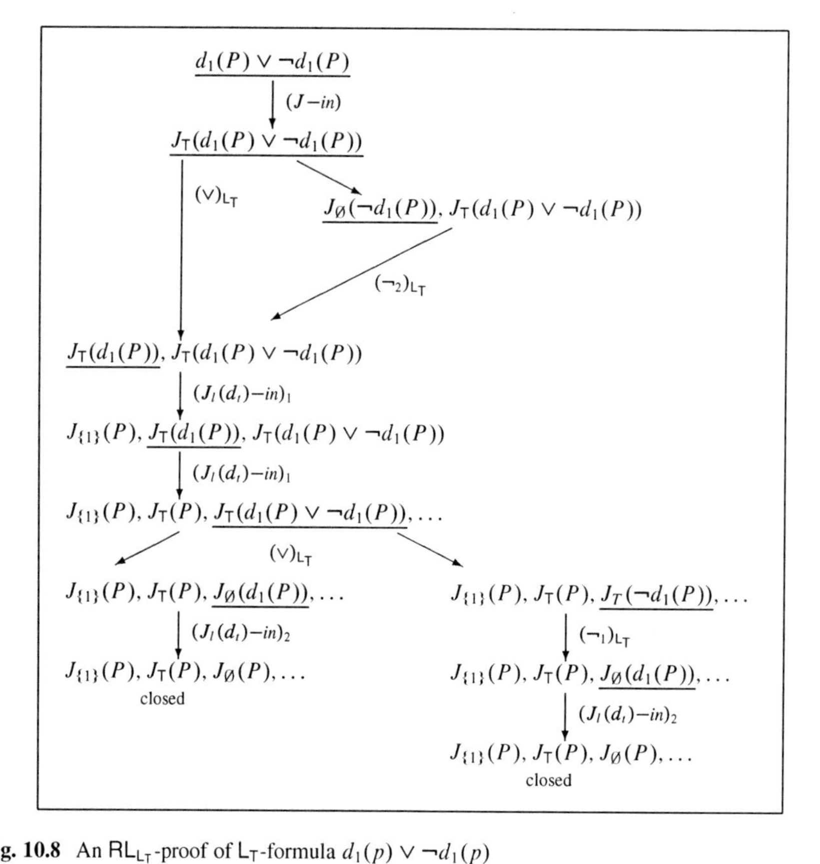

Fig. 10.8 An $\mathsf{RL}_{\mathsf{L_T}}$-proof of $\mathsf{L_T}$-formula $d_1(p) \vee \neg d_1(p)$

Among the rules of $\mathsf{RL}_{\mathsf{L_T}}$-dual tableau are the rules of the form:

$(\vee)_{\mathsf{L_T}}$ $$\frac{J_t(R \vee Q)}{J_{t_1}(R), J_t(R \vee Q) \mid J_{t_2}(Q), J_t(R \vee Q)}$$

for any $t, t_1, t_2 \in \mathcal{SR}_{\mathsf{L_T}}$ such that $t_1 \cup t_2 = t$

$(\neg_1)_{\mathsf{L_T}}$ $\frac{J_{\mathsf{T}}(\neg R)}{J_{\emptyset}(R)}$ $\quad$ $(\neg_2)_{\mathsf{L_T}}$ $\frac{J_{\emptyset}(\neg R)}{J_{\mathsf{T}}(R)}$

These rules follow the scheme $(o_j)_1$ presented in Sect. 10.4.

Let φ be the following $\mathsf{L_T}$-formula:

$$\varphi = d_1(p) \vee \neg d_1(p).$$

The translation of φ into $\mathsf{RL_{L_T}}$-term is:

$$\tau(\varphi) = d_1(P) \vee \neg d_1(P),$$

where $\tau'(p) = P$. Figure 10.8 presents $\mathsf{RL_{L_T}}$-proof of $\tau(\varphi)$, which proves $\mathsf{L_T}$-validity of φ.

Part IV
Relational Reasoning in Logics of Information and Data Analysis

Chapter 11
Dual Tableaux for Information Logics of Plain Frames

11.1 Introduction

Information logics considered here and in the following chapter originated in connection with representation and analysis of data structures known as information systems with incomplete information, introduced in [Lip76], see also [Lip79]. Any such system consists of a collection of objects described in terms of their properties. A property is specified as a pair 'an attribute, a subset of values of this attribute'. Such a form of properties is a manifestation of incompleteness of information. Instead of a single value of an attribute assigned to an object, as is the case in relational database model, here we have a range of values. A disjunctive interpretation of a set of values admits an interpretation that a value is not sufficiently specified, it is only estimated to be in some range. Clearly, also a conjunctive interpretation may be meaningful for some objects. A characterization of objects in terms of attributes and their values induces some relationships among the objects. Typically, they have a form of binary relations and are referred to as information relations derived from an information system. Several families of information relations have been studied in the literature, an extensive catalogue can be found in [DO02]. There are two major classes of these relations: indistinguishability relations and distinguishability relations. The indistinguishability relations reflect degrees of similarity or sameness of objects and distinguishability relations correspond to degrees of dissimilarity or distinctness. Each of these relations is defined in terms of a subset of the attributes of the objects. Thus a distinguishing feature of the relations is that they indicate both which objects are related and also with respect to which of their attributes they are related. In this way the information relations capture a qualitative degree of having a property. Relations of that kind are referred to as relative relations.

Information logics are modal logics where the modal operators are determined by information relations. Depending on a type of the relation, the operators receive various specific, application oriented interpretations. For example, if we consider an indiscernibility relation determined by an attribute i.e., it holds between two objects whenever their sets of values of this attribute are equal, then the necessity

E. Orłowska and J. Golińska-Pilarek, *Dual Tableaux: Foundations, Methodology, Case Studies*, Trends in Logic 36, DOI 10.1007/978-94-007-0005-5_11,

and possibility operators determined by that relation are the lower and the upper approximation operations, respectively, as considered in rough set theory (see [Paw82, Paw91, DO02]).

In this chapter, first, we recall the fundamental notions concerning information systems with incomplete information, information relations, and operators determined by these relations. Next, we develop relational dual tableaux for some typical information logics with modal operators determined by the information relations both from the group of indistinguishability relations and from the group of distinguishability relations. The models of the logics considered in this chapter are based on what are called plain frames where each of the information relations is determined by the whole set of attributes of an information system. In the following chapter we deal with logics of relative frames, consisting of the families of relations determined by all the finite subsets of the set of attributes of an information system.

11.2 Information Systems

In a formal model of an information system with incomplete information, as introduced by Lipski, information systems are collections of information items that have the form of descriptions of some objects in terms of their properties. An *information system* is a structure of the form $S = (\mathbb{OB}, \mathbb{AT}, (\mathbb{VAL}_a)_{a \in \mathbb{AT}}, f)$ where:

- $\mathbb{OB}$ is a non-empty set of objects;
- $\mathbb{AT}$ is a non-empty set of attributes;
- $\mathbb{VAL}_a$ is a non-empty set of values of the attribute a;
- f is a total function $\mathbb{OB} \times \mathbb{AT} \to \bigcup_{a \in \mathbb{AT}} \mathcal{P}(\mathbb{VAL}_a)$ such that for every $(x, a) \in \mathbb{OB} \times \mathbb{AT}$, $f(x, a) \subseteq \mathbb{VAL}_a$; f is referred to as *an information function*.

Usually, instead of $(\mathbb{OB}, \mathbb{AT}, (\mathbb{VAL}_a)_{a \in \mathbb{AT}}, f)$ the more concise notation, namely $(\mathbb{OB}, \mathbb{AT})$, is used. With that short notation, each attribute $a \in \mathbb{AT}$ is considered as a mapping $a \colon \mathbb{OB} \to \mathcal{P}(\mathbb{VAL}_a)$ that assigns subsets of values to objects. An information system $(\mathbb{OB}, \mathbb{AT})$ is *total* (resp. *deterministic*) whenever for every $a \in \mathbb{AT}$ and for every $x \in \mathbb{OB}$, $f(x, a) \neq \emptyset$ (resp. $card(f(x, a)) \leq 1$, in that case x is said to be a *deterministic object*). If an information system is not deterministic, then it is said to be *nondeterministic*. In nondeterministic information systems descriptions of objects are tuples consisting of subsets of values of attributes. Such a representation is also used in symbolic data analysis, see e.g., [Did87, Did88, Pre97], and in rough set-based data analysis, see e.g., [Orł97a, WDB98, WDG00].

Any set $a(x)$ can be viewed as a set of properties of an object x determined by attribute a. Any such property is referred to as *a-property*. Set $\mathbb{VAL}_a \setminus a(x)$ will be referred to as a set of *negative a-properties*. For example, if attribute a is 'colour' and $a(x)$ = green, then x possesses the property of 'being green'; if a is 'languages spoken', and if a person x speaks Polish (Pl), German (D), and French (F), then $a(x) = \{$Pl, D, F$\}$. If both the set of objects and the set of attributes are finite, then we regard such a system as a data table with rows labeled by objects, and columns labeled by attributes; the entry (x, a) contains the value set $a(x)$ of attribute a for object x.

Any information system $S = (\mathbb{OB}, \mathbb{AT})$ contains also some implicit information about relationships among its objects. These relationships are determined by the properties of objects. Usually, they have the form of binary relations. They are referred to as *information relations derived from an information system.* There are two groups of information relations. The relations that reflect various kinds of 'sameness' or 'similarity' of objects are referred to as *indistinguishability relations.* The relations that indicate 'differences' or 'dissimilarity' of objects are referred to as *distinguishability relations.* Below we present a list of the classes of atomic relations that generate a whole family of information relations.

Indistinguishability Relations

Let $S = (\mathbb{OB}, \mathbb{AT})$ be an information system. For every $A \subseteq \mathbb{AT}$ and for all $x, y \in \mathbb{OB}$ we consider the following binary indistinguishability relations on $\mathbb{OB}$:

- The *strong (weak) indiscernibility relation* $ind(A)$ (resp. $wind(A)$): $(x, y) \in ind(A)$ (resp. $(x, y) \in wind(A)$) iff for all (resp. for some) $a \in A$, $a(x) = a(y)$;
- The *strong (weak) similarity relation* $sim(A)$ (resp. $wsim(A)$): $(x, y) \in sim(A)$ (resp. $(x, y) \in wsim(A)$) iff for all (resp. for some) $a \in A$, $a(x) \cap a(y) \neq \emptyset$;
- The *strong (weak) forward inclusion relation* $fin(A)$ (resp. $wfin(A)$): $(x, y) \in fin(A)$ (resp. $(x, y) \in wfin(A)$) iff for all (resp. for some) $a \in A$, $a(x) \subseteq a(y)$;
- The *strong (weak) backward inclusion relation* $bin(A)$ (resp. $wbin(A)$): $(x, y) \in bin(A)$ (resp. $(x, y) \in wbin(A)$) iff for all (resp. for some) $a \in A$, $a(y) \subseteq a(x)$;
- The *strong (weak) negative similarity relation* $nim(A)$ (resp. $wnim(A)$): $(x, y) \in nim(A)$ (resp. $(x, y) \in wnim(A)$) iff for all (resp. for some) $a \in A$, $-a(x) \cap -a(y) \neq \emptyset$, where $-$ is the complement with respect to $\mathbb{VAL}_a$;
- The *strong (weak) incomplementarity relation* $icom(A)$ (resp. $wicom(A)$): $(x, y) \in icom(A)$ (resp. $(x, y) \in wicom(A)$) iff for all (resp. for some) $a \in A$, $a(x) \neq -a(y)$.

If A is a singleton set, then the respective strong and weak relations coincide. Intuitively, two objects are strongly A-indiscernible whenever all of their sets of a-properties determined by the attributes $a \in A$ are the same. Objects are weakly A-indiscernible whenever their properties determined by some members of A are the same. Objects are strongly A-similar (resp. weakly A-similar) whenever all (resp. some) of the sets of their properties determined by the attributes from A are not disjoint, in other words the objects share some properties. Strong (resp. weak) information inclusions hold between the objects whenever their all (resp. some) corresponding sets of properties are included in each other. Strong (resp. weak) negative similarity relation holds between objects whenever they share some negative properties with respect to all (resp. some) attributes. Strong (resp. weak) incomplementarity relation holds between objects whenever a-properties of one object do not coincide with negative a-properties of the other one, for all (resp. some) attributes.

Important applications of the information relations from the indiscernibility group are related to the representation of approximations of subsets of objects in information systems. If $R(A)$ is one of these relations, where A is a subset of $\mathbb{AT}$ and X is a subset of $\mathbb{OB}$, then the *lower* $R(A)$*-approximation of* X, $L_{R(A)}(X)$, and the *upper* $R(A)$*-approximation of* X, $U_{R(A)}(X)$, are defined as follows:

$$L_{R(A)}(X) = \{x \in \mathbb{OB} : \text{ for all } y \in \mathbb{OB}, (x, y) \in R(A) \text{ implies } y \in X\};$$
$$U_{R(A)}(X) = \{x \in \mathbb{OB} : \text{ there exists } y \in \mathbb{OB}, (x, y) \in R(A) \text{ and } y \in X\}.$$

In the rough set theory (see [Paw91]), where a relation $R(A)$ is a strong indiscernibility relation, we obtain the following hierarchy of definability of sets. A subset X of $\mathbb{OB}$ is said to be:

- *A-definable* iff $L_{ind(A)}(X) = X = U_{ind(A)}(X)$;
- *Roughly A-definable* iff $L_{ind(A)}(X) \neq \emptyset$ and $U_{ind(A)}(X) \neq \mathbb{OB}$;
- *Internally A-indefinable* iff $L_{ind(A)}(X) = \emptyset$;
- *Externally A-indefinable* iff $U_{ind(A)}X = \mathbb{OB}$;
- *Totally A-indefinable* iff it is internally and externally A-indefinable.

The other application of the above information relations is related to modeling uncertain knowledge acquired from information about objects collected in an information system. Let X be a subset of $\mathbb{OB}$. The sets of *A-positive* ($POS_A(X)$), *A-borderline* ($BOR_A(X)$), and *A-negative* ($NEG_A(X)$) *instances of* X are as follows:

$$POS_A(X) = L_{ind(A)}(X);$$
$$BOR_A(X) = U_{ind(A)}(X) - L_{ind(A)}(X);$$
$$NEG_A(X) = \mathbb{OB} - U_{ind(A)}(X).$$

The elements of $POS_A(X)$ can be seen as the members of X up to properties from A. The elements of $NEG_A(X)$ are not the members of X up to properties from A.

Knowledge about a set X of objects that can be discovered from an information system can be modelled as $K_A(X) = POS_A(X) \cup NEG_A(X)$. Intuitively, A-knowledge about X consists of those objects that are either A-positive instances of X or A-negative instances of X.

We say that A-knowledge about X is:

- *Complete* if $K_A(X) = \mathbb{OB}$, otherwise *incomplete*;
- *Rough* if $POS_A(X)$, $BOR_A(X)$, and $NEG_A(X)$ are non-empty;
- *Pos-empty* if $POS_A(X) = \emptyset$;
- *Neg-empty* if $NEG_A(X) = \emptyset$;
- *Empty* if it is pos-empty and neg-empty.

Distinguishability Relations

Let $S = (\mathbb{OB}, \mathbb{AT})$ be an information system. For every $A \subseteq \mathbb{AT}$ and for all $x, y \in \mathbb{OB}$ we consider the following binary distinguishability relations on $\mathbb{OB}$:

- The *strong (weak) diversity relation* $div(A)$ (resp. $wdiv(A)$): $(x, y) \in div(A)$ (resp. $(x, y) \in wdiv(A)$) iff for all (resp. for some) $a \in A$, $a(x) \neq a(y)$;
- The *strong (weak) right orthogonality relation* $rort(A)$ (resp. $wrort(A)$): $(x, y) \in rort(A)$ (resp. $(x, y) \in wrort(A)$) iff for all (resp. for some) $a \in A$, $a(x) \subseteq -a(y)$;
- The *strong (weak) left orthogonality relation* $lort(A)$ (resp. $wlort(A)$): $(x, y) \in lort(A)$ (resp. $(x, y) \in wlort(A)$) iff for all (resp. for some) $a \in A$, $-a(x) \subseteq a(y)$;
- The *strong (weak) right negative similarity relation* $rnim(A)$ (respectively $wrnim(A)$): $(x, y) \in rnim(A)$ (resp. $(x, y) \in wrnim(A)$) iff for all (resp. for some) $a \in A$, $a(x) \cap -a(y) \neq \emptyset$;
- The *strong (weak) left negative similarity relation* $lnim(A)$ (respectively $wlnim(A)$): $(x, y) \in lnim(A)$ (resp. $(x, y) \in wlnim(A)$) iff for all (resp. for some) $a \in A$, $-a(x) \cap a(y) \neq \emptyset$;
- The *strong (weak) complementarity relation* $com(A)$ (resp. $wcom(A)$): $(x, y) \in com(A)$ (resp. $(x, y) \in wcom(A)$) iff for all (resp. for some) $a \in A$, $a(x) = -a(y)$.

Intuitively, objects are strongly A-diverse (resp. weakly A-diverse) if all (resp. some) of the sets of their properties determined by members of A are different. The objects are strongly A-right orthogonal (resp. weakly A-right orthogonal) whenever all (resp. some) of the sets of their properties determined by attributes from A are disjoint. The objects are strongly A-left orthogonal (resp. weakly A-left orthogonal) whenever all (resp. some) of their a-properties, for $a \in A$, are exhaustive i.e., $a(x) \cup a(y) = \mathbb{VAL}_a$. Two objects are right or left strongly (resp. weakly) A-negatively similar whenever some properties of one of them are not the properties of the other, for all (resp. some) attributes from A. The objects are strongly (resp. weakly) A-complementary whenever their respective sets of properties are complements of each other, for all (resp. some) attributes from A.

Distinguishability relations can be applied to a non-numerical modelling of degrees of dissimilarity. Diversity relations are applied, among others, in the algorithms for finding cores of sets of attributes. Let an information system $(\mathbb{OB}, \mathbb{AT})$ be given and let A be a subset of $\mathbb{AT}$. We say that an attribute $a \in A$ is *indispensable* in A whenever $ind(A) \neq ind(A-\{a\})$, that is there are some objects such that a is the only attribute from A that can distinguish between them. A *reduct* of A is a minimal subset A' of A such that every $a \in A'$ is indispensable in A' and $ind(A') = ind(A)$. The *core* of A is defined as $CORE(A) = \bigcap\{A' \subseteq \mathbb{AT} : A' \text{ is a reduct of } A\}$. For any pair x, y of objects we define the *discernibility set* $D_{xy} = \{a \in \mathbb{AT} : (x, y) \in div(\{a\})\}$. It is proved in [SR92] that $CORE(A) = \{a \in A : \text{there are } x, y \in \mathbb{OB} \text{ such that } D_{xy} = \{a\}\}$.

In the proposition below some of the properties satisfied by information relations derived from an information system are listed. We recall that a binary relation R on a set U is:

- A weakly reflexive relation whenever for all $x, y \in U$, $(x, y) \in R$ implies $(x, x) \in R$;
- A tolerance relation whenever it is reflexive and symmetric;
- A 3-transitive relation whenever for all $x, y, z, t \in U$, if $(x, z) \in R$, $(z, t) \in R$, and $(t, y) \in R$, then $(x, y) \in R$.

For any property α of R by property co-α of R we mean that $-R$ has the property α.

Proposition 11.2.1. *For every information system $S = (\mathbb{OB}, \mathbb{AT})$, for every $A \subseteq \mathbb{AT}$, the following hold:*

1. *$ind(A)$ is an equivalence relation;*
2. *$sim(A)$ and $nim(A)$ are weakly reflexive and symmetric;*
3. *if S is total, then $sim(A)$ is a tolerance relation;*
4. *$fin(A)$ and $bin(A)$ are reflexive and transitive;*
5. *$icom(A)$ is symmetric and if $A \neq \emptyset$, then $icom(A)$ is reflexive; for every $a \in \mathbb{AT}$, $icom(a)$ is co-3-transitive;*
6. *$wind(A)$ is a tolerance relation and for every $a \in \mathbb{AT}$, $wind(a)$ is transitive;*
7. *$wsim(A)$ is a tolerance relation;*
8. *$wnim(A)$ is weakly reflexive and symmetric;*
9. *$wicom(A)$ is reflexive, symmetric and co-3-transitive;*
10. *$wfin(A)$ and $wbin(A)$ are reflexive; for every $a \in \mathbb{AT}$, $wfin(a)$ and $wbin(a)$ are transitive;*
11. *$div(A)$ is symmetric; if $A \neq \emptyset$, then $div(A)$ is irreflexive; for every $a \in \mathbb{AT}$, $div(a)$ is co-transitive;*
12. *$rort(A)$ is symmetric; if $A \neq \emptyset$, then $rort(A)$ is irreflexive;*
13. *$lort(A)$ is co-weakly reflexive and symmetric;*
14. *$com(A)$ is symmetric and 3-transitive; if $A \neq \emptyset$, then $com(A)$ is irreflexive;*
15. *$rnim(A)$ and $lnim(A)$ are irreflexive, for every $A \neq \emptyset$; for every $a \in \mathbb{AT}$, $rnim(a)$ and $lnim(a)$ are co-transitive;*
16. *$wdiv(A)$ is irreflexive, symmetric and co-transitive;*
17. *$wrort(A)$ is symmetric; if S is total, then $wrort(A)$ is irreflexive;*
18. *$wlort(A)$ is co-weakly reflexive and symmetric;*
19. *$wcom(A)$ is irreflexive and symmetric; for every $a \in \mathbb{AT}$, $wcom(a)$ is 3-transitive;*
20. *$wrnim(A)$ and $wlnim(A)$ are irreflexive and co-transitive.*

The following proposition states some relationships between information relations of different kinds:

Proposition 11.2.2. *For every information system $S = (\mathbb{OB}, \mathbb{AT})$, for every $A \subseteq \mathbb{AT}$, and for all $x, y, z \in \mathbb{OB}$, the following hold:*

1. *$(x, y) \in sim(A)$ and $(x, z) \in fin(A)$ imply $(z, y) \in sim(A)$;*

2. $(x, y) \in ind(A)$ *implies* $(x, y) \in fin(A)$;
3. $(x, y) \in fin(A)$ *and* $(y, x) \in fin(A)$ *imply* $(x, y) \in ind(A)$.

Observe that definitions of information relations include both an information on which objects are related and with respect to which attributes they are related. Relations of that kind are referred to as *relative relations*, they are relative to a subset of attributes. It follows that the formal systems for reasoning about these relations should refer to the structures with relative relations. For that purpose we define a class of *relative frames* which have the form:

$$(U, (R_P^1)_{P \in \mathcal{P}_{fin}(\mathrm{Par})}, \ldots, (R_P^n)_{P \in \mathcal{P}_{fin}(\mathrm{Par})}),$$

where the relations in each family $(R_P^i)_{P \in \mathcal{P}_{fin}(\mathrm{Par})}$ are indexed with finite subsets of a non-empty set Par of parameters. Intuitively, the elements of Par are representations of the attributes of an information system. A *plain frame* is the frame where all the relations are understood as being determined by the whole set of attributes of an information system.

A modal approach to reasoning about incomplete information resulted in various modal systems which are now called information logics. The first logics of that family are defined in [Orł82] published later as [Orł83] and in [Orł84, OP84, Orł85a]. We refer the reader to [DO02] for a comprehensive survey of information logics and to [DG00a, DG00b, DS02b, Bal02, DO07] for some more recent developments. In information logics the elements of the universes of the models are thought of as objects in an information system. This interpretation is quite different from the usual interpretation postulated in modal logics, where the elements of a model represent states (or possible worlds) in which formulas may be true or false.

11.3 Information Logics NIL and IL

The languages of most popular information logics with semantics of plain frames are multimodal languages whose symbols are included in the following pairwise disjoint sets:

- $\mathbb{V}$ – a set of propositional variables possibly including also a propositional constant D interpreted as a set of deterministic objects;
- $\{\leq, \geq, \sigma, \equiv\}$ – a set of relational constants, where $\leq, \geq, \sigma, \equiv$ are the abstract counterparts to the relations of inclusions, similarity, and indiscernibility derived from an information system, respectively;
- $\{\neg, \vee, \wedge, [\leq], [\geq], [\sigma], [\equiv]\}$ – a set of propositional operations.

The set of formulas of a given logic L based on such a language is defined as usual in modal logics (see Sect. 7.3). An L-*frame* is a modal frame of the form $\mathcal{F} = (U, Rel)$, where $Rel \subseteq \{\leq, \geq, \sigma, \equiv\}$. As usual, relations in a frame are denoted with the same symbols as the corresponding relational constants in the language. The relations from Rel are referred to as the accessibility relations. In various information logics the frames satisfy some postulates. Typical conditions on relations in the frames of logics associated with information systems with incomplete information are among the following:

(I1) $\leq = \geq^{-1}$;
(I2) $\leq$ is reflexive and transitive;
(I3) σ is weakly reflexive and symmetric;
(I4) σ is reflexive and symmetric;
(I5) $\equiv$ is an equivalence relation;

For all $x, x', y, y' \in U$,

(I6) If $(x, y) \in \sigma$, $(x, x') \in \leq$ and $(y, y') \in \leq$, then $(x', y') \in \sigma$;
(I7) If $(x, y) \in \sigma$ and $(x, z) \in \leq$, then $(z, y) \in \sigma$;
(I8) If $y \in D$ and $(x, y) \in \leq$, then $x \in D$;
(I9) If $x \in D$ and $(x, y) \in \sigma$, then $(x, y) \in \leq$;
(I10) If $(x, y) \in \equiv$, then $(x, y) \in \leq$;
(I11) If $x, y \in D$ and $(x, y) \in \sigma$, then $(x, y) \in \equiv$;
(I12) If $(x, y) \in \leq$ and $(y, x) \in \leq$, then $(x, y) \in \equiv$;
(I13) If $x \notin D$, then there is $y \in U$ such that $(x, y) \notin \leq$.

We consider two information logics with semantics of plain frames: the logic NIL introduced in [OP84, Vak87] and the logic IL introduced in [Vak89].

The set of relational constants of NIL is $\{\leq, \geq, \sigma\}$. $\mathbb{V}$ consists of propositional variables. The set of propositional operations is $\{\neg, \vee, \wedge, [\leq], [\geq], [\sigma]\}$. The set of NIL-formulas is defined as described in Sect. 7.3.

The NIL-*models* are the structures $(U, \leq, \geq, \sigma, m)$ satisfying the conditions of the definition of models from Sect. 7.3 and conditions (I1), (I2), (I4), and (I6), for all $x, y, z \in U$.

The satisfaction of NIL-formulas by states in a model is defined as in Sect. 7.3, in particular, for formulas built with modal operations we have:

For $T \in \{\leq, \geq, \sigma\}$,

$$\mathcal{M}, s \models [T]\varphi \text{ iff for all } s' \in U, (s, s') \in T \text{ implies } \mathcal{M}, s' \models \varphi.$$

In [Dem00] the following theorem is proved:

Theorem 11.3.1.

1. *The logic* NIL *is decidable;*
2. NIL-*satisfiability is* PSPACE-*complete.*

Moreover, in [Vak87] the following is proved:

Theorem 11.3.2 (Informational representability of NIL). *For every* NIL-*model* $(U, \leq, \geq, \sigma, m)$*, there is a total information system* $\mathcal{S}$ *such that the relations of forward inclusion, backward inclusion, and similarity derived from* $\mathcal{S}$ *coincide with* $\leq, \geq$*, and* σ*, respectively.*

The logic IL is intended to be a tool for reasoning about indiscernibility, similarity, and forward inclusion, and about relationships between them. The set of relational constants of IL is $\{\equiv, \leq, \sigma\}$. $\mathbb{V}$ is a countably infinite set including propositional variables and the propositional constant D which is intuitively interpreted as a set of deterministic objects of an information system. The set of propositional operations is $\{\neg, \vee, \wedge, [\equiv], [\leq], [\sigma]\}$. The set of IL-formulas is defined as usual.

The IL-*models* are the structures of the form $(U, \equiv, \leq, \sigma, D, m)$ satisfying the conditions of the definition of models from Sect. 7.3 and such that $m(D) = D$ and the conditions (I2), (I3), (I5), and (I7), ..., (I13) are satisfied. Satisfaction of IL-formulas by states in a model is defined as in Sect. 7.3.

The following theorem can be found in [DO02].

Theorem 11.3.3.

1. *The logic* IL *is decidable;*
2. IL-*satisfiability problem is* PSPACE-*hard.*

11.4 Relational Formalization of Logics NIL and IL

Let L be a logic with semantics of plain frames. The language of relational logics $\mathsf{RL_L}$ appropriate for expressing L-formulas is $\mathsf{RL}(1, 1')$-language with relational constants representing the accessibility relations from L-models and with propositional constants of L which will be interpreted appropriately as relations (see Sect. 7.4). For the sake of simplicity, we denote these relational constants with the same symbols as in L. An $\mathsf{RL_L}$-structure is of the form $(U, \{T : T \in \mathbb{RC}_{\mathsf{RL_L}} \setminus \{1, 1'\}\}, m)$, where (U, m) is an $\mathsf{RL}(1, 1')$-model and T is a binary relation on U such that $m(T) = T$, for every $T \in \mathbb{RC}_{\mathsf{RL_L}} \setminus \{1, 1'\}$. An $\mathsf{RL_L}$-model is an $\mathsf{RL_L}$-structure that satisfies all the constraints posed on the relational constants in the L-models. The translation of modal formulas of information logics into relational terms is defined as in Sect. 7.4.

More precisely, the language of the relational logic $\mathsf{RL_{NIL}}$ appropriate for a relational representation of logic NIL is $\mathsf{RL}(1, 1')$-language with the set of relational constants $\mathbb{RC}_{\mathsf{RL_{NIL}}} = \{\leq, \geq, \sigma, 1, 1'\}$. An $\mathsf{RL_{NIL}}$-structure is of the form $\mathcal{M} = (U, \leq, \geq, \sigma, m)$, where (U, m) is an $\mathsf{RL}(1, 1')$-model and for every $T \in \{\leq, \geq, \sigma\}$, T is a binary relation on U such that $m(T) = T$. An $\mathsf{RL_{NIL}}$-model is an $\mathsf{RL_{NIL}}$-structure such that the relations $\leq$, $\geq$, and σ satisfy the conditions (I1), (12), (I4), and (I6).

The language of the relational logic $\mathsf{RL_{IL}}$ appropriate for a relational representation of logic IL is $\mathsf{RL}(1, 1')$-language with $\mathbb{RC}_{\mathsf{RL_{IL}}} = \{\leq, \equiv, \sigma, D, 1, 1'\}$. An $\mathsf{RL_{IL}}$-structure is of the form $\mathcal{M} = (U, \leq, \equiv, \sigma, D, m)$, where (U, m) is an $\mathsf{RL}(1, 1')$-model and for every $T \in \{\leq, \equiv, \sigma, D\}$, T is a binary relation on U such that $m(T) = T$. An $\mathsf{RL_{IL}}$-model is an $\mathsf{RL_{IL}}$-structure such that the relations $\leq$, $\equiv$, σ, and D satisfy the conditions (I2), (I3), (I5), (I7), (I10), (12), and the following:

(I8') If $(y, z) \in D$ and $(x, y) \in \leq$, then $(x, z) \in D$;
(I9') If $(x, z) \in D$ and $(x, y) \in \sigma$, then $(x, y) \in \leq$;

(I11') If $(x,z),(y,z) \in D$ and $(x,y) \in \sigma$, then $(x,y) \in \equiv$;
(I13') If $(x,z) \notin D$, then there is $y \in U$ such that $(x,y) \notin \leq$.

Observe that according to the convention established in Sect. 7.4 constant D in $\mathsf{RL_{IL}}$-logic represents a right ideal relation which is a counterpart to the constant D of IL. Note also that in $\mathsf{RL_L}$-models we list explicitly all relations corresponding to accessibility relations from L-models and we denote them with the same symbols as the corresponding constants in the language.

The models of $\mathsf{RL_L}$ such that $1'$ is interpreted as identity are referred to as standard $\mathsf{RL_L}$-models.

By Theorem 7.4.1, we get:

Theorem 11.4.1. *Let L be a logic with semantics of plain frames. Then for every L-formula φ and for all object variables x and y, the following conditions are equivalent:*

1. *φ is L-valid;*
2. *$x\tau(\varphi)y$ is $\mathsf{RL_L}$-valid.*

Dual tableaux for the logics NIL and IL in their relational formalizations are constructed as follows. We add to $\mathsf{RL}(1,1')$-dual tableau the rules corresponding to the constraints on relations that are assumed in the models of these logics. In the following list the rule (rI#) corresponds to the condition (I#), for # $\in \{1,6,7,8',9',10,11',12,13'\}$:

For all object variables x and y,

(wref σ) $\dfrac{x\sigma x}{x\sigma z, x\sigma x}$ z is any object variable

(rI1 $\subseteq$) $\dfrac{x \leq y}{y \geq x, x \leq y}$ (rI1 $\supseteq$) $\dfrac{x \geq y}{y \leq x, x \geq y}$

(rI6) $\dfrac{x\sigma y}{z\sigma t, x\sigma y \mid z \leq x, x\sigma y \mid t \leq y, x\sigma y}$ z, t are any object variables

(rI7) $\dfrac{x\sigma y}{z\sigma y, x\sigma y \mid z \leq x, x\sigma y}$ z is any object variable

(rI8') $\dfrac{xDy}{zDy, xDy \mid x \leq z, xDy}$ z is any object variable

(rI9') $\dfrac{x \leq y}{xDz, x \leq y \mid x\sigma z, x \leq y}$ z is any object variable

(rI10) $\dfrac{x \leq y}{x \equiv y, x \leq y}$ (rI12) $\dfrac{x \equiv y}{x \leq y, x \equiv y \mid y \leq x, x \equiv y}$

(rI11') $\dfrac{x \equiv y}{xDz, x \equiv y \mid yDz, x \equiv y \mid x\sigma y, x \equiv y}$ z is any object variable

(rI13') $\dfrac{xDy}{x \leq z, xDy}$ z is a new object variable

The specific NIL-rules are (ref $\leq$), (tran $\leq$), (ref σ), (sym σ), which are the instances of the corresponding rules presented in Sect. 6.6 (see also Sect. 7.4), and in addition (rI1$\subseteq$), (rI1$\supseteq$), and (rI6).

The specific IL-rules are (ref $\leq$), (tran $\leq$), (wref σ), (sym σ), (ref $\equiv$), (sym $\equiv$), (tran $\equiv$), (rI7), (rI8'), (rI9'), (rI10), (rI11'), (rI12), and (rI13').

As in RL-logic, an $\mathsf{RL_L}$-set is a finite set of $\mathsf{RL_L}$-formulas such that the first-order disjunction of its members is true in all $\mathsf{RL_L}$-models. If $\mathcal{K}$ is a class of $\mathsf{RL_L}$-structures, then the notion of a $\mathcal{K}$-set is defined in a similar way. Correctness of a rule is defined as in the logic RL (see Sects. 2.4 and 2.5).

Theorem 11.4.2 (Correspondence). *Let* L *be an information logic satisfying some of the conditions (I1), ..., (I13) and let* $\mathcal{K}$ *be a class of* $\mathsf{RL_L}$*-structures* $\mathcal{M} = (U, Rel, m)$*, for* $Rel \subseteq \{\leq, \geq, \sigma, \equiv\}$*. Then:*

1. *A relation* $R \in Rel$ *is reflexive (resp. weakly reflexive, symmetric, transitive) if and only if the rule (ref R) (resp. (wref R), (sym R), (tran R)) is* $\mathcal{K}$*-correct;*
2. *Every* $\mathsf{RL_L}$*-structure of* $\mathcal{K}$ *satisfies the condition (I#) iff the rule (rI#) is* $\mathcal{K}$*-correct, where* $\# \in \{1, 6, 7, 8', 9', 10, 11', 12, 13'\}$.

Proof. 1. can be proved in a similar way as Theorem 6.6.1. By way of example, we show 2. for the condition (I6).

Assume that every structure of $\mathcal{K}$ satisfies this condition. Then preservation of validity from the upper set to the bottom sets is obvious. Let X be any finite set of $\mathsf{RL_L}$-formulas. Assume $X_1 = X \cup \{z\sigma t, x\sigma y\}$, $X_2 = X \cup \{z \leq x, x\sigma y\}$, and $X_3 = X \cup \{t \leq y, x\sigma y\}$ are $\mathcal{K}$-sets. Suppose $X \cup \{x\sigma y\}$ is not a $\mathcal{K}$-set, that is there exist an $\mathsf{RL_L}$-structure $\mathcal{M}$ and a valuation v in $\mathcal{M}$ such that $\mathcal{M}, v \not\models x\sigma y$. Since X_1, X_2, X_3 are $\mathcal{K}$-sets, the model $\mathcal{M}$ and the valuation v satisfy $(v(z), v(t)) \in \sigma$, $(v(z), v(x)) \in \leq$, and $(v(t), v(y)) \in \leq$. By the condition (I6), $(v(x), v(y)) \in \sigma$, a contradiction.

Now, assume the rule (rI6) is $\mathcal{K}$-correct. Let $X \stackrel{\text{df}}{=} \{z{-}\sigma t, z{-}\leq x, t{-}\leq y\}$. Then $X \cup \{z\sigma t, x\sigma y\}$, $X \cup \{z \leq x, x\sigma y\}$, and $X \cup \{t \leq y, x\sigma y\}$ are $\mathcal{K}$-sets. Thus, by the assumption, $X \cup \{x\sigma y\}$ is a $\mathcal{K}$-set. Therefore, for every $\mathsf{RL_L}$-structure $\mathcal{M}$ in $\mathcal{K}$ and for every valuation v in $\mathcal{M}$, if $\mathcal{M}, v \models z\sigma t$, $\mathcal{M}, v \models z \leq x$, and $\mathcal{M}, v \models t \leq y$, then $\mathcal{M}, v \models x\sigma y$. □

The above proposition implies that all the specific rules of $\mathsf{RL_L}$-dual tableau are $\mathsf{RL_L}$-correct. Correctness of all the remaining rules can be proved as in $\mathsf{RL}(1, 1')$-dual tableau (see Sects. 2.5 and 2.7), thus we get:

Proposition 11.4.1.

1. *The* NIL*-rules are* $\mathsf{RL_{NIL}}$*-correct;*
2. *The* IL*-rules are* $\mathsf{RL_{IL}}$*-correct.*

It is known that conditions (I12) and (I13) are not expressible in the language of logic IL, hence the completeness proof for its Hilbert-style axiomatization requires a special technique referred to as copying (see [Vak89]). As it is shown above, in the case of relational formalization the rules corresponding to (I12) and (I13) can be explicitly given. They enable us to prove constraints (I12) and (I13), respectively, directly from their representation in the language of $\mathsf{RL}_{\mathsf{IL}}$.

The notions of an RL_{L}-proof tree, a closed branch of such a tree, a closed RL_{L}-proof tree, and RL_{L}-provability are defined as in Sect. 2.4. A branch b of an RL_{L}-proof tree is complete whenever it is closed or it satisfies the completion condition of $\mathsf{RL}(1, 1')$-dual tableau adjusted to the language of RL_{L} and the completion conditions corresponding to the rules that are specific for RL_{L}.

The completion conditions determined by the rules (ref R) for $R \in \{\leq, \sigma, \equiv\}$, (sym R) for $R \in \{\sigma, \equiv\}$, and (tran R) for $R \in \{\leq, \equiv\}$ are the instances of the completion conditions presented in Sect. 6.6.

For all object variables x and y,

Cpl(wref σ) If $x\sigma x \in b$, then for every object variable z, $x\sigma z \in b$, obtained by an application of the rule (wref σ);

Cpl(rI1 $\subseteq$) If $x \leq y \in b$, then $y \geq x \in b$, obtained by an application of the rule (rI1 $\subseteq$);

Cpl(rI1 $\supseteq$) If $x \geq y \in b$, then $y \leq x \in b$, obtained by an application of the rule (rI1 $\supseteq$);

Cpl(rI6) If $x\sigma y \in b$, then for all object variables z and t, either $z\sigma t \in b$ or $z \leq x \in b$ or $t \leq y \in b$, obtained by an application of the rule (rI6);

Cpl(rI7) If $x\sigma y \in b$, then for every object variable z, either $z\sigma y \in b$ or $z \leq x \in b$, obtained by an application of the rule (rI7);

Cpl(rI8') If $xDy \in b$, then for every object variable z, either $zDy \in b$ or $x \leq z \in b$, obtained by an application of the rule (rI8');

Cpl(rI9') If $x \leq y \in b$, then for every object variable z, either $xDz \in b$ or $x\sigma z \in b$, obtained by an application of the rule (rI9');

Cpl(rI10) If $x \leq y \in b$, then $x \equiv y \in b$, obtained by an application of the rule (rI10);

Cpl(rI11') If $x \equiv y \in b$, then for every object variable z, either $xDz \in b$, $yDz \in b$ or $x\sigma y \in b$, obtained by an application of the rule (rI11');

Cpl(rI12) If $x \equiv y \in b$, then either $x \leq y \in b$ or $y \leq x \in b$, obtained by an application of the rule (rI12);

Cpl(rI13) If $xDy \in b$, then for some object variable z, $x \leq z \in b$, obtained by an application of the rule (rI13).

Let L be an information logic. The notions of a complete branch of an RL_{L}-proof tree, a complete RL_{L}-proof tree, and an open branch of an RL_{L}-proof tree are defined as in RL-logic (see Sect. 2.5). In order to prove completeness, we need to define a branch model and to show the three theorems of Table 7.1.

All the rules listed above (see p. 226) guarantee that whenever a branch of an RL_{L}-proof tree contains two formulas one of which is built with an atomic term and the other with its complement, then the branch can be closed. Thus, the closed branch property can be proved as the Proposition 2.8.1.

Let b be an open branch of an $\mathsf{RL_L}$-proof tree. The branch structure has the form $\mathcal{M}^b = (U^b, (\#^b)_{\# \in \{\leq, \geq, \sigma, \equiv, D\}}, m^b)$, where $U^b = \mathbb{OV}_{\mathsf{RL_L}}$, for every relational constant R, $m^b(R) = \{(x, y) \in U^b \times U^b : xRy \notin b\}$, $\#^b = m^b(\#)$ for every $\# \in \{\leq, \geq, \sigma, \equiv, D\}$, and m^b extends to all the compound relational terms as in $\mathsf{RL}(1, 1')$-models. To prove that branch structures satisfy all the conditions that are assumed in the models of a given logic, we employ the corresponding completion conditions.

Proposition 11.4.2 (Branch Model Property). *Let* L *be an information logic that satisfies some of the conditions among (I1), ..., (I13). The branch structure* $\mathcal{M}^b$ *determined by an open branch* b *of an* $\mathsf{RL_L}$*-proof tree is an* $\mathsf{RL_L}$*-model.*

Proof. By way of example, we prove that the branch structure $\mathcal{M}^b = (U^b, \leq^b, \equiv^b, \sigma^b, m^b)$ determined by an open branch b of an $\mathsf{RL_{IL}}$-proof tree satisfies the condition (I12). Assume $(x, y) \in \leq^b$ and $(y, x) \in \leq^b$, that is $x \leq y \notin b$ and $y \leq x \notin b$. Suppose $(x, y) \notin \equiv^b$. Then $x \equiv y \in b$. By the completion condition Cpl(rI12), either $x \leq y \in b$ or $y \leq x \in b$, a contradiction. □

Since the branch models are defined in a standard way, that is for any relational constant R, $m^b(R)$ is defined as in the completeness proof of $\mathsf{RL}(1, 1')$-dual tableau (see Sects. 2.5 and 2.7), the satisfaction in branch model property can be proved as in $\mathsf{RL}(1, 1')$-logic. Thus, we get:

Theorem 11.4.3 (Soundness and Completeness of $\mathsf{RL_{NIL}}$ and $\mathsf{RL_{IL}}$).

1. $\mathsf{RL_{NIL}}$*-dual tableau is sound and complete;*
2. $\mathsf{RL_{IL}}$*-dual tableau is sound and complete.*

Finally, by the above theorem and Theorem 11.4.1, we obtain:

Theorem 11.4.4 (Relational Soundness and Completeness of NIL and IL). *Let* $\mathsf{L} \in \{\mathsf{NIL}, \mathsf{IL}\}$*. Then for every* L*-formula* φ *and for all object variables* x *and* y*, the following conditions are equivalent:*

1. φ *is* L*-valid;*
2. $x\tau(\varphi)y$ *is* $\mathsf{RL_L}$*-provable.*

Example. Let us consider the NIL-formula φ and IL-formula ψ:

$$\varphi = p \to [\leq]\langle\geq\rangle p,$$

$$\psi = [\leq]p \to [\equiv]p.$$

The formula φ (resp. ψ) is true in a modal frame $(U, \leq, \geq)$ (resp. $(U, \leq, \equiv)$) provided that $\leq^{-1} \subseteq \geq$ (resp. $\equiv \subseteq \leq$).

For the sake of simplicity, let us denote $\tau(p)$ by P. According to the translation presented in Sect. 7.4 (see p. 147), the relational representations of these formulas are:

$$\tau(\varphi) = -P \cup -(\leq ; -(\geq ; P)),$$
$$\tau(\psi) = --(\leq ; -P) \cup -(\equiv ; -P).$$

Figure 11.1 presents an $\mathsf{RL}_{\mathsf{NIL}}$-proof of the formula $x\tau(\varphi)y$ which shows NIL-validity of φ, and Fig. 11.2 presents an $\mathsf{RL}_{\mathsf{IL}}$-proof of the formula $x\tau(\psi)y$ that shows IL-validity of ψ.

As in standard modal logics, the relational logic $\mathsf{RL}(1, 1')$ can be used for verification of entailment, model checking in finite models, and verification of satisfaction of a given formula by some objects in a finite model (see Sects. 7.6, 7.7, and 7.8).

Let L be a logic with semantics of plain frames. In order to verify entailment we apply the method presented in Sect. 2.11, that is, first, we translate L-formulas

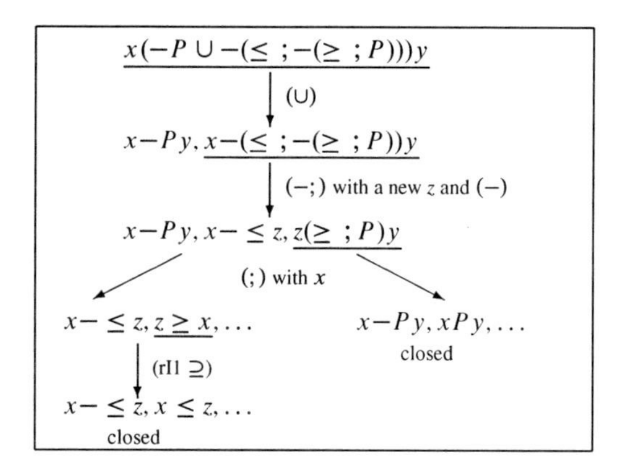

Fig. 11.1 A relational proof of $p \rightarrow [\leq]\langle\geq\rangle p$

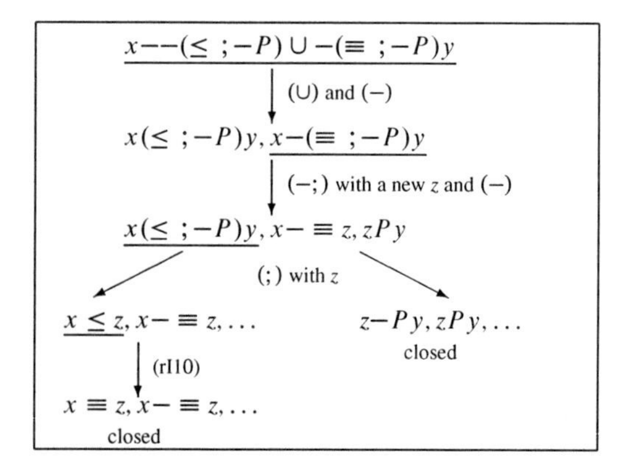

Fig. 11.2 A relational proof of $[\leq]p \rightarrow [\equiv]p$

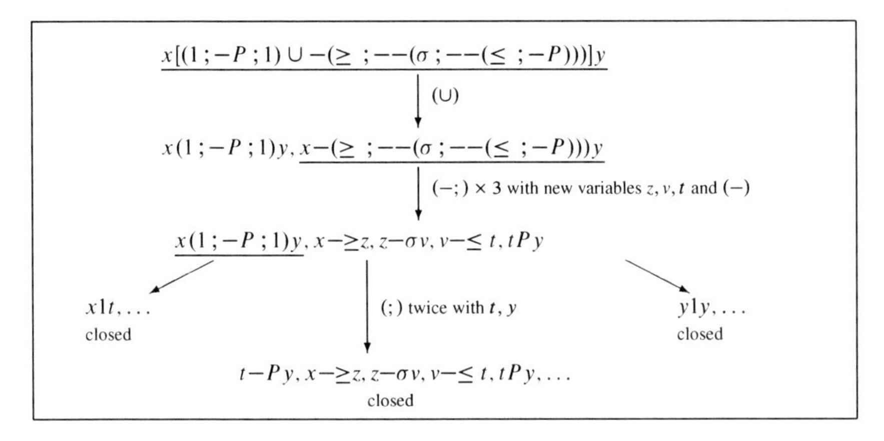

Fig. 11.3 An $\mathsf{RL_{NIL}}$-proof showing that p entails $[\geq][\sigma][\leq]p$

in question into terms of the relational logic $\mathsf{RL_L}$, and then we use the method of verification of entailment for $\mathsf{RL_L}$-logic as it is shown in Sect. 2.11.

For example, in NIL-logic the formula p entails $[\geq][\sigma][\leq]p$. For the sake of simplicity, denote $\tau(p)$ by P. Then:

$$\tau([\geq][\sigma][\leq]p) = -(\geq\,;--(\sigma\,;--(\leq\,;-P))).$$

We need to show that $P = 1$ implies $-(\geq\,;--(\sigma\,;--(\leq\,;-P))) = 1$. According to Proposition 2.2.1 (7.), it suffices to show that the formula:

$$x[(1\,;-P\,;1)\cup-(\geq\,;--(\sigma\,;--(\leq\,;-P)))]y$$

is $\mathsf{RL_{NIL}}$-provable. Figure 11.3 presents an $\mathsf{RL_{NIL}}$-proof of this formula.

11.5 Information Logic CI and Its Relational Formalization

The logic presented in this section and its dual tableau originated in [DO00a]. The language of the logic CI of complementarity and incomplementarity is a multimodal language with symbols from the following pairwise disjoint sets:

- $\mathbb{V}$ – a countable infinite set of propositional variables;
- $\{R, S\}$ – a set of relational constants;
- $\{\neg, \vee, \wedge, [R], [[S]]\}$ – a set of propositional operations.

A CI-frame is a structure (U, R, S) such that:

- U is a non-empty set;
- R is a symmetric and 3-transitive relation on U;

- S is a reflexive relation on U;
- $R \cup S = U \times U$ and $R \cap S = \emptyset$.

Models based on CI-frames are defined as usual (see Sect. 7.3).

Satisfaction of a formula is defined as in Sect. 7.3, that is the clauses for formulas with modal operations are:

- $\mathcal{M}, s \models [R]\varphi$ iff for every $s' \in U$, if $(s, s') \in R$, then $\mathcal{M}, s' \models \varphi$;
- $\mathcal{M}, s \models [[S]]\varphi$ iff for every $s' \in U$, if $\mathcal{M}, s' \models \varphi$, then $(s, s') \in S$.

Observe that although neither irreflexivity of R nor symmetry of S are assumed explicitly, irreflexivity of R is guaranteed by reflexivity of S, and symmetry of S is guaranteed by symmetry of R, since $R = -S$. Irreflexivity of R is not expressible in a modal language with a single accessibility relation. The relational dual tableau for CI will enable us to prove $1' \subseteq -R$. The proof is presented in Fig. 11.5. Symmetry of S is expressible with formula $p \to [[S]]\,[[S]]p$. Its proof is presented in Fig. 11.6.

The language of the relational logic corresponding to CI is $\mathsf{RL}(1, 1')$-language endowed with the relational constants R and S. $\mathsf{RL}_{\mathsf{CI}}$-models are structures of the form $\mathcal{M} = (U, R, S, m)$, where (U, m) is an $\mathsf{RL}(1, 1')$-model and R and S are binary relations on U that provide the interpretation of the corresponding relational constants and satisfy the above conditions assumed in CI-frames.

The translation of CI-formulas into relational terms of the logic $\mathsf{RL}_{\mathsf{CI}}$ is defined as in Sect. 7.4, that is the translation of formulas with modal operations is:

- $\tau([R]\varphi) \stackrel{\mathrm{df}}{=} -(R\,; -\tau(\varphi))$;
- $\tau([[S]]\varphi) \stackrel{\mathrm{df}}{=} -(-S\,; \tau(\varphi))$.

By Theorem 7.4.1, we obtain:

Theorem 11.5.1. *For every* CI*-formula φ and for all object variables x and y, the following conditions are equivalent:*

1. *φ is* CI*-valid;*
2. *$x\tau(\varphi)y$ is* $\mathsf{RL}_{\mathsf{CI}}$*-valid.*

$\mathsf{RL}_{\mathsf{CI}}$-dual tableau includes the rules and axiomatic sets of $\mathsf{RL}(1, 1')$-dual tableau adjusted to $\mathsf{RL}_{\mathsf{CI}}$-language, rules (ref S) and (sym R) which are the instances of the rules presented in Sect. 6.6 and, in addition, it contains the specific rules and axiomatic sets of the following forms:

For all object variables x, y, z, and t,

$$(\text{dis } R, S)\ \frac{}{xRy \mid xSy}$$

$$(\text{3-tran } R)\ \frac{xRy}{xRz, xRy \mid zRt, xRy \mid tRy, xRy} \quad z, t \text{ are any object variables}$$

The rule (dis R, S) is a specialized cut rule. An alternative deterministic form of such rules is discussed in Sect. 25.9.

Specific $\mathsf{RL}_{\mathsf{CI}}$-axiomatic sets are those that include the subset $\{xRy, xSy\}$, for any object variables x and y.

Proposition 11.5.1.

1. *The* $\mathsf{RL}_{\mathsf{CI}}$*-rules are* $\mathsf{RL}_{\mathsf{CI}}$*-correct;*
2. *The* $\mathsf{RL}_{\mathsf{CI}}$*-axiomatic sets are* $\mathsf{RL}_{\mathsf{CI}}$*-sets.*

Proof. The proof of correctness of the rules (ref S), (sym R), and (3-tran R) follows the proof of Theorem 6.6.1. Now, we show correctness of the rule (dis R, S). Let X be a finite set of $\mathsf{RL}_{\mathsf{CI}}$-formulas. The preservation of validity from the upper set to the bottom sets is obvious. Assume that $X \cup \{xRy\}$ and $X \cup \{xSy\}$ are $\mathsf{RL}_{\mathsf{CI}}$-sets. Suppose X is not $\mathsf{RL}_{\mathsf{CI}}$-set. Then there exist an $\mathsf{RL}_{\mathsf{CI}}$-model $\mathcal{M} = (U, R, S, m)$ and a valuation v in $\mathcal{M}$ such that for every $\varphi \in X$, $\mathcal{M}, v \not\models \varphi$. Thus, by the assumption, $\mathcal{M}, v \models xRy$ and $\mathcal{M}, v \models xSy$, hence $(v(x), v(y)) \in R$ and $(v(x), v(y)) \in S$. Therefore, $R \cap S \neq \emptyset$. However, in all $\mathsf{RL}_{\mathsf{CI}}$-models, R and S are disjoint, a contradiction.

Since in every $\mathsf{RL}_{\mathsf{CI}}$-model $\mathcal{M} = (U, R, S, m)$, $R \cup S = U \times U$, for every valuation v in $\mathcal{M}$, $\mathcal{M}, v \models xRy$ or $\mathcal{M}, v \models xSy$, hence $X \cup \{xRy, xSy\}$ is an $\mathsf{RL}_{\mathsf{CI}}$-set, for every set X of $\mathsf{RL}_{\mathsf{CI}}$-formulas. □

An alternative representation of the constraint $R \cup S = U \times U$ can be provided by a rule in the $\mathsf{RL}_{\mathsf{CI}}$-dual tableau. This issue is discussed in Sect. 25.9, see also Sect. 25.6.

The completion conditions corresponding to the specific $\mathsf{RL}_{\mathsf{CI}}$-rules (dis R, S) and (3-tran R) are:

For all object variables x, y, z, and t,

Cpl(dis R, S) Either $xRy \in b$ or $xSy \in b$;
Cpl(3-tran R) If $xRy \in b$, then for all object variables z and t, either $xRz \in b$ or $zRt \in b$ or $tRy \in b$.

It can be proved that the rules specific for $\mathsf{RL}_{\mathsf{CI}}$-dual tableau do not violate the closed branch property.

The branch structure $\mathcal{M}^b = (U^b, R^b, S^b, m^b)$ determined by an open branch of an $\mathsf{RL}_{\mathsf{CI}}$-proof tree is defined as usual, that is $U^b = \mathbb{OV}_{\mathsf{RL}_{\mathsf{CI}}}$, $m^b(T) = \{(x, y) \in U^b \times U^b : xTy \notin b\}$, $T^b = m^b(T)$, for $T \in \{R, S\}$, and m^b extends to all the compound relational terms as in $\mathsf{RL}(1, 1')$-models.

Proposition 11.5.2 (Branch Model Property). *Let b be an open branch of an* $\mathsf{RL}_{\mathsf{CI}}$*-proof tree. Then the branch structure $\mathcal{M}^b = (U^b, R^b, S^b, m^b)$ is an* $\mathsf{RL}_{\mathsf{CI}}$*-model.*

Proof. We need to show that R^b is a symmetric and 3-transitive relation on U^b, S^b is a reflexive relation on U^b, and $R^b \cup S^b = U^b \times U^b$ and $R^b \cap S^b = \emptyset$. Symmetry of R^b and reflexivity of S^b can be proved in a similar way as those properties of relation $m^b(1')$ in the completeness proof of $\mathsf{RL}(1, 1')$-logic in Sect. 2.7. The proof of 3-transitivity of R^b is analogous to the proof of transitivity of $m^b(1')$. Now, suppose that $R^b \cup S^b \neq U^b \times U^b$, that is there are object variables x and y such that $(x, y) \notin R^b$ and $(x, y) \notin S^b$. Then $xRy \in b$ and $xSy \in b$. Since $\{xRy, xSy\}$

is an axiomatic set and all the rules preserve formulas built with atomic relational terms, b is closed, a contradiction. Suppose that $R^b \cap S^b \neq \emptyset$, that is there are object variables x and y such that $(x, y) \in R^b$ and $(x, y) \in S^b$. Then $xRy \notin b$ and $xSy \notin b$. By the completion condition Cpl(dis R, S), for all object variables x and y, either $xRy \in b$ or $xSy \in b$, a contradiction. □

Since the branch model is defined in a standard way, the satisfaction in branch model property can be proved as in $\mathsf{RL}(1, 1')$-logic. Therefore, we obtain:

Theorem 11.5.2 (Soundness and Completeness of $\mathsf{RL}_{\mathsf{CI}}$). *For every $\mathsf{RL}_{\mathsf{CI}}$-formula φ, the following conditions are equivalent:*

1. *φ is $\mathsf{RL}_{\mathsf{CI}}$-valid;*
2. *φ is true in all standard $\mathsf{RL}_{\mathsf{CI}}$-models;*
3. *φ is $\mathsf{RL}_{\mathsf{CI}}$-provable.*

By the above theorem and Theorem 11.5.1, we get:

Theorem 11.5.3 (Relational Soundness and Completeness of CI). *For every CI-formula φ and for all object variables x and y, the following conditions are equivalent:*

1. *φ is CI-valid;*
2. *$x\tau(\varphi)y$ is $\mathsf{RL}_{\mathsf{CI}}$-provable.*

Example. Consider a CI-formula:

$$\varphi = [[S]]\neg p \rightarrow [R]p.$$

This formula is true in a CI-frame (U, R, S) because $R \subseteq -S$. For the sake of simplicity, let us denote $\tau(p)$ by P. Then the relational translation of the formula φ is:

$$\tau(\varphi) = (-S\,; -P) \cup -(R\,; -P).$$

Figure 11.4 presents an $\mathsf{RL}_{\mathsf{CI}}$-proof of the formula $x(-S\,; -P) \cup -(R\,; -P))y$ which shows CI-validity of φ.

Figure 11.5 presents an $\mathsf{RL}_{\mathsf{CI}}$-proof of relational formula $x(-1' \cup -R)y$ which according to Proposition 2.2.1(1.) reflects irreflexivity of relation R.

Now, we consider formula

$$\varphi = p \rightarrow [[S]]\,[[S]]p$$

which reflects symmetry of relation S. Its translation into a relational term is:

$$\tau(\varphi) = -P \cup -(-S\,; -(-S\,; P)).$$

Figure 11.6 depicts an $\mathsf{RL}_{\mathsf{CI}}$-proof of $x\tau(\varphi)y$.

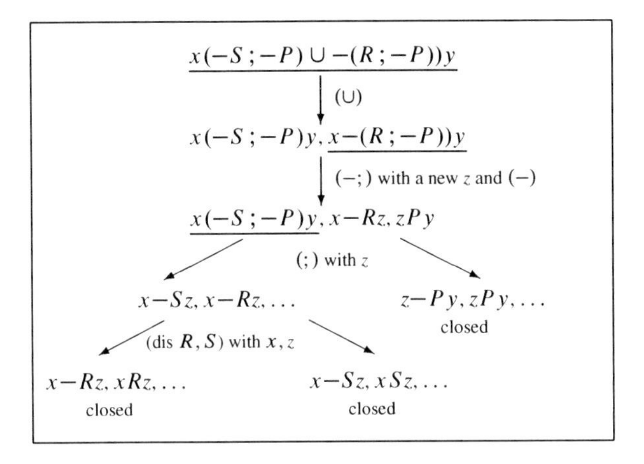

Fig. 11.4 A relational proof of $[\![S]\!]\neg p \rightarrow [R]p$

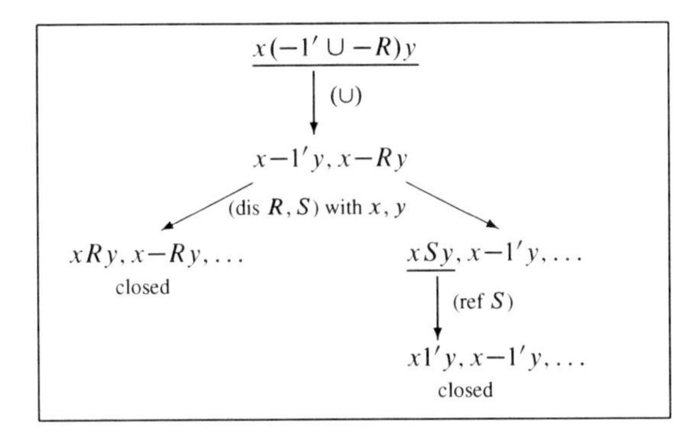

Fig. 11.5 A relational proof of irreflexivity of relation R

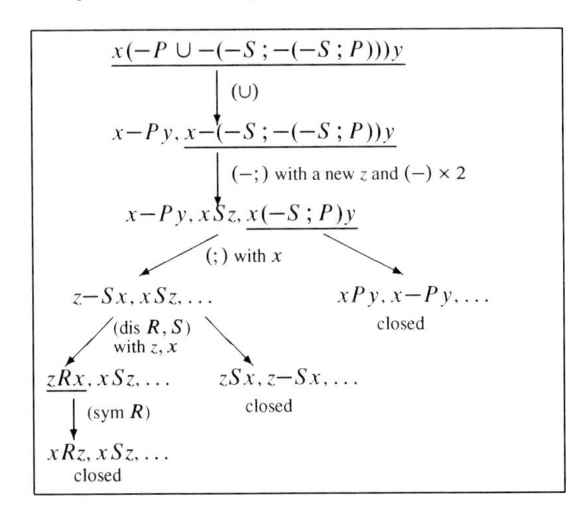

Fig. 11.6 A relational proof of symmetry of relation S

Chapter 12
Dual Tableaux for Information Logics of Relative Frames

12.1 Introduction

In the preceding chapter in Sect. 11.3 we demonstrated that a specification of information relations which would be meaningful for information systems with incomplete information requires an explicit reference to a set of attributes with respect to which the relations are defined. In order to incorporate in a logical formalism the sets of attributes which determine the relations, the notions of a relative relation and a relative frame i.e., the frame whose relations are relative, were introduced in [Orł88], see also [DO02]. More precisely, a relative frame consists of a family, or several families of relations such that the relations within the family are indexed with subsets of a set of parameters, intuitively understood as attributes of an information system. Apart from the ordinary properties of relations such as, for example, symmetry or transitivity, relative relations may have some properties which refer to the family of relations as a whole. These properties are collectively named global properties of relations. For example, one of the typical global property says that a relation indexed with the union of two sets equals intersection of relations indexed with the components of the union. Such an assumption is relevant, among others, when the family consists of equivalence relations. Assuming that the equivalence of objects is established whenever they have the same features corresponding to the set of attributes declared in the definition of this relation, then clearly taking more attributes into account we get a finer granulation of the set of objects than in the case of any smaller number of attributes.

In a full generality, relative frames have families of relations indexed by elements from any level of the powerset hierarchy of a set of parameters. The relative frames of that kind are a convenient tool for specification of hierarchical information such as, for example, a subject classification system. A discussion of these applications of relative frames can be found in [DO02], Chaps. 7–9. Relative frames and their logics are studied, among others, in [DS02b, Bal02, DO07].

In this chapter we confine ourselves to a simple case of indices which are finite subsets of parameters. We present relative versions of the frames considered in the preceding chapter and information logics based on those frames. We construct dual tableaux for these logics focusing on the treatment of global conditions assumed in

E. Orłowska and J. Golińska-Pilarek, *Dual Tableaux: Foundations, Methodology, Case Studies*, Trends in Logic 36, DOI 10.1007/978-94-007-0005-5_12,

the models of the logics. The method of construction of those dual tableaux can be extended to the theories of relative frames determined by a powerset hierarchy of sets of attributes.

12.2 Relative Frames

Let Par be a non-empty set of parameters intuitively interpreted as a set of attributes of an information system. A *relative frame* is a structure:

$$\mathcal{F} = (U, (R^1_P)_{P \in \mathcal{P}_{fin}(\mathrm{Par})}, \ldots, (R^n_P)_{P \in \mathcal{P}_{fin}(\mathrm{Par})}),$$

where relations from $(R^i_P)_{P \in \mathcal{P}_{fin}(\mathrm{Par})}$, $i \in \{1, \ldots, n\}$, are indexed with finite subsets of a non-empty set Par.

Typically, these relations satisfy either of the following postulates:

For all $P, Q \in \mathcal{P}_{fin}(\mathrm{Par})$,

(S1) $R_{P \cup Q} = R_P \cap R_Q$;
(W1) $R_{P \cup Q} = R_P \cup R_Q$.

These conditions are referred to as *global* conditions, because they refer to a family of relations as a whole. The properties of relations listed in Proposition 11.2.1 will be called *local* conditions.

The characterization of a family of relative relations often requires special postulates for the relations indexed with the empty set. The typical condition for the relation $R_\emptyset$ is either of the following:

(S2) $R_\emptyset = U^2$;
(W2) $R_\emptyset = \emptyset$.

Conditions (S1) and (S2) reflect the behaviour of strong relations derived from an information system, and conditions (W1) and (W2) reflect the behaviour of weak relations.

Typically, information logics of relative frames are based on the following classes of frames:

- FS – the class of relative frames in which the families of relative relations satisfy conditions (S1) and (S2). The members of FS are called FS-*frames* or *relative frames with strong relations*;
- FW – the class of relative frames in which the families of relative relations satisfy conditions (W1) and (W2). The members of FW are called FW-*frames* or *relative frames with weak relations.*

Frames with indistinguishability and distinguishability relations listed below provide examples of members of the FS and FW families.

Frames with Indistinguishability Relations:

- FS–IND is the class of FS-frames $(U, (R_P)_{P \subseteq \text{Par}})$ such that for every $p \in \text{Par}$, $R_{\{p\}}$ is an equivalence relation. Consequently, for every $P \subseteq \text{Par}$, R_P is an equivalence relation, since reflexivity, symmetry, and transitivity are preserved under intersection. The members of FS–IND are called *strong indiscernibility frames*;
- FS–SIM is the class of FS-frames $(U, (R_P)_{P \subseteq \text{Par}})$ such that for every $p \in \text{Par}$, $R_{\{p\}}$ is weakly reflexive and symmetric. Consequently, for every $P \subseteq \text{Par}$, R_P is weakly reflexive and symmetric. The members of FS–SIM are called *strong similarity frames*;
- FS–ICOM is the class of FS-frames $(U, (R_P)_{P \subseteq \text{Par}})$ such that for every $p \in \text{Par}$, $R_{\{p\}}$ is reflexive, symmetric, and co-3-transitive. Consequently, for every $P \subseteq \text{Par}$, R_P is reflexive and symmetric. The property of co-3-transitivity is not preserved under intersection. The members of FS–ICOM are called *strong incomplementarity frames*;
- FS–IN is the class of FS-frames $(U, (R_P)_{P \subseteq \text{Par}}, (Q_P)_{P \subseteq \text{Par}})$ with two families of relative relations such that for every $p \in \text{Par}$, $R_{\{p\}}$ and $Q_{\{p\}}$ are reflexive and transitive, and $R_{\{p\}} = Q^{-1}_{\{p\}}$. Consequently, for every $P \subseteq \text{Par}$, R_P and Q_P are reflexive and transitive, and $R_P = Q^{-1}_P$. The members of FS–IN are called *strong inclusion frames*;
- FW–IND is the class of FW-frames $(U, (R_P)_{P \subseteq \text{Par}})$ such that for every $p \in \text{Par}$, $R_{\{p\}}$ is an equivalence relation. Consequently, for every non-empty $P \subseteq \text{Par}$, R_P is reflexive and symmetric. Transitivity is not preserved under union. The members of FW–IND are called *weak indiscernibility frames*;
- FW–SIM is the class of FW-frames $(U, (R_P)_{P \subseteq \text{Par}})$ such that for every $p \in \text{Par}$, $R_{\{p\}}$ is weakly reflexive and symmetric. Consequently, for every $P \subseteq \text{Par}$, R_P is weakly reflexive and symmetric. The members of FW–SIM are called *weak similarity frames*;
- FW–ICOM is the class of FW-frames $(U, (R_P)_{P \subseteq \text{Par}})$ such that for every $p \in \text{Par}$, $R_{\{p\}}$ is reflexive, symmetric, and co-3-transitive. Consequently, for every non-empty $P \subseteq \text{Par}$, R_P is reflexive, symmetric, and co-3-transitive, since all these properties are preserved under union. The members of FW–ICOM are called *weak incomplementarity frames*;
- FW–IN is the class of FW-frames $(U, (R_P)_{P \subseteq \text{Par}}, (Q_P)_{P \subseteq \text{Par}})$ with two families of relative relations such that for every $p \in \text{Par}$, $R_{\{p\}}$ and $Q_{\{p\}}$ are reflexive and transitive, and $R_{\{p\}} = Q^{-1}_{\{p\}}$. Consequently, for every $P \subseteq \text{Par}$, $R_P = Q^{-1}_P$ and if $P \neq \emptyset$, then R_P and Q_P are reflexive. Transitivity is not preserved under union. The members of FW–IN are called *weak inclusion frames*.

Frames with Distinguishability Relations:

- FS–DIV is the class of FS-frames $(U, (R_P)_{P \subseteq \text{Par}})$ such that for every $p \in \text{Par}$, $R_{\{p\}}$ is irreflexive, symmetric, and co-transitive. Consequently, for every

$P \subseteq \mathrm{Par}$, R_P is symmetric, and if $P \neq \emptyset$, then R_P is irreflexive. Co-transitivity is not preserved under intersection. The members of FS–DIV are called *strong diversity frames*;
- FS–RORT is the class of FS-frames $(U, (R_P)_{P \subseteq \mathrm{Par}})$ such that for every $p \in \mathrm{Par}$, $R_{\{p\}}$ is co-weakly reflexive and symmetric. Consequently, for every $P \subseteq \mathrm{Par}$, R_P is co-weakly reflexive and symmetric. The members of FS–RORT are called *strong right orthogonality frames*;
- FS–COM is the class of FS-frames $(U, (R_P)_{P \subseteq \mathrm{Par}})$ such that for every $p \in \mathrm{Par}$, $R_{\{p\}}$ is irreflexive, symmetric, and 3-transitive. Consequently, for every $P \subseteq \mathrm{Par}$, R_P is symmetric and 3-transitive, and if $P \neq \emptyset$, then R_P is irreflexive. The members of FS–COM are called *strong complementarity frames*;
- FW–DIV is the class of FW-frames $(U, (R_P)_{P \subseteq \mathrm{Par}})$ such that for every $p \in \mathrm{Par}$, $R_{\{p\}}$ is irreflexive, symmetric, and co-transitive. Consequently, for every $P \subseteq \mathrm{Par}$, R_P is irreflexive, symmetric, and co-transitive. Co-transitivity is preserved under union. The members of FW–DIV are called *weak diversity frames*;
- FW–RORT is the class of FW-frames $(U, (R_P)_{P \subseteq \mathrm{Par}})$ such that for every $p \in \mathrm{Par}$, $R_{\{p\}}$ is co-weakly reflexive and symmetric. Consequently, for every $P \subseteq \mathrm{Par}$, R_P is co-weakly reflexive and symmetric. The members of FW–RORT are called *weak right orthogonality frames*;
- FW–COM is the class of FW-frames $(U, (R_P)_{P \subseteq \mathrm{Par}})$ such that for every $p \in \mathrm{Par}$, $R_{\{p\}}$ is irreflexive, symmetric, and 3-transitive. Consequently, for every $P \subseteq \mathrm{Par}$, R_P is irreflexive and symmetric. 3-transitivity is not preserved under union. The members of FW–COM are called *weak complementarity frames*.

The languages of information logics with semantics of relative frames, referred to as *Rare-logics*, are the multimodal languages with modal operations determined by relations indexed with finite subsets of a set Par.

12.3 Relational Formalizations of the Logics of Strong and Weak Relative Frames

As a first case study we present logics L_{FS} and L_{FW} based on FS-frames and FW-frames, respectively. The choice of modal operations usually depends on the global conditions assumed in the models. Typically, logic L_{FS} has the sufficiency operations in its language and logic L_{FW} the necessity operations determined by the relational constants. We recall that satisfaction of a formula built with the sufficiency operation, $[[T]]\varphi$, by a state s in a model $\mathcal{M}$ is defined as:

- $\mathcal{M}, s \models [[T]]\varphi$ iff for every $s' \in U$, if $\mathcal{M}, s' \models \varphi$, then $(s, s') \in T$.

Let $\mathsf{L} \in \{\mathsf{L}_{\mathsf{FS}}, \mathsf{L}_{\mathsf{FW}}\}$. An L-*model* is a structure $\mathcal{M} = (U, (R_P)_{P \in \mathcal{P}_{fin}(\mathrm{Par})}, m)$ such that U is a non-empty set, R_P are binary relations on U indexed with finite subsets of a set Par, and m is a meaning function such that:

- $m(p) \subseteq U$, for any propositional variable p;
- $R_\emptyset = \begin{cases} U \times U, & \text{if } \mathsf{L} = \mathsf{L_{FS}}, \\ \emptyset, & \text{if } \mathsf{L} = \mathsf{L_{FW}}; \end{cases}$
- $R_{\{p\}} \subseteq U \times U$, for every $p \in \text{Par}$;
- for all $P, Q \in \mathcal{P}_{fin}(\text{Par})$:

$$R_{P \cup Q} = \begin{cases} R_P \cap R_Q, & \text{if } \mathsf{L} = \mathsf{L_{FS}}, \\ R_P \cup R_Q, & \text{if } \mathsf{L} = \mathsf{L_{FW}}. \end{cases}$$

If a Rare-logic L is based on any of the subclasses of FS or FW of frames listed in Sect. 12.2, then the appropriate local conditions should be assumed in the corresponding L-models.

The relational language corresponding to an L-language is an $\mathsf{RL}(1, 1')$-language endowed with the set $\{R_P\}_{P \in \mathcal{P}_{fin}(\text{Par})}$ of relational constants. The set of relational terms and formulas are defined as in $\mathsf{RL}(1, 1')$-logic (see Sect. 2.3).

The $\mathsf{RL_L}$-models are structures of the form:

$$\mathcal{M} = (U, (R_P)_{P \in \mathcal{P}_{fin}(\text{Par})}, m),$$

where (U, m) is an $\mathsf{RL}(1, 1')$-model and $R_P = m(R_P)$ are binary relations on U satisfying the same conditions as in L-models. As usual, we denote relations in the language and in the models with the same symbols.

The translation of L-formulas into relational terms of the logic $\mathsf{RL_L}$ starts, as usual, with a one-to-one assignment of relational variables to the propositional variables and then the translation τ is defined inductively as in Sect. 7.4. As in classical modal logics, we can prove that for $\mathsf{L} \in \{\mathsf{L_{FS}}, \mathsf{L_{FW}}\}$ and for every L-formula, L-validity is equivalent to $\mathsf{RL_L}$-validity (see Theorem 7.4.1):

Theorem 12.3.1. *For every L-formula φ and for all object variables x and y, the following conditions are equivalent:*

1. *φ is L-valid;*
2. *$x\tau(\varphi)y$ is $\mathsf{RL_L}$-valid.*

Dual tableau systems for Rare-logics are constructed in a similar way as those for information logics of plain frames. Namely, to the $\mathsf{RL}(1, 1')$-dual tableau we add the rules corresponding to the global conditions assumed in a given logic. The rules corresponding to conditions (S1) and (W1) have the following forms:

For all object variables x and y,

$$(\text{rS1}\supseteq)\quad \frac{xR_{P\cup Q}y}{xR_Py \mid xR_Qy} \qquad (\text{rS1}\subseteq)\quad \frac{x{-}R_{P\cup Q}y}{x{-}R_Py,\ x{-}R_Qy}$$

$$(\text{rW1}\supseteq)\quad \frac{xR_{P\cup Q}y}{xR_Py,\ xR_Qy} \qquad (\text{rW1}\subseteq)\quad \frac{x{-}R_{P\cup Q}y}{x{-}R_Py \mid x{-}R_Qy}$$

The rule corresponding to the condition (S2) has the following form:

$$\text{(rS2)} \quad \frac{}{x-R_{\emptyset}y} \qquad x, y \text{ are any object variables}$$

The rule corresponding to the condition (W2) is:

$$\text{(rW2)} \quad \frac{}{xR_{\emptyset}y} \qquad x, y \text{ are any object variables}$$

If some of the local conditions are assumed in L-models, then we add the rules reflecting them. The examples of the rules corresponding to the local conditions can be found in Sect. 11.3.

The specific rules of relational proof systems for the logics $\mathsf{L_{FS}}$ and $\mathsf{L_{FW}}$ are:

- The specific $\mathsf{RL_{L_{FS}}}$-rules: (rS2), (rS1⊇), and (rS1⊆);
- The specific $\mathsf{RL_{L_{FW}}}$-rules: (rW2), (rW1⊇), and (rW1⊆);

The notions of an $\mathsf{RL_{L_{FS}}}$-set, an $\mathsf{RL_{L_{FW}}}$-set, and correctness of a rule are defined as in Sect. 2.4. Then, the following holds:

Proposition 12.3.1.

1. *The specific* $\mathsf{RL_{L_{FS}}}$*-rules are* $\mathsf{RL_{L_{FS}}}$*-correct;*
2. *The specific* $\mathsf{RL_{L_{FW}}}$*-rules are* $\mathsf{RL_{L_{FW}}}$*-correct.*

Proof. By way of example, we prove $\mathsf{RL_{L_{FS}}}$-correctness of the rule (rS2). Let X be any set of $\mathsf{RL_{L_{FS}}}$-formulas. Assume $X \cup \{x-R_{\emptyset}y\}$ is an $\mathsf{RL_{L_{FS}}}$-set. Suppose that X is not an $\mathsf{RL_{L_{FS}}}$-set, that is there exist an $\mathsf{RL_{L_{FS}}}$-model $\mathcal{M}$ and a valuation v in $\mathcal{M}$ such that for every $\varphi \in X$, $\mathcal{M}, v \not\models \varphi$. By the assumption, $\mathcal{M}, v \models x-R_{\emptyset}y$, thus $(v(x), v(y)) \notin R_{\emptyset}$. However, in all $\mathsf{RL_{L_{FS}}}$-models, $R_{\emptyset} = U \times U$, a contradiction. The preservation of validity from the upper set to the bottom set is obvious. □

Let $\mathsf{L} \in \{\mathsf{L_{FS}}, \mathsf{L_{FW}}\}$. The notions of an $\mathsf{RL_L}$-proof tree, a closed branch of such a tree, a closed $\mathsf{RL_L}$-proof tree, and $\mathsf{RL_L}$-provability are defined as in Sect. 2.4.

A branch b of an $\mathsf{RL_L}$-proof tree is complete whenever it is closed or it satisfies the completion conditions of $\mathsf{RL}(1, 1')$-dual tableau adjusted to the language of $\mathsf{RL_L}$ and the following completion conditions corresponding to the rules (rS1⊇), (rS1⊆), (rW1⊇), (rW1⊆), (rS2), and (rW2):

For all object variables x and y,

Cpl(rS1⊇) If $xR_{P\cup Q}y \in b$, then either $xR_Py \in b$ or $xR_Qy \in b$, obtained by an application of the rule (rS1⊇);

Cpl(rS1⊆) If $x-R_{P\cup Q}y \in b$, then both $x-R_Py \in b$ and $x-R_Qy \in b$, obtained by an application of the rule (rS1⊆);

Cpl(rW1⊇) If $xR_{P\cup Q}y \in b$, then both $xR_Py \in b$ and $xR_Qy \in b$, obtained by an application of the rule (rW1⊇);

Cpl(rW1⊆) If $x-R_{P\cup Q}y \in b$, then either $x-R_Py \in b$ or $x-R_Qy \in b$, obtained by an application of the rule (rW1⊆);

Cpl(rS2) For all objects variables x and y, $x - R_{\emptyset} y \in b$, obtained by an application of the rule (rS2);

Cpl(rW2) For all objects variables x and y, $x R_{\emptyset} y \in b$, obtained by an application of the rule (rW2).

The notions of a complete $\mathsf{RL_L}$-proof tree and an open branch of an $\mathsf{RL_L}$-proof tree are defined as in RL-logic (see Sect. 2.5).

The following form of the closed branch property holds:

Fact 12.3.1 (Closed Branch Property). *For every branch b of an $\mathsf{RL_L}$-proof tree, if $x R_{\{p\}} y \in b$ and $x - R_{\{p\}} y \in b$, for $p \in Par$, or $x R_{\emptyset} y \in b$ and $x - R_{\emptyset} y \in b$, then branch b is closed.*

Although, it does not concern relational constants indexed with $P \cup Q$, for $P, Q \subseteq$ Par, it is sufficient for proving satisfaction in branch model property (Proposition 12.3.3). The reason being that the rules for these constants reflect their corresponding definitions in logic L.

Let $\mathsf{L} \in \{\mathsf{L_{FS}}, \mathsf{L_{FW}}\}$. The branch structure determined by an open branch of an $\mathsf{RL_L}$-proof tree is a structure $\mathcal{M}^b = (U^b, (R^b_P)_{P \in \mathcal{P}_{fin}(\mathrm{Par})}, m^b)$ satisfying the following conditions:

- $U^b = \mathbb{OV}_{\mathsf{RL_L}}$;
- $m^b(S) = \{(x, y) \in U^b \times U^b : xSy \notin b\}$, for every $S \in \mathbb{RV}_{\mathsf{RL_L}} \cup \{1, 1'\}$;
- $R^b_{\emptyset} = m^b(R_{\emptyset}) = \{(x, y) \in U^b \times U^b : x R_{\emptyset} y \notin b\}$;
- $R^b_{\{p\}} = m^b(R_{\{p\}}) = \{(x, y) \in U^b \times U^b : x R_{\{p\}} y \notin b\}$, for any $p \in$ Par;
- For all $P, Q \in \mathcal{P}_{fin}$(Par):

$$R^b_{P \cup Q} = m^b(R_{P \cup Q}) = \begin{cases} R^b_P \cap R^b_Q, & \text{if } \mathsf{L} = \mathsf{L_{FS}} \\ R^b_P \cup R^b_Q, & \text{if } \mathsf{L} = \mathsf{L_{FW}}; \end{cases}$$

- m^b extends to all the compound relational terms as in $\mathsf{RL}(1, 1')$-models.

As usual, valuation v^b in $\mathcal{M}^b$ is the identity valuation.

Proposition 12.3.2 (Branch Model Property). *Let $\mathsf{L} \in \{\mathsf{L_{FS}}, \mathsf{L_{FW}}\}$ and let b be a complete branch of an $\mathsf{RL_L}$-proof tree. Then $\mathcal{M}^b$ is an $\mathsf{RL_L}$-model.*

Proof. By way of example, we show that if L-models satisfy the condition (Si) (resp. (Wi)), $i \in \{1, 2\}$, then the branch model satisfies this condition as well. If conditions (S1) or (W1) are assumed, then these properties are satisfied due to the definition of the branch model. Assume that L-models satisfy the condition (S2). By the completion condition Cpl(rS2), for all object variables $x, y \in U^b$, $x - R_{\emptyset} y \in b$. Therefore, for all object variables $x, y \in U^b$, $x R_{\emptyset} y \notin b$, thus $(x, y) \in R^b_{\emptyset}$. Hence, the branch model satisfies the condition (S2). The proof for condition (W2) is similar. All the remaining conditions of $\mathsf{RL_L}$-models are clearly satisfied by $\mathcal{M}^b$. □

If any local condition is assumed in $\mathsf{RL_L}$-models, then we also need to show that $\mathcal{M}^b$ satisfies them.

Next, we show:

Proposition 12.3.3 (Satisfaction in Branch Model Property). *Let* $\mathsf{L} \in \{\mathsf{L_{FS}}, \mathsf{L_{FW}}\}$ *and let* $\mathcal{M}^b$ *be the branch model determined by an open branch* b *of an* $\mathsf{RL_L}$*-proof tree. Then, for every* $\mathsf{RL_L}$*-formula* φ, *if* $\mathcal{M}^b, v^b \models \varphi$, *then* $\varphi \notin b$.

Proof. By way of example, consider logic $\mathsf{L_{FS}}$. Assume that the branch model satisfies a formula $xR_{P\cup Q}y$. By the definition of the branch model, $(x, y) \in R^b_P$ and $(x, y) \in R^b_Q$. Suppose $xR_{P\cup Q}y \in b$. Then, by the completion condition Cpl(rS1$\supseteq$), either $xR_P y \in b$ or $xR_Q y \in b$. By the induction hypothesis, either $(x, y) \notin R^b_P$ or $(x, y) \notin R^b_Q$, a contradiction. □

Thus, we obtain:

Theorem 12.3.2 (Soundness and Completeness of $\mathsf{RL_{L_{FS}}}$ and $\mathsf{RL_{L_{FW}}}$). *Let* $\mathsf{L} \in \{\mathsf{L_{FS}}, \mathsf{L_{FW}}\}$ *and let* φ *be an* $\mathsf{RL_L}$*-formula. Then, the following conditions are equivalent:*

1. φ *is* $\mathsf{RL_L}$*-valid;*
2. φ *is true in all standard* $\mathsf{RL_L}$*-models;*
3. φ *is* $\mathsf{RL_L}$*-provable.*

Recall that an $\mathsf{RL_L}$ model is standard whenever $1'$ is interpreted as the identity. By the above theorem and Theorem 12.3.1, we get:

Theorem 12.3.3 (Relational Soundness and Completeness of Logics $\mathsf{L_{FS}}$ and $\mathsf{L_{FW}}$). *Let* $\mathsf{L} \in \{\mathsf{L_{FS}}, \mathsf{L_{FW}}\}$ *and let* φ *be an* L*-formula. Then for all object variables* x *and* y*, the following conditions are equivalent:*

1. φ *is* L*-valid;*
2. $x\tau(\varphi)y$ *is* $\mathsf{RL_L}$*-provable.*

Example. Consider an $\mathsf{L_{FS}}$-formula φ and an $\mathsf{L_{FW}}$-formula ψ:

$$\varphi = \langle\langle R_{P\cup Q}\rangle\rangle p \rightarrow (\langle\langle R_P\rangle\rangle p \vee \langle\langle R_Q\rangle\rangle p),$$
$$\psi = [R_\emptyset](p \wedge \neg p).$$

The translations of these formulas into relational terms are:

$$\tau(\varphi) = -(-R_{P\cup Q} ; -T) \cup (-R_P ; -T) \cup (-R_Q ; -T),$$
$$\tau(\psi) = -(R_\emptyset ; -(T \cap -T)),$$

where for simplicity of notation $\tau(p) = T$. $\mathsf{L_{FS}}$-validity of φ is equivalent to $\mathsf{RL_{L_{FS}}}$-provability of the formula $x\tau(\varphi)y$. Its $\mathsf{RL_{L_{FS}}}$-proof is presented in Fig. 12.1. $\mathsf{L_{FW}}$-validity of ψ is equivalent to $\mathsf{RL_{L_{FW}}}$-provability of the formula $x\tau(\psi)y$. Its proof is presented in Fig. 12.2.

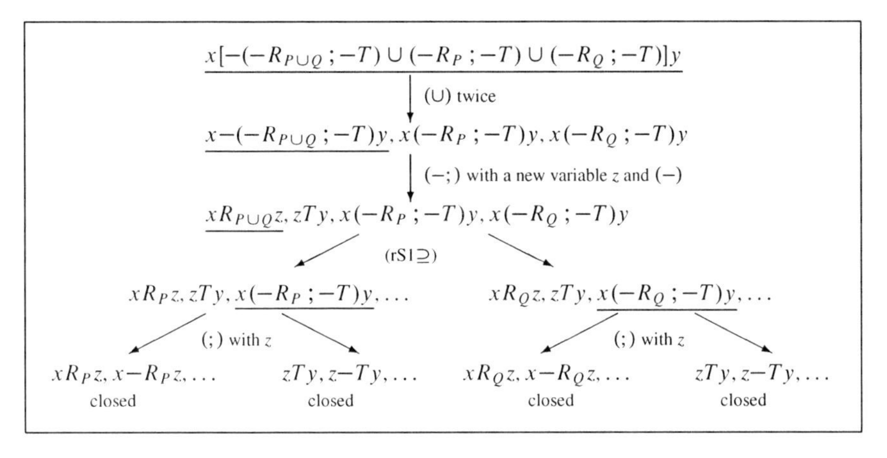

Fig. 12.1 A relational proof of $\mathsf{L_{FS}}$-formula $\langle\langle R_{P\cup Q}\rangle\rangle p \rightarrow (\langle\langle R_P\rangle\rangle p \vee \langle\langle R_Q\rangle\rangle p)$

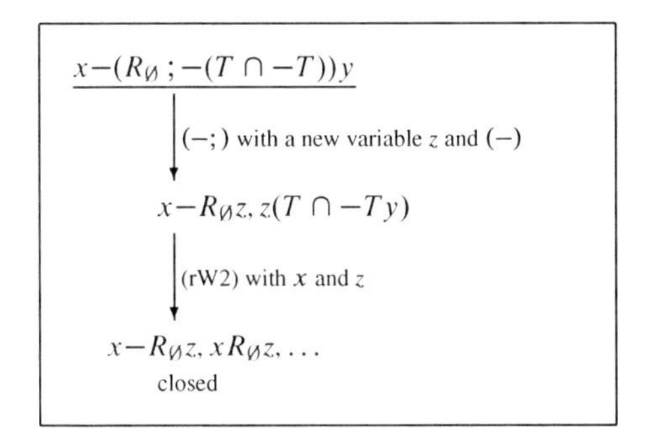

Fig. 12.2 A relational proof of $\mathsf{L_{FW}}$-formula $[R_\emptyset](p \wedge \neg p)$

12.4 Relational Formalization of the Logic Rare-NIL

The second case study of Rare-logics is a relative version of the logic NIL, Rare-NIL. Let Par be a non-empty set of parameters. The vocabulary of the language of Rare-NIL consists of symbols from the following pairwise disjoint sets:

- $\mathbb{V}$ – a countable infinite set of propositional variables;
- $\{\leq_P\}_{P\in\mathcal{P}_{fin}(\mathrm{Par})}$, $\{\geq_P\}_{P\in\mathcal{P}_{fin}(\mathrm{Par})}$, $\{\sigma_P\}_{P\in\mathcal{P}_{fin}(\mathrm{Par})}$ – three families of relational constants;
- $\{\neg, \vee, \wedge\} \cup \{[T] : T$ is a relational constant$\}$ – a set of propositional operations.

A Rare-NIL-model is a structure

$$\mathcal{M} = (U, (\leq_P)_{P\in\mathcal{P}_{fin}(\mathrm{Par})}, (\geq_P)_{P\in\mathcal{P}_{fin}(\mathrm{Par})}, (\sigma_P)_{P\in\mathcal{P}_{fin}(\mathrm{Par})}, m),$$

such that $(U, (\#_P)_{P \in \mathcal{P}_{fin}(\mathrm{Par})}, m)$ is an $\mathsf{L_{FS}}$-model, for every $\# \in \{\leq, \geq, \sigma\}$, and the following conditions are satisfied:

For every $P \in \mathcal{P}_{fin}(\mathrm{Par})$ and for all $x, y, z \in U$,

- $\leq_P = (\geq_P)^{-1}$;
- $\leq_P$ is reflexive and transitive;
- σ_P is reflexive and symmetric;
- if $(x, y) \in \sigma_P$ and $(y, z) \in \leq_P$, then $(x, z) \in \sigma_P$.

Observe that the relations in Rare-NIL-models are constrained in two ways. The three families of relations satisfy the global conditions (S1) and (S2) and the local conditions analogous to the conditions in NIL-models.

In [DO07] it is shown that Rare-NIL-satisfiability (possibly) differs from NIL-satisfiability, namely we have:

Theorem 12.4.1.

1. *Satisfiability problem of Rare-*NIL *is decidable;*
2. *Rare-*NIL *satisfiability is* EXPTIME*-complete.*

The relational language corresponding to Rare-NIL-language is the $\mathsf{RL}(1, 1')$-language endowed with the set $\{r_P\}_{P \in \mathcal{P}_{fin}(\mathrm{Par})}$, $r \in \{\leq, \geq, \sigma\}$ of relational constants. $\mathsf{RL}_{\text{Rare-NIL}}$-models are structures of the form

$$\mathcal{M} = (U, (\leq_P)_{P \in \mathcal{P}_{fin}(\mathrm{Par})}, (\geq_P)_{P \in \mathcal{P}_{fin}(\mathrm{Par})}, (\sigma_P)_{P \in \mathcal{P}_{fin}(\mathrm{Par})}, m),$$

such that (U, m) is an $\mathsf{RL}(1, 1')$-model, $r_P = m(r_P)$, for $r \in \{\leq, \geq, \sigma\}$, are binary relations on U indexed with finite subsets of the set Par, and moreover, relations r_P satisfy all the constraints assumed in Rare-NIL-models.

A dual tableau for Rare-NIL is constructed as in the case of the previous logics. Namely, to the $\mathsf{RL}(1, 1')$-dual tableau we add the rules corresponding to the global and local conditions assumed in this logic. More precisely, we add the following rules:

For $r \in \{\leq, \geq, \sigma\}$, and for all $P, Q \in \mathcal{P}_{fin}(\mathrm{Par})$, the rules of the forms (rS1$\supseteq$), (rS1$\subseteq$), and (rS2) for the relational constants $r_{P \cup Q}$ and $r_\emptyset$, respectively;

For $r \in \{\leq, \sigma\}$, and for every $P \in \mathcal{P}_{fin}(\mathrm{Par})$, the rules (rI1$\subseteq$), (rI1$\supseteq$), (ref R), (tran R) for the relational constants $\leq_P$ and $\geq_P$, and (ref R), (sym R), (rI6) for the relational constants σ_P; these rules can be found in Sect. 11.4.

By Theorem 11.4.2 and Proposition 12.3.1, it can be easily proved that all the specific $\mathsf{RL}_{\text{Rare-NIL}}$-rules are $\mathsf{RL}_{\text{Rare-NIL}}$-correct. Completeness can be proved in a similar way to completeness of $\mathsf{RL}_{\mathsf{NIL}}$-dual tableau and $\mathsf{RL}_{\mathsf{L_{FS}}}$-dual tableau. Thus, we obtain:

Theorem 12.4.2 (Soundness and Completeness of $\mathsf{RL}_{\mathbf{Rare}\text{-}\mathsf{NIL}}$). *For every $\mathsf{RL}_{Rare\text{-}\mathsf{NIL}}$-formula φ, the following conditions are equivalent:*

1. *φ is $\mathsf{RL}_{Rare\text{-}\mathsf{NIL}}$-valid;*
2. *φ is true in all standard $\mathsf{RL}_{Rare\text{-}\mathsf{NIL}}$-models;*
3. *φ is $\mathsf{RL}_{Rare\text{-}\mathsf{NIL}}$-provable.*

By the above theorem and Theorems 11.4.1 and 12.3.1, we have:

Theorem 12.4.3 (Relational Soundness and Completeness of Rare-NIL). *For every Rare-NIL-formula φ and for all object variables x and y, the following conditions are equivalent:*

1. *φ is Rare-NIL-valid;*
2. *$x\tau(\varphi)y$ is $\mathsf{RL}_{Rare\text{-}\mathsf{NIL}}$-provable.*

12.5 Relational Formalization of the Logic Rare-CI

Now, we consider a relational formalization of the logic of relative complementarity and incomplementarity, Rare-CI. The language of this logic is the multimodal language with the symbols from the following pairwise disjoint sets:

- $\mathbb{V}$ – a countable infinite set of propositional variables;
- $\{R_P : P \in \mathcal{P}_{fin}(\mathrm{Par})\} \cup \{S_P : P \in \mathcal{P}_{fin}(\mathrm{Par})\}$ – the set of relational constants;
- $\{\neg, \vee, \wedge\} \cup \{[R_P], [[S_P]] : P \in \mathcal{P}_{fin}(\mathrm{Par})\}$ – the set of propositional operations.

A Rare-CI-model is a structure:

$$\mathcal{M} = (U, (R_P)_{P\in\mathcal{P}_{fin}(\mathrm{Par})}, (S_P)_{P\in\mathcal{P}_{fin}(\mathrm{Par})}, m),$$

such that U is a non-empty set, R_P and S_P are binary relations on U indexed with finite subsets of Par, m is a meaning function such that for every $P \in \mathcal{P}_{fin}(\mathrm{Par})$ the following conditions are satisfied:

- $m(p) \subseteq U$, for every propositional variable p;
- relations R_P satisfy the conditions (S1) and (S2), and relations S_P satisfy the conditions (W1) and (W2);
- R_P are symmetric and 3-transitive; these relations are counterparts to the strong complementarity relations derived from an information system;
- S_P are reflexive; these relations are counterparts to the weak incomplementarity relations derived from an information system;
- $R_P \cup S_P = U \times U$ and $R_P \cap S_P = \emptyset$.

It follows that relations R_P are strong relations and relations S_P are weak relations.

The relational language corresponding to Rare-CI-language is the $\mathsf{RL}(1, 1')$-language endowed with relational constants R_P and S_P, for $P \in \mathcal{P}_{fin}(\mathrm{Par})\}$. $\mathsf{RL}_{\mathrm{Rare}\text{-}\mathsf{CI}}$-models are structures of the form:

$$\mathcal{M} = (U, (R_P)_{P\in\mathcal{P}_{fin}(\mathrm{Par})}, (S_P)_{P\in\mathcal{P}_{fin}(\mathrm{Par})}, m),$$

such that (U, m) is an $\mathsf{RL}(1, 1')$-model, $R_P = m(R_P)$ and $S_P = m(S_P)$ are binary relations on U indexed with finite subsets of a non-empty set Par, and moreover, relations R_P and S_P satisfy all the conditions assumed in Rare-CI-models.

A dual tableau for Rare-CI is constructed from the $\mathsf{RL}(1, 1')$-proof system. We add to it the rules corresponding to the global and local conditions assumed in Rare-CI-logic. Thus, the rules corresponding to the global conditions for relations R_P, $P \in \mathcal{P}_{fin}(\text{Par})$, are the instances of the rules of logic $\mathsf{RL}_{\mathsf{LFS}}$, and the rules for relations S_P, $P \in \mathcal{P}_{fin}(\text{Par})$, are the instances of the rules of logic $\mathsf{RL}_{\mathsf{LFW}}$. The rules reflecting the local conditions posed on relations R_P and S_P are those presented in Sect. 11.5 in the models of logic $\mathsf{RL}_{\mathsf{CI}}$.

$\mathsf{RL}_{\text{Rare-}\mathsf{CI}}$-axiomatic sets are those that include either of the following subsets:

For all object symbols x and y, and for every relational term T,

(Ax1) $\{xTy, x{-}Ty\}$;
(Ax2) $\{xS_Px\}$;
(Ax3) $\{xR_Py, xS_Py\}$.

By Propositions 11.5.1 and 12.3.1, we obtain:

Proposition 12.5.1.

1. *The* $\mathsf{RL}_{Rare\text{-}\mathsf{CI}}$*-rules are* $\mathsf{RL}_{Rare\text{-}\mathsf{CI}}$*-correct;*
2. *The* $\mathsf{RL}_{Rare\text{-}\mathsf{CI}}$*-axiomatic sets are* $\mathsf{RL}_{Rare\text{-}\mathsf{CI}}$*-sets.*

The completion conditions determined by the rules reflecting the global conditions are the instances of those presented in Sect. 12.3. The completion conditions determined by the rules that reflect the local conditions are the instances of those presented in Sect. 11.5.

The completeness proof is based on the same ideas as the completeness proofs of $\mathsf{RL}_{\mathsf{CI}}$, $\mathsf{RL}_{\mathsf{LFS}}$, and $\mathsf{RL}_{\mathsf{LFW}}$ dual tableaux. Namely, by Propositions 11.5.2 and 12.3.2, the branch model property holds. The satisfaction in branch model property is due to Proposition 12.3.3. Finally, by Theorems 11.5.2 and 12.3.2, we get:

Theorem 12.5.1 (Soundness and Completeness of $\mathsf{RL}_{\text{Rare-}\mathsf{CI}}$). *Let φ be an* $\mathsf{RL}_{Rare\text{-}\mathsf{CI}}$*-formula. Then the following conditions are equivalent:*

1. *φ is* $\mathsf{RL}_{Rare\text{-}\mathsf{CI}}$*-valid;*
2. *φ is true in all standard* $\mathsf{RL}_{Rare\text{-}\mathsf{CI}}$*-models;*
3. *φ is* $\mathsf{RL}_{Rare\text{-}\mathsf{CI}}$*-provable.*

By the above theorem and Theorems 11.5.1 and 12.3.1, we have:

Theorem 12.5.2 (Relational Soundness and Completeness of Rare-CI). *For every Rare-*CI*-formula φ and for all object variables x and y, the following conditions are equivalent:*

1. *φ is Rare-*CI*-valid;*
2. *$x\tau(\varphi)y$ is* $\mathsf{RL}_{Rare\text{-}\mathsf{CI}}$*-provable.*

12.6 Relational Formalization of the Logic of Strong Complementarity Frames

Now, we present the relational formalization of the multimodal logic $\mathsf{L}_{\mathsf{FS-COM}}$ based on a family $\mathsf{FS-COM}$ of complementarity frames. The logic $\mathsf{L}_{\mathsf{FS-COM}}$ is defined as the L_{FS}-logic with the following additional semantic conditions: $R_{\{p\}}$ is irreflexive, symmetric, and 3-transitive, for every parameter $p \in \mathrm{Par}$. The same conditions are assumed in the models of the corresponding relational logic $\mathsf{RL}_{\mathsf{FS-COM}}$. A dual tableau for $\mathsf{RL}_{\mathsf{L}_{\mathsf{FS-COM}}}$ includes all the rules and the axiomatic sets of the dual tableau for L_{FS}-logic presented in Sect. 12.3 and in addition the rules reflecting irreflexivity, symmetry, and 3-transitivity, for every relational constant $R_{\{p\}}$. The rule for symmetry can be found in Sect. 6.6 and the rule for 3-transitivity in Sect. 11.5. The rule reflecting irreflexivity of relation R has the following form:

For all object variables x and y,

$$(\text{irref } R) \qquad \frac{}{xRy}$$

Example. It is easy to show that in $\mathsf{L}_{\mathsf{FS-COM}}$-models a relation R_P is symmetric, for every $P \in \mathcal{P}_{fin}(\mathrm{Par})$. Therefore, the formula $x(-R_P \cup R_P^{-1})y$ is valid in all $\mathsf{RL}_{\mathsf{L}_{\mathsf{FS-COM}}}$-models, for every $P \in \mathcal{P}_{fin}(\mathrm{Par})$. By way of example, in Fig. 12.3 we present an $\mathsf{RL}_{\mathsf{L}_{\mathsf{FS-COM}}}$-proof of the formula $x(-R_{\{p\}\cup\{q\}} \cup R^{-1}_{\{p\}\cup\{q\}})y$, for $p, q \in \mathrm{Par}$.

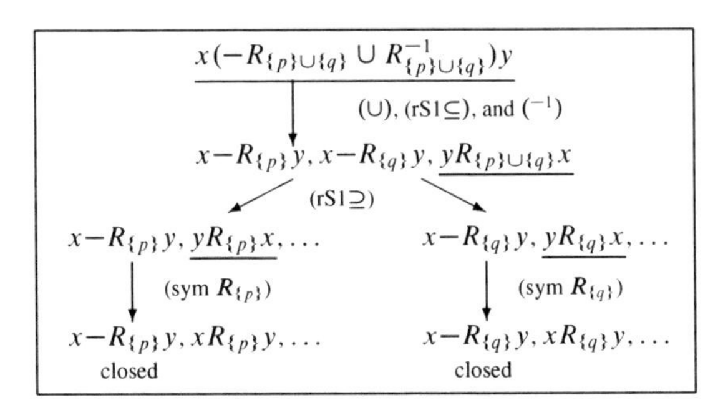

Fig. 12.3 An $\mathsf{RL}_{\mathsf{L}_{\mathsf{FS-COM}}}$-proof of symmetry of the relation $R_{\{p\}\cup\{q\}}$

Chapter 13
Dual Tableau for Formal Concept Analysis

13.1 Introduction

In this chapter we show that an extension of the relational logic $\mathsf{RL}(1, 1')$ can be applied to verification of some logical problems in *formal concept analysis*, FCA, which originated in [Wil82]. A comprehensive exposition of FCA can be found in [GW99]. A basic data structure in formal concept analysis is a *context*. It is a two-sorted relational system consisting of a set whose elements are interpreted as objects, a set of features of these objects and a binary relation which holds between an object and a feature whenever the feature is attributed to the object. In the original formulation of FCA the term 'attribute' is used instead of 'feature', although they are only single-valued attributes. In the more recent developments, FCA distinguishes between single-valued and many-valued attributes. Our terminology is in agreement with that established in relational databases and information systems with incomplete information: an attribute (e.g., color) may have multiple values (e.g., blue, red, etc.), and a feature is a pair 'attribute-value' (e.g., being of blue color).

With the class of contexts a modal logic, referred to as a *context logic*, FCL, can be associated. This logic is based on sufficiency algebras [DO01, DO04] such that the two sufficiency operations of the logic correspond to the mappings of extent and intent determined by a context. Consequently, various problems dealt with in FCA can be represented in logic FCL. We present a relational dual tableau for the logic FCL and we illustrate with examples how it can be applied to the verification of various tasks relevant for FCA. The present chapter is based on [GPO07a]. In [OR08] context algebras associated with the context logic are introduced and a discrete duality for these algebras is presented.

13.2 Basic Notions of Formal Concept Analysis

In this section we recall basic notions of formal concept analysis, i.e., a context, a concept, and the extent and the intent of a concept. Then, we introduce the notion of context frame, and we show how context frames and contexts are related.

E. Orłowska and J. Golińska-Pilarek, *Dual Tableaux: Foundations, Methodology, Case Studies*, Trends in Logic 36, DOI 10.1007/978-94-007-0005-5_13,

A *context* is a triple $\mathcal{C} = (G, M, I)$, where G and M are non-empty sets of *objects* and *features*, respectively, and $I \subseteq G \times M$. The expression gIm is read as 'the object g has the feature m'. A concept is determined by its extent and its intent: the extent consists of all objects belonging to the concept, while the intent is the collection of all features shared by the objects.

For $A \subseteq G$ and $B \subseteq M$ we define:

$$i(A) \stackrel{\text{df}}{=} \{m \in M : \forall g,\ g \in A \text{ implies } gIm\};$$
$$e(B) \stackrel{\text{df}}{=} \{g \in G : \forall m,\ m \in B \text{ implies } gIm\}.$$

Then a *concept* of the context (G, M, I) is defined as a pair (A, B) such that $A \subseteq G$, $B \subseteq M$, $A = e(B)$, and $B = i(A)$. The sets $i(A)$ and $e(B)$ are called the *intent* and the *extent* of the concept (A, B), respectively.

Fact 13.2.1.

1. *$A \subseteq G$ is an extent of some concept iff $A = e(i(A))$;*
2. *The unique concept of which A is an extent is $(A, i(A))$.*

Fact 13.2.2.

1. *$B \subseteq M$ is an intent of some concept iff $B = i(e(B))$;*
2. *The unique concept of which B is an intent is $(e(B), B)$.*

An implication of a context (G, M, I) is a pair of subsets P, Q of M, written as $P \to Q$. An implication $P \to Q$ holds in the context whenever $e(P) \subseteq e(Q)$. Intuitively, it means that each object from G having all the features from P has also the features from Q.

A *context frame* is a structure (X, R, S), where X is a non-empty set, $R, S \subseteq X \times X$, and $S = R^{-1}$. The sufficiency operations determined by a context frame are defined for every $A \subseteq X$ and $T \in \{R, S\}$ as in Sect. 7.3, that is:

$$[[T]]A \stackrel{\text{df}}{=} \{x \in X : \forall y \in X,\ y \in A \text{ implies } xTy\}.$$

Given a context $\mathcal{C} = (G, M, I)$, a *frame derived from* $\mathcal{C}$, $\mathcal{F}_{\mathcal{C}}$, is a structure $\mathcal{F}_{\mathcal{C}} = (X_{\mathcal{C}}, R_{\mathcal{C}}, S_{\mathcal{C}})$ such that $X_{\mathcal{C}} = G \cup M$, $R_{\mathcal{C}} = I$, and $S_{\mathcal{C}} = I^{-1}$.

Proposition 13.2.1. *The mappings of extent and intent determined by a context $\mathcal{C} = (G, M, I)$ are the sufficiency operations determined by the frame $\mathcal{F}_{\mathcal{C}}$, that is:*

1. *$e = [[R_{\mathcal{C}}]]$;*
2. *$i = [[S_{\mathcal{C}}]]$.*

Given a context frame $\mathcal{F} = (X, R, S)$, we define a context $\mathcal{C}_{\mathcal{F}}$ as a structure $(G_{\mathcal{F}}, M_{\mathcal{F}}, I_{\mathcal{F}})$ such that $G_{\mathcal{F}} = Dom(R)$ is the domain of the relation R, $M_{\mathcal{F}} = Rng(R)$ is the range of the relation R, and $I_{\mathcal{F}} = R$.

Proposition 13.2.2. *The sufficiency operations determined by a frame* $\mathcal{F} = (X, R, S)$ *are the operations of extent and intent determined by the context* $\mathcal{C}_{\mathcal{F}}$*, that is:*

1. $[[R]] = e_{\mathcal{F}}$;
2. $[[S]] = i_{\mathcal{F}}$.

Proposition 13.2.3. *If a context* $\mathcal{C} = (G, M, I)$ *satisfies* $G = Dom(I)$ *and* $M = Rng(I)$*, then* $\mathcal{C} = \mathcal{C}_{\mathcal{F}_{\mathcal{C}}}$.

Proposition 13.2.4. *If a context frame* $\mathcal{F} = (X, R, S)$ *satisfies* $X = Dom(R) \cup Rng(R)$*, then* $\mathcal{F} = \mathcal{F}_{\mathcal{C}_{\mathcal{F}}}$.

13.3 Context Logic and Its Dual Tableau

The vocabulary of the language of the modal logic FCL appropriate for reasoning in FCA consists of propositional variables from a non-empty set $\mathbb{V}$, relational constants R and S, and propositional operations of negation ($\neg$), disjunction ($\vee$), conjunction ($\wedge$), and the sufficiency operations $[[R]]$ and $[[S]]$. The set of FCL-formulas is generated from propositional variables with the propositional operations.

An FCL-*model* is a structure of the form $\mathcal{M} = (X, R, S, m)$ such that (X, R, S) is a context frame and a meaning function $m: \mathbb{V} \to X$ provides meanings of propositional variables. The satisfaction of formulas in a model is defined as in modal logics (see Sect. 7.3). We recall that satisfaction of formulas built with sufficiency operations is defined as in Sect. 7.3, that is:

For $T \in \{R, S\}$,

$$\mathcal{M}, s \models [[T]]\varphi \text{ iff for every } s' \in X, \text{ if } \mathcal{M}, s' \models \varphi, \text{ then } sTs'.$$

As usual, an FCL-formula φ is said to be true in the FCL-model $\mathcal{M} = (X, R, S, m)$, $\mathcal{M} \models \varphi$ for short, whenever for every $s \in X$, $\mathcal{M}, s \models \varphi$. An FCL-formula φ is said to be FCL-valid whenever φ is true in all the FCL-models.

By the definition of FCL-models and due to the relationship between contexts and context frames presented in the previous section, we obtain:

Proposition 13.3.1 (Informational Representability).

1. *For every* FCL-*model* $\mathcal{M}$ *there exists a context* $\mathcal{C}$ *such that* $\mathcal{M}$ *is based on a context frame* $\mathcal{F}_{\mathcal{C}}$ *derived from* $\mathcal{C}$;
2. *For every context* $\mathcal{C}$ *there exists an* FCL-*model* $\mathcal{M}$ *such that* $\mathcal{C}$ *is a context derived from the frame which determines* $\mathcal{M}$.

The relational logic $\mathsf{RL}_{\mathsf{FCL}}$ appropriate for expressing FCL-formulas as relational terms is obtained in a standard way, i.e., we expand the language of $\mathsf{RL}(1, 1')$ with relational constants R and S representing the accessibility relations from the FCL-models. The relational terms and formulas are defined as in RL-logic (see Sect. 2.3).

An $\mathsf{RL}_{\mathsf{FCL}}$-structure is of the form $\mathcal{M} = (U, R, S, m)$, where (U, m) is an $\mathsf{RL}(1, 1')$-model and R and S are binary relations on U. An $\mathsf{RL}_{\mathsf{FCL}}$-model is an $\mathsf{RL}_{\mathsf{FCL}}$-structure such that $S = R^{-1}$. As usual, we use the same symbols for constants in the language and for the relations in a structure, and we postulate $m(R) = R$ and $m(S) = S$.

The translation τ of FCL-formulas into relational terms is defined as in standard modal logics (see Sect. 7.4). We recall that the translation of the formulas built with the sufficiency operations is:

$$\tau([[T]]\varphi) \stackrel{\text{df}}{=} -(-T \,;\, \tau(\varphi)), \text{ for } T \in \{R, S\}.$$

The translation is defined so that it preserves validity of formulas. Due to Theorem 7.4.1, we have:

Theorem 13.3.1. *For every* FCL*-formula* φ *and for all object variables* x *and* y*, the following conditions are equivalent:*

1. φ *is* FCL*-valid;*
2. $x\tau(\varphi)y$ *is* $\mathsf{RL}_{\mathsf{FCL}}$*-valid.*

A dual tableau for the logic $\mathsf{RL}_{\mathsf{FCL}}$ is based on the $\mathsf{RL}(1, 1')$-dual tableau. We add to the $\mathsf{RL}(1, 1')$-system the specific rules that reflect the fact $S = R^{-1}$:

For all object symbols x and y,

$$(S)\ \frac{xSy}{yRx, xSy} \qquad\qquad (R)\ \frac{xRy}{ySx, xRy}$$

The alternative forms of the rules above are discussed in Sect. 25.9.

We prove soundness and completeness of $\mathsf{RL}_{\mathsf{FCL}}$ following the method presented in Sect. 7.5. The notions of an $\mathsf{RL}_{\mathsf{FCL}}$-set and correctness of a rule are defined as in Sect. 2.4. The following can be proved in a similar way as Theorem 6.6.1:

Theorem 13.3.2 (Correspondence). *Let* $\mathcal{K}$ *be a class of* $\mathsf{RL}_{\mathsf{FCL}}$*-structures. Then,* $\mathcal{K}$ *is a class of* $\mathsf{RL}_{\mathsf{FCL}}$*-models iff the rules* (S) *and* (R) *are* $\mathcal{K}$*-correct.*

Proof.

$(\rightarrow)$ Let $\mathcal{K}$ be a class of $\mathsf{RL}_{\mathsf{FCL}}$-models, that is in every model of $\mathcal{K}$, $S = R^{-1}$. By way of example, we prove correctness of the rule (S). Let X be a finite set of $\mathsf{RL}_{\mathsf{FCL}}$-formulas. Assume $X \cup \{xSy\}$ is a $\mathcal{K}$-set and suppose $X \cup \{yRx\}$ is not a $\mathcal{K}$-set. Then, there exist an $\mathsf{RL}_{\mathsf{FCL}}$-model $\mathcal{M} = (U, R, S, m)$ in $\mathcal{K}$ and a valuation

v in $\mathcal{M}$ such that $(v(x), v(y)) \in S$ and $(v(y), v(x)) \notin R$. Since $S \subseteq R^{-1}$ holds in $\mathcal{M}$, $(v(y), v(x)) \in R$, a contradiction. The other direction can be proved in a similar way.

($\leftarrow$) Now, let $\mathcal{K}$ be a class of $\mathsf{RL_{FCL}}$-structures. We show that $\mathcal{K}$-correctness of the rules (S) and (R) implies that $S = R^{-1}$ is true in all $\mathcal{K}$-models.

For $S \subseteq R^{-1}$, note that $\{y{-}Sx, ySx\}$ is a $\mathcal{K}$-set. Thus, by correctness of the rule (R), $\{xRy, y{-}Sx\}$ is also a $\mathcal{K}$-set. Therefore, for every $\mathcal{K}$-model and for all $x, y \in U$, if $(x, y) \in S$, then $(x, y) \in R^{-1}$.

For $R^{-1} \subseteq S$, note that $\{y{-}Rx, yRx\}$ is a $\mathcal{K}$-set. Thus, by correctness of the rule (S), $\{xSy, y{-}Rx\}$ is also a $\mathcal{K}$-set. Therefore, for every $\mathcal{K}$-model and for all $x, y \in U$, if $(x, y) \in R^{-1}$, then $(x, y) \in S$. □

The above theorem implies that the rules (S) and (R) are $\mathsf{RL_{FCL}}$-correct. Correctness of the remaining rules can be proved as in $\mathsf{RL}(1, 1')$-dual tableau (see Sects. 2.5 and 2.7), thus we get:

Proposition 13.3.2.

1. *The* $\mathsf{RL_{FCL}}$*-rules are* $\mathsf{RL_{FCL}}$*-correct;*
2. *The* $\mathsf{RL_{FCL}}$*-axiomatic sets are* $\mathsf{RL_{FCL}}$*-sets.*

The notions of an $\mathsf{RL_{FCL}}$-proof tree, a closed branch of such a tree, a closed $\mathsf{RL_{FCL}}$-proof tree, and $\mathsf{RL_{FCL}}$-provability are defined as in Sect. 2.4.

A branch b of an $\mathsf{RL_{FCL}}$-proof tree is complete whenever it is closed or it satisfies the completion conditions of $\mathsf{RL}(1, 1')$-dual tableau adjusted to $\mathsf{RL_{FCL}}$-language and the following which are specific for the $\mathsf{RL_{FCL}}$-dual tableau:

For all object symbols x and y,

Cpl(S) If $xSy \in b$, then $yRx \in b$, obtained by an application of the rule (S);
Cpl(R) If $xRy \in b$, then $ySx \in b$, obtained by an application of the rule (R).

The notions of a complete $\mathsf{RL_{FCL}}$-proof tree and an open branch of an $\mathsf{RL_{FCL}}$-proof tree are defined as in RL-logic (see Sect. 2.5).

Observe that the rules of $\mathsf{RL_{FCL}}$-dual tableau, in particular the specific rules (S) and (R), guarantee that for every branch b of an $\mathsf{RL_{FCL}}$-proof tree, if $xTy \in b$ and $x{-}Ty \in b$ for some atomic relational term T, then there is a node in the branch which contains both of these formulas, and hence branch b is closed. Therefore, the closed branch property holds.

The branch structure $\mathcal{M}^b = (U^b, R^b, S^b, m^b)$ is defined in a standard way (see Sect. 2.6, p. 44), in particular $T^b \stackrel{\text{df}}{=} \{(x, y) \in U^b \times U^b : xTy \notin b\}$ and $m^b(T) = T^b$, for $T \in \{R, S\}$.

Proposition 13.3.3 (Branch Model Property). *Let b be an open branch of an* $\mathsf{RL_{FCL}}$*-proof tree. Then the branch structure $\mathcal{M}^b$ is an* $\mathsf{RL_{FCL}}$*-model.*

Proof. It suffices to show that $S^b = (R^b)^{-1}$. Assume $(x, y) \in S^b$, that is $xSy \notin b$. Suppose $(y, x) \notin R^b$. Then $yRx \in b$ and by the completion condition Cpl(R), $xSy \in b$, a contradiction. The remaining case can be proved in a similar way. □

Since the branch model $\mathcal{M}^b$ is defined in a standard way and the closed branch property holds, the satisfaction in branch model property can be proved as in $\mathsf{RL}(1, 1')$-logic (see Sects. 2.5 and 2.7).

Proposition 13.3.4 (Satisfaction in Branch Model Property). *Let b be an open branch of an $\mathsf{RL_{FCL}}$-proof tree. Then for every $\mathsf{RL_{FCL}}$-formula φ, if $\mathcal{M}^b, v^b \models \varphi$, then $\varphi \notin b$.*

Hence, completeness of $\mathsf{RL_{FCL}}$-dual tableau follows.

Theorem 13.3.3 (Soundness and Completeness of $\mathsf{RL_{FCL}}$). *Let φ be an $\mathsf{RL_{FCL}}$-formula. Then the following conditions are equivalent:*

1. *φ is $\mathsf{RL_{FCL}}$-valid;*
2. *φ is true in all standard $\mathsf{RL_{FCL}}$-models;*
3. *φ is $\mathsf{RL_{FCL}}$-provable.*

Moreover, by Theorems 13.3.1 and 13.3.3, we obtain:

Theorem 13.3.4 (Relational Soundness and Completeness of FCL). *Let φ be an FCL-formula. Then for all object variables x and y, the following conditions are equivalent:*

1. *φ is FCL-valid;*
2. *$x\tau(\varphi)y$ is $\mathsf{RL_{FCL}}$-provable.*

The following example illustrates application of $\mathsf{RL_{FCL}}$-dual tableau to verification of properties of contexts. In this example $\mathsf{RL_{FCL}}$-dual tableau is applied as a validity checker.

Example. Consider the following property of contexts:

(α) For every context $\mathcal{C} = (G, M, I)$ and for any $A \subseteq G$, $A \subseteq e(i(A))$.

Let p be propositional variable representing set A. The representation of (α) in FCL is:

(β) the formula $p \to [[R]][[S]]p$ is FCL-valid.

Denote $\tau(p)$ by P and let x, y be object variables. Relational representation of (β) is:

$x[-P \cup -(-R\,;-(-S\,;P))]y$ is $\mathsf{RL_{FCL}}$-valid.

Figure 13.1 presents an $\mathsf{RL_{FCL}}$-proof of this formula. Recall that in each node of the proof tree we underline the formula to which a rule has been applied during the construction of the proof tree, and we write only those formulas in the nodes which are essential for this construction.

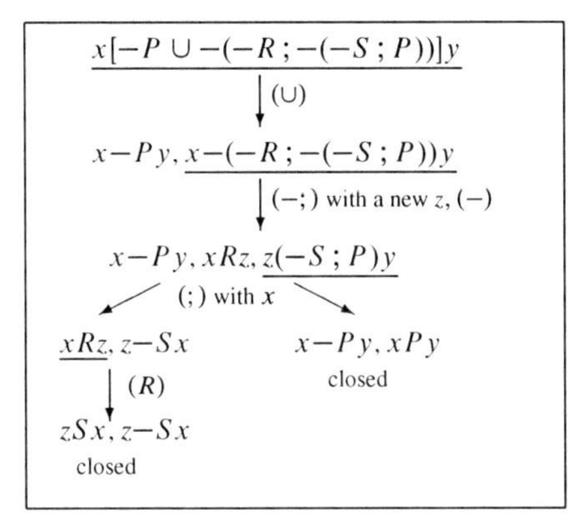

Fig. 13.1 An $\mathsf{RL_{FCL}}$-proof of $A \subseteq e(i(A))$

13.4 Entailment, Model Checking, and Satisfaction in Context Logic

The methods presented in Sects. 7.6–7.8 can be applied to verification of entailment, model checking, and satisfaction in context logic FCL.

Entailment in FCL

In the following example, the relational method of proving entailment presented in Sect. 7.6 is applied to an inference of implications in the contexts.

Example. Consider the following property of contexts:

(α) For every context $\mathcal{C} = (G, M, I)$ and for all sets of features $\mathtt{P}, \mathtt{Q} \subseteq M$, if $\mathtt{P} \rightarrow \mathtt{Q}$ holds in $\mathcal{C}$, then the implication $\mathtt{P} \cup \mathtt{P}' \rightarrow \mathtt{Q}$ holds in $\mathcal{C}$, for every $\mathtt{P}' \subseteq M$.

To represent (α) in FCL, let p, q, and p' be propositional variables representing sets $\mathtt{P}$, $\mathtt{Q}$, and $\mathtt{P}'$, respectively. Then, in view of the definition of implications in a context and due to Proposition 13.3.1, (α) can be represented as:

(β) for every FCL-model $\mathcal{M}$, truth of $[[R]]p \rightarrow [[R]]q$ in $\mathcal{M}$ implies truth of $[[R]](p \vee p') \rightarrow [[R]]q$ in $\mathcal{M}$, for every propositional variable p'.

Denote $\tau(p)$, $\tau(q)$, and $\tau(p')$ by P, Q, and P', respectively. Let x and y be object variables. A relational representation of (β) is:

$x[1\,;-(--(-R\,;P) \cup -(-R\,;Q))\,;1 \cup (--(-R\,;(P \cup P')) \cup -(-R\,;Q))]y$
is $\mathsf{RL_{FCL}}$-valid.

Figure 13.2 presents $\mathsf{RL_{FCL}}$-proof of this formula.

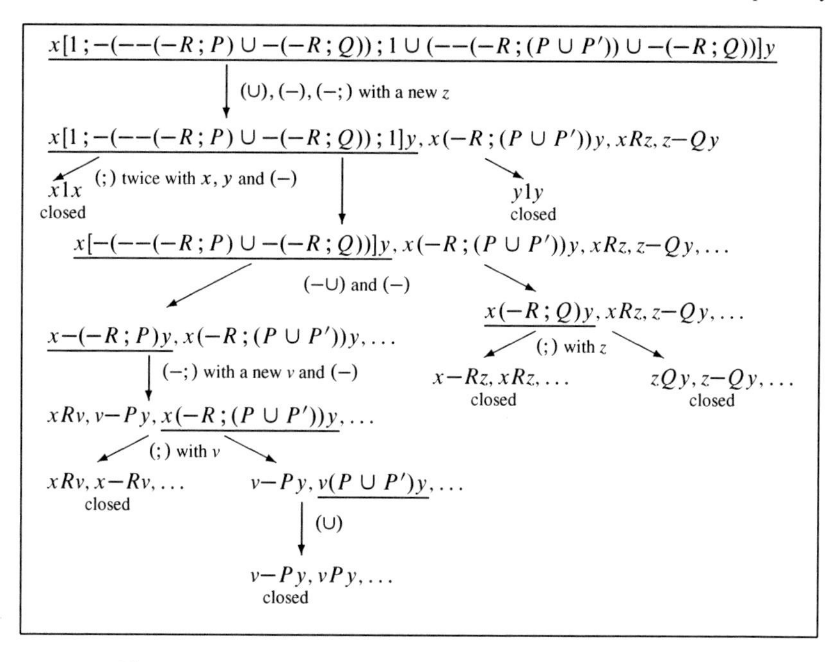

Fig. 13.2 An $\mathsf{RL}_{\mathsf{FCL}}$-proof showing that $\mathtt{P}\rightarrow\mathtt{Q}$ implies $(\mathtt{P}\cup\mathtt{P}')\rightarrow\mathtt{Q}$

Model Checking and Satisfaction in FCL

Assume we are given a finite context and its property such that they can be represented with a formula, say φ, and a finite model, say $\mathcal{M}$, of logic FCL. In order to apply the $\mathsf{RL}_{\mathsf{FCL}}$-dual tableau to the problem of checking whether φ is true in $\mathcal{M}$, we translate this problem to model checking problem in the relational logic $\mathsf{RL}_{\mathsf{FCL}}$ (see Sect. 7.7). We extend the $\mathsf{RL}_{\mathsf{FCL}}$-dual tableau with rules and axiomatic sets which express the given property and a relevant part of the context in question. Then we apply this proof system to verification of the property in the context. Similarly, we can apply the method of verification of satisfaction of FCL-formulas by a valuation in a given finite model. The following example illustrates the main steps of this procedure.

Example. Let $\mathcal{C}=(G,M,I)$ be a context such that $G=\{\mathtt{a},\mathtt{b}\}$, $M=\{\mathtt{m}_1,\mathtt{m}_2\}$, and $I=\{(\mathtt{a},\mathtt{m}_1),(\mathtt{a},\mathtt{m}_2),(\mathtt{b},\mathtt{m}_2)\}$. Consider the following property of this context: (α) if $A=\{\mathtt{a}\}$, then $e(i(A))\subseteq A$. Note that this property does not hold for every subset of G, e.g., it does not hold for the set $\{b\}$. The representation of (α) in FCL is:

(β) $[[R]][[S]]p \rightarrow p$ is true in the model $\mathcal{M}=(X,R,S,m)$ such that $X=\{\mathtt{a},\mathtt{b},\mathtt{m}_1,\mathtt{m}_2\}$, $R=I$, $S=I^{-1}$, and $m(p)=\{\mathtt{a}\}$.

Denote $\tau(p)$ by P and let x, y be object variables. Then, the relational representation of (β) is:

(γ) $\psi \stackrel{\text{df}}{=} x(--(-R\,;-(-S\,;P)) \cup P)y$ is true in $\mathsf{RL_{FCL}}$-model $\mathcal{N} = (U, R, S, n)$ such that $U = X$, $R = n(R)$, $S = n(S)$, and $n(P) = \{(\mathtt{a}, \mathtt{s}) : \mathtt{s} \in U\}$.

For the latter we apply the method of model checking presented in Sect. 7.7. We consider an instance $\mathsf{RL}_{\mathcal{N},\psi}$ of logic $\mathsf{RL_{FCL}}$. The vocabulary of its language consists of a countable infinite set of object variables, the set $\{c_{\mathtt{a}}, c_{\mathtt{b}}, c_{\mathtt{m}_1}, c_{\mathtt{m}_2}\}$ of object constants, the set $\{R, S, P, 1, 1'\}$ of relational constants, and the usual set of relational operations. The $\mathsf{RL}_{\mathcal{N},\psi}$-model is an $\mathsf{RL_{FCL}}$-model $\mathcal{N}' = (U', R, S, n')$ such that:

- $U' = \{\mathtt{a}, \mathtt{b}, \mathtt{m}_1, \mathtt{m}_2\} = X$;
- $n'(c_{\mathtt{a}}) = \mathtt{a}$, $n'(c_{\mathtt{b}}) = \mathtt{b}$, $n'(c_{\mathtt{m}_1}) = \mathtt{m}_1$, and $n'(c_{\mathtt{m}_2}) = \mathtt{m}_2$;
- $R = n'(R) = \{(\mathtt{a}, \mathtt{m}_1), (\mathtt{a}, \mathtt{m}_2), (\mathtt{b}, \mathtt{m}_2)\}$;
- $S = n'(S) = \{(\mathtt{m}_1, \mathtt{a}), (\mathtt{m}_2, \mathtt{a}), (\mathtt{m}_2, \mathtt{b})\}$;
- $n'(P) = n(P) = \{(\mathtt{a}, \mathtt{a}), (\mathtt{a}, \mathtt{b}), (\mathtt{a}, \mathtt{m}_1), (\mathtt{a}, \mathtt{m}_2)\}$.

Thus, truth of ψ in $\mathcal{N}$ is equivalent to its $\mathsf{RL}_{\mathcal{N},\psi}$-validity.

The rules of $\mathsf{RL}_{\mathcal{N},\psi}$-dual tableau specific for the problem (γ) have the following forms:

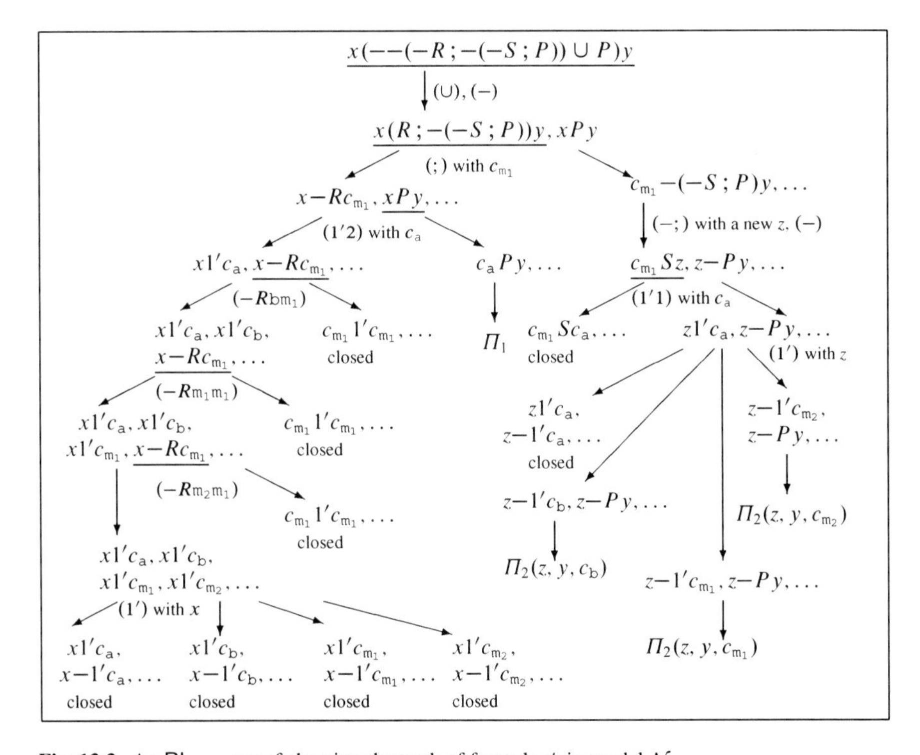

Fig. 13.3 An $\mathsf{RL}_{\mathcal{N},\psi}$-proof showing the truth of formula ψ in model $\mathcal{N}$

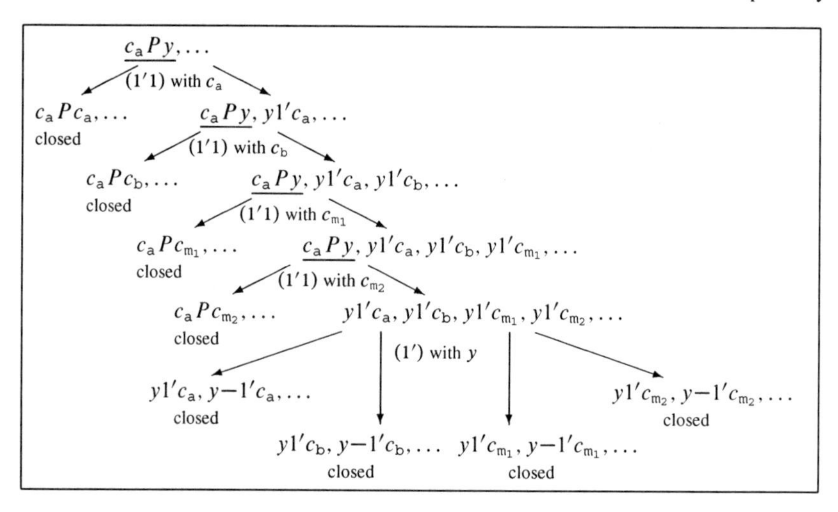

Fig. 13.4 The subtree Π_1

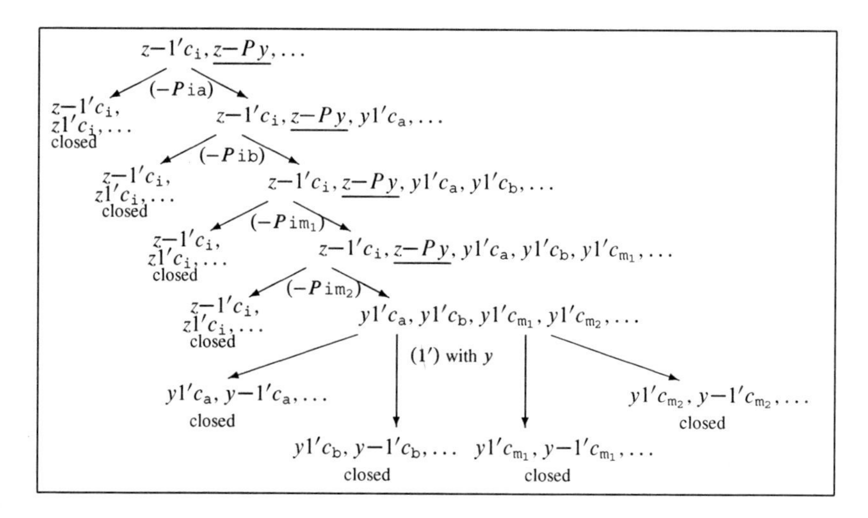

Fig. 13.5 The subtree $\Pi_2(z, y, c_{\mathtt{i}})$, where $\mathtt{i} \in \{\mathtt{b}, \mathtt{m}_1, \mathtt{m}_2\}$

For all object symbols x and y,

$$(-R\mathtt{ij}) \quad \frac{x{-}Ry}{x1'c_{\mathtt{i}}, x{-}Ry \mid y1'c_{\mathtt{j}}, x{-}Ry}$$

for every $(\mathtt{i}, \mathtt{j}) \in U' \times U'$ such that $(\mathtt{i}, \mathtt{j}) \notin R$;

$$(-P\mathtt{ij}) \quad \frac{x{-}Py}{x1'c_{\mathtt{i}}, x{-}Py \mid y1'c_{\mathtt{j}}, x{-}Py}$$

for every $(\mathtt{i}, \mathtt{j}) \in U' \times U'$ such that $(\mathtt{i}, \mathtt{j}) \notin n'(P)$;

$$(-S\mathtt{ij}) \quad \frac{x{-}Sy}{x1'c_{\mathtt{i}}, x{-}Sy \mid y1'c_{\mathtt{j}}, x{-}Sy}$$

for every $(\mathtt{i}, \mathtt{j}) \in U' \times U'$ such that $(\mathtt{i}, \mathtt{j}) \notin S$;

$$(1') \quad \frac{}{x{-}1'c_{\mathtt{a}} \mid x{-}1'c_{\mathtt{b}} \mid x{-}1'c_{\mathtt{m}_1} \mid x{-}1'c_{\mathtt{m}_2}} \qquad x \text{ is any object symbol}$$

$$(\mathtt{i} \neq \mathtt{j}) \quad \frac{}{c_{\mathtt{i}}1'c_{\mathtt{j}}} \qquad \text{for any } \mathtt{i}, \mathtt{j} \in \{\mathtt{a}, \mathtt{b}, \mathtt{m1}, \mathtt{m}_2\}, \mathtt{i} \neq \mathtt{j}$$

The $\mathsf{RL}_{\mathcal{N},\psi}$-axiomatic sets specific for the problem (γ) are:

- $\{c_{\mathtt{i}}Rc_{\mathtt{j}}\}$, for every $(\mathtt{i}, \mathtt{j}) \in \{(\mathtt{a}, \mathtt{m}_1), (\mathtt{a}, \mathtt{m}_2), (\mathtt{b}, \mathtt{m}_2)\}$;
- $\{c_{\mathtt{i}}{-}Rc_{\mathtt{j}}\}$, for every$(\mathtt{i}, \mathtt{j}) \in U \;\; \times U' \setminus \{(\mathtt{a}, \mathtt{m}_1), (\mathtt{a}, \mathtt{m}_2), (\mathtt{b}, \mathtt{m}_2)\}$;
- $\{c_{\mathtt{i}}Pc_{\mathtt{j}}\}$, for every $(\mathtt{i}, \mathtt{j}) \in \{(\mathtt{a}, \mathtt{a}), (\mathtt{a}, \mathtt{b}), (\mathtt{a}, \mathtt{m}_1), (\mathtt{a}, \mathtt{m}_2)\}$;
- $\{c_{\mathtt{i}}{-}Pc_{\mathtt{j}}\}$, for every $(\mathtt{i}, \mathtt{j}) \in U' \times U' \setminus \{(\mathtt{a}, \mathtt{a}), (\mathtt{a}, \mathtt{b}), (\mathtt{a}, \mathtt{m}_1), (\mathtt{a}, \mathtt{m}_2)\}$;
- $\{c_{\mathtt{i}}Sc_{\mathtt{j}}\}$, for every $(\mathtt{i}, \mathtt{j}) \in \{(\mathtt{m}_1, \mathtt{a}), (\mathtt{m}_2, \mathtt{a}), (\mathtt{m}_2, \mathtt{b})\}$;
- $\{c_{\mathtt{i}}{-}Sc_{\mathtt{j}}\}$, for every $(\mathtt{i}, \mathtt{j}) \in U' \times U' \setminus \{(\mathtt{m}_1, \mathtt{a}), (\mathtt{m}_2, \mathtt{a}), (\mathtt{m}_2, \mathtt{b})\}$.

Due to Theorem 7.7.1 the truth of ψ in $\mathcal{N}$ is equivalent to its $\mathsf{RL}_{\mathcal{N},\psi}$-provability. Figure 13.3 presents $\mathsf{RL}_{\mathcal{N},\psi}$-proof of ψ. The subtrees Π_1 and $\Pi_2(z, y, c_{\mathtt{i}})$, $\mathtt{i} \in \{\mathtt{b}, \mathtt{m}_1, \mathtt{m}_2\}$, are presented in Figs. 13.4 and 13.5, respectively.

Chapter 14
Dual Tableau for a Fuzzy Logic

14.1 Introduction

Monoidal triangular norm logic, MTL, was introduced in [EG01]. From the perspective of substructural logics it is the logic of full Lambek calculus endowed with the rules of exchange and weakening, FL_{ew}, with the additional axiom $(\varphi \rightarrow \psi) \vee (\psi \rightarrow \varphi)$ referred to as prelinearity. From the perspective of fuzzy logics it is a logic of left-continuous triangular norms, t-norms for short (see [KM00]).

The t-norms originated in [SS63] (see also [SS83]) to model distances in probabilistic metric spaces. For a modern study of t-norms see [KM00]. A t-norm is a binary operation on the closed real interval $[0, 1]$ which is commutative, associative, non-decreasing in both arguments, and 1 is its neutral element. The classical examples of t-norms which are applied to modelling logical operation of conjunction include Łukasiewicz t-norm $x \odot y = \max(0, x + y - 1)$, product t-norm defined as multiplication of reals, and Gödel t-norm $x \odot y = \min(x, y)$. All those t-norms are continuous. A t-norm $\odot$ is left-continuous if and only if it has a residuum, $\rightarrow$, i.e., the two operations satisfy $z \leq x \rightarrow y$ if and only if $x \odot z \leq y$. The residua of the three t-norms mentioned above play the role of implications: Łukasiewicz implication $x \rightarrow y = \min(1, 1 - x + y)$, Gougen implication $x \rightarrow y = 1$ if $x \leq y$ and y/x otherwise, and Gödel implication $x \rightarrow y = 1$ if $x \leq y$ and y otherwise, respectively.

Algebraic semantics of logic MTL is provided by the class of MTL-algebras. They are abstract counterparts to the standard structures $([0, 1], \leq, \odot, \rightarrow)$, where $\odot$ is a left-continuous t-norm and $\rightarrow$ is its residuum. A completeness of logic MTL with respect to the semantics determined by those standard structures is presented in [JM02]. In [MO02b] a Kripke-style semantics for a first-order version of logic MTL is presented, defined along the lines of semantics for substructural logics presented in [Ono85, Ono93]. In that semantics an algebraic structure is assumed in the universes of the models.

Decidability of fuzzy logics, in particular of MTL and its axiomatic extensions, is discussed in [MOG09]. Decidability of MTL is proved in [Ono].

E. Orłowska and J. Golińska-Pilarek, *Dual Tableaux: Foundations, Methodology, Case Studies*, Trends in Logic 36, DOI 10.1007/978-94-007-0005-5_14,

In this chapter, first, we present MTL-algebras which provide an algebraic semantics of the logic MTL. Next, we present logic MTL with a purely relational semantics developed in [CC08, ORR10]. Based on this semantics a relational dual tableau for the logic is developed (see [GPO10]).

14.2 MTL-Algebras

A t-norm is a binary operation on the closed real interval $\odot\colon [0,1]^2 \to [0,1]$, such that for all $x, y, z \in [0,1]$, the following hold:

- $\odot$ is associative and commutative;
- 1 is the neutral element of $\odot$;
- If $x \leq y$, then $x \odot z \leq y \odot z$.

Let $(x_i)_{i \in J}$ be an indexed family of elements of $[0,1]$. A t-norm is said to be *left-continuous* whenever $(\sup x_i) \odot y = \sup(x_i \odot y)$ for every $y \in [0,1]$.

The abstract structures which capture the properties of left-continuous t-norms are the MTL-algebras. An MTL-algebra is a structure of the form $\mathfrak{A} = (A, \vee, \wedge, \odot, \to, 0, 1)$ such that:

- $(A, \vee, \wedge, 0, 1)$ is a bounded, distributive lattice;
- $(A, \odot, 1)$ is a commutative monoid, i.e., for all $a, b, c \in A$, the following conditions are satisfied:

$$a \odot (b \odot c) = (a \odot b) \odot c,$$
$$a \odot b = b \odot a,$$
$$1 \odot a = a;$$

- $\to$ is a residuum of $\odot$, i.e., $a \odot b \leq c$ iff $a \leq b \to c$, where $\leq$ is the lattice ordering defined by: $a \leq b \overset{\text{df}}{\Longleftrightarrow} a = a \wedge b$ (or equivalently $b = a \vee b$);
- $(a \to b) \vee (b \to a) = 1$.

Observe that monoid $(A, \odot, 1)$ is integral, i.e., its neutral element coincides with the greatest element of the lattice.

14.3 The Logic MTL

The vocabulary of the language of the logic MTL consists of symbols from the following pairwise disjoint sets:

- $\mathbb{V}$ – a countable infinite set of propositional variables;
- $\{0, 1\}$ – the set of propositional constants;
- $\{\vee, \wedge, \odot, \to\}$ – the set of propositional operations of disjunction, conjunction, product, and implication, respectively.

The set of MTL-formulas is the smallest set including $\mathbb{V} \cup \{0, 1\}$ and closed with respect to the propositional operations.

A classical formalization of MTL is based on its algebraic semantics defined in terms of MTL-algebras. Here we present a Kripke-style semantics obtained from a discrete duality for MTL-algebras developed in [ORR10]. This semantics differs from the semantics presented in [MO02b] in that it is a purely relational semantics not assuming any monoid structure on the universes of the models.

An MTL-*model* is a structure $\mathcal{M} = (U, \leq, R, m)$ such that U is a non-empty set, $\leq$ is a reflexive and transitive relation on U, m is a meaning function satisfying the following conditions:

- $m(p) \subseteq U$, for every propositional variable p;
- $m(1) = U$ and $m(0) = \emptyset$;
- (her) If $x \leq y$ and $x \in m(p)$, then $y \in m(p)$, for all $x, y \in U$ and for every propositional variable p.

R is a ternary relation on U satisfying the following conditions, for all $x, y, z, x', y', z', t, w \in U$:

(MTL1) If $(x, y, z) \in R$, $x' \leq x$, $y' \leq y$, and $z \leq z'$, then $(x', y', z') \in R$;
(MTL2) If $(x, y, z) \in R$, then $(y, x, z) \in R$;
(MTL3) If $(x, y, z) \in R$ and $(z, y', z') \in R$, then there exists $u \in U$ such that $(y, y', u) \in R$ and $(x, u, z') \in R$;
(MTL4) There exists $u \in U$ such that $(u, x, x) \in R$;
(MTL5) If $(x, y, z) \in R$, then $y \leq z$;
(MTL6) If $(x, y, z) \in R$ and $(x, t, w) \in R$, then $y \leq w$ or $t \leq z$.

Let $\mathcal{M} = (U, \leq, R, m)$ be an MTL-model and let $s \in U$. Satisfaction of an MTL-formula φ in model $\mathcal{M}$ by s, $\mathcal{M}, s \models \varphi$ for short, is defined as:

- $\mathcal{M}, s \models p$ iff $s \in m(p)$, for every $p \in \mathbb{V} \cup \{0, 1\}$;
- $\mathcal{M}, s \models \varphi \vee \psi$ iff $\mathcal{M}, s \models \varphi$ or $\mathcal{M}, s \models \psi$;
- $\mathcal{M}, s \models \varphi \wedge \psi$ iff $\mathcal{M}, s \models \varphi$ and $\mathcal{M}, s \models \psi$;
- $\mathcal{M}, s \models \varphi \odot \psi$ iff there exist $x, y \in U$ such that $(x, y, s) \in R$ and $\mathcal{M}, x \models \varphi$ and $\mathcal{M}, y \models \psi$;
- $\mathcal{M}, s \models \varphi \rightarrow \psi$ iff for all $x, y \in U$, if $(s, x, y) \in R$ and $\mathcal{M}, x \models \varphi$, then $\mathcal{M}, y \models \psi$.

A formula φ is said to be true in an MTL-model $\mathcal{M}$ if and only if it is satisfied in $\mathcal{M}$ for every $s \in U$. A formula φ is MTL-valid whenever φ is true in all MTL-models.

As shown in [ORR10], algebraic and Kripke-style semantics of MTL are related according to the principle of duality via truth (see [OR07]). In particular, the condition (MTL2) reflects commutativity of $\odot$; the condition (MTL3) corresponds to associativity of $\odot$; (MTL4) expresses the condition $a \leq a \odot 1$, while (MTL5) expresses the condition $1 \odot a \leq a$, (MTL6) reflects prelinearity. It follows that a Hilbert-style axiomatization of MTL corresponding to its algebraic semantics is complete with respect to the Kripke-style semantics.

Since both logic MTL and the relevant logics presented in Chap. 9 are substructural logics, their algebraic semantics is based on residuated lattices. Therefore these logics have the propositional operations $\vee$, $\wedge$, $\odot$, and $\rightarrow$ in common, interpreted as join, meet, product, and its left residuum, respectively. As a consequence the satisfaction relation for formulas built with $\vee$, $\wedge$, $\odot$, and $\rightarrow$, as well as the decomposition rules for these operations and their completion conditions in the dual tableaux, are the same in all these logics. The differences are manifested with specific assumptions on the accessibility relations in the models of the logics which reflect specific assumptions on the operations of the underlying residuated lattices. Similarly, dual tableaux for these logics will differ only in specific rules. There is also a difference in the treatment of negation. Since in logic MTL product is commutative, its left and right residuum coincide, and negation is definable as $\neg a = a \rightarrow 0$, while in logic RLV it is an operation independent of the others.

14.4 Relational Formalization of Logic MTL

To define the relational logic $\mathsf{RL_{MTL}}$ corresponding to MTL-logic we follow the method applied to relevant logics presented in Sect. 9.3. As usual, formulas of $\mathsf{RL_{MTL}}$, interpreted as ternary relations, are intended to represent formulas of the logic MTL and the accessibility relation from its models. The vocabulary of the language of $\mathsf{RL_{MTL}}$ consists of the symbols from the following pairwise disjoint sets:

- $\mathbb{OV}_{\mathsf{RL_{MTL}}}$ – a countable infinite set of object variables;
- $\mathbb{RV}_{\mathsf{RL_{MTL}}}$ – a countable infinite set of ternary relational variables;
- $\{\leq\}$ – the set consisting of binary relational constant;
- $\{R, 1, 0\}$ – the set of ternary relational constants, where 0 is the empty relation and 1 is the universal relation;
- $\{-, \cup, \cap, \odot, \rightarrow\}$ – the set of relational operations.

The set of ternary relational terms is the smallest set including $\mathbb{RV}_{\mathsf{RL_{MTL}}} \cup \{R, 1, 0\}$ and closed on the relational operations. The set of $\mathsf{RL_{MTL}}$-terms consists of ternary relational terms and the relational constant $\leq$. $\mathsf{RL_{MTL}}$-formulas are of the form $T(x, y, z)$ or $x \leq y$, where T is a ternary relational term and x, y, z are object variables.

An $\mathsf{RL_{MTL}}$-model is a structure $\mathcal{M} = (U, \leq, R, m)$ such that U is a non-empty set and the following conditions are satisfied:

- $m(P) = X \times U \times U$, where $X \subseteq U$, for $P \in \mathbb{RV}_{\mathsf{RL_{MTL}}}$;
- $m(1) = U^3$ and $m(0) = \emptyset$;
- $\leq$ is a reflexive and transitive relation on U that provides the interpretation of the relational constant $\leq$;
- (her') If $t \leq x$ and $(t, y, z) \in m(P)$, then $(x, y, z) \in m(P)$, for all $t, x, y, z \in U$ and for every relational variable P;
- R is a ternary relation on U providing the interpretation of the relational constant R and satisfying the conditions (MTL1), ..., (MTL6) of MTL-models;

- m extends to all the relational terms as follows:

$$
\begin{aligned}
&m(-S) = U^3 - m(S),\\
&m(S \cup T) = m(S) \cup m(T),\\
&m(S \cap T) = m(S) \cap m(T),\\
&m(S \odot T) = \{(x, y, z) \in U^3 : \exists t, w \in U, (t, w, x) \in R \,\&\\
&\qquad\qquad (t, y, z) \in m(S) \,\&\, (w, y, z) \in m(T)\},\\
&m(S \to T) = \{(x, y, z) \in U^3 : \forall t, w \in U, \text{ if } (x, t, w) \in R \,\&\\
&\qquad\qquad (t, y, z) \in m(S), \text{ then } (w, y, z) \in m(T)\}.
\end{aligned}
$$

As usual, we use the same symbols for operations and constants in the language and for the corresponding entities in the models. In analogy to binary right ideal relations, the ternary relations on U which are of the form $X \times U \times U$, for some $X \subseteq U$, will be referred to as ideal relations.

Let $\mathcal{M}$ be an $\mathsf{RL_{MTL}}$-model. A valuation in $\mathcal{M}$ is a function assigning elements of U to object variables. An $\mathsf{RL_{MTL}}$-formula $T(x, y, z)$ (resp. $x \leq y$) is satisfied in a model $\mathcal{M}$ by a valuation v whenever $(v(x), v(y), v(z)) \in m(T)$ (resp. $v(x) \leq v(y)$) and it is true in $\mathcal{M}$ whenever it is satisfied in $\mathcal{M}$ by all the valuations. A formula is $\mathsf{RL_{MTL}}$-valid if and only if it is true in all $\mathsf{RL_{MTL}}$-models.

Now, we define a translation function τ from formulas of the logic MTL into ternary relational terms of $\mathsf{RL_{MTL}}$. Let τ': $\mathbb{V} \to \mathbb{RV}_{\mathsf{RL_{MTL}}}$ be a one-to-one mapping assigning relational variables to propositional variables. Then, we define:

- $\tau(0) = 0$ and $\tau(1) = 1$;
- $\tau(p) = \tau'(p)$, for every $p \in \mathbb{V}$;
- $\tau(\varphi \vee \psi) = \tau(\varphi) \cup \tau(\psi)$;
- $\tau(\varphi \wedge \psi) = \tau(\varphi) \cap \tau(\psi)$;
- $\tau(\varphi \odot \psi) = \tau(\varphi) \odot \tau(\psi)$;
- $\tau(\varphi \to \psi) = \tau(\varphi) \to \tau(\psi)$.

As in relevant logics, given an $\mathsf{RL_{MTL}}$-model $\mathcal{M} = (U, \leq, R, m)$, the set of ideal relations on U is closed with respect to the relational operations (see Proposition 9.3.1). Therefore, for every MTL-formula φ and for every $\mathsf{RL_{MTL}}$-model $\mathcal{M} = (U, \leq, R, m)$, $m(\tau(\varphi))$ is an ideal relation.

Proposition 14.4.1. *For every* MTL*-model $\mathcal{M} = (U, \leq, R, m)$ there is an* $\mathsf{RL_{MTL}}$*-model $\mathcal{M}' = (U', \leq', R', m')$ such that for every* MTL*-formula φ and for all $s, t, u \in U$, $\mathcal{M}, s \models \varphi$ iff $(s, t, u) \in m'(\tau(\varphi))$.*

Proof. Let $\mathcal{M} = (U, \leq, R, m)$ be an MTL-model. We define an $\mathsf{RL_{MTL}}$-model $\mathcal{M}' = (U', \leq', R', m')$ as follows:

- $U' = U$, $\leq' = \leq$, and $R' = R$;
- $m'(P) = m(p) \times U^2$, for every $P \in \mathbb{RV}_{\mathsf{RL_{MTL}}} \cup \{0, 1\}$, where p is such that $P = \tau(p)$;
- m' extends to all the compound terms as in $\mathsf{RL_{MTL}}$-models.

Clearly, $\mathcal{M}'$ defined above is an $\mathsf{RL_{MTL}}$-model. We show the required condition by induction on the complexity of formulas.

For $p \in \mathbb{V} \cup \{0, 1\}$ the condition of the proposition holds by definition of m'.

Assume that the condition holds for formulas φ and ψ. By way of example, we prove it for formulas of the form $\varphi \to \psi$. The remaining cases can be proved as in relevant logics.

By the definition, $\mathcal{M}, s \models \varphi \to \psi$ iff for all $x, y \in U$ if $(s, x, y) \in R$ and $\mathcal{M}, x \models \varphi$, then $\mathcal{M}, y \models \psi$ iff by the induction hypothesis, for all $x, y \in U$ if $(s, x, y) \in R'$ and $(x, t, u) \in m'(\tau(\varphi))$, then $(y, t, u) \in m'(\tau(\psi))$ iff $(s, t, u) \in m'(\tau(\varphi) \to \tau(\psi)) = m'(\tau(\varphi \to \psi))$. □

Proposition 14.4.2. *For every* $\mathsf{RL_{MTL}}$*-model* $\mathcal{M}' = (U', \leq', R', m')$ *there is an* MTL*-model* $\mathcal{M} = (U, \leq, R, m)$ *such that for every* MTL*-formula* φ *and for all* $s, t, u \in U$, $\mathcal{M}, s \models \varphi$ *iff* $(s, t, u) \in m'(\tau(\varphi))$.

Proof. The required MTL-model is constructed as follows:

- $U = U'$, $\leq = \leq'$, and $R = R'$;
- $m(p) = \{x \in U :$ there are $y, z \in U$ such that $(x, y, z) \in m(P)\}$, for every $p \in \mathbb{V} \cup \{0, 1\}$, where P is such that $\tau(p) = P$.

The model defined above is an MTL-model and the proof of the proposition is similar to the proof of Proposition 14.4.1. □

Propositions 14.4.1 and 14.4.2 lead to the following theorem:

Theorem 14.4.1. *For every* MTL*-formula* φ *and for all object variables* x, y, *and* z, *the following conditions are equivalent:*

1. φ *is* MTL*-valid;*
2. $\tau(\varphi)(x, y, z)$ *is* $\mathsf{RL_{MTL}}$*-valid.*

A dual tableau for logic $\mathsf{RL_{MTL}}$ consists of the following decomposition rules:

For all object variables x, y, z and for all ternary relational terms S and T,

$$(-)\quad \frac{--S(x, y, z)}{S(x, y, z)}$$

$$(\cup)\quad \frac{(S \cup T)(x, y, z)}{S(x, y, z), T(x, y, z)} \qquad\qquad (-\cup)\quad \frac{-(S \cup T)(x, y, z)}{-S(x, y, z) \mid -T(x, y, z)}$$

$$(\cap)\quad \frac{(S \cap T)(x, y, z)}{S(x, y, z) \mid T(x, y, z)} \qquad\qquad (-\cap)\quad \frac{-(S \cap T)(x, y, z)}{-S(x, y, z), -T(x, y, z)}$$

$$(\odot)\quad \frac{(S \odot T)(x, y, z)}{R(t, w, x), \varphi \mid S(t, y, z), \varphi \mid T(w, y, z), \varphi}$$

t, w are any object variables and $\varphi = (S \odot T)(x, y, z)$

$$(-\odot) \quad \frac{-(S \odot T)(x,y,z)}{-R(t,w,x), -S(t,y,z), -T(w,y,z)}$$

t, w are new object variables such that $t \neq w$

$$(\rightarrow) \quad \frac{(S \rightarrow T)(x,y,z)}{-R(x,t,w), -S(t,y,z), T(w,y,z)}$$

t, w are new object variables such that $t \neq w$

$$(-\rightarrow) \quad \frac{-(S \rightarrow T)(x,y,z)}{R(x,t,w), \varphi \mid S(t,y,z), \varphi \mid -T(w,y,z), \varphi}$$

t, w are any object variables and $\varphi = -(S \rightarrow T)(x,y,z)$

The specific rules are of the form:

For every relational variable P and for all object variables x, y, z, x', y', z', t, and w,

$$(\text{ideal}) \quad \frac{P(x,y,z)}{P(x,t,w), P(x,y,z)} \quad t, w \text{ are any object variables}$$

$$(0) \quad \frac{}{0(x,y,z)} \quad x, y, z \text{ are any object variables}$$

$$(\text{tran} \leq) \quad \frac{x \leq y}{x \leq z, x \leq y \mid z \leq y, x \leq y} \quad z \text{ is any object variable}$$

$$(\text{rher'}) \quad \frac{P(x,y,z)}{t \leq x, P(x,y,z) \mid P(t,y,z), P(x,y,z)} \quad t \text{ is any object variable}$$

$$(\text{rMTL1}) \quad \frac{R(x,y,z)}{R(x',y',z'), \varphi \mid x \leq x', \varphi \mid y \leq y', \varphi \mid z' \leq z, \varphi}$$

x', y', z' are any object variables and $\varphi = R(x,y,z)$

$$(\text{rMTL2}) \quad \frac{R(x,y,z)}{R(y,x,z)}$$

$$(\text{rMTL3}) \quad \frac{}{R(x,y,z) \mid R(z,y',z') \mid -R(y,y',u), -R(x,u,z')}$$

x, y, z, y', z' are any object variables

u is a new object variable such that $\{u\} \cap \{x, y, z, x', y', z'\} = \emptyset$

(rMTL4) $\dfrac{}{-R(u,x,x)}$

x is any object variable,

u is a new object variable such that $u \neq x$

(rMTL5) $\dfrac{y \leq z}{R(x,y,z), y \leq z}$

x is any object variable

(rMTL6) $\dfrac{y \leq w, t \leq z}{R(x,y,z), y \leq w, t \leq z \mid R(x,t,w), y \leq w, t \leq z}$

x is any object variable

The rules (rMTL3) and (rMTL4) are specialized cut rules. An alternative deterministic form of such rules is discussed in Sect. 25.9.

A set of $\mathsf{RL}_{\mathsf{MTL}}$-formulas is said to be an $\mathsf{RL}_{\mathsf{MTL}}$-*axiomatic set* whenever it includes a subset of either of the following forms:

For all object variables x, y and z, and for every relational term S,

(Ax1) $\{x \leq x\}$;
(Ax2) $\{1(x,y,z)\}$;
(Ax3) $\{S(x,y,z), -S(x,y,z)\}$.

The notion of an $\mathsf{RL}_{\mathsf{MTL}}$-set and the notion of $\mathsf{RL}_{\mathsf{MTL}}$-correctness of a rule are defined in a similar way as in Sect. 2.4.

Proposition 14.4.3.

1. *The* $\mathsf{RL}_{\mathsf{MTL}}$*-rules are* $\mathsf{RL}_{\mathsf{MTL}}$*-correct;*
2. *The* $\mathsf{RL}_{\mathsf{MTL}}$*-axiomatic sets are* $\mathsf{RL}_{\mathsf{MTL}}$*-sets.*

Proof. By way of example, we prove correctness of the rules ($\odot$), ($\rightarrow$), and (rMTL4).

($\odot$) Let X be a finite set of $\mathsf{RL}_{\mathsf{MTL}}$-formulas. If $X \cup \{(S \odot T)(x,y,z)\}$ is an $\mathsf{RL}_{\mathsf{MTL}}$-set, then so are sets $X \cup \{R(t,w,x), (S \odot T)(x,y,z)\}$, $X \cup \{S(t,y,z), (S \odot T)(x,y,z)\}$, and $X \cup \{T(w,y,z), (S \odot T)(x,y,z)\}$. Now, assume that $X \cup \{R(t,w,x), (S \odot T)(x,y,z)\}$, $X \cup \{S(t,y,z), (S \odot T)(x,y,z)\}$, and $X \cup \{T(w,y,z), (S \odot T)(x,y,z)\}$ are $\mathsf{RL}_{\mathsf{MTL}}$-sets, and suppose that $X \cup \{(S \odot T)(x,y,z)\}$ is not an $\mathsf{RL}_{\mathsf{MTL}}$-set. Then there exist an $\mathsf{RL}_{\mathsf{MTL}}$-model $\mathcal{M} = (U, \leq, R, m)$ and a valuation v in $\mathcal{M}$ such that for $\varphi \in X \cup \{(S \odot T)(x,y,z)\}$, $\mathcal{M}, v \not\models \varphi$. Therefore, for all $a, b \in U$, $(a, b, v(x)) \notin R$ or $(a, v(y), v(z)) \notin m(S)$ or $(b, v(y), v(z) \notin m(T)$. By the assumption, $(v(t), v(w), v(x)) \in R$, $(v(t), v(y), v(z)) \in m(S)$, and $(v(w), v(y), v(z)) \in m(T)$, a contradiction.

($\rightarrow$) Let X be a finite set of $\mathsf{RL_{MTL}}$-formulas and let t, w be variables that do not occur in X and such that $t \neq w$ and $\{t, w\} \cap \{x, y, z\} = \emptyset$. Assume that $X \cup \{(S \rightarrow T)(x, y, z)\}$ is an $\mathsf{RL_{MTL}}$-set. Suppose $X \cup \{-R(x, t, w), -S(t, y, z), T(w, y, z)\}$ is not an $\mathsf{RL_{MTL}}$-set. Then there exist an $\mathsf{RL_{MTL}}$-model $\mathcal{M} = (U, \leq, R, m)$ and a valuation v in $\mathcal{M}$ such that for every $\varphi \in X \cup \{-R(x, t, w), -S(t, y, z), T(w, y, z)\}$, $\mathcal{M}, v \not\models \varphi$. Therefore, $(v(x), v(t), v(w)) \in R$, $(v(t), v(y), v(z)) \in m(S)$, and $(v(w), v(y), v(z)) \notin m(T)$. Thus, by the definition of $m(S \rightarrow T)$, $(v(x), v(y), v(z)) \notin m(S \rightarrow T)$. However, by the assumption, $(v(x), v(y), v(z)) \in m(S \rightarrow T)$, a contradiction. Now, assume that $X \cup \{-R(x, t, w), -S(t, y, z), T(w, y, z)\}$ is an $\mathsf{RL_{MTL}}$-set. Then, by the assumption on variables t and w, for every $\mathsf{RL_{MTL}}$-model $\mathcal{M} = (U, \leq, R, m)$ and for every valuation v in $\mathcal{M}$ either there exists $\varphi \in X$ such that $\mathcal{M}, v \models \varphi$ or for all $a, b \in U$, if $(v(x), a, b) \in R$ and $(a, v(y), v(z)) \in m(S)$, then $(b, v(y), v(z)) \in m(T)$, in which case $(v(x), v(y), v(z)) \in m(S \rightarrow T)$. Hence, $X \cup \{(S \rightarrow T)(x, y, z)\}$ is an $\mathsf{RL_{MTL}}$-set.

(rMTL4) Let X be a finite set of $\mathsf{RL_{MTL}}$-formulas and let x and u be object variables such that u does not occur in X and $u \neq x$. Clearly, if X is an $\mathsf{RL_{MTL}}$-set, then so is $X \cup \{-R(u, x, x)\}$. Now, assume $X \cup \{-R(u, x, x)\}$ is an $\mathsf{RL_{MTL}}$-set. Then, due to the assumption on the variables x and u, for every $\mathsf{RL_{MTL}}$-model $\mathcal{M} = (U, \leq, R, m)$ and for every valuation v in $\mathcal{M}$ either there is $\varphi \in X$ such that $\mathcal{M}, v \models \varphi$ or for every $a \in U$, $(a, v(x), v(x)) \notin R$. Observe that due to condition (MTL4), there exists $a \in U$ such that $(a, v(x), v(x)) \in R$, hence X is an $\mathsf{RL_{MTL}}$-set.

Correctness of the remaining rules can be proved in a similar way. The easy proof of 2. is omitted. □

The notions of an $\mathsf{RL_{MTL}}$-proof tree, a closed branch of such a tree, a closed $\mathsf{RL_{MTL}}$-proof tree, and $\mathsf{RL_{MTL}}$-provability are defined as in Sect. 2.4.

A branch b of an $\mathsf{RL_{MTL}}$-proof tree is said to be complete whenever it is closed or it satisfies the following completion conditions:

For all object variables x, y, z, x', y', z', t and w, for all ternary relational terms S and T, and for every relational variable P,

Cpl($-$) If $--S(x, y, z) \in b$, then $S(x, y, z) \in b$, obtained by an application of the rule ($-$);

Cpl($\cup$) (resp. Cpl($-\cap$)) If $(S \cup T)(x, y, z) \in b$ (resp. $-(S \cap T)(x, y, z) \in b$), then both $S(x, y, z) \in b$ and $T(x, y, z) \in b$ (resp. both $-S(x, y, z) \in b$ and $-T(x, y, z) \in b$), obtained by an application of the rule ($\cup$) (resp. ($-\cap$));

Cpl($\cap$) (resp. Cpl($-\cup$)) If $(S \cap T)(x, y, z) \in b$ (resp. $-(S \cup T)(x, y, z) \in b$), then either $S(x, y, z) \in b$ or $T(x, y, z) \in b$ (resp. either $-S(x, y, z) \in b$ or $-T(x, y, z) \in b$), obtained by an application of the rule ($\cap$) (resp. ($-\cup$));

Cpl($\odot$) If $(S \odot T)(x, y, z) \in b$, then for all object variables t and w either $R(t, w, x) \in b$ or $S(t, y, z) \in b$ or $T(w, y, z) \in b$, obtained by an application of the rule ($\odot$);

Cpl($-\odot$) If $-(S \odot T)(x, y, z) \in b$, then for some object variables t and w, $-R(t, w, x) \in b$, $-S(t, y, z) \in b$, and $-T(w, y, z) \in b$, obtained by an application of the rule ($-\odot$);

Cpl($\rightarrow$) If $(S \rightarrow T)(x, y, z) \in b$, then for some object variables t and w, $-R(x, t, w) \in b$, $-S(t, y, z) \in b$, and $T(w, y, z) \in b$, obtained by an application of the rule ($\rightarrow$);

Cpl($-\rightarrow$) If $-(S \rightarrow T)(x, y, z) \in b$, then for all object variables t and w either $R(x, t, w) \in b$ or $S(t, y, z) \in b$ or $-T(w, y, z) \in b$, obtained by an application of the rule ($- \rightarrow$);

Cpl(ideal) If $P(x, y, z) \in b$, then for all object variables t and w, $P(x, t, w) \in b$, obtained by an application of the rule (P);

Cpl(0) For all object variables x, y, and z, $0(x, y, z) \in b$, obtained by an application of the rule (0);

Cpl(tran $\leq$) If $x \leq y \in b$, then for every object variable z either $x \leq z \in b$ or $z \leq y \in b$, obtained by an application of the rule (tran $\leq$);

Cpl(her') If $P(x, y, z) \in b$, then for every object variable t either $t \leq x \in b$ or $P(t, y, z) \in b$, obtained by an application of the rule (her');

Cpl(rMTL1) If $R(x, y, z) \in b$, then for all object variables x', y', and z' either $R(x', y', z') \in b$ or $x \leq x' \in b$ or $y \leq y' \in b$ or $z' \leq z \in b$, obtained by an application of the rule (rMTL1);

Cpl(rMTL2) If $R(x, y, z) \in b$, then $R(y, x, z) \in b$, obtained by an application of the rule (rMTL2);

Cpl(rMTL3) For all object variables x, y, z, y', z' either $R(x, y, z) \in b$ or $R(z, y', z') \in b$ or there is an object variable u such that both $-R(y, y', u) \in b$ and $-R(x, u, z') \in b$, obtained by an application of the rule (rMTL3);

Cpl(rMTL4) For every object variable x there exists an object variable u such that $-R(u, x, x) \in b$, obtained by an application of the rule (rMTL4);

Cpl(rMTL5) If $y \leq z \in b$, then for every object variable x, $R(x, y, z) \in b$, obtained by an application of the rule (rMTL5);

Cpl(rMTL6) If $y \leq w \in b$ and $t \leq z \in b$, then for every object variable x either $R(x, y, z) \in b$ or $R(x, t, w) \in b$, obtained by an application of the rule (rMTL6).

The notion of a complete $\mathsf{RL}_{\mathsf{MTL}}$-proof and the notion of an open branch of an $\mathsf{RL}_{\mathsf{MTL}}$-proof tree are defined as in Sect. 2.5.

Although, the rule (rMTL2) does not preserve the formulas of the form $R(x, y, z)$, we can show that the closed branch property holds. The proof is similar to the proof of Proposition 2.8.1.

Let b be an open branch of an $\mathsf{RL}_{\mathsf{MTL}}$-proof tree. We define a branch structure $\mathcal{M}^b = (U^b, \leq^b, R^b, m^b)$ as follows:

- $U^b = \mathbb{OV}_{\mathsf{RL}_{\mathsf{MTL}}}$;
- $\leq^b = \{(x, y) \in (U^b)^2 : x \leq y \notin b\}$ and $m^b(\leq) = \leq^b$;
- $m^b(T) = \{(x, y, z) \in (U^b)^3 : T(x, y, z) \notin b\}$, for every $T \in \mathbb{RV}_{\mathsf{RL}_{\mathsf{MTL}}} \cup \{R, 1, 0\}$;
- m^b extends to all the compound relational terms as in the $\mathsf{RL}_{\mathsf{MTL}}$-models.

Proposition 14.4.4 (Branch Model Property). *Let $\mathcal{M}^b$ be a branch structure determined by an open branch b of an $\mathsf{RL_{MTL}}$-proof tree. Then $\mathcal{M}^b$ is an $\mathsf{RL_{MTL}}$-model.*

Proof. We show that for every relational variable P, $m^b(P) = X \times U \times U$ for some $X \subseteq U$, the relation $\leq^b$ is reflexive and transitive, m^b satisfies the heredity condition, and the relation R^b satisfies the conditions (MTL1), ..., (MTL6).

Let P be a relational variable. Assume that for some $x, y, z \in U^b$, $(x, y, z) \in m^b(P)$. Suppose that for some $t, w \in U^b$, $(x, t, w) \notin m^b(P)$. Then $P(x, t, w) \in b$ and by the completion condition Cpl(ideal), $P(x, y, z) \in b$. Thus, by the definition of m^b, $(x, y, z) \notin m^b(P)$, a contradiction. Hence, $m^b(P)$ is a right ideal relation on U^b.

Since for every object variable x, the set $\{x \leq x\}$ is an axiomatic set, $\leq^b$ is reflexive. Transitivity of $\leq^b$ follows from the completion condition Cpl(tran $\leq$), while the heredity condition follows from the completion condition Cpl(her').

By way of example, we show that R^b satisfies the condition (MTL3). By the completion condition Cpl(rMTL3), for all $x, y, z, y', z' \in U^b$ either $R(x, y, z) \in b$ or $R(z, y', z') \in b$ or there exists $u \in U^b$ such that both $-R(y, y', u) \in b$ and $-R(x, u, z') \in b$. Thus, if $(x, y, z) \in R^b$ and $(z, y', z') \in R^b$, then there exists $u \in U^b$ such that both $(y, y', u) \in R^b$ and $(x, u, z') \in R^b$, hence the condition (MTL3) is satisfied. □

Let v^b be a valuation in $\mathcal{M}^b$ such that $v^b(x) = x$, for every object variable x.

Proposition 14.4.5 (Satisfaction in Branch Model Property). *For every open branch b of an $\mathsf{RL_{MTL}}$-proof tree and for every $\mathsf{RL_{MTL}}$-formula φ, if $\mathcal{M}^b, v^b \models \varphi$, then $\varphi \notin b$.*

Proof. Let φ be an $\mathsf{RL_{MTL}}$-formula. The proof is by induction on the complexity of formulas. If φ is an atomic formula, then the condition of the proposition holds by the definition. Let $\varphi = -S(x, y, z)$ for an atomic term S. If $\mathcal{M}^b, v^b \models -S(x, y, z)$, then $S(x, y, z) \in b$. By the closed branch property, $-S(x, y, z) \notin b$. For binary atomic relational terms the proof is similar. Assume that the condition of the proposition holds for relational terms S and T and their complements. By way of example, we show that it holds for $-(S \to T)$.

Assume $\mathcal{M}^b, v^b \models -(S \to T)(x, y, z)$. Then there are $t, w \in U^b$ such that $(x, t, w) \in R^b$, $(t, y, z) \in m^b(S)$, and $(w, y, z) \notin m^b(T)$. Suppose $-(S \to T)(x, y, z) \in b$. By the completion condition Cpl($-\to$), for all $t, w \in U^b$ either $R(x, t, w) \in b$ or $S(t, y, z) \in b$ or $-T(w, y, z) \in b$, that is either $-R(t, w, x) \notin b$ or $-S(t, y, z) \notin b$ or $T(w, y, z) \notin b$. Therefore, by the induction hypothesis, either $(t, w, x) \notin R^b$ or $(t, y, z) \notin m^b(S)$ or $(w, y, z) \in m^b(T)$, a contradiction.

The remaining cases can be proved in a similar way. □

Hence, completeness of $\mathsf{RL_{MTL}}$-dual tableau follows.

Theorem 14.4.2 (Soundness and Completeness of $\mathsf{RL_{MTL}}$). *Let φ be an $\mathsf{RL_{MTL}}$-formula. Then the following conditions are equivalent:*

1. *φ is $\mathsf{RL_{MTL}}$-valid;*
2. *φ is $\mathsf{RL_{MTL}}$-provable.*

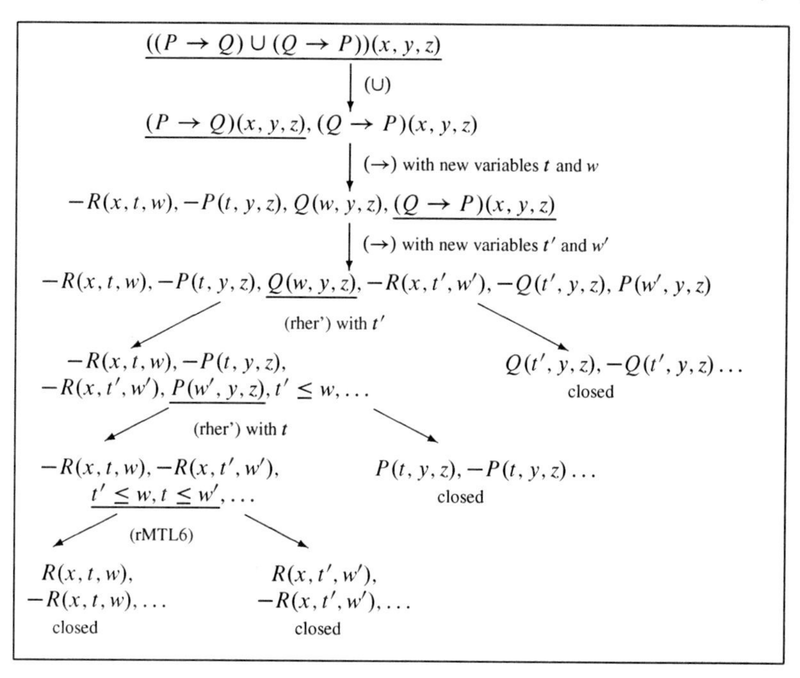

Fig. 14.1 An $\mathsf{RL}_{\mathsf{MTL}}$-proof of the formula $(p \rightarrow q) \vee (q \rightarrow p)$

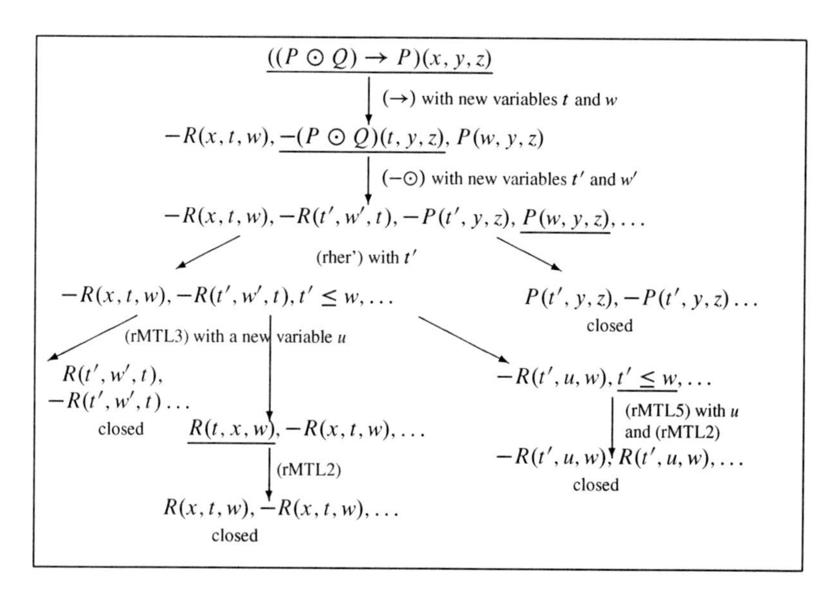

Fig. 14.2 An $\mathsf{RL}_{\mathsf{MTL}}$-proof of the formula $(p \odot q) \rightarrow p$

Furthermore, by Theorems 14.4.1 and 14.4.2, we get:

Theorem 14.4.3 (Relational Soundness and Completeness of MTL). *Let φ be an MTL-formula. Then for all object variables x, y, and z, the following conditions are equivalent:*

1. *φ is MTL-valid;*
2. *$\tau(\varphi)(x, y, z)$ is $\mathsf{RL}_{\mathsf{MTL}}$-provable.*

Example. Consider the following MTL-formulas φ and ψ:

$$\varphi = (p \rightarrow q) \vee (q \rightarrow p), \qquad \psi = (p \odot q) \rightarrow p.$$

The translations of φ and ψ into $\mathsf{RL}_{\mathsf{MTL}}$-terms are:

$$\tau(\varphi) = (P \rightarrow Q) \cup (Q \rightarrow P), \qquad \tau(\psi) = (P \odot Q) \rightarrow P,$$

where $\tau(p) = P$ and $\tau(q) = Q$. MTL-validity of φ and ψ is equivalent to $\mathsf{RL}_{\mathsf{MTL}}$-provability of the formulas $\tau(\varphi)(x, y, z)$ and $\tau(\psi)(x, y, z)$, respectively. Figures 14.1 and 14.2 present their $\mathsf{RL}_{\mathsf{MTL}}$-proofs.

Chapter 15
Dual Tableaux for Logics of Order of Magnitude Reasoning

15.1 Introduction

Order of magnitude reasoning is a reasoning in terms of qualitative ranges of variables instead of their precise values. More precisely, the order of magnitude approach enables us the reasoning in terms of relative magnitudes of variables obtained by comparisons of the sizes of quantities. In a sense, order of magnitude methods of reasoning are situated midway between numerical methods and qualitative formalisms.

The logical approaches to order of magnitude reasoning can be found in [BOO04, BOA05, BMVOA06]. These approaches are based on a system with two landmarks and with relations of comparability and negligibility. The intuitive representation of the underlying models can be illustrated with the linearly ordered set of real numbers, where two landmarks $-\alpha$ and $+\alpha$ are considered.

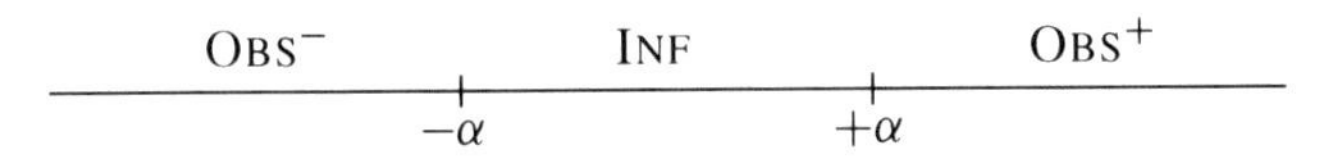

In the picture, $-\alpha$ and $+\alpha$ represent the greatest negative observable and the least positive observable, respectively, partitioning the real line into classes of positive observable numbers, OBS^+, negative observable numbers, OBS^-, and non-observable (also called infinitesimal) numbers, INF. This choice makes sense, in particular, when considering physical metric spaces in which we always have a smallest unit which can be measured; however, it is not possible to identify a least or a greatest non-observable number.

Consider the following example which illustrates a concept of comparability. Assume one aims at specifying the behavior of a device for automatic control of the speed of a car; assume the system has to maintain the speed close to some speed limit v. For practical purposes, any value in an interval $[v-\varepsilon, v+\varepsilon]$ for small ε is admissible. The extreme points of this interval can then be considered as the landmarks $-\alpha$ and $+\alpha$; furthermore, the sets OBS^-, INF, and OBS^+ can be interpreted as SLOW, ADEQUATE, and HIGH speed.

E. Orłowska and J. Golińska-Pilarek, *Dual Tableaux: Foundations, Methodology, Case Studies*, Trends in Logic 36, DOI 10.1007/978-94-007-0005-5_15,

Regarding negligibility, the representation capabilities of a pocket calculator provide an illustrative example of a relation of that kind. In such a device, it is not possible to present any number whose absolute value is less than 10^{-99}. Therefore, it makes sense to consider $-\alpha = -10^{-99}$ and $+\alpha = +10^{-99}$ since any number between -10^{-99} and 10^{-99} cannot be observed or presented. On the other hand, a number x can be said to be negligible with respect to y provided that the difference $y - x$ cannot be distinguished from y. Numerically, and assuming 8+2 (digits and mantissa) display, this amounts to stating that x is negligible with respect to y if and only if $y - x > 10^8$. Furthermore, this example suggests a real-life model in which, for instance, -1000 is negligible with respect to -1. This is even more suggestive if we interpret the numbers as exponents, since 10^{-1000} certainly can be considered negligible with respect to 10^{-1}.

In this chapter we consider a logic OMR of order of magnitude reasoning introduced in [BOO04] and we present a relational dual tableau for this logic based on the developments in [BMVOA06]. In the logic we deal with the comparability and negligibility relations. They determine modal operators which enable us to specify statements about properties of physical systems within the ranges of values of variables. The deduction tool of dual tableau for OMR provides a means for verifying these statements. Decidability of the logic OMR is an open problem.

15.2 A Multimodal Logic of Order of Magnitude Reasoning

The language of the logic OMR of order of magnitude reasoning is a multimodal propositional language with three types of modal operations, each of which is associated with a certain relation: $[<]$ and $[<^{-1}]$ to deal with an ordering of the elements of a system, the operations $[\sqsubset]$ and $[\sqsubset^{-1}]$ to deal with a comparability relation $\sqsubset$, and the operations $[\prec]$ and $[\prec^{-1}]$ to deal with a negligibility relation $\prec$.

The intuitive meaning of the modal operations is as follows:

$[<]\varphi$ – *φ is true for all numbers which are greater than the current one;*
$[<^{-1}]\varphi$ – *φ is true for all numbers which are less than the current one;*
$[\sqsubset]\varphi$ – *φ is true for all numbers which are greater than and comparable with the current one;*
$[\sqsubset^{-1}]\varphi$ – *φ is true for all numbers which are less than and comparable with the current one;*
$[\prec]\varphi$ – *φ is true for all numbers with respect to which the current one is negligible;*
$[\prec^{-1}]\varphi$ – *φ is true for all numbers which are negligible with respect to the current one.*

Formulas of the language are constructed with symbols from the following pairwise disjoint sets:

- $\mathbb{V}$ – a set of propositional variables;
- $\{c^-, c^+\}$ – the set of object constants representing the landmarks;

- $\{<, \sqsubset, \prec\}$ – the set of relational constants representing ordering, comparability, and negligibility, respectively;
- $\{\neg, \vee, \wedge\}$ – the set of classical propositional operations;
- $\{[<], [<^{-1}], [\sqsubset], [\sqsubset^{-1}], [\prec], [\prec^{-1}]\}$ – the set of unary modal operations.

OMR-formulas are generated from the elements of $\mathbb{V} \cup \{c^-, c^+\}$ with the propositional operations.

An OMR-*model* is a structure $\mathcal{M} = (U, <, \sqsubset, \prec, c^-, c^+, m)$, where U is a non-empty set, c^- and c^+ are designated elements of U which, for the sake of simplicity, are denoted with the same symbols as the constants of the language, and m is a meaning function satisfying the following conditions:

- $m(p) \subseteq U$, for every $p \in \mathbb{V}$;
- $<$ is a strict linear ordering on U, that is for all $x, y, z \in U$, the following conditions are satisfied:

  (irref $<$) $(x, x) \notin <$,
  (tran $<$) If $(x, y) \in <$ and $(y, z) \in <$, then $(x, z) \in <$,
  (con $<$) $(x, y) \in <$ or $(y, x) \in <$ or $x = y$;

- $m(c^-) = c^-$, $m(c^+) = c^+$ and $(c^-, c^+) \in <$; then the sets $\textsc{Obs}^+$, $\textsc{Inf}$, and $\textsc{Obs}^-$ are defined by:

  $\textsc{Obs}^- \stackrel{\text{df}}{=} \{x \in U : (x, c^-) \in < \text{ or } x = c^-\}$,
  $\textsc{Inf} \stackrel{\text{df}}{=} \{x \in U : (c^-, x) \in < \text{ and } (x, c^+) \in <\}$,
  $\textsc{Obs}^+ \stackrel{\text{df}}{=} \{x \in U : (c^+, x) \in < \text{ or } c^+ = x\}$;

- $\sqsubset$ is a binary relation on U called the comparability relation:

  $\sqsubset \stackrel{\text{df}}{=} < \cap \big((\textsc{Obs}^- \times \textsc{Obs}^-) \cup (\textsc{Inf} \times \textsc{Inf}) \cup (\textsc{Obs}^+ \times \textsc{Obs}^+)\big)$, that is for any $x, y \in U$, $(x, y) \in \sqsubset$ iff either of the following conditions is satisfied:

  (i$\sqsubset$) $(x, y) \in <$, $((x, c^-) \in <$ or $x = c^-)$, and $((y, c^-) \in <$ or $y = c^-)$,
  (ii$\sqsubset$) $(x, y) \in <$, $(c^-, x) \in <$, $(x, c^+) \in <$, $(c^-, y) \in <$, and $(y, c^+) \in <$,
  (iii$\sqsubset$) $(x, y) \in <$, $(c^+, x) \in <$ or $c^+ = x)$, and $(c^+, y) \in <$ or $c^+ = y)$;

- $\prec$ is a binary relation on U called the negligibility relation and satisfying the following conditions:

  (i$\prec$) $\prec \subseteq <$,
  (ii$\prec$) If $(x, y) \in \prec$ and $(y, z) \in <$, then $(x, z) \in \prec$,
  (iii$\prec$) If $(x, y) \in <$ and $(y, z) \in \prec$, then $(x, z) \in \prec$,
  (iv$\prec$) If $(x, y) \in \prec$, then either $x \notin \textsc{Inf}$ or $y \notin \textsc{Inf}$.

Note that as a consequence of items (ii$\prec$) and (iii$\prec$), the relation $\prec$ is transitive; item (iv$\prec$) states that two non-observable elements cannot be compared by the negligibility relation.

Let φ be an OMR-formula and let $\mathcal{M} = (U, <, \sqsubset, \prec, m)$ be an OMR-model. The satisfaction of φ in $\mathcal{M}$ by a state $s \in U$, $\mathcal{M}, s \models \varphi$ for short, is defined as follows:

- $\mathcal{M}, s \models p$ iff $s \in m(p)$, for every $p \in \mathbb{V}$;
- $\mathcal{M}, s \models c^{\#}$ iff $s = c^{\#}$, for $\# \in \{^-, ^+\}$;
- $\mathcal{M}, s \models \neg\varphi$ iff not $\mathcal{M}, s \models \varphi$;
- $\mathcal{M}, s \models (\varphi \vee \psi)$ iff $\mathcal{M}, s \models \varphi$ or $\mathcal{M}, s \models \psi$;
- $\mathcal{M}, s \models (\varphi \wedge \psi)$ iff $\mathcal{M}, s \models \varphi$ and $\mathcal{M}, s \models \psi$;

For $T \in \{<, \sqsubset, \prec\}$,

- $\mathcal{M}, s \models [T]\varphi$ iff for all $s' \in U$, $(s, s') \in T$ implies $\mathcal{M}, s' \models \varphi$;
- $\mathcal{M}, s \models [T^{-1}]\varphi$ iff for all $s' \in U$, $(s, s') \in T^{-1}$ implies $\mathcal{M}, s' \models \varphi$.

15.3 Dual Tableau for the Logic of Order of Magnitude Reasoning

A relational logic for representation of OMR-formulas, $\mathsf{RL_{OMR}}$, is an instance of the relational logic $\mathsf{RL}_{df}(\mathbb{C})$ presented in Sect. 3.3. The vocabulary of its language consists of symbols from the following pairwise disjoint sets:

- $\mathbb{OV}_{\mathsf{RL_{OMR}}}$ – a set of object variables;
- $\mathbb{OC}_{\mathsf{RL_{OMR}}} = \{c^-, c^+\}$ – the set of object constants corresponding to the distinguished elements from the OMR-models;
- $\mathbb{RV}_{\mathsf{RL_{OMR}}}$ – a set of binary relational variables;
- $\mathbb{RC}_{\mathsf{RL_{OMR}}} = \{1, 1', <, \sqsubset, \prec, C^-, C^+\}$ – the set of relational constants where C^- and C^+ are intended to be point relations representing the landmarks;
- $\{-, \cup, \cap, ;, ^{-1}\}$ – the set of relational operations.

The set of $\mathsf{RL_{OMR}}$-relational terms and the set of $\mathsf{RL_{OMR}}$-formulas are defined in a standard way (see Sect. 3.3).

An $\mathsf{RL_{OMR}}$-*model* is a structure $\mathcal{M} = (U, <, \sqsubset, \prec, c^-, c^+, m)$, where U is a nonempty set, c^- and c^+ are designated elements of U, and m is a meaning function satisfying the following conditions:

- $m(c^-) = c^-$, $m(c^+) = c^+$;
- $m(R) \subseteq U \times U$, for every $R \in \mathbb{RA}_{\mathsf{RL_{OMR}}}$;
- $m(1')$ and $m(1)$ are defined as in $\mathsf{RL}(1, 1')$-models;
- $m(<) = <$ is irreflexive, transitive, and it satisfies the following condition for all $x, y \in U$:

$$\text{(con')} \quad \text{Either } (x, y) \in < \text{ or } (y, x) \in < \text{ or } (x, y) \in m(1');$$

- $(c^-, c^+) \in <$; as in OMR-models, the linearity of $<$ enables us to partition U into the classes $\textsc{Obs}^-$, $\textsc{Obs}^+$, and $\textsc{Inf}$ which are defined as in OMR-models with a difference that equality is replaced by $m(1')$ that may not be the identity (for a discussion of that issue see Sect. 2.7);
- $m(\sqsubset) = \sqsubset = < \cap \big((\mathrm{OBS}^- \times \mathrm{OBS}^-) \cup (\mathrm{INF} \times \mathrm{INF}) \cup (\mathrm{OBS}^+ \times \mathrm{OBS}^+)\big)$, that is for any $x, y \in U$, $(x, y) \in \sqsubset$ iff either of the following conditions is satisfied:

(i'$\sqsubset$) $(x, y) \in <$, $((x, c^-) \in <$ or $(x, c^-) \in m(1'))$, and $((y, c^-) \in <$ or $(y, c^-) \in m(1'))$,
(ii'$\sqsubset$) $(x, y) \in <$, $(c^-, x) \in <$, $(x, c^+) \in <$, $(c^-, y) \in <$, and $(y, c^+) \in <$,
(iii'$\sqsubset$) $(x, y) \in <$, $((c^+, x) \in <$ or $(c^+, x) \in m(1'))$, and $(c^+, y) \in <$ or $(c^+, y) \in m(1'))$;

- $m(\prec) = \prec$ satisfies all the constraints posed on the relation $\prec$ in OMR-models;
- $m(C^\#) = \{x \in U : (x, c^\#) \in m(1')\} \times U$, for $\# \in \{^-, ^+\}$;
- m extends to all the compound relational terms as in RL$(1, 1')$-logic.

As in RL$(1, 1')$-logic, an RL$_{\mathsf{OMR}}$-model in which $m(1')$ is interpreted as the identity is called a standard RL$_{\mathsf{OMR}}$-model. Note that if $m(1')$ is the identity on U, then $m(C^\#) = \{c^\#\} \times U$, $\# \in \{^-, ^+\}$, that is it is a point relation whose domain represents a landmark.

The validity preserving translation assigning relational terms to modal formulas is defined as in the classical modal logic (see Sect. 7.4) with the following additional clauses:

For $\# \in \{^-, ^+\}$,

- $\tau(c^\#) = C^\# ; 1$.

For $T \in \{<, \sqsubset, \prec\}$,

- $\tau([T]\varphi) = -(T ; -\tau(\varphi))$;
- $\tau([T^{-1}]\varphi) = -(T^{-1} ; -\tau(\varphi))$.

By Theorem 7.4.1, in a similar way as in the classical modal logics, we obtain:

Theorem 15.3.1. *For every* OMR*-formula φ and for all object symbols x and y, the following conditions are equivalent:*

- *φ is* OMR*-valid;*
- *$x\tau(\varphi)y$ is* RL$_{\mathsf{OMR}}$*-valid.*

A relational dual tableau for RL$_{\mathsf{OMR}}$ consists of axiomatic sets and decomposition and specific rules of RL$(1, 1')$-dual tableau adjusted to the language of RL$_{\mathsf{OMR}}$ and the rules specific for RL$_{\mathsf{OMR}}$ listed below.

The rule that reflects irreflexivity of $<$ is an instance of the rule (irref R) presented in Sect. 12.6. The rule reflecting transitivity of $<$ is presented in Sect. 6.6 (see also Sect. 7.4). Recall that these rules have the following forms:

For all object symbols x and y,

$$(\text{irref} <)\ \frac{}{x < x} \qquad (\text{tran} <)\ \frac{x < y}{x < z, x < y \mid z < y, x < y}$$

for every object symbol z

The rules for the comparability relation $\sqsubset$ that express its properties have the following forms:

For all object symbols x and y,

$$(\text{r(i'}\sqsubset))\ \frac{x \sqsubset y}{x < y, x \sqsubset y \mid x < c^-, x1'c^-, x \sqsubset y \mid y < c^-, y1'c^-, x \sqsubset y}$$

$$(\text{r(ii'}\sqsubset))\ \frac{x \sqsubset y}{x < y, x \sqsubset y \mid c^- < x, x \sqsubset y \mid x < c^+, x \sqsubset y \mid c^- < y, x \sqsubset y \mid y < c^+, x \sqsubset y}$$

$$(\text{r(iii'}\sqsubset))\ \frac{x \sqsubset y}{x < y, x \sqsubset y \mid c^+ < x, c^+1'x, x \sqsubset y \mid c^+ < y, c^+1'y, x \sqsubset y}$$

$$(-\sqsubset)\ \frac{x - \sqsubset y}{H_1 \mid H_2 \mid H_3}, \text{ where:}$$

$H_0 = \{y < x, x1'y\}$;
$H_1 = H_0 \cup \{c^- < x, c^- < y, x - \sqsubset y\}$;
$H_2 = H_0 \cup \{x < c^-, x1'c^-, c^+ < x, c^+1'x, y < c^-, y1'c^-, c^+ < y, c^+1'y, x - \sqsubset y\}$;
$H_3 = H_0 \cup \{x < c^+, y < c^+, x - \sqsubset y\}$.

The rules for the negligibility relation $\prec$ reflecting its properties have the following forms:

For all object symbols x and y,

$$(\text{r(i}\prec))\ \frac{x < y}{x \prec y, x < y}$$

$$(\text{r(ii}\prec))\ \frac{x \prec y}{x \prec z, x \prec y \mid z < y, x \prec y} \quad \text{for every object symbol } z$$

$$(\text{r(iii}\prec))\ \frac{x \prec y}{x < z, x \prec y \mid z \prec y, x \prec y} \quad \text{for every object symbol } z$$

$$(\text{r(iv}\prec))\ \frac{x < c^-, x1'c^-, c^+ < x, c^+1'x, y < c^-, y1'c^-, c^+ < y, c^+1'y}{x \prec y, x < c^-, x1'c^-, c^+ < x, c^+1'x, y < c^-, y1'c^-, c^+ < y, c^+1'y}$$

The rules for the relational constants $C^\#$, $\# \in \{^-, ^+\}$ are:

For all object symbols x and y,

$$(C^\#)\ \frac{xC^\#y}{x1'c^\#, xC^\#y} \qquad (-C^\#)\ \frac{x-C^\#y}{x-1'c^\#, x-C^\#y}$$

Note that applications of the rules of $\mathsf{RL_{OMR}}$-dual tableau, in particular applications of the specific rules listed above, preserve the formulas built with atomic terms or their complements. Thus, the closed branch property holds.

The specific $\mathsf{RL_{OMR}}$-axiomatic sets are those including either of the following sets:

(Ax1) $\{c^- < c^+\}$;
(Ax2) $\{x < y, y < x, x1'y\}$, for any object symbols x and y.

An alternative representation of the connectivity condition from the OMR-models can be provided by a rule in the $\mathsf{RL_{OMR}}$-dual tableau. This issue is discussed in Sect. 25.9, see also Sect. 25.6.

The notions of an $\mathsf{RL_{OMR}}$-set and $\mathsf{RL_{OMR}}$-correctness of a rule are defined as in Sect. 2.4.

Proposition 15.3.1.

1. *The* $\mathsf{RL_{OMR}}$*-decomposition rules are* $\mathsf{RL_{OMR}}$*-correct;*
2. *The* $\mathsf{RL_{OMR}}$*-specific rules are* $\mathsf{RL_{OMR}}$*-correct;*
3. *The* $\mathsf{RL_{OMR}}$*-axiomatic sets are* $\mathsf{RL_{OMR}}$*-sets.*

Proof. The proof of 1. follows the proof of RL-correctness of the decomposition rules (see Proposition 2.5.1).

For 2., by way of example, we prove $\mathsf{RL_{OMR}}$-correctness of the rule (r(iv$\prec$)). Let X be a finite set of $\mathsf{RL_{OMR}}$-formulas.

(r(iv$\prec$)) Assume that $X \cup \{x \prec y, x < c^-, x1'c^-, c^+ < x, c^+1'x, y < c^-, y1'c^-, c^+ < y, c^+1'y\}$ is an $\mathsf{RL_{OMR}}$-set. Suppose $X \cup \{x < c^-, x1'c^-, c^+ < x, c^+1'x, y < c^-, y1'c^-, c^+ < y, c^+1'y\}$ is not an $\mathsf{RL_{OMR}}$-set. Then there exist an $\mathsf{RL_{OMR}}$-model $\mathcal{M}$ and valuation v in $\mathcal{M}$ such that for every $\varphi \in X$, $\mathcal{M}, v \not\models \varphi$. Hence, by linearity of $<$, all the following conditions are satisfied: $(c^-, v(x)) \in <$, $(v(x), c^+) \in <$, $(c^-, v(y)) \in <$, and $(v(y), c^+) \in <$. These conditions imply that $v(x), v(y) \in \textsc{Inf}$, thus by the condition (iv$\prec$), $(v(x), v(y)) \notin \prec$. However, by the assumption, $(v(x), v(y)) \in \prec$, a contradiction. Preservation of validity from the upper set to the lower set is obvious.

Correctness of the remaining specific rules follows from the corresponding assumption in the $\mathsf{RL_{OMR}}$-models.

For 3., note that for every $\mathsf{RL_{OMR}}$-model $\mathcal{M}$, $(c^-, c^+) \in <$, thus the sets including $\{c^- < c^+\}$ are $\mathsf{RL_{OMR}}$-sets. Similarly, by connectivity of $<$, sets of the form $X \cup \{x < y, x -\!< y, x1'y\}$ are $\mathsf{RL_{OMR}}$-sets. □

The notions of an $\mathsf{RL_{OMR}}$-proof tree, a closed branch of such a tree, a closed $\mathsf{RL_{OMR}}$-proof tree, and $\mathsf{RL_{OMR}}$-provability are defined as in Sect. 2.4.

The completion conditions determined by the rules (irref $<$) and (tran $<$) are the instances of the completion conditions presented in Sects. 12.6 and 6.6, respectively. The completion conditions determined by the rules that are specific for $\mathsf{RL_{OMR}}$-dual tableau have the following forms:

For all object symbols x and y,

Cpl(r(i'$\sqsubset$)) If $x \sqsubset y \in b$, then either $x < y \in b$ or ($x < c^- \in b$ and $x1'c^- \in b$) or ($y < c^- \in b$ and $y1'c^- \in b$), obtained by an application of the rule (r(i'$\sqsubset$));
Cpl(r(ii'$\sqsubset$)) If $x \sqsubset y \in b$, then either $x < y \in b$ or $c^- < x \in b$ or $x < c^+ \in b$ or $c^- < y \in b$ or $y < c^+ \in b$, obtained by an application of the rule (r(ii'$\sqsubset$));
Cpl(r(iii'$\sqsubset$)) If $x \sqsubset y \in b$, then either $x < y \in b$ or ($c^+ < x \in b$ and $c^+1'x \in b$) or ($c^+ < y \in b$ and $c^+1'y \in b$), obtained by an application of the rule (r(iii'$\sqsubset$));
Cpl($-\sqsubset$) If $x{-}\sqsubset y \in b$, then either of the following conditions is satisfied:

- $y < x \in b$, $x1'y \in b$, $c^- < x \in b$, and $c^- < y \in b$,
- $y < x \in b$, $x1'y \in b$, $x < c^- \in b$, $x1'c^- \in b$, $c^+ < x \in b$, $c^+1'x \in b$, $y < c^- \in b$, $y1'c^- \in b$, $c^+ < y \in b$, and $c^+1'y \in b$,
- $y < x \in b$, $x1'y \in b$, $x < c^+ \in b$, and $y < c^+ \in b$,

obtained by an application of the rule ($-\sqsubset$);
Cpl(r(i$\prec$)) If $x < y \in b$, then $x \prec y \in b$, obtained by an application of the rule (r(i$\prec$));
Cpl(r(ii$\prec$)) If $x \prec y \in b$, then for every object symbol z, either $x \prec z \in b$ or $z < y \in b$, obtained by an application of the rule (r(ii$\prec$));
Cpl(r(iii$\prec$)) If $x \prec y \in b$, then for every object symbol z, either $x < z \in b$ or $z \prec y \in b$, obtained by an application of the rule (r(iii$\prec$));
Cpl(r(iv$\prec$)) If $x < c^- \in b$, $x1'c^- \in b$, $c^+ < x \in b$, $c^+1'x \in b$, $y < c^- \in b$, $y1'c^- \in b$, $c^+ < y \in b$, and $c^+1'y \in b$, then $x \prec y \in b$, obtained by an application of the rule (r(iv$\prec$));
Cpl($C^\#$) If $xC^\#y \in b$, then $x1'c^\# \in b$, obtained by an application of the rule ($C^\#$);
Cpl($-C^\#$) If $x{-}C^\#y \in b$, then $x{-}1'c^\# \in b$, obtained by an application of the rule ($-C^\#$).

The notions of a complete branch of an $\mathsf{RL_{OMR}}$-proof tree, a complete $\mathsf{RL_{OMR}}$-proof tree, and an open branch of an $\mathsf{RL_{OMR}}$-proof tree are defined as in RL-logic (see Sect. 2.5).

Let b be an open branch of an $\mathsf{RL_{OMR}}$-proof tree. A branch structure $\mathcal{M}^b = (U^b, <^b, \sqsubset^b, \prec^b, (c^-)^b, (c^+)^b, m^b)$ is defined as follows:

- $U^b = \mathbb{OV}_{\mathsf{RL_{OMR}}} \cup \{c^-, c^+\}$;
- $(c^\#)^b = m^b(c^\#) = c^\#$, for $\# \in \{^-, ^+\}$;
- $m^b(R) = \{(x, y) \in U^b \times U^b : xRy \notin b\}$, for $R \in \mathbb{RA}_{\mathsf{RL_{OMR}}} \setminus \{\sqsubset, C^-, C^+\}$;
- $<^b = m^b(<)$ and $\prec^b = m^b(\prec)$;
- $\sqsubset^b = m^b(\sqsubset)$, $m^b(C^-)$, and $m^b(C^+)$ are defined as in $\mathsf{RL_{OMR}}$-models;
- m^b extends to all the relational terms as in $\mathsf{RL}(1, 1')$-models.

Proposition 15.3.2 (Branch Model Property). *Let b be an open branch of an $\mathsf{RL_{OMR}}$-proof tree. The branch structure $\mathcal{M}^b = (U^b, <^b, \sqsubset^b, \prec^b, (c^-)^b, (c^+)^b, m^b)$ is an $\mathsf{RL_{OMR}}$-model.*

Proof. Obviously, $c^{\#} \in U^b$, for $\# \in \{^-, ^+\}$. Moreover, since $\{c^- < c^+\}$ is an axiomatic set, $c^- < c^+ \notin b$, hence $((c^-)^b, (c^+)^b) \in <^b$.

Now, it suffices to show that $\prec^b$ satisfies all the conditions assumed in $\mathsf{RL_{OMR}}$-models.

By Cpl(irref $<$), for every $x \in U^b$, $x < x \in b$, that is $(x,x) \notin <^b$. Hence, $<^b$ is irreflexive. Assume that for some $x, y, z \in U^b$, $(x,y) \in <^b$, $(y,z) \in <^b$, but $(x,z) \notin <^b$. Then $x < y \notin b$, $y < z \notin b$, and $x < z \in b$. By Cpl(tran $<$) and since $x < z \in b$, either $x < y \in b$ or $y < z \in b$, a contradiction. Therefore, $<^b$ is transitive. Since $\{x < y, y < x, x1'y\}$ is an axiomatic set, for all $x, y \in U^b$, either $x < y \notin b$ or $y < x \notin b$ or $x1'y \notin b$. Thus, for all $x, y \in U^b$, either $(x,y) \in <^b$ or $(y,x) \in <^b$ or $(x,y) \in m^b(1')$. Hence, $<^b$ is linear.

The linearity of $<^b$ enables us to partition U^b into classes $(\mathrm{OBS}^-)^b$, $(\mathrm{OBS}^+)^b$, and INF^b as defined on p. 279. Then, conditions (i$\prec$), (ii$\prec$), and (iii$\prec$) follow directly from the completion conditions Cpl(r(i$\prec$)), Cpl(r(ii$\prec$)), and Cpl(r(iii$\prec$)), respectively.

Now, we prove that $\prec^b$ satisfies the condition (iv$\prec$). Assume that for some $x, y \in U^b$, $(x,y) \in \prec^b$, but $x \in \mathrm{INF}^b$ and $y \in \mathrm{INF}^b$. Due to irreflexivity and transitivity of $<^b$ and extensionality property of $m^b(1')$ (see the proof of Proposition 2.7.5) it is easy to prove that for all $x, y \in U^b$, if $(x,y) \in <^b$, then $(y,x) \notin <^b \cup m^b(1')$. Thus, since $x \in \mathrm{INF}^b$ and $y \in \mathrm{INF}^b$, we obtain $(x, c^-) \notin <^b \cup m^b(1')$, $(c^+, x) \notin <^b \cup m^b(1')$, $(y, c^-) \notin <^b \cup m^b(1')$, and $(c^+, y) \notin <^b \cup m^b(1')$. Thus all the following formulas belong to b: $x < c^-$, $x1'c^-$, $c^+ < x$, $c^+1'x$, $y < c^-$, $y1'c^-$, $c^+ < y$, and $c^+1'y$. Then, by the completion condition Cpl(r(iv$\prec$)), $x \prec y \in b$, so $(x,y) \notin \prec^b$, a contradiction. Therefore, the condition (iv$\prec$) is satisfied. □

A valuation v^b in $\mathcal{M}^b$ is defined in a standard way, that is $v^b(x) = x$, for every $x \in U^b$.

Proposition 15.3.3 (Satisfaction in Branch Model Property). *Let b be an open branch of an $\mathsf{RL_{OMR}}$-proof tree. Then for every $\mathsf{RL_{OMR}}$-formula φ, if $\mathcal{M}^b, v^b \models \varphi$, then $\varphi \notin b$.*

Proof. Let b be an open branch of an $\mathsf{RL_{OMR}}$-proof tree. The proof is by induction on the complexity of formulas. For a formula φ of the form xRy or $x-Ry$, where $R \in \mathbb{RA}_{\mathsf{RL_{OMR}}} \setminus \{\sqsubset, C^-, C^+\}$, the proof is as in RL-logic and it uses the closed branch property (see the proof of Proposition 2.5.5). Now, it suffices to show that the proposition holds for formulas of the form xRy and formulas of the form $x-Ry$, for $R \in \{\sqsubset, C^-, C^+\}$. By way of example, we prove it for $R = -\sqsubset$.

Assume $\mathcal{M}^b, v^b \models x-\sqsubset y$, i.e., $(x,y) \notin \sqsubset^b$. Suppose $x-\sqsubset y \in b$. Then by the completion condition Cpl($-\sqsubset$) and symmetry of $m^b(1')$, either of the following conditions is satisfied:

- $(y,x) \notin <^b \cup m^b(1')$, $(c^-, x) \notin <^b$, and $(c^-, y) \notin <^b$;
- $(y,x) \notin <^b \cup m^b(1')$, $(x, c^-) \notin <^b \cup m^b(1')$, $(c^+, x) \notin <^b \cup m^b(1')$, $(y, c^-) \notin <^b \cup m^b(1')$, and $(c^+, y) \notin <^b \cup m^b(1')$;
- $(y,x) \notin <^b \cup m^b(1')$, $(x, c^+) \notin <^b$, and $(y, c^+) \notin <^b$.

Thus, by the linearity of $<^b$, either of the following conditions is satisfied:

- $(x, y) \in <^b$, $(x, c^-) \in <^b \cup m^b(1')$, and $(y, c^-) \in <^b \cup m^b(1')$;
- $(x, y) \in <^b$, $(c^-, x) \in <^b$, $(x, c^+) \in <^b$, $(c^-, y) \in <^b$, and $(y, c^+) \in <^b$;
- $(x, y) \in <^b$, $(c^+, x) \in <^b \cup m^b(1')$, and $(c^+, y) \in <^b \cup m^b(1')$.

Therefore, $(x, y) \in \sqsubset^b$, a contradiction. For the compound relational terms the proofs are similar to those in RL-logic. □

Finally, we obtain:

Theorem 15.3.2 (Soundness and Completeness of $\mathsf{RL_{OMR}}$). *For every $\mathsf{RL_{OMR}}$-formula φ, the following conditions are equivalent:*

1. *φ is $\mathsf{RL_{OMR}}$-valid;*
2. *φ is true in all standard $\mathsf{RL_{OMR}}$-models;*
3. *φ is $\mathsf{RL_{OMR}}$-provable.*

By the above theorem and Theorem 15.3.1, we have:

Theorem 15.3.3 (Relational Soundness and Completeness of OMR). *For every OMR-formula φ and for all object symbols x and y, the following conditions are equivalent:*

1. *φ is OMR-valid;*
2. *$x\tau(\varphi)y$ is $\mathsf{RL_{OMR}}$-provable.*

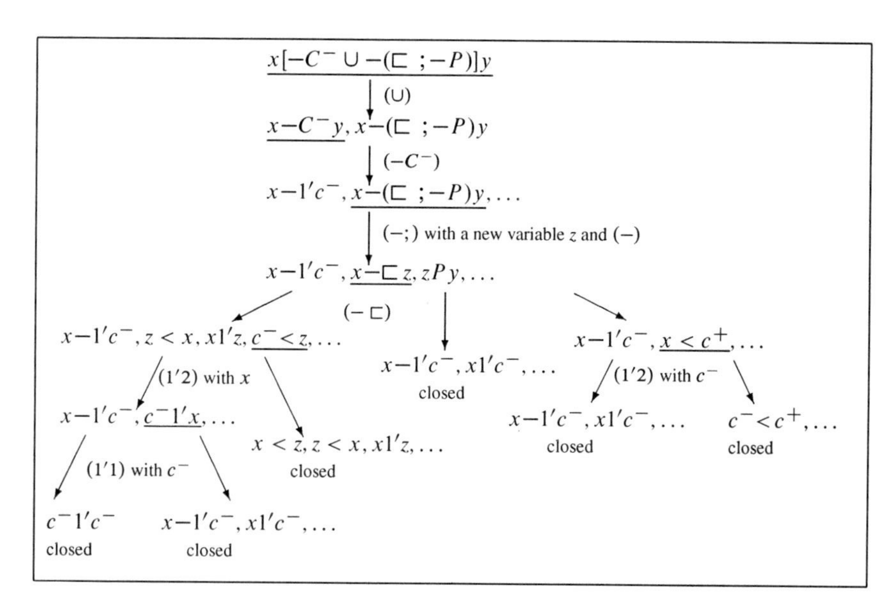

Fig. 15.1 An $\mathsf{RL_{OMR}}$-proof of the formula $\neg c^- \vee [\sqsubset]p$

Example. Consider an OMR-formula:

$$\varphi = \neg c^{-} \vee [\sqsubset]p.$$

The translation of φ to an $\mathsf{RL}_{\mathsf{OMR}}$-relational term is:

$$\tau(\varphi) = -C^{-} \cup -(\sqsubset ;-P),$$

where for simplicity of notation, $\tau(p) = P$. By Theorem 15.3.3, for all object symbols x and y, OMR-validity of φ is equivalent to $\mathsf{RL}_{\mathsf{OMR}}$-provability of the formula $x\tau(\varphi)y$. Figure 15.1 presents an $\mathsf{RL}_{\mathsf{OMR}}$-proof of this formula.

Part V
Relational Reasoning about Time, Space, and Action

Chapter 16
Dual Tableaux for Temporal Logics

16.1 Introduction

The first attempts to create a tense logic are the investigations of A. N. Prior published in [Pri57, Pri67]. Temporal logics are modal logics whose modal operations are determined by two relations on a set of time points expressing earlier-later or past-future relationships between instants of time. The relations are ordering relations possibly satisfying various axioms such as strict ordering, linear ordering, branching time, ordering with or without endpoints, discrete, etc. If P is a relation such that holding of tPt' is interpreted as t precedes t', then the modal operator $\langle P \rangle$ means 'it will at some time be the case that' and $[P]$ is interpreted as 'it will always be the case that'. Similarly, if F is a relation such that tFt' means t follows t', then $\langle F \rangle$ says 'it has at some time been the case that' and $[F]$ is intuitively interpreted as 'it has always been the case that'. Often some other temporal operations are included in languages of temporal logics. The binary operations *Since* and *Until* introduced in [Kam68] and the unary operation *Next*, introduced in [vW65], are among the most popular. If φ and ψ are formulas, then the formula 'φ *Since* ψ' says that ψ has been true since a time point when φ was true. The formula 'φ *Until* ψ' means that there is a future time point at which ψ is true, with φ true at all time points between now and then. The formula '*Next* φ' is meaningful whenever the ordering of time points is discrete. Then it says that φ is true at the immediate successor of a present time moment.

The problem of representing time-varying information and reasoning with such information arises in a wide range of disciplines including logic, computer science, psychology, linguistics, and philosophy. Applications of temporal logic in computer science have been initiated by J. Bubenko in [Bub77] for applications in the theory of databases and by A. Pnueli in [Pnu77] for applications in the theory of programs. Since then temporal logics have become an important issue in artificial intelligence, in the specification and verification of programs, and in reasoning about actions and events. In Chap. 19 we discuss, among others, a temporal logic for the specification and verification of concurrent programs. Surveys of temporal logics and their applications can be found in [Bur79, vB83, vB95, Gol87, AHK02, HR06]. There are many results on decidability of temporal logics. It was proved that the temporal

E. Orłowska and J. Golińska-Pilarek, *Dual Tableaux: Foundations, Methodology, Case Studies*, Trends in Logic 36, DOI 10.1007/978-94-007-0005-5_16,

theories of arbitrary linear orders, of every elementary class of linear orders, of well orders, and of complete orders are decidable. A comprehensive survey of decidability of temporal logics can be found in [BG85].

In this chapter we present relational dual tableaux for a number of point-based temporal structures.

16.2 Basic Temporal Logic

Temporal logics belong to the family of modal logics. In their models the elements of the universes are interpreted as moments of time and the accessibility relation reflects earlier-later relationship. To get access to both past and future moments, in temporal frames we usually include the accessibility relation and its converse, denoted by P and F, respectively. Modal operations determined by these relational constants refer to past and future moments of time, respectively. The logics with these operations as the only modal operations are referred to as the *standard temporal logics*. We obtain various classes of standard temporal logics by assuming specific properties of the accessibility relations.

The common language of standard temporal logics is a modal language as defined in Sect. 7.3 with relational constants F and P. In this chapter we consider the basic temporal logic TL whose models are structures of the form $\mathcal{M} = (U, F, P, m)$ as defined in Sect. 7.3 (see p. 146) such that F and P are transitive relations on U satisfying $F = P^{-1}$. Relations F and P may satisfy some additional conditions, for example linearity or density, as presented in the next section.

The language of temporal logics with operations determined by relations F and P is more expressive than the ordinary modal language with operations determined by one of these relations, for example, continuity of a strict linear ordering is expressible only in the presence of both past and future operations.

In order to obtain a relational representation of the logic TL, we follow the method presented in Sect. 7.4. We define logic $\mathsf{RL}_{\mathsf{TL}}$ whose language is appropriate for expressing formulas of TL as relations. It is an $\mathsf{RL}(1, 1')$-language such that $\mathbb{RC}_{\mathsf{RL}_{\mathsf{TL}}} = \{1, 1', F, P\}$ and $\mathbb{OC}_{\mathsf{RL}_{\mathsf{TL}}} = \emptyset$. An $\mathsf{RL}_{\mathsf{TL}}$-model is a structure $\mathcal{M} = (U, F, P, m)$ such that (U, m) is an $\mathsf{RL}(1, 1')$-model, F and P are binary relations on U that provide the interpretation of relational constants F and P, respectively, and satisfy all the properties of relations F and P that are assumed in TL-models. The translation of TL-formulas into relational terms is defined as in Sect. 7.4 (see p. 147). Due to Theorem 7.4.1, we have:

Theorem 16.2.1. *For every* TL*-formula* φ *and for all object variables* x *and* y*, the following conditions are equivalent:*

1. φ *is* TL*-valid;*
2. $x\tau(\varphi)y$ *is* $\mathsf{RL}_{\mathsf{TL}}$*-valid.*

$\mathsf{RL_{TL}}$-dual tableau is an extension of $\mathsf{RL}(1, 1')$-dual tableau with the specific rules that reflect properties of the relations F and P. The rule reflecting transitivity is presented in Sect. 6.6 (see also Sect. 7.4). We recall that this rule has the following form:

For all object variables x and y and for every $R \in \{F, P\}$,

$$(\text{tran } R) \quad \frac{xRy}{xRz, xRy \mid zRy, xRy} \quad z \text{ is any object variable}$$

The rules that reflect the conditions $P^{-1} \subseteq F$ and $F \subseteq P^{-1}$ have the following forms, respectively:

For all object variables x and y,

$$(F) \quad \frac{xFy}{yPx, xFy} \qquad (P) \quad \frac{xPy}{yFx, xPy}$$

Alternative forms of rules (F) and (P) are discussed in Sect. 25.9.

The notions of an $\mathsf{RL_{TL}}$-set and correctness of a rule are defined as in Sect. 2.4. Following the proof of correspondence for the $\mathsf{RL_{FCL}}$-rules (S) and (R) in Theorem 13.3.2, it can be shown that the rules (F) and (P) are correct if and only if the conditions $P^{-1} \subseteq F$ and $F \subseteq P^{-1}$, respectively, hold in all $\mathsf{RL_{TL}}$-structures. Thus, based on correctness of $\mathsf{RL}(1, 1')$-rules and on correctness of the rule (tran R) proved in Sect. 6.6, we get:

Proposition 16.2.1.

1. *The* $\mathsf{RL_{TL}}$*-rules are* $\mathsf{RL_{TL}}$*-correct;*
2. *The* $\mathsf{RL_{TL}}$*-axiomatic sets are* $\mathsf{RL_{TL}}$*-sets.*

The notions of an $\mathsf{RL_{TL}}$-proof tree, a closed branch of such a tree, a closed $\mathsf{RL_{TL}}$-proof tree, and $\mathsf{RL_{TL}}$-provability are defined as in Sect. 2.4.

A branch b of an $\mathsf{RL_{TL}}$-proof tree is complete whenever it is closed or it satisfies the completion conditions of $\mathsf{RL}(1, 1')$-dual tableau adjusted to $\mathsf{RL_{TL}}$-language and the following conditions specific for the $\mathsf{RL_{TL}}$-dual tableau:

For all object variables x and y and for every $R \in \{F, P\}$,

Cpl(tran R) If $xRy \in b$, then for every object variable z, either $xRz \in b$ or $zRy \in b$, obtained by an application of the rule (tran R);
Cpl(F) If $xFy \in b$, then $yPx \in b$, obtained by an application of the rule (F);
Cpl(P) If $xPy \in b$, then $yFx \in b$, obtained by an application of the rule (P).

The rules of $\mathsf{RL_{TL}}$-dual tableau, in particular the specific rules listed above, guarantee that for every branch b of an $\mathsf{RL_{TL}}$-proof tree, if $xTy \in b$ and $x{-}Ty \in b$, for some atomic relational term T, then there is a node in the branch which contains both of these formulas, which implies that branch b is closed. Therefore, the closed branch property holds.

The notions of a complete $\mathsf{RL_{TL}}$-proof tree and an open branch of an $\mathsf{RL_{TL}}$-proof tree are defined as in RL-logic (see Sect. 2.5).

Let b be an open branch of an $\mathsf{RL_{TL}}$-proof tree. The branch structure $\mathcal{M}^b = (U^b, F^b, P^b, m^b)$ is defined in a standard way (see Sect. 2.6), i.e., (U^b, m^b) is an $\mathsf{RL}(1,1')$-branch model and $m^b(R) = \{(x,y) \in U^b \times U^b : xRy \notin b\}$, for every $R \in \{F, P\}$. Following the proof of the branch model property in the completeness proof of $\mathsf{RL_{EQ}}$-dual tableau and $\mathsf{RL_{FCL}}$-dual tableau (see Sects. 6.6 and 13.3, respectively), the branch model property for $\mathsf{RL_{TL}}$-dual tableau can be proved. Since the branch model $\mathcal{M}^b$ is defined in a standard way and the closed branch property holds, the satisfaction in branch model property can be proved as in $\mathsf{RL}(1,1')$-logic (see Sects. 2.5 and 2.7). Hence, completeness of $\mathsf{RL_{TL}}$-dual tableau follows.

Theorem 16.2.2 (Soundness and Completeness of $\mathsf{RL_{TL}}$). *Let φ be an $\mathsf{RL_{TL}}$-formula. Then the following conditions are equivalent:*

1. *φ is $\mathsf{RL_{TL}}$-valid;*
2. *φ is true in all standard $\mathsf{RL_{TL}}$-models;*
3. *φ is $\mathsf{RL_{TL}}$-provable.*

The above theorem and Theorem 16.2.1 imply:

Theorem 16.2.3 (Relational Soundness and Completeness of TL). *Let φ be a TL-formula. Then for all object variables x and y, the following conditions are equivalent:*

1. *φ is TL-valid;*
2. *$x\tau(\varphi)y$ is $\mathsf{RL_{TL}}$-provable.*

16.3 Semantic Restrictions on Basic Temporal Logic

Various temporal logics are obtained from TL by assuming some properties of the time ordering F and P. Let $R \in \{F, P\}$. The following conditions on R are among the most typical:

For all x, y, and z,

- R is irreflexive: $(x,x) \notin R$;
- R is serial: there exists y such that $(x,y) \in R$;
- R is unbound from below: R^{-1} is serial;
- R is discrete: $(x,y) \in R$ implies (1) there exists z such that $(x,z) \in R$ and for all t if $(x,t) \in R$, then $(z,t) \in R$, and (2) there exists z such that $(z,y) \in R$ and for all t if $(t,y) \in R$, then $(t,z) \in R$;
- R is weakly connected: $(x,y) \in R$ and $(x,z) \in R$ imply $(y,z) \in R$ or $(z,y) \in R$ or $y = z$;
- R is connected: either $(x,y) \in R$ or $(y,x) \in R$ or $x = y$;
- R is dense: $(x,y) \in R$ implies there exists z such that $(x,z) \in R$ and $(z,y) \in R$;
- R is weakly directed: $(x,y) \in R$ and $(x,z) \in R$ imply there exists t such that $(y,t) \in R$ and $(z,t) \in R$;
- R is Euclidean: $(x,y) \in R$ and $(x,z) \in R$ imply $(y,z) \in R$;
- R is partially functional: $(x,y) \in R$ and $(x,z) \in R$ imply $y = z$;
- R is functional: there exists exactly one y such that $(x,y) \in R$.

These properties of temporal ordering are of great importance in temporal reasoning. The adequate modelling of time scale should guarantee that any time moment does not precede itself. It seems that the most appropriate model for linear time scales is a strict total order which is irreflexive and connected. In several applications we need to distinguish between discrete and dense time scales, for example to model execution of computer programs, a discrete time is appropriate.

In what follows, any logic whose models are TL-models such that the relation R possibly satisfies some constraints from the above list is referred to as a standard temporal logic and is denoted by $\mathsf{L_{TL}}$. It is always specified semantically in terms of a class of models.

The relational logic for an $\mathsf{L_{TL}}$-logic is defined as in the previous section with the minor change of notation. We will write R and R^{-1} instead of F and P (or P and F), respectively, both in the language, in the structures, and in the models. An $\mathsf{RL_{L_{TL}}}$-structure is an $\mathsf{RL_{TL}}$-model $\mathcal{M} = (U, R, R^{-1}, m)$. An $\mathsf{RL_{L_{TL}}}$-model is an $\mathsf{RL_{L_{TL}}}$-structure such that the relation R satisfies all the conditions assumed in the $\mathsf{L_{TL}}$-models.

As usual, we can prove the following:

Theorem 16.3.1. *Let $\mathsf{L_{TL}}$ be a standard temporal logic. Then for every $\mathsf{L_{TL}}$-formula φ and for all object variables x and y, the following conditions are equivalent:*

1. *φ is $\mathsf{L_{TL}}$-valid;*
2. *$x\tau(\varphi)y$ is $\mathsf{RL_{L_{TL}}}$-valid.*

$\mathsf{RL_{L_{TL}}}$-dual tableaux are extensions of $\mathsf{RL_{TL}}$-dual tableau with the specific rules and/or axiomatic sets that reflect properties of the relation R. Below we present the rules and axiomatic sets that reflect the properties of relations from the above list.

Connectivity of R leads to the following axiomatic set:

(Ax) $\{xRy, yRx, x1'y\}$.

Alternatively, connectivity can be expressed with a rule (see Sect. 25.9).

The rule that reflects irreflexivity can be found in Sect. 12.6. We recall that this rule has the following form:

(irref R) $\dfrac{}{xRx}$ x is any object variable

The rules for the remaining properties have the following forms:

(ser R) $\dfrac{}{x{-}(R\,;1)x}$ x is any object variable

(un R) $\dfrac{}{x{-}(R^{-1}\,;1)x}$ x is any object variable

(dis$_1$ R) $\dfrac{}{xRy \mid x{-}({-}(R\,;{-}R^{-1})\,;R^{-1})x}$ x, y are any object variables (dis$_2$ R) $\dfrac{}{xRy \mid y{-}(-(R^{-1} ; -R) ; R)y}$ x, y are any object variables

(wcon R) $\dfrac{yRz, zRy, y1'z}{xRy, yRz, zRy, y1'z \mid xRz, yRz, zRy, y1'z}$

x is any object variable

(den R) $\dfrac{}{xRy \mid x{-}(R ; R)y}$ x, y are any object variables

(wdir R) $\dfrac{}{xRy \mid xRz \mid y{-}(R ; R^{-1})z}$ x, y, z are any object variables

(Euc R) $\dfrac{yRz}{xRy, yRz \mid xRz, yRz}$ x is any object variable

(pfun R) $\dfrac{y1'z}{xRy, y1'z \mid xRz, y1'z}$ x is any object variable

(fun R) $\dfrac{}{xRy, x{-}(R ; 1)x \mid xRz, x{-}(R ; 1)x \mid y{-}1'z, x{-}(R ; 1)x}$

x, y, z are any object variables

Many of these rules have the form of a specialized cut rule. Some of them could be replaced by rules with a non-empty premise, and then an ordinary cut rule must be introduced to the dual tableau in question. This issue is discussed in detail in Sect. 25.9.

The rules (ser R), (dis$_1$ R), (dis$_2$ R), (wcon R), (den R), (wdir R), (Euc R), (pfun R), (fun R), and (un R) reflect that the relation R is serial, discrete, weakly connected, dense, weakly directed, Euclidean, partially functional, functional, and unbound from below, respectively. Introduction of any of these rules to the $\mathsf{RL}_{\mathsf{L}_{\mathsf{TL}}}$-dual tableau does not violate the closed branch property which can be proved as in Proposition 2.5.3.

Let L_{TL} be a standard temporal logic and let $\mathcal{K}$ be a class of L_{TL}-structures. The notion of a $\mathcal{K}$-set and the notion of $\mathcal{K}$-correctness are defined as in Sect. 2.4.

Theorem 16.3.2 (Correspondence). *Let* L_{TL} *be a standard temporal logic and let* $\mathcal{K}$ *be a class of* $\mathsf{RL}_{\mathsf{L}_{\mathsf{TL}}}$*-structures. A condition (c) is true in the class* $\mathcal{K}$ *iff its corresponding rule(s) is(are)* $\mathcal{K}$*-correct.*

Proof. By way of example, we prove the statement for a logic where relation R is dense and for a logic where R is discrete.

(den R) Let L_{TL} be a logic where relation R is dense. Assume that R is dense in every $\mathcal{K}$-structure. Let X be a finite set of $\mathsf{RL}_{\mathsf{L}_{\mathsf{TL}}}$-formulas. Preservation of validity from the upper set to the lower sets is obvious. Now, assume that $X \cup \{xRy\}$ and $X \cup \{x{-}(R ; R)y\}$ are $\mathcal{K}$-sets and suppose X is not a $\mathcal{K}$-set. Then, by the assumption, there exist a $\mathcal{K}$-structure $\mathcal{M}$ and a valuation v in $\mathcal{M}$ such that $(v(x), v(y)) \in R$

and for every $z \in U$ either $(v(x), z) \notin R$ or $(z, v(y)) \notin R$, a contradiction with density of R. Hence, the rule (den R) is $\mathcal{K}$-correct. Now, assume the rule (den R) is $\mathcal{K}$-correct. Let $X \stackrel{\text{df}}{=} \{x{-}Ry, x(R\,;R)y\}$. Then $X \cup \{xRy\}$ and $X \cup \{x{-}(R\,;R)y\}$ are $\mathcal{K}$-sets. Thus, by the assumption, $\{x{-}Ry, x(R\,;R)y\}$ is a $\mathcal{K}$-set, which means that for every $\mathcal{K}$-structure $\mathcal{M}$ and for all $x, y \in U$ if $(x, y) \in R$, then there exists $z \in U$ such that $(x, z) \in R$ and $(z, y) \in R$. Hence, R is dense.

(dis R) Let $\mathsf{L_{TL}}$ be a logic where relation R is discrete. Assume that R is discrete in every $\mathcal{K}$-structure. Let X be a finite set of $\mathsf{RL_{L_{TL}}}$-formulas. Assume that $X \cup \{xRy\}$ and $X \cup \{x{-}({-}(R\,;{-}R^{-1})\,;R^{-1})x\}$ are $\mathcal{K}$-sets and suppose X is not a $\mathcal{K}$-set. Then, by the assumption, there exist a $\mathcal{K}$-structure $\mathcal{M}$ and a valuation v in $\mathcal{M}$ such that $(v(x), v(y)) \in R$ and for all $z \in U$ either $(v(x), z) \notin R$ or there exists $t \in U$ such that $(v(x), t) \in R$ and $(z, t) \notin R$, which contradicts discreteness of R. Preservation of validity from the upper set to the lower sets is obvious. Therefore, the rule (dis_1 R) is $\mathcal{K}$-correct. Correctness of the rule (dis_2 R) can be proved in a similar way. Now, assume that the rules (dis_1 R) and (dis_2 R) are $\mathcal{K}$-correct. We show that R is discrete in every $\mathcal{K}$-structure. Let $X \stackrel{\text{df}}{=} \{x{-}Ry, x({-}(R\,;{-}R^{-1})\,;R^{-1})x\}$. Then, in every $\mathcal{K}$-structure the sets $X \cup \{xRy\}$ and $X \cup \{x{-}({-}(R\,;{-}R^{-1})\,;R^{-1})x\}$ are $\mathcal{K}$-sets. Thus, by $\mathcal{K}$-correctness of the rule (dis_1 R), X is a $\mathcal{K}$-set, that is for all $x, y \in U$, if $(x, y) \in R$, then there exists $z \in U$ such that $(x, z) \in R$ and for all $t \in U$, if $(x, t) \in R$, then $(z, t) \in R$. Therefore, the part (1) of discreteness holds. In a similar way, correctness of the rule (dis_2 R) implies part (2) of discreteness. □

The above theorem leads to the following:

Proposition 16.3.1. *Let* $\mathsf{L_{TL}}$ *be a standard temporal logic. Then:*

1. *The* $\mathsf{RL_{L_{TL}}}$*-rules are* $\mathsf{RL_{L_{TL}}}$*-correct;*
2. *The* $\mathsf{RL_{L_{TL}}}$*-axiomatic sets are* $\mathsf{RL_{L_{TL}}}$*-sets.*

A branch b of an $\mathsf{RL_{L_{TL}}}$-proof tree is complete whenever it is closed or it satisfies the completion conditions of $\mathsf{RL_{L_{TL}}}$-dual tableau and the following conditions specific for the $\mathsf{RL_{L_{TL}}}$-dual tableau:

For all object variables x, y, z,

Cpl(irref R) $xRx \in b$, obtained by an application of the rule (irref R);
Cpl(ser R) $x{-}(R\,;1)x \in b$, obtained by an application of the rule (ser R);
Cpl(un R) $x{-}(R^{-1}\,;1)x \in b$, obtained by an application of the rule (un R);
Cpl(dis_1 R) Either $xRy \in b$ or $x{-}({-}(R\,;{-}R^{-1})\,;R^{-1})x \in b$, obtained by an application of the rule (dis_1 R);
Cpl(dis_2 R) Either $xRy \in b$ or $y{-}({-}(R^{-1}\,;{-}R)\,;R)y \in b$, obtained by an application of the rule (dis_2 R);
Cpl(wcon R) If $yRz \in b$, $zRy \in b$, and $y1'z \in b$, then either $xRy \in b$ or $xRz \in b$, obtained by an application of the rule (wcon R);
Cpl(den R) Either $xRy \in b$ or $x{-}(R\,;R)y \in b$, obtained by an application of the rule (den R);

Cpl(wdir R) Either $xRy \in b$ or $xRz \in b$ or $y-(R;R^{-1})z \in b$, obtained by an application of the rule (wdir R);

Cpl(Euc R) If $yRz \in b$, then either $xRy \in b$ or $xRz \in b$, obtained by an application of the rule (Euc R);

Cpl(pfun R) If $y1'z \in b$, then for every object variable x, either $xRy \in b$ or $xRz \in b$, obtained by an application of the rule (pfun R);

Cpl(fun R) $x-(R;1)x \in b$ and either $xRy \in b$ or $xRz \in b$ or $y-1'z \in b$, obtained by an application of the rule (fun R).

The notions of a complete $\mathsf{RL}_{\mathsf{L}_{\mathsf{TL}}}$-proof tree and an open branch of an $\mathsf{RL}_{\mathsf{L}_{\mathsf{TL}}}$-proof tree are defined as in RL-logic (see Sect. 2.5).

Let b be an open branch of an $\mathsf{RL}_{\mathsf{L}_{\mathsf{TL}}}$-proof tree. The branch structure $\mathcal{M}^b = (U^b, R^b, (R^{-1})^b, m^b)$ is defined as in the completeness proof of $\mathsf{RL}_{\mathsf{TL}}$-dual tableau, in particular $R^b = m^b(R) \stackrel{\text{df}}{=} \{(x,y) \in U^b \times U^b : xRy \notin b\}$ and $(R^{-1})^b \stackrel{\text{df}}{=} (R^b)^{-1}$.

Proposition 16.3.2 (Branch Model Property). *Let L_{TL} be a standard temporal logic. For every open branch b of an $\mathsf{RL}_{\mathsf{L}_{\mathsf{TL}}}$-proof tree, the branch structure $\mathcal{M}^b = (U^b, R^b, (R^{-1})^b, m^b)$ is an $\mathsf{RL}_{\mathsf{L}_{\mathsf{TL}}}$-model.*

Proof. It suffices to show that if a condition is true in all models of a logic $\mathsf{RL}_{\mathsf{L}_{\mathsf{TL}}}$, then it is true in $\mathcal{M}^b$.

By way of example, we show that this holds for a logic L_{TL} whose models have a discrete relation R. Then, the dual tableau for $\mathsf{RL}_{\mathsf{L}_{\mathsf{TL}}}$ contains rules (dis$_1$ R) and (dis$_2$ R). By the completion condition Cpl(dis$_1$ R), for all $x, y \in U^b$, either $xRy \in b$ or $x-(-(R;-R^{-1});R^{-1})x \in b$. Thus, by the completion conditions Cpl(−;), Cpl(−), and Cpl($-^{-1}$), for all $x, y \in U^b$, either $xRy \in b$ or for some $z \in U^b$, $x(R;-R^{-1})z \in b$ and $x-Rz \in b$. Thus, by the completion conditions Cpl(;) and Cpl($-^{-1}$), if $(x,y) \in R^b$, then there exists $z \in U^b$ such that $(x,z) \in R^b$ and if $(x,t) \in R^b$, then $(z,t) \in R^b$. On the other hand, by the completion condition Cpl(dis$_2$ R), it can be proved that if $(x,y) \in R^b$, then there exists $z \in U^b$ such that $(z,y) \in R^b$ and if $(t,y) \in R^b$, then $(t,z) \in R^b$. Therefore, R^b is discrete. □

The satisfaction in branch model property can be proved as in $\mathsf{RL}_{\mathsf{TL}}$-logic. Therefore, we get:

Theorem 16.3.3 (Soundness and Completeness of $\mathsf{RL}_{\mathsf{L}_{\mathsf{TL}}}$). *Let L_{TL} be a standard temporal logic and let φ be an $\mathsf{RL}_{\mathsf{L}_{\mathsf{TL}}}$-formula. Then the following conditions are equivalent:*

1. *φ is $\mathsf{RL}_{\mathsf{L}_{\mathsf{TL}}}$-valid;*
2. *φ is true in all standard $\mathsf{RL}_{\mathsf{L}_{\mathsf{TL}}}$-models;*
3. *φ is $\mathsf{RL}_{\mathsf{L}_{\mathsf{TL}}}$-provable.*

The above theorem and Theorem 16.3.1 imply:

Theorem 16.3.4 (Relational Soundness and Completeness of $\mathsf{L_{TL}}$). *Let $\mathsf{L_{TL}}$ be a standard temporal logic and let φ be an $\mathsf{L_{TL}}$-formula. Then for all object variables x and y, the following conditions are equivalent:*

1. *φ is $\mathsf{L_{TL}}$-valid;*
2. *$x\tau(\varphi)y$ is $\mathsf{RL_{L_{TL}}}$-provable.*

Example. Let $\mathsf{L^1_{TL}}$ be a standard temporal logic whose models have an Euclidean relation R. Figure 16.1 presents an $\mathsf{RL_{L^1_{TL}}}$-proof of the formula:

$$\varphi = \langle R\rangle p \to [R]\langle R\rangle p,$$

which reflects this property.

Let $\mathsf{L^2_{TL}}$ be a standard temporal logic whose models have a dense relation R. Figure 16.2 presents an $\mathsf{RL_{L^2_{TL}}}$-proof of the formula

$$\psi = [R][R]p \to [R]p.$$

For simplicity, we write P instead of $\tau(p)$. The relational translation of φ and ψ are:

$$\tau(\varphi) = -(R\,;P) \cup -(R\,;-(R\,;P)),$$
$$\tau(\psi) = --(R\,;--(R\,;-P)) \cup -(R\,;-P).$$

In the following sections we present some signature extensions of the basic temporal logic.

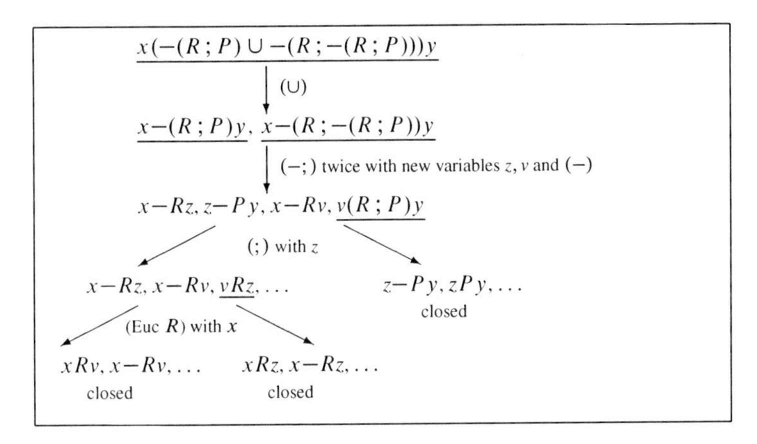

Fig. 16.1 An $\mathsf{RL_{L^1_{TL}}}$-proof of the formula $\langle R\rangle p \to [R]\langle R\rangle p$

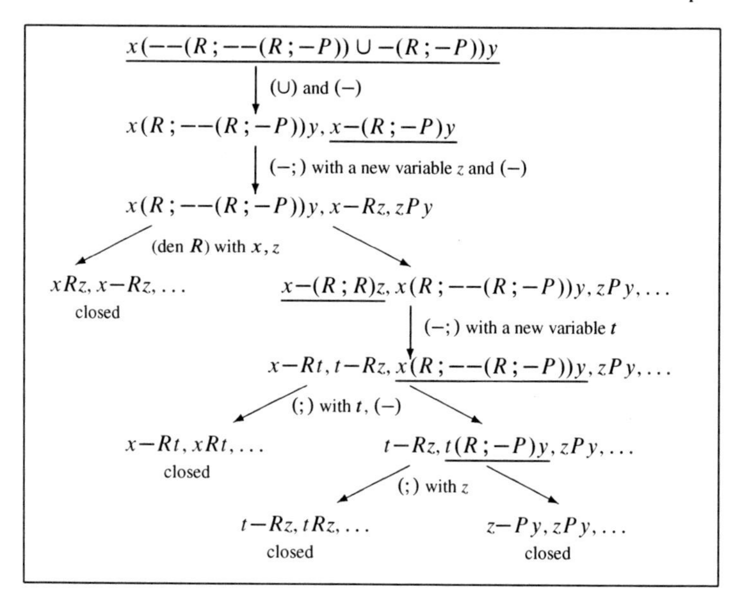

Fig. 16.2 An $\mathsf{RL}_{\mathsf{L}^2_{\mathsf{TL}}}$-proof of the formula $[R][R]p \rightarrow [R]p$

16.4 Temporal Logics with *Since* and *Until*

In languages of temporal logics we often admit binary operations *Since* and *Until* with the following semantics. Let R be a temporal ordering, then:

- $\mathcal{M}, s \models \varphi\, Since\, \psi$ iff there exists $s' \in U$ such that $(s', s) \in R$ and $\mathcal{M}, s' \models \psi$ and for all $u \in U$, if $(s', u) \in R$ and $(u, s) \in R$, then $\mathcal{M}, u \models \varphi$;
- $\mathcal{M}, s \models \varphi\, Until\, \psi$ iff there exists $s' \in U$ such that $(s, s') \in R$ and $\mathcal{M}, s' \models \psi$ and for all $u \in U$, if $(s, u) \in R$ and $(u, s') \in R$, then $\mathcal{M}, u \models \varphi$.

A formula φ *Since* ψ means that there is a past moment s' at which ψ is satisfied and at all moments between s' and now φ is satisfied. Similarly, φ *Until* ψ says that there is a future moment s' at which ψ is satisfied and φ is satisfied at all moments between now and s'. In the presence of *Until* the *next-state operation*, *Next*, is definable in the modal language:

- $\mathcal{M}, s \models Next\, \varphi$ iff $\mathcal{M}, s \models (\varphi \wedge \neg\varphi)Until\, \varphi$.

Let $\mathsf{TL}_{\mathsf{SU}}$ denote a temporal logic with a time ordering R and operations *Since*, *Until*, and *Next*. To define a relational representation of $\mathsf{TL}_{\mathsf{SU}}$-formulas, we extend the set of relational terms of TL by admitting relational counterparts of logical operations *Since*, *Until*, and *Next* among the relational operations. For the sake of simplicity, they are denoted in the same way as the respective propositional operations. Namely, the vocabulary of the language of logic $\mathsf{RL}_{\mathsf{TL}_{\mathsf{SU}}}$ is the $\mathsf{RL}(1, 1')$-language such that:

- $\mathbb{OC}_{\mathsf{RL}_{\mathsf{TL}_{\mathsf{SU}}}} = \emptyset$;
- $\mathbb{RC}_{\mathsf{RL}_{\mathsf{TL}_{\mathsf{SU}}}} = \{1, 1', R\}$, where R is the relational constant representing the time ordering;
- $\{Since, Until, Next\}$ is included in the set of relational operations.

The set of relational terms is obtained from $\mathbb{RV}_{\mathsf{RL}_{\mathsf{TL}_{\mathsf{SU}}}} \cup \{1, 1', R\}$ by making its closure with respect to the standard relational operations and the operations *Since*, *Until*, and *Next*.

$\mathsf{RL}_{\mathsf{TL}_{\mathsf{SU}}}$-structures are of the form $\mathcal{M} = (U, R, R^{-1}, m)$, where (U, m) is an $\mathsf{RL}(1, 1')$-model, $R = m(R)$ is a binary relation on U satisfying all the conditions assumed in $\mathsf{TL}_{\mathsf{SU}}$-logic, and the relational operations *Since*, *Until*, and *Next* are interpreted as operations on binary relations on U. $\mathsf{RL}_{\mathsf{TL}_{\mathsf{SU}}}$-models are $\mathsf{RL}_{\mathsf{TL}_{\mathsf{SU}}}$-structures $\mathcal{M} = (U, R, R^{-1}, m)$ such that the relational operations *Since*, *Until*, and *Next* are interpreted as follows. Let T, Q be relational terms, then:

- $m(T\ Since\ Q) \stackrel{\text{df}}{=} \{(x, y) : \exists t[(t, x) \in R \wedge (t, y) \in m(Q) \wedge \forall u\,((t, u) \in R \wedge (u, x) \in R \rightarrow (u, u) \in m(T))]\}$;
- $m(T\ Until\ Q) \stackrel{\text{df}}{=} \{(x, y) : \exists t[(x, t) \in R \wedge (t, y) \in m(Q) \wedge \forall u\,((x, u) \in R \wedge (u, t) \in R \rightarrow (u, u) \in m(T))]\}$;
- $m(Next\ T) \stackrel{\text{df}}{=} \{(x, y) : \exists t[(x, t) \in R \wedge (t, y) \in m(T) \wedge \neg\exists u\,((x, u) \in R \wedge (u, t) \in R)]\}$.

This interpretation of the operations *Since*, *Until*, and *Next* is motivated with the role that they play in the representation of $\mathsf{TL}_{\mathsf{SU}}$-formulas which are interpreted as right ideal relations (see translation function defined on p. 302 and Theorem 16.4.1).

The next proposition shows that operations *Since*, *Until*, and *Next* are definable in the logic $\mathsf{RL}(1, 1')$. It is due to the fact that their relational definitions involve implicitly the information that the relations to which they apply are meant to be right ideal relations. For the reasons of readability, we will identify symbols of the language with the corresponding entities in the models, if it does not lead to a confusion. In particular, we will omit the symbol of a meaning function.

Proposition 16.4.1. *For every $\mathsf{RL}_{\mathsf{TL}_{\mathsf{SU}}}$-model $\mathcal{M} = (U, R, R^{-1}, m)$ and for all relations T and Q on U the following hold:*

1. $T\ Since\ Q = (R \cap -(R ; (-T \cap 1') ; R))^{-1} ; Q$;
2. $T\ Until\ Q = (R \cap -(R^{-1} ; (-T \cap 1') ; R^{-1})^{-1}) ; Q$;
3. $Next\ T = (R \cap -(R ; R)) ; T$.

Proof. Let $\mathcal{M} = (U, R, R^{-1}, m)$ be an $\mathsf{RL}_{\mathsf{TL}_{\mathsf{SU}}}$-model, let T and Q be relations on U, and let $x, y \in U$. First, note that the following holds: $(*)$ $(t, x) \in R ; (-T \cap 1') ; R$ iff there exists $u \in U$ such that $(t, u) \in R$ and $(u, x) \in R$ and $(u, u) \notin T$. Indeed, $(t, x) \in R ; (-T \cap 1') ; R$ iff there exist $u, v \in U$ such that $(t, v) \in R$, $(v, u) \notin T$,

$(v,u) \in 1'$, and $(u,x) \in R$. Therefore, if the left side of $(*)$ holds, then it can be easily proved, by the extensionality property of $\mathsf{RL}(1,1')$-models (see Sect. 2.7), that the right side of $(*)$ also holds. Conversely, if the right side of $(*)$ holds, then taking $v := u$ the left side also holds.

To prove 1., note that $(x,y) \in (R \cap -(R;(-T \cap 1');R))^{-1};Q)$ iff there exists $t \in U$ such that $(t,x) \in R$ and $(t,y) \in Q$ and $(t,x) \notin R;(-T \cap 1');R$. By $(*)$, the latter is equivalent to: there exists $t \in U$ such that $(t,x) \in R$ and $(t,y) \in Q$ and for all $u \in U$ if $(t,u) \in R$ and $(u,x) \in R$, then $(u,u) \in T$, which is equivalent to $(x,y) \in T$ *Since* Q.

2. and 3. can be proved in a similar way. □

The translation of $\mathsf{TL_{SU}}$-formulas into relational terms is defined by an extension of the function τ defined in Sect. 7.4 with the following clauses:

- $\tau(\varphi\ Since\ \psi) = \tau(\varphi)\ Since\ \tau(\psi)$;
- $\tau(\varphi\ Until\ \psi) = \tau(\varphi)\ Until\ \tau(\psi)$;
- $\tau(Next\ \varphi) = Next\ \tau(\varphi)$.

In view of Proposition 7.4.1, it is easy to check that the relational terms, obtained from temporal formulas built with operations *Since*, *Until*, and *Next*, represent right ideal relations. Therefore, the following can be proved in a similar way as Theorem 7.4.1:

Theorem 16.4.1. *For every* $\mathsf{TL_{SU}}$*-formula* φ *and for all object variables* x *and* y*, the following conditions are equivalent:*

1. φ *is* $\mathsf{TL_{SU}}$*-valid;*
2. $x\tau(\varphi)y$ *is* $\mathsf{RL_{TL_{SU}}}$*-valid.*

Proof. It suffices to show that Propositions 7.4.2 and 7.4.3 are true for all $\mathsf{TL_{SU}}$-formulas. For that purpose, we need to show:

(1) For every $\mathsf{TL_{SU}}$-model $\mathcal{M} = (U,R,R^{-1},m)$ there exists an $\mathsf{RL_{TL_{SU}}}$-model $\mathcal{M}' = (U,R,R^{-1},m')$ with the same universe and the same relation R as those in $\mathcal{M}$, and such that for all $s,s' \in U$ and for every $\mathsf{TL_{SU}}$-formula φ of the form $\psi\ Since\ \vartheta$, $\psi\ Until\ \vartheta$, and $Next\ \psi$, the following holds:

$$(*)\ \mathcal{M},s \models \varphi \text{ iff } (s,s') \in m'(\tau(\varphi));$$

(2) For every standard $\mathsf{RL_{TL_{SU}}}$-model $\mathcal{M}' = (U,R,R^{-1},m')$ there exists a $\mathsf{TL_{SU}}$-model $\mathcal{M} = (U,R,R^{-1},m)$ with the same universe and the same relation R as those in $\mathcal{M}'$, and such that for all $s,s' \in U$ and for every $\mathsf{TL_{SU}}$-formula φ of the form $\psi\ Since\ \vartheta$, $\psi\ Until\ \vartheta$, and $Next\ \psi$, $(*)$ holds.

Then, the rest of the proof is similar to the proof of Theorem 7.4.1.

By way of example, we prove (1). Let $\mathcal{M} = (U,R,R^{-1},m)$ be a $\mathsf{TL_{SU}}$-model. Then we define $\mathsf{RL_{TL_{SU}}}$-model $\mathcal{M}' = (U,R,R^{-1},m')$ as in the proof of Proposition 7.4.2, namely:

- $m'(1) = U \times U$;
- $m'(1')$ is the identity on U;
- $m'(\tau(p)) = \{(x,y) \in U \times U : x \in m(p)\}$, for every propositional variable p;

- $m'(R) = R$;
- m' extends to all the compound terms as in $\mathsf{RL}_{\mathsf{TL_{SU}}}$-models.

Let $\varphi = \psi\ \mathit{Since}\ \vartheta$ and let $s, s' \in U$. Assume that $\mathcal{M}, s \models \varphi$, that is there exists $t \in U$ such that $(t, s) \in R$, $\mathcal{M}, t \models \vartheta$, and for all $u \in U$, if $(t, u) \in R$ and $(u, s) \in R$, then $\mathcal{M}, u \models \psi$. Then, by the induction hypothesis, this is equivalent to: there exists $t \in U$ such that $(t, s) \in R$ and $(t, s') \in m'(\tau(\vartheta))$ and for all $u \in U$, if $(t, u) \in R$ and $(u, s) \in R$, then $(u, s') \in m'(\tau(\psi))$. Since $\tau(\psi)$ is a right ideal relation, by Proposition 7.4.1, $(u, s') \in m'(\tau(\psi))$ iff $(u, u) \in m'(\tau(\psi))$. Hence, $(s, s') \in m'(\tau(\psi) \mathit{Since}\, \tau(\vartheta))$.

In a similar way we can prove that $(*)$ holds for the formulas of the form $\psi\ \mathit{Until}\ \vartheta$ and $\mathit{Next}\ \psi$. □

$\mathsf{RL}_{\mathsf{TL_{SU}}}$-dual tableau is an extension of $\mathsf{RL}(1, 1')$-dual tableau with the rules and the axiomatic sets reflecting properties of the relation R and with the rules corresponding to the new relational operations *Since*, *Until*, and *Next*. These rules have the following forms:

For all object variables x, y and for all relational terms T and Q,

$$(\mathit{Since})\quad \frac{x(T\ \mathit{Since}\ Q)y}{tRx, K \mid tQy, K \mid t{-}(R\,;(-T \cap 1')\,;R)x, K}$$

$K = x(T\ \mathit{Since}\ Q)y$, t is any object variable

$$(-\mathit{Since})\quad \frac{x{-}(T\ \mathit{Since}\ Q)y}{t{-}Rx, t{-}Qy, t(R\,;(-T \cap 1')\,;R)x}$$

t is a new object variable

$$(\mathit{Until})\quad \frac{x(T\ \mathit{Until}\ Q)y}{xRt, K \mid tQy, K \mid t{-}(R^{-1}\,;(-T \cap 1')\,;R^{-1})x, K}$$

$K = x(T\ \mathit{Until}\ Q)y$, t is any object variable

$$(-\mathit{Until})\quad \frac{x{-}(T\ \mathit{Until}\ Q)y}{x{-}Rt, t{-}Qy, t(R^{-1}\,;(-T \cap 1')\,;R^{-1})x}$$

t is a new object variable

$$(\mathit{Next})\quad \frac{x(\mathit{Next}\ T)y}{xRt, x(\mathit{Next}\ T)y \mid tTy, x(\mathit{Next}\ T)y \mid x{-}(R\,;R)t, x(\mathit{Next}\ T)y}$$

t is any object variable

$$(-\mathit{Next})\quad \frac{x{-}(\mathit{Next}\ T)y}{x{-}Rt, t{-}Ty, x(R\,;R)t}$$

t is a new object variable

Theorem 16.4.2 (Correspondence). *Let $\mathcal{K}$ be a class of $\mathsf{RL}_{\mathsf{TL}_{\mathsf{SU}}}$-structures. Then $\mathcal{K}$ is a class of $\mathsf{RL}_{\mathsf{TL}_{\mathsf{SU}}}$-models iff the rules (#) and (−#) are $\mathcal{K}$-correct for every $\# \in \{Since, Until, Next\}$.*

Proof. ($\rightarrow$) Assume that $\mathcal{K}$ is a class of $\mathsf{RL}_{\mathsf{TL}_{\mathsf{SU}}}$-models. We need to show that for every $\# \in \{Since, Until, Next\}$, the rules (#) and (−#) are $\mathcal{K}$-correct. By way of example, we show it for the operation *Since*.

Let X be a finite set of $\mathsf{RL}_{\mathsf{TL}_{\mathsf{SU}}}$-formulas. Clearly, if the upper set of formulas in the rule (*Since*) is a $\mathcal{K}$-set, then all the lower sets of formulas in the rule are also $\mathcal{K}$-sets. Now, assume that $X \cup \{tRx, x(T\ Since\ Q)y\}$, $X \cup \{tQy, x(T\ Since\ Q)y\}$, and $X \cup \{t{-}(R\,;(-T \cap 1')\,;R)x, x(T\ Since\ Q)y\}$ are $\mathcal{K}$-sets and suppose that $X \cup \{x(T\ Since\ Q)y\}$ is not a $\mathcal{K}$-set. Then there exist an $\mathsf{RL}_{\mathsf{TL}_{\mathsf{SU}}}$-model $\mathcal{M} = (U, R, R^{-1}, m)$ in $\mathcal{K}$ and a valuation v in $\mathcal{M}$ such that $(v(x), v(y)) \notin m(T\ Since\ Q)$. By the assumption, there exists $t \in U$ such that $(v(t), v(x)) \in R$ and $(v(t), v(y)) \in m(Q)$ and $(v(t), v(x)) \notin m(R\,;(-T \cap 1')\,;R)$. By Proposition 16.4.1, $(v(x), v(y)) \in m(T\ Since\ Q)$, a contradiction.

Now, we prove that the rule (−*Since*) is $\mathcal{K}$-correct. Let X be a finite set of $\mathsf{RL}_{\mathsf{TL}_{\mathsf{SU}}}$-formulas. Let t be a variable that does not occur in X and let $x, y \neq t$. Assume that $X \cup \{t{-}Rx, t{-}Qy, t(R\,;(-T \cap 1')\,;R)x\}$ is a $\mathcal{K}$-set and suppose that $X \cup \{x{-}(T\ Since\ Q)y\}$ is not a $\mathcal{K}$-set. Then there exist an $\mathsf{RL}_{\mathsf{TL}_{\mathsf{SU}}}$-model $\mathcal{M} = (U, R, R^{-1}, m)$ in $\mathcal{K}$ and a valuation v in $\mathcal{M}$ such that $(v(x), v(y)) \in m(T\ Since\ Q)$. By the assumption and since t does not occur in $X \cup \{x{-}(T\ Since\ Q)y\}$, for all $t \in U$ either $(t, v(x)) \notin R$ or $(t, v(y)) \notin m(Q)$ or $(t, v(x)) \in m(R\,;(-T \cap 1')\,;R)$. By Proposition 16.4.1, $(v(x), v(y)) \notin m(T\ Since\ Q)$, a contradiction.

($\leftarrow$) Let $\mathcal{K}$ be a class of $\mathsf{RL}_{\mathsf{TL}_{\mathsf{SU}}}$-structures. Assume that for every $\# \in \{Since, Until, Next\}$ the rules (#) and (−#) are $\mathcal{K}$-correct. We show that $\mathcal{K}$ is a class of $\mathsf{RL}_{\mathsf{TL}_{\mathsf{SU}}}$-models, that is that the meaning of the operations *Since*, *Until*, and *Next* is as in $\mathsf{RL}_{\mathsf{TL}_{\mathsf{SU}}}$-models defined on p. 301. By way of example, we show it for *Until*. By Proposition 16.4.1, we need to show that $\mathcal{K}$-correctness of the rules (*Until*) and (−*Until*) implies $m(T\ Until\ Q) = m((R \cap -(R^{-1}\,;(-T \cap 1')\,;R^{-1}))^{-1}\,;Q)$.

($\supseteq$) Let $X \stackrel{\mathrm{df}}{=} \{x{-}Rt, t{-}Qy, t(R^{-1}\,;(-T \cap 1')\,;R^{-1})x\}$. Then, clearly all the sets $X \cup \{xRt, x(T\ Until\ Q)y\}$, $X \cup \{tQy, x(T\ Until\ Q)y\}$, and $X \cup \{t{-}(R^{-1}\,;(-T \cap 1')\,;R^{-1})x, x(T\ Until\ Q)y\}$ are $\mathcal{K}$-sets. By $\mathcal{K}$-correctness of the rule (*Until*), $X \cup \{x(T\ Until\ Q)y\}$ is a $\mathcal{K}$-set, which means that if $(x, y) \in m((R \cap -(R^{-1}\,;(-T \cap 1')\,;R^{-1}))^{-1}\,;Q)$, then $(x, y) \in m(T\ Until\ Q)$.

($\subseteq$) Let $X \stackrel{\mathrm{df}}{=} \{x(R \cap -(R^{-1}\,;(-T \cap 1')\,;R^{-1}))^{-1}\,;Qy\}$ and let $x, y \neq t$. Then, $X \cup \{x{-}Rt, t{-}Qy, t(R^{-1}\,;(-T \cap 1')\,;R^{-1})x\}$ is a $\mathcal{K}$-set. Thus, by $\mathcal{K}$-correctness of the rule (−*Until*), $X \cup \{x{-}(T\ Until\ Q)y\}$ is a $\mathcal{K}$-set, which means that if $(x, y) \in m(T\ Until\ Q)$, then $(x, y) \in m((R \cap -(R^{-1}\,;(-T \cap 1')\,;R^{-1}))^{-1}\,;Q)$. □

By the above theorem and since correctness of the remaining rules can be proved as in $\mathsf{RL}_{\mathsf{L}_{\mathsf{TL}}}$-logics (see Proposition 16.2.1 and Theorem 16.3.2), we have:

Proposition 16.4.2.

1. *The $\mathsf{RL}_{\mathsf{TL}_{\mathsf{SU}}}$-rules are $\mathsf{RL}_{\mathsf{TL}_{\mathsf{SU}}}$-correct;*
2. *The $\mathsf{RL}_{\mathsf{TL}_{\mathsf{SU}}}$-axiomatic sets are $\mathsf{RL}_{\mathsf{TL}_{\mathsf{SU}}}$-sets.*

The completion conditions determined by the rules for operations *Since*, *Until*, and *Next* are:

For all object variables x, y and for all relational terms T and Q,

Cpl(*Since*) If $x(T\ Since\ Q)y \in b$, then for every object variable t, either $tRx \in b$ or $tQy \in b$ or $t-(R\,;(-T \cap 1')\,;R)x \in b$, obtained by an application of the rule (*Since*);

Cpl($-$*Since*) If $x-(T\ Since\ Q)y \in b$, then for some object variable t, $t-Rx \in b$ and $t-Qy \in b$ and $t(R\,;(-T \cap 1')\,;R)x \in b$, obtained by an application of the rule ($-$*Since*);

Cpl(*Until*) If $x(T\ Until\ Q)y \in b$, then for every object variable t, either $xRt \in b$ or $tQy \in b$ or $t-(R^{-1}\,;(-T \cap 1')\,;R^{-1})x \in b$, obtained by an application of the rule (*Until*);

Cpl($-$*Until*) If $x-(T\ Until\ Q)y \in b$, then for some object variable t, $x-Rt \in b$ and $t-Qy \in b$ and $t(R^{-1}\,;(-T \cap 1')\,;R^{-1})x \in b$, obtained by an application of the rule ($-$*Until*);

Cpl(*Next*) If $x(Next\ T)y \in b$, then for every object variable t, either $xRt \in b$ or $tTy \in b$ or $x-(R\,;R)t \in b$, obtained by an application of the rule (*Next*);

Cpl($-$*Next*) If $x-(Next\ T)y \in b$, then for some variable t, $x-Rt \in b$ and $t-Ty \in b$ and $x(R\,;R)t \in b$, obtained by an application of the rule ($-$*Next*).

As in $\mathsf{RL}_{\mathsf{LTL}}$-logics, every branch that contains formulas xTy and $x-Ty$, for some atomic relational term T, is closed. Thus, the closed branch property holds.

Let b be an open branch of an $\mathsf{RL}_{\mathsf{TL_{SU}}}$-proof tree. The branch structure $\mathcal{M}^b = (U^b, R^b, (R^{-1})^b, m^b)$ is defined in a standard way, namely:

- $U^b = \mathbb{OV}_{\mathsf{RL_{TL_{SU}}}}$;
- $R^b = m^b(R)$ and $(R^{-1})^b = (R^b)^{-1}$;
- $m^b(P) = \{(x, y) \in U^b \times U^b : xPy \notin b\}$, for every atomic relational term P;
- m^b extends to all the compound relational terms as in $\mathsf{RL}_{\mathsf{TL_{SU}}}$-models.

It follows from this definition that $\mathcal{M}^b$ is an $\mathsf{RL}_{\mathsf{TL_{SU}}}$-model, so the branch model property is satisfied. Actually, we only need to prove that the time ordering R^b satisfies all the conditions assumed in $\mathsf{RL}_{\mathsf{TL_{SU}}}$-models. This can be done as in the completeness proof of $\mathsf{RL}_{\mathsf{TL_L}}$-dual tableaux presented in the previous section.

Proposition 16.4.3 (Satisfaction in Branch Model Property). *Let b be an open branch of an $\mathsf{RL}_{\mathsf{TL_{SU}}}$-proof tree. Then for every $\mathsf{RL}_{\mathsf{TL_{SU}}}$-formula φ, $\mathcal{M}^b, v^b \models \varphi$ implies $\varphi \notin b$.*

Proof. Let b be an open branch of an $\mathsf{RL}_{\mathsf{TL_{SU}}}$-proof tree. The proof is by induction on the complexity of formulas. If φ is of the form xTy or $x-Ty$ for some atomic relational term T, we prove the above condition as in RL-logic (see the proof of Proposition 2.5.5). Then, we show that the condition holds for the compound relational terms. By way of example, we show it for *Next*.

Assume $\mathcal{M}^b, v^b \models x(Next\ T)y$, Then, by Proposition 16.4.1, $(x, y) \in m^b((R \cap -(R\,;R))\,;T)$. Suppose $x(Next\ T)y \in b$. Then, by the completion condition

Cpl(*Next*), for every $t \in U^b$, either $xRt \in b$ or $tTy \in b$ or $x-(R\,;R)t \in b$. By the completion condition Cpl(−;), for every $t \in U^b$, either $xRt \in b$ or $tTy \in b$ or for some $u \in U^b$ both $x-Ru \in b$ and $u-Rt \in b$. By the induction hypothesis, for every $t \in U^b$, either $(x,t) \notin R^b$ or $(t,y) \notin m^b(T)$ or for some $u \in U^b$ both $(x,u) \in R^b$ and $(u,t) \in R^b$. Therefore, $(x,y) \notin m^b((R \cap -(R\,;R))\,;T)$, a contradiction.

Assume $\mathcal{M}^b, v^b \models x-(Next\,T)y$, Then, $(x,y) \notin m^b((R \cap -(R\,;R))\,;T)$. Suppose $x-(Next\,T)y \in b$. Then, by the completion condition Cpl(−*Next*), for some $t \in U^b$ the following hold: $xRt \notin b$, $tTy \notin b$, and $x(R\,;R)t \in b$. By the completion condition Cpl(;), there exists $t \in U^b$ such that $xRt \notin b$ and $tTy \notin b$ and for all $u \in U^b$, either $xRu \in b$ or $uRt \in b$. Thus, by the induction hypothesis, there exists $t \in U^b$ such that $(x,t) \in R^b$ and $(t,y) \in m^b(T)$ and for all $u \in U^b$, either $(x,u) \notin R^b$ or $(u,t) \notin R^b$. Therefore, $(x,y) \in m^b((R \cap -(R\,;R))\,;T)$, a contradiction. □

Therefore, we have:

Theorem 16.4.3 (Soundness and Completeness of $\mathsf{RL}_{\mathsf{TL}_{\mathsf{SU}}}$). *For every $\mathsf{RL}_{\mathsf{TL}_{\mathsf{SU}}}$-formula φ, the following conditions are equivalent:*

1. *φ is $\mathsf{RL}_{\mathsf{TL}_{\mathsf{SU}}}$-valid;*
2. *φ is true in all standard $\mathsf{RL}_{\mathsf{TL}_{\mathsf{SU}}}$-models;*
3. *φ is $\mathsf{RL}_{\mathsf{TL}_{\mathsf{SU}}}$-provable.*

The theorem above and Theorem 16.4.1 imply:

Theorem 16.4.4 (Relational Soundness and Completeness of $\mathsf{TL}_{\mathsf{SU}}$). *For every $\mathsf{TL}_{\mathsf{SU}}$-formula φ and for all object variables x and y, the following conditions are equivalent:*

1. *φ is $\mathsf{TL}_{\mathsf{SU}}$-valid;*
2. *$x\tau(\varphi)y$ is $\mathsf{RL}_{\mathsf{TL}_{\mathsf{SU}}}$-provable.*

Example. Let φ be the following formula:

$$\varphi = (p \wedge \langle R\rangle(p \wedge [R^{-1}\rangle\neg q)] \rightarrow \neg q\;Until\;p.$$

Its relational translation is:

$$\tau(\varphi) = -((P\,;1) \cap (R\,;((P\,;1) \cap -(R^{-1}\,;--(Q\,;1))))) \cup (-(Q\,;1)Until(P\,;1)),$$

where $\tau(p) = P\,;1$ and $\tau(q) = Q\,;1$. $\mathsf{TL}_{\mathsf{SU}}$-validity of φ is equivalent to $\mathsf{RL}_{\mathsf{TL}_{\mathsf{SU}}}$-provability of the formula $x\tau(\varphi)y$. Figure 16.3 presents an $\mathsf{RL}_{\mathsf{TL}_{\mathsf{SU}}}$-proof of φ.

16.5 Standard Temporal Logics with Nominals

Temporal logics with nominals were considered by Arthur Prior [Pri67] and Robert Bull [Bul70] in the late 1960s. Nominals are propositional constants interpreted as singleton sets. In computer science nominals were introduced to the dynamic logic

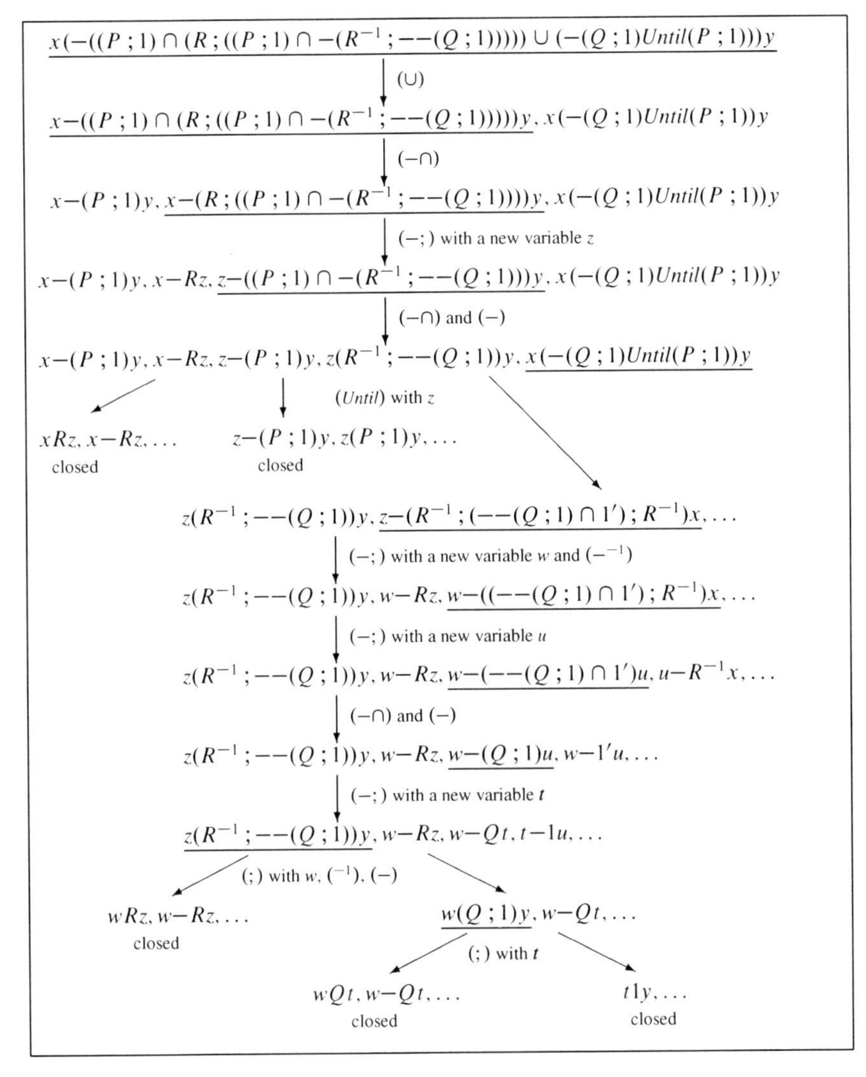

Fig. 16.3 An $\mathsf{RL}_{\mathsf{TL}_{\mathsf{SU}}}$-proof of $[p \wedge \langle R\rangle(p \wedge [R^{-1}]\neg q)] \rightarrow \neg q\ Until\ p$

in [Pas84] and then studied in [PT85]. In temporal languages both nominals and propositional variables are considered as atomic formulas.

Given a model $\mathcal{M} = (U, R, R^{-1}, m)$, we define the satisfaction of nominals:

- $\mathcal{M}, s \models c$ iff $m(c) = \{s\}$.

An extensive study of nominals can be found in [Bla90].

Nominals increase drastically expressiveness of modal languages. Below we list examples of classes of relations which are not definable in a modal language with a single accessibility relation, unless the language contains the nominals (see e.g., [Bla90]).

- R is irreflexive: $c \rightarrow \neg\langle R\rangle c$;
- R is antisymmetric: $c \rightarrow [R](\langle R\rangle c \rightarrow c)$;
- R is directed: $\langle R\rangle\langle R^{-1}\rangle c$;
- R is connected: $\langle R^{-1}\rangle c \vee c \vee \langle R\rangle c$;
- R is discrete: $c \rightarrow (\langle R\rangle(\varphi \vee \neg\varphi) \rightarrow \langle R\rangle[R^{-1}][R^{-1}]\neg c)$.

Let a logic $\mathsf{TL}(\mathbb{C})$ be obtained from the temporal logic TL by extending its language with a set $\mathbb{C}$ of nominals. Its models are TL-models $\mathcal{M} = (U, F, P, m)$ such that $m(c) \in U$, for every $c \in \mathbb{C}$ (see Sect. 7.3). In analogy to logics considered in Sect. 16.3, we consider standard temporal logics with nominals, $\mathsf{L}_{\mathsf{TL}(\mathbb{C})}$, based on $\mathsf{TL}(\mathbb{C})$. The corresponding relational logics are based on the relational logic with point relations introduced with axioms. The logic is defined in Sect. 3.2.

Let $\mathsf{L}_{\mathsf{TL}(\mathbb{C})}$ be a standard temporal logic with the set $\mathbb{C}$ of nominals. With every nominal $c \in \mathbb{C}$, we associate a relational constant C_c. Then, a relational logic appropriate for expressing $\mathsf{L}_{\mathsf{TL}(\mathbb{C})}$-formulas, $\mathsf{RL}_{\mathsf{L}_{\mathsf{TL}(\mathbb{C})}}$, is based on logic $\mathsf{RL}_{ax}(\{C_c : c \in \mathbb{C}\})$ which is an instance of the logics considered in Sect. 3.2. The set of relational constants of $\mathsf{RL}_{\mathsf{L}_{\mathsf{TL}(\mathbb{C})}}$ is $\{1, 1', R\} \cup \{C_c : c \in \mathbb{C}\}$, where R represents the time ordering and $\{C_c : c \in \mathbb{C}\}$ is the set of relational constants representing nominals from $\mathbb{C}$.

$\mathsf{RL}_{\mathsf{L}_{\mathsf{TL}(\mathbb{C})}}$-models are structures $\mathcal{M} = (U, R, R^{-1}, m)$ such that (U, m) is an $\mathsf{RL}_{ax}(\{C_c : c \in \mathbb{C}\})$-model, as defined in Sect. 3.2, and $R = m(R)$ is the relation on U that satisfies all the conditions assumed in $\mathsf{L}_{\mathsf{TL}(\mathbb{C})}$-models.

The translation of $\mathsf{L}_{\mathsf{TL}(\mathbb{C})}$-formulas into relational terms is defined as in Sect. 7.4 (p. 147), that is $\tau(c) = C_c ; 1$. As usual, the translation is defined so that it preserves the validity of formulas. Due to Theorem 7.4.1, we have:

Theorem 16.5.1. *Let $\mathsf{L}_{\mathsf{TL}(\mathbb{C})}$ be a standard temporal logic with nominals. For every $\mathsf{L}_{\mathsf{TL}(\mathbb{C})}$-formula φ and for all object variables x and y, the following conditions are equivalent:*

1. *φ is $\mathsf{L}_{\mathsf{TL}(\mathbb{C})}$-valid;*
2. *$x\tau(\varphi)y$ is $\mathsf{RL}_{\mathsf{L}_{\mathsf{TL}(\mathbb{C})}}$-valid.*

$\mathsf{RL}_{\mathsf{L}_{\mathsf{TL}(\mathbb{C})}}$-dual tableau is an extension of $\mathsf{RL}_{ax}(\{C_c : c \in \mathbb{C}\})$-dual tableau with the rules reflecting the properties of the time ordering R. We recall that the specific rules of $\mathsf{RL}_{ax}(\{C_c : c \in \mathbb{C}\})$-dual tableau are (see Sect. 3.2):

For all object symbols x and y and for every $c \in \mathbb{C}$,

$$(C1)\quad \frac{}{z{-}C_c t} \qquad z, t \text{ are new object variables and } z \neq t$$

$$(C2)\quad \frac{xC_c y}{xC_c z,\ xC_c y} \qquad z \text{ is any object symbol}$$

$$(C3)\quad \frac{x1'y}{xC_c z,\ x1'y \mid yC_c z,\ x1'y} \qquad z \text{ is any object symbol}$$

Soundness and completeness of $\mathsf{RL}_{\mathsf{L}_{\mathsf{TL}(\mathbb{C})}}$-dual tableau follow from soundness and completeness of $\mathsf{RL}_{ax}(\{C_c : c \in \mathbb{C}\})$-dual tableau (see Theorem 3.2.1) and $\mathsf{RL}_{\mathsf{L}_{\mathsf{TL}}}$-dual tableaux for standard temporal logics (see Theorem 16.3.3). Hence, we have:

Theorem 16.5.2 (Soundness and Completeness of $\mathsf{RL}_{\mathsf{L}_{\mathsf{TL}(\mathbb{C})}}$). *Let $\mathsf{L}_{\mathsf{TL}(\mathbb{C})}$ be a standard temporal logic with nominals. Then, for every $\mathsf{RL}_{\mathsf{L}_{\mathsf{TL}(\mathbb{C})}}$-formula φ, the following conditions are equivalent:*

1. *φ is $\mathsf{RL}_{\mathsf{L}_{\mathsf{TL}(\mathbb{C})}}$-valid;*
2. *φ is true in all standard $\mathsf{RL}_{\mathsf{L}_{\mathsf{TL}(\mathbb{C})}}$-models;*
3. *φ is $\mathsf{RL}_{\mathsf{L}_{\mathsf{TL}(\mathbb{C})}}$-provable.*

By the above and Theorem 16.5.1, we obtain:

Theorem 16.5.3 (Relational Soundness and Completeness of $\mathsf{L}_{\mathsf{TL}(\mathbb{C})}$). *Let $\mathsf{L}_{\mathsf{TL}(\mathbb{C})}$ be a temporal logic with nominals. Then, for every $\mathsf{L}_{\mathsf{TL}(\mathbb{C})}$-formula φ and for all object variables x and y, the following conditions are equivalent:*

1. *φ is $\mathsf{L}_{\mathsf{TL}(\mathbb{C})}$-valid;*
2. *$x(\tau(\varphi))y$ is $\mathsf{RL}_{\mathsf{L}_{\mathsf{TL}(\mathbb{C})}}$-provable.*

Example. The formula:

$$\varphi = c \rightarrow \neg\langle R\rangle c$$

defines irreflexivity of relation R, hence it is true in all $\mathsf{L}_{\mathsf{TL}(\mathbb{C})}$-structures in which R is irreflexive. The translation of φ is:

$$\tau(\varphi) = -(C_c ; 1) \cup -(R ; (C_c ; 1)),$$

where $\tau(c) = C_c ; 1$. Then, φ is valid in all irreflexive $\mathsf{L}_{\mathsf{TL}(\mathbb{C})}$-structures iff $x\tau(\varphi)y$ is provable in $\mathsf{RL}_{\mathsf{L}_{\mathsf{TL}(\mathbb{C})}}$-dual tableau with the rule (irref R) presented in Sect. 16.3. Figure 16.4 presents an $\mathsf{RL}_{\mathsf{L}_{\mathsf{TL}(\mathbb{C})}}$-proof of $x\tau(\varphi)y$ which shows that φ is true in all irreflexive $\mathsf{L}_{\mathsf{TL}(\mathbb{C})}$-structures.

The formula:

$$\psi = \langle R^{-1}\rangle c \vee c \vee \langle R\rangle c$$

defines connectivity of relation R, thus ψ is true in all $\mathsf{L}_{\mathsf{TL}(\mathbb{C})}$-structures in which R is connected. The translation of ψ is:

$$\tau(\psi) = (R^{-1} ; (C_c ; 1)) \cup (C_c ; 1) \cup (R ; (C_c ; 1)),$$

where $\tau(c) = C_c ; 1$. Then, validity of ψ in all connected $\mathsf{L}_{\mathsf{TL}(\mathbb{C})}$-structures is equivalent to $\mathsf{RL}_{\mathsf{L}_{\mathsf{TL}(\mathbb{C})}}$-provability of $x\tau(\psi)y$. $\mathsf{RL}_{\mathsf{L}_{\mathsf{TL}(\mathbb{C})}}$-dual tableau includes axiomatic sets (Ax), presented in Sect. 16.3 (p. 295), that reflect connectivity of R. Figure 16.5 presents an $\mathsf{RL}_{\mathsf{L}_{\mathsf{TL}(\mathbb{C})}}$-proof of $x\tau(\psi)y$.

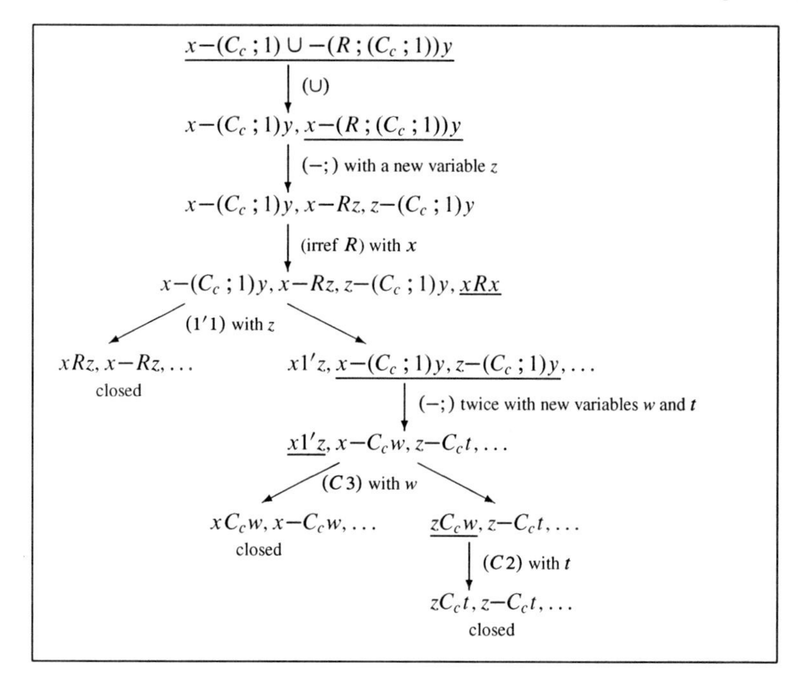

Fig. 16.4 An $\mathsf{RL}_{\mathsf{L}_{\mathsf{TL}(\mathbb{C})}}$-proof showing that $c \rightarrow \neg\langle R\rangle c$ is true in all irreflexive $\mathsf{L}_{\mathsf{TL}(\mathbb{C})}$-structures

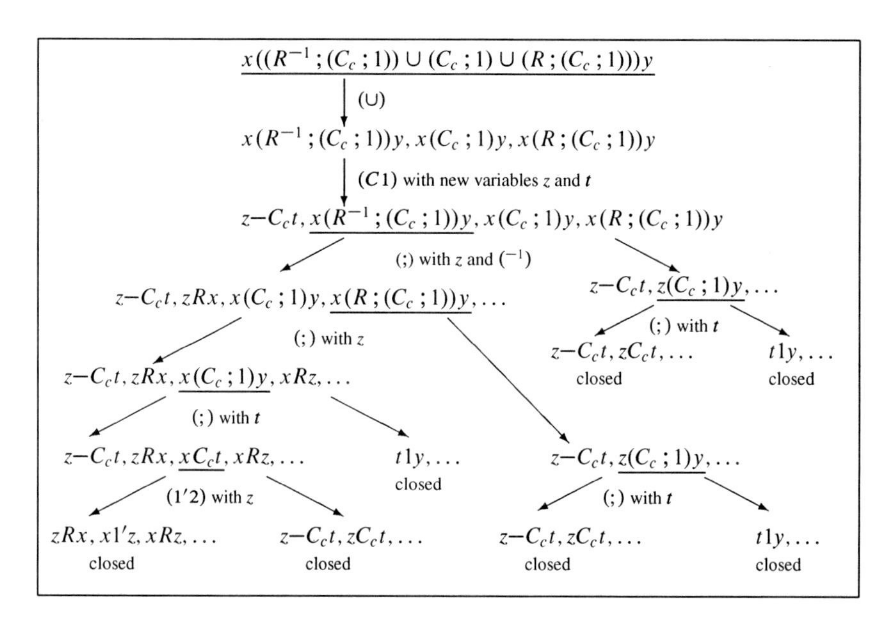

Fig. 16.5 An $\mathsf{RL}_{\mathsf{L}_{\mathsf{TL}(\mathbb{C})}}$-proof showing that $\langle R^{-1}\rangle c \vee c \vee \langle R\rangle c$ is true in all connected $\mathsf{L}_{\mathsf{TL}(\mathbb{C})}$-structures

16.6 Temporal Information Logics

Temporal information logic TIL considered in the present section was developed to provide a means of reasoning in temporal databases where properties of objects change with the lapse of time. In logic TIL we intend to represent data that have the form of a description of time varying properties of objects. For example, we are interested in such attributes as Height, Temperature, Blood Pressure, usually at given moments of time. Further, their change in a given time interval may be of essential importance too. For that purpose, we include in information systems a parameter which represents the moment to which an information about values of attributes applies.

By a *dynamic information system* (see [Orł82]) we mean a system of the form $\mathcal{S} = (\mathbb{OB}, \mathbb{T}, R, \mathbb{AT}, \{\mathbb{VAL}_a : a \in \mathbb{AT}\}, f)$, where $\mathbb{OB}$ is a non-empty set of objects, $\mathbb{T}$ is a non-empty set of moments of time, R is an ordering on a set $\mathbb{T}$, $\mathbb{AT}$ is a non-empty set of attributes, $\mathbb{VAL}_a$, for $a \in \mathbb{AT}$, is a non-empty set of values of attribute a, f is a function $f\colon \mathbb{OB} \times \mathbb{T} \times \mathbb{AT} \to \bigcup\{\mathbb{VAL}_a : a \in \mathbb{AT}\}$, such that $f(x,t,a) \in \mathbb{VAL}_a$, for all $x \in \mathbb{OB}$, $t \in \mathbb{T}$, and $a \in \mathbb{AT}$.

As an example, consider Table 16.1 containing partial results of photoelectric observations of stars, presented in the Astrophysical Journal.

The table can be treated as a dynamic information system such that the set $\mathbb{OB}$ of objects consists of the stars, that is $\mathbb{OB} = \{$*S Canis Minoris, R Caneri, R Leonis, T Centauri*$\}$, the set $\mathbb{T}$ of moments of time consists of non-negative real numbers representing Julian Days given in the second column of the table, relation R is the natural order in the set of real numbers restricted to the set $\mathbb{T}$, the set $\mathbb{AT}$ of attributes consists of two wavelength regions of spectrum, $\mathbb{AT} = \{visual(V), blue{-}visual(B{-}V)\}$, the set $\mathbb{VAL}$ of values of attributes consists of the magnitudes of a star in the given wavelength regions.

The language of logic TIL is a language of basic temporal logic with specific atomic formulas. An atomic piece of information in an information system is a statement of the form: an object x assumes a value v of an attribute a. Hence, instead of propositional variables, we admit structured atomic formulas built with syntactic components of three types. Let $\mathbb{OV}$, $\mathbb{AV}$, and $\mathbb{AVV}$ be sets of object variables, attribute variables, and attribute value variables, respectively. They are arbitrary, pairwise disjoint, countable sets. Then, the atomic formulas of the language are

Table 16.1 A dynamic information system

	JD	V	B-V
S Cmi	1688.788	11.12	1.97
	1798.538	9.28	1.76
R Cnc	1719.750	3.38	1.47
	1800.558	9.51	2.02
R Leo	1688.821	6.27	1.62
	1833.481	9.91	2.87
T Cen	1687.826	6.05	1.44
	1717.816	6.12	1.73

of the form (o, a, v), for $o \in \mathbb{OV}$, $a \in \mathbb{AV}$, and $v \in \mathbb{AVV}$. The compound formulas are built from atomic formulas with the usual propositional operations of temporal logics.

We define semantics of the logic TIL by means of notion of a model determined by a dynamic information system. By a TIL-model we mean any pair $\mathcal{M} = (\mathcal{S}, m)$, where $\mathcal{S} = (\mathbb{OB}, \mathbb{T}, R, \mathbb{AT}, \{\mathbb{VAL}_a : a \in \mathbb{AT}\}, f)$ is a dynamic information system, and m is a meaning function which assigns objects to object variables, attributes to attribute variables, and values of attributes to attribute value variables: $m(o) \in \mathbb{OB}$, $m(a) \in \mathbb{AT}$, $m(v) \in \bigcup\{\mathbb{VAL}_a : a \in \mathbb{AT}\}$. We define satisfaction of formulas in a moment of time in the usual way (see Sect. 7.3), with the exception that for atomic formulas we have:

- $\mathcal{M}, t \models (o, a, v)$ iff $f(m(o), t, m(a)) = m(v)$.

The relational logic $\mathsf{RL}_{\mathsf{TIL}}$ corresponding to the logic TIL is similar to the relational logic for the basic temporal logic. The minor difference is that the relational variables are indexed with triples of the form (o, a, v). Models of the relational logic for TIL are determined by dynamic information systems in the same way as the respective TIL-models. As usual, the translation starts with a one-to-one assignment of relational variables $P_{(o,a,v)}$ to the atomic TIL-formulas (o, a, v). Let τ' be such an assignment. Then the translation τ of TIL-formulas into $\mathsf{RL}_{\mathsf{TIL}}$-terms is defined as in Sect. 7.4 with the following clause for atomic formulas: $\tau(o, a, v) \stackrel{\text{df}}{=} P_{(o,a,v)} ; 1$.

In view of Proposition 7.4.1, it is easy to check that the relational terms obtained from TIL-formulas represent right ideal relations. Therefore, the following can be proved in a similar way as Theorem 7.4.1:

Theorem 16.6.1. *For every* TIL*-formula φ and for all object variables x and y, the following conditions are equivalent:*

1. *φ is* TIL*-valid;*
2. *$x\tau(\varphi)y$ is* $\mathsf{RL}_{\mathsf{TIL}}$*-valid.*

Proof. It suffices to show that Propositions 7.4.2 and 7.4.3 are true for all TIL-formulas. Thus, we need to show that for every TIL-model $\mathcal{M}$ there exists a standard $\mathsf{RL}_{\mathsf{TIL}}$-model $\mathcal{M}'$ that satisfies the same TIL-formulas as model $\mathcal{M}$, and that for every $\mathsf{RL}_{\mathsf{TIL}}$-model $\mathcal{M}'$ there exists a TIL-model $\mathcal{M}$ that satisfies the same TIL-formulas as $\mathcal{M}'$. If a TIL-model $\mathcal{M}$ is given, then the model $\mathcal{M}'$ is defined as a standard $\mathsf{RL}_{\mathsf{TIL}}$-model such that the interpretation of a relational variable $P_{(o,a,v)}$ is a right ideal relation whose domain is the set of the form $\{x : f(m(o), x, m(a)) = m(v)\}$. If an $\mathsf{RL}_{\mathsf{TIL}}$-model $\mathcal{M}'$ is given, then the model $\mathcal{M}$ is defined as a TIL-model such that the interpretation of an atomic formula (o, a, v) is the domain of the relation $P_{(o,a,v)}$. The rest of the proof is similar to the proof of Theorem 7.4.1. □

$\mathsf{RL}_{\mathsf{TIL}}$-dual tableau is an $\mathsf{RL}_{\mathsf{TL_L}}$-dual tableau adjusted to the $\mathsf{RL}_{\mathsf{TIL}}$-language and extended with the rule of the following form:

For all object variables x and y,

$$(\mathsf{TIL})\ \frac{xP_{(o,a,v)}y}{xP_{(o',a',v')}y, xP_{(o,a,v)}y \mid xP_{(o',a',v)}y, xP_{(o,a,v)}y \mid xP_{(o,a,v')}y, xP_{(o,a,v)}y}$$

for any $o' \in \mathbb{OV}$, $a' \in \mathbb{AV}$, and $v' \in \mathbb{AVV}$

The rule (TIL) reflects the following property of relations $P_{(o,a,v)}$: in an underlying information system, given an object, a moment of time, and an attribute, the function f assigns a unique value of the attribute to this triple.

The completion condition determined by this rule is:

For all object variables x and y,

Cpl(TIL) If $xP_{(o,a,v)}y \in b$, then for all $o' \in \mathbb{OV}$, $a' \in \mathbb{AV}$, and $v' \in \mathbb{AVV}$, either $xP_{(o',a',v')}y \in b$ or $xP_{(o',a',v)}y \in b$ or $xP_{(o,a,v')}y \in b$.

Proofs of all the propositions needed for proving soundness and completeness of $\mathsf{RL_{TIL}}$ follow the analogous proofs in the logics $\mathsf{RL_{TL}}$ and $\mathsf{RL_{L_{TL}}}$ presented in the previous sections of this chapter. Thus, we have:

Theorem 16.6.2 (Soundness and Completeness of $\mathsf{RL_{TIL}}$). *For every $\mathsf{RL_{TIL}}$-formula φ, the following conditions are equivalent:*

1. *φ is $\mathsf{RL_{TIL}}$-valid;*
2. *φ is true in all standard $\mathsf{RL_{TIL}}$-models;*
3. *φ is $\mathsf{RL_{TIL}}$-provable.*

By the above and Theorem 16.6.1, we get:

Theorem 16.6.3 (Relational Soundness and Completeness of TIL). *For every TIL-formula φ and for all object variables x and y, the following conditions are equivalent:*

1. *φ is TIL-valid;*
2. *$x(\tau(\varphi))y$ is $\mathsf{RL_{TIL}}$-provable.*

Chapter 17
Dual Tableaux for Interval Temporal Logics

17.1 Introduction

The representation of time by means of intervals rather than points was initiated in philosophical logic, see e.g., [Hum79, Röp80, Bur82, vB83]. In computer science the interval structure of time was adopted by James Allen [All83] for use in solving some artificial intelligence problems such as planning and by Ben Moszkowski [Mos83] for reasoning about periods of time found in a formal description of hardware and software systems. Since then interval temporal logics have been extensively studied both in logic and computer science, and they are successfully applied to the specification and verification of properties of real time systems.

A calculus of time intervals in a linearly ordered time structure was introduced by Allen in [All83]. In [LM87] (see also [LM94]), this calculus was presented and studied as a relation algebra, called an interval algebra. In the literature various propositional and first-order interval temporal logics have been proposed; a comprehensive survey can be found in [GMS04]. The most popular propositional logics are Halpern and Shoham's HS [HS91, Ven90], Venema's CDT logic [GMS03a, GMSS06, Ven91], Moszkowski's Propositional Interval Temporal Logic PITL (see [Mos83]), and Goranko, Montanari, and Sciavicco's Propositional Neighborhood Logic PNL [BM05, BMS07, GMS03b].

Propositional interval temporal logics are very expressive. It is known that both HS and CDT are strictly more expressive than every point-based temporal logic on linear orders: they enable us to express properties of pairs of time points rather than single time points. In a linearly ordered set 13 different binary relations between intervals are possible (see [All83]): *equals* $(1')$, *ends* (E), *during* (D), *begins* (B), *overlaps* (O), *meets* (M), *precedes* (P) together with their converses (see Table 17.1). Propositional interval temporal logics are usually characterized by modalities of the form $\langle R \rangle$ and $\langle R^{-1} \rangle$, where R is any of these relations.

In this chapter we present relational dual tableaux for several interval temporal logics: for the Halpern–Shoham logic, for some of its axiomatic and signature extensions, and also for its proper fragment without the converses of the interval relations. The extensions are chosen so that they reflect some characteristics of intervals or properties of the ordering of time points. The content of this chapter is based on [BGPO06].

E. Orłowska and J. Golińska-Pilarek, *Dual Tableaux: Foundations, Methodology, Case Studies*, Trends in Logic 36, DOI 10.1007/978-94-007-0005-5_17,

Table 17.1 Allen's interval relations

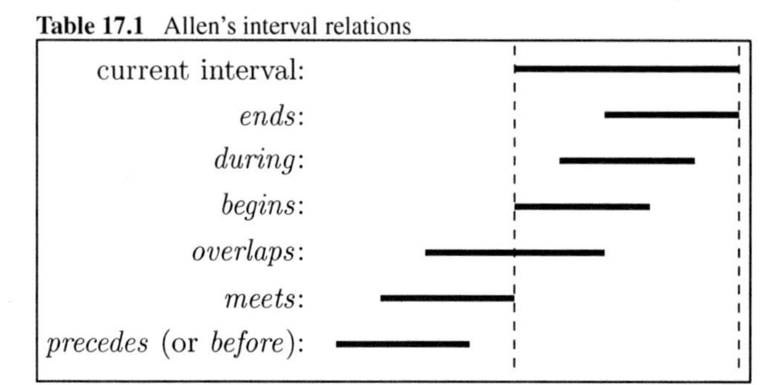

17.2 Halpern–Shoham Logic

Halpern–Shoham logic, HS, is a propositional interval logic characterized by four temporal modalities that correspond to Allen's relations begins and ends, and their converses [HS91, Ven90]. These four modalities suffice to define all unary modalities corresponding to Allen's relations. Hence, the logic HS is the most expressive interval temporal logic. It is undecidable as shown in [HS91]. HS-language is a modal language with four basic modal operations $\langle B\rangle$, $\langle E\rangle$, $\langle B^{-1}\rangle$, and $\langle E^{-1}\rangle$. The other propositional operations, such as $\wedge$, $\rightarrow$, and the propositional constants $\top$ (true) and $\bot$ (false), as well as the necessity modalities $[B]$, $[E]$, $[B^{-1}]$, and $[E^{-1}]$ are defined as in Sect. 7.3.

Given a strict linear ordering $(D, <)$, a *non-strict interval* on D is a pair $[c, d]$ such that $c \leq d$. We denote the set of all non-strict intervals on D by $\mathbb{I}(D)^+$. An HS-*model* is a structure $\mathcal{M}^+ = (D, \mathbb{I}(D)^+, <, m)$, where $(D, <)$ is a strict linear ordering and m is a meaning function assigning a set of intervals to every propositional variable $p \in \mathbb{V}$, with the intuition that p is true in these intervals. The semantics of HS is defined in terms of the satisfaction relation $\models$ as follows. Let $\mathcal{M}^+ = (D, \mathbb{I}(D)^+, <, m)$ be an HS-model and let $[c, d] \in \mathbb{I}(D)^+$:

- $\mathcal{M}^+, [c, d] \models p$ iff $[c, d] \in m(p)$, for every propositional variable $p \in \mathbb{V}$;
- $\mathcal{M}^+, [c, d] \models \neg\varphi$ iff $\mathcal{M}^+, [c, d] \not\models \varphi$;
- $\mathcal{M}^+, [c, d] \models \varphi \vee \psi$ iff $\mathcal{M}^+, [c, d] \models \varphi$ or $\mathcal{M}^+, [c, d] \models \psi$;
- $\mathcal{M}^+, [c, d] \models \langle B\rangle\varphi$ iff $\exists c' \in D$ such that $c' < d$ and $\mathcal{M}^+, [c, c'] \models \varphi$;
- $\mathcal{M}^+, [c, d] \models \langle E\rangle\varphi$ iff $\exists c' \in D$ such that $c < c'$ and $\mathcal{M}^+, [c', d] \models \varphi$;
- $\mathcal{M}^+, [c, d] \models \langle B^{-1}\rangle\varphi$ iff $\exists c' \in D$ such that $d < c'$ and $\mathcal{M}^+, [c, c'] \models \varphi$;
- $\mathcal{M}^+, [c, d] \models \langle E^{-1}\rangle\varphi$ iff $\exists c' \in D$ such that $c' < c$ and $\mathcal{M}^+, [c', d] \models \varphi$.

The notions of the truth and HS-validity are defined as usual.

Note that $\mathbb{I}(D)^+$ includes also intervals of the form $[c, c]$. They are called *point intervals*. Since point intervals have no intervals that begin and/or end them, they can be distinguished by the formulas $[B]\bot$ and $[E]\bot$. Namely, $[B]\bot$ is satisfied in a model $\mathcal{M}^+$ by $[c, d]$ iff d is the beginning of this interval. Similarly, $[E]\bot$ is satisfied in a model $\mathcal{M}^+$ by $[c, d]$ iff c is the end point of this interval. This allows us to define two derived operations, $[BP]$ and $[EP]$, that express properties holding on the begin point and on the end point of the current interval, respectively:

$$[BP]\varphi \stackrel{\text{df}}{=} ([B]\bot \wedge \varphi) \vee \langle B\rangle([B]\bot \wedge \varphi);$$
$$[EP]\varphi \stackrel{\text{df}}{=} ([E]\bot \wedge \varphi) \vee \langle E\rangle([E]\bot \wedge \varphi).$$

In the presence of point intervals, the modalities corresponding to the other Allen's relations are definable in logic HS:

$$\langle D\rangle\varphi \stackrel{\text{df}}{=} \langle B\rangle\langle E\rangle\varphi; \qquad \langle D^{-1}\rangle\varphi \stackrel{\text{df}}{=} \langle B^{-1}\rangle\langle E^{-1}\rangle\varphi;$$
$$\langle O\rangle\varphi \stackrel{\text{df}}{=} \langle B\rangle\langle E^{-1}\rangle\varphi; \qquad \langle O^{-1}\rangle\varphi \stackrel{\text{df}}{=} \langle B^{-1}\rangle\langle E\rangle\varphi;$$
$$\langle M\rangle\varphi \stackrel{\text{df}}{=} [BP]\langle E^{-1}\rangle\varphi; \qquad \langle M^{-1}\rangle\varphi \stackrel{\text{df}}{=} [EP]\langle B^{-1}\rangle\varphi;$$
$$\langle P\rangle\varphi \stackrel{\text{df}}{=} \langle M\rangle\langle M\rangle\varphi; \qquad \langle P^{-1}\rangle\varphi \stackrel{\text{df}}{=} \langle M^{-1}\rangle\langle M^{-1}\rangle\varphi.$$

17.3 Relational Logic for Halpern–Shoham Logic

In this section we define a relational logic $\mathsf{RL_{HS}}$ associated with logic HS. The vocabulary of the language of the logic $\mathsf{RL_{HS}}$ consists of the pairwise disjoint sets listed below:

- $\mathbb{IV} = \{i, j, k, \dots\}$ – a countable infinite set of interval variables;
- $\mathbb{PV} = \{i_1, i_2 : i \in \mathbb{IV}\}$ – a countable infinite set of point variables; since intervals are meant to be certain pairs of points, to every interval variable i we associate two point variables denoted i_1 and i_2, with the intuition that $i = [i_1, i_2]$;
- $\mathbb{IRV}$ – a countable infinite set of interval relational variables;
- $\mathbb{PRC} = \{1', <\}$ – the set of point relational constants;
- $\mathbb{IRC} = \{1, B, E\}$ – the set of interval relational constants;
- $\{-, \cup, \cap, ;, ^{-1}\}$ – the set of relational operations.

The constant $<$ is intended to represent an ordering on the set of time points. The set of *point relational terms*, $\mathbb{PRT}$, is the smallest set of expressions that includes $\mathbb{PRC}$ and is closed with respect to the relational operations. The set of *interval relational terms*, $\mathbb{IRT}$, is the smallest set of expressions that includes $\mathbb{IRA} = \mathbb{IRV} \cup \mathbb{IRC}$ and is closed with respect to the relational operations. The set of point relational formulas consists of expressions of the form xRy, where $x, y \in \mathbb{PV}$ and $R \in \mathbb{PRT}$. The set of interval relational formulas consists of expressions of the form iRj, where $i, j \in \mathbb{IV}$ and $R \in \mathbb{IRT}$. $\mathsf{RL_{HS}}$-formulas are point formulas or interval formulas. R is said to be an atomic relational term whenever $R \in \mathbb{PRC} \cup \mathbb{IRA}$. A formula xRy is said to be atomic whenever R is an atomic relational term.

An $\mathsf{RL_{HS}}$-model is a structure $\mathcal{M} = (U, \mathbb{I}(U)^+, <, B, E, m)$, where U and $\mathbb{I}(U)^+$ are non-empty sets and $m\colon \mathbb{PRT} \cup \mathbb{IRT} \to \mathcal{P}(U \times U) \cup \mathcal{P}(\mathbb{I}(U)^+ \times \mathbb{I}(U)^+)$

is a meaning function which assigns binary relations on $U \times U$ to point relational terms and binary relations on $\mathbb{I}(U)^+ \times \mathbb{I}(U^+)$ to interval relational terms as follows:

- $m(1')$ is the relation on U defined as in $\mathsf{RL}(1, 1')$-models;
- $< = m(<)$ is a strict linear ordering on U, that is for all $c, d, e \in U$, the following hold:

  (irref $<$) $(c, c) \notin <$,
  (tran $<$) if $(c, d) \in <$ and $(d, e) \in <$, then $(c, e) \in <$,
  (con $<$) $(c, d) \in <$ or $(d, c) \in <$ or $(c, d) \in m(1')$;

- m extends to all the compound relational terms $R \in \mathbb{PRT}$ as in RL-models;
- $\mathbb{I}(U)^+ = \{[c, d] \in U \times U : (c, d) \in < \cup\, m(1')\}$;
- $m(1) = \mathbb{I}(U)^+ \times \mathbb{I}(U)^+$;
- $B = m(B) = \{([c, d], [c', d']) \in m(1) : (c, c') \in m(1')$ and $(d', d) \in <\}$;
- $E = m(E) = \{([c, d], [c', d']) \in m(1) : (c, c') \in <$ and $(d, d') \in m(1')\}$;
- m extends to all the compound interval relational terms $R \in \mathbb{IRT}$ in a similar way as in RL-models with the appropriate understanding of relational operations as the operations on $\mathbb{I}(U)^+$. Accordingly, $m(-R) = (\mathbb{I}(U)^+ \times (\mathbb{I}(U)^+) \setminus m(R)$.

$\mathsf{RL_{HS}}$-models such that $m(1')$ is the identity on U are referred to as standard $\mathsf{RL_{HS}}$-models. A valuation in an $\mathsf{RL_{HS}}$-model $\mathcal{M} = (U, \mathbb{I}(U)^+, <, B, E, m)$ is any function v: $\mathbb{PV} \cup \mathbb{IV} \to U \cup \mathbb{I}(U)^+$ such that:

- If $x \in \mathbb{PV}$, then $v(x) \in U$;
- If $i \in \mathbb{IV}$, then $v(i) = [v(i_1), v(i_2)] \in \mathbb{I}(U)^+$.

An $\mathsf{RL_{HS}}$-formula xRy is said to be satisfied in a model $\mathcal{M}$ by a valuation v, $\mathcal{M}, v \models xRy$ for short, whenever $(v(x), v(y)) \in m(R)$. A formula is true in $\mathcal{M}$ whenever it is satisfied in $\mathcal{M}$ by every valuation v. A formula is $\mathsf{RL_{HS}}$-valid whenever it is true in all $\mathsf{RL_{HS}}$-models.

17.4 Translation of Halpern–Shoham Logic into a Relational Logic

In this section we define the translation of HS-formulas into relational terms of the logic $\mathsf{RL_{HS}}$ that enables us to represent the validity problem of an HS-formula as the validity problem of the corresponding relational formula. Let τ' be a one-to-one assignment of interval relational variables to the propositional variables. Then the translation τ that maps HS-formulas to $\mathsf{RL_{HS}}$-relational terms is defined as follows:

- $\tau(p) = \tau'(p)\,; 1$, for every propositional variable $p \in \mathbb{V}$;
- $\tau(\neg\varphi) = -\tau(\varphi)$;
- $\tau(\varphi \vee \psi) = \tau(\varphi) \cup \tau(\psi)$;
- $\tau(\langle R\rangle\varphi) = R\,; \tau(\varphi)$, for every $R \in \{B, E, B^{-1}, E^{-1}\}$.

Proposition 17.4.1. *For every* HS*-model* $\mathcal{M}^+$ *and for every* HS*-formula* φ*, there exists a standard* $\mathsf{RL_{HS}}$*-model* $\mathcal{M}$ *such that for all interval variables* i *and* j*,* φ *is true in* $\mathcal{M}^+$ *iff* $i\,\tau(\varphi)\,j$ *is true in* $\mathcal{M}$*.*

Proof. Let φ be an HS-formula and let $\mathcal{M}^+ = (D, \mathbb{I}(D)^+, <, m)$ be an HS-model. Then the corresponding standard $\mathsf{RL_{HS}}$-model $\mathcal{M} = (U, \mathbb{I}(U)^+, <', B, E, m')$ is defined as follows:

- $U = D$;
- $\mathbb{I}(U)^+ = \mathbb{I}(D)^+$;
- $<' = <$;
- $m'(1) = \mathbb{I}(U)^+ \times \mathbb{I}(U)^+$ and $m'(1') = Id_U$;
- $m'(P) = \{([c,d],[c',d']) \in m'(1) : [c,d] \in m(p)\}$, for every interval relational variable P such that $\tau'(p) = P$;
- $B = \{([c,d],[c',d']) \in m'(1) : c = c' \text{ and } d' < d\}$;
- $E = \{([c,d],[c',d']) \in m'(1) : c < c' \text{ and } d' = d\}$.

Given a valuation v in a model $\mathcal{M}^+$, we show by induction on the complexity of φ that for all interval variables i and j, the following holds:

$$\mathcal{M}^+, v(i) \models \varphi \text{ iff } \mathcal{M}, v \models i\,\tau(\varphi)\,j.$$

From that, we can conclude that φ is true in $\mathcal{M}^+$ iff $i\,\tau(\varphi)\,j$ is true in $\mathcal{M}$. By way of example, we prove the required condition for the formulas of the form: $\psi \vee \vartheta$, $\langle B\rangle\psi$, and $\langle E^{-1}\rangle\psi$.

If $\varphi = \psi \vee \vartheta$, then $\mathcal{M}^+, v(i) \models \psi \vee \vartheta$ iff $\mathcal{M}^+, v(i) \models \psi$ or $\mathcal{M}^+, v(i) \models \vartheta$ iff, by the induction hypothesis, $\mathcal{M}, v \models i\,\tau(\psi)\,j$ or $\mathcal{M}, v \models i\,\tau(\vartheta)\,j$ iff $\mathcal{M}, v \models i(\tau(\psi) \cup \tau(\vartheta))j$ iff $\mathcal{M}, v \models i\,\tau(\psi \vee \vartheta)\,j$.

If $\varphi = \langle B\rangle\psi$, then $\mathcal{M}^+, v(i) \models \langle B\rangle\psi$ iff there exists $c' < v(i_2)$ such that $\mathcal{M}^+, [v(i_1), c'] \models \psi$ iff, by the induction hypothesis and by the definition of $\mathcal{M}$, $(v(i), [v(i_1), c']) \in B$ and $([v(i_1), c'], [v(j_1), v(j_2)]) \in m'(\tau(\psi))$ iff $\mathcal{M}, v \models i(B\,;\tau(\psi))j$ iff $\mathcal{M}, v \models i\,\tau(\langle B\rangle\psi)\,j$.

Finally, if $\varphi = \langle E^{-1}\rangle\psi$, then $\mathcal{M}^+, v(i) \models \langle E^{-1}\rangle\psi$ iff there exists $c' < v(i_1)$ such that $\mathcal{M}^+, [c', v(i_2)] \models \psi$ iff, by the induction hypothesis and by the definition of $\mathcal{M}$, $(v(i), [c', v(i_2)]) \in E^{-1}$ and $([c', v(i_2)], [v(j_1), v(j_2)]) \in m'(\tau(\psi))$ iff $\mathcal{M}, v \models i(E^{-1}\,;\tau(\psi))j$ iff $\mathcal{M}, v \models i\,\tau(\langle E^{-1}\rangle\psi)\,j$. □

The following proposition can be proved in a similar way:

Proposition 17.4.2. *For every standard* $\mathsf{RL_{HS}}$*-model* $\mathcal{M}$ *and for every* HS*-formula* φ*, there exists an* HS*-model* $\mathcal{M}^+$ *such that for all interval variables* i *and* j*,* φ *is true in* $\mathcal{M}^+$ *iff* $i\,\tau(\varphi)\,j$ *is true in* $\mathcal{M}$*.*

From the above propositions we obtain:

Theorem 17.4.1. *For every* HS*-formula* φ *and for all interval variables* i *and* j*, the following conditions are equivalent:*

1. φ *is* HS*-valid;*
2. $i\,\tau(\varphi)\,j$ *is true in all standard* $\mathsf{RL_{HS}}$*-models.*

17.5 Dual Tableau for Halpern–Shoham Logic

In this section we present a dual tableau for the logic $\mathsf{RL_{HS}}$ and we show how it can be used for verification of validity in HS.

$\mathsf{RL_{HS}}$-dual tableau contains the decomposition and specific rules of $\mathsf{RL}(1,1')$-dual tableau (see Sect. 2.7) adjusted to $\mathsf{RL_{HS}}$-language. In particular, the rules $(\cup)$, $(\cap)$, $(-\cup)$, $(-\cap)$, $(-)$, $(^{-1})$, $(-^{-1})$, $(;)$, and $(-;)$ are assumed both for point relations and for interval relations, i.e., either x, y, z appearing in the rules are point variables and R, S are point relational terms or x, y, z are interval variables and R, S are interval relational terms. The rules $(1'1)$ and $(1'2)$ are assumed for point relational constants, i.e., x and y appearing in the rules are point variables and R is a point relational constant. Furthermore, $\mathsf{RL_{HS}}$-dual tableau contains the rules of the following forms:

Decomposition Rules from Interval Relations to Point Relations:

For all $i, j, k \in \mathbb{IV}$ and for every $R \in \mathbb{IRA}$,

$$(R_1)\ \frac{iRj}{i_1 1' k_1, iRj \mid i_2 1' k_2, iRj \mid kRj, iRj}$$

$$(R_2)\ \frac{iRj}{j_1 1' k_1, iRj \mid j_2 1' k_2, iRj \mid iRk, iRj}$$

For all $i, j \in \mathbb{IV}$,

$$(B)\ \frac{iBj}{i_1 1' j_1, iBj \mid j_2 < i_2, iBj} \qquad (-B)\ \frac{i{-}Bj}{i_1{-}1' j_1, j_2{-}{<} i_2, i{-}Bj}$$

$$(E)\ \frac{iEj}{i_2 1' j_2, iEj \mid i_1 < j_1, iEj} \qquad (-E)\ \frac{i{-}Ej}{i_2{-}1' j_2, i_1{-}{<} j_1, i{-}Ej}$$

Specific Rules:

For all $x, y, z \in \mathbb{PV}$,

$$(\text{irref} <)\ \frac{}{x < x}$$

$$(\text{tran} <)\ \frac{x < y}{x < z, x < y \mid z < y, x < y}$$

An $\mathsf{RL_{HS}}$-axiomatic set is a set including a subset of either of the following forms:

For all $x, y \in \mathbb{PV}$, $i, j \in \mathbb{IV}$, $R \in \mathbb{PRT}$, and $T \in \mathbb{IRT}$,

(Ax1) $\{xRy, x{-}Ry\}$;
(Ax1') $\{iTj, i{-}Tj\}$;
(Ax2) $\{x1'x\}$;
(Ax3) $\{x < y, x1'y, y < x\}$;
(Ax4) $\{i1j\}$;
(Ax5) $\{i_1 < i_2, i_1 1' i_2\}$.

The axiomatic sets of the form (Ax3) and (Ax5) can be replaced by a rule as discussed in Sect. 25.9.

Observe that any application of the rules of $\mathsf{RL_{HS}}$-dual tableau, in particular an application of the specific rules listed above, preserves the formulas built with atomic terms or their complements. Thus, the closed branch property holds.

The notion of an $\mathsf{RL_{HS}}$-set is defined in a similar way as in the relational logics of classical algebras of binary relations (see Sect. 2.4), that is a finite set of $\mathsf{RL_{HS}}$-formulas is an $\mathsf{RL_{HS}}$-set whenever the first-order disjunction of its members is valid in every $\mathsf{RL_{HS}}$-model. Also $\mathsf{RL_{HS}}$-correctness is defined in a similar way as in Sect. 2.4.

Proposition 17.5.1.

1. *The* $\mathsf{RL_{HS}}$*-rules are* $\mathsf{RL_{HS}}$*-correct;*
2. *The* $\mathsf{RL_{HS}}$*-axiomatic sets are* $\mathsf{RL_{HS}}$*-sets.*

Proof. By way of example, we show that the rules (B) and $(-E)$ are $\mathsf{RL_{HS}}$-correct. The proofs of the correctness of the remaining rules are similar. Let X be any finite set of $\mathsf{RL_{HS}}$-formulas.

(B) It is easy to see that if $X \cup \{iBj\}$ is an $\mathsf{RL_{HS}}$-set, then so are $X \cup \{i_1 1' j_1, iBj\}$ and $X \cup \{j_2 < i_2, iBj\}$. Now, assume that $X \cup \{i_1 1' j_1, iBj\}$ and $X \cup \{j_2 < i_2, iBj\}$ are $\mathsf{RL_{HS}}$-sets, while $X \cup \{iBj\}$ is not an $\mathsf{RL_{HS}}$-set. Then there exist an $\mathsf{RL_{HS}}$-model $\mathcal{M}$ and a valuation v in $\mathcal{M}$ such that $\mathcal{M}, v \not\models iBj$. However, by the assumption, $\mathcal{M}, v \models i_1 1' j_1$ and $\mathcal{M}, v \models j_2 < i_2$, that is $(v(i_1), v(j_1)) \in m(1')$, and $(v(j_2), v(i_2)) \in <$. By the definition of B, we obtain $(v(i), v(j)) \in B$, thus $\mathcal{M}, v \models iBj$, a contradiction.

The proof of correctness of the rule $(-E)$ is analogous. Namely, if $X \cup \{i{-}Ej\}$ is an $\mathsf{RL_{HS}}$-set, then so is $X \cup \{i_2{-}1' j_2, i_1{-}{<}j_1, i{-}Ej\}$. Now, assume that $X \cup \{i_2{-}1' j_2, i_1{-}{<}j_1, i{-}Ej\}$ is an $\mathsf{RL_{HS}}$-set, while $X \cup \{i{-}Ej\}$ is not an $\mathsf{RL_{HS}}$-set. Then there exist an $\mathsf{RL_{HS}}$-model $\mathcal{M}$ and a valuation v in $\mathcal{M}$ such that $\mathcal{M}, v \not\models i{-}Ej$. However, by the assumption, $\mathcal{M}, v \models i_2{-}1' j_2$ or $\mathcal{M}, v \models i_1{-}{<}\, j_1$, that is $(v(i_2), v(j_2)) \notin m(1')$ or $(v(i_1), v(j_1)) \notin <$. By the definition of E, $(v(i), v(j)) \notin E$, thus $\mathcal{M}, v \models i{-}Ej$, a contradiction. □

The notions of an $\mathsf{RL_{HS}}$-proof tree, a closed branch of such a tree, a closed $\mathsf{RL_{HS}}$-proof tree, and $\mathsf{RL_{HS}}$-provability are defined as in Sect. 2.4.

A branch b of an $\mathsf{RL_{HS}}$-proof tree is complete whenever it is closed or it satisfies the usual completion conditions determined by the rules $(\cup)$, $(-\cup)$, $(\cap)$, $(-\cap)$, $(-)$, $(^{-1})$, $(-^{-1})$, $(;)$, and $(-;)$ listed in Sect. 2.5 adapted both for point relations and for interval relations and, in addition:

For all $x, y \in \mathbb{PV}$ and $R \in \mathbb{PRC}$,

Cpl(1′1) If $xRy \in b$ then, for every $z \in \mathbb{PV}$, $xRz \in b$ or $y1'z \in b$, obtained by an application of the rule (1′1);

Cpl(1′2) If $xRy \in b$ then, for every $z \in \mathbb{PV}$, $z1'x \in b$ or $zRy \in b$, obtained by an application of the rule (1′2);

For all $x, y \in \mathbb{PV}$,

Cpl(irref $<$) $x < x \in b$, obtained by an application of the rule (irref $<$);
Cpl(tran $<$) If $x < y \in b$, then for every $z \in \mathbb{PV}$, $x < z \in b$ or $z < y \in b$, obtained by an application of the rule (tran $<$);

For all $i, j \in \mathbb{IV}$,

Cpl(R_1) If $iRj \in b$, then for every $k \in \mathbb{IV}$ either $i_1 1' k_1 \in b$, $i_2 1' k_2 \in b$, or $kRj \in b$, obtained by an application of the rule (R_1);
Cpl(R_2) If $iRj \in b$, then for every $k \in \mathbb{IV}$ either $j_1 1' k_1 \in b$, $j_2 1' k_2 \in b$, or $iRk \in b$, obtained by an application of the rule (R_2);
Cpl(B) If $iBj \in b$, then either $i_1 1' j_1 \in b$ or $j_2 < i_2 \in b$, obtained by an application of the rule (B);
Cpl($-B$) If $i-Bj \in b$, then $i_1 - 1' j_1$, $j_2 - < i_2 \in b$, obtained by an application of the rule ($-B$);
Cpl(E) If $iEj \in b$, then either $i_2 1' j_2 \in b$ or $i_1 < j_1 \in b$, obtained by an application of the rule (E);
Cpl($-E$) If $i-Ej \in b$, then $i_2 - 1' j_2$, $i_1 - < j_1 \in b$, obtained by an application of the rule ($-E$).

The notions of a complete $\mathsf{RL}_{\mathsf{HS}}$-proof tree and an open branch of an $\mathsf{RL}_{\mathsf{HS}}$-proof tree are defined as in RL-logic (see Sect. 2.5).

Let b be an open branch of an $\mathsf{RL}_{\mathsf{HS}}$-proof tree. The branch structure $\mathcal{M}^b = (U^b, \mathbb{I}(U^b)^+, <^b, B^b, E^b, m^b)$ is defined as follows:

- $U^b = \mathbb{PV}$;
- $m^b(R) = \{(x, y) \in U^b \times U^b : xRy \notin b\}$, for every $R \in \mathbb{PRC}$;
- m^b extends to all the compound relational terms $R \in \mathbb{PRT}$ as in $\mathsf{RL}_{\mathsf{HS}}$-models;
- $\mathbb{I}(U^b)^+ = \{[c, d] \in U^b \times U^b : (c, d) \in <^b \cup m^b(1')\}$;
- $m^b(R) = \{(i, j) \in \mathbb{I}(U^b)^+ \times \mathbb{I}(U^b)^+ : iRj \notin b\}$, for every $R \in \mathbb{IRV}$;
- $m^b(1) = \mathbb{I}(U^b)^+ \times \mathbb{I}(U^b)^+$;
- $B^b = m^b(B) = \{([c, d], [c', d']) \in \mathbb{I}(U^b)^+ \times \mathbb{I}(U^b)^+ : (c, c') \in m^b(1')$ and $(d', d) \in <^b\}$;
- $E^b = m^b(E) = \{([c, d], [c', d']) \in \mathbb{I}(U^b)^+ \times \mathbb{I}(U^b)^+ : (c, c') \in <^b$ and $(d, d') \in m^b(1')\}$;
- m^b extends to all the compound interval relational terms as in $\mathsf{RL}_{\mathsf{HS}}$-models.

Proposition 17.5.2 (Branch Model Property). *Let b be an open branch of an $\mathsf{RL}_{\mathsf{HS}}$-proof tree. The branch structure $\mathcal{M}^b$ is an $\mathsf{RL}_{\mathsf{HS}}$-model.*

Proof. We show that $<^b$ satisfies the conditions (irref $<$), (tran $<$), and (con $<$). All the remaining conditions are satisfied by the definition of the branch structure.

By the completion condition Cpl(irref $<$), for every $x \in U^b$, we have $x < x \in b$, which means that $(x, x) \notin <^b$ for every $x \in U^b$, therefore $<^b$ is irreflexive. To prove transitivity, assume $(x, y) \in <^b$ and $(y, z) \in <^b$, that is $x < y \notin b$ and $y < z \notin b$. Suppose $(x, z) \notin <^b$. Then $x < z \in b$. By the completion condition Cpl(tran $<$), $x < y \in b$ or $y < z \in b$, a contradiction. Therefore $<^b$ satisfies the condition (tran $<$). Since b is open, for all $x, y \in U^b$, $x < y \notin b$ or $y < x \notin b$ or $x1'y \notin b$.

Otherwise, since the rules preserve formulas built with atomic relational terms or their complements, all of these formulas eventually appear in a node of b and then b would be closed. Thus, $(x, y) \in <^b$ or $(y, x) \in <^b$ or $(x, y) \in m^b(1')$, therefore $<^b$ satisfies the condition (con $<$). □

Given a branch model $\mathcal{M}^b = (U^b, \mathbb{I}(U^b)^+, <^b, B^b, E^b, m^b)$, let $v^b\colon \mathbb{PV} \cup \mathbb{IV} \to U^b \cup \mathbb{I}(U^b)^+$ be a function such that $v^b(x) = x$, for every $x \in \mathbb{PV}$, and $v(i) = [i_1, i_2]$, for every $i \in \mathbb{IV}$.

Proposition 17.5.3. *Let b be an open branch of an* $\mathsf{RL_{HS}}$*-proof tree. Then the function v^b is a valuation in the branch model $\mathcal{M}^b$.*

Proof. By the definition of v^b, if $x \in \mathbb{PV}$ then $v^b(x) \in U^b$, and if $i \in \mathbb{IV}$ then $v^b(i) = [v^b(i_1), v^b(i_2)]$. It remains to show that for every $i \in \mathbb{IV}$, $(v^b(i_1), v^b(i_2)) \in <^b \cup m^b(1')$. Suppose that there exists $i \in \mathbb{IV}$ such that $(v^b(i_1), v^b(i_2)) \notin <^b \cup m^b(1')$. Then $(v^b(i_1), v^b(i_2)) \notin <^b$ and $(v^b(i_1), v^b(i_2)) \notin m^b(1')$. By the definition of m^b, this implies that $i_1 < i_2 \in b$ and $i_1 1' i_2 \in b$. Due to (Ax5), b is closed, a contradiction. □

Proposition 17.5.4 (Satisfaction in Branch Model Property). *Let b be an open branch of an* $\mathsf{RL_{HS}}$*-proof tree. Then for every* $\mathsf{RL_{HS}}$*-formula φ, if $\mathcal{M}^b, v^b \models \varphi$, then $\varphi \notin b$.*

Proof. Let φ be an $\mathsf{RL_{HS}}$-formula. If φ is a point formula, than the proposition can be proved as in the logic $\mathsf{RL}(1, 1')$ (see Sects. 2.5 and 2.7). Let $\varphi = iRj$ be an interval formula. The proof is by induction on the complexity of R. If R is an interval relational variable or its complement, the required condition can be proved as in Proposition 2.5.5. Now, we show that the proposition holds for interval relational constants and their complements.

For $R = 1$, it holds trivially, since every set including $i1j$ is axiomatic.

Let $R = B$. Assume $(i, j) \in B^b$, that is $(i_1, j_1) \in m^b(1')$ and $(j_2, i_2) \in <^b$. Then $i_1 1' j_1 \notin b$ and $j_2 < i_2 \notin b$. Suppose $iBj \in b$. By the completion condition Cpl(B), either $i_1 1' j_1 \in b$ or $j_2 < i_2 \in b$, a contradiction.

Similarly, for $R = -B$ (resp. $E, -E$) we use the completion conditions Cpl($-B$) (resp. Cpl(E), Cpl($-E$)).

Therefore, the proposition holds for all formulas built with an atomic term or its complement. For the formulas built with compound terms it can be proved in a similar way as in RL-logic (see Sect. 2.5). □

Following the general method of proving completeness presented in Sect. 2.6, the above propositions imply:

Theorem 17.5.1 (Soundness and Completeness of $\mathsf{RL_{HS}}$). *For every* $\mathsf{RL_{HS}}$*-formula φ, the following conditions are equivalent:*

1. *φ is* $\mathsf{RL_{HS}}$*-valid;*
2. *φ is true in all standard* $\mathsf{RL_{HS}}$*-models;*
3. *φ is* $\mathsf{RL_{HS}}$*-provable.*

Finally, due to the above theorem and Theorem 17.4.1, we get:

Theorem 17.5.2 (Relational Soundness and Completeness of HS). *For every* HS*-formula* φ *and for all interval variables* i *and* j*, the following conditions are equivalent:*

1. φ *is* HS*-valid;*
2. $i\,\tau(\varphi)\,j$ *is* $\mathsf{RL}_{\mathsf{HS}}$*-provable.*

Example. Consider the HS-formula:

$$\varphi = \langle B\rangle\langle B\rangle p \rightarrow \langle B\rangle p,$$

which reflects transitivity of relation B. The translation of φ into an $\mathsf{RL}_{\mathsf{HS}}$-term is:

$$\tau(\varphi) = -(B\,;(B\,;P)) \cup (B\,;P),$$

where for simplicity $\tau(p)$ is denoted by P. By Theorem 17.4.1, $\mathsf{RL}_{\mathsf{HS}}$-dual tableau can be used for verification of validity of φ. Figure 17.1 presents an $\mathsf{RL}_{\mathsf{HS}}$-proof of the relational formula $i\,\tau(\varphi)\,j$ from which HS-validity of φ follows.

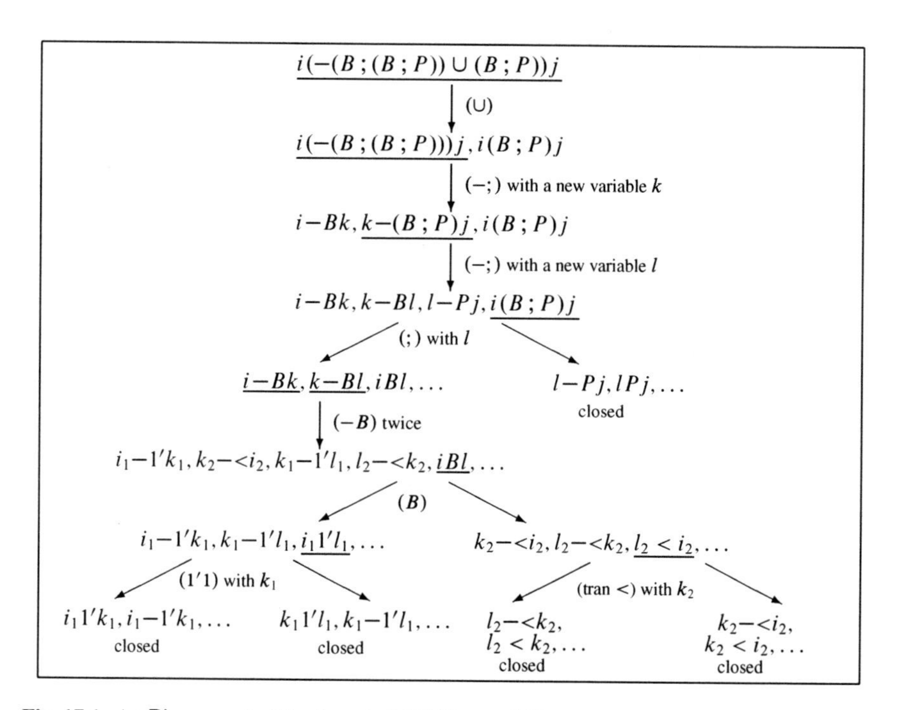

Fig. 17.1 An $\mathsf{RL}_{\mathsf{HS}}$-proof of the formula $\langle B\rangle\langle B\rangle p \rightarrow \langle B\rangle p$

17.6 Dual Tableaux for Other Interval Temporal Logics

In this section we exploit the modularity of the relational approach, and we show how to extend HS-dual tableau to cope with other interval relations and other meaningful temporal domains.

Logics Based on Strict Intervals

In the previous sections we considered the non-strict semantics of HS where, given a strict ordering $(D, <)$, the set of non-strict intervals $\mathbb{I}(D)^+$ is defined as the set of all $[c, d]$ such that $c \leq d$. As a consequence, the set of intervals includes the point intervals of the form $[c, c]$. In the literature an alternative semantics for interval logics is considered, namely the *strict semantics*, where the point intervals are excluded. Given a strict ordering $(D, <)$, a *strict interval* is a pair $[c, d]$ where $c < d$. The set of all strict intervals on D is denoted by $\mathbb{I}(D)^-$. The models based on strict intervals are defined in a way analogous to the non-strict case.

In what follows, we show how to modify the dual tableau for $\mathsf{RL_{HS}}$ in the case of the strict semantics. To this end, we define the relational logic $\mathsf{RL}^-_{\mathsf{HS}}$ (strict $\mathsf{RL_{HS}}$), having the same syntax as non-strict $\mathsf{RL_{HS}}$, but a different semantics.

An $\mathsf{RL}^-_{\mathsf{HS}}$-model is a structure $\mathcal{M}^- = (U, \mathbb{I}(U)^-, <, B, E, m)$, where (U, m) is an $\mathsf{RL}(1, 1')$-model, $<$, B, and E are defined as in $\mathsf{RL_{HS}}$-models, $\mathbb{I}(U)^- = \{[c, d] \in U \times U : (c, d) \in <\}$ and $m(1) = \mathbb{I}(U)^- \times \mathbb{I}(U)^-$. An $\mathsf{RL}^-_{\mathsf{HS}}$-valuation is any function $v: \mathbb{PV} \cup \mathbb{IV} \to \mathcal{P}(U) \cup \mathcal{P}(\mathbb{I}(U)^- \times \mathbb{I}(U)^-)$ such that:

- If $x \in \mathbb{PV}$, then $v(x) \in U$;
- If $i \in \mathbb{IV}$, then $v(i) = [v(i_1), v(i_2)] \in \mathbb{I}(U)^-$.

The notions of satisfaction and validity of a formula are defined as in logic $\mathsf{RL_{HS}}$.

A dual tableau for $\mathsf{RL}^-_{\mathsf{HS}}$ is obtained from the $\mathsf{RL_{HS}}$-dual tableau by replacing the axiomatic set (Ax5) with:

(Ax5$^-$) $i_1 < i_2$, for $i \in \mathbb{IV}$.

$\mathsf{RL}^-_{\mathsf{HS}}$-dual tableau is sound and complete; it can be proved in a similar way as for $\mathsf{RL_{HS}}$-dual tableau.

Other Interval Temporal Logics

Now, we show how to modify the relational logic $\mathsf{RL_{HS}}$ and its dual tableau to obtain a relational logic and a corresponding dual tableau for any interval logic based on relations chosen from the 13 Allen's relations. Let $I \subseteq \{E, E^{-1}, D, D^{-1}, B, B^{-1}, O, O^{-1}, M, M^{-1}, P, P^{-1}\}$. A language of a logic L_I is obtained from the HS-language by replacing the set of relational constants $\{B, E, B^{-1}, E^{-1}\}$ with the set I and the set of modal operations with $\{\langle R\rangle : R \in I\}$. Given an interval logic L_I, the corresponding relational logic $\mathsf{RL}_{\mathsf{L}_I}$ differs from $\mathsf{RL_{HS}}$ only in the choice of the set of interval relational constants, that is $\mathbb{IRC} = \{1\} \cup I$. Models of $\mathsf{RL}_{\mathsf{L}_I}$ are defined as those of $\mathsf{RL_{HS}}$ with the assumption that the semantics of the relational constants from I is defined in accordance with the semantics of the chosen primitive interval relations. A validity preserving translation τ of L_I-formulas into $\mathsf{RL}_{\mathsf{L}_I}$-formulas is obtained from the translation presented in Sect. 17.4 by assuming $R \in I$.

A dual tableau for $\mathsf{RL}_{\mathsf{L}_I}$ can be obtained from $\mathsf{RL}_{\mathsf{HS}}$-dual tableau (in the case of the non-strict semantics for intervals) or from $\mathsf{RL}_{\mathsf{HS}}^{-}$-dual tableau (in the case of the strict semantics), by replacing the rules (B), (E), $(-B)$, and $(-E)$ with the rules appropriate for the choice of basic relations. The rules for the relations D, M, P, and O have the following forms:

For all $i, j \in \mathbb{IV}$,

$$(D) \quad \frac{iDj}{i_1 < j_1, iDj \mid j_2 < i_2, iDj} \qquad (-D) \quad \frac{i{-}Dj}{i_1{-}{<}j_1, j_2{-}{<}i_2, i{-}Dj}$$

$$(M) \quad \frac{iMj}{j_2 1' i_1, iMj} \qquad (-M) \quad \frac{i{-}Mj}{j_2{-}1' i_1, i{-}Mj}$$

$$(P) \quad \frac{iPj}{j_2 < i_1, iPj} \qquad (-P) \quad \frac{i{-}Pj}{j_2{-}{<}i_1, i{-}Pj}$$

$$(O) \quad \frac{iOj}{j_1 < i_1, iOj \mid i_1 < j_2, iOj \mid j_2 < i_2, iOj}$$

$$(-O) \quad \frac{i{-}Oj}{j_1{-}{<}i_1, i_1 < j_2, j_2 < i_2, i{-}Oj}$$

The rules presented in Sect. 17.5 allow us to easily adapt $\mathsf{RL}_{\mathsf{HS}}$-dual tableau to any propositional interval temporal logic that is a proper fragment of HS. Here we show two examples of such a modification.

The logic with relations B and B^{-1} and the logic with relations E and E^{-1} are decidable (see [GMS04]). The logic BE is obtained from HS by deleting the operations $\langle B^{-1}\rangle$ and $\langle E^{-1}\rangle$. It was studied in [Lod00], where its undecidability was proved. Since BE does not have modal operations determined by converses of relations B and E, the relational logic $\mathsf{RL}_{\mathsf{BE}}$ appropriate for representation of BE-formulas is the logic $\mathsf{RL}_{\mathsf{HS}}$ without the converse operation $^{-1}$. A dual tableau for $\mathsf{RL}_{\mathsf{BE}}$ can be obtained from that of $\mathsf{RL}_{\mathsf{HS}}$ by deleting the rules $(^{-1})$ and $(-^{-1})$.

Interval logics based on the relation *meet*, M, and its converse are usually called *neighborhood logics*. First-order neighborhood logics were first introduced and studied in [CH97]. Their propositional variant, called *Propositional Neighborhood Logic*, PNL, was proposed and investigated recently in [GMS03b].

In [GMS03b] the authors studied logic PNL both with the non-strict and strict semantics over linear orderings. Let PNL^{+} and PNL^{-} be the respective logics. The relational logic $\mathsf{RL}_{\mathsf{PNL}^{+}}$ is obtained from logic $\mathsf{RL}_{\mathsf{HS}}$ by taking the interval relational constant M in place of B and E. A dual tableau for $\mathsf{RL}_{\mathsf{PNL}^{+}}$ can be obtained from that of $\mathsf{RL}_{\mathsf{HS}}$ by replacing the rules (B), $(-B)$, (E), and $(-E)$ with the rules (M) and $(-M)$. In the case of the strict semantics, the relational logic $\mathsf{RL}_{\mathsf{PNL}^{-}}$ can be obtained from $\mathsf{RL}_{\mathsf{HS}}^{-}$ in the same way.

In all the relational systems RL_{L} presented above, the strict ordering $<$ is considered to be linear, without any further assumption. We can modify this constraint by requiring that the order of point intervals in the models is, for example, unbounded from below, serial, dense, or discrete. Then, in order to obtain a dual tableau for such a restriction, we replace the rules and axiomatic sets that reflect strict linearity of $<$ by the rules corresponding to the appropriate properties of $<$ assumed in the models, as presented in Sect. 16.3.

Chapter 18
Dual Tableaux for Spatial Reasoning

18.1 Introduction

Qualitative spatial reasoning is concerned with the qualitative aspects of representation and reasoning about spatial entities as opposed to one-dimensional situations. It aims to express non-numerical relationships among spatial regions. The basic concepts in most of these theories are 'part of' and 'connection' relations which are the typical examples of a more general notion of a 'contact' relation.

The formalization of 'part of' relationship goes back to the mereological systems of Stanisław Leśniewski developed from 1916 onwards [Leś16, Leś29, Leś31]. Next, based on an earlier work of de Laguna [dL22], Whitehead [Whi29] introduced a kind of connection relation as a basic relation between regions. His system includes Leśniewski's mereology. Later, Grzegorczyk [Grz60] and Clarke [Cla81] presented an axiomatization of Whitehead's relation.

One of the more recent spatial theories is the Region Connection Calculus, RCC, introduced in [RCC92]. The primitive concept of this theory is a binary relation of 'connection' between regions. Intuitively, in terms of points incident in regions, two regions are connected whenever they share a common point. With the connection relation a set of the other spatial relations is defined. These relations describe different connections between regions such as being externally connected, partially overlapping, being a tangential part of, and so on. In the theory RCC all meaningful degrees of connection are formally defined. The fundamental structure considered in the theory is a Boolean algebra endowed with a contact relation satisfying certain axioms.

Spatial theories having as a basic concept a proximity relation, first considered in [Efr52], are investigated in [NW70, DV07, VDDB02, VDB01], among others. The proximity relation is a binary relation between subsets of a non-empty set, which holds between the two sets whenever, intuitively, they are near in some sense. The proximity relation satisfies axioms which are among the typical axioms of the connection relation. The spatial intuition of proximity is also formalized in terms of nearness spaces as presented in [BdRV01]. A natural counterpart to the proximity and nearness spaces are the apartness spaces introduced in [BD03]. In [DO08] the apartness algebras and apartness frames are defined which lead to an intuitionistic

E. Orłowska and J. Golińska-Pilarek, *Dual Tableaux: Foundations, Methodology, Case Studies*, Trends in Logic 36, DOI 10.1007/978-94-007-0005-5_18,

logic with a sufficiency operation. Consequently, a dual tableau for such a logic can be developed based on the dual tableau for intuitionistic logic presented in Sect. 8.2 and on the developments of Sect. 11.5 where the rules for a sufficiency operation are discussed.

Qualitative spatial reasoning from a philosophical perspective is discussed in [Sim87] and from a relation algebraic perspective in [Dün05].

In this chapter we consider four types of spatial theories: theories based on a plain contact relation, theories based on Boolean algebras additionally equipped with a contact relation, theories of region connection calculus, and theories based on a proximity relation. All of these theories, except for those from the first group that are based on logic $\mathsf{RL}(1, 1')$, are presented as first-order theories of appropriate classes of relations, following the developments in [Dün01b, DWM99, DOW01, DWM01, Ste00, VDDB02]. This enables us to construct their proof systems in a uniform way.

18.2 Dual Tableaux for Spatial Theories Based on a Plain Contact Relation

Typically, the most basic notion of theories for spatial reasoning is that of a contact relation [Cla81, Dün01b]. In this section we consider theories of the class of relation algebras generated by a binary relation C interpreted as a *contact relation.* This class is denoted by CRA. Each of these theories has some specific axioms that characterize relation C. One of them has the following axioms (see [DO00b]):

(C1) C is reflexive;
(C2) C is symmetric;
(C3) If $C(a) = C(b)$, then $a = b$, where $C(x) = \{y : xCy\}$.

Axiom (C3) is referred to as the extensionality axiom.

In terms of a contact relation, several mereological relations can be defined. In Table 18.1 we list some typical relations of that kind.

Table 18.1 Mereological relations definable from the contact relation C

$P = -(C\,;-C)$	part of
$PP = P \cap -1'$	proper part of
$O = P^{-1}\,; P$	overlap
$PO = O \cap -(P \cup P^{-1})$	partial overlap
$EC = C \cap -O$	external contact
$TPP = PP \cap (EC\,; EC)$	tangential proper part
$NTPP = PP \cap -TPP$	non-tangential proper part
$DC = -C$	disconnected
$DR = -O$	discrete

The language of the relational logic $\mathsf{RL}_{\mathsf{CRA}}$, adequate for reasoning in relation algebras generated by a contact relation, is $\mathsf{RL}(1, 1')$-language with:

- $\mathbb{RV}_{\mathsf{RL}_{\mathsf{CRA}}} = \emptyset$;
- $\mathbb{RC}_{\mathsf{RL}_{\mathsf{CRA}}} = \{1, 1', C\}$.

An $\mathsf{RL}_{\mathsf{CRA}}$-model is a structure $\mathcal{M} = (U, C, m)$ such that (U, m) is an $\mathsf{RL}(1, 1')$-model, C is a reflexive and symmetric relation on U such that $m(C) = C$ and, in addition:

(ext C) If $\{z \in U : (x, z) \in C\} = \{z \in U : (y, z) \in C\}$, then $(x, y) \in m(1')$.

Note that the above condition is a counterpart to the extensionality axiom (C3) although the meaning of $1'$ may not be the identity. $\mathsf{RL}_{\mathsf{CRA}}$-models in which $1'$ is interpreted as the identity relation are referred to as standard models.

An $\mathsf{RL}_{\mathsf{CRA}}$-dual tableau consists of the rules and the axiomatic sets of the logic $\mathsf{RL}(1, 1')$ and the following specific rules reflecting the axioms of contact relation C.

For any object symbols x and y,

$$(\text{ref } C)\ \frac{xCy}{x1'y, xCy} \qquad (\text{sym } C)\ \frac{xCy}{yCx}$$

$$(\text{ext } C)\ \frac{}{x{-}Cz, yCz \,|\, y{-}Cz, xCz \,|\, x{-}1'y}$$

x and y are any object symbols, z is a new variable, and $z \neq x, y$

We can prove soundness and completeness of $\mathsf{RL}_{\mathsf{CRA}}$ following the method developed for $\mathsf{RL}(1, 1')$-logic. As usual, $\mathsf{RL}_{\mathsf{CRA}}$-sets of formulas are defined as in Sect. 2.4, i.e., the first-order disjunction of their members is valid. Correctness of a rule is defined as in Sect. 2.4.

Proposition 18.2.1. *The rules (ref C), (sym C), and (ext C) are* $\mathsf{RL}_{\mathsf{CRA}}$*-correct.*

Proof. The rules (ref C) and (sym C) reflect reflexivity and symmetry of the relation C, respectively, as it is shown in Theorem 6.6.1. Now, we prove correctness of the rule (ext C). Let X be a finite set of $\mathsf{RL}_{\mathsf{CRA}}$-formulas. Let x, y be any object symbols and let z be an object variable that does not occur in X and such that $z \neq x, y$. Clearly, if X is an $\mathsf{RL}_{\mathsf{CRA}}$-set, then so are $X \cup \{x{-}Cz, yCz\}$, $X \cup \{y{-}Cz, xCz\}$, and $X \cup \{x{-}1'y\}$. Now, assume that $X \cup \{x{-}Cz, yCz\}$, $X \cup \{y{-}Cz, xCz\}$, and $X \cup \{x{-}1'y\}$ are $\mathsf{RL}_{\mathsf{CRA}}$-sets. Suppose X is not an $\mathsf{RL}_{\mathsf{CRA}}$-set, that is there exist an $\mathsf{RL}_{\mathsf{CRA}}$-model $\mathcal{M} = (U, C, m)$ and a valuation v in $\mathcal{M}$ such that $\mathcal{M}, v \not\models \varphi$, for every $\varphi \in X$. It follows from the assumption that model $\mathcal{M}$ and valuation v satisfy:

$\mathcal{M}, v \models x{-}Cz$ or $\mathcal{M}, v \models yCz$;
$\mathcal{M}, v \models y{-}Cz$ or $\mathcal{M}, v \models xCz$;
$\mathcal{M}, v \models x{-}1'y$.

By the assumption on variable z, for every $a \in U$ the following hold:

If $(v(x), a) \in C$, then $(v(y), a) \in C$;
If $(v(y), a) \in C$, then $(v(x), a) \in C$;
$(v(x), v(y)) \notin m(1')$.

Thus $\{z \in U : (v(x), z) \in C\} = \{z \in U : (v(y), z) \in C\}$ and $(v(x), v(y)) \notin m(1')$. However, the condition (ext C) implies $(v(x), v(y)) \in m(1')$, a contradiction. □

By the above proposition and since correctness of all the remaining rules can be proved as in $\mathsf{RL}(1, 1')$-logic (see Sects. 2.5 and 2.7), we get:

Proposition 18.2.2.

1. *The* $\mathsf{RL}_{\mathsf{CRA}}$*-rules are* $\mathsf{RL}_{\mathsf{CRA}}$*-correct;*
2. *The* $\mathsf{RL}_{\mathsf{CRA}}$*-axiomatic sets are* $\mathsf{RL}_{\mathsf{CRA}}$*-sets.*

The notions of an $\mathsf{RL}_{\mathsf{CRA}}$-proof tree, a closed branch of such a tree, a closed $\mathsf{RL}_{\mathsf{CRA}}$-proof tree, and $\mathsf{RL}_{\mathsf{CRA}}$-provability are defined as in Sect. 2.4.

A branch b of an $\mathsf{RL}_{\mathsf{CRA}}$-proof tree is complete whenever it is closed or it satisfies the completion conditions of $\mathsf{RL}(1, 1')$-dual tableau and the following conditions specific for the $\mathsf{RL}_{\mathsf{CRA}}$-dual tableau:

For any object symbols x and y,

Cpl(ref C) If $xCy \in b$, then $x1'y \in b$, obtained by an application of the rule (ref C);
Cpl(sym C) If $xCy \in b$, then $yCx \in b$, obtained by an application of the rule (sym C);
Cpl(ext C) Either $x{-}1'y \in b$ or there exists an object variable z such that either ($x{-}Cz \in b$ and $yCz \in b$) or ($y{-}Cz \in b$ and $xCz \in b$), obtained by an application of the rule (ext C).

The notions of a complete $\mathsf{RL}_{\mathsf{CRA}}$-proof tree and an open branch of an $\mathsf{RL}_{\mathsf{CRA}}$-proof tree are defined as in RL-logic (see Sect. 2.5). Observe that the rules of $\mathsf{RL}_{\mathsf{CRA}}$-dual tableau, in particular the specific rules (ref C), (sym C), and (ext C), guarantee that whenever a branch of an $\mathsf{RL}_{\mathsf{CRA}}$-proof tree contains formulas xRy and $x{-}Ry$, for an atomic term R, then the branch can be closed. Thus, the closed branch property can be proved as in Proposition 2.8.1.

The branch structure $\mathcal{M}^b = (U^b, C^b, m^b)$ is defined in the standard way (see Sect. 2.6, p. 44), in particular $m^b(C) = C^b = \{(x, y) \in U^b \times U^b : xCy \notin b\}$.

Proposition 18.2.3 (Branch Model Property). *Let b be an open branch of an* $\mathsf{RL}_{\mathsf{CRA}}$*-proof tree. Then, the branch structure $\mathcal{M}^b = (U^b, C^b, m^b)$ is an* $\mathsf{RL}_{\mathsf{CRA}}$*-model.*

Proof. It suffices to show that C^b is reflexive, symmetric, and it satisfies the condition (ext C). By the completion conditions Cpl(ref C) and Cpl(sym C), it easily follows that C^b is reflexive and symmetric. To prove the extensionality property of C^b, let $x, y \in U^b$ be such that $\{z \in U^b : (x, z) \in C^b\} = \{z \in U^b : (y, z) \in C^b\}$.

By the completion condition Cpl(ext C), either $x{-}1'y \in b$ or there exists an object variable z, such that either ($x{-}Cz \in b$ and $yCz \in b$) or ($y{-}Cz \in b$ and $xCz \in b$). If $x{-}1'y \in b$, then $x1'y \notin b$, since otherwise, by the closed branch property, b would be closed. Hence, $(x,y) \in m^b(1')$. If for some $z \in U^b$, $x{-}Cz \in b$ and $yCz \in b$, then $(x,z) \in C^b$ and $(y,z) \notin C^b$. Thus, $\{z \in U^b : (x,z) \in C^b\} \neq \{z \in U^b : (y,z) \in C^b\}$, a contradiction. If there is $z \in U^b$ such that $y{-}Cz \in b$ and $xCz \in b$, then $(y,z) \in C^b$ and $(x,z) \notin C^b$, which also contradicts the assumption. □

Since $m^b(C)$ is defined in a standard way, the satisfaction in branch model property can be proved as in $\mathsf{RL}(1,1')$-logic (see Sects. 2.5 and 2.7). Hence, completeness of $\mathsf{RL_{CRA}}$-dual tableau follows.

Theorem 18.2.1 (Soundness and Completeness of $\mathsf{RL_{CRA}}$). *For every $\mathsf{RL_{CRA}}$-formula φ, the following conditions are equivalent:*

1. *φ is $\mathsf{RL_{CRA}}$-valid;*
2. *φ is true in all standard $\mathsf{RL_{CRA}}$-models;*
3. *φ is $\mathsf{RL_{CRA}}$-provable.*

In order to reason directly about the mereological relations definable from the contact relation C, we can develop the relational logic $\mathsf{RL_{Mer}}$ which is obtained from $\mathsf{RL_{CRA}}$ by adding to its language the set of relational constants $\mathsf{Mer} = \{P, PP, O, PO, EC, TPP, NTPP, DC, DR\}$. An $\mathsf{RL_{Mer}}$-structure is a system $\mathcal{M} = (U, C, \{\# : \# \in \mathsf{Mer}\}, m)$, where (U, C, m) is an $\mathsf{RL_{CRA}}$-model and $m(\#) = \# \subseteq U \times U$, for every relational constant $\# \in \mathsf{Mer}$. An $\mathsf{RL_{Mer}}$-model is an $\mathsf{RL_{Mer}}$-structure $\mathcal{M} = (U, C, \{\# : \# \in \mathsf{Mer}\}, m)$ such that relations from Mer are defined by conditions given in Table 18.1. We slightly abuse the notation here by identifying symbols of the language with the corresponding entities in the models.

A dual tableau for $\mathsf{RL_{Mer}}$ consists of the rules and the axiomatic sets of $\mathsf{RL_{CRA}}$-dual tableau, and the following specific rules that reflect the definitions of mereological relations in terms of the contact relation C:

For all object symbols x and y,

$$(P)\ \frac{xPy}{x{-}Cz, zCy} \qquad (-P)\ \frac{x{-}Py}{xCz, x{-}Py \mid z{-}Cy, x{-}Py}$$

z is a new object variable $\qquad$ z is any object symbol

$$(PP)\ \frac{xPPy}{xPy, xPPy \mid x{-}1'y, xPPy} \qquad (-PP)\ \frac{x{-}PPy}{x{-}Py, x1'y, x{-}PPy}$$

$$(O)\ \frac{xOy}{zPx, xOy \mid zPy, xOy} \qquad (-O)\ \frac{x{-}Oy}{z{-}Px, z{-}Py}$$

z is any object symbol $\qquad$ z is a new object variable

$$(PO)\quad \frac{xPOy}{xOy, xPOy \mid x{-}Py, xPOy \mid y{-}Px, xPOy}$$

$$(-PO)\quad \frac{x{-}POy}{x{-}Oy, xPy, yPx, x{-}POy}$$

$$(EC)\quad \frac{xECy}{xCy, xECy \mid x{-}Oy, xECy} \qquad (-EC)\quad \frac{x{-}ECy}{x{-}Cy, xOy, x{-}ECy}$$

$$(TPP)\quad \frac{xTPPy}{xPPy, xTPPy \mid xECz, xTPPy \mid zECy, xTPPy}$$

z is any object symbol

$$(-TPP)\quad \frac{x{-}TPPy}{x{-}PPy, x{-}ECz, z{-}ECy} \qquad z \text{ is a new object variable}$$

$$(NTPP)\quad \frac{xNTTPy}{xPPy, xNTPPy \mid x{-}TPPy, xNTPPy}$$

$$(-NTPP)\quad \frac{x{-}NTPPy}{x{-}PPy, xTPPy, x{-}NTPPy}$$

$$(DC)\quad \frac{xDCy}{x{-}Cy, xDCy} \qquad (-DC)\quad \frac{x{-}DCy}{xCy, x{-}DCy}$$

$$(DR)\quad \frac{xDRy}{x{-}Oy, xDRy} \qquad (-DR)\quad \frac{x{-}DRy}{xOy, x{-}DRy}$$

The alternative forms of the rules above are discussed in Sect. 25.9.

Let $\mathcal{K}$ be a class of $\mathsf{RL_{Mer}}$-structures. The notion of a $\mathcal{K}$-set and the notion of $\mathcal{K}$-correctness of a rule are defined as in Sect. 2.4.

Theorem 18.2.2 (Correspondence). *Let $\mathcal{K}$ be a class of $\mathsf{RL_{Mer}}$-structures and let $\# \in \mathsf{Mer} = \{P, O, PO, PP, EC, TPP, NTPP, DC, DR\}$. Then, the following conditions are equivalent:*

1. *$\mathcal{K}$ is the class of $\mathsf{RL_{Mer}}$-models;*
2. *For every $\# \in \mathsf{Mer}$, the rules $(\#)$ and $(-\#)$ are $\mathcal{K}$-correct.*

Proof.

(1. $\rightarrow$ 2.) Assume that $\mathcal{K}$ is the class of $\mathsf{RL_{Mer}}$-models. By way of example, we prove correctness of the rules (P) and $(-P)$.

(P) Let X be a finite set of $\mathsf{RL_{Mer}}$-formulas and let z be a variable that does not occur in X and $z \neq x, y$. Assume $X \cup \{xPy\}$ is an $\mathsf{RL_{Mer}}$-set and suppose that $X \cup \{x{-}Cz, zCy\}$ is not an $\mathsf{RL_{Mer}}$-set. It follows that there is an $\mathsf{RL_{Mer}}$-model $\mathcal{M}$ and a valuation v in $\mathcal{M}$ such that for every $\varphi \in X$, $\mathcal{M}, v \not\models \varphi$

and, moreover, $(v(x), v(z)) \in C$ and $(v(z), v(y)) \notin C$, that is $(v(x), v(y)) \in (C ; -C)$. However, by the assumption, $(v(x), v(y)) \in P$, which by the condition $P \subseteq -(C ; -C)$ implies $(v(x), v(y)) \notin (C ; -C)$, a contradiction. Now, assume that $X \cup \{x{-}Cz, zCy\}$ is an $\mathsf{RL_{Mer}}$-set. Then, by the assumption on variable z, for every $\mathsf{RL_{Mer}}$-model $\mathcal{M}$ and for every valuation v in $\mathcal{M}$, either there exists $\varphi \in X$ such that $\mathcal{M}, v \models \varphi$ or for every $z \in U$, $(v(x), z) \notin C$ or $(z, v(y)) \in C$, that is $(v(x), v(y)) \notin (C ; -C)$. Suppose $X \cup \{xPy\}$ is not an $\mathsf{RL_{Mer}}$-set. Then, there exist an $\mathsf{RL_{Mer}}$-model $\mathcal{M}$ and a valuation v such that $(v(x), v(y)) \notin P$. Thus, by the condition $-P \subseteq (C ; -C)$, $(v(x), v(y)) \in (C ; -C)$, a contradiction. Therefore, the rule (P) is $\mathsf{RL_{Mer}}$-correct.

$(-P)$ Let X be a finite set of $\mathsf{RL_{Mer}}$-formulas. Clearly, if $X \cup \{x{-}Py\}$ is an $\mathsf{RL_{Mer}}$-set, then so are $X \cup \{xCz, x{-}Py\}$ and $X \cup \{z{-}Cy, x{-}Py\}$. Now, assume that $X \cup \{xCz, x{-}Py\}$ and $X \cup \{z{-}Cy, x{-}Py\}$ are $\mathsf{RL_{Mer}}$-sets. Suppose $X \cup \{x{-}Py\}$ is not an $\mathsf{RL_{Mer}}$-set. Then, there exist an $\mathsf{RL_{Mer}}$-model $\mathcal{M}$ and a valuation v in $\mathcal{M}$ such that for every $\varphi \in X$, $\mathcal{M}, v \not\models \varphi$ and $(v(x), v(y)) \in P$. It follows that for every $w \in U$, either $(v(x), w) \notin C$ or $(w, v(y)) \in C$. However, by the assumption there exists $w \in U$ such that $(v(x), w) \in C$ and $(w, v(y)) \notin C$, a contradiction. Therefore, the rule $(-P)$ is $\mathsf{RL_{Mer}}$-correct.

(2. $\rightarrow$ 1.) Assume that for every $\# \in \mathsf{Mer}$, the rules $(\#)$ and $(-\#)$ are $\mathcal{K}$-correct. We need to show that relations from Mer satisfy equations given in Table 18.1. By way of example, we show that $P = -(C ; -C)$.

$(\subseteq)$ By correctness of the rule $(-P)$, for every finite set X of $\mathsf{RL_{Mer}}$-formulas, $X \cup \{x{-}Py\}$ is a $\mathcal{K}$-set iff $X \cup \{xCz, x{-}Py\}$ and $X \cup \{z{-}Cy, x{-}Py\}$ are $\mathcal{K}$-sets. Let $X \stackrel{\mathrm{df}}{=} \{x{-}Cz, zCy\}$. By the assumption, $\{x{-}Py, x{-}Cz, zCy\}$ is a $\mathcal{K}$-set. Thus, for every $\mathcal{K}$-structure $\mathcal{M}$ and for every valuation v in $\mathcal{M}$, if $(v(x), v(z)) \in C$ and $(v(z), v(y)) \notin C$, then $(v(x), v(y)) \notin P$. Thus, if $(v(x), v(y)) \in (C ; -C)$, then $(v(x), v(y)) \notin P$, and hence $P \subseteq -(C ; -C)$.

$(\supseteq)$ By correctness of the rule (P), for every finite set X of $\mathsf{RL_{Mer}}$-formulas and for every variable z that does not occur in X and such that $z \neq x, y$, $X \cup \{xPy\}$ is a $\mathcal{K}$-set iff $X \cup \{x{-}Cz, zCy\}$ is a $\mathcal{K}$-set. Let $X \stackrel{\mathrm{df}}{=} \{x(C ; -C)y\}$. Then, by the assumption on variable z, $X \cup \{x{-}Cz, zCy\}$ is a $\mathcal{K}$-set. Thus, by $\mathcal{K}$-correctness of the rule $(-P)$, $\{xPy, x(C ; -C)y\}$ is a $\mathcal{K}$-set. Hence, $-(C ; -C) \subseteq P$. □

The above theorem implies:

Proposition 18.2.4.

1. *The* $\mathsf{RL_{Mer}}$*-rules are* $\mathsf{RL_{Mer}}$*-correct;*
2. *The* $\mathsf{RL_{Mer}}$*-axiomatic sets are* $\mathsf{RL_{Mer}}$*-sets.*

In order to prove completeness, we need to define the completion conditions corresponding to the rules specific for $\mathsf{RL_{Mer}}$-dual tableau, i.e., the rules (P), $(-P)$, (O), $(-O)$, (PP), $(-PP)$, (EC), $(-EC)$, (DR), and $(-DR)$.

For all object symbols x and y,

Cpl(P) If $xPy \in b$, then for some object variable z, both $x{-}Cz \in b$ and $zCy \in b$, obtained by an application of the rule (P);

Cpl($-P$) If $x{-}Py \in b$, then for every object symbol z, either $xCz \in b$ or $z{-}Cy \in b$, obtained by an application of the rule ($-P$);

Cpl(PP) If $xPPy \in b$, then either $xPy \in b$ or $x{-}1'y \in b$, obtained by an application of the rule (PP);

Cpl($-PP$) If $x{-}PPy \in b$, then both $x{-}Py \in b$ and $x1'y \in b$, obtained by an application of the rule ($-PP$);

Cpl(O) If $xOy \in b$, then for every object symbol z, either $zPx \in b$ or $zPy \in b$, obtained by an application of the rule (O);

Cpl($-O$) If $x{-}Oy \in b$, then for some object variable z, both $z{-}Px \in b$ and $z{-}Py \in b$, obtained by an application of the rule ($-O$);

Cpl(PO) If $xPOy \in b$, then either $xOy \in b$ or $x{-}Py \in b$ or $y{-}Px \in b$, obtained by an application of the rule (PO);

Cpl($-PO$) If $x{-}POy \in b$, then $x{-}Oy \in b$ and $xPy \in b$ and $yPx \in b$, obtained by an application of the rule ($-PO$);

Cpl(EC) If $xECy \in b$, then either $xCy \in b$ or $x{-}Oy \in b$, obtained by an application of the rule (EC);

Cpl($-EC$) If $x{-}ECy \in b$, then both $x{-}Cy \in b$ and $xOy \in b$, obtained by an application of the rule ($-EC$);

Cpl(TPP) If $xTPPy \in b$, then either $xPPy \in b$ or for every object symbol z, $xECz \in b$ or $zECy \in b$, obtained by an application of the rule (TPP);

Cpl($-TPP$) If $x{-}TPPy \in b$, then $xPPy \in b$ and for some object variable z, both $x{-}ECz \in b$ and $z{-}ECy \in b$, obtained by an application of the rule ($-TPP$);

Cpl($NTPP$) If $xNTPPy \in b$, then either $xPPy \in b$ or $x{-}TPPy \in b$, obtained by an application of the rule ($NTPP$);

Cpl($-NTPP$) If $x{-}NTPPy \in b$, then both $x{-}PPy \in b$ and $xTPPy \in b$, obtained by an application of the rule ($-NTPP$);

Cpl(DC) If $xDCy \in b$, then $x{-}Cy \in b$, obtained by an application of the rule (DC);

Cpl($-DC$) If $x{-}DCy \in b$, then $xCy \in b$, obtained by an application of the rule ($-DC$);

Cpl(DR) If $xDRy \in b$, then $x{-}Oy \in b$, obtained by an application of the rule (DR);

Cpl($-DR$) If $x{-}DRy \in b$, then $xOy \in b$, obtained by an application of the rule ($-DR$).

The notions of a complete $\mathsf{RL}_{\mathsf{Mer}}$-proof tree and an open branch of an $\mathsf{RL}_{\mathsf{Mer}}$-proof tree are defined as in RL-logic (see Sect. 2.5).

The following form of the closed branch property holds:

Fact 18.2.1 (Closed Branch Property). *For every branch of an* $\mathsf{RL}_{\mathsf{Mer}}$*-proof tree, if* xRy *and* $x{-}Ry$*, for an atomic* $R \in \mathsf{Mer} \setminus \{P, -O, -TPP\}$*, belong to the branch, then the branch is closed.*

Although it does not concern the relational constants P, $-O$, and $-TPP$, it is sufficient for proving satisfaction in branch model property, see Proposition 18.2.5. The reason is that the rules for these constants reflect their corresponding definitions in logic $\mathsf{RL}_{\mathsf{Mer}}$.

Let b be an open branch of an $\mathsf{RL_{Mer}}$-proof tree. The branch structure $\mathcal{M}^b = (U^b, C^b, \{\#^b : \# \in \mathsf{Mer}\}, m^b)$ is defined as in the completeness proof of $\mathsf{RL_{CRA}}$-dual tableau with the following additional clauses:

- For every $\# \in \mathsf{Mer}$, $\#^b$ is defined according to Table 18.1; for example, $O^b \stackrel{\text{df}}{=} (P^{-1})^b ; P^b$.
- $m^b(\#) = \#^b$.

By the above definition, $\mathcal{M}^b$ is an $\mathsf{RL_{Mer}}$-model, hence the branch model property holds. Now, we prove the satisfaction in branch model property.

Proposition 18.2.5 (Satisfaction in Branch Model Property). *Let b be an open branch of an $\mathsf{RL_{Mer}}$-proof tree. Then, for every $\mathsf{RL_{Mer}}$-formula φ, if $\mathcal{M}^b, v^b \models \varphi$, then $\varphi \notin b$.*

Proof. We need to show that the proposition holds for every relational constant # and its complement, for $\# \in \mathsf{Mer}$. By way of example, we prove the required condition for the relational constant O and for $-EC$.

$(-O)$ Assume that $\mathcal{M}^b$ and v^b satisfy a formula $x{-}Oy$, that is $(x, y) \notin O^b$. By the definition of O^b, for every $z \in U^b$, either $(z, x) \notin P^b$ or $(z, y) \notin P^b$. Suppose $x{-}Oy \in b$. By the completion condition Cpl$(-O)$, there exists $z \in U^b$ such that $z{-}Px \in b$ and $z{-}Py \in b$. By the completion condition Cpl$(-P)$, there exists $z \in U^b$ such that for every $u \in U^b$, either $zCu \in b$ or $u{-}Cx \in b$, and for every $t \in U^b$, either $zCt \in b$ or $t{-}Cy \in b$. Thus, by the induction hypothesis, there exists $z \in U^b$ such that for every $u \in U^b$, either $(z, u) \notin C^b$ or $(u, x) \in C^b$, and for every $t \in U^b$, either $(z, t) \notin C^b$ or $(t, y) \in C^b$. Therefore, there exists $z \in U^b$ such that $(z, x) \notin (C^b ; -C^b)$ and $(z, y) \notin (C^b ; -C^b)$. Hence, $(z, x) \in P^b$ and $(z, y) \in P^b$, a contradiction.

$(-EC)$ Assume $\mathcal{M}^b, v^b \models x{-}ECy$, that is $(x, y) \notin EC^b$. Then, either $(x, y) \notin C^b$ or $(x, y) \in O^b$. Suppose $x{-}ECy \in b$. By the completion condition Cpl$(-EC)$, $x{-}Cy \in b$ and $xOy \in b$. By the induction hypothesis, $(x, y) \in C^b$ and $(x, y) \notin O^b$, a contradiction.

The proofs of the remaining cases are similar. □

Finally, we obtain:

Theorem 18.2.3 (Soundness and Completeness of $\mathsf{RL_{Mer}}$). *For every $\mathsf{RL_{Mer}}$-formula φ, the following conditions are equivalent:*

1. *φ is $\mathsf{RL_{Mer}}$-valid;*
2. *φ is true in all standard $\mathsf{RL_{Mer}}$-models;*
3. *φ is $\mathsf{RL_{Mer}}$-provable.*

Example. In Figs. 18.1–18.3 we present examples of $\mathsf{RL_{Mer}}$-proofs of some properties of mereological relations. We show that $P \cap DC = \emptyset$ by showing that $-P \cup -DC$ is the universal relation (Fig. 18.1). Then we show that $(PP ; PP) \subseteq PP$ by constructing an $\mathsf{RL_{Mer}}$-proof of the formula $x(-(PP ; PP) \cup PP)y$ (Fig. 18.2). Finally, we prove that P is antisymmetric, i.e., $P \cap P^{-1} \subseteq 1'$, by constructing an $\mathsf{RL_{Mer}}$-proof of the formula $x(-(P \cap P^{-1}) \cup 1')y$ (Fig. 18.3).

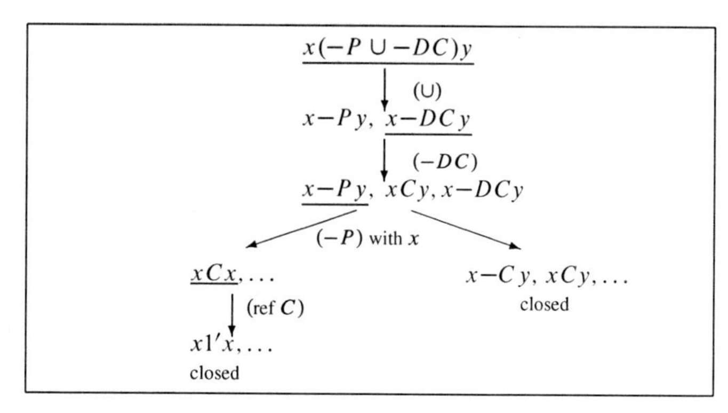

Fig. 18.1 An $\mathsf{RL}_{\mathsf{Mer}}$-proof of $P \cap DC = \emptyset$

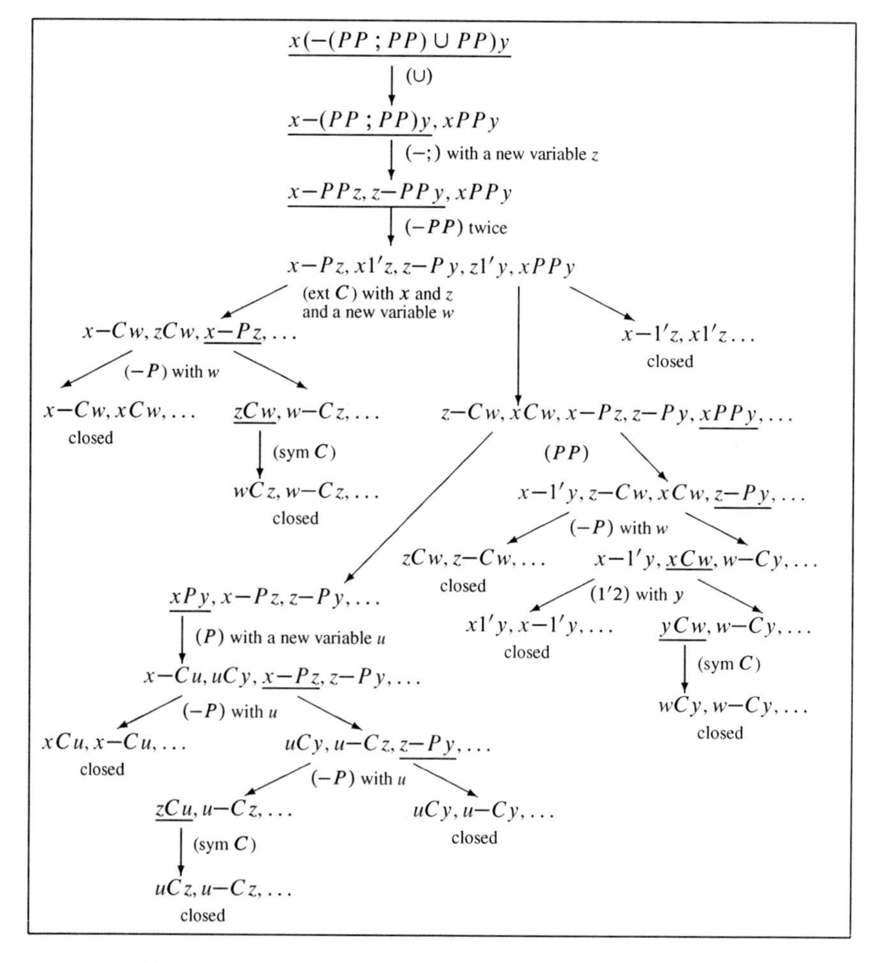

Fig. 18.2 An $\mathsf{RL}_{\mathsf{Mer}}$-proof of $PP\,;PP \subseteq PP$

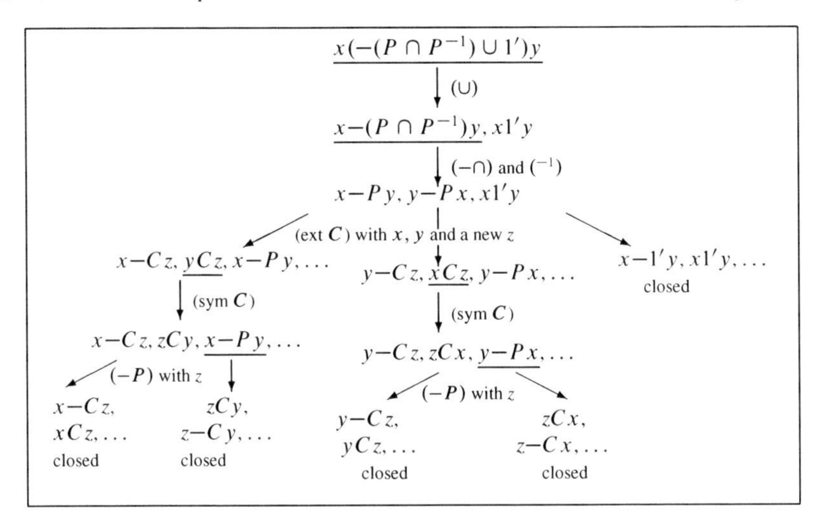

Fig. 18.3 An $\mathsf{RL}_{\mathsf{Mer}}$-proof of antisymmetry of the 'part of' relation P

18.3 Dual Tableaux for Spatial Theories Based on a Contact Relation on a Boolean Algebra

In this section we present a first-order theory of Boolean algebras with a contact relation. Such structures are considered, for example, in [Ste00, DW08].

The language of the first-order theory of Boolean algebras, $\mathsf{F}_{\mathsf{BA}fun}$, is a first-order language with the identity and function symbols (cf. Sect. 1.2) representing the Boolean operations from $fun = \{-, \vee, \wedge\}$ and the Boolean constants 0 and 1. The language of the first-order theory of Boolean algebras with a contact relation, $\mathsf{F}_{\mathsf{BAC}fun}$, is an $\mathsf{F}_{\mathsf{BA}fun}$-language endowed with the relational constant C. $\mathsf{F}_{\mathsf{BAC}fun}$-models are Boolean algebras with a contact relation C satisfying some of the following conditions.

Let $(U, -, \vee, \wedge, 0, 1)$ be a Boolean algebra, let C be a binary relation on U, let $a, b, c \in U$, and let $\leq$ be the Boolean ordering on U.

(C0) If aCb, then $a, b \neq 0$;
(C1) If $a \neq 0$, then aCa.

Sometimes, a stronger axiom is postulated instead of (C1):

(C2) If $a \wedge b \neq 0$, then aCb.

Further axioms include:

(C3) C is symmetric;
(C4) C is reflexive;
(C5) If aCb and $b \leq c$, then aCc;
(C6) If $aC(b \vee c)$, then aCb or aCc;
(C7) If $C(a) \subseteq C(b)$, then $a \leq b$.

Observe that (C7) follows from (C3) and (C5). The other typical axioms are:

(C8) If not aCb, then there is c such that not aCc and not $(-c)Cb$;
(C9) If $a, b \neq 0$ and $a \vee b = 1$, then aCb;
(C10) If $a \neq 1$, then there exists $b \neq 0$ such that not bCa.

In the presence of the axioms (C0), ..., (C6) axiom (C9) is equivalent to:

(C11) If $a \neq 0, 1$, then $aC(-a)$.

The primitive notions of $\mathsf{F}_{\mathsf{BA}fun}$ and $\mathsf{F}_{\mathsf{BAC}fun}$ can be equivalently presented in purely relational terms by elimination of function symbols. The binary Boolean operations $\vee$ and $\wedge$ are represented as ternary relations $R_\vee$ and $R_\wedge$, respectively, and the Boolean complement – as a binary relation R_-. Relations R_-, $R_\vee$, $R_\wedge$ are intended to reflect the following intuitions:

$R_-(x, y)$ iff $y = -x$;
$R_\vee(x, y, z)$ iff $z = x \vee y$;
$R_\wedge(x, y, z)$ iff $z = x \wedge y$.

Then the formalism appropriate for reasoning in the spatial theories involving these relations is a first-order theory of binary relations C and R_-, and ternary relations $R_\vee$ and $R_\wedge$. It will be denoted by $\mathsf{F}_{\mathsf{BAC}}$. Similarly, F_{BA} denotes a first-order language, without function symbols, of the theory of Boolean algebras.

An F_{BA}-model is a structure $\mathcal{M} = (U, R_-, R_\vee, R_\wedge, 0, 1, m)$ such that (U, m) is an F-model defined in Sect. 1.2, R_- is a binary relation on U interpreting the relational constant R_-, $R_\vee$ and $R_\wedge$ are ternary relations on U interpreting the relational constants $R_\vee$ and $R_\wedge$, respectively, and $0, 1$ are distinguished elements of U interpreting object constants from the language. An F_{BA}-model is standard whenever the interpretation of $=$ is the identity. As usual, we slightly abuse the notation using the same symbols for the relational constants in the language and for the corresponding relations in the models.

A valuation in a model $\mathcal{M} = (U, R_-, R_\vee, R_\wedge, 0, 1, m)$ is a function $v: \mathbb{OV}_{\mathsf{F}_{\mathsf{BA}}} \cup \{0, 1\} \rightarrow U \cup \{0, 1\}$ such that $v(1) = 1$ and $v(0) = 0$. The satisfaction relation, truth in a model, and validity are defined as in F-logic in Sect. 1.2.

F_{BA}-models are assumed to satisfy the following semantic conditions reflecting the role of the axioms of Boolean algebras:

For $\# \in \{\vee, \wedge\}$ and for all x, y, z,

(ReBA0) There is z such that $R_-(x, z)$;
(ReBA1) $R_-(z, x)$ and $R_-(z, y)$ imply $x = y$;
(ReBA2#) There is z such that $R_\#(x, y, z)$;
(ReBA3#) $R_\#(z, t, x)$ and $R_\#(z, t, y)$ imply $x = y$.

The above conditions encode the fact that R_- and $R_\#$, for $\# \in \{\vee, \wedge\}$, are functions.

(ReBA4#) $R_\#(x, y, z)$ iff $R_\#(y, x, z)$;
(ReBA5#) There are t, u such that $R_\#(x, y, t)$ and $R_\#(t, z, u)$ iff there are t', u' such that $R_\#(y, z, t')$ and $R_\#(x, t', u')$;

(ReBA6) There are t, u such that $R_\vee(x, y, t)$ and $R_\wedge(t, z, u)$ iff there are t', u' such that $R_\wedge(x, z, t')$ and $R_\wedge(y, z, u')$ and $R_\vee(t', u', u)$;

(ReBA7) There are t, u such that $R_\wedge(x, y, t)$ and $R_\vee(t, z, u)$ iff there are t', u' such that $R_\vee(x, z, t')$ and $R_\vee(y, z, u')$ and $R_\wedge(t', u', u)$.

These conditions encode the properties of join and meet, namely commutativity (ReBA4), associativity (ReBA5), and distributivity ((ReBA6) and (ReBA7)).

Boolean constants 0 and 1 are characterized with the conditions:

(ReBA8) $R_\vee(x, 0, x)$;
(ReBA9) $R_\wedge(x, 1, x)$;
(ReBA10) There is y such that $R_-(x, y)$, $R_\vee(x, y, 1)$, and $R_\wedge(x, y, 0)$.

$\mathsf{F}_{\mathsf{BAC}}$-models are the structures $\mathcal{M} = (U, R_-, R_\vee, R_\wedge, C, 0, 1, m)$ such that $(U, R_-, R_\vee, R_\wedge, 0, 1, m)$ is an F_{BA}-model and C is a binary relation on U. The relation C may satisfy some of the conditions (ReC0), ..., (ReC11) which correspond to the conditions (C0), ..., (C11), respectively.

Let $x, y, z, t \in U$. Then:

(ReC0) If $C(x, y)$, then $x \neq 0$ and $y \neq 0$;
(ReC1) If $x \neq 0$, then $C(x, x)$;
(ReC2) If $R_\wedge(x, y, 0)$, then $C(x, y)$;
(ReC3) If $C(x, y)$, then $C(y, x)$;
(ReC4) If $x = y$, then $C(x, y)$;
(ReC5) If $C(x, z)$ and $R_\vee(z, y, y)$, then $C(x, y)$;
(ReC6) If $C(x, t)$ and $R_\vee(y, z, t)$, then $C(x, y)$ or $C(x, z)$;
(ReC7) If for all $z \in U$, $C(x, z)$ implies $C(y, z)$, then $R_\vee(x, y, y)$;
(ReC8) If $C(x, y)$, then there are z, t such that not $C(x, z)$ and $R_-(z, t)$ and not $C(t, y)$;
(ReC9) If $x \neq 0$, $y \neq 0$, and $R_\vee(x, y, 1)$, then $C(x, y)$;
(ReC10) If $x \neq 1$, then there exists z such that $z \neq 0$ and $\neg C(z, x)$;
(ReC11) If $x \neq 0$ and $x \neq 1$, then there exists z such that $C(x, z)$ and $R_-(x, z)$.

Observe that $R_\vee(x, y, y)$ iff $x \leq y$.

Theorem 18.3.1. *For every* $\mathsf{F}_{\mathsf{BAC}fun}$*-formula* φ*, there is an* $\mathsf{F}_{\mathsf{BAC}}$*-formula* φ' *such that* φ *is true in all* $\mathsf{F}_{\mathsf{BAC}fun}$*-models satisfying some of the axioms from* {*(C0), ..., (C11)*} *iff* φ' *is true in all* $\mathsf{F}_{\mathsf{BAC}}$*-models satisfying the corresponding conditions from* {*(ReC0), ..., (ReC11)*}.

In order to prove the above theorem, it suffices to show: (1) for every $\mathsf{F}_{\mathsf{BAC}fun}$-model $\mathcal{M}$ there exists an $\mathsf{F}_{\mathsf{BAC}}$-model $\mathcal{M}'$ such that $(*)$ φ is true in $\mathcal{M}$ iff φ' is true in $\mathcal{M}'$, and (2) for every standard $\mathsf{F}_{\mathsf{BAC}}$-model $\mathcal{M}'$ there exists an $\mathsf{F}_{\mathsf{BAC}fun}$-model $\mathcal{M}$ such that $(*)$ holds. We construct the models $\mathcal{M}$ and $\mathcal{M}'$ so that they satisfy: (Ci) is true in $\mathcal{M}$ iff (ReCi) is true in $\mathcal{M}'$, for $i \in \{0, \ldots, 11\}$, and the same holds for the axioms of Boolean algebras and the conditions (ReBA0), (ReBA1), (ReBA2#), (ReBA3#), (ReBA4#), (ReBA5#), for $\# \in \{\vee, \wedge\}$, and (ReBA7), ..., (ReBA10).

$\mathsf{F}_{\mathsf{BAC}}$-dual tableau contains the axiomatic sets and the rules of F-dual tableau adjusted to $\mathsf{F}_{\mathsf{BAC}}$-language (see Sect. 1.3). In the rules ($\exists$) and ($\neg\forall$), z is any object variable or constant, and in the rules ($\forall$) and ($\neg\exists$), z is a new variable. Furthermore, we admit specific rules that reflect properties of relational constants that are specific for $\mathsf{F}_{\mathsf{BAC}}$.

The rules and the axiomatic sets reflecting the properties of relations presented above have the following forms:

For $\# \in \{\vee, \wedge\}$ and for all object symbols x and y,

$$(\text{rReBA0}) \quad \frac{}{\neg R_{-}(x,z)} \quad z \text{ is a new object variable and } z \neq x$$

$$(\text{rReBA1}) \quad \frac{x = y}{R_{-}(z,x), x = y \mid R_{-}(z,y), x = y}$$

z is any object symbol

$$(\text{rReBA2\#}) \quad \frac{}{\neg R_{\#}(x,y,z)} \quad z \text{ is a new object variable and } z \neq x, y$$

$$(\text{rReBA3\#}) \quad \frac{x = y}{R_{\#}(z,t,x), x = y \mid R_{\#}(z,t,y), x = y}$$

z, t are any object symbols

$$(\text{rReBA4\#}) \quad \frac{R_{\#}(x,y,z)}{R_{\#}(y,x,z)} \quad z \text{ is any object symbol}$$

The rule corresponding to the implication from left to right of the condition (ReBA5#) has the following form:

$$(\text{rReBA5\#}(\rightarrow)) \quad \frac{}{R_{\#}(x,y,t) \mid R_{\#}(t,z,u) \mid \neg R_{\#}(y,z,t'), \neg R_{\#}(x,t',u')}$$

x, y, z, t, u are any object symbols, t', u' are new object variables such that $t' \neq u'$ and $\{t', u'\} \cap \{x, y, z, t, u\} = \emptyset$

The implication from right to left is reflected by the rule of the form:

$$(\text{rReBA5\#}(\leftarrow)) \quad \frac{}{R_{\#}(y,z,t') \mid R_{\#}(x,t',u') \mid \neg R_{\#}(x,y,t), \neg R_{\#}(t,z,u)}$$

x, y, z, t', u' are any object symbols, t, u are new object variables such that $t \neq u$ and $\{t, u\} \cap \{x, y, z, t', u'\} = \emptyset$

Similarly, the rules corresponding to the conditions (ReBA6) and (ReBA7) have the following forms:

(rReBA6(→)) $\dfrac{}{R_\vee(x,y,t) \mid R_\wedge(t,z,u) \mid \neg R_\wedge(x,z,t'), \neg R_\wedge(y,z,u'), \neg R_\vee(t',u',u)}$

x, y, z, t, u are any object symbols, t', u' are new object variables such that $t' \neq u'$ and $\{t', u'\} \cap \{x, y, z, t, u\} = \emptyset$

(rReBA6(←)) $\dfrac{}{R_\wedge(x,z,t') \mid R_\wedge(y,z,u') \mid R_\vee(t',u',u) \mid \neg R_\vee(x,y,t), \neg R_\wedge(t,z,u)}$

x, y, z, t', u' are any object symbols, t, u are new object variables such that $t \neq u$ and $\{t, u\} \cap \{x, y, z, t', u'\} = \emptyset$

(rReBA7(→)) $\dfrac{}{R_\wedge(x,y,t) \mid R_\vee(t,z,u) \mid \neg R_\vee(x,z,t'), \neg R_\vee(y,z,u'), \neg R_\wedge(t',u',u)}$

x, y, z, t, u are any object symbols, t', u' are new object variables such that $t' \neq u'$ and $\{t', u'\} \cap \{x, y, z, t, u\} = \emptyset$

(rReBA7(←)) $\dfrac{}{R_\vee(x,z,t') \mid R_\vee(y,z,u') \mid R_\wedge(t',u',u) \mid \neg R_\wedge(x,y,t), \neg R_\vee(t,z,u)}$

x, y, z, t', u' are any object symbols, t, u are new object variables such that $t \neq u$ and $\{t, u\} \cap \{x, y, z, t', u'\} = \emptyset$.

The conditions (ReBA8) and (ReBA9) lead to the following axiomatic sets:

(AxReBA8) $\{R_\vee(x,0,x)\}$, (AxReBA9) $\{R_\wedge(x,1,x)\}$.

The condition (ReBA10) is reflected by the rule of the form:

(rReBA10) $\dfrac{}{\neg R_-(x,y), \neg R_\vee(x,y,1), \neg R_\wedge(x,y,0)}$

x is any object symbol, y is a new object variable and $y \neq x$

The rules corresponding to the axioms characterizing the conditions (ReCi) have the following forms:

For all object symbols x and y,

(rReC0) $\dfrac{}{C(x,y) \mid x = 0, y = 0}$

(rReC1) $\dfrac{C(x,x)}{x \neq 0, C(x,x)}$ (rReC2) $\dfrac{C(x,y)}{\neg R_\wedge(x,y,0), C(x,y)}$

(rReC3) $\dfrac{C(x,y)}{C(y,x)}$ (rReC4) $\dfrac{C(x,y)}{x = y, C(x,y)}$

$$\text{(rReC5)} \quad \frac{C(x,y)}{C(x,z), C(x,y) \mid R_{\vee}(z,y,y), C(x,y)} \quad z \text{ is any object symbol}$$

$$\text{(rReC6)} \quad \frac{C(x,y), C(x,z)}{C(x,t), C(x,y), C(x,z) \mid R_{\vee}(y,z,t), C(x,y), C(x,z)}$$

z, t are any object symbols

$$\text{(rReC7)} \quad \frac{R_{\vee}(x,y,y)}{\neg C(x,z), C(y,z), R_{\vee}(x,y,y)} \quad z \text{ is a new object variable}$$

$$\text{(rReC8)} \quad \frac{}{C(x,y) \mid C(x,z), \neg R_{-}(z,t), C(t,y)}$$

z, t are new object variables such that $z \neq t$ and $\{z,t\} \cap \{x,y\} = \emptyset$

$$\text{(rReC9)} \quad \frac{C(x,y)}{x \neq 0, C(x,y) \mid y \neq 0, C(x,y) \mid R_{\vee}(x,y,1), C(x,y)}$$

$$\text{(rReC10)} \quad \frac{}{x \neq 1 \mid z = 0, C(z,x)}$$

z is a new object variable such that $z \neq x$

$$\text{(rReC11)} \quad \frac{}{x \neq 0 \mid x \neq 1 \mid \neg C(x,z), \neg R_{-}(x,z)}$$

z is a new object variable such that $z \neq x$

$\mathsf{F}_{\mathsf{BAC}}$-dual tableau includes specialized cut rules. As mentioned in Sect. 16.3, in the presence of the ordinary cut rule they could be replaced by the rules with a non-empty premise (see Sect. 25.9).

Let $\mathcal{K}$ be a class of $\mathsf{F}_{\mathsf{BAC}}$-models. The notions of a $\mathcal{K}$-set and $\mathcal{K}$-correctness of a rule are defined in a similar way as in the logic F (see Sect. 1.3). Recall, that a finite set X of $\mathsf{F}_{\mathsf{BAC}}$-formulas is a $\mathcal{K}$-set whenever the disjunction of its formulas is $\mathcal{K}$-valid.

Theorem 18.3.2 (Correspondence). *Let $\mathcal{K}$ be a class of $\mathsf{F}_{\mathsf{BAC}}$-models. Then for every $i \in \{0, \ldots, 11\}$, the following conditions are equivalent:*

1. *(ReCi) is true in all models of $\mathcal{K}$;*
2. *The rule (rReCi) is $\mathcal{K}$-correct.*

Proof. By way of example, we prove the statement for the condition (ReC10).

Let $\mathcal{K}$ be a class of $\mathsf{F}_{\mathsf{BAC}}$-models satisfying the condition (ReC10). Let X be a finite set of $\mathsf{F}_{\mathsf{BAC}}$-formulas. Let z be a variable that does not occur in X and such that

$z \neq x$. Assume that $X \cup \{x \neq 1\}$ and $X \cup \{z = 0, C(z,x)\}$ are $\mathcal{K}$-sets. Then, by the assumption on variable z, for every $\mathcal{K}$-model $\mathcal{M}$ and for every valuation v in $\mathcal{M}$, either there exists $\varphi \in X$ such that $\mathcal{M}, v \models \varphi$ or both $v(x) \neq 1$ and for every $z \in U$, either $z = 0$ or $C(z, v(x))$. Suppose X is not a $\mathcal{K}$-set. Then, by the assumption, there exist a $\mathcal{K}$-model $\mathcal{M}$ and a valuation v in $\mathcal{M}$ such that $v(x) \neq 1$ and for every $z \in U$, either $z = 0$ or $C(z, v(x))$. Since $v(x) \neq 1$, by the condition (ReC8), there exists $z \in U$ such that $z \neq 0$ and $\neg C(z, v(x))$, a contradiction. Therefore, the rule (rReC10) is $\mathcal{K}$-correct.

Assume the rule (rReC10) is $\mathcal{K}$-correct. Let $X \stackrel{\text{df}}{=} \{x = 1, \exists t(t \neq 0 \wedge C(t,x))\}$ and let z be a variable such that $z \neq x$ and $z \neq t$. Then, $X \cup \{x \neq 1\}$ and $X \cup \{z = 0, C(z,x)\}$ are $\mathcal{K}$-sets. Thus, by the assumption, X is a $\mathcal{K}$-set, that is in every $\mathcal{K}$-model $\mathcal{M}$, either $x = 1$ or there exists $t \in U$ such that $t \neq 0$ and $C(t,x)$, which means that the condition (ReC10) holds.

The proofs of the remaining cases are similar. □

From now on, throughout this section, by an $\mathsf{F_{BAC}}$-model we mean a model $(U, R_-, R_\vee, R_\wedge, C, 0, 1, m)$ satisfying all the conditions (ReBA0), (ReBA1), (ReBA2#), ..., (ReBA5#), for # $\in \{\vee, \wedge\}$, (ReBA6), ..., (ReBA10), and all the conditions (ReC0), ..., (ReC11). Theorem 18.3.2 implies:

Proposition 18.3.1.

1. *The* $\mathsf{F_{BAC}}$*-rules are* $\mathsf{F_{BAC}}$*-correct;*
2. *The* $\mathsf{F_{BAC}}$*-axiomatic sets are* $\mathsf{F_{BAC}}$*-sets.*

The notions of an $\mathsf{F_{BAC}}$-proof tree, a closed branch of such a tree, a closed $\mathsf{F_{BAC}}$-proof tree, and $\mathsf{F_{BAC}}$-provability are defined as in the logic F (see Sect. 1.3).

A branch b of an $\mathsf{F_{BAC}}$-proof tree is complete whenever it is closed or it satisfies the completion conditions of the dual tableau for the logic F (see Sect. 1.3) and the completion conditions corresponding to the specific rules of the $\mathsf{F_{BAC}}$-dual tableau:

For every # $\in \{\vee, \wedge\}$,

Cpl(rReBA0) For every object symbol x, there exists an object variable z such that $\neg R_-(x,z) \in b$, obtained by an application of the rule (rReBA0);

Cpl(rReBA1) For all object symbols x and y, if $x = y \in b$, then for every object symbol z, either $R_-(z,x) \in b$ or $R_-(z,y) \in b$, obtained by an application of the rule (rReBA1);

Cpl(rReBA2#) For all object symbols x and y, there exists an object variable z such that $\neg R_\#(x,y,z) \in b$, obtained by an application of the rule (rReBA2#);

Cpl(rReBA3#) For all object symbols x and y, if $x = y \in b$, then for all object symbols z and t, either $R_\#(z,t,x) \in b$ or $R_\#(z,t,y) \in b$, obtained by an application of the rule (rReBA3#);

Cpl(rReBA4#) For all object symbols x, y, and z, if $R_\#(x,y,z) \in b$, then $R_\#(y,x,z) \in b$, obtained by an application of the rule (rReBA4#);

Cpl(rReBA5#(→)) For all object symbols x, y, z, t, u, either $R_\#(x,y,t) \in b$ or $R_\#(t,z,u) \in b$ or for some object variables t' and u' both $\neg R_\#(y,z,t') \in b$ and $\neg R_\#(x,t',u') \in b$, obtained by an application of the rule (rReBA5#(→));

Cpl(rReBA5#(←)) For all object symbols x, y, z, t', u', either $R_{\#}(y, z, t') \in b$ or $R_{\#}(x, t', u') \in b$ or for some object variables t and u both $\neg R_{\#}(x, y, t) \in b$ and $\neg R_{\#}(t, z, u) \in b$, obtained by an application of the rule (rReBA5#(←));

Cpl(rReBA6(→)) For all object symbols x, y, z, t, u, either $R_{\vee}(x, y, t) \in b$ or $R_{\wedge}(t, z, u) \in b$ or there are object variables t', u' such that $\neg R_{\wedge}(x, z, t') \in b$ and $\neg R_{\wedge}(y, z, u') \in b$ and $\neg R_{\vee}(t', u', u)$, obtained by an application of the rule (rReBA6(→));

Cpl(rReBA6(←)) For all object symbols x, y, z, t', u', either $R_{\wedge}(x, z, t') \in b$ or $R_{\wedge}(y, z, u') \in b$ or $R_{\vee}(t', u', u)$ or there are object variables t and u such that both $\neg R_{\vee}(x, y, t) \in b$ and $\neg R_{\wedge}(t, z, u) \in b$, obtained by an application of the rule (rReBA6(←));

Cpl(rReBA7(→)) For all object symbols x, y, z, t, u, either $R_{\wedge}(x, y, t) \in b$ or $R_{\vee}(t, z, u) \in b$ or there are object variables t', u' such that $\neg R_{\vee}(x, z, t') \in b$ and $\neg R_{\vee}(y, z, u') \in b$ and $\neg R_{\wedge}(t', u', u)$, obtained by an application of the rule (rReBA7(→));

Cpl(rReBA7(←)) For all object symbols x, y, z, t', u', either $R_{\vee}(x, z, t') \in b$ or $R_{\vee}(y, z, u') \in b$ or $R_{\wedge}(t', u', u)$ or there are object variables t and u such that both $\neg R_{\wedge}(x, y, t) \in b$ and $\neg R_{\vee}(t, z, u) \in b$, obtained by an application of the rule (rReBA7(←));

Cpl(rReBA10) For every object symbol x, there exists an object variable y such that $\neg R_{-}(x, y) \in b$ and $\neg R_{\vee}(x, y, 1) \in b$, and $\neg R_{\wedge}(x, y, 0)$, obtained by an application of the rule (rReBA10);

Cpl(rReC0) For all object symbols x and y, either $C(x, y) \in b$ or both $x = 0 \in b$ and $y = 0 \in b$, obtained by an application of the rule (rReC0);

Cpl(rReC1) For all object symbols x and y, if $C(x, x) \in b$, then $x \neq 0 \in b$, obtained by an application of the rule (rReC1);

Cpl(rReC2) For all object symbols x, y, if $C(x, y) \in b$, then $\neg R_{\wedge}(x, y, 0) \in b$, obtained by an application of the rule (rReC2);

Cpl(rReC3) For all object symbols x and y, if $C(x, y) \in b$, then $C(y, x) \in b$, obtained by an application of the rule (rReC3);

Cpl(rReC4) For all object symbols x and y, if $C(x, y) \in b$, then $x = y \in b$, obtained by an application of the rule (rReC4);

Cpl(rReC5) For all object symbols x and y, if $C(x, y) \in b$, then for every object symbol z, either $C(x, z) \in b$ or $R_{\vee}(z, y, y) \in b$, obtained by an application of the rule (rReC5);

Cpl(rReC6) For all object symbols x, y, and z, if $C(x, y) \in b$ and $C(x, z) \in b$, then for every object symbol t, either $C(x, t) \in b$ or $R_{\vee}(y, z, t) \in b$, obtained by an application of the rule (rReC6);

Cpl(rReC7) For all object symbols x and y, if $R_{\vee}(x, y, y) \in b$, then for some object variable z both $\neg C(x, z) \in b$ and $C(y, z) \in b$, obtained by an application of the rule (rReC7);

Cpl(rReC8) For all object symbols x and y either $C(x, y) \in b$ or there are object variables z and t such that $C(x, z) \in b$ and $\neg R_{-}(z, t) \in b$, and $C(t, y) \in b$, obtained by an application of the rule (rReC8);

Cpl(rReC9) For all object symbols x and y, if $C(x, y) \in b$, then either $x \neq 0 \in b$ or $y \neq 0$ or $R_{\vee}(x, y, 1) \in b$, obtained by an application of the rule (rReC9);

Cpl(rReC10) For every object symbol x, either $x \neq 1 \in b$ or for some object variable z, both $z = 0 \in b$ and $C(z, x) \in b$, obtained by an application of the rule (rReC10);

Cpl(rReC11) For every object symbol x, either $x \neq 0 \in b$ or $x \neq 1 \in b$ or for some object variable z, both $\neg C(x, z) \in b$ and $\neg R_{-}(x, z) \in b$, obtained by an application of the rule (rReC11).

The notions of a complete $\mathsf{F_{BAC}}$-proof tree and an open branch of an $\mathsf{F_{BAC}}$-proof tree are defined as in F-logic (see Sect. 1.3). Observe that all the rules considered in this section guarantee that whenever a branch of an $\mathsf{F_{BAC}}$-proof tree contains an atomic $\mathsf{F_{BAC}}$-formula φ and its negation, then the branch can be closed, which enables us to prove the closed branch property.

Let b be an open branch of an $\mathsf{F_{BAC}}$-proof tree. The branch structure $\mathcal{M}^b = (U^b, R_{-}^b, R_{\vee}^b, R_{\wedge}^b, C^b, 0^b, 1^b, m^b)$ is defined as follows:

- $U^b = \mathbb{OS}_{\mathsf{F_{BAC}}}$;
- $0^b = m^b(0) = 0$ and $1^b = m^b(1) = 1$;
- $R^b = m^b(R) = \{(x, y) \in (U^b)^2 : R(x, y) \notin b\}$, for $R \in \{R_{-}, C, =\}$;
- $Q^b = m^b(Q) = \{(x, y, z) \in (U^b)^3 : Q(x, y, z) \notin b\}$, for $Q \in \{R_{\vee}, R_{\wedge}\}$.

Proposition 18.3.2 (Branch Model Property). *Let b be an open branch of an $\mathsf{F_{BAC}}$-proof tree. Then the branch structure $\mathcal{M}^b = (U^b, R_{-}^b, R_{\vee}^b, R_{\wedge}^b, C^b, 0^b, 1^b, m^b)$ is an $\mathsf{F_{BAC}}$-model.*

Proof. Let b be an open branch of an $\mathsf{F_{BAC}}$-proof tree. It suffices to show that $\mathcal{M}^b$ satisfies the conditions (ReBA0), (ReBA1), (ReBA2#), ..., (ReBA5#), for $\# \in \{\vee, \wedge\}$, (ReBA6), ..., (ReBA10), (ReC0), ..., (ReC11). By way of example, we show that $\mathcal{M}^b$ satisfies the conditions (ReBA2#) and (ReC7).

(ReBA2#) Note that by the completion condition Cpl(rReBA2#), for all $x, y \in U^b$ there exists $z \in U^b$ such that $\neg R_{\#}(x, y, z) \in b$. Thus, $R_{\#}(x, y, z) \notin b$. Hence, for all $x, y \in U^b$ there exists $z \in U^b$ such that $R_{\#}^b(x, y, z)$.

(ReC7) Let $x, y \in U^b$. Assume that for all $z \in U^b$, $C^b(x, z)$ implies $C^b(y, z)$, that is for all object symbols z, either $C(x, z) \in b$ or $C(y, z) \notin b$. Suppose $R_{\vee}^b(x, y, y)$ is not satisfied in $\mathcal{M}^b$. Then, $R_{\vee}(x, y, y) \in b$, and by the completion condition Cpl(rReC7), there exists $z \in U^b$ such that both $\neg C(x, z) \in b$ and $C(y, z) \in b$, a contradiction.

The proofs of the remaining conditions are similar. In each case we use the corresponding completion condition. ◻

Since the interpretation of atomic formulas in the branch model is defined in the standard way, i.e., as in the completeness proof of F-dual tableau (see Sect. 1.3), the satisfaction in branch model property holds and the completeness of $\mathsf{F_{BAC}}$-dual tableau can be proved as in the logics F and $\mathsf{RL}(1, 1')$.

Theorem 18.3.3 (Soundness and Completeness of $\mathsf{F_{BAC}}$). *Let φ be an $\mathsf{F_{BAC}}$-formula. Then the following conditions are equivalent:*

1. *φ is $\mathsf{F_{BAC}}$-valid;*
2. *φ is true in all standard $\mathsf{F_{BAC}}$-models;*
3. *φ is $\mathsf{F_{BAC}}$-provable.*

18.4 Dual Tableau for Region Connection Calculus

The first-order theory of region connection calculus (see [RCC92]), $\mathsf{F_{RCC}}_{fun}$, is obtained from $\mathsf{F_{BA}}_{fun}$ by endowing its language with a relational constant C representing a contact relation, relational constants DC, EC, PO, TPP, $NTPP$, TPP^{-1}, and $NTPP^{-1}$ representing the mereological relations defined in Sect. 18.2, and the relational constant EQ representing equality of regions. The mereological relations provide a partition of the universe of regions, that is they are pairwise disjoint and their union is the universal relation. Such a theory is, in fact, based on the contact relation (see Table 18.1). A relation algebra generated by these relations is investigated in [LM94]. In [DSW01] it is shown that the basic RCC relations cannot be the atoms of any relation algebra.

The models of $\mathsf{F_{RCC}}_{fun}$ satisfy the conditions (RCC1), (RCC2), and (RCC3) that coincide with (C3), (C4), and (C6), respectively, and the following:

(RCC4) $aC1$;
(RCC5) $b \neq 1$ implies: $[aC(-b)$ iff $a-NTPPb]$ and $[aO(-b)$ iff $a-Pb]$;
(RCC6) $aC(b \wedge c)$ iff there exists d such that dPb, dPc, and aCd;
(RCC7) $a \wedge b \neq 0$ iff aOb;
(RCC8) $\mathrm{syq}(C, C) \subseteq 1'$,

where syq is the operation of symmetric quotient defined as:

$$\mathrm{syq}(R, P) \stackrel{\mathrm{df}}{=} (R \backslash P) \cap (R^{-1}/P^{-1}),$$

and operations $\backslash$ and $/$ of the right and left residual of the composition of relations, respectively, are defined by the following conditions, for all relations P and R:

$$P \backslash R \stackrel{\mathrm{df}}{=} -(-R\,;P^{-1}),$$

$$P/R \stackrel{\mathrm{df}}{=} -(R^{-1}\,;-P).$$

By $\mathsf{F_{RCC}}$ we mean the theory obtained from $\mathsf{F_{RCC}}_{fun}$ by elimination of function symbols. $\mathsf{F_{RCC}}$-structures are $\mathsf{F_{BA}}$-models with binary relations C, DC, EC, PO, TPP, and $NTPP$ that interpret the specific relational constants of the region connection calculus according to their definitions in Table 18.1.

$\mathsf{F_{RCC}}$-models are the $\mathsf{F_{RCC}}$-structures that satisfy the conditions which are function-free representations of the RCC-axioms. (ReRCC1), (ReRCC2), and (ReRCC3) coincide with (ReC3), (ReC4), and (ReC6), respectively.

(ReRCC4) $C(x, 1)$;
(ReRCC5) $y \neq 1$ implies:

(1) There exists z such that $C(x, z)$ and $R_{-}(y, z)$ iff $\neg NTTP(x, y)$,
(2) There exists z such that $O(x, z)$ and $R_{-}(y, z)$ iff $\neg P(x, y)$;

(ReRCC6) There is t such that $C(x, t)$ and $R_{\wedge}(y, z, t)$ iff there is w such that $P(w, y)$, $P(w, z)$, and $C(x, w)$;
(ReRCC7) There is z such that $R_{\wedge}(x, y, z)$ and $z \neq 0$ iff $O(x, y)$;
(ReRCC8) If for all z $C(x, z)$ or not $C(z, y)$ and for all w not $C(x, w)$ or $C(w, y)$, then $x = y$.

The following theorem states the relationship between the theory of region connection calculus and the $\mathsf{F}_{\mathsf{RCC}}$-models.

Theorem 18.4.1. *For every $\mathsf{F}_{\mathsf{RCC}fun}$-formula φ there exists an $\mathsf{F}_{\mathsf{RCC}}$-formula φ' such that φ is true in all algebras of* RCC *iff φ' is true in all $\mathsf{F}_{\mathsf{RCC}}$-models.*

$\mathsf{F}_{\mathsf{RCC}}$-dual tableau is obtained from the F-dual tableau by endowing it with the specific rules and the axiomatic sets reflecting the conditions (ReBA0), (ReBA1), (ReBA2#), ..., (ReBA5#), for $\# \in \{\vee, \wedge\}$, (ReBA6), ..., (ReBA10), and the rules (rReRCC1), (rReRCC2), (rReRCC3) which coincide with (rReC3), (rReC4), and (rReC6), respectively. All these rules and axiomatic sets are presented in the previous section. $\mathsf{F}_{\mathsf{RCC}}$-dual tableau contains also the rules (#) and (−#), for every $\# \in \mathsf{Mer}$, presented in Sect. 18.2 with a minor syntactic transformation which adapts the rules to their first-order presentation. Namely, instead of xRy and $x{-}Ry$ we write $R(x, y)$ and $\neg R(x, y)$, respectively, and instead of $1'$ we write $=$. Moreover, in the rules which introduce arbitrary variable, now we allow any object symbol, i.e., an object variable or 0 or 1. We also include the axiomatic sets and the specific rules that reflect conditions (ReRCC5), ..., (ReRCC8).

Condition (ReRCC4) leads to the following axiomatic set:

(AxReRCC4) $\{C(x, 1)\}$, for any object symbol x.

The two implications of the first part of (ReRCC5) have the following rules:

(r1ReRCC5(→)) $\dfrac{}{y \neq 1 \mid C(x, z) \mid R_{-}(y, z) \mid NTTP(x, y)}$

x, y, z are any object symbols

(r1ReRCC5(←)) $\dfrac{}{y \neq 1 \mid \neg NTTP(x, y) \mid \neg C(x, z), \neg R_{-}(y, z)}$

x, y are any object symbols
z is a new object variable, $z \neq x, y$

The rules for the second part of (ReRCC5) are:

$$(\text{r2ReRCC5}(\rightarrow)) \quad \frac{}{y \neq 1 \mid O(x,z) \mid R_{-}(y,z) \mid P(x,y)}$$

x, y, z are any object symbols

$$(\text{r2ReRCC5}(\leftarrow)) \quad \frac{}{y \neq 1 \mid \neg P(x,y) \mid \neg O(x,z), \neg R_{-}(y,z)}$$

x, y are any object symbols
z is a new object variable, $z \neq x, y$

The remaining rules are:

$$(\text{rReRCC6}(\rightarrow)) \quad \frac{}{C(x,t) \mid R_{\wedge}(y,z,t) \mid \neg P(w,y), \neg P(w,z), \neg C(x,w)}$$

x, y, z, t are any object symbols
w is a new object variable, $w \neq x, y, z, t$

$$(\text{rReRCC6}(\leftarrow)) \quad \frac{}{P(w,y) \mid P(w,z) \mid C(x,w) \mid \neg C(x,t), \neg R_{\wedge}(y,z,t)}$$

x, y, z, w are any object symbols
t is a new object variable, $t \neq x, y, z, w$

$$(\text{rReRCC7}(\rightarrow)) \quad \frac{}{R_{\wedge}(x,y,z) \mid z \neq 0 \mid \neg O(x,y)}$$

x, y, z are any object symbols

$$(\text{rReRCC7}(\leftarrow)) \quad \frac{}{O(x,y) \mid \neg R_{\wedge}(x,y,z), z = 0}$$

x, y are any object symbols
z is a new object variable, $z \neq x, y$

$$(\text{rReRCC8}) \quad \frac{x = y}{C(x,z), \neg C(z,y), x = y \mid \neg C(x,w), C(w,y), x = y}$$

x, y are any object symbols
z, w are new object variables such that $\{z, w\} \cap \{x, y\} = \emptyset$

The rules above are specialized cut rules. A discussion of the alternative deterministic forms of specialized cut rules can be found in Sect. 25.9.

Theorem 18.4.2 (Correspondence). *Let $\mathcal{K}$ be a class of $\mathsf{F}_{\mathsf{RCC}}$-structures. Then, the following hold:*

1. *The condition (ReRCC4) is valid in $\mathcal{K}$ iff the axiomatic sets (AxReRCC4) are $\mathcal{K}$-sets;*
2. *The condition (ReRCC5) is valid in $\mathcal{K}$ iff for every $i \in \{1,2\}$, the rules (riReRCC5($\rightarrow$)) and (riReRCC5($\leftarrow$)) are $\mathcal{K}$-correct;*

3. *For every $\mathcal{C} \in \{$(ReRCC6), (ReRCC7)$\}$, the condition $\mathcal{C}$ is valid in $\mathcal{K}$ iff the rules $(r\mathcal{C}(\rightarrow))$ and $(r\mathcal{C}(\leftarrow))$ are $\mathcal{K}$-correct;*
4. *The condition (ReRCC8) is valid in $\mathcal{K}$ iff the rule (rReRCC8) is $\mathcal{K}$-correct.*

Proof. By way of example, we prove the proposition for the condition (ReRCC8).

Assume (ReRCC8) is true in all $\mathcal{K}$-structures. We show that the rule (rReRCC8) is $\mathcal{K}$-correct. Let X be a finite set of $\mathsf{F}_{\mathsf{RCC}}$-formulas, let x, y be object symbols, and let z and w be variables that do not occur in X and such that $z \neq w$ and $\{z, w\} \cap \{x, y\} = \emptyset$. Assume that $X \cup \{C(x,z), \neg C(z,y), x = y\}$ and $X \cup \{\neg C(x,w), C(w,y), x = y\}$ are $\mathcal{K}$-sets. Suppose $X \cup \{x = y\}$ is not a $\mathcal{K}$-set. Then, there exist a $\mathcal{K}$-structure $\mathcal{M}$ and a valuation v in $\mathcal{M}$ such that $v(x) \neq v(y)$ and, due to the assumption on variables z and w, for all $z \in U$, either $C(v(x), z)$ or not $C(z, v(y))$ and for all $w \in U$ either not $C(v(x), w)$ or $C(w, v(y))$. Thus, by the condition (ReRCC8), $v(x) = v(y)$, a contradiction. Therefore, the rule (rReRCC8) is $\mathcal{K}$-correct.

Now, assume that the rule (rReRCC8) is $\mathcal{K}$-correct. Let $X \stackrel{\text{df}}{=} \{\exists z'(\neg C(x,z') \wedge C(z'y)), \exists w'(C(x,w') \wedge \neg C(w',y))\}$. Let z, w be object variables such that $z \neq w$ and $\{z,w\} \cap \{x,y,z',w'\} = \emptyset$. Then, $X \cup \{C(x,z), \neg C(z,y), x = y\}$ and $X \cup \{\neg C(x,w), C(w,y), x = y\}$ are $\mathcal{K}$-sets. Thus, by the assumption, $X \cup \{x = y\}$ is also $\mathcal{K}$-set, which means that for every $\mathcal{K}$-structure $\mathcal{M}$, if $x \neq y$ is true in $\mathcal{M}$, then either there exists $z' \in U$ such that not $C(x,z')$ and $C(z',y)$ or there exists $w' \in U$ such that $C(x,w')$ and not $C(w',y)$. Hence, the condition (ReRCC8) holds. □

The above theorem implies:

Proposition 18.4.1.

1. *The $\mathsf{F}_{\mathsf{RCC}}$-rules are $\mathsf{F}_{\mathsf{RCC}}$-correct;*
2. *The $\mathsf{F}_{\mathsf{RCC}}$-axiomatic sets are $\mathsf{F}_{\mathsf{RCC}}$-sets.*

The completion conditions determined by the rules specific for $\mathsf{F}_{\mathsf{RCC}}$ are:

Cpl(r1ReRCC5($\rightarrow$)) For all object symbols x, y, and z, either $y \neq 1 \in b$ or $NTTP(x,y) \in b$ or $C(x,z) \in b$ or $R_{-}(y,z) \in b$, obtained by an application of the rule (r1ReRCC5($\rightarrow$));

Cpl(r1ReRCC5($\leftarrow$)) For all object symbols x and y, either $y \neq 1 \in b$ or $\neg NTTP(x,y) \in b$ or for some object variable z both $\neg C(x,z) \in b$ and $\neg R_{-}(y,z) \in b$, obtained by an application of the rule (r1ReRCC5($\leftarrow$));

Cpl(r2ReRCC5($\rightarrow$)) For all object symbols x, y, and z, either $y \neq 1 \in b$ or $P(x,y) \in b$ or $O(x,z) \in b$ or $R_{-}(y,z) \in b$, obtained by an application of the rule (r2ReRCC5($\rightarrow$));

Cpl(r2ReRCC5($\leftarrow$)) For all object symbols x and y, either $y \neq 1 \in b$ or $\neg P(x,y) \in b$ or there exists an object variable z such that both $\neg O(x,z) \in b$ and $\neg R_{-}(y,z) \in b$, obtained by an application of the rule (r2ReRCC5($\leftarrow$));

Cpl(rReRCC6($\rightarrow$)) For all object symbols x, y, z, and t, either $C(x,t) \in b$ or $R_{\wedge}(y,z,t) \in b$ or there exists an object variable w such that $\neg P(w,y) \in b$, $\neg P(w,z) \in b$, and $\neg C(x,w) \in b$, obtained by an application of the rule (rReRCC6($\rightarrow$));

Cpl(rReRCC6(←)) For all object symbols x, y, z, and w, either $P(w, y) \in b$ or $P(w, z) \in b$ or $C(x, w) \in b$, or for some object variable t both $\neg C(x, t) \in b$ and $\neg R_\wedge(y, z, t) \in b$, obtained by an application of the rule (rReRCC6(←));

Cpl(rReRCC7(→)) For all object symbols x, y, and z, either $\neg O(x, y) \in b$ or $R_\wedge(x, y, z) \in b$ or $z \neq 0$, obtained by an application of the rule (rReRCC7(→));

Cpl(rReRCC7(←)) For all object symbols x and y, either $O(x, y) \in b$ or for some object variable z, both $\neg R_\wedge(x, y, z) \in b$ and $z = 0 \in b$, obtained by an application of the rule (rReRCC7(←));

Cpl(rReRCC8) For all object symbols x and y, if $x = y \in b$, then for some object variables z and w, either both $C(x, z) \in b$ and $\neg C(z, y) \in b$ or both $\neg C(x, w) \in b$ and $C(w, y) \in b$, obtained by an application of the rule (rReRCC8).

The rules of $\mathsf{F_{RCC}}$-dual tableau guarantee that whenever a branch of an $\mathsf{F_{RCC}}$-proof tree contains an atomic $\mathsf{F_{RCC}}$-formula φ and its negation, then the branch can be closed. Thus, the closed branch property can be proved.

Let b be an open branch of an $\mathsf{F_{RCC}}$-proof tree. The branch structure $\mathcal{M}^b = (U^b, R^b_-, R^b_\vee, R^b_\wedge, C^b, \{\#^b : \# \in \mathsf{Mer}\}, 0^b, 1^b, m^b)$ is defined as follows:

- $U^b = \mathbb{OS}_{\mathsf{F_{RCC}}}$;
- $0^b = m^b(0) = 0$ and $1^b = m^b(1) = 1$;
- $R^b = m^b(R) = \{(x, y) \in (U^b)^2 : R(x, y) \notin b\}$, for $R \in \{R_-, C, =\}$;
- $Q^b = m^b(Q) = \{(x, y, z) \in (U^b)^3 : Q(x, y, z) \notin b\}$, for $Q \in \{R_\vee, R_\wedge\}$;
- For every $\# \in \mathsf{Mer}$, $\#^b$ is defined according to Table 18.1; for example, $P^b \stackrel{\text{df}}{=} -(C^b ; -C^b)$;
- $m^b(\#) = \#^b$.

Proposition 18.4.2 (Branch Model Property). *Let b be an open branch of an $\mathsf{F_{RCC}}$-proof tree. Then, the branch structure $\mathcal{M}^b$ is an $\mathsf{F_{RCC}}$-model.*

Proof. Let b be an open branch of an $\mathsf{F_{RCC}}$-proof tree. It suffices to show that $\mathcal{M}^b$ satisfies the conditions (ReRCC1), ..., (ReRCC8). By way of example, we show that $\mathcal{M}^b$ satisfies the condition (ReRCC6).

(→) By the completion condition Cpl(rReRCC6(→)), for all $x, y, z, t \in U^b$, either $C(x, t) \in b$ or $R_\wedge(y, z, t) \in b$ or there exists an object variable w such that $\neg P(w, y) \in b$ and $\neg P(w, z) \in b$, and $\neg C(x, w) \in b$. Thus, by the definition of $\mathcal{M}^b$ and the completion condition Cpl($\neg P$) (see the proof of Proposition 18.2.5), either not $C^b(x, t)$ or not $R^b_\wedge(y, z, t)$ or there exists $w \in U^b$ such that $P^b(w, y)$ and $P^b(w, z)$, and $C^b(x, w)$. Hence, if $C^b(x, t)$ and $R^b_\wedge(y, z, t)$, then there exists $w \in U^b$ such that $P^b(w, y)$ and $P^b(w, z)$, and $C^b(x, w)$.

(←) By the completion condition Cpl(rReRCC6(←)), for all $x, y, z, w \in U^b$, either $P(w, y) \in b$ or $P(w, z) \in b$ or $C(x, w) \in b$ or there exists an object variable t such that both $\neg C(x, t) \in b$ and $\neg R_\wedge(y, z, t) \in b$. By the definition of $\mathcal{M}^b$ and the completion condition Cpl(P) (see Proposition 18.2.5), if $P^b(w, y)$, $P^b(w, z)$, and $C^b(x, w)$ hold, then there exists $t \in U^b$ such that $C^b(x, t)$ and $R^b_\wedge(y, z, t)$. Therefore, $\mathcal{M}^b$ satisfies the condition (ReRCC6).

The proofs of the remaining conditions are similar. □

The satisfaction in branch model property for $\mathsf{F}_{\mathsf{RCC}}$ follows from Proposition 18.2.5. In this way, we get:

Theorem 18.4.3 (Soundness and Completeness of $\mathsf{F}_{\mathsf{RCC}}$). *Let φ be an $\mathsf{F}_{\mathsf{RCC}}$-formula. Then, the following conditions are equivalent:*

1. *φ is $\mathsf{F}_{\mathsf{RCC}}$-valid;*
2. *φ is true in all standard $\mathsf{F}_{\mathsf{RCC}}$-models;*
3. *φ is $\mathsf{F}_{\mathsf{RCC}}$-provable.*

Example. Consider the following property holding in RCC:

$$aOb \text{ implies } 0 \lneq (a \wedge b).$$

This property is represented by the following $\mathsf{F}_{\mathsf{RCC}}$-formula:

$$\varphi = \forall x \forall y[\neg O(x,y) \vee \exists w(R_{\wedge}(x,y,w) \wedge \neg R_{\vee}(w,0,0) \wedge R_{\vee}(0,w,w))].$$

Figure 18.4 presents an $\mathsf{F}_{\mathsf{RCC}}$-proof of φ which shows that the property in question holds in RCC.

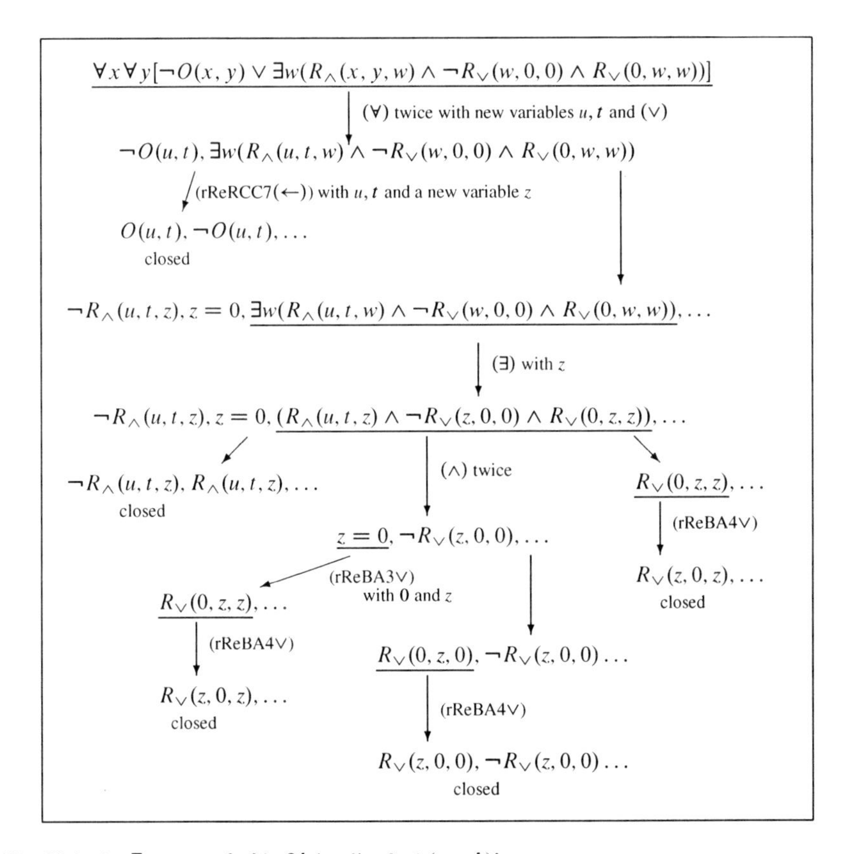

Fig. 18.4 An $\mathsf{F}_{\mathsf{RCC}}$-proof of '$aOb$ implies $0 \lneq (a \wedge b)$'

18.5 Dual Tableaux for Spatial Theories of Proximity Relation

A proximity relation is a binary relation δ on a Boolean algebra satisfying the following axioms:

(δ1) $a\delta b$ implies $a, b \neq 0$;
(δ2) $c\delta(a \vee b)$ iff $c\delta a$ or $c\delta b$;
(δ3) $(a \vee b)\delta c$ iff $a\delta c$ or $b\delta c$.

Development of a dual tableau for the theory of proximity follows the lines of Sect. 18.3. We consider the first-order theory $\mathsf{F}_{\mathsf{BA}\delta}$ of binary relations δ and R_-, and ternary relations $R_\vee$ and $R_\wedge$.

$\mathsf{F}_{\mathsf{BA}\delta}$-models are structures $\mathcal{M} = (U, R_-, R_\vee, R_\wedge, \delta, 0, 1, m)$ satisfying conditions (ReBA0), (ReBA1), (ReBA2#), ..., (ReBA5#), for # $\in \{\vee, \wedge\}$, (ReBA6), ..., (ReBA10) presented in Sect. 18.3, and for all $x, y, t \in U$:

(Reδ1) $\delta(x, y)$ implies $x \neq 0$ and $y \neq 0$;
(Reδ2) There exists z such that $R_\vee(x, y, z)$ and $\delta(t, z)$ iff $\delta(t, x)$ or $\delta(t, y)$;
(Reδ3) There exists z such that $R_\vee(x, y, z)$ and $\delta(z, t)$ iff $\delta(x, t)$ or $\delta(y, t)$.

$\mathsf{F}_{\mathsf{BA}\delta}$-dual tableau is the F-dual tableau endowed with the specific rules and axiomatic sets corresponding to the conditions (ReBA0), (ReBA1), (ReBA2#), ..., (ReBA5#), for # $\in \{\vee, \wedge\}$, (ReBA6), ..., (ReBA10) presented in Sect. 18.3, and the specific rules corresponding to the conditions (Reδ1), (Reδ2), (Reδ3). The latter rules have the following forms:

For all object symbols x, y, and t,

(rReδ1) $$\frac{}{\delta(x, y) \mid x = 0, y = 0}$$

(rReδ2($\rightarrow$)) $$\frac{\delta(t, x), \delta(t, y)}{R_\vee(x, y, z), \delta(t, x), \delta(t, y) \mid \delta(t, z), \delta(t, x), \delta(t, y)}$$

z is any object symbol

(rReδ2($\leftarrow$)) $$\frac{}{\neg R_\vee(x, y, z), \neg\delta(t, z) \mid \delta(t, x), \delta(t, y)}$$

z is a new object variable such that $z \neq x, y, t$

(rReδ3($\rightarrow$)) $$\frac{\delta(x, t), \delta(y, t)}{R_\vee(x, y, z), \delta(x, t), \delta(y, t) \mid \delta(z, t), \delta(x, t), \delta(y, t)}$$

z is any object symbol

(rReδ3($\leftarrow$)) $$\frac{}{\neg R_\vee(x, y, z), \neg\delta(z, t) \mid \delta(x, t), \delta(y, t)}$$

z is a new object variable, $z \neq x, y, t$

Alternative forms of rules (rReδ1), (rReδ2($\leftarrow$)), and (rReδ3($\leftarrow$)) are discussed in Sect. 25.9.

The following can be proved:

Proposition 18.5.1.

1. *The* $\mathsf{F_{BA\delta}}$*-rules are* $\mathsf{F_{BA\delta}}$*-correct;*
2. *The* $\mathsf{F_{BA\delta}}$*-axiomatic sets are* $\mathsf{F_{BA\delta}}$*-sets.*

The completion conditions determined by the rules (rReδ1), (rReδ2($\rightarrow$)), (rReδ2($\leftarrow$)), (rReδ3($\rightarrow$)), and (rReδ3($\leftarrow$)) are:

For all object symbols x, y, and t,

Cpl(rReδ1) Either $\delta(x,y) \in b$ or both $x = 0 \in b$ and $y = 0 \in b$, obtained by an application of the rule (rReδ1);

Cpl(rReδ2($\rightarrow$)) If $\delta(t,x) \in b$ and $\delta(t,y) \in b$, then for every object symbol z, either $R_{\vee}(x,y,z) \in b$ or $\delta(t,z) \in b$, obtained by an application of the rule (rReδ2($\rightarrow$));

Cpl(rReδ2($\leftarrow$)) Either both $\delta(t,x) \in b$ and $\delta(t,y) \in b$ or for some object variable z both $\neg R_{\vee}(x,y,z) \in b$ and $\neg\delta(t,z) \in b$, obtained by an application of the rule (rReδ2($\leftarrow$));

Cpl(rReδ3($\rightarrow$)) If $\delta(x,t) \in b$ and $\delta(y,t) \in b$, then for every object symbol z, either $R_{\vee}(x,y,z) \in b$ or $\delta(z,t) \in b$, obtained by an application of the rule (rReδ3($\rightarrow$));

Cpl(rReδ3($\leftarrow$)) Either both $\delta(x,t) \in b$ and $\delta(y,t) \in b$ or for some object variable z both $\neg R_{\vee}(x,y,z) \in b$ and $\neg\delta(z,t) \in b$, obtained by an application of the rule (rReδ3($\leftarrow$)).

The rules of $\mathsf{F_{BA\delta}}$-dual tableau guarantee that whenever a branch of an $\mathsf{F_{BA\delta}}$-proof tree contains an atomic $\mathsf{F_{BA\delta}}$-formula φ and its negation, then the branch can be closed. Thus, the closed branch property can be proved.

The branch structure is defined as in the completeness proof of $\mathsf{F_{BAC}}$-dual tableau (see p. 347), where C is replaced by δ.

Proposition 18.5.2 (Branch Model Property). *Let b be an open branch of an* $\mathsf{F_{BA\delta}}$*-proof tree. Then, the branch structure* $\mathcal{M}^b = (U^b, R^b_{-}, R^b_{\vee}, R^b_{\wedge}, \delta^b, 0^b, 1^b, m^b)$ *is an* $\mathsf{F_{BA\delta}}$*-model.*

Proof. It suffices to show that $\mathcal{M}^b$ satisfies the conditions (Reδ1), (Reδ2), (Reδ3). By way of example, we show that $\mathcal{M}^b$ satisfies the condition (Reδ2).

Assume that there exists $z \in U^b$ such that $R^b_{\vee}(x,y,z)$ and $\delta^b(t,z)$, that is $R_{\vee}(x,y,z) \notin b$ and $\delta(t,z) \notin b$. Suppose $\delta^b(t,x)$ and $\delta^b(t,y)$ do not hold in $\mathcal{M}^b$. Then, $\delta(t,x) \in b$ and $\delta(t,y) \in b$. By the completion condition Cpl(rReδ2($\rightarrow$)), for every $z \in U^b$, either $R_{\vee}(x,y,z) \in b$ or $\delta(t,z) \in b$, a contradiction.

On the other hand, by the completion condition Cpl(rReδ2($\leftarrow$)), either $\delta(t,x) \in b$ or $\delta(t,y) \in b$ or for some $z \in U^b$ both $\neg R_{\vee}(x,y,z) \in b$ and $\neg\delta(t,z) \in b$. Thus, if $\delta^b(t,x)$ and $\delta^b(t,y)$, then there exists $z \in U^b$ such that $R^b_{\vee}(x,y,z)$ and $\delta^b(t,z) \in b$. Therefore, the condition (Reδ2) is true in $\mathcal{M}^b$. □

Since the interpretation of atomic formulas in the branch model is defined in the standard way, the satisfaction in branch model property can be proved as for $\mathsf{F_{BAC}}$-dual tableau. Thus, we obtain:

Theorem 18.5.1 (Soundness and Completeness of $\mathsf{F_{BA\delta}}$). *For every $\mathsf{F_{BA\delta}}$-formula φ, the following conditions are equivalent:*

1. *φ is $\mathsf{F_{BA\delta}}$-valid;*
2. *φ is true in all standard $\mathsf{F_{BA\delta}}$-models;*
3. *φ is $\mathsf{F_{BA\delta}}$-provable.*

Various types of proximity relations, defined by adding specific axioms to the basic axioms ($\delta 1$), ($\delta 2$), and ($\delta 3$), are considered in the literature. Some of these axioms are among the following:

($\delta 4$) $a\delta b$ implies $a = b$ (separated proximity);
($\delta 5$) If not $a\delta b$, then there exists c such that not $a\delta c$ and not $(-c)\delta b$ (Efremovic proximity);
($\delta 6$) δ is symmetric;
($\delta 7$) $a \wedge b \neq 0$ implies $a\delta b$.

In order to reason about a proximity relation δ satisfying additional conditions ($\delta 4$), ..., ($\delta 7$), we develop logics $\mathsf{F_{BA(\delta i)}}$, $i \in \{4, \ldots, 7\}$, based on the logic $\mathsf{F_{BA\delta}}$. $\mathsf{F_{BA(\delta i)}}$-models are $\mathsf{F_{BA\delta}}$-models that satisfy the condition (Reδi), for all $x, y \in U$:

(Re$\delta 4$) $\delta(x, y)$ implies $x = y$;
(Re$\delta 5$) If not $\delta(x, y)$, then there are $z, t \in U$ such that not $\delta(x, z)$ and $R_{-}(z, t)$ and not $\delta(t, y)$;
(Re$\delta 6$) $\delta(x, y)$ implies $\delta(y, x)$;
(Re$\delta 7$) If there exists $z \in U$ such that $R_{\wedge}(x, y, z)$ and $z \neq 0$, then $\delta(x, y)$.

A dual tableau for logic $\mathsf{F_{BA(\delta i)}}$ is $\mathsf{F_{BA\delta}}$-dual tableau endowed with the specific rule corresponding to the condition (Reδi), for $i \in \{4, \ldots, 7\}$. The rules have the following forms:

For all object symbols x and y,

(rRe$\delta 4$) $$\frac{x = y}{\delta(x, y), x = y}$$

(rRe$\delta 5$) $$\frac{\delta(x, y)}{\delta(x, z), \neg R_{-}(z, t), \delta(t, y), \delta(x, y)}$$

z, t are new object variables, $z \neq t$ and $\{z, t\} \cap \{x, y\} = \emptyset$

(rRe$\delta 6$) $$\frac{\delta(x, y)}{\delta(y, x)}$$

(rReδ7) $$\frac{\delta(x,y)}{R_{\wedge}(x,y,z), \delta(x,y) \mid z \neq 0, \delta(x,y)}$$

z is any object symbol

Theorem 18.5.2 (Correspondence). *Let $\mathcal{K}$ be a class of $\mathsf{F}_{\mathsf{BA}\delta}$-models. Then for every $i \in \{4,\ldots,7\}$, the following conditions are equivalent:*

1. *The condition (Reδi) is true in all $\mathcal{K}$-models;*
2. *The rule (rReδi) is $\mathcal{K}$-correct.*

Proof. By way of example, we prove the theorem for (Reδ5).

Assume that the condition (Reδ5) is true in all $\mathcal{K}$-models. Let X be a finite set of $\mathsf{F}_{\mathsf{BA}\delta}$-formulas. Let x, y be object symbols and let z, t be object variables that do not occur in X and such that $z \neq t$, $\{z,t\} \cap \{x,y\} = \emptyset$. Assume $X \cup \{\delta(x,z), \neg R_{-}(z,t), \delta(t,y), \delta(x,y)\}$ is a $\mathcal{K}$-set and suppose $X \cup \{\delta(x,y)\}$ is not a $\mathcal{K}$-set. Then, there exist a $\mathcal{K}$-model $\mathcal{M}$ and a valuation v in $\mathcal{M}$ such that $\mathcal{M}, v \not\models \delta(x,y)$. By the assumption on variables z and t, for all $z, t \in U$, either $\delta(v(x), z)$ or not $R_{-}(z,t)$ or $\delta(t, v(y))$. Then, by the condition (Reδ5), $\mathcal{M}, v \models \delta(x,y)$, a contradiction. Therefore, the rule (rReδ5) is $\mathcal{K}$-correct.

Assume that the rule (rReδ5) is $\mathcal{K}$-correct. Let $X \stackrel{\mathrm{df}}{=} \{\exists z' \exists t' (\neg\delta(x,z') \wedge R_{-}(z',t') \wedge \neg\delta(t',y))\}$. Let z, t be object variables such that $z \neq t$ and $\{z,t\} \cap \{x,y,z',t'\} = \emptyset$. Then, $X \cup \{\delta(x,z), \neg R_{-}(z,t), \delta(t,y), \delta(x,y)\}$ is a $\mathcal{K}$-set, so by the assumption $X \cup \{\delta(x,y)\}$ is also a $\mathcal{K}$-set. Therefore, the following is true in every $\mathcal{K}$-model: if not $\delta(x,y)$, then there exist z' and t' such that not $\delta(x,z')$ and $R_{-}(z',t')$ and not $\delta(t',y)$. Hence, the condition (Reδ5) is true in all $\mathcal{K}$-models. □

The above theorem implies:

Proposition 18.5.3. *Let $i \in \{4,\ldots,7\}$. Then:*

1. *The $\mathsf{F}_{\mathsf{BA}(\delta i)}$-rules are $\mathsf{F}_{\mathsf{BA}(\delta i)}$-correct;*
2. *The $\mathsf{F}_{\mathsf{BA}(\delta i)}$-axiomatic sets are $\mathsf{F}_{\mathsf{BA}(\delta i)}$-sets.*

The completion conditions determined by the rules (rReδ4), ..., (rReδ7) are:

For all object symbols x and y,

Cpl(rReδ4) If $x = y \in b$, then $\delta(x,y) \in b$, obtained by an application of the rule (rReδ4);

Cpl(rReδ5) If $\delta(x,y) \in b$, then there are object variables z and t such that $\delta(x,z) \in b$ and $\neg R_{-}(z,t) \in b$, and $\delta(t,y) \in b$, obtained by an application of the rule (rReδ5);

Cpl(rReδ6) If $\delta(x,y) \in b$, then $\delta(y,x) \in b$, obtained by an application of the rule (rReδ6);

Cpl(rReδ7) If $\delta(x,y) \in b$, then for every object symbol z, either $R_{\wedge}(x,y,z) \in b$ or $z \neq 0 \in b$, obtained by an application of the rule (rReδ1).

The closed branch property, the branch model property and the satisfaction in branch model property can be proved as for $\mathsf{F}_{\mathsf{BA}\delta}$-dual tableau. Hence, we get:

Theorem 18.5.3 (Soundness and Completeness of $\mathsf{F}_{\mathsf{BA}(\delta i)}$). *Let $i \in \{4, \ldots, 7\}$. For every $\mathsf{F}_{\mathsf{BA}\delta}$-formula φ, the following conditions are equivalent:*

1. *φ is $\mathsf{F}_{\mathsf{BA}(\delta i)}$-valid;*
2. *φ is true in all standard $\mathsf{F}_{\mathsf{BA}(\delta i)}$-models;*
3. *φ is $\mathsf{F}_{\mathsf{BA}(\delta i)}$-provable.*

Chapter 19
Dual Tableaux for Logics of Programs

19.1 Introduction

Dynamic logic is a framework suitable for specification and verification of dynamic properties of systems. It is a multimodal logic with the modal operations of necessity and possibility determined by binary relations understood as state transition relations or input-output relations associated with computer programs. The logic evolved from the early ideas of J. Yanov [Yan59], E. Engeler [Eng67], and C. Hoare [Hoa69], through the algorithmic logic of A. Salwicki [Sal70, MS87], and finally it was formulated by V. Pratt [Pra76] in the form which emphasizes a modal nature of interactions between programs and assertions (see also [FL79]). A number of interesting variants of propositional dynamic logic, PDL, obtained by restricting or extending it in various ways, have been studied including deterministic PDL, automata PDL, programs with restrictions on allowable tests, programs with complementation and intersection, programs with a converse operation, and logics with well-foundness and halting predicates; for a comprehensive survey see [HKT00].

In this chapter we present, first, a relational dual tableau for logic PDL. It includes an infinitary rule reflecting the behaviour of an iteration operation modelled in the logic with the reflexive and transitive closure of a relation. Next, we consider an extension of PDL with the operations of postspecification and prespecification of a program modelled with the residuals of the composition of programs (see [Orł93]). Furthermore, following [DO96] a dual tableau is presented for PDL endowed with the demonic union and the demonic composition operations relevant for nondeterministic programs. Some versions of the iteration operation adequate for nondeterministic programs are also discussed in this section. A relational model of demonic nondeterministic programs is considered in [Ngu91]. Finally, a logic of event structures is presented and its relational dual tableau is developed following [Orł95]. Event structures considered here are based on the developments in [Win80] and [Win86], and the logic of event structures is presented in [Pen88]. All the logics considered in this chapter are decidable.

E. Orłowska and J. Golińska-Pilarek, *Dual Tableaux: Foundations, Methodology, Case Studies*, Trends in Logic 36, DOI 10.1007/978-94-007-0005-5_19,

19.2 Relational Formalization of Propositional Dynamic Logic

The formulas of a propositional dynamic logic, PDL, are generated with the operations relevant for binary relations and for modal formulas, respectively.

The vocabulary of the language of PDL consists of symbols from the following pairwise disjoint sets:

- $\mathbb{V}$ – a countably infinite set of propositional variables;
- $\mathbb{RC}_{\mathsf{PDL}}$ – a countably infinite set of relational constants interpreted as atomic programs;
- $\{\cup, ;, ?, ^*\}$ – the set of relational operations, where $\cup$ is interpreted as a nondeterministic choice, ; is interpreted as a sequential composition of programs, ? is the test operation, and * is interpreted as a nondeterministic iteration;
- $\{\neg, \vee, \wedge, \rightarrow, [\,], \langle\rangle\}$ – the set of propositional operations of negation, disjunction, conjunction, implication, necessity, and possibility respectively.

The set of PDL-relational terms interpreted as compound programs and the set of PDL-formulas are the smallest sets containing $\mathbb{RC}_{\mathsf{PDL}}$ and $\mathbb{V}$, respectively, and satisfying the following conditions:

- If S and T are PDL-relational terms, then so are $S \cup T$, $S ; T$, and T^*;
- If φ is a PDL-formula, then $\varphi?$ is a PDL-relational term;
- If φ and ψ are PDL-formulas, then so are $\neg\varphi, \varphi \vee \psi, \varphi \wedge \psi$, and $\varphi \rightarrow \psi$;
- If φ is a PDL-formula and T is a PDL-relational term, then $[T]\varphi$ and $\langle T\rangle\varphi$ are PDL-formulas.

Let R be a binary relation on a set U. Then, we define:

R^0 is the identity relation on U,
$R^{i+1} \stackrel{\text{df}}{=} (R^i ; R)$,
$R^* \stackrel{\text{df}}{=} \bigcup_{i \geq 0} R^i$.

A PDL-*model* is a structure $\mathcal{M} = (U, \{r_R\}_{R \in \mathbb{RC}_{\mathsf{PDL}}}, m)$, where U is a non-empty set interpreted as a set of states of a computation, r_R are binary relations on U, and m is a meaning function satisfying the following conditions:

- $m(p) \subseteq U$ for $p \in \mathbb{V}$,
- $m(R) = r_R$ for $R \in \mathbb{RC}_{\mathsf{PDL}}$,

m extends to all the compound PDL-relational terms:

- $m(T^*) = m(T)^* = \bigcup_{i \geq 0} m(T)^i$,
- $m(S \cup T) = m(S) \cup m(T)$,
- $m(S ; T) = m(S) ; m(T)$,
- $m(\varphi?) = \{(s, s) \in U \times U : \mathcal{M}, s \models \varphi\}$.

The satisfaction of a PDL-formula φ in a PDL-model $\mathcal{M}$ by a state $s \in U$, $\mathcal{M}, s \models \varphi$, is defined as:

- $\mathcal{M}, s \models p$ iff $s \in m(p)$, for $p \in \mathbb{V}$;
- $\mathcal{M}, s \models \neg\varphi$ iff $\mathcal{M}, s \not\models \varphi$;
- $\mathcal{M}, s \models \varphi \vee \psi$ iff $\mathcal{M}, s \models \varphi$ or $\mathcal{M}, s \models \psi$;
- $\mathcal{M}, s \models \varphi \wedge \psi$ iff $\mathcal{M}, s \models \varphi$ and $\mathcal{M}, s \models \psi$;
- $\mathcal{M}, s \models \varphi \rightarrow \psi$ iff $\mathcal{M}, s \models \neg\varphi \vee \psi$;

and for every PDL-relational term T,

- $\mathcal{M}, s \models [T]\varphi$ iff for all $s' \in U$, if $(s, s') \in m(T)$, then $\mathcal{M}, s' \models \varphi$;
- $\mathcal{M}, s \models \langle T \rangle\varphi$ iff there exists $s' \in U$ such that $(s, s') \in m(T)$ and $\mathcal{M}, s' \models \varphi$.

A PDL-formula φ is true in a PDL-model $\mathcal{M}$ whenever it is satisfied in $\mathcal{M}$ by every $s \in U$, and it is PDL-valid if and only if it is true in all PDL-models.

Intuitively, $(s, s') \in m(R)$ means that there exists a computation of program R starting in the state s and terminating in the state s'. Program $S \cup T$ performs S or T nondeterministically; program $S\,;T$ performs first S and then T. Expression $\varphi?$ is a command to continue if φ is true, and fail otherwise. Program T^* performs T zero or more times sequentially.

Theorem 19.2.1.

1. PDL *has the finite model property;*
2. *Model checking problem for* PDL *is* P*-complete and solvable in linear time;*
3. *Satisfiability problem for* PDL *is* EXPTIME*-complete.*

The proofs of the above theorem can be found in [FL79, Pra78, Pra79, HKT00, KP81, Seg77].

The language of the relational logic $\mathsf{RL}_{\mathsf{PDL}}$ appropriate for expressing PDL-formulas is an extension of $\mathsf{RL}(1, 1')$-language such that:

- $\mathbb{RC}_{\mathsf{RL}_{\mathsf{PDL}}} = \{1, 1'\} \cup \mathbb{RC}_{\mathsf{PDL}}$;
- The set of relational operations includes the operation *.

The set of $\mathsf{RL}_{\mathsf{PDL}}$-terms is the smallest set containing relational variables and constants and closed on all the relational operations. As usual, $\mathsf{RL}_{\mathsf{PDL}}$-formulas are of the form xTy, where T is an $\mathsf{RL}_{\mathsf{PDL}}$-relational term and x, y are object variables.

An $\mathsf{RL}_{\mathsf{PDL}}$-model is a structure $\mathcal{M} = (U, \{r_R\}_{R \in \mathbb{RC}_{\mathsf{PDL}}}, m)$ such that (U, m) is an $\mathsf{RL}(1, 1')$-model that satisfies:

- $m(R) = r_R$, for every $R \in \mathbb{RC}_{\mathsf{PDL}}$;
- $m(T^*) = m(T)^*$, for every relational term T, where the operation $*$ is defined as in PDL-models.

Let $v\colon \mathbb{OV}_{\mathsf{RL_{PDL}}} \to U$ be a valuation in an $\mathsf{RL_{PDL}}$-model $\mathcal{M}$ and let φ be an $\mathsf{RL_{PDL}}$-formula. The satisfaction of φ in $\mathcal{M}$ by v, the truth of φ in $\mathcal{M}$, and $\mathsf{RL_{PDL}}$-validity of φ are defined as in Sect. 2.3.

Let τ' be a one-to-one mapping that assigns relational variables to propositional variables. We extend τ' to relational constants of PDL in such a way that $\tau'(R) = R$, for every $R \in \mathbb{RC}_{\mathsf{PDL}}$. The translation τ of PDL-terms and PDL-formulas into $\mathsf{RL_{PDL}}$-relational terms is defined as follows:

- $\tau(R) = \tau'(R)$, for $R \in \mathbb{RC}_{\mathsf{PDL}}$;

and for all relational terms T and S,

- $\tau(T^*) = \tau(T)^*$,
- $\tau(T \cup S) = \tau(T) \cup \tau(S)$,
- $\tau(T\,;S) = \tau(T)\,;\tau(S)$,
- $\tau(\varphi?) = 1' \cap \tau(\varphi)$,
- $\tau(p) = \tau'(p)\,;1$, for $p \in \mathbb{V}$,
- $\tau(\varphi \vee \psi) = \tau(\varphi) \cup \tau(\psi)$,
- $\tau(\varphi \wedge \psi) = \tau(\varphi) \cap \tau(\psi)$,
- $\tau(\varphi \to \psi) = \tau(\neg\varphi \vee \psi)$,
- $\tau(\langle T\rangle\varphi) = \tau(T)\,;\tau(\varphi)$,
- $\tau([T]\varphi) = -(\tau(T)\,;-\tau(\varphi))$.

Relational terms obtained from formulas of the dynamic logic include both declarative information and procedural information provided by these formulas. The declarative part which represents static facts about a domain is represented by means of a Boolean reduct of algebras of relations, and the procedural part, which is intended to model dynamics of the domain, requires the relational operations. In the relational terms which represent the formulas after the translation the two types of information receive a uniform representation and the process of reasoning about both statics and dynamics, and about relationships between them can be performed within the framework of a single uniform formalism.

As in the standard modal logics (see Theorem 7.4.1), we have:

Theorem 19.2.2. *For every* PDL*-formula and for all object variables x and y, the following conditions are equivalent:*

1. *φ is* PDL*-valid;*
2. *$x\tau(\varphi)y$ is* $\mathsf{RL_{PDL}}$*-valid.*

Proof. It suffices to show that Propositions 7.4.2 and 7.4.3 are true for all PDL-formulas. Thus, we need to show:

(1) For every PDL-model $\mathcal{M} = (U, \{r_R\}_{R\in\mathbb{RC}_{\mathsf{PDL}}}, m)$, there exists an $\mathsf{RL_{PDL}}$-model $\mathcal{M}' = (U, \{r_R\}_{R\in\mathbb{RC}_{\mathsf{PDL}}}, m')$ with the same universe and the same relations r_R, $R \in \mathbb{RC}_{\mathsf{PDL}}$, as those in $\mathcal{M}$ and such that for all $s, s' \in U$, for every PDL-formula φ, and for every PDL-term T, the following hold:

(a) $(s, s') \in m(T)$ iff $(s, s') \in m'(\tau(T))$,

(b) $\mathcal{M}, s \models \varphi$ iff $(s, s') \in m'(\tau(\varphi))$.

(2) For every standard $\mathsf{RL}_{\mathsf{PDL}}$-model $\mathcal{M}' = (U, \{r_R\}_{R \in \mathbb{RC}_{\mathsf{PDL}}}, m')$, there exists a PDL-model $\mathcal{M} = (U, \{r_R\}_{R \in \mathbb{RC}_{\mathsf{PDL}}}, m)$ with the same universe and the same relations r_R, $R \in \mathbb{RC}_{\mathsf{PDL}}$, as those in $\mathcal{M}'$, such that (a) and (b) hold.

By way of example, we prove (1). Let $\mathcal{M} = (U, \{r_R\}_{R \in \mathbb{RC}_{\mathsf{PDL}}}, m)$ be a PDL-model. We define an $\mathsf{RL}_{\mathsf{PDL}}$-model $\mathcal{M}' = (U, \{r_R\}_{R \in \mathbb{RC}_{\mathsf{PDL}}}, m')$ in a similar way as in the proof of Proposition 7.4.2, namely:

- $m'(1) = U \times U$;
- $m'(1')$ is the identity on U;
- $m'(\tau(p)) = \{(x, y) \in U \times U : x \in m(p)\}$, for every propositional variable p;
- $m'(R) = r_R$, for every $R \in \mathbb{RC}_{\mathsf{PDL}}$;
- m' extends to all the compound terms as in $\mathsf{RL}_{\mathsf{PDL}}$-models.

The proof of (a) and (b) is by mutual induction. By the definition of $\mathcal{M}'$, (a) holds for every PDL-relational constant. For atomic PDL-formulas, (b) can be proved as in the standard modal logics (see Proposition 7.4.2). Now, assume that (a) and (b) hold for PDL-relational terms S and T and PDL-formulas φ and ψ. We need to show that condition (a) holds for T^*, $S \cup T$, $S\,;T$, $\varphi?$, and the condition (b) holds for formulas $\varphi \vee \psi$, $\varphi \wedge \psi$, $\varphi \rightarrow \psi$, $\langle T \rangle \varphi$, and $[T]\varphi$.

By way of example, we prove (a) for T^* and $\varphi?$. Let $s, s' \in U$. Then:

$(s, s') \in m(T^*)$ iff there exists $i \geq 0$ such that $(s, s') \in m(T)^i$ iff $s = s'$ or there exist $i > 0$ and $t_1, \ldots, t_i \in U$ such that $s = t_1$, $(t_1, t_2) \in m(T)$, $\ldots$, $(t_i, s') \in m(T)$ iff, by the induction hypothesis, $(s, s') \in m'(1')$ or there exists $i > 0$ and there are $t_1, \ldots, t_i \in U$ such that $(s, t_1) \in m'(1')$, $(t_1, t_2) \in m'(\tau(T))$, $\ldots$, $(t_i, s') \in m'(\tau(T))$ iff $(s, s') \in m'(\tau(T))^*$ iff $(s, s') \in m'(\tau(T^*))$.

$(s, s') \in m(\varphi?)$ iff $s = s'$ and $\mathcal{M}, s \models \varphi$ iff, by the induction hypothesis, $(s, s') \in m'(1')$ and $(s, s') \in m'(\tau(\varphi))$ iff $(s, s') \in m'(1' \cap \tau(\varphi))$ iff $(s, s') \in m'(\tau(\varphi?))$.

The condition (b) can be proved in a similar way as in Proposition 7.4.2. By way of example, we show it for a formula of the form $\langle T \rangle \varphi$. Let $s, s' \in U$. Then:

$\mathcal{M}, s \models \langle T \rangle \varphi$ iff there exists $t \in U$ such that $(s, t) \in m(T)$ and $\mathcal{M}, t \models \varphi$ iff, by the induction hypothesis, there exists $t \in U$ such that $(s, t) \in m'(\tau(T))$ and $(t, s') \in m'(\tau(\varphi))$ iff $(s, s') \in m'(\tau(\langle T \rangle \varphi))$. □

$\mathsf{RL}_{\mathsf{PDL}}$-dual tableau is an extension of $\mathsf{RL}(1, 1')$-dual tableau with the decomposition rules for the operation * and its complement:

For all object symbols x, y and for every relational term T,

$$(*)\ \frac{xT^*y}{xT^iy,\ xT^*y} \qquad (-*)\ \frac{x{-}(T^*)y}{x{-}(T^0)y \mid \ldots \mid x{-}(T^i)y \mid \ldots}$$

for any $i \geq 0$ where $T^0 = 1'$, $T^{i+1} = T\,; T^i$

Observe that the rule $(-*)$ is an infinitary rule.

As usual, an $\mathsf{RL}_{\mathsf{PDL}}$-set is a finite set of $\mathsf{RL}_{\mathsf{PDL}}$-formulas such that the first-order disjunction of its members is true in all $\mathsf{RL}_{\mathsf{PDL}}$-models. Correctness of finitary rules is defined in a similar way as in the relational logics of classical algebras of binary relations (see Sect. 2.4). Correctness of an infinitely branching rule is a natural extension of the definition of correctness of finitary rules, namely, a rule $\frac{\Phi}{\Phi_1 \mid \ldots \mid \Phi_n \mid \ldots}$ is $\mathsf{RL_{PDL}}$-correct whenever for every set X of $\mathsf{RL_{PDL}}$-formulas, $X \cup \Phi$ is an $\mathsf{RL_{PDL}}$-set if and only if for every $i \in \omega$, $X \cup \Phi_i$ is an $\mathsf{RL_{PDL}}$-set.

Proposition 19.2.1.

1. *The* $\mathsf{RL_{PDL}}$*-rules are* $\mathsf{RL_{PDL}}$*-correct;*
2. *The* $\mathsf{RL_{PDL}}$*-axiomatic sets are* $\mathsf{RL_{PDL}}$*-sets.*

Proof. By way of example, we show correctness of the rules (*) and (−*). Let X be a finite set of $\mathsf{RL_{PDL}}$-formulas.

(*) Clearly, if $X \cup \{xT^*y\}$ is an $\mathsf{RL_{PDL}}$-set, then so are $X \cup \{xT^iy, xT^*y\}$, for every $i \geq 0$. Let $i \geq 0$. Assume that $X \cup \{xT^iy, xT^*y\}$ is an $\mathsf{RL_{PDL}}$-set and suppose that $X \cup \{xT^*y\}$ is not an $\mathsf{RL_{PDL}}$-set. Then there exist an $\mathsf{RL_{PDL}}$-model $\mathcal{M}$ and a valuation v in $\mathcal{M}$ such that for every $\varphi \in X$, $\mathcal{M}, v \not\models \varphi$ and $\mathcal{M}, v \not\models xT^*y$. By the assumption, $\mathcal{M}, v \models xT^iy$, hence $(v(x), v(y)) \in m(T^i)$, for some $i \geq 0$. Thus, by the definition of $m(T^*)$, $(v(x), v(y)) \in m(T^*)$, so $\mathcal{M}, v \models xT^*y$, a contradiction. Therefore, the rule (*) is $\mathsf{RL_{PDL}}$-correct.

(−*) Assume $X \cup \{x{-}(T^*)y\}$ is an $\mathsf{RL_{PDL}}$-set. Then for every $\mathsf{RL_{PDL}}$-model $\mathcal{M}$ and for every valuation v in $\mathcal{M}$, either there exists $\varphi \in X$ such that $\mathcal{M}, v \models \varphi$ or for every $i \geq 0$, $\mathcal{M}, v \models x{-}(T^i)y$. Thus, for every $\mathsf{RL_{PDL}}$-model $\mathcal{M}$, for every valuation v in $\mathcal{M}$, and for every $i \geq 0$, either there exists $\varphi \in X$ such that $\mathcal{M}, v \models \varphi$ or $\mathcal{M}, v \models x{-}(T^i)y$. Hence, for every $i \geq 0$, $X \cup \{x{-}(T^i)y\}$ is an $\mathsf{RL_{PDL}}$-set. Now, assume that for every $i \geq 0$, $X \cup \{x{-}(T^i)y\}$ is an $\mathsf{RL_{PDL}}$-set. Suppose $X \cup \{x{-}(T^*)y\}$ is not an $\mathsf{RL_{PDL}}$-set, that is there exist an $\mathsf{RL_{PDL}}$-model $\mathcal{M}$ and a valuation v in $\mathcal{M}$ such that for every $\varphi \in X$ $\mathcal{M}, v \not\models \varphi$ and $\mathcal{M}, v \models xT^*y$. However, by the assumption, for every $i \geq 0$, $\mathcal{M}, v \models x{-}(T^i)y$, hence $\mathcal{M}, v \models x{-}(T^*)y$, a contradiction.

Correctness of the remaining rules can be proved as in the relational logics of classical algebras of binary relations in Sect. 2.5. □

An $\mathsf{RL_{PDL}}$-proof tree for φ is defined as in logic RL (see Sect. 2.4); however, here it is not necessarily finitely branching tree. The notions of a closed branch of an $\mathsf{RL_{PDL}}$-proof tree, a closed $\mathsf{RL_{PDL}}$-proof tree, and $\mathsf{RL_{PDL}}$-provability are defined as in Sect. 2.4.

A branch b of an $\mathsf{RL_{PDL}}$-proof tree is complete whenever it is closed or it satisfies the completion conditions of $\mathsf{RL}(1, 1')$-dual tableau and in addition the completion conditions which are specific for the $\mathsf{RL_{PDL}}$-dual tableau:

For all object symbols x, y and for every relational term T,

Cpl(*) If $xT^*y \in b$, then for every $i \geq 0$, $xT^iy \in b$, obtained by an application of the rule (*);

Cpl(−*) If $x{-}(T^*)y \in b$, then for some $i \geq 0$, $x{-}(T^i)y \in b$, obtained by an application of the rule (−*).

The notions of a complete $\mathsf{RL_{PDL}}$-proof tree and an open branch of an $\mathsf{RL_{PDL}}$-proof tree are defined as in RL-logic (see Sect. 2.5). Observe that any application of the rules of $\mathsf{RL_{PDL}}$-dual tableau, in particular an application of the rules (*) and (−*), preserves the formulas built with atomic terms or their complements. Thus, the closed branch property holds.

Let b be an open branch of an $\mathsf{RL_{PDL}}$-proof tree. The branch structure $\mathcal{M}^b = (U^b, \{r_R^b\}_{R \in \mathbb{RC}_{\mathsf{PDL}}}, m^b)$ is defined in a similar way as in $\mathsf{RL}(1, 1')$-logic, that is:

- $U^b = \mathbb{OS}_{\mathsf{RL_{PDL}}}$;
- $r_R^b = m^b(R) = \{(x, y) \in U^b \times U^b : xRy \notin b\}$, for every $R \in \mathbb{RC}_{\mathsf{RL_{PDL}}}$;
- m^b extends to all the compound relational terms as in $\mathsf{RL_{PDL}}$-models.

Clearly, the branch model property holds, that is $\mathcal{M}^b$ is an $\mathsf{RL_{PDL}}$-model.

Proposition 19.2.2 (Satisfaction in Branch Model Property). *Let b be an open branch of an $\mathsf{RL_{PDL}}$-proof tree. Then for every $\mathsf{RL_{PDL}}$-formula φ, if $\mathcal{M}^b, v^b \models \varphi$, then $\varphi \notin b$.*

Proof. The proof is by induction on the complexity of formulas. The atomic case can be proved as in Sect. 2.5, due to the closed branch property. By way of example, we prove the proposition for the formulas built with T^* or its complement.

First, we prove:

(1) If the proposition holds for a formula built with a relational term T (resp. $-T$), then for every $i \geq 0$, it holds for the formula built with T^i (resp. $-(T^i)$).

The proof is by induction on i. Assume $\mathcal{M}^b, v^b \models xT^0y$, that is $(x, y) \in m^b(1')$. Directly from the definition of $\mathcal{M}^b$, $x1'y \notin b$. Assume $\mathcal{M}^b, v^b \models x{-}(T^0)y$, that is $(x, y) \notin m^b(1')$. Thus $x1'y \in b$, hence $x{-}1'y \notin b$, by the closed branch property. Assume the proposition holds for $k \geq 0$. We prove that it holds for $k + 1$. Let $\mathcal{M}^b, v^b \models xT^{k+1}y$. Then there exists $z \in U^b$ such that $(x, z) \in m^b(T)$ and $(z, y) \in m^b(T^k)$. Suppose $xT^{k+1}y \in b$. Then by the completion condition Cpl(;), for every $z \in U^b$, either $xTz \in b$ or $zT^ky \in b$. By the induction hypothesis, for every $z \in U^b$, either $(x, z) \notin m^b(T)$ or $(z, y) \notin m^b(T^k)$, a contradiction. Let $\mathcal{M}^b, v^b \models x{-}(T^{k+1})y$. Then for all $z \in U^b$, either $(x, z) \notin m^b(T)$ or $(z, y) \notin m^b(T^k)$. Suppose $x{-}(T^{k+1})y \in b$. Then by the completion condition Cpl(−;), for some $z \in U^b$, $x{-}Tz \in b$ and $z{-}(T^k)y \in b$. By the induction hypothesis, for some $z \in U^b$, $(x, z) \in m^b(T)$ and $(z, y) \in m^b(T^k)$, a contradiction.

Now, we prove that the proposition holds for formulas built with terms of the form T^* and $-(T^*)$.

Assume $\mathcal{M}^b, v^b \models xT^*y$. Then $(v(x), v(y)) \in m^b(T^*)$, that is there is $i \geq 0$ such that $(v(x), v(y)) \in m^b(T^i)$. Suppose $xT^*y \in b$. Then, by the completion condition Cpl(*), for every $j \geq 0$, $xT^jy \in b$. Thus, by (1), $(v(x), v(y)) \notin m^b(T^i)$, a contradiction.

Assume $\mathcal{M}^b, v^b \models x{-}(T^*)y$. Then $(v(x), v(y)) \notin m^b(T^*)$, so for all $i \geq 0$, $(v(x), v(y)) \notin m^b(T^i)$. Suppose $x{-}(T^*)y \in b$. Then, by the completion condition Cpl(−*), for some $i \geq 0$, $x{-}(T^i)y \in b$. Thus, by (1), $(v(x), v(y)) \subset m^b(T^i)$, a contradiction. □

Following the general method of proving soundness and completeness described in Sect. 2.6 (p. 44), we obtain:

Theorem 19.2.3 (Soundness and Completeness of $\mathsf{RL}_{\mathsf{PDL}}$). *For every $\mathsf{RL}_{\mathsf{PDL}}$-formula φ, the following conditions are equivalent:*

1. *φ is $\mathsf{RL}_{\mathsf{PDL}}$-valid;*
2. *φ is true in all standard $\mathsf{RL}_{\mathsf{PDL}}$-models;*
3. *φ is $\mathsf{RL}_{\mathsf{PDL}}$-provable.*

The above theorem and Theorem 19.2.2 imply:

Theorem 19.2.4 (Relational Soundness and Completeness of PDL). *For every PDL-formula φ and for all object variables x and y, the following conditions are equivalent:*

1. *φ is PDL-valid;*
2. *$x\tau(\varphi)y$ is $\mathsf{RL}_{\mathsf{PDL}}$-provable.*

Example. Consider the following PDL-formulas:

$$\varphi = [R^*]p \rightarrow (p \wedge [R]p),$$

$$\psi = (p \wedge [R^*](p \rightarrow [R]p)) \rightarrow [R^*]p.$$

Translations of these formulas into $\mathsf{RL}_{\mathsf{PDL}}$-relational terms are:

$$\tau(\varphi) = --(R^* ; -P) \cup (P \cap -(R ; -P)),$$
$$\tau(\psi) = -(P \cap -(R^* ; -(-P \cup -(R ; -P)))) \cup -(R^* ; -P),$$

respectively, where for simplicity $\tau(p) = P$. Figures 19.1 and 19.2 present $\mathsf{RL}_{\mathsf{PDL}}$-proofs of $x\tau(\varphi)y$ and $x\tau(\psi)y$, respectively, which, by Theorem 19.2.4, show PDL-validity of φ and ψ, respectively.

19.3 Relational Formalization of Dynamic Logic with Program Specifications

In this section we consider the propositional dynamic logic with program specifications, PDLS, which is an extension of PDL-logic with program constructors that cover a wide spectrum of command formation rules from various programming languages.

The language of PDLS-logic is a PDL-language endowed with the propositional operation *if-then-else* and with the relational operations \ and / representing the weakest prespecification and the weakest postspecification, respectively.

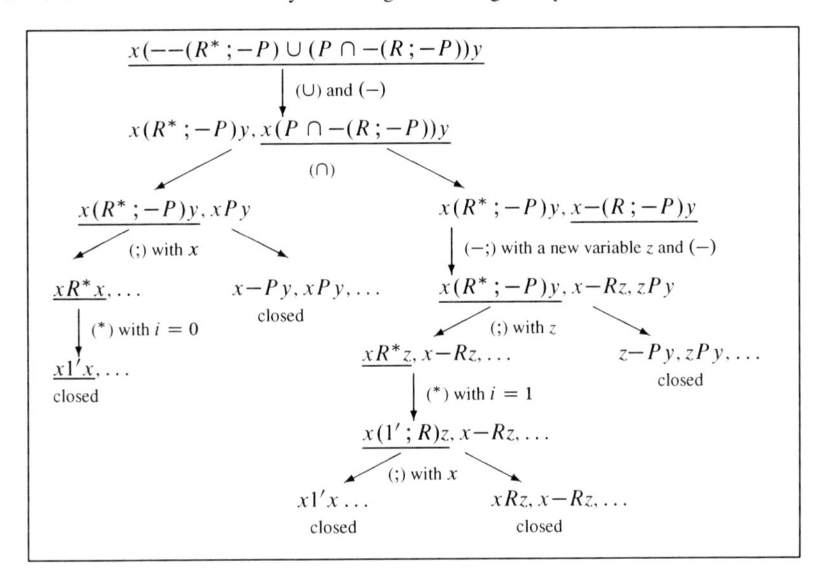

Fig. 19.1 An $\mathsf{RL_{PDL}}$-proof of the formula $[R^*]p \rightarrow (p \wedge [R]p)$

The set of PDLS-relational terms and the set of PDLS-formulas are defined as in PDL-logic (see Sect. 19.2) with the following additional clauses:

- $S\backslash T$ and S/T are PDLS-relational terms, for all PDLS-terms S and T;
- *If* φ *then* S *else* T is a PDLS-relational term, for every PDLS-formula φ and for all PDLS-terms S and T.

A PDLS-model is a PDL-model $\mathcal{M} = (U, \{r_R\}_{R \in \mathbb{RC}_{\mathsf{PDLS}}}, m)$ such that the following conditions are satisfied for all PDLS-relational terms S, T and for every PDLS-formula φ:

- $m(S\backslash T) = m(S)\backslash m(T)$;
- $m(S/T) = m(S)/m(T)$;
- $m(\mathit{if}\, \varphi\, \mathit{then}\, S\, \mathit{else}\, T) = m(\varphi?) ; m(S) \cup m((\neg\varphi)?) ; m(T)$.

We recall that the operations $\backslash$ and $/$ of the right and left residual of the composition of relations, respectively, are defined as in Sect. 18.4:

$$S\backslash T \stackrel{\text{df}}{=} -(-T ; S^{-1});$$
$$S/T \stackrel{\text{df}}{=} -(T^{-1} ; -S).$$

Intuitively, $S\backslash T$ is the weakest prespecification of S to achieve T, that is the greatest program such that $(S\backslash T) ; S \subseteq T$. Similarly, S/T is the weakest postspecification

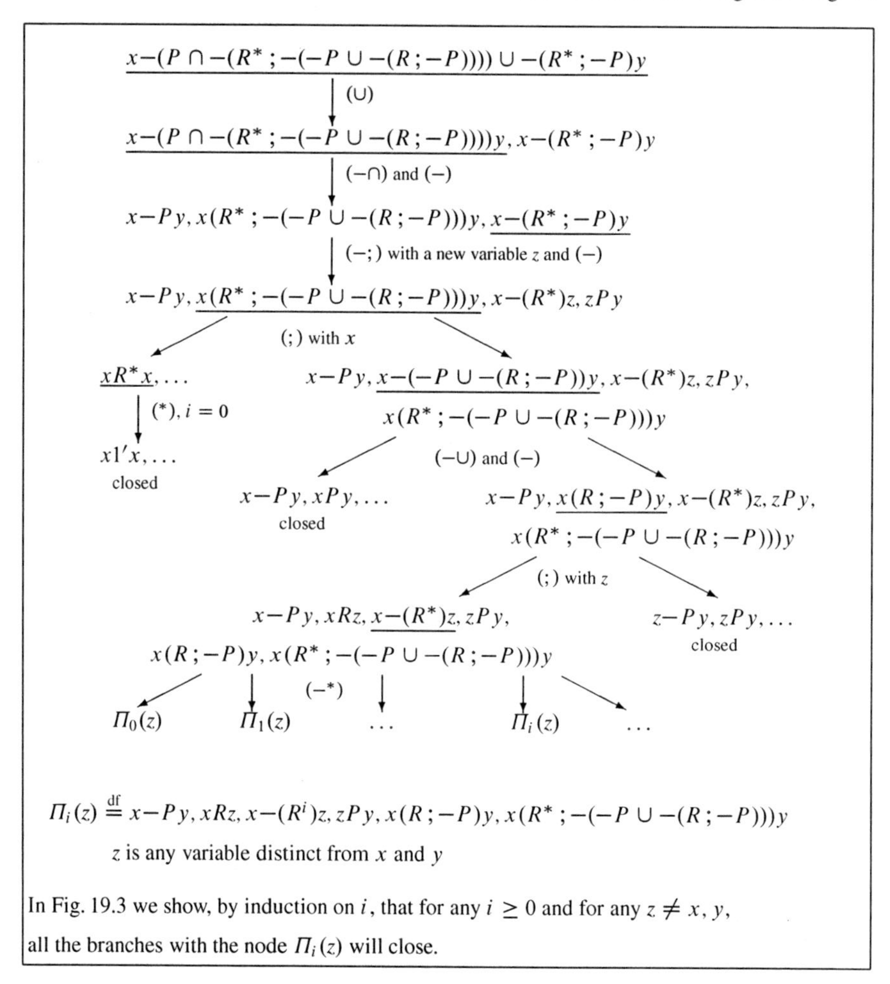

Fig. 19.2 An $\mathsf{RL}_{\mathsf{PDL}}$-proof of the formula $(p \wedge [R^*](p \rightarrow [R]p)) \rightarrow [R^*]p$

of T to achieve S, that is the greatest program such that $T\,;(S/T) \subseteq S$. Program *if* φ *then* S *else* T performs S whenever an input state satisfies φ, and performs T otherwise.

The satisfaction, truth in a model, and validity of a PDLS-formula are defined as in PDL-logic.

The relational logic appropriate for expressing PDLS-formulas is an $\mathsf{RL}_{\mathsf{PDL}}$-logic defined in Sect. 19.2. The translation of PDLS-formulas into PDL-relational terms is defined as in the previous section and extended with the clauses:

- $\tau(S \backslash T) \stackrel{\text{df}}{=} -(-\tau(T)\,;\tau(S)^{-1})$;
- $\tau(S/T) \stackrel{\text{df}}{=} -(\tau(T)^{-1}\,;-\tau(S))$;
- $\tau(\mathit{if}\,\varphi\;\mathit{then}\;S\;\mathit{else}\;T) \stackrel{\text{df}}{=} \tau(\varphi?)\,;\tau(S) \cup \tau((\neg\varphi)?)\,;\tau(T)$.

For $i = 0$ we have:

$$x{-}Py,\ xRz,\ x{-}1'z,\ \underline{zPy},\ x(R\,;-P)y,\ x(R^*\,;-(-P \cup -(R\,;-P)))y$$

(1′2) with x

$\underline{z1'x},\ x{-}1'z, \ldots$; $xPy,\ x{-}Py, \ldots$ closed

(1′1) with z

$z1'z, \ldots$ closed ; $x1'z,\ x{-}1'z, \ldots$ closed

For $i = 1$ we have:

$$x{-}Py,\ xRz,\ \underline{x{-}(1'\,;R)z},\ zPy,\ x(R\,;-P)y,\ x(R^*\,;-(-P \cup -(R\,;-P)))y$$

(−;) with a new variable t

$$\underline{xRz},\ x{-}1't,\ t{-}Rz, \ldots$$

(1′2) with t

$x1't,\ x{-}1't, \ldots$ closed ; $tRz,\ t{-}Rz, \ldots$ closed

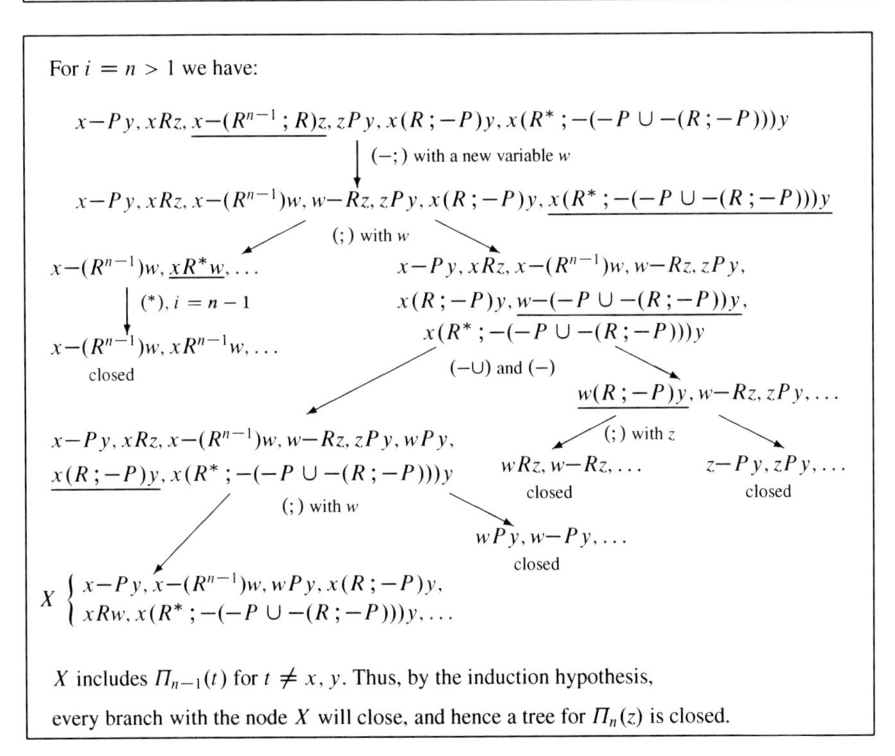

Fig. 19.3 The inductive proof of $\Pi_i(z)$

Following the proof of Theorem 19.2.2, we can prove that the translation τ preserves validity of formulas:

Theorem 19.3.1. *For every* PDLS*-formula and for all object variables x and y, the following conditions are equivalent:*

1. φ *is* PDLS*-valid;*
2. $x\tau(\varphi)y$ *is* $\mathsf{RL}_{\mathsf{PDL}}$*-valid.*

Proof. As in the proof of Theorem 19.2.2, it suffices to show:

(1) For every PDLS-model $\mathcal{M} = (U, \{r_R\}_{R\in\mathbb{RC}_{\mathsf{PDLS}}}, m)$, there exists an $\mathsf{RL}_{\mathsf{PDL}}$-model $\mathcal{M}' = (U, \{r_R\}_{R\in\mathbb{RC}_{\mathsf{PDLS}}}, m')$ with the same universe and the same relations r_R, $R \in \mathbb{RC}_{\mathsf{PDLS}}$, as those in the model $\mathcal{M}$ and such that for all $s, s' \in U$, for every PDLS-formula φ, and for every PDLS-term T, the following hold:

 (a) $(s, s') \in m(T)$ iff $(s, s') \in m'(\tau(T))$,
 (b) $\mathcal{M}, s \models \varphi$ iff $(s, s') \in m'(\tau(\varphi))$.

(2) For every standard $\mathsf{RL}_{\mathsf{PDL}}$-model $\mathcal{M}' = (U, \{r_R\}_{R\in\mathbb{RC}_{\mathsf{PDLS}}}, m')$, there exists a PDLS-model $\mathcal{M} = (U, \{r_R\}_{R\in\mathbb{RC}_{\mathsf{PDLS}}}, m)$ with the same universe and the same relations r_R, $R \in \mathbb{RC}_{\mathsf{PDLS}}$, as those in the model $\mathcal{M}'$ and such that for all $s, s' \in U$, for every PDLS-formula φ, and for every PDLS-term T, conditions (a) and (b) hold.

By way of example, we prove (2). Let $\mathcal{M}' = (U, \{r_R\}_{R\in\mathbb{RC}_{\mathsf{PDLS}}}, m')$ be a standard $\mathsf{RL}_{\mathsf{PDL}}$-model. Then we define a PDLS-model $\mathcal{M} = (U, \{r_R\}_{R\in\mathbb{RC}_{\mathsf{PDLS}}}, m)$ in a similar way as in the proof of Proposition 7.4.3, namely:

- $m(p) = \{x \in U : \exists y \in U,\ (x, y) \in m'(\tau(p))\}$, for every propositional variable p;
- $m(R) = r_R$, for every $R \in \mathbb{RC}_{\mathsf{PDLS}}$;
- m extends to all the compound PDLS-terms as in PDLS-models.

As in the proof of Theorem 19.2.2, the proof of (a) and (b) is by mutual induction. For the PDLS-constants, the condition (a) holds by the definition of $\mathcal{M}$. For propositional variables, the condition (b) can be proved in a similar way as in the standard modal logics (see Propositions 7.4.2 and 7.4.3). Now, assume that (a) and (b) hold for PDLS-relational terms S and T and a PDLS-formula φ. By way of example, we show that (a) holds for $S\backslash T$. Let $s, s' \in U$. Then:

$(s, s') \in m(S\backslash T)$ iff for every $u \in U$, either $(s, u) \in m(T)$ or $(s', u) \notin m(S)$ iff, by the induction hypothesis, for every $u \in U$ either, $(s, u) \in m'(\tau(T))$ or $(s', u) \notin m'(\tau(S))$ iff $(s, s') \in -m'(-\tau(T) ; \tau(S)^{-1})$ iff $(s, s') \in m'(\tau(S\backslash T))$.□

The above theorem implies that a dual tableau for PDLS-logic is just the PDL-dual tableau. Therefore, we have:

Theorem 19.3.2 (Relational Soundness and Completeness of PDLS). *For every PDLS-formula φ and for all object variables x and y, the following conditions are equivalent:*

1. *φ is PDLS-valid;*
2. *$x\tau(\varphi)y$ is $\mathsf{RL}_{\mathsf{PDL}}$-provable.*

19.4 Relational Formalization of Logics of Demonic Nondeterministic Programs

The main motivation for introducing demonic program constructors can be expressed as a principle that possible nontermination implies definite nontermination. To incorporate this principle in the relational semantics of programs, the classical operations of nondeterministic choice and the sequential composition should be modified appropriately:

- If two commands of a nondeterministic program α can be executed nondeterministically, and if an execution of one of them does not terminate, then the execution of α does not terminate.
- If a sequence of commands of a nondeterministic program α can be executed sequentially, and if an execution of one of them does not terminate, then the execution of α does not terminate.

The above postulates reflect Murphy's law: 'If it can go wrong, it will'. Nondeterminism modelled according to these postulates is referred to as demonic nondeterminism, as opposed to angelic nondeterminism [BZ86].

The calculus of binary relations with operations of demonic union ($||$) and demonic composition ($;;$) of relations was studied in [Ngu91]. An extension of the calculus with the demonic iteration operation $d(*)$ has been proposed in [DO96]. The demonic iteration is motivated by the following postulate:

> If a command of a nondeterministic program α is executed nondeterministically any finite number of times and if one of these executions does not terminate, then the execution of α does not terminate.

In this section we consider the propositional logic of demonic nondeterministic programs, PDL^d, which deals with the three demonic operations. The language of the logic PDL^d is a PDL-language endowed with the relational operations $||$, $;;$, and $d(*)$ interpreted as the program operations of demonic union, demonic composition, and demonic iteration, respectively.

The set of PDL^d-relational terms and the set of PDL^d-formulas are defined in a similar way as in PDL-logic (see p. 360) with the following additional clause:

- If R and S are PDL^d-relational terms, then so are $R \,||\, S$, $R \,;;\, S$, and $R^{d(*)}$.

Let R and S be binary relations on U. The demonic union, demonic composition, and demonic iteration are defined by:

$$R \,||\, S \stackrel{\text{df}}{=} \{(x,y) : \exists t\ (x,t) \in R \text{ and } \exists t'\ (x,t') \in S \text{ and } (x,y) \in R \cup S\},$$
$$R\,;;\,S \stackrel{\text{df}}{=} \{(x,y) : \forall z\,((x,z) \in R \text{ implies } \exists t\,(z,t) \in S) \text{ and } (x,y) \in R\,;S\},$$
$$R^{d(*)} \stackrel{\text{df}}{=} ||_{i \in \omega} R^{d(i)}, \text{ where}$$
$$R^{d(0)} \stackrel{\text{df}}{=} 1' \text{ and } R^{d(i+1)} \stackrel{\text{df}}{=} R\,;;\,R^{d(i)}.$$

Since the demonic composition is associative, $R^{d(i+1)} = R^{d(i)}\,;;\,R$, for every $i \geq 0$. Moreover, if the relations R and S are serial, i.e., $(R\,;1) = 1$ and $(S\,;1) = 1$, then the demonic union (resp. composition) collapses to the classical union (resp. composition).

Proposition 19.4.1. *Let R and S be relations on a set U. Then:*

1. $R \,||\, S = (R \cup S) \cap (R\,;(U \times U)) \cap (S\,;(U \times U))$,
2. $R\,;;\,S = (R\,;S) \cap -(R\,;-(S\,;(U \times U)))$,
3. $(x,y) \in R^{d(*)}$ *iff* $(x,y) \in R^*$ *and for all* $i \geq 0$, $(x,y) \in (R^{d(i)}\,;(U \times U))$.

Items 1. and 2. follow from the definition of the operations $||$ and $;;$, respectively. The inductive proof of 3. can be found in [DO96].

A PDL^d-*model* is a PDL-model $\mathcal{M} = (U, \{r\}_{R \in \mathbb{RC}_{\mathsf{PDL}^d}}, m)$ such that for all PDL^d-relational terms S and T, the following conditions are satisfied:

- $m(R \,||\, S) = m(R) \,||\, m(S)$,
- $m(R\,;;\,S) = m(R)\,;;\,m(S)$,
- $m(R^{d(*)}) = m(R)^{d(*)}$.

As usual, we use the same symbols for the operations in the language and for the corresponding operations in the semantics.

The satisfaction, truth in a model, and validity of a PDL^d-formula are defined as in PDL-logic.

In [DO96] the following is proved:

Theorem 19.4.1.

The logic PDL^d *has the finite model property.*

It follows that the logic PDL^d is decidable.

The language of the relational logic $\mathsf{RL}_{\mathsf{PDL}^d}$ appropriate for expressing PDL^d-formulas is the language of $\mathsf{RL}_{\mathsf{PDL}}$-logic defined in Sect. 19.2, endowed with the binary relational operations $||$ and $;;$, and the unary operation $d(*)$. The set of $\mathsf{RL}_{\mathsf{PDL}^d}$-terms is the smallest set containing all the atomic relational terms and closed on all the relational operations. $\mathsf{RL}_{\mathsf{PDL}^d}$-formulas are defined as usual (see Sect. 2.3).

An $\mathsf{RL}_{\mathsf{PDL}^d}$-model is an $\mathsf{RL}_{\mathsf{PDL}}$-model $\mathcal{M} = (U, \{r_R\}_{R \in \mathbb{RC}_{\mathsf{PDL}^d}}, m)$ such that m extends to all $\mathsf{RL}_{\mathsf{PDL}^d}$-relational terms built with the operations $||$, $;;$, and $^{d(*)}$ as in

PDL^d-models. The satisfaction, truth, and validity of $\mathsf{RL}_{\mathsf{PDL}^d}$-formulas are defined in a standard way (see Sect. 2.3).

The translation of PDL^d-formulas into PDL-relational terms is defined as in the logic PDL (see Sect. 19.2, p. 362) with the following additional clauses for formulas built with demonic operations:

- $\tau(R \,||\, S) = \tau(R) \,||\, \tau(S)$,
- $\tau(R\,;;\,S) = \tau(R)\,;;\,\tau(S)$,
- $\tau(R^{d(*)}) = \tau(R)^{d(*)}$.

Translation function τ is defined so that it preserves validity of formulas.

Theorem 19.4.2. *For every* PDL^d*-formula and for all object variables* x *and* y*, the following conditions are equivalent:*

1. φ *is* PDL^d*-valid;*
2. $x\tau(\varphi)y$ *is* $\mathsf{RL}_{\mathsf{PDL}^d}$*-valid.*

The above theorem can be proved in a similar way as Theorem 19.2.2.

A dual tableau for the logic $\mathsf{RL}_{\mathsf{PDL}^d}$ is an extension of $\mathsf{RL}_{\mathsf{PDL}}$-dual tableau with the decomposition rules for formulas built with demonic operations. These rules have the following forms:

For all relational terms R, S and for all object symbols x and y,

$$(||)\quad \frac{x(R \,||\, S)y}{xRz, x(R \,||\, S)y \mid xSt, x(R \,||\, S)y \mid xRy, xSy, x(R \,||\, S)y}$$

z, t are any object symbols

$$(-||)\quad \frac{x{-}(R \,||\, S)y}{x{-}Rz, x{-}St, x{-}Ry \mid x{-}Rz, x{-}St, x{-}Sy}$$

z, t are new object variables such that $z \neq t$

$$(;;)\quad \frac{x(R\,;;\,S)y}{x{-}Rz, z(S\,;1)y \mid x(R\,;S)y} \qquad z \text{ is a new object variable}$$

$$(-;;)\quad \frac{x{-}(R\,;;\,S)y}{x{-}Rz, z{-}Sy, x(R\,;-(S\,;1))y} \qquad z \text{ is a new object variable}$$

$$({}^{d(*)})\quad \frac{xR^{d(*)}y}{xR^*y \mid x(R^{d(0)}\,;1)y \mid \ldots \mid x(R^{d(i)}\,;1)y \mid \ldots} \qquad i \in \omega$$

$$(-^{d(*)})\quad \frac{x{-}(R^{d(*)})y}{x{-}(R^*)y, x{-}(R^{d(i)}\,;1)y, x{-}(R^{d(*)})y} \qquad \text{for any } i \in \omega$$

The notions of an $\mathsf{RL}_{\mathsf{PDL}^d}$-set and correctness of $\mathsf{RL}_{\mathsf{PDL}^d}$-rule are defined as in $\mathsf{RL}_{\mathsf{PDL}}$-logic (see Sect. 19.2).

Proposition 19.4.2.

1. *The* $\mathsf{RL}_{\mathsf{PDL}^d}$*-rules are* $\mathsf{RL}_{\mathsf{PDL}^d}$*-correct;*
2. *The* $\mathsf{RL}_{\mathsf{PDL}^d}$*-axiomatic sets are* $\mathsf{RL}_{\mathsf{PDL}^d}$*-sets.*

Proof. In view of Proposition 19.2.1, we need to prove correctness of the rules specific for $\mathsf{RL}_{\mathsf{PDL}}$-dual tableau. By way of example, we show correctness of the rules $(;;)$ and $(-^{d(*)})$. Let X be a finite set of $\mathsf{RL}_{\mathsf{PDL}^d}$-formulas.

$(;;)$ Let z be an object variable which does not occur in X and $z \neq x, y$. Assume $X \cup \{x(R\,;;S)y\}$ is an $\mathsf{RL}_{\mathsf{PDL}^d}$-set and suppose that $X \cup \{x{-}Rz, zS\,;1y\}$ or $X \cup \{x(R\,;S)y\}$ is not an $\mathsf{RL}_{\mathsf{PDL}^d}$-set. Then either there exist an $\mathsf{RL}_{\mathsf{PDL}^d}$-model $\mathcal{M}$ and a valuation v in $\mathcal{M}$ such that for every $\varphi \in X$, $\mathcal{M}, v \not\models \varphi$ and $(v(x), v(y)) \in m(R\,;-(S\,;1))$ or there exist an $\mathsf{RL}_{\mathsf{PDL}^d}$-model $\mathcal{M}'$ and a valuation v' in $\mathcal{M}'$ such that for every $\varphi \in X$, $\mathcal{M}', v' \not\models \varphi$ and $(v'(x), v'(y)) \notin m'(R\,;S)$. However, by the assumption and by Proposition 19.4.1(2.), for every $\mathsf{RL}_{\mathsf{PDL}^d}$-model $\mathcal{M}$ and for every valuation v in $\mathcal{M}$, either there exists $\varphi \in X$ such that $\mathcal{M}, v \models \varphi$ or both $(v(x), v(y)) \in m(R\,;S)$ and $(v(x), v(y)) \notin m(R\,;-(S\,;1))$, a contradiction.

Now, assume that $X \cup \{x{-}Rz, zS\,;1y\}$ and $X \cup \{x(R\,;S)y\}$ are $\mathsf{RL}_{\mathsf{PDL}^d}$-sets. By the assumption on variable z, for every $\mathsf{RL}_{\mathsf{PDL}^d}$-model $\mathcal{M}$ and for every valuation v in $\mathcal{M}$, either there exists $\varphi \in X$ such that $\mathcal{M}, v \models \varphi$ or both $(v(x), v(y)) \in m(R\,;S)$ and for every $z \in U$, if $(v(x), v(z)) \in m(R)$, then $(v(z), v(y)) \in (S\,;1)$. Hence, by Proposition 19.4.1(2.), $X \cup \{x(R\,;;S)y\}$ is an $\mathsf{RL}_{\mathsf{PDL}^d}$-set. Therefore, the rule $(;;)$ is $\mathsf{RL}_{\mathsf{PDL}^d}$-correct.

$(-^{d(*)})$ Clearly, if $X \cup \{x{-}(R^{d(*)})y\}$ is an $\mathsf{RL}_{\mathsf{PDL}^d}$-set, then so are $X \cup \{x{-}(R^*)y, x{-}(R^{d(i)}\,;1)y, x{-}(R^{d(*)})y\}$, for every $i \geq 0$. Let $i \geq 0$. Assume that $X \cup \{x{-}(R^*)y, x{-}(R^{d(i)}\,;1)y, x{-}(R^{d(*)})y\}$ is an $\mathsf{RL}_{\mathsf{PDL}^d}$-set and suppose that $X \cup \{x{-}(R^{d(*)})y\}$ is not an $\mathsf{RL}_{\mathsf{PDL}^d}$-set. Then, there exist an $\mathsf{RL}_{\mathsf{PDL}^d}$-model $\mathcal{M}$ and a valuation v in $\mathcal{M}$ such that for every $\varphi \in X$, $\mathcal{M}, v \not\models \varphi$ and $(v(x), v(y)) \in m(R^{d(*)})$. Then, by the assumption, $(v(x), v(y)) \notin m(R^*)$ or $(v(x), v(y)) \notin m(R^{d(i)}\,;1)$. Thus, by Proposition 19.4.1(3.), $(v(x), v(y)) \notin m(R^{d(*)})$, a contradiction. □

The notion of an $\mathsf{RL}_{\mathsf{PDL}^d}$-proof tree is defined as in Sect. 19.2 for $\mathsf{RL}_{\mathsf{PDL}}$-logic and the notions of a closed branch, a closed $\mathsf{RL}_{\mathsf{PDL}^d}$-proof tree, and $\mathsf{RL}_{\mathsf{PDL}^d}$-provability are defined as in Sect. 2.4.

A branch b of an $\mathsf{RL}_{\mathsf{PDL}^d}$-proof tree is complete whenever it is closed or it satisfies the completion conditions of $\mathsf{RL}_{\mathsf{PDL}}$-dual tableau and the completion conditions specific for the $\mathsf{RL}_{\mathsf{PDL}^d}$-dual tableau:

For all object symbols x, y and for all relational terms R, S,

Cpl($||$) If $x(R\,||\,S)y \in b$, then for all object symbols z and t, either $xRz \in b$ or $xSt \in b$ or both $xRy \in b$ and $xSy \in b$, obtained by an application of the rule $(||)$;

Cpl($-||$) If $x{-}(R\,||\,S)y \in b$, then there are object variables z and t such that $x{-}Rz \in b$ and $x{-}St \in b$ and either $x{-}Ry \in b$ or $x{-}Sy \in b$, obtained by an application of the rule $(-||)$;

Cpl(; ;) If $x(R\,;;S)y \in b$, then either $x(R\,;S)y \in b$ or there exists an object variable z such that both $x{-}Rz \in b$ and $z(S\,;1)y \in b$, obtained by an application of the rule (; ;);

Cpl(−; ;) If $x{-}(R\,;;S)y \in b$, then $x(R\,;-(S\,;1))y \in b$ and there exists an object variable z such that $x{-}Rz \in b$ and $z{-}Sy \in b$, obtained by an application of the rule (−; ;);

Cpl($^{d(*)}$) If $xR^{d(*)}y \in b$, then either $xR^*y \in b$ or there exists $i \geq 0$ such that $x(R^{d(i)}\,;1)y \in b$, obtained by an application of the rule ($^{d(*)}$);

Cpl($-^{d(*)}$) If $x{-}(R^{d(*)})y \in b$, then $x{-}(R^*)y \in b$ and for every $i \geq 0$, $x{-}(R^{d(i)}\,;1)y \in b$, obtained by an application of the rule ($-^{d(*)}$).

The notions of a complete $\mathsf{RL}_{\mathsf{PDL}^d}$-proof tree and an open branch of an $\mathsf{RL}_{\mathsf{PDL}^d}$-proof tree are defined as in Sect. 2.5. Observe that the PDL^d-rules, in particular the decomposition rules for the operations $||$, ; ;, $^{d(*)}$ and the rules for their complements, preserve formulas built with atomic relational terms or their complements. Thus, the closed branch property holds.

Let b be an open branch of an $\mathsf{RL}_{\mathsf{PDL}^d}$-proof tree. The branch structure $\mathcal{M}^b = (U^b, \{r_R^b\}_{R \in \mathbb{RC}_{\mathsf{PDL}^d}}, m^b)$ is defined as in the proof of completeness of $\mathsf{RL}_{\mathsf{PDL}}$-dual tableau with the assumption that m^b extends to all the $\mathsf{RL}_{\mathsf{PDL}^d}$-terms built with the operations $||$, ; ;, and $^{d(*)}$ as in $\mathsf{RL}_{\mathsf{PDL}^d}$-models. Thus, for every open branch b of an $\mathsf{RL}_{\mathsf{PDL}^d}$-proof tree, $\mathcal{M}^b$ is an $\mathsf{RL}_{\mathsf{PDL}^d}$-model, hence the branch model property holds.

Proposition 19.4.3 (Satisfaction in Branch Model Property). *Let b be an open branch of an $\mathsf{RL}_{\mathsf{PDL}^d}$-proof tree. Then for every $\mathsf{RL}_{\mathsf{PDL}^d}$-formula φ, if $\mathcal{M}^b, v^b \models \varphi$, then $\varphi \notin b$.*

Proof. The proof is by induction on the complexity of formulas. By way of example, we show the proposition for formulas built with terms of the form $R\,;;S$.

Assume $\mathcal{M}^b, v^b \models x(R\,;;S)y$. Then, by Proposition 19.4.1(2.), $(v(x), v(y)) \notin m^b(R\,;-(S\,;1))$ and $(v(x), v(y)) \in m^b(R\,;S)$. Suppose $x(R\,;;S)y \in b$. By the completion condition Cpl(; ;), either $x(R\,;S)y \in b$ or there exists $z \in U^b$ such that $x{-}Rz \in b$ and $z(S\,;1)y \in b$. Hence, by the completion condition Cpl(;) and due to the induction hypothesis, either $(x,y) \notin m^b(R\,;S)$ or there exists $z \in U^b$ such that $(x,z) \in m^b(R)$ and $(z,y) \notin m^b(S\,;1)$, a contradiction. □

Therefore, we have:

Theorem 19.4.3 (Soundness and Completeness of $\mathsf{RL}_{\mathsf{PDL}^d}$). *For every $\mathsf{RL}_{\mathsf{PDL}^d}$-formula φ, the following conditions are equivalent:*

1. *φ is $\mathsf{RL}_{\mathsf{PDL}^d}$-valid;*
2. *φ is true in all standard $\mathsf{RL}_{\mathsf{PDL}^d}$-models;*
3. *φ is $\mathsf{RL}_{\mathsf{PDL}^d}$-provable.*

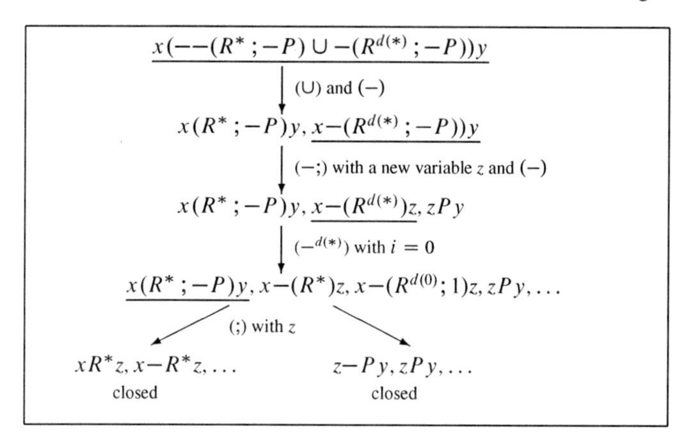

Fig. 19.4 An $\mathsf{RL}_{\mathsf{PDL}^d}$-proof of the formula $[R^*]p \to [R^{d(*)}]p$

Theorem 19.4.4 (Relational Soundness and Completeness of PDL^d). *For every PDL^d-formula φ and for all object variables x and y, the following conditions are equivalent:*

1. *φ is PDL^d-valid;*
2. *$x\tau(\varphi)y$ is $\mathsf{RL}_{\mathsf{PDL}^d}$-provable.*

Example. Consider the following PDL^d-formula:

$$\varphi = [R^*]p \to [R^{d(*)}]p.$$

Its translation into an $\mathsf{RL}_{\mathsf{PDL}^d}$-relational term is:

$$\tau(\varphi) = --(R^* ; -P) \cup -(R^{d(*)} ; -P),$$

where for simplicity $\tau(p) = P$. Figure 19.4 presents an $\mathsf{RL}_{\mathsf{PDL}^d}$-proof of $x\tau(\varphi)y$ which, by Theorem 19.4.4, shows PDL^d-validity of φ.

19.5 Relational Formalization of Event Structure Logics

The notion of an event structure was introduced in [Win80] and investigated in [NPW81, Win86, LT87], among others. A logic ESL for event structures was introduced and studied in [Pen88]. An event structure is intended to model behavior of a distributed system in terms of a set of event occurrences, a causality relation, which is a partial order in a set of event occurrences, and a conflict relation, which determines forbidden successions of events. Two event occurrences which are neither comparable with respect to the causality relation nor in conflict, may occur concurrently.

An *event structure* is a system $\mathcal{ES} = (E, csl, cnf)$ such that E is a non-empty set of events, csl and cnf are binary relations on E called *causality relation* and *conflict relation*, respectively, such that the following conditions are satisfied:

(E1) csl is reflexive, antisymmetric, and transitive;
(E2) cnf is irreflexive and symmetric;
(E3) $cnf\,;csl \subseteq cnf$.

Condition (E3) is referred to as *conflict inheritance*. Conditions (E2) and (E3) imply that relations csl and cnf are disjoint.

Let X be a subset of E. We define:

X is *conflict free* iff for all $e, e' \in E$, if $e \in X$ and $(e, e') \in cnf$, then $e' \notin X$;
X is *backward closed* iff for all $e, e' \in E$, if $e \in X$ and $(e', e) \in csl$, then $e' \in X$;
X is a *run in structure* $\mathcal{ES}$ iff X is a maximal (with respect to inclusion) conflict free and backward closed subset of E.

In [Pen88] the following is proved:

Proposition 19.5.1. *Let $\mathcal{ES} = (E, csl, cnf)$ be an event structure and let $X \subseteq E$. Then the following conditions are equivalent:*

1. *X is a run in $\mathcal{ES}$;*
2. *$x \in X$ iff for every $y \in E$, $(x, y) \in cnf$ implies $y \notin X$.*

The language of logic ESL is a modal language (see Sect. 7.3) with relational constants csl, csl^{-1}, and cnf and with the propositional constant run. An ESL-model is a system $\mathcal{M} = (U, csl, cnf, run)$ such that (U, csl, cnf) is an event structure, and $run \subseteq U$ is a run in that structure. The satisfaction of formulas in ESL-models is defined as usual (see Sect. 7.3), with the following clause for the propositional constant run:

- $\mathcal{M}, s \models run$ iff $s \in run$.

Truth in a model and validity of an ESL-formula are defined as in Sect. 7.3.

In [Pen88] the following has been proved:

Theorem 19.5.1.

1. *Logic* ESL *does not have the finite model property;*
2. *Logic* ESL *is decidable.*

The language of the relational logic $\mathsf{RL}_{\mathsf{ESL}}$ appropriate for expressing ESL-formulas is an $\mathsf{RL}(1, 1')$-language such that its set of relational constants is

$\{csl, cnf, run\}$. An $\mathsf{RL}_{\mathsf{ESL}}$-model is a structure (U, csl, cnf, run, m) such that (U, m) is an $\mathsf{RL}(1, 1')$-model, (U, csl, cnf) is an event structure, and $run = X \times U$, for some $X \subseteq U$ such that X is a run in this event structure.

Proposition 19.5.1 implies:

Proposition 19.5.2. *(U, csl, cnf, run, m) is an $\mathsf{RL}_{\mathsf{ESL}}$-model iff (U, csl, cnf) is an event structure and $run = -(cnf\ ; run)$.*

The translation of ESL-formulas into relational terms of $\mathsf{RL}_{\mathsf{ESL}}$-logic is defined as in the standard modal logics with the additional clause for the propositional constant run:

- $\tau(run) \stackrel{\text{df}}{=} run$.

As in the standard modal logics (see Theorem 7.4.1), we have:

Theorem 19.5.2. *For every ESL-formula and for all object variables x and y, the following conditions are equivalent:*

1. *φ is ESL-valid;*
2. *$x\tau(\varphi)y$ is $\mathsf{RL}_{\mathsf{ESL}}$-valid.*

$\mathsf{RL}_{\mathsf{ESL}}$-dual tableau is an extension of $\mathsf{RL}(1, 1')$-dual tableau with the rules reflecting all the constraints posed on the accessibility relations in ESL-models: reflexivity, antisymmetry, and transitivity of causality relation, irreflexivity and symmetry of conflict relation, conflict inheritance, and properties of run. The rules (ref csl) and (tran csl) reflecting reflexivity and transitivity of csl, respectively, and the rule (sym cnf) reflecting symmetry of cnf are analogous to those presented in Sect. 6.6 (see also Sect. 7.4). We recall that these rules have the following forms:

For all object symbols x and y,

$$(\text{ref } csl)\ \frac{xcsly}{x1'y, xcsly} \qquad (\text{sym } cnf)\ \frac{xcnfy}{ycnfx}$$

$$(\text{tran } csl)\ \frac{xcsly}{xcslz, xcsly \mid zcsly, xcsly} \qquad z \text{ is any object symbol}$$

The rule (irref cnf) that reflects irreflexivity of cnf is analogous to the rule (irref R_p) presented in Sect. 12.6.

For every object symbol x,

$$(\text{irref } cnf)\ \frac{}{xcnfx}$$

The remaining rules specific for $\mathsf{RL}_{\mathsf{ESL}}$-dual tableau have the following forms:

For all object symbols x and y,

$$(\text{asym } csl)\ \frac{x1'y}{xcsly, x1'y \mid ycslx, x1'y}$$

(inh) $\dfrac{xcnfy}{xcnfz, xcnfy \mid zcsly, xcnfy}$ z is any object symbol

(run_1) $\dfrac{xruny}{x{-}cnfz, z{-}runy}$ z is a new object variable

(run_2) $\dfrac{x{-}runy}{xcnfz, x{-}runy \mid zruny, x{-}runy}$ z is any object symbol

(run_3) $\dfrac{}{xruny \mid x{-}runy}$

(run_4) $\dfrac{xruny}{xrunz, xruny}$ z is any object variable

Proposition 19.5.3.

1. *The* $\mathsf{RL}_{\mathsf{ESL}}$*-rules are* $\mathsf{RL}_{\mathsf{ESL}}$*-correct;*
2. *The* $\mathsf{RL}_{\mathsf{ESL}}$*-axiomatic sets are* $\mathsf{RL}_{\mathsf{ESL}}$*-sets.*

Proof. We show correctness of the rules that are specific for $\mathsf{RL}_{\mathsf{ESL}}$-dual tableau. Correctness of the rules (ref csl), (asym csl), and (tran csl) follows from reflexivity, asymmetry, and transitivity of the relation csl, respectively. Correctness of the rules (irref cnf) and (sym cnf) follows from irreflexivity and symmetry of the relation cnf, respectively. Correctness of these rules can be proved in a similar way as in Theorem 6.6.1. The rule (inh) reflects the condition $cnf\,;csl \subseteq cnf$. The rules (run_1) and (run_2) reflect the condition $run = -(cnf\,;run)$. The rule (run_3) is a cut-like rule needed in the proof of completeness. The rule (run_4) reflects that the relation run is a right ideal relation. □

The notions of an $\mathsf{RL}_{\mathsf{ESL}}$-proof tree, a closed branch of such a tree, a closed $\mathsf{RL}_{\mathsf{ESL}}$-proof tree, and $\mathsf{RL}_{\mathsf{ESL}}$-provability are defined as in Sect. 2.4.

A branch b of an $\mathsf{RL}_{\mathsf{ESL}}$-proof tree is complete whenever it is closed or it satisfies the completion conditions of $\mathsf{RL}(1, 1')$-dual tableau, the completion conditions determined by the rules (ref csl), (tran csl), (sym cnf), and (irref cnf) which are the instances of the completion conditions presented in Sects. 6.6 and 12.6, and the following completion conditions specific for the $\mathsf{RL}_{\mathsf{ESL}}$-dual tableau:

For all object symbols x and y,

Cpl(asym csl) If $x1'y \in b$, then either $xcsly \in b$ or $ycslx \in b$, obtained by an application of the rule (asym csl);

Cpl(inh) If $xcnfy \in b$, then for every object symbol z, either $xcnfz \in b$ or $zcsly \in b$, obtained by an application of the rule (inh);

Cpl(run_1) If $xruny \in b$, then for some object variable z both $x{-}cnfz \in b$ and $z{-}runy \in b$, obtained by an application of the rule (run_1);

Cpl(run_2) If $x{-}runy \in b$, then for every object symbol z, either $xcnfz \in b$ or $zruny \in b$, obtained by an application of the rule (run_2);

Cpl(run_3) Either $xruny \in b$ or $x{-}runy \in b$, obtained by an application of the rule (run_3);

Cpl(run_4) If $xruny \in b$, then for every object variable z, $xrunz \in b$, obtained by an application of the rule (run_4).

The notions of a complete $\mathsf{RL_{ESL}}$-proof tree and an open branch of an $\mathsf{RL_{ESL}}$-proof tree are defined as in RL-logic (see Sect. 2.5). As in Proposition 2.8.1, it can be proved that for every branch of an $\mathsf{RL_{ESL}}$-proof tree, if xRy and $x{-}Ry$ for an atomic term R belong to the branch, then the branch can be closed. Thus, the closed branch property holds.

Let b be an open branch of an $\mathsf{RL_{ESL}}$-proof tree. The branch structure $\mathcal{M}^b = (U^b, csl^b, cnf^b, run^b, m^b)$ is defined in a standard way (see Sect. 2.6), in particular $T^b \stackrel{\text{df}}{=} \{(x,y) \in U^b \times U^b : xTy \notin b\}$ and $m^b(T) \stackrel{\text{df}}{=} T^b$, for every $T \in \{csl, cnf, run\}$.

Proposition 19.5.4 (Branch Model Property). *For every open branch b of an $\mathsf{RL_{ESL}}$-proof tree, $\mathcal{M}^b$ is an $\mathsf{RL_{ESL}}$-model.*

Proof. By way of example, we show that run^b satisfies all the required properties. By the completion condition Cpl(run_4), run^b is a right ideal relation. Hence, by Proposition 19.5.2, it suffices to show that $run^b = -(cnf^b ; run^b)$.

For $run^b \subseteq -(cnf^b ; run^b)$, assume that $(x,y) \in run^b$. Then, $xruny \notin b$. By the completion condition Cpl(run_3), $x{-}runy \in b$. Thus, by the completion condition Cpl(run_2), for every $z \in U^b$, either $xcnfz \in b$ or $zruny \in b$. Suppose $(x,y) \notin -(cnf^b ; run^b)$. Then, there exists $z \in U^b$ such that $xcnf \notin b$ and $zruny \notin b$, a contradiction. The other inclusion can be proved in a similar way. □

Since the branch model $\mathcal{M}^b$ is defined in a standard way and the closed branch property holds, the satisfaction in branch model property can be proved as in $\mathsf{RL}(1,1')$-logic (see Sects. 2.5 and 2.7). Hence, completeness of $\mathsf{RL_{ESL}}$-dual tableau follows:

Theorem 19.5.3 (Soundness and Completeness of $\mathsf{RL_{ESL}}$). *For every $\mathsf{RL_{ESL}}$-formula φ, the following conditions are equivalent:*

1. *φ is $\mathsf{RL_{ESL}}$-valid;*
2. *φ is true in all standard $\mathsf{RL_{ESL}}$-models;*
3. *φ is $\mathsf{RL_{ESL}}$-provable.*

By Theorems 19.5.2 and 19.5.3, we get:

Theorem 19.5.4 (Relational Soundness and Completeness of ESL). *For every ESL-formula φ and for all object variables x and y, the following conditions are equivalent:*

1. *φ is ESL-valid;*
2. *$x\tau(\varphi)y$ is $\mathsf{RL_{ESL}}$-provable.*

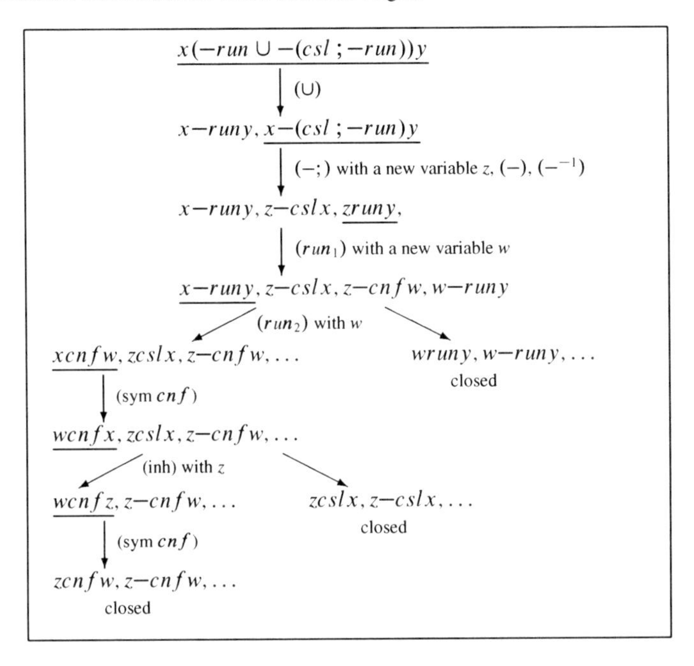

Fig. 19.5 An $\mathsf{RL}_{\mathsf{ESL}}$-proof of the formula $run \rightarrow [csl^{-1}]run$

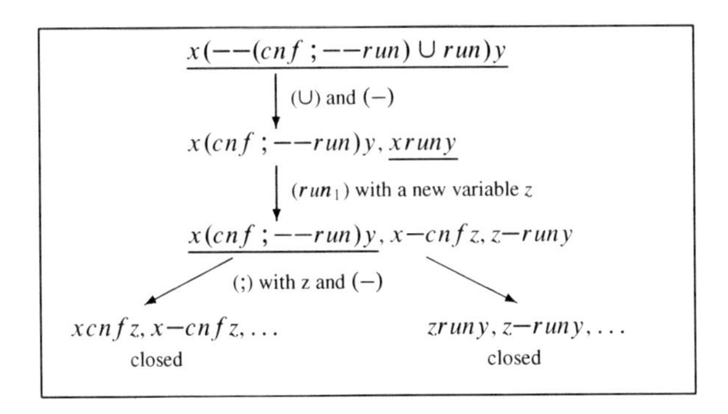

Fig. 19.6 An $\mathsf{RL}_{\mathsf{ESL}}$-proof of the formula $[cnf]\neg run \rightarrow run$

Example. Consider the following ESL-formulas:

$$\varphi = run \rightarrow [csl^{-1}]run,$$

$$\psi = [cnf]\neg run \rightarrow run.$$

The translations of these formulas into relational terms are:

$$\tau(\varphi) = -run \cup -(csl\, ; -run),$$
$$\tau(\psi) = --(cnf\, ; --run) \cup run.$$

Figures 19.5 and 19.6 present $\mathsf{RL}_{\mathsf{ESL}}$-proofs of $x\tau(\varphi)y$ and $x\tau(\psi)y$, respectively. By Theorem 19.5.4 formulas φ and ψ are ESL-valid.

Part VI
Beyond Relational Theories

Chapter 20
Dual Tableaux for Threshold Logics

20.1 Introduction

Threshold logics are a tool for specification, design, and verification of switching circuits constructed from electronic gates. One of the first ideas in this direction can be found in [PM60]; see also [Der65]. Threshold logics developed as a circuit design paradigm alternative to the classical Boolean logic. A threshold element is a generalization of a conventional gate. A single threshold element can represent a number of switching functions obtained through various combinations of weights and a threshold. The concept of threshold provides a representation of a level and switching states below or above the level. A survey of the development and applications of threshold logics can be found in [BQA03, Mur71].

In this chapter we consider a class of first-order threshold logics such that the weights and the thresholds are elements of a commutative group. We develop dual tableaux for this class of logics and we prove their completeness. Next, we show that the standard threshold logic based on the additive group of integers is mutually interpretable with the classical first-order logic. In the above mentioned applications the propositional calculus of this standard threshold logic is used. The dual tableaux presented in this chapter originated in [Orł74, Orł76]. Their generalization to threshold logics based on arbitrary groups was developed in [Cie80].

20.2 Threshold Logics

Let G be the set of elements of the non-zero commutative group $(G, +, 0)$ with the relation $<$ of linear ordering satisfying the following conditions:

For all $a, b, c, d \in G$,

(1g) If $a < b$ and $c \leq d$, then $a + c < b + d$;
(2g) If $a < b$, then $-b < -a$.

Since G is non-zero and by (2g), we can distinguish an element $g_0 \in G$ such that $g_0 > 0$.

E. Orłowska and J. Golińska-Pilarek, *Dual Tableaux: Foundations, Methodology, Case Studies*, Trends in Logic 36, DOI 10.1007/978-94-007-0005-5_20,

Let an element 1 be defined as:

$$(3g)\quad 1 \stackrel{\text{df}}{=} \begin{cases} \min\{g \in G : g > 0\} & \text{if this minimum exists} \\ g_0 & \text{otherwise.} \end{cases}$$

By this definition, we have:

$$(4g)\quad 1 > 0.$$

The terms and formulas of the threshold logic, TL_G, determined by a group G are constructed with the symbols from the following pairwise disjoint sets:

- $\mathbb{OV}_{\mathsf{TL}_G}$ – a countable set of individual variables;
- $\mathbb{F}_{\mathsf{TL}_G}$ – a countable set of function symbols;
- $\mathbb{P}_{\mathsf{TL}_G}$ – a countable set of predicate symbols;
- G – the set of elements of the group;
- $\{\forall, \exists\}$ – the set of quantifiers.

The set $\mathbb{T}_{\mathsf{TL}_G}$ of terms is defined in the usual way as in the first-order logic (see Sect. 5.5). The set of formulas of logic TL_G is the smallest set that satisfies the following conditions:

- If P is a k-ary predicate symbol, $k \geq 1$, and $\tau_1, \ldots, \tau_k \in \mathbb{T}_{\mathsf{TL}_G}$, then $P(\tau_1, \ldots, \tau_k)$ is a TL_G-formula; formulas of that form are referred to as atomic formulas;
- If $\varphi_1, \ldots, \varphi_l$ are TL_G-formulas and $n_1, \ldots, n_l \in G \setminus \{0\}$ and $t \in G$, then $(n_1, \ldots, n_l, t)(\varphi_1, \ldots, \varphi_l)$ is a TL_G-formula, for $l \geq 1$;
- If $\varphi(x)$, $x \in \mathbb{OV}_{\mathsf{TL}_G}$, is a TL_G-formula, then $\forall x\varphi(x)$ and $\exists x\varphi(x)$ are TL_G-formulas.

In the formula $(n_1, \ldots, n_l, t)(\varphi_1, \ldots, \varphi_l)$ the elements $n_1, \ldots, n_l$ are called the *weights* of the formulas $\varphi_1, \ldots, \varphi_l$, respectively. The element t is called its *threshold*. Strings $(n_1, \ldots, n_l, t)$ play the role of propositional operations.

A TL_G*-model* is a triple $\mathcal{M} = (U, G, m)$ such that:

- U is a non-empty set;
- m is a meaning function which assigns:
  - k-ary relations on U to k-ary predicate symbols, that is $m(P) \subseteq U^k$, for every k-ary predicate symbol P, $k \geq 1$;
  - k-ary functions on U to k-ary function symbols, that is $m(f)\colon U^k \to U$, for every k-ary function symbol f, $k \geq 1$.

Any function $v\colon \mathbb{OV}_{\mathsf{TL}_G} \to U$ is a valuation in a TL_G-model $\mathcal{M} = (U, G, m)$. The meaning function m extends to all terms so that for any term τ, $m(\tau)$ is a function from the set of valuations into U:

- If $\tau = x$, for $x \in \mathbb{OV}_{\mathsf{TL}_G}$, then $m(\tau)(v) = v(x)$;
- If $\tau = f(\tau_1, \ldots, \tau_k)$, then $m(\tau)(v) = m(f)(m(\tau_1)(v), \ldots, m(\tau_k)(v))$, for every k-ary function symbol f, $k \geq 1$.

Let $\mathcal{M} = (U, G, m)$ be a TL_G-model and let v be a valuation in $\mathcal{M}$. The satisfaction of a TL_G-formula φ in model $\mathcal{M}$ by valuation v, $\mathcal{M}, v \models \varphi$, is defined inductively as:

- $\mathcal{M}, v \models P(\tau_1, \ldots, \tau_k)$ iff $(m(\tau_1)(v), \ldots, m(\tau_k)(v)) \in m(P)$;
- $\mathcal{M}, v \models (n_1, \ldots, n_k, t)(\varphi_1, \ldots, \varphi_k)$ iff $\sum_{i=1}^{k} h_{\mathcal{M},v}(n_i, \varphi_i) < t$,

  where $h_{\mathcal{M},v}$ is the function assigning elements of G to pairs (n, φ), for $n \in G$ and a TL_G-formula φ:

$$h_{\mathcal{M},v}(n, \varphi) \stackrel{\text{df}}{=} \begin{cases} n & \text{if } \mathcal{M}, v \models \varphi \\ 0 & \text{otherwise;} \end{cases}$$

- $\mathcal{M}, v \models \forall x \varphi(x)$ iff for every valuation v' in $\mathcal{M}$ such that v and v' coincide on $\mathbb{OV}_{\mathsf{TL}_G} \setminus \{x\}$, $\mathcal{M}, v' \models \varphi(x)$;
- $\mathcal{M}, v \models \exists x \varphi(x)$ iff for some valuation v' as above, $\mathcal{M}, v' \models \varphi(x)$.

It follows that $\mathcal{M}, v \not\models (n_1, \ldots, n_k, t)(\varphi_1, \ldots, \varphi_k)$ iff $\sum_{i=1}^{k} h_{\mathcal{M},v}(n_i, \varphi_i) \geq t$.

A TL_G-formula φ is true in a TL_G-model $\mathcal{M}$ if and only if $\mathcal{M}, v \models \varphi$ for every valuation v in $\mathcal{M}$, and it is TL_G-valid whenever it is true in all TL_G-models. A formula φ is unsatisfiable if and only if for every TL_G-model $\mathcal{M}$ and for every valuation v in $\mathcal{M}$, $\mathcal{M}, v \not\models \varphi$.

Proposition 20.2.1. *Let φ be a TL_G-formula, let $\mathcal{M}$ be a TL_G-model, and let v be a valuation in $\mathcal{M}$. Then:*

1. *If $n \geq t$ and $t \leq 0$, then $(n, t)\varphi$ is unsatisfiable;*
2. *$\mathcal{M}, v \models \varphi$ iff $\mathcal{M}, v \not\models (1, 1)\varphi$.*

Proof. For 1., let $n \geq t$ and $t \leq 0$, and let φ be a TL_G-formula. Observe that for every TL_G-model $\mathcal{M}$ and for every valuation v in $\mathcal{M}$, $\mathcal{M}, v \models (n, t)\varphi$ iff $h_{\mathcal{M},v}(n, \varphi) < t$. The assumptions of the proposition enable us to conclude that for every TL_G-model $\mathcal{M}$ and for every valuation v in $\mathcal{M}$, $h_{\mathcal{M},v}(n, \varphi) \geq t$, hence $\mathcal{M}, v \not\models (n, t)\varphi$. Therefore $(n, t)\varphi$ is unsatisfiable.

For 2., assume that $\mathcal{M}, v \models (1, 1)\varphi$. Then $h_{\mathcal{M},v}(1, \varphi) < 1$, which means that $h_{\mathcal{M},v}(1, \varphi) = 0$, hence $\mathcal{M}, v \not\models \varphi$. Assume that $\mathcal{M}, v \not\models (1, 1)\varphi$. Then $h_{\mathcal{M},v}(1, \varphi) \geq 1$, that is $h_{\mathcal{M},v}(1, \varphi) = 1$, hence $\mathcal{M}, v \models \varphi$, which completes the proof. □

Note that the above proposition implies that the operation $(1, 1)$ behaves like the classical negation.

20.3 Dual Tableaux for Threshold Logics

A dual tableau for the logic TL_G consists of the rules of the following forms:
For all TL_G-formulas $\varphi, \varphi_1, \ldots, \varphi_k$, $k \geq 1$, and for all $n_1, \ldots, n_k, n, t \in G$,

(TL1) $$\frac{(n_1, \ldots, n_k, t)(\varphi_1, \ldots, \varphi_k)}{\varphi_k, \psi \mid (n_1, \ldots, n_{k-1}, t - n_k)(\varphi_1, \ldots, \varphi_{k-1}), \psi}$$
where $k > 1$, $n_k < 0$, and $\psi = (n_1, \ldots, n_{k-1}, t)(\varphi_1, \ldots, \varphi_{k-1})$

(TL2) $$\frac{(n_1, \ldots, n_k, t)(\varphi_1, \ldots, \varphi_k)}{(1,1)\varphi_k, \psi \mid (n_1, \ldots, n_{k-1}, t)(\varphi_1, \ldots, \varphi_{k-1}), \psi}$$
where $k > 1$, $n_k > 0$, and $\psi = (n_1, \ldots, n_{k-1}, t - n_k)(\varphi_1, \ldots, \varphi_{k-1})$

(TL3) $$\frac{(n,t)(\varphi)}{\varphi} \quad \text{for } n < t \text{ and } t \leq 0$$

(TL4) $$\frac{(n,t)(\varphi)}{(1,1)(\varphi)} \quad \text{for } 0 < t \leq n \text{ and not } (n = t = 1)$$

(TL5) $$\frac{(1,1)((n_1, \ldots, n_k, t)(\varphi_1, \ldots, \varphi_k))}{(-n_1, \ldots, -n_k, \varepsilon(n_1, \ldots, n_k, t) - t)(\varphi_1, \ldots, \varphi_k)}$$

where $\varepsilon(n_1, \ldots, n_k, t) = \min(\{1\} \cup S(n_1, \ldots, n_k, t))$ and
$S(n_1, \ldots, n_k, t) = \{g \in G : g > 0 \text{ and } g = t - \sum_{i=1}^{k} \chi_Q(i) \text{ for } Q \subseteq \{1, \ldots, k\}\}$,
where

$$\chi_Q(i) \stackrel{\text{df}}{=} \begin{cases} n_i & \text{for } i \in Q \\ 0 & \text{for } i \notin Q \end{cases} \qquad \text{for every } Q \subseteq \{1, \ldots, k\}.$$

(TL6) $$\frac{(1,1)(\forall x \varphi(x))}{\exists x (1,1)(\varphi(x))}$$

(TL7) $$\frac{(1,1)(\exists x \varphi(x))}{\forall x (1,1)(\varphi(x))}$$

(TL8) $$\frac{\forall x \varphi(x)}{\varphi(z)} \quad z \text{ is a new individual variable}$$

(TL9) $$\frac{\exists x \varphi(x)}{\varphi(\tau), \exists x \varphi(x)} \quad \tau \text{ is any term}$$

A finite set X of TL-formulas is said to be TL-axiomatic whenever it includes either of the subsets of the following forms:

For any TL_G-formula φ,

(Ax1) $\{\varphi, (1,1)\varphi\}$;
(Ax2) $\{(n,t)\varphi\}$, for $n < t, t > 0$.

The notions of a TL_G-set and TL_G-correctness of a rule are defined in a similar way as in F-logic in Sect. 1.3.

Proposition 20.3.1.

1. *The* TL_G*-axiomatic sets are* TL_G*-sets;*
2. *The* TL_G*- rules are* TL_G*-correct.*

Proof.

1. Let φ be a TL_G-formula, let $\mathcal{M}$ be a TL_G-model, and let v be a valuation in $\mathcal{M}$. Note that if $\mathcal{M}, v \not\models \varphi$, then $h_{\mathcal{M},v}(1,\varphi) = 0 < 1$, hence $\mathcal{M}, v \models (1,1)\varphi$. Therefore for every TL_G-model and for every valuation v in $\mathcal{M}$, $\mathcal{M}, v \models \varphi$ or $\mathcal{M}, v \models (1,1)\varphi$, hence every finite set X of TL_G-formulas including φ and $(1,1)\varphi$ is a TL_G-set. Let $n > t$ and $t > 0$. Then $h_{\mathcal{M},v}(n,\varphi) < t$, for every TL_G-model $\mathcal{M}$, for every valuation v in $\mathcal{M}$, and for every TL_G-formula φ. Therefore $\mathcal{M}, v \models (n,t)\varphi$, hence $\{(n,t)\varphi\}$ is a TL_G-set.

2. By way of example, we prove correctness of the rules ($\mathsf{TL}1$), ($\mathsf{TL}3$), and ($\mathsf{TL}5$). Let X be a finite set of TL_G-formulas.

($\mathsf{TL}1$) Let $k > 1$ and $n_k < 0$. Then, $h_{\mathcal{M},v}(n_k, \varphi_k) \leq 0$ for every TL_G-model $\mathcal{M}$ and for every valuation v in $\mathcal{M}$. Assume $X \cup \{(n_1, \ldots, n_k, t)\ (\varphi_1, \ldots, \varphi_k)\}$ is a TL_G-set, that is for every TL_G-model $\mathcal{M}$ and for every valuation v in $\mathcal{M}$, either there exists $\vartheta \in X$ such that $\mathcal{M}, v \models \vartheta$ or $\sum_{i=1}^{k} h_{\mathcal{M},v}(n_i, \varphi_i) < t$. Let $\mathcal{M}$ be a TL_G-model and let v be a valuation in $\mathcal{M}$. If $\mathcal{M}, v \models \varphi_k$, then $h_{\mathcal{M},v}(n_k, \varphi_k) = n_k < 0$, hence $\sum_{i=1}^{k-1} h_{\mathcal{M},v}(n_i, \varphi_i) < t - n_k$. Therefore, $\mathcal{M}, v \models \varphi_k$ and $\mathcal{M}, v \models (n_1, \ldots, n_{k-1}, t - n_k)(\varphi_1, \ldots, \varphi_k)$. If $\mathcal{M}, v \not\models \varphi_k$, then $h_{\mathcal{M},v}(n_k, \varphi_k) = 0$, hence $\sum_{i=1}^{k-1} h_{\mathcal{M},v}(n_i, \varphi_i) < t$. Thus, either there exists $\vartheta \in X$ such that $\mathcal{M}, v \models \vartheta$ or both $\mathcal{M}, v \models \varphi_k$ and $\mathcal{M}, v \models (n_1, \ldots, n_{k-1}, t)(\varphi_1, \ldots, \varphi_k)$. Therefore, sets $X \cup \{\varphi_k, \psi\}$ and $X \cup \{(n_1, \ldots, n_{k-1}, t - n_k)(\varphi_1, \ldots, \varphi_{k-1}), \psi\}$ are TL_G-sets, where $\psi = (n_1, \ldots, n_{k-1}, t)(\varphi_1, \ldots, \varphi_{k-1})$.

Assume that $X \cup \{\varphi_k, \psi\}$ and $X \cup \{(n_1, \ldots, n_{k-1}, t - n_k)(\varphi_1, \ldots, \varphi_{k-1}), \psi\}$ are TL_G-sets, where $\psi = (n_1, \ldots, n_{k-1}, t)(\varphi_1, \ldots, \varphi_{k-1})$. Suppose that $X \cup \{(n_1, \ldots, n_k, t)\ (\varphi_1, \ldots, \varphi_k)\}$ is not a TL_G-set, that is there exist a TL_G-model $\mathcal{M}$ and a valuation v in $\mathcal{M}$ such that $\sum_{i=1}^{k} h_{\mathcal{M},v}(n_i, \varphi_i) \geq t$. Suppose that $\mathcal{M}, v \models \varphi_k$ which implies $h_{\mathcal{M},v}(n_k, \varphi_k) = n_k < 0$. If $\mathcal{M}, v \models (n_1, \ldots, n_{k-1}, t - n_k)$ $(\varphi_1, \ldots, \varphi_{k-1})$, then $\sum_{i=1}^{k-1} h_{\mathcal{M},v}(n_i, \varphi_i) < t - n_k$. Thus:

$$\sum_{i=1}^{k} h_{\mathcal{M},v}(n_i, \varphi_i) = \sum_{i=1}^{k-1} h_{\mathcal{M},v}(n_i, \varphi_i) + h_{\mathcal{M},v}(n_k, \varphi_k) < (t - n_k) + n_k = t,$$

a contradiction.

If $\mathcal{M}, v \not\models (n_1, \ldots, n_{k-1}, t - n_k)(\varphi_1, \ldots, \varphi_{k-1})$, then by the assumption $\mathcal{M}, v \models \psi$. Therefore, we obtain $\sum_{i=1}^{k-1} h_{\mathcal{M},v}(n_i, \varphi_i) < t$, which implies $\sum_{i=1}^{k} h_{\mathcal{M},v}(n_i, \varphi_i) < t + n_k < t$, a contradiction.

If $\mathcal{M}, v \not\models \varphi_k$, then by the assumption $h_{\mathcal{M},v}(n_k, \varphi_k) = 0$ and $\mathcal{M}, v \models \psi$. Then $\sum_{i=1}^{k} h_{\mathcal{M},v}(n_i, \varphi_i) < t + 0 = t$, a contradiction. Hence, the rule (TL1) is TL_G-correct.

(TL3) Let $n < t$ and $t \leq 0$. Assume $X \cup \{(n,t)(\varphi)\}$ is a TL_G-set. Then for every TL_G-model $\mathcal{M}$ and for every valuation v in $\mathcal{M}$, either there exists $\vartheta \in X$ such that $\mathcal{M}, v \models \vartheta$ or $h_{\mathcal{M},v}(n, \varphi) < t$. Suppose $X \cup \{\varphi\}$ is not a TL_G-set. Then there exist a TL_G-model $\mathcal{M}$ and a valuation v in $\mathcal{M}$ such that $\mathcal{M}, v \not\models \varphi$. Thus $h_{\mathcal{M},v}(n, \varphi) = 0$. However, by the assumption, model $\mathcal{M}$ and valuation v satisfy $h_{\mathcal{M},v}(n, \varphi) < t$. Hence $t > 0$, a contradiction. Now, assume $X \cup \{\varphi\}$ is a TL_G-set, that is for every TL_G-model $\mathcal{M}$ and for every valuation v in $\mathcal{M}$, either there exists $\vartheta \in X$ such that $\mathcal{M}, v \models \vartheta$ or $\mathcal{M}, v \models \varphi$. Suppose $X \cup \{(n,t)(\varphi)\}$ is not a TL_G-set. Then there exist a TL_G-model $\mathcal{M}$ and a valuation v in $\mathcal{M}$ such that $h_{\mathcal{M},v}(n, \varphi) \geq t$. By the assumption, $h_{\mathcal{M},v}(n, \varphi) = n$, which implies $n \geq t$, a contradiction. Hence the rule (TL3) is TL_G-correct.

(TL5) For simplicity, in the following we will write ε instead of $\varepsilon(n_1, \ldots, n_k, t)$ and S instead of $S(n_1, \ldots, n_k, t)$. Note that $\varepsilon > 0$. Assume that $X \cup \{(1,1)((n_1, \ldots, n_k, t)(\varphi_1, \ldots, \varphi_k))\}$ is a TL_G-set. Then for every TL_G-model $\mathcal{M}$ and for every valuation v in $\mathcal{M}$, either there exists $\vartheta \in X$ such that $\mathcal{M}, v \models \vartheta$ or $\mathcal{M}, v \models (1,1)((n_1, \ldots, n_k, t)(\varphi_1, \ldots, \varphi_k))$, which implies that $(n_1, \ldots, n_k, t)(\varphi_1, \ldots, \varphi_k)$ is not satisfied in $\mathcal{M}$ by valuation v, and hence $\sum_{i=1}^{k} h_{\mathcal{M},v}(n_i, \varphi_i) \geq t$. Suppose that $X \cup \{(-n_1, \ldots, -n_k, \varepsilon - t)(\varphi_1, \ldots, \varphi_k)\}$ is not a TL_G-set. Then there exist a TL_G-model $\mathcal{M}$ and a valuation v in $\mathcal{M}$ such that $\mathcal{M}, v \not\models (-n_1, \ldots, -n_k, \varepsilon - t)(\varphi_1, \ldots, \varphi_k)$, that is $\sum_{i=1}^{k} h_{\mathcal{M},v}(-n_i, \varphi_i) \geq \varepsilon - t$. Therefore $\sum_{i=1}^{k} -h_{\mathcal{M},v}(n_i, \varphi_i) \geq \varepsilon - t$. By the assumption, $\sum_{i=1}^{k} -h_{\mathcal{M},v}(n_i, \varphi_i) \leq -t$. Thus $\varepsilon \geq 0$, a contradiction.

Now, assume $X \cup \{(-n_1, \ldots, -n_k, \varepsilon - t)(\varphi_1, \ldots, \varphi_k)\}$ is a TL_G-set. Note that the formula $(-n_1, \ldots, -n_k, \varepsilon - t)(\varphi_1, \ldots, \varphi_k))$ is satisfied in a TL_G-model $\mathcal{M}$ by a valuation v in $\mathcal{M}$ whenever $\sum_{i=1}^{k} h_{\mathcal{M},v}(-n_i, \varphi_i) < \varepsilon - t$, that is $t - \sum_{i=1}^{k} h_{\mathcal{M},v}(n_i, \varphi_i) < \varepsilon$. Suppose $X \cup \{(1,1)((n_1, \ldots, n_k, t)(\varphi_1, \ldots, \varphi_k))\}$ is not a TL_G-set. Then there exist a TL_G-model $\mathcal{M}$ and a valuation v in $\mathcal{M}$ such that $\mathcal{M}, v \not\models (1,1)((n_1, \ldots, n_k, t)(\varphi_1, \ldots, \varphi_k))$, which implies that $\mathcal{M}, v \models (n_1, \ldots, n_k, t)(\varphi_1, \ldots, \varphi_k)$. Thus, $\sum_{i=1}^{k} h_{\mathcal{M},v}(n_i, \varphi_i) < t$, so $t - \sum_{i=1}^{k} h_{\mathcal{M},v}(n_i, \varphi_i) > 0$. Then, by the definition of S, $t - \sum_{i=1}^{k} h_{\mathcal{M},v}(n_i, \varphi_i) \in S$, and hence $t - \sum_{i=1}^{k} h_{\mathcal{M},v}(n_i, \varphi_i) \geq \varepsilon$. However, by the assumption, $\mathcal{M}, v \models (-n_1, \ldots, -n_k, \varepsilon - t)(\varphi_1, \ldots, \varphi_k))$, thus $t - \sum_{i=1}^{k} h_{\mathcal{M},v}(n_i, \varphi_i) < \varepsilon$, a contradiction.

Correctness of rules (TL6) and (TL7) follows easily from Proposition 20.2.1. The proofs of correctness of the remaining rules are similar. □

A TL_G-proof tree for a TL_G-formula φ is a finitely branching tree defined as in Sect. 1.3. The notions of a closed branch of a TL_G-proof tree, a closed TL_G-proof tree, and TL_G-provability are defined as in Sect. 1.3. The rules guarantee that whenever φ and $(1,1)\varphi$, for an atomic formula φ, belong to a branch b of a TL_G-proof tree, then they eventually will appear in the same node of b. Hence, we obtain:

Proposition 20.3.2 (Closed Branch Property). *For every branch b of a TL_G-proof tree, if both $\varphi \in b$ and $(1,1)\varphi \in b$, for some atomic formula φ, then branch b is closed.*

A branch b of a TL_G-proof tree is complete whenever it is closed or it satisfies the following completion conditions:

For all TL_G-formulas $\varphi, \varphi_1, \ldots, \varphi_k$, $k \geq 1$, and for all $n_1, \ldots, n_k, n, t \in G$,

Cpl(TL1) If $(n_1, \ldots, n_k, t)(\varphi_1, \ldots, \varphi_k) \in b$, for $k > 1$ and $n_k < 0$, then either $\varphi_k \in b$ and $(n_1, \ldots, n_{k-1}, t)(\varphi_1, \ldots, \varphi_{k-1}) \in b$ or $(n_1, \ldots, n_{k-1}, t - n_k)(\varphi_1, \ldots, \varphi_{k-1}) \in b$ and $(n_1, \ldots, n_{k-1}, t)(\varphi_1, \ldots, \varphi_{k-1}) \in b$, obtained by an application of the rule (TL1);
Cpl(TL2) If $(n_1, \ldots, n_k, t)(\varphi_1, \ldots, \varphi_k) \in b$, for $k > 1$ and $n_k > 0$, then either $(1,1)(\varphi_k) \in b$ and $(n_1, \ldots, n_{k-1}, t - n_k)(\varphi_1, \ldots, \varphi_{k-1}) \in b$ or $(n_1, \ldots, n_{k-1}, t - n_k)(\varphi_1, \ldots, \varphi_{k-1}) \in b$ and $(n_1, \ldots, n_{k-1}, t)(\varphi_1, \ldots, \varphi_{k-1}) \in b$, obtained by an application of the rule (TL2);
Cpl(TL3) If $(n,t)(\varphi) \in b$, for $n < t$ and $t \leq 0$, then $\varphi \in b$, obtained by an application of the rule (TL3);
Cpl(TL4) If $(n,t)(\varphi) \in b$, for $n \geq t, t > 0$, and not $(n = t = 1)$, then $(1,1)(\varphi) \in b$, obtained by an application of the rule (TL4);
Cpl(TL5) If $(1,1)((n_1, \ldots, n_k, t)(\varphi_1, \ldots, \varphi_k)) \in b$, then $(-n_1, \ldots, -n_k, \varepsilon(n_1, \ldots, n_k, t) - 1)(\varphi_1, \ldots, \varphi_k) \in b$, obtained by an application of the rule (TL5);
Cpl(TL6) If $(1,1)(\forall x \varphi(x)) \in b$, then $\exists x(1,1)(\varphi(x)) \in b$, obtained by an application of the rule (TL6);
Cpl(TL7) If $(1,1)(\exists x \varphi(x)) \in b$, then $\forall x(1,1)(\varphi(x)) \in b$, obtained by an application of the rule (TL7);
Cpl(TL8) If $\forall x \varphi(x) \in b$, then for some individual variable z, $\varphi(z) \in b$, obtained by an application of the rule (TL8);
Cpl(TL9) If $\exists x \varphi(x) \in b$, then for every term τ, $\varphi(\tau) \in b$, obtained by an application of the rule (TL9).

The notion of a complete TL_G-proof tree and the notion of an open branch of a TL_G-proof tree are defined as in F-logic (see Sect. 1.3).

The branch structure is a structure $\mathcal{M}^b = (U^b, G, m^b)$ satisfying the following conditions:

- $U^b = \mathbb{T}_{\mathsf{TL}_G}$;
- m^b is a meaning function such that:
  - $m^b(P) = \{(\tau_1, \ldots, \tau_k) \in U^b \times \ldots \times U^b : P(\tau_1, \ldots, \tau_k) \notin b\}$, for every k-ary predicate symbol P, $k \geq 1$;
  - $m^b(f) = f$, for every k-ary function symbol f, $k \geq 1$;
- m^b extends to all terms as in TL_G-models.

It is easy to see that the structure defined above is a TL_G-model, and hence the branch model property holds.

Let $v^b \colon \mathbb{OV}_{\mathsf{TL}_G} \to U^b$ be defined as $v^b(x) = x$, for every $x \in \mathbb{OV}_{\mathsf{TL}_G}$. It follows that $m^b(\tau)(v^b) = \tau$, for every term τ.

Proposition 20.3.3 (Satisfaction in Branch Model Property). *Let $\mathcal{M}^b$ be a TL_G-model determined by an open branch b of a TL_G-proof tree. For every TL_G-formula φ, if $\mathcal{M}^b, v^b \models \varphi$, then $\varphi \notin b$.*

Proof. The proof is by induction on the complexity of formulas. For formulas of the form $\varphi = P(\tau_1, \ldots, \tau_k)$ the theorem holds by the definition of $\mathcal{M}^b$.

Let $\varphi = (1,1)(P(\tau_1, \ldots, \tau_k))$. Assume that $\mathcal{M}^b, v^b \models (1,1)(P(\tau_1, \ldots, \tau_k))$. Then $h_{\mathcal{M}^b,v^b}(1, P(\tau_1, \ldots, \tau_k)) < 1$, hence $\mathcal{M}^b, v^b \not\models P(\tau_1, \ldots, \tau_k)$, for otherwise $h_{\mathcal{M}^b,v^b}(1, P(\tau_1, \ldots, \tau_k)) = 1$. Therefore, by the definition of satisfaction, $(\tau_1, \ldots, \tau_k) \notin m^b(P)$. By the definition of $\mathcal{M}^b$, $P(\tau_1, \ldots, \tau_k) \in b$. Hence, $(1,1)(P(\tau_1, \ldots, \tau_k)) \notin b$, for otherwise, by the closed branch property, b would be closed.

Let $\varphi = (n_1, \ldots, n_k, t)(\varphi_1, \ldots, \varphi_k)$, where $k > 1$ and $n_k < 0$. Assume $\mathcal{M}^b, v^b \models \varphi$, which implies $\sum_{i=1}^{k} h_{\mathcal{M}^b,v^b}(n_i, \varphi_i) < t$. Suppose that $\varphi \in b$. Then, by the completion condition Cpl($\mathsf{TL}1$), either $\varphi_k \in b$ and $(n_1, \ldots, n_{k-1}, t)(\varphi_1, \ldots, \varphi_{k-1}) \in b$ or $(n_1, \ldots, n_{k-1}, t - n_k)(\varphi_1, \ldots, \varphi_{k-1}) \in b$ and $(n_1, \ldots, n_{k-1}, t)(\varphi_1, \ldots, \varphi_{k-1}) \in b$. Consider the first case. Then, by the induction hypothesis, we get $\mathcal{M}^b, v^b \not\models \varphi_k$ and $\mathcal{M}^b, v^b \not\models (n_1, \ldots, n_{k-1}, t)(\varphi_1, \ldots, \varphi_{k-1})$. Therefore, $h_{\mathcal{M}^b,v^b}(n_k, \varphi_k) = 0$ and $\sum_{i=1}^{k-1} h_{\mathcal{M}^b,v^b}(n_i, \varphi_i) \geq t$. Hence $\sum_{i=1}^{k} h_{\mathcal{M}^b,v^b}(n_i, \varphi_i) \geq t$, a contradiction. Now, consider the second case. Then $\mathcal{M}^b, v^b \not\models (n_1, \ldots, n_{k-1}, t - n_k)(\varphi_1, \ldots, \varphi_k)$ and $\mathcal{M}^b, v^b \not\models (n_1, \ldots, n_{k-1}, t)(\varphi_1, \ldots, \varphi_{k-1})$. Thus, we obtain $\sum_{i=1}^{k-1} h_{\mathcal{M}^b,v^b}(n_i, \varphi_i) \geq t$ and $\sum_{i=1}^{k-1} h_{\mathcal{M}^b,v^b}(n_i, \varphi_i) \geq t - n_k$, which implies $\sum_{i=1}^{k} h_{\mathcal{M}^b,v^b}(n_i, \varphi_i) \geq t$, a contradiction.

Let $\varphi = (n,t)(\psi)$, $n \geq t > 0$, and not $(n = t = 1)$. Assume $\mathcal{M}^b, v^b \models \varphi$. Then $h_{\mathcal{M}^b,v^b}(n, \psi) < t$. This implies $\mathcal{M}^b, v^b \not\models \psi$, for otherwise $n < t$. Suppose $\varphi \in b$. By the completion condition Cpl($\mathsf{TL}4$), $(1,1)(\psi) \in b$. By the induction hypothesis, $\mathcal{M}^b, v^b \not\models (1,1)(\psi)$. So $h_{\mathcal{M}^b,v^b}(1, \psi) \geq 1$. Since $\mathcal{M}^b, v^b \not\models \psi$, $h_{\mathcal{M}^b,v^b}(1, \psi) = 0 \geq 1$, a contradiction.

Let $\varphi = (1,1)(\forall x \psi(x))$. Assume $\mathcal{M}^b, v^b \models \varphi$. Then $h_{\mathcal{M}^b,v^b}(1, \forall x \psi(x)) < 1$, hence $\mathcal{M}^b, v^b \not\models \forall x \psi(x)$. Suppose $\varphi \in b$. By the completion condition Cpl($\mathsf{TL}6$), $\exists x (1,1)(\psi(x)) \in b$. By the induction hypothesis, $\mathcal{M}^b, v^b \not\models \exists x (1,1)(\psi(x))$. Since $\mathcal{M}^b, v^b \not\models \forall x \psi(x)$, there exists $\tau \in U^b$ such that $\mathcal{M}^b, v' \not\models \psi(x)$, where $v'(x) = \tau$ and $v'(z) = v^b(z)$, for every $z \neq x$. Hence, by Proposition 20.2.1, $\mathcal{M}^b, v' \models (1,1)\psi(x)$. Therefore, $\mathcal{M}^b, v^b \models \exists x (1,1)(\psi(x))$, a contradiction.

The proofs of the remaining cases are similar. □

Proposition 20.3.4. *Let φ be a TL_G-formula. If φ is TL_G-valid, then there is a closed TL_G-proof tree for it.*

Proof. Assume φ is TL_G-valid. Suppose there is no any closed TL_G-proof tree for φ. Then there exists a complete TL_G-proof tree for φ with an open branch, say b. By Proposition 20.3.3, since $\varphi \in b$, $\mathcal{M}^b, v^b \not\models \varphi$. Since $\mathcal{M}^b$ is a TL_G-model, φ is not TL_G-valid, a contradiction □

Propositions 20.3.1 and 20.3.4 yield:

Theorem 20.3.1 (Soundness and Completeness of TL_G). *Let φ be a TL_G-formula. Then the following conditions are equivalent:*

1. *φ is TL_G-valid;*
2. *φ is TL_G-provable.*

Observe that the reduct of TL_G-dual tableau consisting of the rules $\mathsf{TL}1$, $\mathsf{TL}2$, $\mathsf{TL}3$, $\mathsf{TL}4$, and $\mathsf{TL}5$ is a decision procedure for the propositional threshold logic.

20.4 Mutual Interpretability of a Threshold Logic and Classical First-Order Logic

Let Z be an additive group of integers. Let TL_Z and FOL be a threshold logic and the first-order logic as defined in Sect. 5.5, respectively, with the same sets of predicate symbols and function symbols. Then threshold logic TL_Z and the first-order logic FOL are mutually interpretable.

Theorem 20.4.1.

1. *There exists a translation γ of terms and formulas of logic FOL into terms and formulas of logic TL_Z, respectively, such that for every FOL-formula φ, φ is FOL-valid iff $\gamma(\varphi)$ is TL_Z-valid;*
2. *There exists a translation γ' of terms and formulas of logic TL_Z into terms and formulas of logic FOL, respectively, such that for every TL_Z-formula φ, φ is TL_Z-valid iff $\gamma'(\varphi)$ is FOL-valid.*

Proof.

1. Let γ be the mapping from the set of terms and formulas of logic FOL into the set of terms and formulas of logic TL_Z defined as:

- $\gamma(\tau) = \tau$, for any term τ of logic FOL;
- $\gamma(P(\tau_1, \ldots, \tau_k)) = P(\tau_1, \ldots, \tau_k)$, for all terms $\tau_1, \ldots, \tau_k$ and for every k-ary predicate symbol $P \in \mathbb{P}_{\mathsf{FOL}}$, $k \geq 1$;
- If φ and ψ are FOL-formulas, then:

$$\begin{aligned}
&\gamma(\neg\varphi) = (1, 1)(\gamma(\varphi));\\
&\gamma(\varphi \vee \psi) = (-1, -1, 0)(\gamma(\varphi), \gamma(\psi));\\
&\gamma(\varphi \wedge \psi) = (-1, -1, -1)(\gamma(\varphi), \gamma(\psi));\\
&\gamma(\varphi \rightarrow \psi) = (1, -1, 1)(\gamma(\varphi), \gamma(\psi));\\
&\gamma(\exists x\varphi) = \exists x(\gamma(\varphi));\\
&\gamma(\forall x\varphi) = \forall x(\gamma(\varphi)).
\end{aligned}$$

Now, it suffices to show that for every FOL-formula φ, the following hold:

(a) For every FOL-model $\mathcal{M}$ and for every valuation v in $\mathcal{M}$, there exist a TL_Z-model $\mathcal{M}'$ and a valuation v' in $\mathcal{M}'$ such that $\mathcal{M}, v \models \varphi$ iff $\mathcal{M}', v' \models \gamma(\varphi)$;
(b) For every TL_Z-model $\mathcal{M}'$ and for every valuation v' in $\mathcal{M}'$, there exist an FOL-model $\mathcal{M}$ and a valuation v in $\mathcal{M}$ such that $\mathcal{M}, v \models \varphi$ iff $\mathcal{M}', v' \models \gamma(\varphi)$.

For (a), let $\mathcal{M} = (U, m)$ be an FOL-model and let v be a valuation in $\mathcal{M}$. We construct a TL_Z-model $\mathcal{M}' = (U', G, m')$ setting $U' = U$, $m'(P) = m(P)$, and $m'(f) = m(f)$, for every predicate symbol P and for every function symbol f. Let v' be a valuation in $\mathcal{M}'$ such that $v' = v$. Then we prove the required condition by induction on the complexity of formulas. Directly from the definition of γ, it holds for the atomic formulas.

Let $\varphi = \neg\psi$. Assume $\mathcal{M}, v \models \varphi$. Then $\mathcal{M}, v \not\models \psi$. By the induction hypothesis $\mathcal{M}', v' \not\models \gamma(\psi)$. Thus $h_{\mathcal{M}',v'}(1, \gamma(\psi)) = 0$, and hence $\mathcal{M}', v' \models (1,1)(\gamma(\psi))$. Assume that $\mathcal{M}', v' \models \gamma(\varphi)$, that is $h_{\mathcal{M}',v'}(1, \gamma(\psi)) < 1$. Thus $h_{\mathcal{M}',v'}(1, \gamma(\psi)) = 0$, and hence $\mathcal{M}', v' \not\models \gamma(\psi)$. Suppose $\mathcal{M}, v \not\models \varphi$. Then by the induction hypothesis, $\mathcal{M}, v \not\models \psi$, hence $\mathcal{M}, v \models \varphi$.

Let $\varphi = \psi \vee \vartheta$. Assume $\mathcal{M}, v \models \varphi$. Then $\mathcal{M}, v \models \psi$ or $\mathcal{M}, v \models \vartheta$. By the induction hypothesis, $\mathcal{M}', v' \models \gamma(\psi)$ or $\mathcal{M}', v' \models \gamma(\vartheta)$. Thus $h_{\mathcal{M}',v'}(-1, \gamma(\psi)) + h_{\mathcal{M}',v'}(-1, \gamma(\vartheta)) < -1 < 0$, hence $\mathcal{M}', v' \models (-1,-1,0)(\gamma(\psi), \gamma(\vartheta))$. Therefore, $\mathcal{M}', v' \models \gamma(\varphi)$. Assume $\mathcal{M}', v' \models \gamma(\varphi)$. Then $h_{\mathcal{M}',v'}(-1, \gamma(\psi)) + h_{\mathcal{M}',v'}(-1, \gamma(\vartheta)) < 0$, hence $\mathcal{M}', v' \models \gamma(\psi)$ or $\mathcal{M}', v' \models \gamma(\vartheta)$. By the induction hypothesis, $\mathcal{M}, v \models \psi$ or $\mathcal{M}, v \models \vartheta$. Thus $\mathcal{M}, v \models \varphi$.

The proofs of the remaining cases are similar.

For (b), let $\mathcal{M}' = (U', G, m')$ be a TL_Z-model and let v' be a valuation in $\mathcal{M}'$. Then let $\mathcal{M} = (U, m)$ be a FOL-model such that $U = U'$, $m(P) = m'(P)$, and $m(f) = m'(f)$, for every predicate symbol P and for every function symbol f. Let v be a valuation in $\mathcal{M}$ such that $v = v'$. Then the required condition can be proved by induction on the complexity of formulas in a similar way as in (a).

2. Let γ' be the mapping from the set of terms and formulas of logic TL_Z into the set of terms and formulas of logic FOL, respectively, defined as follows:

- $\gamma'(\tau) = \tau$, for any term τ of logic TL_Z;
- $\gamma'(P(\tau_1, \ldots, \tau_k)) = P(\tau_1, \ldots, \tau_k)$, for all terms $\tau_1, \ldots, \tau_k$ and for every k-ary predicate symbol $P \in \mathbb{P}_{\mathsf{TL}_Z}$, $k \geq 1$;
- If φ is a TL_Z-formula, then:

$$\gamma'((n,t)(\varphi)) = \neg\gamma'(\varphi), \text{ if } 0 < t \leq n;$$
$$\gamma'((n,t)(\varphi)) = \gamma'(\varphi), \text{ if } n < t \leq 0;$$
$$\gamma'((n,t)(\varphi)) = \varphi \wedge \neg\varphi, \text{ if } n > 0 \text{ and } t < 0 \text{ or } n < 0 \text{ and } t > 0;$$
$$\gamma'((n,t)(\varphi)) = \varphi \vee \neg\varphi, \text{ if } 0 < n < t;$$
$$\gamma'(\exists x\varphi) = \exists x(\gamma'(\varphi));$$
$$\gamma'(\forall x\varphi) = \forall x(\gamma'(\varphi));$$

- If $n_1, \ldots, n_k \in Z \setminus \{0\}, t \in Z$, and $\varphi_1, \ldots, \varphi_k$ are TL_Z-formulas, $k > 0$, then

  $\gamma'((n_1, \ldots, n_k, t)(\varphi_1, \ldots, \varphi_k)) = \gamma'((n_1, \ldots, n_{k-1}, t - n_k)(\varphi_1, \ldots, \varphi_{k-1})) \vee (\neg\gamma'(\varphi_k) \wedge \gamma'((n_1, \ldots, n_{k-1}, t)(\varphi_1, \ldots, \varphi_{k-1})))$, for $n_k > 0$;

  $\gamma'((n_1, \ldots, n_k, t)(\varphi_1, \ldots, \varphi_k)) = \gamma'((n_1, \ldots, n_{k-1}, t)(\varphi_1, \ldots, \varphi_{k-1})) \vee (\gamma'(\varphi_k) \wedge \gamma'((n_1, \ldots, n_{k-1}, t - n_k)(\varphi_1, \ldots, \varphi_{k-1})))$, for $n_k < 0$.

The preservation of validity by γ' can be proved in a similar way as in the case of γ. □

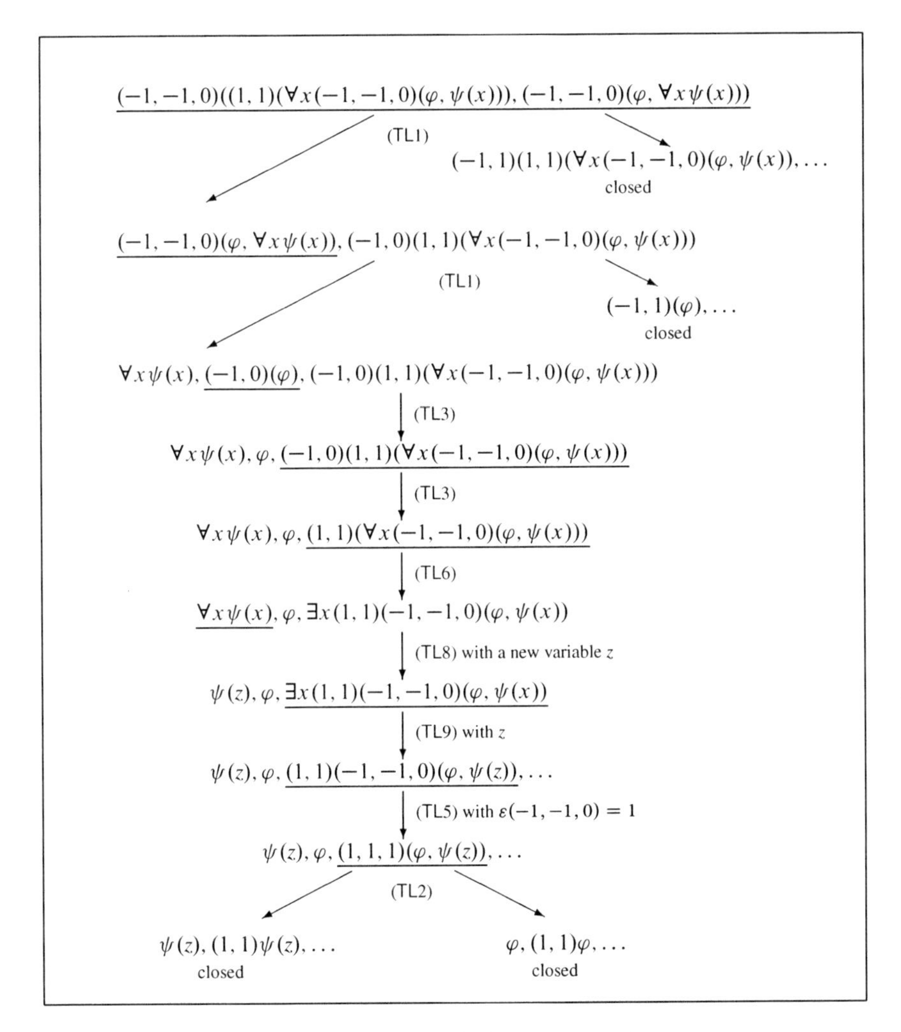

Fig. 20.1 A TL_Z-proof of the formula $\forall x(\varphi \vee \psi(x)) \rightarrow (\varphi \vee \forall x\psi(x))$

Example. Consider the following FOL-formula:

$$\forall x(\varphi \vee \psi(x)) \rightarrow (\varphi \vee \forall x \psi(x)).$$

Its equivalent form is:

$$\alpha = \neg \forall x(\varphi \vee \psi(x)) \vee (\varphi \vee \forall x \psi(x)).$$

By Theorem 20.4.1 (1.), FOL-validity of α is equivalent to TL_Z-provability of $\gamma(\alpha)$, where the translation γ is defined in the proof of Theorem 20.4.1. For the sake of simplicity, we write φ and $\psi(x)$ instead of $\gamma(\varphi)$ and $\gamma(\psi(x))$, respectively:

$$\gamma(\alpha) = (-1, -1, 0)((1, 1)(\forall x(-1, -1, 0)(\varphi, \psi(x))), (-1, -1, 0)(\varphi, \forall x \psi(x))).$$

Figure 20.1 presents a TL_Z-proof of the formula $\gamma(\alpha)$.

Chapter 21
Signed Dual Tableau for Gödel–Dummett Logic

21.1 Introduction

In 1933 Kurt Gödel [Göd33] introduced a family of finitely many-valued propositional logics. His goal was to show that the intuitionistic logic cannot be characterized by a finite matrix. Dummett in [Dum59] generalized them to infinite-valued logics and presented their complete Hilbert-style axiomatization. It consists of the axioms of the intuitionistic propositional logic and the linearity axiom $(\varphi \rightarrow \psi) \vee (\psi \rightarrow \varphi)$. It is known that the set of tautologies of these logics is the same for any infinite set of truth values. Kripke-style semantics for these logics is determined by the intuitionistic Kripke models which are linearly ordered. The logic LC is an intersection of the sets of tautologies of all finite-valued Gödel logics. Gödel–Dummett logics have many applications both in logic and in computer science. Logic LC is employed in the investigations of the provability logic of the intuitionistic arithmetic [Vis82] and relevant logics [DM71]. It is applied to the foundations of logic programming [Pea99] and also it is considered as one of the most important fuzzy logics [Háj98].

This chapter presents a signed dual tableau decision procedure for logic LC developed in [AK01]. Two signs **T** (True) and **F** (False) are used. There is an analogy between signed dual tableaux and relational dual tableaux. In relational dual tableaux there are decomposition rules determined by all the relational operations and by their Boolean complements. In signed dual tableaux every operation of a logic in question determines two rules applicable to formulas built with that operation and endowed with signs **T** and **F**, respectively. They are the counterparts to the positive and complemented occurrence of the term in a relational formula which is a translation of the formula from the original language of the logic. The method of proving completeness of signed dual tableaux is also analogous to the completeness proofs of relational dual tableaux. It involves constructing a branch structure and proving the branch model property and the satisfaction in branch model property. However, in the case of signed formulas the satisfaction in branch model property involves two cases, namely, satisfaction of formulas signed with **T** and **F**, respectively. Following this method, completeness for the LC-dual tableau is proved in the present chapter.

E. Orłowska and J. Golińska-Pilarek, *Dual Tableaux: Foundations, Methodology, Case Studies*, Trends in Logic 36, DOI 10.1007/978-94-007-0005-5_21,

21.2 Gödel–Dummett Logic

The language of the logic LC is that of the intuitionistic logic, that is its vocabulary consists of symbols from the following pairwise disjoint sets:

- $\mathbb{V}$ – a countable set of propositional variables;
- $\{\vee, \wedge, \rightarrow\}$ – the set of propositional operations;
- $\{\bot\}$ – the set consisting of the propositional constant interpreted as falsity.

The set of LC-formulas is the smallest set including $\mathbb{V} \cup \{\bot\}$ and closed with respect to all the propositional operations.

An LC-*model* is a structure:

$$\mathcal{M} = (N \cup \{\omega\}, \leq, v)$$

such that N is the set of natural numbers, $\omega \notin N$, ω is the greatest element of $N \cup \{\omega\}$, $\leq$ is the natural ordering of N, and v: $\mathbb{V} \cup \{\bot\} \rightarrow N \cup \{\omega\}$ is a valuation such that:

- $v(p) \in N \cup \{\omega\}$, for every propositional variable p;
- $v(\bot) = 0$;

Valuation v extends to all the LC-formulas as follows:

- $v(\varphi \vee \psi) = \max(v(\varphi), v(\psi))$;
- $v(\varphi \wedge \psi) = \min(v(\varphi), v(\psi))$;
- $v(\varphi \rightarrow \psi) = \begin{cases} \omega & \text{if } v(\varphi) \leq v(\psi) \\ v(\psi) & \text{otherwise} \end{cases}$

As usual in the intuitionistic logic, negation is definable as $\neg\varphi \stackrel{\text{df}}{=} \varphi \rightarrow \bot$.

An LC-formula φ is said to be true in an LC-model $\mathcal{M} = (N \cup \{\omega\}, \leq, v)$, $\mathcal{M} \models \varphi$, whenever $v(\varphi) = \omega$, and it is LC-valid whenever it is true in all LC-models.

21.3 Signed Dual Tableau Decision Procedure for Gödel–Dummett Logic

Rules of the dual tableau for Gödel–Dummett logic apply to *signed formulas* which are obtained from LC-formulas by prefixing them with the signs **T** or **F**. We extend valuations in LC-models to signed formulas as follows:

- $v(\mathbf{T}(\varphi)) = \omega$ iff $v(\varphi) = \omega$;
- $v(\mathbf{F}(\varphi)) = \omega$ iff $v(\varphi) \neq \omega$.

Decomposition rules have the following forms:

For all LC-formulas $\varphi, \psi, \varphi_1, \varphi_2, \psi_1$ and ψ_2,

$$(\mathbf{T}\vee) \quad \frac{\mathbf{T}(\varphi \vee \psi)}{\mathbf{T}(\varphi), \mathbf{T}(\psi)} \qquad (\mathbf{F}\vee) \quad \frac{\mathbf{F}(\varphi \vee \psi)}{\mathbf{F}(\varphi) \mid \mathbf{F}(\psi)}$$

$$(\mathbf{T}\wedge) \quad \frac{\mathbf{T}(\varphi \wedge \psi)}{\mathbf{T}(\varphi) \mid \mathbf{T}(\psi)} \qquad (\mathbf{F}\wedge) \quad \frac{\mathbf{F}(\varphi \wedge \psi)}{\mathbf{F}(\varphi), \mathbf{F}(\psi)}$$

$$(\mathbf{T}\vee \rightarrow) \quad \frac{\mathbf{T}((\varphi_1 \vee \varphi_2) \rightarrow \psi)}{\mathbf{T}(\varphi_1 \rightarrow \psi) \mid \mathbf{T}(\varphi_2 \rightarrow \psi)} \qquad (\mathbf{F}\vee \rightarrow) \quad \frac{\mathbf{F}((\varphi_1 \vee \varphi_2) \rightarrow \psi)}{\mathbf{F}(\varphi_1 \rightarrow \psi), \mathbf{F}(\varphi_2 \rightarrow \psi)}$$

$$(\mathbf{T} \rightarrow \vee) \quad \frac{\mathbf{T}(\varphi \rightarrow (\psi_1 \vee \psi_2))}{\mathbf{T}(\varphi \rightarrow \psi_1), \mathbf{T}(\varphi \rightarrow \psi_2)} \qquad (\mathbf{F} \rightarrow \vee) \quad \frac{\mathbf{F}(\varphi \rightarrow (\psi_1 \vee \psi_2))}{\mathbf{F}(\varphi \rightarrow \psi_1) \mid \mathbf{F}(\varphi \rightarrow \psi_2)}$$

$$(\mathbf{T}\wedge \rightarrow) \quad \frac{\mathbf{T}((\varphi_1 \wedge \varphi_2) \rightarrow \psi)}{\mathbf{T}(\varphi_1 \rightarrow \psi), \mathbf{T}(\varphi_2 \rightarrow \psi)} \qquad (\mathbf{F}\wedge \rightarrow) \quad \frac{\mathbf{F}((\varphi_1 \wedge \varphi_2) \rightarrow \psi)}{\mathbf{F}(\varphi_1 \rightarrow \psi) \mid \mathbf{F}(\varphi_2 \rightarrow \psi)}$$

$$(\mathbf{T} \rightarrow \wedge) \quad \frac{\mathbf{T}(\varphi \rightarrow (\psi_1 \wedge \psi_2))}{\mathbf{T}(\varphi \rightarrow \psi_1) \mid \mathbf{T}(\varphi \rightarrow \psi_2)} \qquad (\mathbf{F} \rightarrow \wedge) \quad \frac{\mathbf{F}(\varphi \rightarrow (\psi_1 \wedge \psi_2))}{\mathbf{F}(\varphi \rightarrow \psi_1), \mathbf{F}(\varphi \rightarrow \psi_2)}$$

$$(\mathbf{T} \rightarrow (\rightarrow)) \quad \frac{\mathbf{T}(\varphi \rightarrow (\psi_1 \rightarrow \psi_2))}{\mathbf{T}(\psi_1 \rightarrow \psi_2), \mathbf{T}(\varphi \rightarrow \psi_2)}$$

$$(\mathbf{F} \rightarrow (\rightarrow)) \quad \frac{\mathbf{F}(\varphi \rightarrow (\psi_1 \rightarrow \psi_2))}{\mathbf{F}(\psi_1 \rightarrow \psi_2) \mid \mathbf{F}(\varphi \rightarrow \psi_2)}$$

$$(\mathbf{T}(\rightarrow) \rightarrow) \quad \frac{\mathbf{T}((\varphi_1 \rightarrow \varphi_2) \rightarrow \psi)}{\mathbf{T}(\varphi_2 \rightarrow \psi) \mid \mathbf{T}(\psi), \mathbf{F}(\varphi_1 \rightarrow \varphi_2)}$$

$$(\mathbf{F}(\rightarrow) \rightarrow) \quad \frac{\mathbf{F}((\varphi_1 \rightarrow \varphi_2) \rightarrow \psi)}{\mathbf{T}(\varphi_1 \rightarrow \varphi_2), \mathbf{F}(\varphi_2 \rightarrow \psi) \mid \mathbf{F}(\psi)}$$

The specific rules have the following forms:

For all propositional variables p, q, and r,

$$(\text{tran}) \quad \frac{\mathbf{F}(p \rightarrow q), \mathbf{F}(q \rightarrow r)}{\mathbf{F}(p \rightarrow r), \mathbf{F}(p \rightarrow q), \mathbf{F}(q \rightarrow r)}$$

$$(\text{lmax}) \quad \frac{\mathbf{F}(p \rightarrow q), \mathbf{F}(p)}{\mathbf{F}(q), \mathbf{F}(p \rightarrow q), \mathbf{F}(p)} \qquad (\text{rmax}) \quad \frac{\mathbf{T}(p \rightarrow q)}{\mathbf{T}(q), \mathbf{T}(p \rightarrow q)}$$

$$(\text{lin}) \quad \frac{\mathbf{T}(p \rightarrow q)}{\mathbf{F}(q \rightarrow p), \mathbf{T}(p \rightarrow q)} \qquad (\text{min}\,\bot) \quad \frac{\mathbf{F}(p \rightarrow \bot)}{\mathbf{F}(\bot \rightarrow p), \mathbf{F}(p \rightarrow \bot)}$$

A finite set X of signed LC-formulas is said to be LC-axiomatic whenever it includes a subset of either of the following forms:

(Ax1) $\{\mathbf{F}(\bot)\}$;
(Ax2) $\{\mathbf{T}(\varphi), \mathbf{F}(\varphi)\}$, for any LC-formula φ.

A finite set $\{\varphi_1, \ldots, \varphi_n\}$ of signed LC-formulas is said to be an LC-set whenever for every LC-model $\mathcal{M}$ there exists $i \in \{1, \ldots, n\}$ such that φ_i is true in $\mathcal{M}$. LC-correctness of a rule is defined as in F-logic in Sect. 1.3.

Proposition 21.3.1.

1. *The* LC*-decomposition rules are* LC*-correct;*
2. *The* LC*-specific rules are* LC*-correct;*
3. *The* LC*-axiomatic sets are* LC*-set.*

Proof. Let X be a finite set of LC-formulas. By way of example, we prove correctness of the rules ($\mathbf{T}\vee$), ($\mathbf{F}\vee \rightarrow$), ($\mathbf{T} \rightarrow \wedge$), ($\mathbf{T}(\rightarrow) \rightarrow$), (tran), (rmax), and (lin).

($\mathbf{T}\vee$) Assume $X \cup \{\mathbf{T}(\varphi \vee \psi)\}$ is an LC-set. Then, for every LC-model $\mathcal{M}$, either there exists $\vartheta \in X$ such that $\mathcal{M} \models \vartheta$ or $\mathcal{M} \models \mathbf{T}(\varphi \vee \psi)$, that is $v(\varphi \vee \psi) = \omega$. Thus, for every LC-model $\mathcal{M}$, either $\mathcal{M} \models \vartheta$ or $v(\varphi) = \omega$ or $v(\psi) = \omega$, which means that either $\mathcal{M} \models \vartheta$ or $\mathcal{M} \models \mathbf{T}(\varphi)$ or $\mathcal{M} \models \mathbf{T}(\psi)$, and hence $X \cup \{\mathbf{T}(\varphi), \mathbf{T}(\psi)\}$ is an LC-set. Assume $X \cup \{\mathbf{T}(\varphi), \mathbf{T}(\psi)\}$ is an LC-set. Then, for every LC-model $\mathcal{M}$, either there exists $\vartheta \in X$ such that $\mathcal{M} \models \vartheta$ or $\mathcal{M} \models \mathbf{T}(\varphi)$ or $\mathcal{M} \models \mathbf{T}(\psi)$, which means that either $\mathcal{M} \models \vartheta$ or $v(\varphi) = \omega$ or $v(\psi) = \omega$. Thus, either $\mathcal{M} \models \vartheta$ or $\max(v(\varphi), v(\psi)) = \omega$, so either $\mathcal{M} \models \vartheta$ or $\mathcal{M} \models \mathbf{T}(\varphi \vee \psi)$. Hence, $X \cup \{\mathbf{T}(\varphi \vee \psi)\}$ is an LC-set.

($\mathbf{F}\vee \rightarrow$) Since for every LC-model $\mathcal{M}$, $v((\varphi_1 \vee \varphi_2) \rightarrow \psi) \neq \omega$ iff $v(\varphi_1) > v(\psi)$ or $v(\varphi_2) > v(\psi)$, correctness of the rule ($\mathbf{F}\vee \rightarrow$) follows.

($\mathbf{T} \rightarrow \wedge$) Correctness of the rule ($\mathbf{T} \rightarrow \wedge$) follows from the property: for every LC-model $\mathcal{M}$, $v(\varphi \rightarrow (\psi_1 \wedge \psi_2)) \neq \omega$ iff $v(\varphi) \leq v(\psi_1)$ and $v(\varphi) \leq v(\psi_2)$.

($\mathbf{T}(\rightarrow) \rightarrow$) Assume $X \cup \{\mathbf{T}((\varphi_1 \rightarrow \varphi_2) \rightarrow \psi)\}$ is an LC-set. Then, for every LC-model $\mathcal{M}$, either there exists $\vartheta \in X$ such that $\mathcal{M} \models \vartheta$ or $v(\varphi_1 \rightarrow \varphi_2) \leq v(\psi)$, which implies $v(\varphi_2) \leq v(\psi)$. Suppose that either $X \cup \{\mathbf{T}(\varphi_2 \rightarrow \psi)\}$ or $X \cup \{\mathbf{T}(\psi), \mathbf{F}(\varphi_1 \rightarrow \varphi_2)\}$ is not an LC-set. The former implies that there exists an LC-model $\mathcal{M}$ such that $v(\varphi_2) > v(\psi)$, a contradiction. The latter implies that there exists an LC-model $\mathcal{M}$ such that $v(\psi) \neq \omega$ and $v(\varphi_1 \rightarrow \varphi_2) = \omega$. Hence, $v(\varphi_1 \rightarrow \varphi_2) > v(\psi)$, a contradiction. Assume $X \cup \{\mathbf{T}(\varphi_2 \rightarrow \psi)\}$ and $X \cup \{\mathbf{T}(\psi), \mathbf{F}(\varphi_1 \rightarrow \varphi_2)\}$ are LC-sets. Then, for every LC-model $\mathcal{M}$, either there exists $\vartheta \in X$ such that $\mathcal{M} \models \vartheta$ or:

$$(*)\ v(\varphi_2) \leq v(\psi) \text{ and } (v(\psi) = \omega \text{ or } v(\varphi_1 \rightarrow \varphi_2) \neq \omega).$$

Let $\mathcal{M}$ be an LC-model. If $v(\varphi_1 \rightarrow \varphi_2) = \omega$, then by $(*)$ we get $v(\psi) = \omega$, and thus $v(\varphi_1 \rightarrow \varphi_2) \leq v(\psi)$. Hence, $X \cup \{\mathbf{T}((\varphi_1 \rightarrow \varphi_2) \rightarrow \psi)\}$ is an LC-set. If $v(\varphi_1 \rightarrow \varphi_2) \neq \omega$, then $v(\varphi_1 \rightarrow \varphi_2) = v(\varphi_2)$. By $(*)$, we get $v(\varphi_1 \rightarrow \varphi_2) = v(\varphi_2) \leq v(\psi)$. Hence, $X \cup \{\mathbf{T}((\varphi_1 \rightarrow \varphi_2) \rightarrow \psi)\}$ is an LC-set.

(tran) Assume $X \cup \{\mathbf{F}(p \to q), \mathbf{F}(q \to r), \mathbf{F}(p \to r)\}$ is an LC-set. Then for every LC-model $\mathcal{M}$, either there exists $\vartheta \in X$ such that $\mathcal{M} \models \vartheta$ or $v(p) > v(q)$ or $v(q) > v(r)$ or $v(p) > v(r)$. Suppose $X \cup \{\mathbf{F}(p \to q), \mathbf{F}(q \to r)\}$ is not an LC-set. Then there exists an LC-model $\mathcal{M}$ such that $v(p) \leq v(q)$ and $v(q) \leq v(r)$. This implies $v(p) \leq v(r)$, a contradiction. Clearly, if $X \cup \{\mathbf{F}(p \to q), \mathbf{F}(q \to r)\}$ is an LC-set, then so is $X \cup \{\mathbf{F}(p \to q), \mathbf{F}(q \to r), \mathbf{F}(p \to r)\}$.

Correctness of the rules (rmax) and (lin) follow from the following properties of LC-models, respectively:

$$v(p) \leq v(q) \text{ iff } v(p) \leq v(q) \text{ or } v(q) = \omega;$$

$$v(p) \leq v(q) \text{ iff } v(p) \leq v(q) \text{ or } v(q) > v(p).$$

The proofs of correctness of the remaining rules are similar.

For 3., note that in every LC-model $\mathcal{M}$, $v(\bot) = 0$, so $v(\mathbf{F}(\bot)) = \omega$. Therefore $\mathcal{M} \models \mathbf{F}(\bot)$, hence $X \cup \{\mathbf{F}(\bot)\}$ is an LC-set. Since in every LC-model $\mathcal{M}$, either $v(\varphi) = \omega$ or $v(\varphi) \neq \omega$, we get $\mathcal{M} \models \mathbf{T}(\varphi)$ or $\mathcal{M} \models \mathbf{F}(\varphi)$, and hence $X \cup \{\mathbf{T}(\varphi), \mathbf{F}(\varphi)\}$ is an LC-set. □

An application of a rule $\frac{\Phi}{\Phi_1}$ (resp. $\frac{\Phi}{\Phi_1 \mid \Phi_2}$) to a finite set X of signed LC-formulas is said to be *essential* whenever the result of an application of the rule to X includes at least one formula which does not appear in X, that is $(X \setminus \Phi) \cup \Phi_i \neq X$, for some $i \in \{1, 2\}$.

Let φ be an LC-formula. An LC-proof tree for φ is a tree with the following properties:

- The formula $\mathbf{T}(\varphi)$ is at the root of this tree;
- Each node except the root is obtained by an essential application of an LC-rule to its predecessor node;
- A node does not have successors whenever it is an LC-axiomatic set of formulas or no rule introducing any new signed formula is applicable to its signed formulas.

Observe that every LC-proof tree is finite.

As usual, a branch of an LC-proof tree is said to be closed whenever it ends with an axiomatic set of formulas. An LC-proof tree for φ is said to be an LC-proof of φ whenever all of its branches are closed. A formula φ is LC-provable whenever there exists an LC-proof for it.

A branch b of an LC-proof tree is complete whenever it is closed or it satisfies the following completion conditions:

For all LC-formulas $\varphi, \psi, \varphi_1, \varphi_2, \psi_1$ and ψ_2,

Cpl($\mathbf{T}\vee$) (resp. Cpl($\mathbf{F}\wedge$)) If $\mathbf{T}(\varphi \vee \psi) \in b$ (resp. $\mathbf{F}(\varphi \wedge \psi) \in b$), then both $\mathbf{T}(\varphi) \in b$ and $\mathbf{T}(\psi) \in b$ (resp. $\mathbf{F}(\varphi) \in b$ and $\mathbf{F}(\psi) \in b$), obtained by an application of the rule ($\mathbf{T}\vee$) (resp. ($\mathbf{F}\wedge$));

Cpl($\mathbf{T}\wedge$) (resp. Cpl($\mathbf{F}\vee$)) If $\mathbf{T}(\varphi \wedge \psi) \in b$ (resp. $\mathbf{F}(\varphi \vee \psi) \in b$), then either $\mathbf{T}(\varphi) \in b$ or $\mathbf{T}(\psi) \in b$ (resp. $\mathbf{F}(\varphi) \in b$ or $\mathbf{F}(\psi) \in b$), obtained by an application of the rule ($\mathbf{T}\wedge$) (resp. ($\mathbf{F}\vee$));

Cpl($\mathbf{T}\vee\rightarrow$) (resp. Cpl($\mathbf{F}\wedge\rightarrow$)) If $\mathbf{T}((\varphi_1 \vee \varphi_2) \rightarrow \psi)) \in b$ (resp. $\mathbf{F}((\varphi_1 \wedge \varphi_2) \rightarrow \psi) \in b$), then either $\mathbf{T}(\varphi_1 \rightarrow \psi) \in b$ or $\mathbf{T}(\varphi_2 \rightarrow \psi) \in b$ (resp. $\mathbf{F}(\varphi_1 \rightarrow \psi) \in b$ or $\mathbf{F}(\varphi_2 \rightarrow \psi) \in b$), obtained by an application of the rule ($\mathbf{T}\vee\rightarrow$) (resp. ($\mathbf{F}\wedge\rightarrow$));

Cpl($\mathbf{T}\rightarrow\vee$) (resp. Cpl($\mathbf{F}\rightarrow\wedge$)) If $\mathbf{T}(\varphi \rightarrow (\psi_1 \vee \psi_2)) \in b$ (resp. $\mathbf{F}(\varphi \rightarrow (\psi_1 \wedge \psi_2)) \in b$), then both $\mathbf{T}(\varphi \rightarrow \psi_1) \in b$ and $\mathbf{T}(\varphi \rightarrow \psi_2) \in b$ (resp. $\mathbf{F}(\varphi \rightarrow \psi_1) \in b$ and $\mathbf{F}(\varphi \rightarrow \psi_2) \in b$), obtained by an application of the rule ($\mathbf{T}\rightarrow\vee$) (resp. ($\mathbf{F}\rightarrow\wedge$));

Cpl($\mathbf{T}\wedge\rightarrow$) (resp. Cpl($\mathbf{F}\vee\rightarrow$)) If $\mathbf{T}((\varphi_1 \wedge \varphi_2) \rightarrow \psi) \in b$ (resp. $\mathbf{F}((\varphi_1 \vee \varphi_2) \rightarrow \psi) \in b$), then both $\mathbf{T}(\varphi_1 \rightarrow \psi) \in b$ and $\mathbf{T}(\varphi_2 \rightarrow \psi) \in b$ (resp. $\mathbf{F}(\varphi_1 \rightarrow \psi) \in b$ and $\mathbf{F}(\varphi_2 \rightarrow \psi) \in b$), obtained by an application of the rule ($\mathbf{T}\wedge\rightarrow$) (resp. ($\mathbf{F}\vee\rightarrow$));

Cpl($\mathbf{T}\rightarrow\wedge$) (resp. Cpl($\mathbf{F}\rightarrow\vee$)) If $\mathbf{T}(\varphi \rightarrow (\psi_1 \wedge \psi_2)) \in b$ (resp. $\mathbf{F}(\varphi \rightarrow (\psi_1 \vee \psi_2)) \in b$), then either $\mathbf{T}(\varphi \rightarrow \psi_1) \in b$ or $\mathbf{T}(\varphi \rightarrow \psi_2) \in b$ (resp. $\mathbf{F}(\varphi \rightarrow \psi_1) \in b$ or $\mathbf{F}(\varphi \rightarrow \psi_2) \in b$), obtained by an application of the rule ($\mathbf{T}\rightarrow\wedge$) (resp. ($\mathbf{F}\rightarrow\vee$));

Cpl($\mathbf{T}\rightarrow(\rightarrow)$) If $\mathbf{T}(\varphi \rightarrow (\psi_1 \rightarrow \psi_2)) \in b$, then both $\mathbf{T}(\psi_1 \rightarrow \psi_2) \in b$ and $\mathbf{T}(\varphi \rightarrow \psi_2) \in b$, obtained by an application of the rule ($\mathbf{T}\rightarrow(\rightarrow)$);

Cpl($\mathbf{F}\rightarrow(\rightarrow)$) If $\mathbf{F}(\varphi \rightarrow (\psi_1 \rightarrow \psi_2)) \in b$, then either $\mathbf{F}(\psi_1 \rightarrow \psi_2) \in b$ or $\mathbf{T}(\varphi \rightarrow \psi_2) \in b$, obtained by an application of the rule ($\mathbf{F}\rightarrow(\rightarrow)$);

Cpl($\mathbf{T}(\rightarrow)\rightarrow$) If $\mathbf{T}((\varphi_1 \rightarrow \varphi_2) \rightarrow \psi) \in b$, then either $\mathbf{T}(\varphi_2 \rightarrow \psi) \in b$ or both $\mathbf{T}(\psi) \in b$ and $\mathbf{F}(\varphi_1 \rightarrow \varphi_2) \in b$, obtained by an application of the rule ($\mathbf{T}(\rightarrow)\rightarrow$);

Cpl($\mathbf{F}(\rightarrow)\rightarrow$) If $\mathbf{F}((\varphi_1 \rightarrow \varphi_2) \rightarrow \psi) \in b$, then either both $\mathbf{T}(\varphi_1 \rightarrow \varphi_2) \in b$ and $\mathbf{F}(\varphi 2 \rightarrow \psi) \in b$ or $\mathbf{F}(\psi) \in b$, obtained by an application of the rule ($\mathbf{F}(\rightarrow)\rightarrow$);

For all propositional variables p, q, and r,

Cpl(tran) If $\mathbf{F}(p \rightarrow q) \in b$ and $\mathbf{F}(q \rightarrow r) \in b$, then $\mathbf{F}(p \rightarrow r) \in b$, obtained by an application of the rule (tran);

Cpl(lmax) If $\mathbf{F}(p \rightarrow q) \in b$ and $\mathbf{F}(p) \in b$, then $\mathbf{F}(q) \in b$, obtained by an application of the rule (lmax);

Cpl(rmax) If $\mathbf{T}(p \rightarrow q) \in b$, then $\mathbf{T}(q) \in b$, obtained by an application of the rule (rmax);

Cpl(lin) If $\mathbf{T}(p \rightarrow q) \in b$, then $\mathbf{F}(q \rightarrow p) \in b$, obtained by an application of the rule (lin);

Cpl(min $\bot$) If $\mathbf{F}(p \rightarrow \bot) \in b$, then $\mathbf{F}(\bot \rightarrow p) \in b$, obtained by an application of the rule (min $\bot$).

The notions of a complete LC-proof tree and an open branch of an LC-proof tree are defined as in F-logic (see Sect. 1.3). Note that the rules of LC-dual tableau guarantee that if b is a complete branch of an LC-proof tree and both $\mathbf{T}(p \rightarrow q) \in b$ and $\mathbf{F}(p \rightarrow q) \in b$, then there is a node in b containing these two formulas. Therefore, we obtain:

Proposition 21.3.2 (Closed Branch Property). *Let b be an open branch of an LC-proof tree. Then for all $p, q \in \mathbb{V} \cup \{\bot\}$, $\mathbf{T}(p \rightarrow q) \in b$ iff $\mathbf{F}(p \rightarrow q) \notin b$.*

Proposition 21.3.3. *For every open branch b of an* LC*-proof tree, there are no $p_1, \ldots, p_n \in \mathbb{V} \cup \{\bot\}$, $n \geq 1$, such that all the following conditions are satisfied:*

- $p_1 = p_n$;
- *For every $i \in \{1, \ldots, n-1\}$, $\mathbf{F}(p_i \to p_{i+1}) \in b$ or $\mathbf{T}(p_{i+1} \to p_i) \in b$;*
- *For some $i, j \in \{1, \ldots, n\}$, $\mathbf{T}(p_i \to p_j) \in b$.*

Proof. Suppose there are $p_1, \ldots, p_n \in \mathbb{V} \cup \{\bot\}$ such that $p_1 = p_n$, and for every $i \in \{1, \ldots, n-1\}$, $\mathbf{F}(p_i \to p_{i+1}) \in b$ or $\mathbf{T}(p_{i+1} \to p_i) \in b$, and for some $i, j \in \{1, \ldots, n\}$, $\mathbf{T}(p_i \to p_j) \in b$. Then, by the completion condition Cpl(lin), for every $i \in \{1, \ldots, n-1\}$, $\mathbf{F}(p_i \to p_{i+1}) \in b$, and by the completion condition Cpl(tran), for all $i, j \in \{1, \ldots, n\}$, $\mathbf{F}(p_i \to p_j) \in b$. Since for some $i, j \in \{1, \ldots, n\}$, $\mathbf{T}(p_i \to p_j) \in b$, there exists $i, j \in \{1, \ldots, n\}$ such that $\mathbf{F}(p_i \to p_j) \in b$ and $\mathbf{T}(p_i \to p_j) \in b$, which contradicts Proposition 21.3.2. □

Let $q_1, \ldots, q_l, p \in \mathbb{V} \cup \{\bot\}$, $l \geq 1$, and let $n \geq 0$. A sequence $q_1, \ldots, q_l$ is said to be *n-sequence for p* whenever the following conditions are satisfied:

- $p = q_l$;
- For every $i \in \{1, \ldots, l-1\}$, $\mathbf{F}(q_i \to q_{i+1}) \in b$ or $\mathbf{T}(q_{i+1} \to q_i) \in b$;
- For n different i's, $\mathbf{T}(q_{i+1} \to q_i) \in b$.

By Proposition 21.3.3, for every $p \in \mathbb{V} \cup \{\bot\}$, there exists a maximal n for which there is an n-sequence for p.

Let b be an open branch of an LC-proof tree. The branch structure is a system $\mathcal{M}^b = (N \cup \{\omega\}, \leq, v^b)$ such that for every $p \in \mathbb{V} \cup \{\bot\}$, $v^b(p) = \omega$ if $\mathbf{F}(p) \in b$, otherwise $v^b(p) = \max\{n : \text{there exists } n\text{-sequence for } p\}$. Valuation v^b extends to all the LC-formulas as in LC-models.

Proposition 21.3.4 (Branch Model Property). *For every open branch b of an* LC*-proof tree, $\mathcal{M}^b$ is an* LC*-model.*

Proof. We will show that $v^b(\bot) = 0$. Since b is an open branch, $\mathbf{F}(\bot) \notin b$. Therefore, $v^b(\bot) \neq \omega$. Now, let $q_1, \ldots, q_l$ be an n-sequence for $\bot$. Since for every $i \in \{1, \ldots, l-1\}$, $\mathbf{F}(q_i \to q_{i+1}) \in b$ or $\mathbf{T}(q_{i+1} \to q_i) \in b$, by the completion condition Cpl(lin), for every $i \in \{1, \ldots, l-1\}$, $\mathbf{F}(q_i \to q_{i+1}) \in b$. By the completion conditions Cpl(min $\bot$) and Cpl(tran), for all $i, j \in \{1, \ldots, l\}$, $\mathbf{F}(q_i \to q_j) \in b$. By Proposition 21.3.3, the only n for which an n-sequence for $\bot$ exists is $n = 0$. Hence, $v^b(\bot) = 0$. □

Proposition 21.3.5 (Satisfaction in Branch Model Property). *For every open branch b and for every* LC*-formula φ, the following hold:*

1. *If $\mathbf{T}(\varphi) \in b$, then $\mathcal{M}^b \not\models \varphi$;*
2. *If $\mathbf{F}(\varphi) \in b$, then $\mathcal{M}^b \models \varphi$.*

Proof. The proof is by induction on the complexity of formulas.

Let $\varphi = p \in \mathbb{V}$. Then, $\mathbf{F}(p) \in b$ iff $v^b(p) = \omega$ iff $\mathcal{M}^b \models p$. Since b is an open branch, if $\mathbf{F}(p) \in b$, then $\mathbf{T}(p) \notin b$. Therefore, if $\mathcal{M}^b \models p$, then $\mathbf{F}(p) \in b$, and hence $\mathbf{T}(p) \notin b$.

Let $\varphi = (p \to q)$. Assume $\mathbf{T}(p \to q) \in b$. By the completion condition Cpl(lin), $\mathbf{F}(q \to p) \in b$. By the completion condition Cpl(rmax), $\mathbf{T}(q) \in b$, and by the induction hypothesis $v^b(q) \neq \omega$. Let $q_1, \ldots, q_l$, $l \geq 1$, be an n-sequence for q. Note that for every $i \in \{1, \ldots, l\}$, $q_i \neq p$, because otherwise $\mathbf{F}(p \to q) \in b$ and hence b would be closed. On the other hand, since $\mathbf{F}(q \to p) \in b$ and $\mathbf{T}(p \to q) \in b$, a sequence $q_1, \ldots, q_l, p$ is an $n+1$-sequence for p. Therefore $v^b(p) > v^b(q)$, and hence $\mathcal{M}^b \not\models \varphi$. Now, assume $\mathbf{F}(p \to q) \in b$. If $\mathbf{F}(p) \in b$, then by the completion condition Cpl(lmax), we get $\mathbf{F}(q) \in b$, so $v^b(p) = v^b(q) = \omega$. Therefore, $\mathcal{M}^b \models \varphi$. If $\mathbf{F}(p) \notin b$, then $v^b(p) \neq \omega$. Let $p_1, \ldots, p_l$ be an n-sequence for p. Then $p_1, \ldots, p_l, q$ is an n-sequence for q, hence $v^b(p) \leq v^b(q)$. Therefore, $\mathcal{M}^b \models \varphi$.

Now, we prove 1. or 2. for some exemplary formulas.

Let $\varphi = (\varphi_1 \wedge \varphi_2)$. Assume $\mathbf{T}(\varphi_1 \wedge \varphi_2) \in b$. By the completion condition Cpl($\mathbf{T}\wedge$), either $\mathbf{T}(\varphi_1) \in b$ or $\mathbf{T}(\varphi_2) \in b$. By the induction hypothesis, $v^b(\varphi_1) \neq \omega$ or $v^b(\varphi_2) \neq \omega$. Thus, $\min(v^b(\varphi_1), v^b(\varphi_2)) \neq \omega$, and hence $\mathcal{M}^b \not\models \varphi_1 \wedge \varphi_2$.

Let $\varphi = ((\varphi_1 \vee \varphi_2) \to \psi)$. Assume that $\mathbf{F}(\varphi) \in b$. By the completion condition Cpl($\mathbf{F}\vee \to$), both $\mathbf{F}(\varphi_1 \to \psi) \in b$ and $\mathbf{F}(\varphi_2 \to \psi) \in b$. By the induction hypothesis, $v^b(\varphi_1) \leq v^b(\psi)$ and $v^b(\varphi_2) \leq v^b(\psi)$. Thus, $\max(v^b(\varphi_1), v^b(\varphi_2)) \leq v^b(\psi)$, and hence $\mathcal{M}^b \models \varphi$.

Let $\varphi = (\vartheta \to (\psi_1 \wedge \psi_2))$. Assume that $\mathbf{T}(\varphi) \in b$. By the completion condition Cpl($\mathbf{T} \to \wedge$), either $\mathbf{T}(\vartheta \to \psi_1) \in b$ or $\mathbf{T}(\vartheta \to \psi_2) \in b$. By the induction hypothesis, $v^b(\vartheta) > v^b(\psi_1)$ or $v^b(\vartheta) > v^b(\psi_2)$. Thus, $v^b(\vartheta) > \max(v^b(\psi_1), v^b(\psi_2))$, and hence $\mathcal{M}^b \not\models \varphi$.

Let $\varphi = (\vartheta \to (\psi_1 \to \psi_2))$. Assume that $\mathbf{F}(\varphi) \in b$. By the completion condition Cpl($\mathbf{F} \to (\to)$), either $\mathbf{F}(\psi_1 \to \psi_2) \in b$ or $\mathbf{F}(\vartheta \to \psi_2) \in b$. By the induction hypothesis, $v^b(\psi_1 \to \psi_2) = \omega$ or $v^b(\vartheta) \leq v^b(\psi_2)$. Note that $v^b(\varphi)$ equals ω or $v^b(\psi_2)$. Therefore $v^b(\vartheta) \leq v^b(\psi_1 \to \psi_2)$, and hence $\mathcal{M}^b \models \varphi$.

Let $\varphi = ((\varphi_1 \to \varphi_2) \to \psi)$. Assume that $\mathbf{F}(\varphi) \in b$. Then, by the completion condition Cpl($\mathbf{F}(\to) \to$), either both $\mathbf{T}(\varphi_1 \to \varphi_2) \in b$ and $\mathbf{F}(\varphi_2 \to \psi) \in b$ or $\mathbf{F}(\psi) \in b$. If both $\mathbf{T}(\varphi_1 \to \varphi_2) \in b$ and $\mathbf{F}(\varphi_2 \to \psi) \in b$, then by the induction hypothesis, $\mathcal{M}^b \not\models \varphi_1 \to \varphi_2$ and $\mathcal{M}^b \models \varphi_2 \to \psi$. Thus, $v^b(\varphi_1 \to \varphi_2) = v^b(\varphi_2)$ and $v^b(\varphi_2) \leq v^b(\psi)$. Therefore, $v^b(\varphi_1 \to \varphi_2) \leq v^b(\psi)$, and hence $\mathcal{M}^b \models \varphi$. If $\mathbf{F}(\psi) \in b$, then by the induction hypothesis, $\mathcal{M}^b \models \psi$. Thus, $v^b(\psi) = t$. Therefore, $v^b(\varphi_1 \to \varphi_2) \leq v^b(\psi)$, and hence $\mathcal{M}^b \models \varphi$.

The proofs of the remaining cases are similar. □

Proposition 21.3.6. *Let φ be an* LC*-formula. If φ is* LC*-valid, then φ is* LC*-provable.*

Proof. Assume φ is LC-valid. Suppose there is no any closed LC-proof tree for φ. Then there exists a complete LC-proof tree for φ with an open branch, say b. Since $\mathbf{T}(\varphi) \in b$, by Proposition 21.3.4, $\mathcal{M}^b \not\models \varphi$. Hence, by Proposition 21.3.5 φ is not LC-valid, a contradiction. □

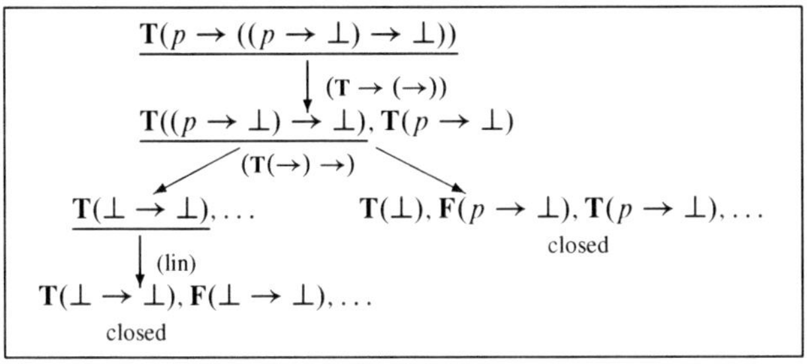

$\mathbf{T}(p \to ((p \to \bot) \to \bot))$

(T → (→))

$\mathbf{T}((p \to \bot) \to \bot), \mathbf{T}(p \to \bot)$

(T(→) →)

$\mathbf{T}(\bot \to \bot), \ldots$ | $\mathbf{T}(\bot), \mathbf{F}(p \to \bot), \mathbf{T}(p \to \bot), \ldots$ closed

(lin)

$\mathbf{T}(\bot \to \bot), \mathbf{F}(\bot \to \bot), \ldots$ closed

Fig. 21.1 An LC-proof of the formula $p \to ((p \to \bot) \to \bot)$

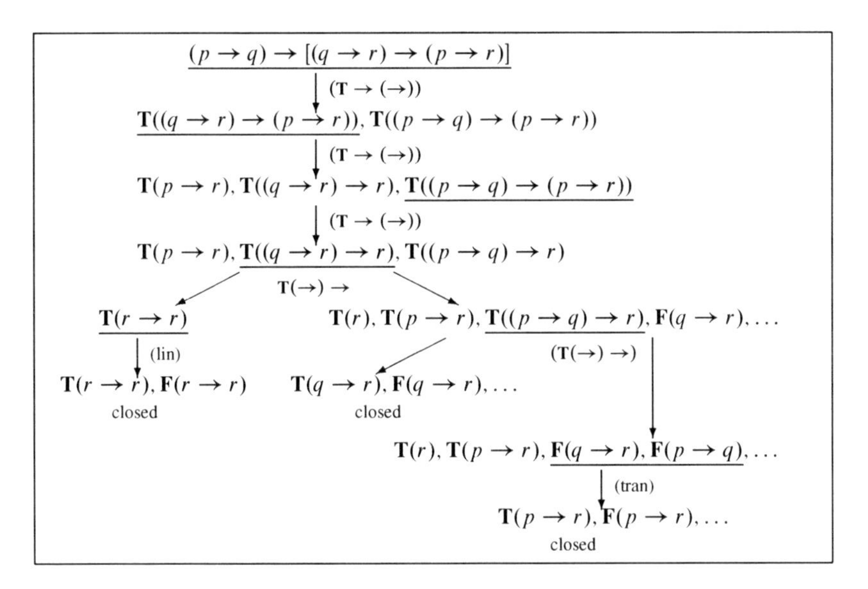

Fig. 21.2 An LC-proof of the formula $(p \to q) \to [(q \to r) \to (p \to r)]$

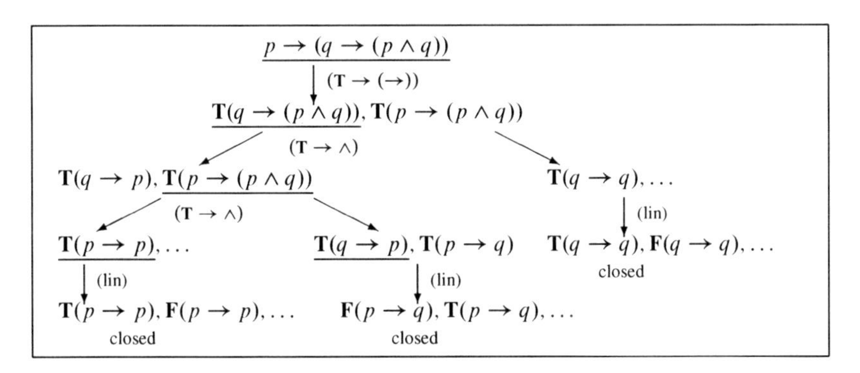

Fig. 21.3 An LC-proof of the formula $p \to (q \to (p \land q))$

By Propositions 21.3.1 and 21.3.6, we get:

Theorem 21.3.1 (Soundness and Completeness of LC). *For every LC-formula φ, the following conditions are equivalent:*

1. *φ is LC-valid;*
2. *φ is LC-provable.*

Since every LC-proof tree is finite and the rules of LC-dual tableau preserve and reflect validity of LC-formulas, LC-dual tableau is a decision procedure for the logic LC.

Example. Consider the following LC-formulas:

$$\varphi = p \to ((p \to \bot) \to \bot),$$
$$\psi = (p \to q) \to [(q \to r) \to (p \to r)],$$
$$\vartheta = p \to (q \to (p \wedge q)).$$

Their LC-proofs are presented in Figs. 21.1–21.3, respectively.

Chapter 22
Dual Tableaux for First-Order Post Logics

22.1 Introduction

Emil Post's doctoral dissertation [Pos21] included a description of an n-valued, functionally complete algebra, for a finite $n \geq 2$. The notion of Post algebra was introduced in [Ros42]. The first algebraic formulation of Post algebras with a family of unary disjoint operations was presented in [Eps60]. In [Rou69, Rou70] an equivalent formulation of the class of Post algebras was given where monotone operations instead of disjoint operations were used. It became a starting point of extensive research and since then various generalizations of Post algebras have been developed, see e.g., [Ras73, Ras85, ER90, ER91]. Post algebras were also applied to a multiple-valued formalization of some logics of programs, see [Ras94].

In this chapter we present a dual tableau for the class of first-order Post logics based on Post algebras of order $n \geq 2$, along the lines of [Sal72, Orł85b].

22.2 Post Algebras of Order n

A Post algebra of order $n \geq 2$ is a structure of the form:

$$\mathfrak{P}_n = (P, -, \vee, \wedge, \rightarrow, d_1, \ldots, d_{n-1}, e_0, \ldots, e_{n-1}),$$

where for all $a, b \in P$ the following conditions are satisfied:

(P1) $(P, \vee, \wedge)$ is a distributive lattice;
(P2) $e_0, \ldots, e_{n-1}$ are distinguished elements of P such that e_0 is the smallest element and e_{n-1} is the greatest element of the lattice;
(P3) $(P, -, \vee, \wedge, \rightarrow, e_0, e_{n-1})$ is a Heyting algebra;
(P4) $d_i(a \vee b) = d_i a \vee d_i b$;
(P5) $d_i(a \wedge b) = d_i a \wedge d_i b$;
(P6) $d_i(a \rightarrow b) = (d_1 a \rightarrow d_1 b) \wedge \ldots \wedge (d_i a \rightarrow d_i b)$;
(P7) $d_i(-a) = -d_1 a$;
(P8) $d_i d_j a = d_j a$;

E. Orłowska and J. Golińska-Pilarek, *Dual Tableaux: Foundations, Methodology, Case Studies*, Trends in Logic 36, DOI 10.1007/978-94-007-0005-5_22,

(P9) $d_i e_j = e_{n-1}$ if $i \leq j$, and $d_i e_j = e_0$ if $i > j$;
(P10) $a = (d_1 a \wedge e_1) \vee \ldots \vee (d_{n-1} a \wedge e_{n-1})$;
(P11) $d_1 a \vee -d_1 a = e_{n-1}$.

Proposition 22.2.1. *Let $\leq$ be the ordering in the lattice $(P, \vee, \wedge)$. Then in any Post algebra of order n, the following hold for all $i, j \in \{1, \ldots, n-1\}$ and for all $a, b \in P$:*

1. $e_0 \leq e_1 \leq \ldots \leq e_{n-1}$;
2. $d_i a \leq d_j a$, *for* $j \leq i$;
3. *Operations d_i are monotone, i.e., if $a \leq b$, then $d_i a \leq d_i b$;*
4. *The set B_P of elements of the form $d_i a$ is closed with respect to operations $-, \vee$, and $\wedge$ and the algebra $\mathfrak{B}_{\mathfrak{P}_n} = (B_P, -, \vee, \wedge)$ is a Boolean algebra;*
5. $d_i(a \to b) = \bigwedge_{k=1}^{i} (-d_k a \vee d_k b)$;
6. $-d_i(a \to b) = d_1 a \wedge \bigwedge_{k=1}^{i-1} (d_{k+1} a \vee -d_k b) \wedge -d_i b$.

Infinite meets and joins in $\mathfrak{P}_n$ are denoted by $\bigcap^{\mathfrak{P}_n}$ and $\bigcup^{\mathfrak{P}_n}$, respectively. Similarly, $\bigcap^{\mathfrak{B}_{\mathfrak{P}_n}}$ and $\bigcup^{\mathfrak{B}_{\mathfrak{P}_n}}$ denote the infinite meet and join, respectively, in the Boolean algebra $\mathfrak{B}_{\mathfrak{P}_n}$ determined by $\mathfrak{P}_n$. In [Eps60] the following was proved:

Proposition 22.2.2. *For every element $a \in \mathfrak{P}_n$ and for any indexed family $\{a_t\}_{t \in T}$ of elements of $\mathfrak{P}_n$, the infinite meets and joins satisfy the following conditions:*

1. $a = \bigcup_{t \in T}^{\mathfrak{P}_n} a_t$ *iff* $d_i a = \bigcup_{t \in T}^{\mathfrak{B}_{\mathfrak{P}_n}} d_i a_t$;
2. $a = \bigcap_{t \in T}^{\mathfrak{P}_n} a_t$ *iff* $d_i a = \bigcap_{t \in T}^{\mathfrak{B}_{\mathfrak{P}_n}} d_i a_t$.

The disjoint operations c_i, $i \in \{0, \ldots, n-1\}$, introduced in [Eps60], can be defined in terms of the monotone operations d_i as:

$$c_0 a \stackrel{\text{df}}{=} -d_1 a = -a;$$

$$c_i a \stackrel{\text{df}}{=} d_i a \wedge -d_{i+1} a, \text{ for } i \in \{1, \ldots, n-2\};$$

$$c_{n-1} a \stackrel{\text{df}}{=} d_{n-1} a.$$

Then, $c_i a \wedge c_j a = e_0$, for all $i, j \in \{0, \ldots, n-1\}$ such that $i \neq j$. Furthermore, $c_i(e_j) = e_{n-1}$ if $i = j$, and otherwise $c_i(e_j) = e_0$.

22.3 First-Order *n*-Valued Post Logic

The language of an n-valued Post logic, P_n, is a first-order language whose formulas are constructed with the symbols from the following pairwise disjoint sets:

- $\mathbb{OV}_{\mathsf{P}_n}$ – a countable infinite set of individual (object) variables;
- $\{E_0, \ldots, E_{n-1}\}$ – the set of propositional constants;

- $\mathbb{P}^k_{\mathsf{P}_n}$ – a countable set of predicate symbols, where $k \geq 1$;
- $\{\neg, D_1, \ldots, D_{n-1}\}$ – the set of unary propositional operations;
- $\{\vee, \wedge, \rightarrow\}$ – the set of binary propositional operations;
- $\{\forall, \exists\}$ – the set of quantifiers.

As usual, we slightly abuse the notation using the symbols $\neg, \vee, \wedge, \rightarrow$ both for the operations in Post algebras and in the language of the logic.

Atomic P_n-formulas are of the form E_i, for $i \in \{0, \ldots, n-1\}$, or $R(x_1, \ldots, x_k)$, where $x_1, \ldots, x_k \in \mathbb{OV}_{\mathsf{P}_n}$ and R is a k-ary predicate symbol, $k \geq 1$. The set of P_n-formulas is the smallest set including the set of atomic formulas and closed with respect to propositional operations and quantifiers.

Algebraic semantics of P_n-language is provided by the class of complete Post algebras of order n, $\mathfrak{P}_n = (P, -, \vee, \wedge, \rightarrow, d_1, \ldots, d_{n-1}, e_0, \ldots, e_{n-1})$. Elements $e_0, \ldots, e_{n-1}$ of $\mathfrak{P}_n$ play the role of truth values. Propositional operations correspond to the respective algebraic operations, and quantifiers $\forall$ and $\exists$ correspond to infinite meet and join in $\mathfrak{P}_n$, respectively. Intuitively, a formula $D_i\varphi$ says that the value of φ is not less than e_i. A P_n*-model* is a structure $\mathcal{M} = (U, \mathfrak{P}_n, m)$, where U is a non-empty set, $\mathfrak{P}_n$ is a complete Post algebra of order n, and m is a meaning function such that:

- $m(E_i) = e_i$, for $i \in \{0, \ldots, n-1\}$;
- $m(R) \in \{e_0, \ldots, e_{n-1}\}^{U^k}$, for every k-ary predicate symbol $R \in \mathbb{P}_{\mathsf{P}_n}$, $k \geq 1$.

Thus, m assigns functions from U^k into $\{e_0, \ldots, e_{n-1}\}$ to k-ary predicate symbols. It follows that the meaning of a k-ary predicate is an n-ary characteristic function of a k-ary relation.

Let $\mathcal{M}$ be a P_n-model. A valuation in $\mathcal{M}$ is a function $v\colon \mathbb{OV}_{\mathsf{P}_n} \rightarrow U$ assigning elements of the universe to individual variables. Given a P_n-model $\mathcal{M}$ and a valuation v in $\mathcal{M}$, we define function $val_{\mathcal{M},v}$ that assigns elements of algebra $\mathfrak{P}_n$ to formulas:

- $val_{\mathcal{M},v}(E_i) = e_i$, for every $i \in \{0, \ldots, n-1\}$;
- $val_{\mathcal{M},v}(R(x_1, \ldots, x_k)) = m(R)(v(x_1), \ldots, v(x_k))$, for every k-ary predicate symbol $R \in \mathbb{P}_{\mathsf{P}_n}$, $k \geq 1$, and for all $x_1, \ldots, x_k \in \mathbb{OV}_{\mathsf{P}_n}$;
- $val_{\mathcal{M},v}(\neg\varphi) = -val_{\mathcal{M},v}(\varphi)$;
- $val_{\mathcal{M},v}(\varphi \vee \psi) = val_{\mathcal{M},v}(\varphi) \vee val_{\mathcal{M},v}(\psi)$;
- $val_{\mathcal{M},v}(\varphi \wedge \psi) = val_{\mathcal{M},v}(\varphi) \wedge val_{\mathcal{M},v}(\psi)$;
- $val_{\mathcal{M},v}(\varphi \rightarrow \psi) = val_{\mathcal{M},v}(\varphi) \rightarrow val_{\mathcal{M},v}(\psi)$;
- $val_{\mathcal{M},v}(D_i\varphi) = d_i val_{\mathcal{M},v}(\varphi)$;
- $val_{\mathcal{M},v}(\forall x\varphi(x)) = \bigcap^{\mathfrak{P}_n}_{u \in U} val_{\mathcal{M},v_u}(\varphi(x))$;
- $val_{\mathcal{M},v}(\exists x\varphi(x)) = \bigcup^{\mathfrak{P}_n}_{u \in U} val_{\mathcal{M},v_u}(\varphi(x))$;

where v_u is the valuation in $\mathcal{M}$ such that $v_u(x) = u$ and $v_u(z) = v(z)$, for all $z \neq x$.

The function $val_{\mathcal{M},v}$ extends to finite sets of formulas. Let $X = \{\varphi_1, \ldots, \varphi_l\}$, $l \leq 1$, be a finite set of P_n-formulas. We define:

$$val_{\mathcal{M},v}(X) \stackrel{\text{df}}{=} val_{\mathcal{M},v}(\varphi_1 \vee \ldots \vee \varphi_l).$$

A P_n-formula φ is said to be true in a model $\mathcal{M}$ whenever for every valuation v in $\mathcal{M}$, $val_{\mathcal{M},v}(\varphi) = e_{n-1}$, and it is said to be e_s*-valid*, $s \in \{1, \ldots, n-1\}$, if for every P_n-model $\mathcal{M}$ and for every valuation v in $\mathcal{M}$, $val_{\mathcal{M},v}(\varphi) \geq e_s$. A formula φ is P_n-valid whenever it is e_{n-1}-valid. The definition of semantics leads to the following observation:

Proposition 22.3.1. *For every P_n-formula φ and for every $s \in \{1, \ldots, n-1\}$, the following conditions are equivalent:*

1. *φ is e_s-valid;*
2. *$D_s\varphi$ is P_n-valid.*

22.4 Dual Tableaux for Post Logics

Dual tableaux for Post logics consist of the decomposition rules of the following forms:

For all P_n-formulas φ and ψ, for every individual variable x, and for all $i, j \in \{1, \ldots, n-1\}$,

Decomposition Rules for Propositional Operations

$$(\vee)\ \frac{D_i(\varphi \vee \psi)}{D_i\varphi, D_i\psi} \qquad (\neg\vee)\ \frac{\neg D_i(\varphi \vee \psi)}{\neg D_i\varphi \mid \neg D_i\psi}$$

$$(\wedge)\ \frac{D_i(\varphi \wedge \psi)}{D_i\varphi \mid D_i\psi} \qquad (\neg\wedge)\ \frac{\neg D_i(\varphi \wedge \psi)}{\neg D_i\varphi, \neg D_i\psi}$$

$$(\rightarrow)\ \frac{D_i(\varphi \rightarrow \psi)}{\neg D_1\varphi, D_1\psi \mid \ldots \mid \neg D_i\varphi, D_i\psi}$$

$$(\neg\rightarrow)\ \frac{\neg D_i(\varphi \rightarrow \psi)}{D_1\varphi \mid D_2\varphi, \neg D_1\psi \mid \ldots \mid \neg D_i\varphi, \neg D_{i-1}\psi \mid \neg D_i\psi}$$

$$(\neg)\ \frac{D_i(\neg\varphi)}{\neg D_1\varphi} \qquad (\neg\neg)\ \frac{\neg D_i(\neg\varphi)}{D_1\varphi}$$

$$(ij)\ \frac{D_iD_j(\varphi)}{D_j\varphi} \qquad (\neg ij)\ \frac{\neg D_iD_j(\varphi)}{\neg D_j\varphi}$$

Decomposition Rules for Quantifiers

$$(\forall)\quad \frac{D_i \forall x\varphi(x)}{D_i\varphi(z)} \quad z \text{ is a new individual variable}$$

$$(\neg\forall)\quad \frac{\neg D_i \forall x\varphi(x)}{\neg D_i\varphi(z), \neg D_i \forall x\varphi(x)} \quad z \text{ is any individual variable}$$

$$(\exists)\quad \frac{D_i \exists x\varphi(x)}{D_i\varphi(z), D_i \exists x\varphi(x)} \quad z \text{ is any individual variable}$$

$$(\neg\exists)\quad \frac{\neg D_i \exists x\varphi(x)}{\neg D_i\varphi(z)} \quad z \text{ is a new individual variable}$$

We observe that any application of the rules of P_n-dual tableau preserves the atomic formulas and their negations.

A set of P_n-formulas is said to be P_n-axiomatic whenever it includes either of the following sets of formulas:

For every P_n-formula φ,

(Ax1) $\{D_i(E_j)\}$, for $1 \leq i \leq j$;
(Ax2) $\{\neg D_i(E_j)\}$, for $i > j \geq 0$;
(Ax3) $\{D_i\varphi, \neg D_j\varphi\}$, for $1 \leq i \leq j$.

Proposition 22.4.1. *Let $X = \{\varphi_1, \ldots, \varphi_l\}$, $l \geq 1$, be a finite set of P_n-formulas, let $\mathcal{M}$ be a P_n-model, and let v be a valuation in $\mathcal{M}$. Then, for every P_n-decomposition rule for a propositional operation of the form $\frac{\Phi}{\Phi_1 | \ldots | \Phi_t}$, $t \geq 1$, the following holds:*

$$val_{\mathcal{M},v}(X) \vee val_{\mathcal{M},v}(\Phi) = \bigwedge_{i=1}^{t} (val_{\mathcal{M},v}(X) \vee val_{\mathcal{M},v}(\Phi_i)).$$

Proof. By way of example, we prove the statement for the rules $(\neg\vee)$, $(\rightarrow)$, $(\neg ij)$.

$(\neg\vee)$ By axioms (P1) and (P4) of Post algebras and since $\mathfrak{B}_{\mathfrak{P}_n}$ is a Boolean algebra, we obtain:

$$\begin{aligned}
&(val_{\mathcal{M},v}(X) \vee val_{\mathcal{M},v}(\neg D_i\varphi)) \wedge (val_{\mathcal{M},v}(X) \vee val_{\mathcal{M},v}(\neg D_i\psi)) \\
&= val_{\mathcal{M},v}(X) \vee (val_{\mathcal{M},v}(\neg D_i\varphi) \wedge val_{\mathcal{M},v}(\neg D_i\psi)) \\
&= val_{\mathcal{M},v}(X) \vee (-d_i val_{\mathcal{M},v}(\varphi) \wedge -d_i val_{\mathcal{M},v}(\psi)) \\
&= val_{\mathcal{M},v}(X) \vee -(d_i val_{\mathcal{M},v}(\varphi) \vee d_i val_{\mathcal{M},v}(\psi)) \\
&= val_{\mathcal{M},v}(X) \vee -d_i val_{\mathcal{M},v}(\varphi \vee \psi) \\
&= val_{\mathcal{M},v}(X) \vee val_{\mathcal{M},v}(\neg D_i(\varphi \vee \psi)).
\end{aligned}$$

($\rightarrow$) By the axiom (P1) and the condition 5. of Proposition 22.2.1, we have:

$$
\begin{aligned}
& val_{\mathcal{M},v}(X) \vee val_{\mathcal{M},v}(D_i(\varphi \rightarrow \psi)) \\
&= val_{\mathcal{M},v}(X) \vee \bigwedge_{k=1}^{i} (-d_k val_{\mathcal{M},v}(\varphi) \vee d_k val_{\mathcal{M},v}(\psi)) \\
&= \bigwedge_{k=1}^{i} (val_{\mathcal{M},v}(X) \vee -d_k val_{\mathcal{M},v}(\varphi) \vee d_k val_{\mathcal{M},v}(\psi)).
\end{aligned}
$$

($\neg ij$) By axiom (P8) we have:

$$
\begin{aligned}
val_{\mathcal{M},v}(X) \vee val_{\mathcal{M},v}(\neg D_i D_j \varphi) &= val_{\mathcal{M},v}(X) \vee -d_j val_{\mathcal{M},v}(\varphi) \\
&= val_{\mathcal{M},v}(X) \vee val_{\mathcal{M},v}(\neg D_j \varphi).
\end{aligned}
$$

The proofs for the remaining rules are similar. □

Proposition 22.2.1 (1.) and (4.) imply:

Proposition 22.4.2. *Let φ and ψ be P_n-formulas and let $i, j \in \{1, \ldots, n-1\}$. Then, for every P_n-model $\mathcal{M}$ and for every valuation v in $\mathcal{M}$, the following hold:*

1. *$val_{\mathcal{M},v}(D_i\varphi \vee D_i\psi) = e_{n-1}$ iff $val_{\mathcal{M},v}(D_i\varphi) = e_{n-1}$ or $val_{\mathcal{M},v}(D_i\psi) = e_{n-1}$;*
2. *If $i \leq j$, then $val_{\mathcal{M},v}(D_i\varphi) \vee val_{\mathcal{M},v}(\neg D_j\varphi) = e_{n-1}$.*

Proof. For 1., note that for every $i \in \{1, \ldots, n-1\}$ and for every formula φ, either $val_{\mathcal{M},v}(D_i\varphi) = e_{n-1}$ or $val_{\mathcal{M},v}(D_i\varphi) = e_0$, hence by the definition of semantics, 1. follows.

For 2., observe that:

$$
val_{\mathcal{M},v}(D_i\varphi) \vee val_{\mathcal{M},v}(\neg D_j\varphi) = d_i(val_{\mathcal{M},v}(\varphi)) \vee -d_j(val_{\mathcal{M},v}(\varphi)).
$$

If $val_{\mathcal{M},v}(\varphi) = e_k$, for some $k \in \{1, \ldots, n\}$ such that $1 \leq k < i \leq j$, then $d_i(val_{\mathcal{M},v}(\varphi)) \vee -d_j(val_{\mathcal{M},v}(\varphi)) = d_i e_k \vee (-d_j e_k) = e_0 \vee (-e_0) = e_0 \vee e_{n-1} = e_{n-1}$. If $val_{\mathcal{M},v}(\varphi) = e_k$ for some $k \in \{1, \ldots, n\}$ such that $1 \leq i \leq k < j$, then $d_i(val_{\mathcal{M},v}(\varphi)) \vee -d_j(val_{\mathcal{M},v}(\varphi)) = d_i e_k \vee (-d_j e_k) = e_{n-1} \vee (-e_0) = e_{n-1} \vee e_{n-1} = e_{n-1}$. If $val_{\mathcal{M},v}(\varphi) = e_k$ for some $k \in \{1, \ldots, n\}$ such that $1 \leq i \leq j \leq k$, then $d_i(val_{\mathcal{M},v}(\varphi)) \vee -d_j(val_{\mathcal{M},v}(\varphi)) = d_i e_k \vee (-d_j e_k) = e_{n-1} \vee (-e_{n-1}) = e_{n-1} \vee e_0 = e_{n-1}$. □

As usual, a P_n-set is a finite set of P_n-formulas such that the disjunction of its members is true in all P_n-models. P_n-correctness of a rule is defined in a similar way as in the logic F (see Sect. 1.3), i.e., a rule is P_n-correct whenever it preserves and reflects P_n-validity.

Proposition 22.4.3.

1. *The* P_n*-rules are* P_n*-correct;*
2. *The* P_n*-axiomatic sets are* P_n*-sets.*

Proof. For 1., observe that correctness of decomposition rules for propositional operations follows from Proposition 22.4.1. Correctness of decomposition rules for quantifiers follows from Proposition 22.2.2 and Proposition 22.4.2(1.).

2. follows from Proposition 22.4.2(2.). □

In order to prove e_s-validity of a P_n-formula φ, we built a P_n-decomposition tree for the formula $D_s\varphi$. As usual, each node of the tree includes all the formulas of its predecessor node, possibly except for those which have been transformed by a rule. A node of the tree does not have successors whenever its set of formulas includes a P_n-axiomatic subset or none of the rules is applicable to it. The notions of a closed branch of a P_n-proof tree, a closed P_n-proof tree, and P_n-provability are defined as in Sect. 1.3.

Proposition 22.4.4 (Closed Branch Property). *Let* φ *be an atomic* P_n*-formula and let* $1 \leq i \leq j$*. For every branch* b *of a* P_n*-proof tree, if* $D_i\varphi \in b$ *and* $D_j\varphi \in b$*, then* b *is closed.*

Proof. Let φ be an atomic P_n-formula, let $1 \leq i \leq j$, and let b be a branch of a P_n-proof tree. Observe that the rules of P_n-dual tableau, in particular the specific rules, guarantee that if formulas $D_i\varphi$ and $D_j\varphi$ belong to the branch b, then there is a node of that branch which includes both of them. Thus, branch b contains an axiomatic set of formulas, hence it is closed. □

A branch b of a P_n-proof tree is complete whenever it is closed or it satisfies the following completion conditions:

For all P_n-formulas φ and ψ, for every individual variable x, and for all $i, j \in \{1, \ldots, n-1\}$,

Cpl($\vee$) (resp. Cpl($\neg\vee$)) If $D_i(\varphi \vee \psi) \in b$ (resp. $\neg D_i(\varphi \vee \psi) \in b$), then both $D_i\varphi \in b$ and $D_i\psi \in b$ (resp. $\neg D_i\varphi \in b$ and $\neg D_i\psi \in b$), obtained by an application of the rule ($\vee$) (resp. ($\neg\vee$));

Cpl($\wedge$) (resp. Cpl($\neg\wedge$)) If $D_i(\varphi \wedge \psi) \in b$ (resp. $\neg D_i(\varphi \wedge \psi) \in b$), then either $D_i\varphi \in b$ or $D_i\psi \in b$ (resp. $\neg D_i\varphi \in b$ or $\neg D_i\psi \in b$), obtained by an application of the rule ($\wedge$) (resp. ($\neg\wedge$));

Cpl($\rightarrow$) If $D_i(\varphi \rightarrow \psi) \in b$, then there exists $k \in \{1, \ldots, i\}$ such that both $\neg D_k\varphi \in b$ and $D_k\psi \in b$, obtained by an application of the rule ($\rightarrow$);

Cpl($\neg\rightarrow$) If $\neg D_i(\varphi \rightarrow \psi) \in b$, then either $D_1\varphi \in b$ or $\neg D_i\psi \in b$ or there exists $k \in \{2, \ldots, i\}$ such that both $D_k\varphi \in b$ and $\neg D_{k-1}\psi \in b$, obtained by an application of the rule ($\neg\rightarrow$);

Cpl($\neg$) If $D_i(\neg\varphi) \in b$, then $\neg D_1\varphi \in b$, obtained by an application of the rule ($\neg$);

Cpl($\neg\neg$) If $\neg D_i(\neg\varphi) \in b$, then $D_1\varphi \in b$, obtained by an application of the rule ($\neg\neg$);

Cpl(ij) If $D_i D_j(\varphi) \in b$, then $D_j \varphi \in b$, obtained by an application of the rule (ij);
Cpl($\neg ij$) If $\neg D_i D_j(\varphi) \in b$, then $\neg D_j \varphi \in b$, obtained by an application of the rule $(\neg ij)$;
Cpl($\forall$) (resp. Cpl($\neg\exists$)) If $D_i \forall x \varphi(x) \in b$ (resp. $\neg D_i \exists x \varphi(x) \in b$), then for some individual variable z, $D_i \varphi(z) \in b$ (resp. $\neg D_i \varphi(z) \in b$), obtained by an application of the rule ($\forall$) (resp. $\neg\exists$));
Cpl($\exists$) (resp. Cpl($\neg\forall$)) If $D_i \exists x \varphi(x) \in b$ (resp. $\neg D_i \forall x \varphi(x) \in b$), then for every individual variable z, $D_i \varphi(z) \in b$ (resp. $\neg D_i \varphi(z) \in b$), obtained by an application of the rule ($\exists$) (resp. ($\neg\forall$)).

The notions of a complete P_n-proof tree and an open branch of a P_n-proof tree are defined as in F-logic (see Sect. 1.3).

Let b be an open branch of a P_n-proof tree. We define a branch structure $\mathcal{M}^b = (U^b, \mathfrak{P}_n, m^b)$ as follows:

- $U^b = \mathbb{OV}_{\mathsf{P}_n}$;
- $m^b(E_i) = e_i$;
- For all $x_1, \ldots, x_k \in \mathbb{OV}_{\mathsf{P}_n}$ and for every k-ary predicate symbol $R \in \mathbb{P}_{\mathsf{P}_n}$, $k \geq 1$:

$$m^b(R)(x_1, \ldots, x_k) = \begin{cases} e_{i-1} & \text{if } i \text{ is the smallest element of } \{1, \ldots, n-1\} \\ & \text{such that } D_i R(x_1, \ldots, x_k) \in b \\ e_{n-1} & \text{if for all } i < n,\ D_i R(x_1, \ldots, x_k) \notin b. \end{cases}$$

Clearly, a branch structure $\mathcal{M}^b$ is a P_n-model. Therefore, the branch model property holds. Let v^b be the identity valuation in $\mathcal{M}^b$. Now, we show the satisfaction in branch model property:

Proposition 22.4.5 (Satisfaction in Branch Model Property). *Let b be an open branch of a P_n-proof tree. Then, for every P_n-formula φ and for every $i \in \{1, \ldots, n-1\}$, the following hold:*

1. *If $D_i \varphi \in b$, then $val_{\mathcal{M}^b, v^b}(D_i \varphi) < e_{n-1}$;*
2. *If $\neg D_i \varphi \in b$, then $val_{\mathcal{M}^b, v^b}(\neg D_i \varphi) < e_{n-1}$.*

Proof. The proof is by induction on the complexity of formulas. First, we prove that 1. and 2. hold for atomic P_n-formulas.

If $D_i(E_j) \in b$, then $i > j$, since otherwise b would be closed. Thus, $val_{\mathcal{M}^b, v^b}(D_i(E_j)) = e_0 < e_{n-1}$. Assume that $D_i R(x_1, \ldots, x_k) \in b$, for some predicate symbol R and for some individual variables $x_1, \ldots, x_k$. Then, we have $val_{\mathcal{M}^b, v^b}(D_i R(x_1, \ldots, x_k)) = d_i val_{\mathcal{M}^b, v^b}(R(x_1, \ldots, x_k))$ and, by the definition of $m^b(R)$, $val_{\mathcal{M}^b, v^b}(R(x_1, \ldots, x_k)) = e_j$, for some $j < i$. Thus, $d_i val_{\mathcal{M}^b, v^b}(R(x_1, \ldots, x_k)) = d_i e_j = e_0 < e_{n-1}$.

If $\neg D_i(E_j) \in b$, then $i \leq j$, since otherwise b would be closed. Thus, $val_{\mathcal{M}^b, v^b}(\neg D_i(E_j)) = -e_{n-1} = e_0 < e_{n-1}$. Assume $\neg D_i R(x_1, \ldots, x_k) \in b$, for some predicate symbol R and for some individual variables $x_1, \ldots, x_k$. Then, by the closed branch property, for all $j \leq i$, $D_j R(x_1, \ldots, x_k) \not\in b$. Thus, by the definition of $m^b(R)$, $val_{\mathcal{M}^b, v^b}(R(x_1, \ldots, x_k)) = e_j$, for some $j \geq i$. Hence, $val_{\mathcal{M}^b, v^b}(\neg D_i R(x_1, \ldots, x_k)) = -d_i e_j = -e_{n-1} = e_0 < e_{n-1}$.

Now, we prove that 1. and 2. hold for negations of atomic P_n-formulas. Let φ be an atomic P_n-formula.

Observe that $val_{\mathcal{M}^b, v^b}(D_i(\neg\varphi)) = -d_1 val_{\mathcal{M}^b, v^b}(\varphi) = val_{\mathcal{M}^b, v^b}(\neg D_1 \varphi)$, due to axiom (P7). Assume that $D_i(\neg\varphi) \in b$. Then, by the completion condition Cpl($\neg$), $\neg D_1 \varphi \in b$. By the induction hypothesis, $val_{\mathcal{M}^b, v^b}(\neg D_1 \varphi) < e_{n-1}$, therefore $val_{\mathcal{M}^b, v^b}(D_i(\neg\varphi)) < e_{n-1}$. The statement 2. can be proved in a similar way.

Now, we prove that 1. and 2. hold for compound P_n-formulas. Assume that 1. and 2. hold for formulas ψ, ψ_1, ψ_2, and their negations.

Let $\varphi = D_j \psi$. Assume $D_i D_j \psi \in b$. Then, by the completion condition Cpl(ij), $D_j \psi \in b$. By the induction hypothesis, $val_{\mathcal{M}^b, v^b}(D_j \psi) < e_{n-1}$. By axiom (P8) of Post algebras, $val_{\mathcal{M}^b, v^b}(D_i D_j \psi) = val_{\mathcal{M}^b, v^b}(D_j \psi) < e_{n-1}$.

Let $\varphi = \psi_1 \wedge \psi_2$. Assume $D_i(\psi_1 \wedge \psi_2) \in b$. Then, by the completion condition Cpl($\wedge$), either $D_i \psi_1 \in b$ or $D_i \psi_2 \in b$. By the induction hypothesis, $val_{\mathcal{M}^b, v^b}(D_i(\psi_1)) < e_{n-1}$ or $val_{\mathcal{M}^b, v^b}(D_i \psi_2) < e_{n-1}$. Suppose that $val_{\mathcal{M}^b, v^b}(D_i(\psi_1 \wedge \psi_2)) = e_{n-1}$. Then, $val_{\mathcal{M}^b, v^b}(D_i \psi_1) = e_{n-1}$ and $val_{\mathcal{M}^b, v^b}(D_i \psi_2) = e_{n-1}$, a contradiction.

Let $\varphi = \psi_1 \rightarrow \psi_2$. Assume $D_i(\psi_1 \rightarrow \psi_2) \in b$. Then, by the completion condition Cpl($\rightarrow$), there exists $k \in \{1, \ldots, i\}$ such that both $\neg D_k \psi_1 \in b$ and $D_k \psi_2 \in b$. By the induction hypothesis, $-d_k val_{\mathcal{M}^b, v^b}(\psi_1) < e_{n-1}$ and $d_k val_{\mathcal{M}^b, v^b}(\psi_2) < e_{n-1}$. Thus, $-d_k val_{\mathcal{M}^b, v^b}(\psi_1) \vee d_k val_{\mathcal{M}^b, v^b}(\psi_2) < e_{n-1}$. Suppose that $val_{\mathcal{M}^b, v^b}(D_i(\psi_1 \rightarrow \psi_2)) = e_{n-1}$. Note that the following holds: $val_{\mathcal{M}^b, v^b}(D_i(\psi_1 \rightarrow \psi_2)) = \bigwedge_{k=1}^{i}(d_k val_{\mathcal{M}^b, v^b}(\psi_1) \rightarrow d_k val_{\mathcal{M}^b, v^b}(\psi_2)) = e_{n-1}$ iff for every $k \in \{1, \ldots, i\}$, $d_k val_{\mathcal{M}^b, v^b}(\psi_1) \rightarrow d_k val_{\mathcal{M}^b, v^b}(\psi_2) = e_{n-1}$. Hence, by Proposition 22.2.1(5.), $-d_k val_{\mathcal{M}^b, v^b}(\psi_1) \vee d_k val_{\mathcal{M}^b, v^b}(\psi_2) = e_{n-1}$, a contradiction.

Let $\varphi = \forall x \psi(x)$. Assume $D_i(\forall x \psi(x)) \in b$. Then, by the completion condition Cpl($\forall$), for some individual variable z, $D_i \psi(z) \in b$. Thus, by the induction hypothesis, $d_i val_{\mathcal{M}^b, v^b}(\psi(z)) < e_{n-1}$. Suppose $val_{\mathcal{M}^b, v^b}(D_i(\forall x \psi(x))) = e_{n-1}$. Then, by the axiom (P9) and Proposition 22.2.2, for every individual variable z, $d_i val_{\mathcal{M}^b, v^b}(\psi(z)) = e_{n-1}$, a contradiction.

The proofs of the remaining cases are similar. □

Theorem 22.4.1 (Soundness and Completeness of P_n). *For every P_n-formula φ and for every $s \in \{1, \ldots, n-1\}$, the following conditions are equivalent:*

1. *φ is e_s-valid;*
2. *$D_s \varphi$ is P_n-provable.*

Proof. Assume φ is e_s-valid. Thus, by Proposition 22.3.1, $D_s \varphi$ is P_n-valid. Then, it must exist a closed P_n-decomposition tree for $D_s \varphi$, since otherwise there is a tree with an open branch, and then, by Proposition 22.4.5(1.), $D_s \varphi$ would not be P_n-valid. Hence, $D_s \varphi$ is P_n-provable. Now, assume that $D_s \varphi$ is P_n-provable, that is there exists a closed P_n-decomposition tree for $D_s\varphi$. Then, Proposition 22.4.3, enables us to prove that $D_s\varphi$ is P_n-valid. Thus, by Proposition 22.3.1, φ is e_s-valid. □

Observe that the reduct of P_n-dual tableau consisting of the decomposition rules for propositional operations is a decision procedure for the propositional Post logic.

Example. Consider the following P_n-formula:

$$\chi = (\varphi \to \psi) \vee (\psi \to \varphi).$$

In order to prove P_n-validity of χ, we show that $D_{n-1}\chi$ is P_n-provable. Figure 22.1 presents its P_n-proof. In this tree all branches end with the sets of the following form: $H(i,j) = \{\neg D_i\varphi, D_i\psi, \neg D_j\psi, D_j\varphi\}$, for $i, j \in \{1, \dots, n-1\}$. It is easy to prove that $H(i,j)$ is a P_n-axiomatic set, for all $i, j \in \{1, \dots, n-1\}$. Indeed, if $i \leq j$, then $\{D_i\psi, \neg D_j\psi\}$ is axiomatic, and if $i > j$, then $\{D_j\varphi, \neg D_i\varphi\}$ is axiomatic.

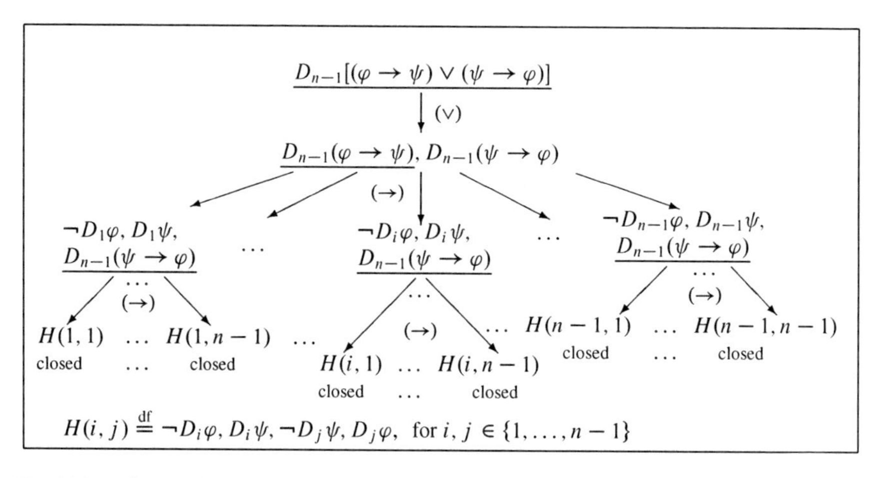

Fig. 22.1 A P_n-proof of the formula $(\varphi \to \psi) \vee (\psi \to \varphi)$

Chapter 23
Dual Tableau for Propositional Logic with Identity

23.1 Introduction

In this chapter we consider propositional logic with identity, referred to as SCI (Sentential Calculus with Identity) introduced in [Sus68]. It is two-valued as the classical logic, but it rejects the main assumption of Frege's philosophy that the meaning of a sentence is its logical value. Non-fregean logics are based on the principle that *denotations of sentences of a given language are different from their truth values*. SCI is obtained from the classical propositional logic by endowing its language with an operation of *identity*, $\equiv$, and the axioms which say that formula $\varphi \equiv \psi$ is interpreted as 'φ has the same denotation as ψ'. Identity axioms together with two-valuedness imply that the set of denotations of sentences has at least two elements. Any other assumptions about the range of sentences or properties of the identity operation lead to axiomatic extensions of SCI. In general, the identity operation is different from the equivalence operation, that is two sentences with the same truth values may have different denotations. If we add $(\varphi \leftrightarrow \psi) \equiv (\varphi \equiv \psi)$ to the set of SCI axioms, then we obtain the classical propositional logic, where the identity and equivalence operations are indistinguishable. In this way the Fregean axiom can be formulated in SCI. Some extensions of SCI are known to correspond to modal logics S4 and S5 and to the three-valued Łukasiewicz logic (see [Sus71a]). Decidability of the logic SCI is proved in [Sus71c].

Non-fregean logics were an inspiration for some other logical systems. In the paper [BS73] it is indicated that Lindenbaum algebras, obtained by the Tarski–Lindenbaum method and further developed by Rasiowa and Sikorski, are too weak for studying some logical systems. For example, propositional logics with identity and the first-order non-fregean logics are not algebraizable in the Rasiowa–Sikorski style. This fact inspired an introduction of abstract logics in [BS73], aimed at generalizing of the concept of a logical system. Many ideas from the paper [BS73] have been studied within the theory of abstract algebraic logics.

Basic definitions and main results concerning non-fregean logics can be found in [Sus71b, Sus71c, BS72, Sus73, Sus72, GPH05], among others.

E. Orłowska and J. Golińska-Pilarek, *Dual Tableaux: Foundations, Methodology, Case Studies*, Trends in Logic 36, DOI 10.1007/978-94-007-0005-5_23,

23.2 A Propositional Logic with Identity

The vocabulary of the language of the non-fregean propositional logic, SCI, consists of the symbols from the following pairwise disjoint sets:

- $\mathbb{V}$ – a countable infinite set of propositional variables;
- $\{\neg, \vee, \wedge, \rightarrow, \leftrightarrow, \equiv\}$ – the set of propositional operations of negation $\neg$, disjunction $\vee$, conjunction $\wedge$, implication $\rightarrow$, equivalence $\leftrightarrow$, and identity $\equiv$.

The set of SCI-formulas is the smallest set including $\mathbb{V}$ and closed with respect to all the propositional operations.

An SCI-*model* is a structure $\mathcal{M} = (U, \sim, \sqcup, \sqcap, \Rightarrow, \Leftrightarrow, \circ, D)$, where U is a non-empty set, D is any non-empty subset of U, and $\sim, \sqcup, \sqcap, \Rightarrow, \Leftrightarrow, \circ$ are operations on U with arities 1, 2, 2, 2, 2, 2, respectively, such that:

For all $a, b \in U$,

(SCI1) $\sim a \in D$ iff $a \notin D$;
(SCI2) $a \sqcup b \in D$ iff $a \in D$ or $b \in D$;
(SCI3) $a \sqcap b \in D$ iff $a \in D$ and $b \in D$;
(SCI4) $a \Rightarrow b \in D$ iff $a \notin D$ or $b \in D$;
(SCI5) $a \Leftrightarrow b \in D$ iff $a \in D$ iff $b \in D$;
(SCI6) $a \circ b \in D$ iff $a = b$.

Let $\mathcal{M}$ be an SCI-model. A valuation in $\mathcal{M}$ is any mapping $v\colon \mathbb{V} \rightarrow U$. A valuation v extends homomorphically to all the formulas:

$v(\neg\varphi) = \sim v(\varphi)$;
$v(\varphi \vee \psi) = v(\varphi) \sqcup v(\psi)$;
$v(\varphi \wedge \psi) = v(\varphi) \sqcap v(\psi)$;
$v(\varphi \rightarrow \psi) = v(\varphi) \Rightarrow v(\psi)$;
$v(\varphi \leftrightarrow \psi) = v(\varphi) \Leftrightarrow v(\psi)$;
$v(\varphi \equiv \psi) = v(\varphi) \circ v(\psi)$.

Let v be a valuation in an SCI-model $\mathcal{M}$. An SCI-formula φ is satisfied by v in $\mathcal{M}$, $\mathcal{M}, v \models \varphi$, whenever $v(\varphi) \in D$. An SCI-formula φ is true in $\mathcal{M}$ if it is satisfied by all valuations in $\mathcal{M}$. A formula is SCI-valid if it is true in all SCI-models.

The logic SCI is two valued. We may define the logical value of a formula φ in a model $\mathcal{M}$ as:

$$val_{\mathcal{M}}(\varphi) \stackrel{\text{df}}{=} \begin{cases} \text{true} & \text{if for every } v \text{ in } \mathcal{M}, v(\varphi) \in D \\ \text{false} & \text{otherwise.} \end{cases}$$

The following proposition shows that SCI is extensional in the sense that any subformula ψ of an SCI-formula φ can be replaced with another formula ϑ such that its denotation is the same as ψ without affecting the denotation of φ.

Proposition 23.2.1. *Let $\mathcal{M}$ be an* SCI*-model, let v be a valuation in $\mathcal{M}$, let φ be an* SCI*-formula containing a subformula ψ, and let φ' be the result of replacing some occurrences of ψ in φ by a formula ϑ. Then, $\mathcal{M}, v \models \psi \equiv \vartheta$ implies $\mathcal{M}, v \models \varphi \equiv \varphi'$.*

Proof. The proof is by induction on the complexity of formulas. Let φ be a propositional variable p and let ϑ be an SCI-formula. Then p is the only subformula of φ and, clearly, if $v(p) = v(\vartheta)$, then the proposition holds. In what follows $\varphi(\psi)$ denotes a formula φ with a subformula ψ, ϑ denotes any formula such that $v(\psi) = v(\vartheta)$, and φ' denotes a formula resulting from φ by replacing some occurrences of ψ with ϑ.

Let $\varphi(\psi) = \neg\phi$. Then ψ is a subformula of ϕ and $v(\neg\phi(\psi)) = {\sim}v(\phi(\psi))$. By the induction hypothesis, $v(\phi(\psi)) = v(\phi(\vartheta))$, hence ${\sim}v(\phi(\psi)) = {\sim}v(\phi(\vartheta))$. Therefore $v(\varphi) = v(\varphi')$.

Let $\varphi(\psi) = (\phi_1 \vee \phi_2)$. Then ψ is a subformula of ϕ_1 or ϕ_2. Without loss of generality, we may assume that ψ is a subformula of ϕ_1. Then, $v(\phi_1(\psi) \vee \phi_2) = v(\phi_1(\psi)) \sqcup v(\phi_2)$. By the induction hypothesis, $v(\phi_1(\psi)) = v(\phi_1(\vartheta))$, hence $v(\phi_1(\psi)) \sqcup v(\phi_2) = v(\phi_1(\vartheta)) \sqcup v(\phi_2)$. Therefore, $v(\varphi) = v(\varphi')$.

Let $\varphi(\psi) = (\phi_1 \equiv \phi_2)$. Then, ψ is a subformula of ϕ_1 or ϕ_2. Without loss of generality, we may assume that ψ is a subformula of ϕ_1. Then, $v(\phi_1(\psi) \equiv \phi_2) = v(\phi_1(\psi)) \circ v(\phi_2)$. By the induction hypothesis, $v(\phi_1(\psi)) = v(\phi_1(\vartheta))$, hence $v(\phi_1(\psi)) \circ v(\phi_2) = v(\phi_1(\vartheta)) \circ v(\phi_2) = v(\phi_1(\vartheta) \equiv \phi_2)$. Therefore $v(\varphi) = v(\varphi')$.

The proofs of the remaining cases are similar. □

A Hilbert-style axiomatization of SCI consists of the axioms of the classical propositional logic PC, which characterize the operations $\neg, \vee, \wedge, \rightarrow, \leftrightarrow$, and the following axioms for the identity operation $\equiv$:

($\equiv_1$) $\varphi \equiv \varphi$;
($\equiv_2$) $(\varphi \equiv \psi) \rightarrow (\neg\varphi \equiv \neg\psi)$;
($\equiv_3$) $(\varphi \equiv \psi) \rightarrow (\varphi \rightarrow \psi)$;
($\equiv_4$) $[(\varphi \equiv \psi) \wedge (\vartheta \equiv \xi)] \rightarrow [(\varphi \# \vartheta) \equiv (\psi \# \xi)]$, for $\# \in \{\vee, \wedge, \rightarrow, \leftrightarrow, \equiv\}$.

The only rule of inference is modus ponens. It can be shown that all the SCI-axioms are true in every SCI-model. It is known that they provide a complete axiomatization of logic SCI.

Fact 23.2.1. *For every* PC*-formula φ, the following conditions are equivalent:*

1. *φ is* PC*-valid;*
2. *φ is* SCI*-valid.*

Note also that the reduct $(U, \sim, \sqcup, \sqcap)$ of an SCI-model is not necessarily a Boolean algebra, for example $a \sqcap b = b \sqcap a$ is not true in all SCI-models. Consider an SCI-model $\mathcal{M} = (U, \sim, \sqcup, \sqcap, \Rightarrow, \Leftrightarrow, \circ, D)$, where $U = \{0, 1, 2\}$, $D = \{1, 2\}$, and the operations $\sim, \sqcup, \sqcap, \Rightarrow, \Leftrightarrow, \circ$ are defined by:

$$\sim a \stackrel{\text{df}}{=} \begin{cases} 0 & \text{if } a \neq 0 \\ 1 & \text{otherwise} \end{cases}$$

$$a \sqcup b \stackrel{\text{df}}{=} \begin{cases} 0 & \text{if } a = 0 \text{ and } b = 0 \\ 1 & \text{otherwise} \end{cases}$$

$$a \sqcap b \stackrel{\text{df}}{=} \begin{cases} 0 & \text{if } a = 0 \text{ or } b = 0 \\ 1 & \text{if } b = 2 \text{ and } a \neq 0 \\ 2 & \text{otherwise} \end{cases}$$

$$a \Rightarrow b \stackrel{\text{df}}{=} \begin{cases} 0 & \text{if } a \neq 0 \text{ and } b = 0 \\ 1 & \text{otherwise} \end{cases}$$

$$a \Leftrightarrow b \stackrel{\text{df}}{=} \begin{cases} 0 & \text{if } a \neq 0, b = 0 \text{ or } a = 0, b \neq 0 \\ 1 & \text{otherwise} \end{cases}$$

$$a \circ b \stackrel{\text{df}}{=} \begin{cases} 0 & \text{if } a \neq b \\ a & \text{if } a = b \text{ and } a \neq 0 \\ 1 & \text{otherwise.} \end{cases}$$

This structure is an SCI-model. Indeed, the following hold:

$$\sim a \in D \text{ iff } a = 0 \text{ iff } a \notin D;$$
$$a \sqcup b \in D \text{ iff } a \neq 0 \text{ or } b \neq 0 \text{ iff } a \in D \text{ or } b \in D;$$
$$a \sqcap b \in D \text{ iff } a \neq 0 \text{ and } b \neq 0 \text{ iff } a \in D \text{ and } b \in D;$$
$$a \Rightarrow b \in D \text{ iff } a = 0 \text{ or } b \neq 0 \text{ iff } a \notin D \text{ or } b \in D;$$
$$a \Leftrightarrow b \in D \text{ iff either } a = b = 0 \text{ or } a \neq 0 \neq b \text{ iff } a \in D \text{ iff } b \in D;$$
$$a \circ b \in D \text{ iff } a = b.$$

However, we have $2 \sqcap 1 = 2$, while $1 \sqcap 2 = 1$. Hence, $a \sqcap b = b \sqcap a$ is not true in this model.

23.3 Axiomatic Extensions of the Propositional Logic with Identity

The class of all different SCI-theories is uncountable. Therefore, the question of natural extensions of SCI arises. Let X be a set of SCI-formulas. The axiomatic extension of SCI, SCI^X, is the logic obtained from SCI by adding formulas of X to SCI-axioms. There are three natural and extensively studied axiomatic extensions of SCI, the logics $\mathsf{SCI}^{\mathsf{B}}$, $\mathsf{SCI}^{\mathsf{T}}$, and $\mathsf{SCI}^{\mathsf{H}}$.

Logic $\mathsf{SCI}^{\mathsf{B}}$

The specific axioms of this logic are:

(B1) $[(\varphi \wedge \psi) \vee \vartheta] \equiv [(\psi \vee \vartheta) \wedge (\varphi \vee \vartheta)]$;
(B2) $[(\varphi \vee \psi) \wedge \vartheta] \equiv [(\psi \wedge \vartheta) \vee (\varphi \wedge \vartheta)]$;
(B3) $[\varphi \vee (\psi \wedge \neg\psi)] \equiv \varphi$;
(B4) $[\varphi \wedge (\psi \vee \neg\psi)] \equiv \varphi$;
(B5) $(\varphi \rightarrow \psi) \equiv (\neg\varphi \vee \psi)$;
(B6) $(\varphi \leftrightarrow \psi) \equiv [(\varphi \rightarrow \psi) \wedge (\psi \rightarrow \varphi)]$.

An $\mathsf{SCI}^{\mathsf{B}}$-model is an SCI-model $\mathcal{M} = (U, \sim, \sqcup, \sqcap, \Rightarrow, \Leftrightarrow, \circ, D)$ such that D is an ultrafilter on $(U, \sim, \sqcup, \sqcap)$ and for all $a, b \in U$, $a \circ b \in D$ iff $a = b$. Any axiomatic extension of SCI which includes $\mathsf{SCI}^{\mathsf{B}}$ is referred to as a Boolean SCI-logic.

In [Sus71a] the following was proved:

Theorem 23.3.1. *For every* SCI*-formula* φ*, the following conditions are equivalent:*

1. φ *is true in all* $\mathsf{SCI}^{\mathsf{B}}$*-models;*
2. φ *is provable in* $\mathsf{SCI}^{\mathsf{B}}$.

Logic $\mathsf{SCI}^{\mathsf{T}}$

The logic $\mathsf{SCI}^{\mathsf{T}}$ is an extension of $\mathsf{SCI}^{\mathsf{B}}$ with the following axiom:

(T) $\varphi \equiv \psi$, for all formulas φ and ψ such that $\varphi \leftrightarrow \psi$ is provable in SCI.

This logic has many interesting properties (see [Sus71a]). Below we list some of them:

Proposition 23.3.1.

1. *The set of all* $\mathsf{SCI}^{\mathsf{T}}$*-provable formulas is the smallest set of* $\mathsf{SCI}^{\mathsf{B}}$*-provable formulas closed on the Gödel rule:*

$$(G) \quad \frac{\varphi, \psi}{\varphi \equiv \psi}$$

2. *The set of all* $\mathsf{SCI}^{\mathsf{T}}$*-provable formulas is the smallest set of* SCI*-provable formulas closed on the quasi-Fregean rule:*

$$(QF) \quad \frac{(\varphi \leftrightarrow \psi)}{\varphi \equiv \psi}.$$

An $\mathsf{SCI}^{\mathsf{T}}$-model is an $\mathsf{SCI}^{\mathsf{B}}$-model $\mathcal{M} = (U, \sim, \sqcup, \sqcap, \Rightarrow, \Leftrightarrow, \circ, D)$ such that:

For all $a, b, c, d, e \in U$,

- $a \circ a = e \sqcup \sim e$;
- $[(a \circ b) \Rightarrow (a \Leftrightarrow b)] = e \sqcup \sim e$;
- $[[(a \circ b) \sqcap (c \circ d)] \Rightarrow [(a \# c) \circ (b \# d)]] = e \sqcup \sim e$, for $\# \in \{\sqcup, \sqcap, \circ\}$.

In [Sus71c] the following was proved:

Theorem 23.3.2. *For every* SCI*-formula* φ*, the following conditions are equivalent:*

1. φ *is true in all* SCI^T*-models;*
2. φ *is provable in* SCI^T.

Furthermore, the following was observed:

Theorem 23.3.3.

1. *Let* $\mathcal{M} = (U, \sim, \sqcup, \sqcap, \Rightarrow, \Leftrightarrow, \circ, D)$ *be an* SCI^T*-model. Then, the structure* $(U, \sim, \sqcup, \sqcap, I)$*, where a unary operation* I *on* U *is defined as* $I(a) \stackrel{\text{df}}{=} a \circ (a \sqcup \sim a)$*, for every* $a \in U$*, is a topological Boolean algebra, i.e.,* I *is an interior operation;*
2. *Let* $\mathcal{T} = (U, \sim, \sqcup, \sqcap, I)$ *be a topological Boolean algebra. Then, the structure* $\mathcal{M} = (U, \sim, \sqcup, \sqcap, \Rightarrow, \Leftrightarrow, \circ, D)$ *such that* $\Rightarrow$ *and* $\Leftrightarrow$ *are operations on* U *satisfying the conditions of* SCI*-models and, in addition, for all* $a, b \in U$*,* $a \circ b \stackrel{\text{df}}{=} I(a \Leftrightarrow b)$ *and* D *is an ultrafilter on* U *such that* $a \circ b \in D$ *iff* $a = b$*, is an* SCI^T*-model.*

Logic SCI^H

The logic SCI^H is an extension of SCI^B with the following axioms:

(H1) $1 \equiv (\varphi \vee \neg\varphi)$;
(H2) $0 \equiv (\varphi \wedge \neg\varphi)$;
(H3) $(\varphi \equiv \psi) \equiv [(\varphi \equiv \psi) \equiv 1]$;
(H4) $\neg(\varphi \equiv \psi) \equiv [(\varphi \equiv \psi) \equiv 0]$,

where 1 and 0 are propositional constants defined as $p \vee \neg p$ and $p \wedge \neg p$, respectively.

An SCI^H-model is an SCI^B-model $\mathcal{M} = (U, \sim, \sqcup, \sqcap, \Rightarrow, \Leftrightarrow, \circ, D)$ such that 1 and 0 are the greatest and the smallest element, respectively, of the Boolean algebra $(U, \sim, \sqcup, \sqcap)$, and the following is satisfied:

$$a \circ b = \begin{cases} 1 & \text{if } a = b \\ 0 & \text{otherwise.} \end{cases}$$

It is easy to see that the operation I defined as $I(a) \stackrel{\text{df}}{=} a \circ 1$ has the property:

$$I(a) = \begin{cases} 1 & \text{if } a = 1 \\ 0 & \text{otherwise.} \end{cases}$$

Theorem 23.3.4.

1. *Let* $\mathcal{M} = (U, \sim, \sqcup, \sqcap, \Rightarrow, \Leftrightarrow, \circ, D)$ *be an* $\mathsf{SCI}^{\mathsf{H}}$*-model. Then, the structure* $(U, \sim, \sqcup, \sqcap, I)$*, where* I *is an interior operation on* U *defined as above, is a topological Boolean algebra with only two open elements;*
2. *Let* $\mathcal{H} = (U, \sim, \sqcup, \sqcap, I)$ *be a topological Boolean algebra with only two open elements. Then, the structure* $\mathcal{M} = (U, \sim, \sqcup, \sqcap, \Rightarrow, \Leftrightarrow, \circ, D)$ *such that* $\Rightarrow$ *and* $\Leftrightarrow$ *are operations on* U *satisfying the conditions of* SCI*-models and, in addition, for all* $a, b \in U$, $a \circ b \stackrel{\text{df}}{=} \begin{cases} 1 & \text{if } I(a) = I(b) \\ 0 & \text{otherwise} \end{cases}$*, and* D *is an ultrafilter on* U *such that* $a \circ b \in D$ *iff* $a = b$*, is an* $\mathsf{SCI}^{\mathsf{H}}$*-model.*

Theorem 23.3.5. *For every* SCI*-formula* φ*, the following conditions are equivalent:*

1. φ *is true in all* $\mathsf{SCI}^{\mathsf{H}}$*-models;*
2. φ *is provable in* $\mathsf{SCI}^{\mathsf{H}}$*.*

There are some relationships between logics $\mathsf{SCI}^{\mathsf{T}}$ and $\mathsf{SCI}^{\mathsf{H}}$ and the modal logics S4 and S5, respectively (see Sect. 7.3).

Let σ be a mapping from the set of SCI-formulas into the set of modal formulas defined inductively as follows:

- $\sigma(p) = p$, for every propositional variable p;
- $\sigma(\varphi) = \varphi$, if $\equiv$ does not occur in φ;
- $\sigma(\varphi \equiv \psi) = [R](\sigma(\varphi) \leftrightarrow \sigma(\psi))$, where R is an accessibility relation of modal logics.

The following is known (see [Sus71a]):

Proposition 23.3.2. *For every* SCI*-formula* φ*, the following hold:*

1. φ *is true in all* $\mathsf{SCI}^{\mathsf{T}}$*-models iff* $\sigma(\varphi)$ *is* S4*-valid;*
2. φ *is true in all* $\mathsf{SCI}^{\mathsf{H}}$*-models iff* $\sigma(\varphi)$ *is* S5*-valid.*

Now, consider a mapping σ' from the set of modal formulas into the set of SCI-formulas. The function σ' is defined inductively as follows:

- $\sigma'(p) = p$, for every propositional variable p;
- $\sigma'(\varphi) = \varphi$, if $[R]$ does not occur in φ;
- $\sigma'([R]\varphi) = (\sigma'(\varphi) \equiv (\sigma'(\varphi) \vee \neg\sigma'(\varphi)))$.

The following was proved in [Sus71a]:

Proposition 23.3.3. *For every modal formula* φ*, the following hold:*

1. φ *is* S4*-valid iff* $\sigma'(\varphi)$ *is true in all* $\mathsf{SCI}^{\mathsf{T}}$*-models;*
2. φ *is* S5*-valid iff* $\sigma'(\varphi)$ *is true in all* $\mathsf{SCI}^{\mathsf{H}}$*-models.*

23.4 Dual Tableau for the Propositional Logic with Identity

A dual tableau for the logic SCI, developed in [GP07], consists of decomposition rules $(\vee)$, $(\wedge)$, $(\neg\vee)$, $(\neg\wedge)$, $(\rightarrow)$, $(\neg\rightarrow)$, $(\leftrightarrow)$, $(\neg\leftrightarrow)$, $(\neg)$ of F-dual tableau (see Sect. 1.3) adjusted to the SCI-language and, in addition, the specific rule:

$$(\equiv)\ \frac{\varphi(\psi)}{\psi \equiv \vartheta, \varphi(\psi) \mid \varphi(\vartheta), \varphi(\psi)}$$

where φ and ϑ are SCI-formulas, ψ is a subformula of φ, and $\varphi(\vartheta)$ is obtained from $\varphi(\psi)$ by replacing some occurrences of ψ with ϑ.

Observe that any application of the rules of SCI-dual tableau, in particular an application of the specific rule $(\equiv)$, preserves atomic formulas and their negations. Thus, the closed branch property holds.

A finite set of formulas is SCI-axiomatic whenever it includes either of the sets of the following forms:

For any SCI-formula φ,

(Ax1) $\{\varphi \equiv \varphi\}$;
(Ax2) $\{\varphi, \neg\varphi\}$.

A finite set X of SCI-formulas is said to be an SCI-set whenever for every SCI-model $\mathcal{M}$ and for every valuation v in $\mathcal{M}$ there exists $\varphi \in X$ such that $\mathcal{M}, v \models \varphi$. Correctness of a rule is defined in a similar way as in F-logic in Sect. 1.3.

Proposition 23.4.1.

1. *The* SCI-*rules are* SCI-*correct;*
2. *The* SCI-*axiomatic sets are* SCI-*sets.*

Proof. By way of example, we prove correctness of the rule $(\equiv)$. Let X be a finite set of SCI-formulas and let $\varphi(\psi)$ be an SCI-formula. Clearly, if $X \cup \{\varphi(\psi)\}$ is an SCI-set, then so are $X \cup \{\psi \equiv \vartheta, \varphi(\psi)\}$ and $X \cup \{\varphi(\vartheta), \varphi(\psi)\}$. Assume that $X \cup \{\psi \equiv \vartheta, \varphi(\psi)\}$ and $X \cup \{\varphi(\vartheta), \varphi(\psi)\}$ are SCI-sets. Suppose $X \cup \{\varphi(\psi)\}$ is not an SCI-set. Then there exist an SCI-model $\mathcal{M}$ and a valuation v in $\mathcal{M}$ such that for every formula $\chi \in X \cup \{\varphi(\psi)\}$, $\mathcal{M}, v \not\models \chi$. By the assumption, $\mathcal{M}, v \models \psi \equiv \vartheta$ and $\mathcal{M}, v \models \varphi(\vartheta)$, that is $v(\psi) = v(\vartheta)$ and $v(\varphi(\vartheta)) \in D$. Hence, by Proposition 23.2.1, $v(\varphi(\psi)) \in D$. Thus $\mathcal{M}, v \models \varphi(\psi)$, a contradiction. □

The notions of an SCI-proof tree, a closed branch of such a tree, a closed SCI-proof tree, and SCI-provability are defined in a similar way as in Sect. 1.3.

A branch b of an SCI-proof tree is complete whenever it satisfies the completion conditions that correspond to decomposition rules (see Sect. 1.3) and the completion condition that correspond to the rule $(\equiv)$ specific for SCI-dual tableau:

Cpl$(\equiv)$ If $\varphi \in b$ and ψ is a subformula of φ, then for every SCI-formula ϑ, either $\psi \equiv \vartheta \in b$ or $\varphi(\vartheta) \in b$, obtained by an application of the rule $(\equiv)$.

The notions of a complete SCI-proof tree and an open branch of an SCI-proof tree are defined as in Sect. 1.3.

We define inductively a *depth* of SCI-formulas as:

For all SCI-formulas φ and ψ,

- $d(p) = d(\varphi \equiv \psi) = 0$, for every $p \in \mathbb{V}$;
- $d(\neg\varphi) = d(\varphi) + 1$;
- $d(\varphi \vee \psi) = d(\varphi \wedge \psi) = d(\varphi \rightarrow \psi) = d(\varphi \leftrightarrow \psi) = \max(d(\varphi), d(\psi)) + 1$.

Let $\mathbb{FR}_{\mathsf{SCI}}$ be a set of all SCI-formulas and let $n \geq 0$. By $\mathbb{FR}^n_{\mathsf{SCI}}$ we denote the set of all SCI-formulas of the depth n.

Let b be an open branch of an SCI-proof tree. We define a branch structure $\mathcal{M}^b = (U^b, \sim^b, \sqcup^b, \sqcap^b, \Rightarrow^b, \Leftrightarrow^b, \circ^b, D^b)$ as:

- $U^b = \mathbb{FR}_{\mathsf{SCI}}$;
- $D^b = \bigcup_{n\in\omega} D^b_n$, where:

$$D^b_0 = \{\psi \in \mathbb{FR}^0_{\mathsf{SCI}} : \psi \notin b\},$$
$$D^b_{n+1} = X_1 \cup \ldots \cup X_5, \text{ where:}$$
$$X_1 = \{\neg\psi \in \mathbb{FR}^{n+1}_{\mathsf{SCI}} : \psi \notin D^b_n\},$$
$$X_2 = \{\psi \vee \theta \in \mathbb{FR}^{n+1}_{\mathsf{SCI}} : \psi \in \textstyle\bigcup_{k\leq n} D^b_k \text{ or } \theta \in \bigcup_{k\leq n} D^b_k\},$$
$$X_3 = \{\psi \wedge \theta \in \mathbb{FR}^{n+1}_{\mathsf{SCI}} : \psi, \theta \in \textstyle\bigcup_{k\leq n} D^b_k\};$$
$$X_4 = \{\psi \rightarrow \theta \in \mathbb{FR}^{n+1}_{\mathsf{SCI}} : \psi \notin \textstyle\bigcup_{k<n} D^b_k \text{ or } \theta \in \bigcup_{k\leq n} D^b_k\};$$
$$X_5 = \{\psi \leftrightarrow \theta \in \mathbb{FR}^{n+1}_{\mathsf{SCI}} : \psi, \theta \in \textstyle\bigcup_{k\leq n} D^b_k \text{ or } \psi, \theta \notin \bigcup_{k\leq n} D^b_k\};$$

- $\sim^b \psi = \neg\psi$;
- $\psi \sqcup^b \theta = (\psi \vee \theta)$;
- $\psi \sqcap^b \theta = (\psi \wedge \theta)$;
- $\psi \Rightarrow^b \theta = (\psi \rightarrow \theta)$;
- $\psi \Leftrightarrow^b \theta = (\psi \leftrightarrow \theta)$;
- $\psi \circ^b \theta = (\psi \equiv \theta)$.

Fact 23.4.1. *Let ψ be an* SCI*-formula and let $d(\psi) = n$, for some $n \geq 0$. Then, $\psi \in D^b$ iff $\psi \in D^b_n$.*

Let v^b be a valuation in $\mathcal{M}^b$ such that $v^b(p) = p$, for all $p \in \mathbb{V}$. By the definition of $\mathcal{M}^b$, $v^b(\varphi) = \varphi$, for every SCI-formula φ.

Proposition 23.4.2. *Let b be an open branch of an* SCI*-proof tree. Then, for every* SCI*-formula ψ, if $\psi \in D^b$, then $\psi \notin b$.*

Proof. The proof is by induction on the depth of formulas. For formulas of the depth 0, the proposition holds by the definition of the set D^b. Let $\psi = \neg\theta$, for some formula θ such that $d(\theta) = 0$. Assume $\psi \in D^b_1$. By the definition of D^b, $\theta \notin D^b_0$, thus $\theta \in b$. Hence, $\neg\theta \notin b$, and so $\psi \notin b$.

Suppose the proposition holds for all formulas of depth not greater than n and their negations. Assume $d(\psi) = n + 1$ and $\psi \in D^b_{n+1}$.

Let $\psi = \theta \vee \chi$ for some formulas θ and χ such that $\max(d(\theta), d(\chi)) = n$. Since $\psi \in D^b_{n+1}$, $\theta \in \bigcup_{k \leq n} D^b_k$ or $\chi \in \bigcup_{k \leq n} D^b_k$. By the induction hypothesis, $\theta \notin b$ or $\chi \notin b$. Suppose $\psi \in b$. The completion condition Cpl($\vee$) implies both $\psi \in b$ and $\theta \in b$, a contradiction.

Let $\psi = \neg\neg\chi$. Then $\neg\chi \notin D^b_n$. Suppose $\psi \in b$. By the completion condition Cpl($\neg$), $\chi \in b$. By the induction hypothesis, $\chi \notin D^b_{n-1}$. By the definition of the set D^b, if a formula χ of the depth $n-1$ satisfies $\chi \notin D^b_{n-1}$, then $\neg\chi \in D^b_n$, a contradiction.

Let $\psi = \neg(\theta \vee \chi)$. Then $(\theta \vee \chi) \notin D^b_n$. Suppose $\neg(\theta \vee \chi) \in b$. By the completion condition Cpl($\neg\vee$), either $\neg\theta \in b$ or $\neg\chi \in b$. By the induction hypothesis, either $\neg\theta \notin D^b$ or $\neg\chi \notin D^b$. Therefore, either $\theta \in D^b$ or $\chi \in D^b$. So by the construction of the set D^b, we have $(\theta \vee \chi) \in D^b_n$, a contradiction.

The proofs of the remaining cases are similar. □

Let us define the relation $R_\circ$ on the set of SCI-formulas as:

$$(\psi, \theta) \in R_\circ \overset{\text{df}}{\Longleftrightarrow} (\psi \circ \theta) \in D^b.$$

Proposition 23.4.3. *For every open branch b of an SCI-proof tree, $R_\circ$ is an equivalence relation on the set U^b.*

Proof. If for some $\psi \in U^b$, $(\psi, \psi) \notin R_\circ$, then $\psi \equiv \psi \in b$, which would mean that b is closed, a contradiction. Let $(\psi, \theta) \in R_\circ$ and suppose that $(\theta, \psi) \notin R_\circ$. Then $(\psi \equiv \theta) \notin b$ and $(\theta \equiv \psi) \in b$. By the completion condition Cpl($\equiv$), $(\psi \equiv \theta) \in b$ or $(\theta \equiv \theta) \in b$. The first case contradicts $(\psi \equiv \theta) \notin b$, the second one implies that the branch is closed, a contradiction. Let $(\psi, \theta) \in R_\circ$, $(\theta, \chi) \in R_\circ$, and suppose that $(\psi, \chi) \notin R_\circ$. Then, $(\psi \equiv \theta) \notin b$, $(\theta \equiv \chi) \notin b$, and $(\psi \equiv \chi) \in b$. By the completion condition Cpl($\equiv$), either $(\psi \equiv \theta) \in b$ or $(\theta \equiv \chi) \in b$, a contradiction. □

Let b be an open branch of an SCI-proof tree. We define the quotient structure $\mathcal{M}^b_q = (U^b_q, \sim^b_q, \sqcup^b_q, \sqcap^b_q, \circ^b_q, D^b_q)$ as:

- $U^b_q = \{\|\psi\| : \psi \in U^b\}$, where $\|\psi\|$ is the equivalence class of $R_\circ$ generated by ψ;
- $D^b_q = \{\|\psi\| : \psi \in D^b\}$;
- $\sim^b_q \|\psi\| = \|\sim^b \psi\|$;
- $\|\psi\| \sqcup^b_q \|\theta\| = \|\psi \sqcup^b \theta\|$;
- $\|\psi\| \sqcap^b_q \|\theta\| = \|\psi \sqcap^b \theta\|$;
- $\|\psi\| \Rightarrow^b_q \|\theta\| = \|\psi \Rightarrow^b \theta\|$;
- $\|\psi\| \Leftrightarrow^b_q \|\theta\| = \|\psi \Leftrightarrow^b \theta\|$;
- $\|\psi\| \circ^b_q \|\theta\| = \|\psi \circ^b \theta\|$.

Let v^b_q be a valuation such that $v^b_q(p) = \|p\|$, for every $p \in \mathbb{V}$.

Proposition 23.4.4 (Branch Model Property).

1. *The structure* $\mathcal{M}_q^b$ *is an* SCI*-model;*
2. *For every* SCI*-formula* φ, $v^b(\varphi) \in D^b$ *iff* $v_q^b(\varphi) \in D_q^b$.

Proof. We show that the model $\mathcal{M}_q^b$ satisfies all the conditions of SCI-models. D_q^b is a non-empty subset of U_q^b, since D^b is a non-empty subset of U^b. Indeed, D^b is non-empty, since for every SCI-formula ψ, a formula $\psi \equiv \psi \notin b$, hence $\psi \equiv \psi \in D^b$.

Let $\psi, \theta \in U^b$ and let $\max(d(\psi), d(\varphi)) = n$, $n \geq 0$. Then, the following hold:

$\sim^b \psi \in D^b$ iff $\neg\psi \in D^b$ iff for some n, $\psi \notin D_n^b$ iff $\psi \notin D^b$;
$\psi \sqcup^b \theta \in D^b$ iff $(\psi \vee \theta) \in D^b$ iff $\psi \in \bigcup_{k \leq n} D_k^b$ or $\theta \in \bigcup_{k \leq n} D_k^b$ iff $\psi \in D^b$ or $\theta \in D^b$;
$\psi \sqcap^b \theta \in D^b$ iff $(\psi \wedge \theta) \in D^b$ iff $\psi, \theta \in \bigcup_{k \leq n} D_k^b$ iff $\psi \in D^b$ and $\theta \in D^b$;
$\psi \Rightarrow^b \theta \in D^b$ iff $(\psi \rightarrow \theta) \in D^b$ iff $\psi \notin \bigcup_{k<n} D_k^b$ or $\theta \in \bigcup_{k \leq n} D_k^b$ iff $\psi \notin D^b$ or $\theta \in D^b$;
$\psi \Leftrightarrow^b \theta \in D^b$ iff $(\psi \leftrightarrow \theta) \in D^b$ iff $\psi, \theta \in \bigcup_{k \leq n} D_k^b$ or $\psi, \theta \notin \bigcup_{k \leq n} D_k^b$ iff $\psi \in D^b$ iff $\theta \in D^b$.

The above properties together with the definition of $\mathcal{M}_q^b$ and Proposition 23.4.3 imply:

$\sim_q^b \|\psi\| \in D_q^b$ iff $\sim^b \psi \in D^b$ iff $\psi \notin D^b$ iff $\|\psi\| \notin D_q^b$;
$\|\psi\| \sqcup_q^b \|\theta\| \in D_q^b$ iff $\psi \in D^b$ or $\theta \in D^b$ iff $\|\psi\| \in D_q^b$ or $\|\theta\| \in D_q^b$;
$\|\psi\| \sqcap_q^b \|\theta\| \in D_q^b$ iff $\psi \in D^b$ and $\theta \in D^b$ iff $\|\psi\| \in D_q^b$ and $\|\theta\| \in D_q^b$;
$\|\psi\| \Rightarrow_q^b \|\theta\| \in D_q^b$ iff $\psi \notin D^b$ or $\theta \in D^b$ iff $\|\psi\| \notin D_q^b$ or $\|\theta\| \in D_q^b$;
$\|\psi\| \Leftrightarrow_q^b \|\theta\| \in D_q^b$ iff $\psi \notin D^b$ iff $\theta \in D^b$ iff $\|\psi\| \in D_q^b$ iff $\|\theta\| \in D_q^b$;
$\|\psi\| \circ_q^b \|\theta\| \in D_q^b$ iff $\|\psi \circ^b \theta\| \in D_q^b$ iff $\psi \circ^b \theta \in D^b$ iff $(\psi, \theta) \in R_\circ$ iff $\|\psi\| = \|\theta\|$.

Thus, $\mathcal{M}_q^b$ is an SCI-model.
2. follows directly from the definition of D_q^b. □

By Propositions 23.4.2 and 23.4.4, we have:

Proposition 23.4.5. *Let* b *be an open branch of an* SCI*-proof tree. Then, for every* SCI*-formula* φ, *if* $\mathcal{M}_q^b, v_q^b \models \varphi$, *then* $\varphi \notin b$.

Theorem 23.4.1 (Soundness and Completeness of SCI). *Let* φ *be an* SCI*-formula. Then the following conditions are equivalent:*

1. φ *is* SCI*-valid;*
2. φ *is* SCI*-provable.*

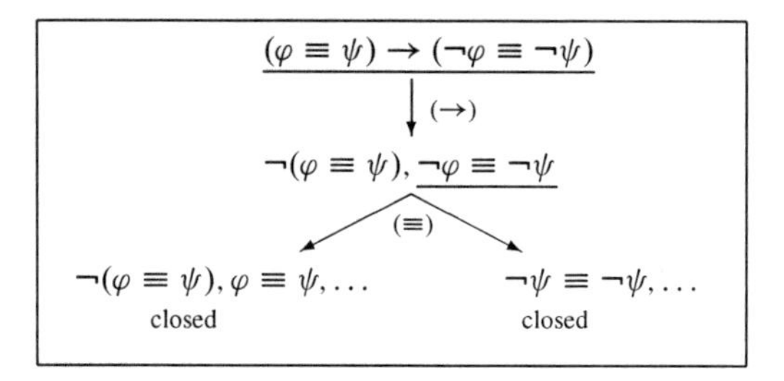

Fig. 23.1 An SCI-proof of SCI-axiom $(\equiv_2)$

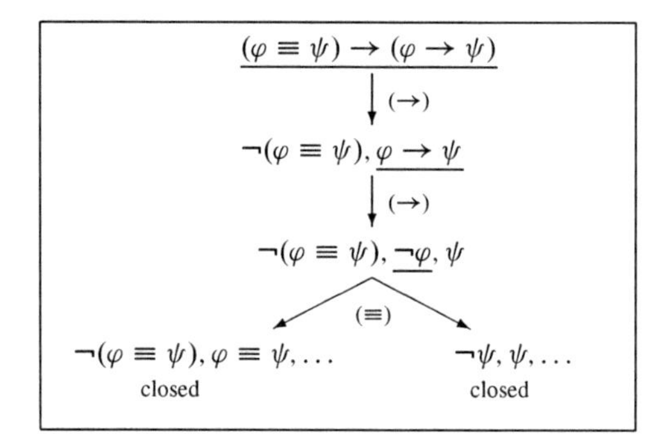

Fig. 23.2 An SCI-proof of SCI-axiom $(\equiv_3)$

Proof. The implication 1. $\rightarrow$ 2. holds by Proposition 23.4.1. Now, assume that φ is SCI-valid. Suppose there is no any closed SCI-proof tree for φ. Then, there exists a complete SCI-proof tree for φ with an open branch, say b. Since $\varphi \in b$, by Proposition 23.4.5, we get $\mathcal{M}_q^b \not\models \varphi$. Hence, by Proposition 23.4.4, φ is not SCI-valid, a contradiction. □

Example. We show that all SCI-axioms characterizing $\equiv$ are SCI-provable. Clearly, the axiom $(\equiv_1)$ of the form $\varphi \equiv \varphi$ is SCI-provable, since $\{\varphi \equiv \varphi\}$ is an SCI-axiomatic set. In Figs. 23.1 and 23.2 we present SCI-proofs of axiom $(\equiv_2)$ and axiom $(\equiv_3)$, respectively.

Figure 23.3 presents an SCI-proof of $(\equiv_4)$, for any $\# = \{\vee, \wedge, \rightarrow, \leftrightarrow, \equiv\}$.

23.5 Dual Tableaux for Axiomatic Extensions of the Propositional Logic with Identity

In Sect. 23.3 axiomatic extensions $\mathsf{SCI}^{\mathsf{T}}$ and $\mathsf{SCI}^{\mathsf{H}}$ of the logic SCI were presented. By Proposition 23.3.2, an SCI-formula φ is true in all $\mathsf{SCI}^{\mathsf{T}}$-models (resp. $\mathsf{SCI}^{\mathsf{H}}$-models) if and only if $\sigma(\varphi)$ is S4-valid (resp. S5-valid), where σ is the translation of SCI-formulas into modal formulas defined in Sect. 23.3.

On the other hand, in Sects. 7.4 and 7.5 we showed that S4-validity (resp. S5-validity) of a modal formula ψ is equivalent to $\mathsf{RL}_{\mathsf{S4}}$-provability (resp. $\mathsf{RL}_{\mathsf{S5}}$-

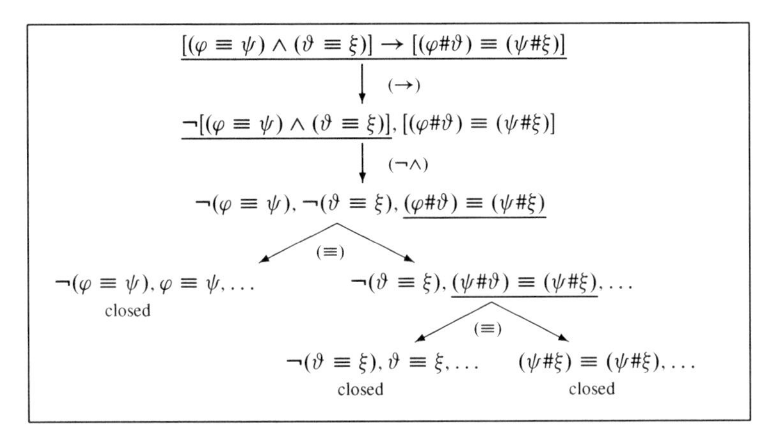

Fig. 23.3 An SCI-proof of SCI-axiom $(\equiv_4)$

provability) of the translation of ψ, $\tau(\psi)$, into a relational formula defined in Sect. 7.4 (see Theorem 7.4.1, p. 147).

We can define a translation of SCI-formulas into relational terms of standard modal logics by $\chi(\varphi) \stackrel{\text{df}}{=} \tau(\sigma(\varphi))$. Let χ' be a one-to-one assignment of relational variables to the propositional variables. Then, the translation χ of SCI-formulas satisfies:

- $\chi(p) = \chi'(p) ; 1$, for any propositional variable $p \in \mathbb{V}$;
- $\chi(\neg\varphi) = -\chi(\varphi)$;
- $\chi(\varphi \vee \psi) = \chi(\varphi) \cup \chi(\psi)$;
- $\chi(\varphi \wedge \psi) = \chi(\varphi) \cap \chi(\psi)$;
- $\chi(\varphi \rightarrow \psi) = -\chi(\varphi) \cup \chi(\psi)$;
- $\chi(\varphi \leftrightarrow \psi) = (-\chi(\varphi) \cup \chi(\psi)) \cap (-\chi(\psi) \cup \chi(\varphi)))$;
- $\chi(\varphi \equiv \psi) = -(R ; -\chi(\varphi \leftrightarrow \psi))$.

By Proposition 23.3.2 and Theorem 7.4.1, an SCI-formula φ is true in all $\mathsf{SCI}^{\mathsf{T}}$-models (resp. $\mathsf{SCI}^{\mathsf{H}}$-models) iff $\chi(\varphi)$ is $\mathsf{RL}_{\mathsf{S4}}$-provable (resp. $\mathsf{RL}_{\mathsf{S5}}$-provable).

Theorem 23.5.1. *For every* SCI*-formula φ and for all object variables x and y, the following conditions are equivalent:*

1. *φ is true in all* $\mathsf{SCI}^{\mathsf{T}}$*-models (resp.* $\mathsf{SCI}^{\mathsf{H}}$*-models);*
2. *$x\chi(\varphi)y$ is* $\mathsf{RL}_{\mathsf{S4}}$*-provable (resp.* $\mathsf{RL}_{\mathsf{S5}}$*-provable).*

It follows that $\mathsf{RL}_{\mathsf{S4}}$-dual tableau (resp. $\mathsf{RL}_{\mathsf{S5}}$-dual tableau) can be used to verify $\mathsf{SCI}^{\mathsf{T}}$-validity (resp. $\mathsf{SCI}^{\mathsf{H}}$-validity) of SCI-formulas.

Example. Consider SCI-formulas φ and ψ:

$$\varphi = (p \vee \neg p) \equiv (q \vee \neg q); \qquad \psi = (p \wedge \neg p) \equiv (q \wedge \neg q).$$

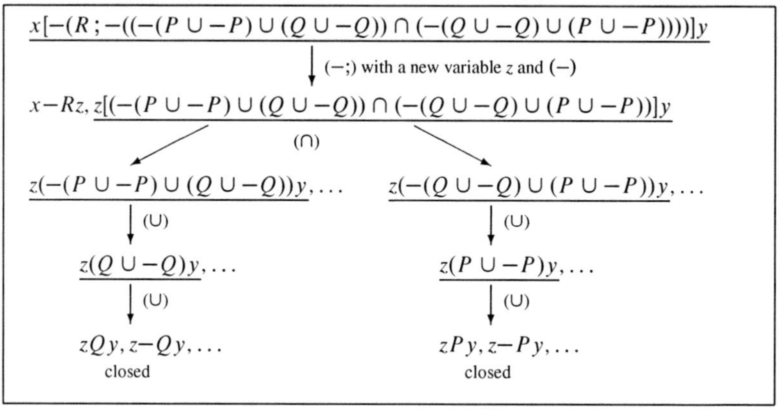

x[−(R ; −((−(P ∪ −P) ∪ (Q ∪ −Q)) ∩ (−(Q ∪ −Q) ∪ (P ∪ −P))))]y
(−;) with a new variable z and (−)
x−Rz, z[(−(P ∪ −P) ∪ (Q ∪ −Q)) ∩ (−(Q ∪ −Q) ∪ (P ∪ −P))]y
(∩)
z(−(P ∪ −P) ∪ (Q ∪ −Q))y, . . .
z(−(Q ∪ −Q) ∪ (P ∪ −P))y, . . .
(∪)
(∪)
z(Q ∪ −Q)y, . . .
z(P ∪ −P)y, . . .
(∪)
(∪)
zQy, z−Qy, . . .
closed
zPy, z−Py, . . .
closed

Fig. 23.4 An RL_{S5}-proof of $\mathsf{SCI}^{\mathsf{H}}$-validity of SCI-formula $(p \vee \neg p) \equiv (q \vee \neg q)$

x[−(R ; −((−(P ∩ −P) ∪ (Q ∩ −Q)) ∩ (−(Q ∩ −Q) ∪ (P ∩ −P))))]y
(−;) with a new variable z and (−)
x−Rz, z[(−(P ∩ −P) ∪ (Q ∩ −Q)) ∩ (−(Q ∩ −Q) ∪ (P ∩ −P))]y
(∩)
z(−(P ∩ −P) ∪ (Q ∩ −Q))y, . . .
z(−(Q ∩ −Q) ∪ (P ∩ −P))y, . . .
(∪)
(∪)
z−(P ∩ −P)y, . . .
z−(Q ∩ −Q)y, . . .
(−∩) and (−)
(−∩) and (−)
z−Py, zPy, . . .
closed
z−Qy, zQy, . . .
closed

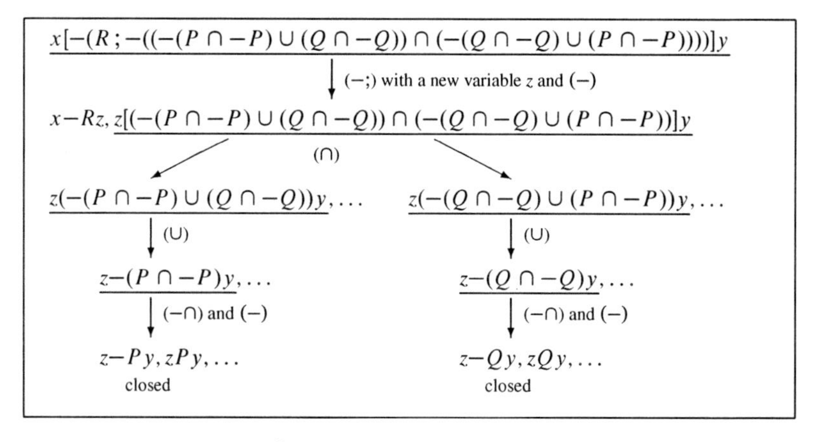

Fig. 23.5 An RL_{S5}-proof of $\mathsf{SCI}^{\mathsf{H}}$-validity of SCI-formula $(p \wedge \neg p) \equiv (q \wedge \neg q)$

These formulas are not SCI-valid. Indeed, let $\mathcal{M}=(U, \sim, \sqcup, \sqcap, \Rightarrow, \Leftrightarrow, \circ, D)$, be an SCI-model such that $U=\{0,1,2,3\}$, $D=\{2,3\}$, and the operations $\sim, \sqcup, \sqcap, \Rightarrow$, $\Leftrightarrow, \circ$ are defined as:

$$\sim a \stackrel{\text{df}}{=} \begin{cases} 0 & \text{if } a = 2 \\ 1 & \text{if } a = 3 \\ 2 & \text{if } a = 0 \\ 3 & \text{if } a = 1 \end{cases} \qquad a \circ b \stackrel{\text{df}}{=} \begin{cases} 0 & \text{if } a \neq b \\ 3 & \text{otherwise} \end{cases}$$

$$a \sqcup b \stackrel{\text{df}}{=} \max(a,b) \qquad a \sqcap b \stackrel{\text{df}}{=} \min(a,b)$$

$$a \Rightarrow b \stackrel{\text{df}}{=} \max(\sim a, b) \qquad a \Leftrightarrow b \stackrel{\text{df}}{=} \min(\max(\sim a, b), \max(\sim b, a)).$$

This structure is an SCI-model. Let v be a valuation in $\mathcal{M}$ such that $v(p) = 0$ and $v(q) = 3$. Then:

$$v(p \vee \neg p) = 2 \text{ and } v(p \wedge \neg p) = 0;$$
$$v(q \vee \neg q) = 3 \text{ and } v(q \wedge \neg q) = 1.$$

Therefore, $v((p \vee \neg p) \equiv (q \vee \neg q)) = 0$ and $v((p \wedge \neg p) \equiv (q \wedge \neg q)) = 0$. Hence, φ and ψ are not true in $\mathcal{M}$. However, by Theorem 23.3.4, formulas φ and ψ are true in all $\mathsf{SCI}^{\mathsf{H}}$-models. The translations of these formulas into relational terms are:

$$\chi(\varphi) \stackrel{\text{df}}{=} -(R\,;-((-(P \cup -P) \cup (Q \cup -Q)) \cap (-(Q \cup -Q) \cup (P \cup -P))));$$

$$\chi(\psi) \stackrel{\text{df}}{=} -(R\,;-((-(P \cap -P) \cup (Q \cap -Q)) \cap (-(Q \cap -Q) \cup (P \cap -P)))),$$

where $\chi'(p) = P$ and $\chi'(q) = Q$. Figures 23.4 and 23.5 present an $\mathsf{RL}_{\mathsf{S5}}$-proof of $\chi(\varphi)$ and $\chi(\psi)$, respectively, which by Theorem 23.5.1 show that φ and ψ are true in all $\mathsf{SCI}^{\mathsf{H}}$-models.

Chapter 24
Dual Tableaux for Logics of Conditional Decisions

24.1 Introduction

The roots of the logics and algebras of conditional decisions can be traced back to the work by Claude Elwood Shanon in [Sha38] and his information theory [Sha48].

Logics of conditional decisions are the formalisms for the specification of $n \geq 2$ alternative decisions that depend on a condition which holds in a degree. In the binary case the formulas of the logic may be viewed as binary decision trees or *if-then-else* statements which represent branching in algorithms; they are the basic statements in most of the programming languages. The paper [McC63] is one of the foundations of that research. Binary decision diagrams were introduced in [Lee59] as a compact representation of binary decision trees used for a representation of Boolean functions; see also [Ake78]. They are extensively applied to the synthesis of combinatorial circuits and in the implementation and formal verification of systems. Interest in multiple-valued branching was inspired by observations in [Dij68].

A propositional logic of binary decisions presented in this chapter was introduced in [EO67], and it was extended to the class of $\forall\exists$-formulas of classical first-order logic in [Orł69], see also [PS95]. The algebras of conditional decisions named as B-actions were investigated in [Ber91]. A further study of these algebras can be found, e.g., in [Man93, Sto98].

In this chapter we present, first, a logic of conditional decisions, where the conditions are specified in the form of formulas of the classical propositional logic. In this case a condition may hold or not. We present a dual tableau decision procedure for this logic following [EO67]. The rules of the system satisfy stronger semantic property than the usual correctness, namely they preserve values of formulas under a valuation. We also present an algebraic version of the logic with an equational axiomatization which can be derived directly from the dual tableau rules. We show a relationship of the logic with the algebras of B-action introduced in [Ber91] and with the axiomatization of *if-then-else* presented in [BT83, GM87].

In high level programming languages three-valued conditions are often used, where a third value intuitively corresponds to the situation, when testing of the condition does not halt, and then the *if-then-else* statement diverges. Therefore, we extend the binary case to a family of n-ary, $n \geq 2$, logics of conditional decisions,

E. Orłowska and J. Golińska-Pilarek, *Dual Tableaux: Foundations, Methodology, Case Studies*, Trends in Logic 36, DOI 10.1007/978-94-007-0005-5_24,

where the conditions are specified as formulas of the Rosser–Turquette logic discussed in Sect. 10.2, with an operation of implication added to the language. We present a dual tableau decision procedure for that class of logics.

24.2 Logic of Conditional Decisions and Its Dual Tableau Decision Procedure

The formulas of a logic of conditional decisions, LCD, are built from expressions representing conditions and decisions. Conditions are PC-formulas (see Sect. 7.2) endowed with propositional constants 0 and 1 which represent elements of a two-element Boolean algebra. We will refer to them as PC-formulas as well. LCD-formulas represent decisions. They are defined by:

- 0 and 1 are LCD-formulas;
- If φ is a condition and α_0, α_1 are LCD-formulas, then $\varphi\alpha_0\alpha_1$ is an LCD-formula.

It follows that the symbols of the LCD-language are those of PC and 0 and 1.

The set of subformulas of an LCD-formula α, $Sub(\alpha)$, is defined as:

- If $\alpha = 0$ or $\alpha = 1$, then $Sub(\alpha) = \{\alpha\}$;
- If $\alpha = \varphi\alpha_0\alpha_1$, then $Sub(\alpha) = \{\alpha\} \cup Sub(\alpha_0) \cup Sub(\alpha_1)$.

Hence, $Sub(\alpha)$ consists of all the decisions occurring in α.

An LCD-*model* is a PC-model $\mathcal{M} = (\{0, 1\}, v)$ such that $\{0, 1\}$ is a two-element Boolean algebra and valuation v is extended to LCD-formulas as:

- $v(0) = 0$ and $v(1) = 1$;
- $v(\varphi\alpha_0\alpha_1) = \begin{cases} v(\alpha_0) & \text{if } v(\varphi) = 0 \\ v(\alpha_1) & \text{if } v(\varphi) = 1. \end{cases}$

As usual, we use symbols 0 and 1 both in the language and in the models.

An LCD-formula α is said to be true in $\mathcal{M} = (\{0, 1\}, v)$ whenever $v(\alpha) = 1$, and it is LCD-valid if it is true in all LCD-models.

Rules of LCD-dual tableau are of the form $\frac{\Sigma_1\alpha\Sigma_2}{\Sigma_1\beta\Sigma_2}$, where Σ_1 and Σ_2 are strings of symbols of the LCD-language such that $\Sigma_1\alpha\Sigma_2$ and $\Sigma_1\beta\Sigma_2$ are LCD-formulas and α and β are their subformulas, respectively. An application of such a rule to a formula γ of the form $\Sigma_1\alpha\Sigma_2$ results in a formula obtained from γ by replacing an occurrence of α indicated in the string $\Sigma_1\alpha\Sigma_2$ with formula β.

Fact 24.2.1. *Let* $\frac{\Sigma_1\alpha\Sigma_2}{\Sigma_1\beta\Sigma_2}$ *be a rule as above. Then, for every* LCD-*model* $\mathcal{M} = (\{0, 1\}, v)$, *if* $v(\alpha) = v(\beta)$, *then* $v(\Sigma_1\alpha\Sigma_2) = v(\Sigma_1\beta\Sigma_2)$.

Let φ and ψ be conditions and let α_0, α_1 be LCD-formulas. Decomposition rules of LCD-dual tableau are:

$$(\vee)\ \frac{\Sigma_1(\varphi \vee \psi)\alpha_0\alpha_1\Sigma_2}{\Sigma_1\varphi\psi\alpha_0\alpha_1\alpha_1\Sigma_2} \qquad (\wedge)\ \frac{\Sigma_1(\varphi \wedge \psi)\alpha_0\alpha_1\Sigma_2}{\Sigma_1\varphi\alpha_0\psi\alpha_0\alpha_1\Sigma_2}$$

$$(\rightarrow)\ \frac{\Sigma_1(\varphi \rightarrow \psi)\alpha_0\alpha_1\Sigma_2}{\Sigma_1\varphi\alpha_1\psi\alpha_0\alpha_1\Sigma_2} \qquad (\neg)\ \frac{\Sigma_1(\neg\varphi)\alpha_0\alpha_1\Sigma_2}{\Sigma_1\varphi\alpha_1\alpha_0\Sigma_2}$$

Let φ be a condition, let p be a propositional variable, let $\alpha_0, \alpha_1, \beta$ be LCD-formulas, and let $\Sigma_3 p\alpha_0\alpha_1\Sigma_4$ be an LCD-formula such that $p\alpha_0\alpha_1$ is its subformula. Simplification rules of LCD-dual tableau are:

$$(0)\ \frac{\Sigma_1 0\alpha_0\alpha_1\Sigma_2}{\Sigma_1\alpha_0\Sigma_2} \qquad (1)\ \frac{\Sigma_1 1\alpha_0\alpha_1\Sigma_2}{\Sigma_1\alpha_1\Sigma_2}$$

$$(S1)\ \frac{\Sigma_1\varphi\alpha\alpha\Sigma_2}{\Sigma_1\alpha\Sigma_2}$$

$$(S2)\ \frac{\Sigma_1(p\Sigma_3 p\alpha_0\alpha_1\Sigma_4\beta)\Sigma_2}{\Sigma_1(p\Sigma_3\alpha_0\Sigma_4\beta)\Sigma_2} \qquad (S3)\ \frac{\Sigma_1(p\beta\Sigma_3 p\alpha_0\alpha_1\Sigma_4)\Sigma_2}{\Sigma_1(p\beta\Sigma_3\alpha_1\Sigma_4)\Sigma_2}$$

The rules satisfy the stronger semantic property than that of preserving and reflecting validity, which typically holds in the dual tableaux presented in the book. Namely, they preserve values of formulas under a valuation. More precisely, we have:

Proposition 24.2.1. *Let $\mathcal{M} = (\{0,1\}, v)$ be an LCD-model. Then, for every LCD-rule $\frac{\Sigma_1\alpha\Sigma_2}{\Sigma_1\beta\Sigma_2}$, $v(\Sigma_1\alpha\Sigma_2) = v(\Sigma_1\beta\Sigma_2)$.*

Proof. By way of example, we prove the statement for the rules: $(\vee)$, $(S1)$, and $(S2)$. Recall that, by Fact 24.2.1, for all LCD-formulas $\Sigma_1\alpha\Sigma_2$ and $\Sigma_1\beta\Sigma_2$ such that α and β are their subformulas, respectively, if $v(\alpha) = v(\beta)$, then $v(\Sigma_1\alpha\Sigma_2) = v(\Sigma_1\beta\Sigma_2)$. Therefore, in order to prove the statement, we need to show that $v(\alpha) = v(\beta)$.

$(\vee)$ By the definition of valuation:

$$v(\varphi(\psi\alpha_0\alpha_1)\alpha_1) = \begin{cases} v(\psi\alpha_0\alpha_1) & \text{if } v(\varphi) = 0 \\ v(\alpha_1) & \text{if } v(\varphi) = 1 \end{cases}$$

$$\text{where } v(\psi\alpha_0\alpha_1) = \begin{cases} v(\alpha_0) & \text{if } v(\psi) = 0 \\ v(\alpha_1) & \text{if } v(\psi) = 1 \end{cases}$$

Thus, we obtain:

$$v(\varphi(\psi\alpha_0\alpha_1)\alpha_1) = \begin{cases} v(\alpha_0) & \text{if } v(\varphi) = v(\psi) = 0 \\ v(\alpha_1) & \text{if } v(\varphi) = 1 \text{ or } v(\psi) = 1 \end{cases} = v((\varphi \vee \psi)\alpha_0\alpha_1).$$

$(S1)$ Directly from the definition of valuation we have: $v(\varphi\alpha\alpha) = v(\alpha)$.

$(S2)$ In view of Fact 24.2.1, we may assume that Σ_3 and Σ_4 are empty. If $v(p) = 0$, then $v(p(p\alpha_0\alpha_1)\beta) = v(p\alpha_0\alpha_1) = v(\alpha_0)$. Similarly, if $v(p) = 1$, then $v(p(p\alpha_0\alpha_1)\beta) = v(\alpha_1)$. Hence, $v(p(p\alpha_0\alpha_1)\beta) = v(p\alpha_0\alpha_1)$. □

An LCD-*proof sequence* for an LCD-formula α is a sequence of LCD-formulas with the following properties:

- The first element of the sequence is α;
- Each formula of the sequence except the first one is obtained from its predecessor formula by an application of an LCD-rule;
- A formula of the sequence does not have a successor whenever none of the rules is applicable to it.

It is easy to see that every LCD-proof sequence is finite. An LCD-formula α is said to be LCD-provable whenever there exists an LCD-proof sequence for α such that the last formula of the sequence is 1; such a sequence is then referred to as an LCD-proof of α.

Now, we define some notions that will be useful in the completeness proof.

Operation free LCD-*formulas* are defined as follows:

- $0, 1$ are operation free formulas;
- $\varphi\alpha_0\alpha_1$ is an operation free formula if and only if φ is a propositional variable and α_0, α_1 are operation free formulas.

An operation free formula $p\alpha_0\alpha_1$ has a *unique condition* whenever p occurs neither in α_0 nor in α_1.

An LCD-formula α is said to be *indecomposable* whenever it satisfies the following conditions:

- α is an operation free formula;
- $p11$ and $p00$ are not subformulas of α, for any propositional variable p;
- Every subformula of α has a unique condition.

Note that a formula $p\alpha_0\alpha_1$ is indecomposable if and only if both α_0 and α_1 are indecomposable and if $\alpha_i \in \{1, 0\}$, then $\alpha_i \neq \alpha_j$, for all $i, j \in \{0, 1\}$ such that $i \neq j$.

Proposition 24.2.2. *For every indecomposable* LCD-*formula* α *such that* $\alpha \neq 1$, *there exists an* LCD-*model* $\mathcal{M} = (\{0, 1\}, v)$ *such that* $v(\alpha) = 0$.

Proof. The proof is by the induction on the complexity of α. If $\alpha = 0$, then the proposition holds. Let $\alpha = p0\beta$, for some propositional variable p and for some indecomposable formula β such that p does not occur in β and $\beta \neq 0$. Let v be any valuation such that $v(p) = 0$. Then $v(p0\beta) = 0$, hence the proposition holds. Similarly, if $\alpha = p\beta 0$, then we take a valuation such that $v(p) = 1$. Now, we show that if the proposition holds for indecomposable formulas $\alpha_0 \neq 1$ and $\alpha_1 \neq 1$ such that p occurs neither in α_0 nor in α_1, then it holds for $p\alpha_0\alpha_1$. By the induction hypothesis, there is a valuation v_0 such that $v_0(\alpha_0) = 0$. We define the required valuation v as: $v(p) = 0$ and $v(q) = v_0(q)$, for every propositional variable q

occurring in α_0. Then, $v(p\alpha_0\alpha_1) = v_0(\alpha_0) = 0$. If $\alpha_0 = 1$, then $\alpha_1 \neq 1$. Thus, by the induction hypothesis, there is a valuation v_1 such that $v_1(\alpha_1) = 0$. Then, we define $v(p) = 1$ and $v(q) = v_1(q)$, for every propositional variable q occurring in α_1. Hence, $v(p\alpha_0\alpha_1) = v_1(\alpha_1) = 0$. If $\alpha_1 = 1$, then $\alpha_0 \neq 1$, and we define valuation v in a way analogous to the previous case. □

Proposition 24.2.3. *For every* LCD*-formula* α*, there exists an* LCD*-proof sequence such that its last formula is indecomposable.*

Proof. Let α be an LCD-formula. Let β be the last formula in an LCD-proof sequence for α. Then none of the rules is applicable to β. Clearly, if $\beta = 0$ or $\beta = 1$, then the proposition holds. Let $\beta = \varphi\alpha_0\alpha_1$. If β is not an operation free formula, then $\varphi \in \{0, 1\}$ or either its condition or a condition of some of its subformula is a compound PC-formula. But then either of the rules $(\vee)$, $(\wedge)$, $(\rightarrow)$, $(\neg)$, (0), or (1) can be applied to β, a contradiction. If $p11$ or $p00$ is a subformula of β, for some propositional variable p, then the rule $(S1)$ can be applied to β, a contradiction. If some of the subformulas of β does not have a unique condition, then the rule $(S2)$ or $(S3)$ can be applied to β, a contradiction. Therefore, β must be an indecomposable LCD-formula. □

Theorem 24.2.1 (Soundness and Completeness of LCD**).** *For every* LCD*-formula* α*, the following conditions are equivalent:*

1. α *is* LCD*-valid;*
2. α *is* LCD*-provable.*

Proof. (1. $\rightarrow$ 2.) Let α be LCD-valid. Suppose α is not LCD-provable, that is there is no any LCD-proof sequence for α which ends with 1. By Proposition 24.2.3, there exists an LCD-proof sequence for α such that its last formula, say β, is indecomposable. By Proposition 24.2.2, there exists an LCD-model $\mathcal{M} = (\{0, 1\}, v)$ such that $v(\beta) = 0$. By Proposition 24.2.1, $v(\alpha) = 0$, a contradiction.

(2. $\rightarrow$ 1.) Now, assume that α is LCD-provable, that is there exists an LCD-proof sequence for α such that its last formula is 1. By Proposition 24.2.1, for every LCD-model $\mathcal{M} = (\{0, 1\}, v)$, $v(\alpha) = 1$. □

As a corollary, we obtain:

Theorem 24.2.2. *For every* PC*-formula* φ*, the following conditions are equivalent:*

1. φ *is* PC*-valid;*
2. $\varphi 01$ *is* LCD*-provable.*

Thus, since every LCD-proof tree is finite and the rules of LCD-dual tableau preserve and reflect validity of LCD-formulas, LCD-dual tableau is a decision procedure for the logic LCD as well as for the logic PC.

Example. We consider an LCD-formula:

$$\alpha = pp\neg(q \wedge \neg q)01\neg(\neg q \vee q)101,$$

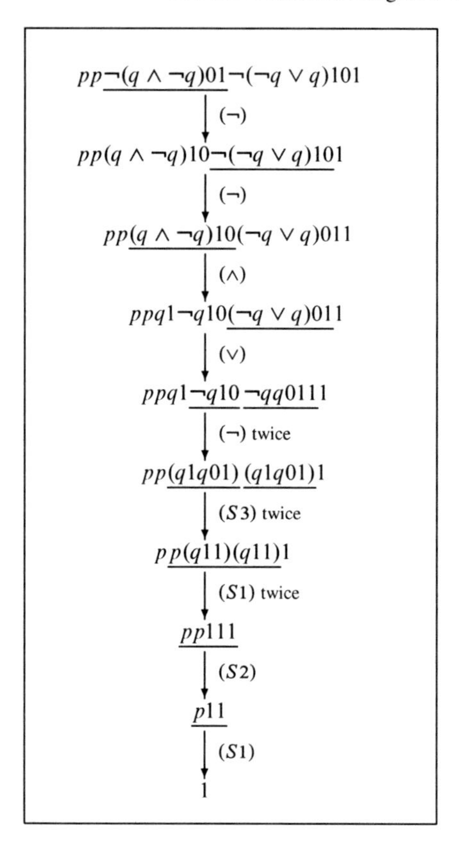

Fig. 24.1 An LCD-proof of LCD-formula $pp\neg(q \wedge \neg q)01\neg(\neg q \vee q)101$

and a PC-formula:

$$\varphi = (p \wedge \neg(p \vee q)) \rightarrow q.$$

Figures 24.1 and 24.2 present LCD-proofs of α and $\varphi 01$, respectively. Due to Theorems 24.2.1 and 24.2.2, respectively, α is LCD-valid and φ is PC-valid.

24.3 Algebras of Conditional Decisions

The logic LCD of conditional decisions determines in a natural way the class ACD of algebras of conditional decisions of the form: $\mathfrak{A} = (\mathfrak{B}, A, f)$, where $\mathfrak{B} = (B, -, +, \cdot, 0, 1)$ is a Boolean algebra, A is a non-empty set, and $f: B \times A \times A \rightarrow A$ is a ternary operation satisfying the following axioms:

For all $b, b_1, b_2 \in B$ and for all $a, a_0, a_1, a_2 \in A$,

- $f(0, a_0, a_1) = a_0$;
- $f(1, a_0, a_1) = a_1$;

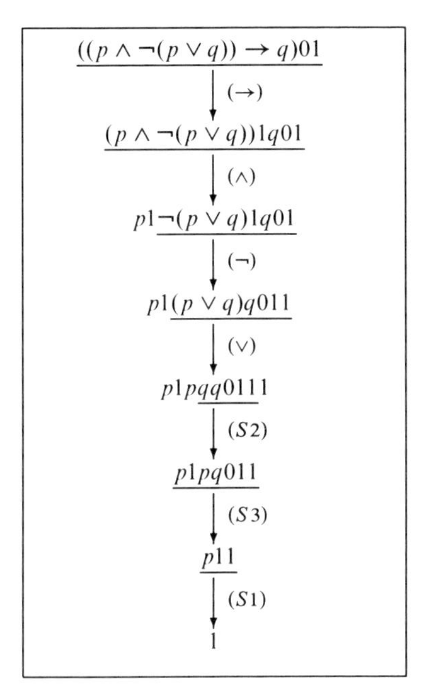

Fig. 24.2 An LCD-proof of PC-formula $(p \wedge \neg(p \vee q)) \to q$

- $f(-b, a_0, a_1) = f(b, a_1, a_0)$;
- $f(b_1 + b_2, a_0, a_1) = f(b_1, f(b_2, a_0, a_1), a_1)$;
- $f(b_1 \cdot b_2, a_0, a_1) = f(b_1, a_1, f(b_2, a_0, a_1))$;
- $f(b, f(b, a_0, a_1), a_2) = f(b, a_0, a_2)$;
- $f(b, a_0, f(b, a_1, a_2)) = f(b, a_0, a_2)$;
- $f(b, a, a) = a$.

It is easy to see that the terms $f(b, a_0, a_1)$ of $\mathfrak{A}$ are the direct counterparts to the LCD-formulas and the axioms correspond to the rules of LCD-dual tableau.

Alternatively, we may define the algebras of the form $(A, (f_b)_{b \in \mathfrak{B}})$ with a family of binary operations $f_b \colon A \times A \to A$ on a non-empty set A, each of which is determined by an element of a Boolean algebra $\mathfrak{B}$. For simplicity, we may write $b(a_0, a_1)$ instead of $f_b(a_0, a_1)$. In [Ber91] these algebras are referred to as *algebras with* $\mathfrak{B}$*-action*, see also [Sto98].

It is easy to see that terms $f(b, a_0, a_1)$ of the language of an LCD-algebra may be understood as the LCD-formulas, where Boolean terms b play the role of conditions and elements a_0 and a_1 of set A play the role of decisions. With this reformulation in mind, the following theorem shows a relationship between LCD-logic and ACD-algebras.

Theorem 24.3.1. *For all $b, b' \in B$ and for all $a_0, a_1, a'_0, a'_1 \in A$, if there exists an LCD-proof sequence starting with the formula ba_0a_1 and ending with the formula $b'a'_0a'_1$, then the equation $f(b, a_0, a_1) = f(b', a'_0, a'_1)$ is true in all ACD-algebras.*

In order to get the converse theorem, we have to add one more rule to the LCD-rules:

$$(c) \qquad \frac{\varphi\psi\alpha_0\alpha_1\psi\alpha'_0\alpha'_1}{\psi\varphi\alpha_0\alpha'_0\varphi\alpha_1\alpha'_1}$$

It is easy to verify that rule (c) satisfies the condition of Proposition 24.2.1.

Example. Consider an ACD-equation:

$$f(b \cdot c + -b \cdot d, a_0, a_1) = f(b, f(d, a_0, a_1), f(c, a_0, a_1)).$$

The terms:

$$t_0 \stackrel{\text{df}}{=} f(b \cdot c + -b \cdot d, a_0, a_1) \quad \text{and} \quad t_1 \stackrel{\text{df}}{=} f(b, f(d, a_0, a_1), f(c, a_0, a_1))$$

correspond to formulas:

$$\alpha_0 \stackrel{\text{df}}{=} [(b \wedge c) \vee (\neg b \wedge d)]a_0a_1 \quad \text{and} \quad \alpha_1 \stackrel{\text{df}}{=} b(da_0a_1)(ca_0a_1),$$

respectively. Figure 24.3 presents an LCD-proof starting with α_0 and ending with α_1. By Theorem 24.3.1, it shows that $t_1 = t_2$ is true in every ACD-algebra.

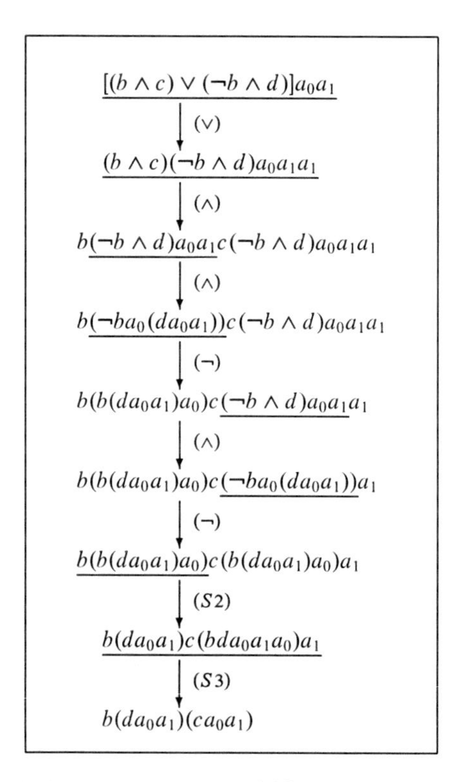

Fig. 24.3 An LCD-proof of an equation true in all ACD-algebras

In [BT83] an axiomatization of a more general class $\mathfrak{C}$ of one-sorted algebras of *if-then-else* is presented. The algebras are of the form $\mathfrak{A} = (A, \mathsf{T}, \mathsf{F}, \bot, f)$, where A is a non-empty set, T, F, and $\bot$ are distinguished elements of $\mathfrak{A}$, and f is a ternary operation $f\colon A^3 \to A$ defined by:

$$f(x, y, z) = \begin{cases} z & \text{if } x = \mathsf{F} \\ y & \text{if } x = \mathsf{T} \\ \bot & \text{otherwise.} \end{cases}$$

Viewing T and F as Boolean constants 1 and 0, respectively, this operation represents *if-then-else* statements, where verification of the condition may not halt. In [BT83] a set Ax of equational axioms is presented from which all the equations true in all the algebras of $\mathfrak{C}$ can be proved.

The set Ax of axioms consists of:

- $f(\mathsf{T}, x, y) = x$;
- $f(\mathsf{F}, x, y) = y$;
- $f(x, x, y) = f(x, \mathsf{T}, y)$;
- $f(x, y, x) = f(x, y, \mathsf{F})$;
- $f(\bot, x, y) = \bot$;
- $f(x, \bot, \bot) = \bot$;
- $f(x, f(y, z, u), f(y, v, w)) = f(y, f(x, z, v), f(x, u, w))$;
- $f(x, f(x, y, z), w) = f(x, y, w)$;
- $f(x, y, f(x, z, w)) = f(x, y, w)$.

The following completeness theorem is proved in [BT83]:

Theorem 24.3.2. *For all terms t and t' of the language of $\mathfrak{C}$, the following conditions are equivalent:*

1. *$t = t'$ is true in all algebras of $\mathfrak{C}$;*
2. *$t = t'$ is provable from Ax with equational deduction.*

It is easy to see that a decision logic based on the algebras of class $\mathfrak{C}$ can be defined in the style of LCD-logic of Sect. 24.2.

24.4 Relational Interpretation of the Logic of Conditional Decisions

In this section we show that $\mathsf{RL}(1, 1')$-dual tableau, presented in Sect. 2.7, can be used for proving validity of LCD-formulas. First, we define a translation of LCD-formulas into relational terms of the logic $\mathsf{RL}(1, 1')$. Let τ' be a one-to-one assignment of relational variables to the propositional variables. Then the translation τ is defined inductively as:

- $\tau(p) = \tau'(p)\,;1$, for every propositional variable p;
- $\tau(1) = 1$ (the universal relation);
- $\tau(0) = -1$ (the empty relation);

Let φ and ψ be LCD-conditions. Then:

- $\tau(\varphi \vee \psi) = \tau(\varphi) \cup \tau(\psi)$;
- $\tau(\varphi \wedge \psi) = \tau(\varphi) \cap \tau(\psi)$;
- $\tau(\varphi \rightarrow \psi) = -\tau(\varphi) \cup \tau(\psi)$;
- $\tau(\neg\varphi) = -\tau(\varphi)$;

Let α_0 and α_1 be LCD-formulas and let φ be an LCD-condition. Then:

- $\tau(\varphi\alpha_0\alpha_1) = (-\tau(\varphi) \cap 1')\,;\tau(\alpha_0) \cup (\tau(\varphi) \cap 1')\,;\tau(\alpha_1)$.

The following can be easily proved:

Fact 24.4.1. *If $\Sigma_1\alpha\Sigma_2$ and $\Sigma_1\beta\Sigma_2$ are* LCD*-formulas such that α and β are their subformulas, then for every* RL$(1,1')$*-model $\mathcal{M} = (U,m)$, if $m(\tau(\alpha)) = m(\tau(\beta))$, then $m(\tau(\Sigma_1\alpha\Sigma_2)) = m(\tau(\Sigma_1\beta\Sigma_2))$.*

Proposition 24.4.1. *For every* LCD*-rule $\frac{\Sigma_1\alpha\Sigma_2}{\Sigma_1\beta\Sigma_2}$ and for every* RL$(1,1')$*-model $\mathcal{M} = (U,m)$, $m(\tau(\Sigma_1\alpha\Sigma_2)) = m(\tau(\Sigma_1\beta\Sigma_2))$.*

Proof. By way of example, we show the required equality for the rules $(\vee)$ and (1). In view of Fact 24.4.1, it is sufficient to show that $m(\tau((\varphi \vee \psi)\alpha_0\alpha_1)) = m(\tau(\varphi(\psi\alpha_0\alpha_1)\alpha_1))$ and $m(\tau(1\alpha_0\alpha_1)) = m(\tau(\alpha_1))$.

Let $\tau(\varphi) = R_\varphi$, $\tau(\psi) = R_\psi$, $\tau(\alpha_0) = R_{\alpha_0}$, and $\tau(\alpha_1) = R_{\alpha_1}$. For simplicity of presentation, we will identify $R_\varphi, R_\psi, R_{\alpha_0}, R_{\alpha_1}$ with their interpretations in the model.

First, recall that the following are true for all binary relations R, S, and T:

(0) $(R \cup S) \cap T = (R \cap T) \cup (S \cap T)$ and $-(R \cup S) \cap T = (-R \cap T) \cap (-S \cap T)$;
(1) $(R \cup S)\,;T = R\,;T \cup S\,;T$;
(2) $(R \cap 1') \cup (S \cap 1') = (R \cap 1') \cup ((-R \cap 1') \cap (S \cap 1'))$;
(3) $((R \cap 1') \cap (S \cap 1'))\,;T = (R \cap 1')\,;((S \cap 1')\,;T)$.

Now, we have:

$m(\tau((\varphi \vee \psi)\alpha_0\alpha_1)) =$

$(-(R_\varphi \cup R_\psi) \cap 1')\,;R_{\alpha_0} \cup ((R_\varphi \cup R_\psi) \cap 1')\,;R_{\alpha_1} \stackrel{(0)}{=}$

$((-R_\varphi \cap 1') \cap (-R_\psi \cap 1'))\,;R_{\alpha_0} \cup ((R_\varphi \cap 1') \cup (R_\psi \cap 1'))\,;R_{\alpha_1} \stackrel{(2)}{=}$

$((-R_\varphi \cap 1') \cap (-R_\psi \cap 1'))\,;R_{\alpha_0} \cup ((R_\varphi \cap 1') \cup ((-R_\varphi \cap 1') \cap (R_\psi \cap 1')))\,;R_{\alpha_1} \stackrel{(1)}{=}$

$((-R_\varphi \cap 1') \cap (-R_\psi \cap 1'))\,;R_{\alpha_0} \cup ((-R_\varphi \cap 1') \cap (R_\psi \cap 1'))\,;R_{\alpha_1} \cup (R_\varphi \cap 1')\,;R_{\alpha_1} \stackrel{(3)}{=}$

$((-R_\varphi \cap 1')\,;((-R_\psi \cap 1')\,;R_{\alpha_0}) \cup (-R_\varphi \cap 1')\,;((R_\psi \cap 1')\,;R_{\alpha_1}) \cup (R_\varphi \cap 1')\,;R_{\alpha_1} \stackrel{(1)}{=}$
$(-R_\varphi \cap 1')\,;((-R_\psi \cap 1')\,;R_{\alpha_0} \cup (R_\psi \cap 1')\,;R_{\alpha_1}) \cup (R_\varphi \cap 1')\,;R_{\alpha_1} =$
$m(\tau(\varphi(\psi\alpha_0\alpha_1)\alpha_1))$.

Hence, the rule $(\vee)$ preserves the meaning of the translations of its formulas. The same holds for the rule (1):

$$m(\tau(1\alpha_0\alpha_1)) = (-1 \cap 1') ; R_{\alpha_0} \cup (1 \cap 1') ; R_{\alpha_1} = 1' ; R_{\alpha_1} = R_{\alpha_1} = m(\tau(\alpha_1)).$$

The proofs for the other LCD-rules are similar. □

Theorem 24.4.1. *For every* LCD*-formula* α *and for all object variables* x *and* y*, the following conditions are equivalent:*

1. α *is* LCD*-valid;*
2. $x\tau(\alpha)y$ *is* RL$(1, 1')$*-valid.*

Proof. (1. → 2.) Assume α is LCD-valid. Then, by Theorem 24.2.1, there exists an LCD-proof sequence for α such that its last formula is 1. Recall that $m(\tau(1))$ is the universal relation in every RL$(1, 1')$-model. Therefore, since by Proposition 24.4.1 the LCD-rules preserve the meaning of the corresponding relational terms, $m(\tau(\alpha))$ is the universal relation in every RL$(1, 1')$-model. Hence, $x\tau(\alpha)y$ is RL$(1, 1')$-valid.

(2. → 1.) Assume $x\tau(\alpha)y$ is RL$(1, 1')$-valid and suppose that α is not LCD-valid. Then, there exists an LCD-model $\mathcal{M} = (\{0, 1\}, v)$ such that $v(\alpha) = 0$. Let $\mathcal{M}' = (U, m)$ be a standard RL$(1, 1')$-model such that:

- $U = \{0, 1\}$;
- $m(1) = U \times U$ and $m(1') = Id_U$;

  For every propositional variable p,

- $m(\tau(p)) = \begin{cases} U \times U & \text{if } v(p) = 1 \\ \emptyset & \text{if } v(p) = 0. \end{cases}$

We show that the latter extends to all the LCD-conditions, i.e., for every condition φ, we have:

$$(a) \quad m(\tau(\varphi)) = \begin{cases} U \times U & \text{if } v(\varphi) = 1 \\ \emptyset & \text{if } v(\varphi) = 0. \end{cases}$$

The proof of (a) is by induction on the complexity of conditions.

Clearly, (a) holds for $\varphi = i$, $i \in \{0, 1\}$, and for $\varphi = p$, for any propositional variable p.

Let $\varphi = \neg p$. Then $v(\varphi) = 1$ iff $v(p) = 0$ iff $m(\tau(p)) = \emptyset$ iff $-m(\tau(p)) = U \times U$ iff $m(\tau(\varphi)) = U \times U$.

Let $\varphi = \psi \vee \vartheta$. Then $v(\varphi) = 1$ iff $v(\psi) = 1$ or $v(\vartheta) = 1$ iff, by the induction hypothesis, $m(\tau(\psi)) = U \times U$ or $m(\tau(\vartheta)) = U \times U$ iff $m(\tau(\varphi)) = U \times U$.

The remaining cases can be proved in a similar way.

Now, we prove, that for every LCD-formula β, we have:

$$(b) \quad \mathcal{M} \models \beta \text{ iff } \mathcal{M}' \models x\tau(\beta)y.$$

The proof is by induction on the complexity of formulas. If $\beta = i$, $i \in \{0, 1\}$, then condition (b) holds. Let $\beta = \varphi\beta_0\beta_1$ and let us assume that (b) holds for β_0 and β_1.

Assume $v(\beta) = 1$. If $v(\varphi) = 0$, then $v(\beta_0) = 1$. By condition (a), $m(\tau(\varphi)) = \emptyset$, so $-m(\tau(\varphi)) \cap m(1') = 1'$. By the induction hypothesis, $m(\tau(\beta_0)) = U \times U$, so $(-m(\tau(\varphi)) \cap m(1')) ; m(\tau(\beta_0)) = U \times U$. Hence, $m(\tau(\beta)) = U \times U$. If $v(\varphi) = 1$, then $v(\beta_1) = 1$. By condition (a), $m(\tau(\varphi)) = U \times U$, so $m(\tau(\varphi)) \cap 1' = 1'$. By the induction hypothesis, $m(\tau(\beta_1)) = U \times U$, so $(m(\tau(\varphi)) \cap m(1')) ; m(\tau(\beta_1)) = U \times U$. Hence, $m(\tau(\beta)) = U \times U$.

Assume $m(\tau(\beta)) = U \times U$. Since $m(\tau(\varphi))$ is either the universal relation or the empty relation and $m(\tau(\beta_0))$ and $m(\tau(\beta_1))$ are right ideal relations, $m(\tau(\beta)) = U \times U$ iff either $(-m(\tau(\varphi)) \cap m(1')) ; m(\tau(\beta_0)) = U \times U$ or $(m(\tau(\varphi)) \cap m(1')) ; m(\tau(\beta_1)) = U \times U$. If $(-m(\tau(\varphi)) \cap m(1')) ; m(\tau(\beta_0)) = U \times U$, then $m(\tau(\varphi)) = \emptyset$ and $m(\tau(\beta_0)) = U \times U$. Thus, by condition (a) and by the induction hypothesis, $v(\varphi) = 0$ and $v(\beta_0) = 1$, hence $v(\varphi) = 1$. The proof of the remaining case is similar.

Now, since $\mathcal{M} \not\models \alpha$, by condition (b) $\mathcal{M}' \not\models x\tau(\alpha)y$. However, this contradicts the assumption of $\mathsf{RL}(1, 1')$-validity of $x\tau(\alpha)y$. □

Thus, the dual tableau of the logic $\mathsf{RL}(1, 1')$ can be used for verification of validity of LCD-formulas.

24.5 Logics of Conditional Decisions of Order n and Their Dual Tableau Decision Procedures

In this section we present a logic of conditional decisions of order n, LCD_n, for $n \geq 2$, where the conditions are specified as formulas of the Rosser–Turquette logic RT discussed in Sect. 10.2, with an operation of implication added to the language.

LCD_n-symbols are those of logic RT together with the constants $0, \ldots, n-1$. As in LCD-logic, the formulas of the logic LCD_n are built from expressions representing conditions and decisions. Conditions are generated from propositional variables and propositional constants $0, \ldots, n-1$ with the operations $\vee$, $\wedge$, $\rightarrow$, and J_i, $i \in \{0, \ldots, n-1\}$. We will identify propositional constants with their interpretations as natural numbers in the models.

Formulas are defined by:

- $0, \ldots, n-1$ are LCD_n-formulas;
- If φ is an LCD_n-condition and $\alpha_0, \ldots, \alpha_{n-1}$ are LCD_n-formulas, then $\varphi\alpha_0 \ldots \alpha_{n-1}$ is an LCD_n-formula.

The set of subformulas of an LCD_n-formula α is defined in a similar way as in LCD-logic (see Sect. 24.2).

An LCD_n-*model* is a structure $\mathcal{M} = (\{0, \ldots, n-1\}, v)$ such that v is a valuation of propositional variables into elements of $\{0, \ldots, n-1\}$ and it extends to all the conditions and to the LCD_n-formulas as:

- $v(i) = i$, for every $i \in \{0, \ldots, n-1\}$;
- $v(\varphi \vee \psi) = \max(v(\varphi), v(\psi))$;
- $v(\varphi \wedge \psi) = \min(v(\varphi), v(\psi))$;
- $v(\varphi \rightarrow \psi) = \begin{cases} 0 & \text{if } v(\varphi) > v(\psi) \\ n-1 & \text{if } v(\varphi) \leq v(\psi); \end{cases}$
- $v(J_i\varphi) = \begin{cases} 0 & \text{if } v(\varphi) \neq i \\ n-1 & \text{if } v(\varphi) = i \end{cases}$ for every $i \in \{0, \ldots, n-1\}$;
- $v(\varphi\alpha_0 \ldots \alpha_{n-1}) = v(\alpha_i)$, if $v(\varphi) = i$, for $i \in \{0, \ldots, n-1\}$.

Thus, an LCD_n-formula represents n alternatives each of which holds depending an a degree of satisfaction of the condition in the formula.

An LCD_n-formula α is (n, s)-valid, $0 < s \leq n-1$, whenever for every LCD_n-model $\mathcal{M} = (\{0, \ldots, n-1\}, v)$ $v(\alpha) \geq s$. A formula which is $(n, n-1)$-valid is referred to as an LCD_n-valid formula. Observe that if $n = 2$, then $v(J_0\varphi) = 1 - v(\varphi)$ and $v(J_1\varphi) = v(\varphi)$, so logic LCD_2 coincides with logic LCD presented in Sect. 24.2.

As in the LCD-dual tableau, the rules of an LCD_n-dual tableau are of the form $\frac{\Sigma_1\alpha\Sigma_2}{\Sigma_1\beta\Sigma_2}$, where Σ_1 and Σ_2 are strings of symbols of the LCD_n-language such that $\Sigma_1\alpha\Sigma_2$ and $\Sigma_1\beta\Sigma_2$ are LCD_n-formulas and α and β are their subformulas, respectively. The following holds:

Fact 24.5.1. *Let* $\frac{\Sigma_1\alpha\Sigma_2}{\Sigma_1\beta\Sigma_2}$ *be an* LCD_n*-rule. Then for every* LCD_n*-model* $\mathcal{M} = (\{0, \ldots, n-1\}, v)$*, if* $v(\alpha) = v(\beta)$*, then* $v(\Sigma_1\alpha\Sigma_2) = v(\Sigma_1\beta\Sigma_2)$.

Let φ and ψ be LCD_n-conditions and let $\alpha_0, \ldots, \alpha_{n-1}$ be LCD_n-formulas. In what follows $C_{j=0}^{k}\alpha_j$ denotes the concatenation of formulas $\alpha_0 \ldots \alpha_k$.

Decomposition rules of LCD_n-dual tableau have the following forms:

$$(\vee)\ \frac{\Sigma_1(\varphi \vee \psi)C_{j=0}^{n-1}\alpha_j\,\Sigma_2}{\Sigma_1\varphi C_{j=0}^{n-1}(\psi C_{k=0}^{n-1}\alpha_{\max(j,k)})\Sigma_2} \qquad (\wedge)\ \frac{\Sigma_1(\varphi \wedge \psi)C_{j=0}^{n-1}\alpha_j\,\Sigma_2}{\Sigma_1\varphi C_{j=0}^{n-1}(\psi C_{k=0}^{n-1}\alpha_{\min(j,k)})\Sigma_2}$$

$$(\rightarrow)\ \frac{\Sigma_1(\varphi \rightarrow \psi)C_{j=0}^{n-1}\alpha_j\,\Sigma_2}{\Sigma_1\varphi C_{j=0}^{n-1}(\psi C_{k=0}^{n-1}\beta_k^j)\Sigma_2} \quad \text{where } \beta_k^j = \begin{cases} \alpha_0 & \text{if } k < j \\ \alpha_{n-1} & \text{if } k \geq j \end{cases}$$

$$(J_i)\ \frac{\Sigma_1 J_i\varphi C_{j=0}^{n-1}\alpha_j\,\Sigma_2}{\Sigma_1\varphi C_{j=0}^{n-1}\gamma_i^j\,\Sigma_2} \quad \text{where } \gamma_i^j = \begin{cases} \alpha_0 & \text{if } i \neq j \\ \alpha_{n-1} & \text{if } i = j \end{cases}$$

Let φ be an LCD_n-condition, let p be a propositional variable, let $\alpha_0, \ldots, \alpha_{n-1}, \beta_0, \ldots, \beta_{n-1}$ be LCD_n-formulas, and let $\Sigma_3 p\, C_{k=0}^{n-1}\beta_k\,\Sigma_4$ be an LCD_n-formula such that $p\, C_{k=0}^{n-1}\beta_k$ is its subformula. Let $i \in \{0, \ldots, n-1\}$.

Simplification rules of LCD_n-dual tableau are:

$$(i)\ \frac{\Sigma_1 i C_{j=0}^{n-1}\alpha_j \Sigma_2}{\Sigma_1 \alpha_i \Sigma_2}$$

$$(S1)\ \frac{\Sigma_1 \varphi C_{j=0}^{n-1}\alpha_j \Sigma_2}{\Sigma_1 \alpha \Sigma_2} \qquad \text{where } \alpha_j = \alpha \text{ for all } j \in \{0,\dots,n-1\}$$

$$(S2)\ \frac{\Sigma_1 (p\, C_{j=0}^{l-1}\alpha_j (\Sigma_3 p\, C_{k=0}^{n-1}\beta_k \Sigma_4) C_{j=l+1}^{n-1}\alpha_j)\Sigma_2}{\Sigma_1 (p\, C_{j=0}^{l-1}\alpha_j \Sigma_3 \beta_l \Sigma_4 C_{j=l+1}^{n-1}\alpha_j)\Sigma_2} \qquad l \in \{0,\dots,n-1\}$$

Observe that if $l = 0$, then the rule $(S2)$ has the form:

$$\frac{\Sigma_1 (p\, (\Sigma_3 p\, C_{k=0}^{n-1}\beta_k \Sigma_4) C_{j=1}^{n-1}\alpha_j)\Sigma_2}{\Sigma_1 (p\, \Sigma_3 \beta_0 \Sigma_4 C_{j=1}^{n-1}\alpha_j)\Sigma_2}$$

Similarly, if $l = n-1$, then the rule $(S2)$ has the form:

$$\frac{\Sigma_1 (p\, C_{j=0}^{n-2}\alpha_j (\Sigma_3 p\, C_{k=0}^{n-1}\beta_k \Sigma_4)\Sigma_2}{\Sigma_1 (p\, C_{j=0}^{n-2}\alpha_j \Sigma_3 \beta_{n-1} \Sigma_4)\Sigma_2}$$

Proposition 24.5.1. *Let $\mathcal{M} = (\{0,\dots,n-1\}, v)$ be an LCD_n-model. Then for every LCD_n-rule $\frac{\Sigma_1\alpha\Sigma_2}{\Sigma_1\beta\Sigma_2}$, $v(\Sigma_1\alpha\Sigma_2) = v(\Sigma_1\beta\Sigma_2)$.*

Proof. By way of example, we prove the statement for the rules $(\vee)$ and $(S2)$. Due to Fact 24.5.1, it suffices to show $v(\alpha) = v(\beta)$.

$(\vee)$ Without loss of generality, we may assume that $v(\varphi) \leq v(\psi)$. Let $v(\varphi) = i$ and let $v(\psi) = l$, for $i, l \in \{0,\dots,n-1\}$. By the definition of valuation:

$$v((\varphi \vee \psi)\alpha_0 \dots \alpha_{n-1}) = v(\alpha_l).$$

On the other hand:

$$v(\varphi C_{j=0}^{n-1}(\psi C_{k=0}^{n-1}\alpha_{\max(j,k)})) = v(\psi C_{k=0}^{n-1}\alpha_{\max(i,k)}) = v(\alpha_l).$$

Hence, $v((\varphi \vee \psi)\alpha_0 \dots \alpha_{n-1}) = v(\varphi C_{j=0}^{n-1}(\psi C_{k=0}^{n-1}\alpha_{\max(j,k)}))$.

$(S2)$ In view of Proposition 24.5.1, we may assume that Σ_3 and Σ_4 are empty. Assume that $v(p) = s \neq l$. Then, $v(p\, C_{j=0}^{l-1}\alpha_j (p\, C_{k=0}^{n-1}\beta_k) C_{j=l+1}^{n-1}\alpha_j) = v(\alpha_s) = v(p\, C_{j=0}^{l-1}\alpha_j \beta_j C_{j=l+1}^{n-1}\alpha_j)$. On the other hand, if $v(p) = s = l$, then we have: $v(p\, C_{j=0}^{l-1}\alpha_j (p\, C_{k=0}^{n-1}\beta_k) C_{j=l+1}^{n-1}\alpha_j) = v(p\, C_{k=0}^{n-1}\beta_k) = v(\beta_l)$ which is equal to $v(p\, C_{j=0}^{l-1}\alpha_j \beta_l C_{j=l+1}^{n-1}\alpha_j)$. Hence, $v(p\, C_{j=0}^{l-1}\alpha_j (p\, C_{k=0}^{n-1}\beta_k) C_{j=l+1}^{n-1}\alpha_j) = v(p\, C_{j=0}^{l-1}\alpha_j \beta_l C_{j=l+1}^{n-1}\alpha_j)$. □

An LCD_n-proof sequence for an LCD_n-formula is defined as in LCD-logic in Sect. 24.2. An LCD_n-formula α is said to be LCD_n-provable whenever there exists an LCD_n-proof sequence for α such that the last formula of the sequence is $n-1$. Such a proof sequence is then referred to as an LCD_n-proof of α.

Completeness of LCD_n-dual tableau can be proved in a similar way as completeness of LCD-dual tableau. The notions which are used in the completeness proof are direct extensions of the corresponding notions for logic LCD.

Operation free LCD_n*-formulas* are defined as:

- $0, \ldots, n-1$ are operation free formulas;
- $\varphi\alpha_0 \ldots \alpha_{n-1}$ is an operation free formula if and only if φ is a propositional variable and $\alpha_0, \ldots, \alpha_{n-1}$ are operation free formulas.

An operation free formula $p\alpha_0 \ldots \alpha_{n-1}$ has a *unique condition* whenever p does not occur in α_i, for any $i \in \{0, \ldots, n-1\}$.

An LCD_n-formula α is said to be *indecomposable* whenever it satisfies the following conditions:

- α is an operation free formula;
- $p\underbrace{i \ldots i}_{n \text{ times}}$ is not a subformula of α, for any propositional variable p and $i \in \{0, \ldots, n-1\}$;
- Every subformula of α has a unique condition.

Note that if a formula $p\alpha_0 \ldots \alpha_{n-1}$ is indecomposable, then every α_i is indecomposable and, moreover, if $\alpha_i \in \{0, \ldots, n-1\}$, then there exists $j \in \{0, \ldots, n-1\} \setminus \{i\}$ such that $\alpha_i \neq \alpha_j$.

Proposition 24.5.2. *For every indecomposable* LCD_n*-formula* α *such that* $\alpha \neq n-1$, *there exists an* LCD_n*-model* $\mathcal{M} = (\{0, \ldots, n-1\}, v)$ *such that* $v(\alpha) \neq n-1$.

Proof. The proof is by induction on the complexity of indecomposable LCD_n-formulas. If $\alpha = i$, for some $i \in \{0, \ldots, n-2\}$, then the proposition holds. Let $\alpha = pi_0 \ldots i_{n-1}$, for some propositional variable p and $i_0, \ldots, i_{n-1} \in \{0, \ldots, n-1\}$. Since α is indecomposable, there exists $k \in \{0, \ldots, n-1\}$ such that $i_k \neq n-1$. Let v be any valuation such that $v(p) = i_k$. Then $v(pi_0 \ldots i_{n-1}) = i_k \neq n-1$, hence the proposition holds. Now, consider an indecomposable formula $p\alpha_0 \ldots \alpha_{n-1}$ such that p does not occur in α_i for any $i \in \{0, \ldots, n-1\}$. Clearly, there exists $i \in \{0, \ldots, n-1\}$ such that $\alpha_i \neq n-1$, for otherwise formula $p\alpha_0 \ldots \alpha_{n-1}$ is not indecomposable. By the induction hypothesis, there is a valuation v' such that $v'(\alpha_i) \neq n-1$. Consider a valuation v satisfying: $v(p) = i$ and $v(q) = v'(q)$, for every propositional variable q occurring in α_i. Then, $v(p\alpha_0 \ldots \alpha_{n-1}) = v'(\alpha_i) \neq n-1$, and hence the proposition holds. □

Proposition 24.5.3. *For every* LCD_n*-formula* α, *there exists an* LCD_n*-proof sequence such that its last formula is indecomposable.*

Proof. Let α be an LCD_n-formula. Let β be the last formula in an LCD_n-proof sequence for α. Then none of the rules is applicable to β. Clearly, if $\beta = i$ for some $i \in \{0,\ldots,n-1\}$, then the proposition holds. Let $\beta = \varphi\alpha_0\ldots\alpha_{n-1}$. If β is not an operation free formula, then $\varphi \in \{0,\ldots,n-1\}$ or its condition or a condition of some of its subformula is a compound LCD_n-formula. But then either of the rules $(\vee)$, $(\wedge)$, $(\rightarrow)$, (J_i), or (i) can be applied to β, a contradiction. If $pi\ldots i$ is a subformula of β, for some propositional variable p and $i \in \{0,\ldots,n-1\}$, then the rule $(S1)$ can be applied to β, a contradiction. If some of the subformulas of β does not have a unique condition, then the rule $(S2)$ can be applied to β, a contradiction. Therefore, β must be an indecomposable LCD_n-formula. □

Theorem 24.5.1 (Soundness and Completeness of LCD_n). *For every LCD_n-formula α, the following conditions are equivalent:*

1. *α is LCD_n-valid;*
2. *α is LCD_n-provable.*

Proof. (1. $\rightarrow$ 2.) Let α be LCD_n-valid. Suppose α is not LCD_n-provable, that is there is no any LCD_n-proof sequence for α that ends with $n-1$. Then, by Proposition 24.5.3, there exists an LCD-proof sequence for α such that its last formula, say β, is indecomposable. By Proposition 24.5.2, there exists an LCD_n-model $\mathcal{M} = (\{0,\ldots,n-1\}, v)$ such that $v(\beta) \neq n-1$. By Proposition 24.5.1, $v(\alpha) \neq n-1$, a contradiction.

(2. $\rightarrow$ 1.) Now, assume that α is LCD_n-provable, that is there exists an LCD_n-proof sequence for α such that its last formula is $n-1$. By Proposition 24.5.1, for every LCD_n-model $\mathcal{M} = (\{0,\ldots,n-1\}, v)$, $v(\alpha) = n-1$. □

As in the case of LCD-dual tableau, we conclude that LCD_n-dual tableau is a decision procedure for the logic LCD_n. Note also that ACD-algebras presented in Sect. 24.3 can be generalized to the algebras of order n on the basis of logics LCD_n.

Observe that for every LCD_n-formula φ, $v(\varphi 01\ldots n-1) = v(\varphi)$. Therefore, (n,s)-validity of φ in the Rosser–Turquette logic RT is equivalent to (n,s)-validity of $\varphi 01\ldots n-1$ in logic LCD_n, $n \geq 1$. Hence, by Theorem 24.5.1, the following holds:

Theorem 24.5.2. *For every* RT*-formula φ, the following conditions are equivalent:*

1. *φ is* RT*-valid;*
2. *$\varphi 01\ldots n-1$ is LCD_n-provable.*

Now, we show how to obtain an (n,s)-dual tableau which will enable us verification of (n,s)-validity of LCD_n-formulas, $0 < s \leq n-1$. An indecomposable LCD_n-formula α of the form $\Sigma_1(pi_0\ldots i_{n-1})\Sigma_2$, where Σ_1 and Σ_2 are strings of symbols of the LCD_n-language, $pi_0\ldots i_{n-1}$ is a subformula of α, and $i_k \in \{0,\ldots,n-1\}$, for every $k \in \{0,\ldots,n-1\}$, will be referred to as a *simple formula*. It follows that simple formulas are those indecomposable formulas which are not propositional constants.

The (n, s)-dual tableau includes all the rules of LCD_n-dual tableau and the rule of the form:

For every simple formula $\Sigma_1(pi_0 \dots i_{n-1})\Sigma_2$,

$$(S3) \quad \frac{\Sigma_1(pi_0 \dots i_{n-1})\Sigma_2}{\Sigma_1 i_k \Sigma_2} \qquad i_k = \min\{i_0, \dots, i_{n-1}\}.$$

Proposition 24.5.4. *Let $0 < s \leq n-1$. Rule $(S3)$ preserves and reflects (n, s)-validity, i.e., the following conditions are equivalent:*

1. *$\Sigma_1(pi_0 \dots i_{n-1})\Sigma_2$ is (n, s)-valid;*
2. *$\Sigma_1 i_k \Sigma_2$ is (n, s)-valid, where $i_k = \min\{i_0, \dots, i_{n-1}\}$.*

Proof. First, observe that if α is an indecomposable LCD_n-formula, then it is (n, s)-valid iff all of its subformulas are (n, s)-valid. Note also that if $i_k \neq 0$, then $pi_0 \dots i_{n-1}$ is i_k-valid. Let us assume that $\Sigma_1(pi_0 \dots i_{n-1})\Sigma_2$ is (n, s)-valid. Then, $pi_0 \dots i_{n-1}$ is (n, s)-valid, so $i_k \geq s$. Hence, $\Sigma_1 i_k \Sigma_2$ is (n, s)-valid. Now, assume that $\Sigma_1 i_k \Sigma_2$ is (n, s)-valid. Then, $i_k \geq s$, which implies (n, s)-validity of $pi_0 \dots i_{n-1}$. Hence, $\Sigma_1(pi_0 \dots i_{n-1})\Sigma_2$ is (n, s)-valid. □

Observe that the rule $(S3)$ does not have the property of preserving values of its premise and conclusion in a model under a valuation, as the rules of LCD_n-dual tableau do. For example, consider an LCD_4-model $\mathcal{M} = (\{0, \dots, 3\}, v)$ such that $v(p) = 0$. Then $v(p2111) = 2$, while $v(1) = 1$. However, by Proposition 24.5.4, formula $p2111$ is $(3, 1)$-valid.

Propositions 24.5.1 and 24.5.4 imply:

Proposition 24.5.5. *The (n, s)-rules preserve and reflect (n, s)-validity.*

The notion of an (n, s)-proof sequence is defined in an analogous way as in LCD_n-dual tableau. Observe that the last formula of an (n, s)-proof sequence is $i \in \{0, \dots, n-1\}$. Indeed, if none of the rules of LCD_n-dual tableau can be applied to a formula α in a sequence, then either $\alpha \in \{0, \dots, n-1\}$ or α is a simple formula to which the rule $(S3)$ can be applied. As the result of an application of the rule $(S3)$ we obtain a formula α' which is either $i \in \{0, \dots, n-1\}$ or a simple formula or $p\underbrace{i \dots i}_{n \text{ times}}$ is a subformula of α' for some propositional variable p and for some $i \in \{0, \dots, n-1\}$. If the latter holds, we apply to α' the rule $(S1)$. If α' is a simple formula, then we apply the rule $(S3)$. In both cases we obtain either $i \in \{0, \dots, n-1\}$ or a simple formula or a formula with a subformula $p\underbrace{i \dots i}_{n \text{ times}}$. It is easy to see that this process of applications of (n, s)-rules is finite and the last formula in the sequence is $i \in \{0, \dots, n-1\}$.

Let $0 < s \leq n-1$. An LCD_n-formula α is said to be (n, s)-*provable* whenever there exists an (n, s)-proof sequence for α such that the last formula of the sequence is $i \in \{1, \dots, n-1\}$ such that $i \geq s$.

Theorem 24.5.3. *For every* LCD_n*-formula* α *and for every* $s \in \{1, \ldots, n-1\}$*, the following conditions are equivalent:*

1. α *is* (n, s)*-valid;*
2. α *is* (n, s)*-provable.*

Proof. Let α be an LCD_n-formula. Consider an (n, s)-proof sequence for α. Such a sequence is finite and its last formula is $i \in \{0, \ldots, n-1\}$. If α is (n, s)-valid, then by Proposition 24.5.5, the last formula, say i, is (n, s)-valid, so $i \geq s$. Hence, α is (n, s)-provable. On the other hand, if α is (n, s)-provable, then the last formula in its proof sequence is i such that $i \geq s$, which is an (n, s)-valid formula. Therefore, by Proposition 24.5.5, α is (n, s)-valid. □

Example. Consider the logic of conditional decisions of order 3. The rules of LCD_3-dual tableau are as follows. Let φ and ψ be conditions and let $\alpha_0, \alpha_1, \alpha_2$ be LCD_3-formulas. Decomposition rules of LCD_3-dual tableau are instances of the decomposition rules of LCD_n-dual tableau with $n = 3$:

$$(\vee)\ \frac{\Sigma_1(\varphi \vee \psi)\alpha_0\alpha_1\alpha_2\Sigma_2}{\Sigma_1\varphi((\psi\alpha_0\alpha_1\alpha_2)(\psi\alpha_1\alpha_1\alpha_2)\alpha_2)\Sigma_2}$$

$$(\wedge)\ \frac{\Sigma_1(\varphi \wedge \psi)\alpha_0\alpha_1\alpha_2\Sigma_2}{\Sigma_1\varphi(\alpha_0(\psi\alpha_0\alpha_1\alpha_1)(\psi\alpha_0\alpha_1\alpha_2))\Sigma_2}$$

$$(\rightarrow)\ \frac{\Sigma_1(\varphi \rightarrow \psi)\alpha_0\alpha_1\alpha_2\Sigma_2}{\Sigma_1\varphi(\alpha_2(\psi\alpha_0\alpha_2\alpha_2)(\psi\alpha_0\alpha_0\alpha_2))\Sigma_2}$$

$$(J_0)\ \frac{\Sigma_1 J_0\varphi\alpha_0\alpha_1\alpha_2\Sigma_2}{\Sigma_1\varphi\alpha_2\alpha_0\alpha_0\Sigma_2} \quad (J_1)\ \frac{\Sigma_1 J_1\varphi\alpha_0\alpha_1\alpha_2\Sigma_2}{\Sigma_1\varphi\alpha_0\alpha_2\alpha_0\Sigma_2} \quad (J_2)\ \frac{\Sigma_1 J_2\varphi\alpha_0\alpha_1\alpha_2\Sigma_2}{\Sigma_1\varphi\alpha_0\alpha_0\alpha_2\Sigma_2}$$

Let φ be an LCD_3-condition, let p be a propositional variable, let $\alpha, \alpha_0, \alpha_1, \alpha_2, \beta_0, \beta_1, \beta_2$ be LCD_3-formulas, and let $\Sigma_3 p\beta_0\beta_1\beta_2\Sigma_4$ be an LCD_3-formula such that $p\beta_0\beta_1\beta_2$ is its subformula. Simplification rules of LCD_3-dual tableau are instances of LCD_n-simplification rules with $n = 3$:

$$(0)\ \frac{\Sigma_1 0\alpha_0\alpha_1\alpha_2\Sigma_2}{\Sigma_1\alpha_0\Sigma_2} \quad (1)\ \frac{\Sigma_1 1\alpha_0\alpha_1\alpha_2\Sigma_2}{\Sigma_1\alpha_1\Sigma_2} \quad (2)\ \frac{\Sigma_1 2\alpha_0\alpha_1\alpha_2\Sigma_2}{\Sigma_1\alpha_2\Sigma_2}$$

$$(S1)\ \frac{\Sigma_1\varphi\alpha\alpha\alpha\Sigma_2}{\Sigma_1\alpha\Sigma_2}$$

$$(S2) \text{ for } l = 0 \qquad \frac{\Sigma_1(p(\Sigma_3 p\beta_0\beta_1\beta_2\Sigma_4)\alpha_1\alpha_2)\Sigma_2}{\Sigma_1(p\Sigma_3\beta_0\Sigma_4\alpha_1\alpha_2)\Sigma_2}$$

$$(S2) \text{ for } l = 1 \qquad \frac{\Sigma_1(p\alpha_0(\Sigma_3 p\beta_0\beta_1\beta_2\Sigma_4)\alpha_2)\Sigma_2}{\Sigma_1(p\alpha_0\Sigma_3\beta_1\Sigma_4\alpha_2)\Sigma_2}$$

(S2) for $l = 2$ $$\frac{\Sigma_1(p\alpha_0\alpha_1(\Sigma_3 p\beta_0\beta_1\beta_2\Sigma_4))\Sigma_2}{\Sigma_1(p\alpha_0\alpha_1\Sigma_3\beta_2\Sigma_4)\Sigma_2}$$

(S3) $$\frac{\Sigma_1(pi_0i_1i_2)\Sigma_2}{\Sigma_1 i_k\Sigma_2} \qquad i_k = \min\{i_0, i_1, i_2\}$$

Let φ be the formula of the three-valued Rosser–Turquette logic:

$$\varphi = J_0(p \vee q) \rightarrow (J_0 p \vee J_0 q).$$

Figure 24.4 presents an LCD_3-proof of $\varphi 012$, which in view of Theorem 24.5.2 shows that φ is valid in the three-valued Rosser–Turquette logic.

Consider an LCD_3-formula:

$$\psi = J_1(q \vee p)(p121)21.$$

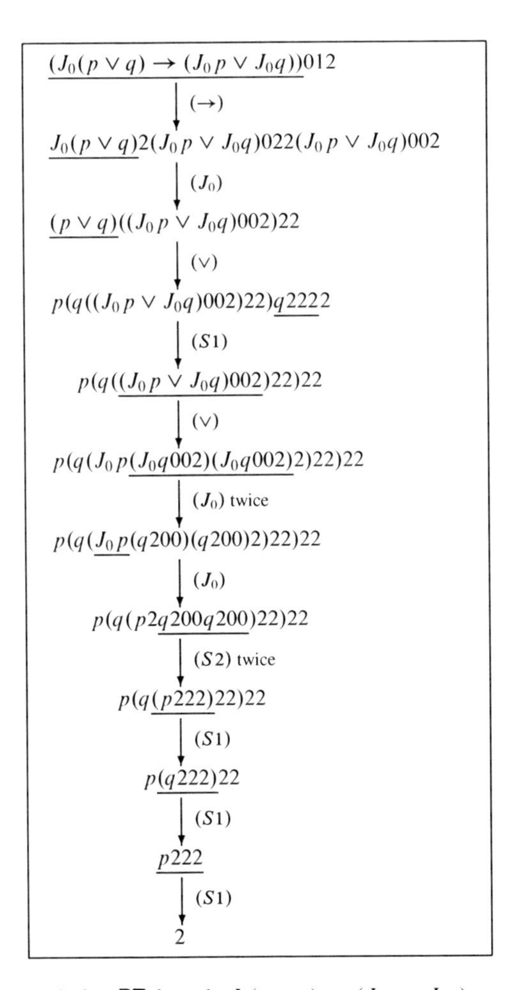

Fig. 24.4 An LCD_3-proof of an RT-formula $J_0(p \vee q) \rightarrow (J_0 p \vee J_0 q)$

It is easy to check that this formula is not LCD_3-valid. However, ψ is $(3,1)$-provable, hence, by Theorem 24.5.3, it is $(3,1)$-valid. Figure 24.5 presents its $(3,1)$-proof.

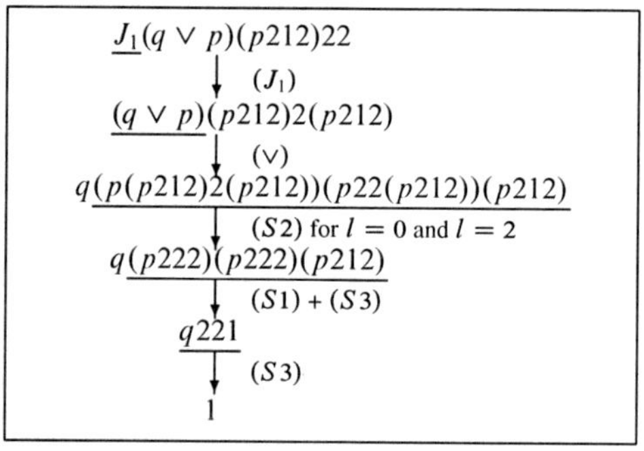

$$\underline{J_1}(q \vee p)(p212)22$$
$$\downarrow (J_1)$$
$$\underline{(q \vee p)}(p212)2(p212)$$
$$\downarrow (\vee)$$
$$q\underline{(p(p212)2(p212))(p22(p212))(p212)}$$
$$\downarrow (S2) \text{ for } l = 0 \text{ and } l = 2$$
$$q\underline{(p222)(p222)(p212)}$$
$$\downarrow (S1) + (S3)$$
$$\underline{q221}$$
$$\downarrow (S3)$$
$$1$$

Fig. 24.5 A $(3,1)$-proof of an LCD_3-formula $J_1(q \vee p)(p121)21$

Part VII
Conclusion

Chapter 25
Methodological Principles of Dual Tableaux

25.1 Introduction

Dual tableaux for theories considered in the book operate either on a relational language or a first-order language associated with the theories.

In this chapter we discuss general principles of defining a relational logic for a given theory, the methods of construction of dual tableaux for relational logics and first-order logics, and the method of proving completeness of dual tableaux.

In Sect. 25.2 the major steps leading to relational formalization of theories in relation to their presentation are listed. In Sect. 25.3 a general relational logic is presented such that most of the theories considered in the book are interpretable in some instance of that logic. In Sect. 25.4 the problem of translation between relational languages and first-order languages is discussed. In Sect. 25.5 a general format of dual tableaux is presented. In Sects. 25.6 and 25.7 a correspondence theory for dual tableaux is developed. The methods of constructing dual tableau rules reflecting constraints on the models of relational logics and definitions of relational operations are presented. In Sect. 25.8 a method of proving completeness of dual tableaux is discussed. The method is applied to all the dual tableau systems for relational logics considered in the book. In Sect. 25.9 a variety of forms of dual tableau rules is brought into attention. Depending on implementation requirements some of them may be more efficient than the others. In Sect. 25.10 the existing implementations of dual tableaux are briefly presented. In Sect. 25.11 some directions for developing decision procedures are outlined. The contents of Sects. 25.5–25.8 are based on [MO02a].

Dual tableaux not considered in this book include the systems for: a fragment of set theory presented in [OOP04]; some substructural logics, different from the logics dealt with in Chaps. 9 and 14, presented in [Mac97, Mac98, Mac99]; some information logics with semantics of relative frames presented in [Kon87, Kon97, DO02], Chaps. 7–9; a logic of nondeterministic program specifications presented in [BK99]; the implication problem for association rules presented in [Mac01]; the many-valued modal logics presented in [KO01].

E. Orłowska and J. Golińska-Pilarek, *Dual Tableaux: Foundations, Methodology, Case Studies*, Trends in Logic 36, DOI 10.1007/978-94-007-0005-5_25,

25.2 Theories Interpreted Relationally

A relational interpretation of a theory consists in providing relational semantics and/or relational proof system for the theory. In this section we outline the successive steps leading to the construction of relational proof systems for the theories presented in this book. The construction depends in an essential way on a presentation of the theory in question. In each case the goal is to associate with a given theory T a relational theory T' such that T is interpretable in T'. Following [TMR68] interpretability of T in T' means that there is a truth preserving translation from the language of T into the language of T'.

Case 1

A theory is presented in the form of a class of algebras of relations, say T. Then the steps leading to a relational proof system for T are:

- The construction of a relational logic RL_{T} such that T is interpretable in RL_{T}.
- The construction of a proof system for RL_{T}.

The algebras and their relational logics presented in Part II are examples of this procedure.

Case 2

A theory T is presented in the form of a class of algebras (not necessarily algebras of relations). Then the development of a relational proof system for T consists of:

- The construction of a class of relational systems $Re(\mathsf{T})$ for T such that the first-order theory of T is interpretable in the first-order theory of $Re(\mathsf{T})$, $\mathsf{F}_{Re(\mathsf{T})}$.
- The construction of a proof system for $\mathsf{F}_{Re(\mathsf{T})}$.

Relational proof systems obtained in this way are presented in Chap. 18.

Case 3

A theory has the form of a non-classical logic, say L. Then we obtain a relational proof system for L following the steps below:

- The construction of a relational logic RL_{L} such that L is interpretable in RL_{L}.
- The construction of a proof system for RL_{L}.

Relational proof systems obtained in this way are presented in Parts III, IV, and Chaps. 16, 17, and 19.

The main step in this construction consists of defining validity preserving translation function τ from formulas of the logic L into relational terms of the logic RL_{L} and proving the following property:

Translation preserves and reflects validity
For every L-formula φ, if $\mathsf{RL_L}$ is a logic of n-ary relations, then for all object variables $x_1, \ldots, x_n$, the following conditions are equivalent:

1. φ is L-valid;
2. $\tau(\varphi)(x_1, \ldots, x_n)$ is $\mathsf{RL_L}$-valid.

Relational logics enable us to represent within a uniform formalism the three basic components of any logical system: syntax, semantics, and deduction.

Syntax: With the formal language of a logic L there is associated a language of relational terms.

Semantics and model theory: With a logic L there is associated a class of relational models for L, and in these models the formulas from L are interpreted as relations. Each relational model determines an algebra of relations. The class of algebras derived from the underlying class of relational models for L provides a relation-algebraic semantics for L.

Proof theory: With a logic L there is associated a relational logic for L, $\mathsf{RL_L}$, such that its proof system provides a deduction method for L.

In relational representation of formulas we articulate explicitly information about both their syntactic structure and semantic satisfaction conditions. Generally speaking, formulas are represented as terms over a class of algebras of relations, and validity of a formula is reflected by an equation over this class. Each of the propositional operations becomes a relational operation and in this way an original syntactic form of formulas is preserved. Semantic information about a formula which is provided by a satisfaction condition consists of the two basic parts: first, we say which states satisfy the subformulas of the given formula, and second, how those states are related to each other via the accessibility relation. Those two ingredients of semantic information are interrelated and inseparable. In relational representation of formulas the terms representing accessibility relations appear explicitly in the respective relational terms corresponding to the formulas. They become the arguments of the relational operations in a term in the same way as the other of its subterms, obtained from subformulas of the given formula. In this way in the relational term corresponding to a formula both syntactic and semantic information about the formula are integrated into a single information item.

One of the advantages of the relational representation of formulas is that we gain compositionality. In most of the non-classical logics their formulas that are built with intensional propositional operations, for example with modal or temporal operations, are not compositional. The meaning of a compound formula is not necessarily a function of the meanings of its subformulas. In the relational formalism the counterparts of these operations become compositional. In several applications of logics there is a need to distinguish between information about static facts and dynamic transitions between states in a domain which a given logic is intended to model. In relational logics these two types of information interact in a uniform framework.

Relational proof theory enables us to build proof systems for non-classical logics in a systematic modular way. First, deduction rules applicable to relational terms built with Boolean operations are defined, they are the common relational core of the logics. These rules constitute a basis of all the relational proof systems. Next, for any particular logic some specific rules are designed and adjoined to the core set of rules. Hence, we need not implement each deduction system from scratch, we should only extend the core system with a module corresponding to a specific part of a logic under consideration.

From the algebraic perspective relational formalization of non-classical logics leads to what might be called non-classical algebras of relations. In these algebras the relational operations are admitted which are relational counterparts of intensional propositional operations. They are not always expressible in terms of the standard relational operations. Examples of such operations can be found in Sects. 9.3 and 16.4.

25.3 Relational Logics

Among the languages considered in this book, the most expressive relational language is the language of typed relations presented in Sect. 6.2. However, since the majority of theories considered in the book are based on simpler languages, in this section we present a family of relational languages which includes the languages of those theories as its instances. Only few languages dealt with in the book are not captured in that family. In Chap. 5 the language of the relational logic of fork algebras is expanded with object terms so that an internal structure of objects can be reflected in the syntax (see Sect. 5.3). In the language presented in Chap. 6 the relational symbols are interpreted as heterogenous relations and it is reflected in the syntax through introduction of types (see Sect. 6.2). Finally, in Chap. 10 the semantics of the relational language for presentation of many-valued logics is based on models with a family of meaning functions determined by the range of truth values assumed in the logic under consideration. None of these three features is attributed to the language of this section. All the remaining relational languages presented in the book are instances of this language.

Expressions of relational languages are constructed from symbols of the following pairwise disjoint sets:

- $\mathbb{OV}$ – a countable infinite set of *object variables*;
- $\mathbb{OC}$ – a countable (possibly empty) set of *object constants*;
- $\mathbb{RV}$ – a countable (possibly empty) set of *relational variables* representing relations of various finite arities;
- $\mathbb{RC}$ – a countable (possibly empty) set of *relational constants* representing relations of various finite arities;
- A set of relational operations, including set operations $\cup$, $\cap$, and $-$.

Relational constants and object constants are usually those from the languages of the theories in question. Some relational operations may act on relations of different arities. Examples of such operations can be found in Chap. 6. Set operations always act on relations of the same arities.

Sets $\mathbb{OS} = \mathbb{OV} \cup \mathbb{OC}$ and $\mathbb{RA} = \mathbb{RV} \cup \mathbb{RC}$ are called the set of *object symbols* and the set of *atomic relational terms*, respectively. The set of *relational terms*, $\mathbb{RT}$, is the smallest set including atomic relational terms and closed with respect to the relational operations. We say that a term is n-ary whenever it represents an n-ary relation. The set of *relational formulas* consists of expressions of the form $T(\overline{x})$, where T is a relational term and $\overline{x}$ is a finite sequence of object symbols.

Let L be a relational logic based on a relational language. If X is any of the syntactic categories defined above, then we write X_{L} for its instance in L. Moreover, if Y is a term or a formula of logic L, then $X_{\mathsf{L}}(Y)$ denotes the set of symbols of the category X_{L} appearing in Y. For example, $\mathbb{OC}_{\mathsf{L}}(Y)$ is the set of object constants in Y.

An L-*structure* is a pair $\mathcal{M} = (U, m)$, where U is a non-empty set and m is a meaning function which:

- Assigns elements of U to object constants, so that for each object constant c, $m(c) \in U$;
- Assigns n-ary relations on U, $n \geq 1$, to the atomic relational terms, so that for each n-ary atomic relational term R, $m(R) \subseteq U^n$;
- Assigns relational operations on U to the symbols of relational operations of the language and preserves $\cup$, $\cap$, and $-$, i.e., $m(T \cup T') = m(T) \cup m(T')$, $m(T \cap T') = m(T) \cap m(T')$, $m(-T) = U^n \setminus m(T)$, for all n-ary relational terms T and T';
- Extends to compound relational terms as follows: for each k-ary relational operation symbol # and for all terms T_i, $i = 1, \ldots, k$, $m(\#(T_1, \ldots, T_k)) = m(\#)(m(T_1), \ldots, m(T_k))$.

Usually, the same symbols are used for the operations in the syntax and for their assignment under m.

Given a logic L, L-*models* are L-structures such that the meaning function m satisfies all the conditions specific for L. The conditions concern interpretation of relational operations specific for a given logic and/or interpretation of specific relational and/or object constants of the logic. For example, in a logic of binary relations we may assume that relational variables represent equivalence relations. Often these constraints are expressed in a first-order language and the operations are first-order definable.

Most often relational variables and relational constants in a language are of the same arity. In few cases they are of the different arities: in Chap. 9 there is a ternary relational constant and two quaternary relational constants; in Chap. 6 relational variables are of any finite arity.

A *valuation* in an L-structure $\mathcal{M}$ is a function $v\colon \mathbb{OS} \to U$ assigning elements of U to object symbols such that $v(c) = m(c)$, for every object constant c.

If $\overline{x} = x_1, \ldots, x_n$ is a sequence of n object symbols, then we write $v(\overline{x})$ to denote the sequence $v(x_1), \ldots, v(x_n)$. We also define $\mathrm{set}(\overline{x}) \stackrel{\mathrm{df}}{=} \{x_1, \ldots, x_n\}$.

Satisfaction of an L-formula in an L-model $\mathcal{M}$ by a valuation v is defined as:

$$\mathcal{M}, v \models T(\overline{x}) \text{ iff } v(\overline{x}) \in m(T).$$

The relational formula $T(\overline{x})$ is *true* in $\mathcal{M}$ if and only if it is satisfied by all valuations in $\mathcal{M}$. Consequently, if set$(\overline{x}) \subseteq \mathbb{OV}_{\mathsf{L}}$, then $T(\overline{x})$ is true in $\mathcal{M}$ whenever $m(T) = U^n$. The formula $T(\overline{x})$ is L-valid whenever it is true in every L-model.

25.4 Relational Languages Versus First-Order Languages

There is a natural relationship between relational languages and the first-order languages. If we assume that the set of object variables (resp. object constants) in the relational language is the set of individual variables (resp. individual constants) in the first-order language, and the set of atomic relational terms coincides with the set of predicate symbols, then there is a natural translation τ from relational formulas into first-order formulas: if R is a symbol of n-ary relation and $x_1, \ldots, x_n$ are object symbols, then $\tau(R(x_1, \ldots, x_n)) = R(x_1, \ldots, x_n)$. If all the relational operations are first-order definable, then the image under τ of a compound term is obtained by replacing the relational operations by their first-order definitions. For example, in logic RL presented in Sect. 2.5 the translation of terms built with the operation of composition, ;, is:

$$\tau((R\,;P)(x,y)) = \exists z(\tau(R(x,z)) \wedge \tau(P(z,y))),$$

where R and P are terms of RL denoting binary relations, z is an object variable, and x, y are object symbols.

Every relational model $\mathcal{M}$ determines a first-order model $\mathcal{M}^\tau$. The models share the universe and the meaning functions in both models act on object constants and atomic relational terms in the same way. In model $\mathcal{M}$ the extension of meaning function m to all the relational terms is determined by its action on relational operations, while in $\mathcal{M}^\tau$ the extension of m on the compound formulas is determined by its standard action on propositional operations and quantifiers. Furthermore, every valuation in $\mathcal{M}$ is at the same time a valuation in $\mathcal{M}^\tau$.

Lemma 25.4.1. *Let* L *be a relational logic. For every* L*-formula* $T(\overline{x})$*, for every* L*-model* $\mathcal{M}$*, and for every valuation* v *in* $\mathcal{M}$*:*

$$\mathcal{M}, v \models T(\overline{x}) \textit{ iff } \mathcal{M}^\tau, v \models \tau(T(\overline{x})).$$

In this way every logic of relations with first-order definable relational operations determines a first-order theory.

By a *relational literal* we mean an atomic relational term or a complemented atomic relational term. For the sake of simplicity, any expression $R(\overline{x})$, such that R

is a relational literal and $\overline{x}$ is a sequence of object symbols, is also referred to as a literal. If a literal A is $R(\overline{x})$, where R is a relational literal, then we denote the literal $-R(\overline{x})$ by $\neg A$ as in the first-order languages. A literal is positive (resp. negative) if it is of the form $R(\overline{x})$ (resp. $-R(\overline{x})$) for some atomic relational term R and for a sequence $\overline{x}$ of object symbols.

On the other hand, some first-order formulas are definable in relational languages. Let $\varphi = \psi(R_1, \dots, R_m)(\overline{x}, \overline{z})$ be a first-order formula built with predicates R_i such that the elements of set$(\overline{x})$ are the free variables in φ, the elements of set$(\overline{z})$ are the bound variables in φ, and set$(\overline{x}) \cap$ set$(\overline{z}) = \emptyset$. If there is a relational logic L and an L-term T such that for every L-model $\mathcal{M}$ and for every valuation v in $\mathcal{M}$, $\mathcal{M}^\tau, v \models \varphi$ iff $\mathcal{M}, v \models T(\overline{x})$, then φ is said to be L-*definable*.

Example. Let $\varphi = \exists y(xRy)$. The formula φ is definable in the logic RL (see Sect. 2.5) by $x(R\,;(R \cup -R))x$. Indeed, for every RL-model $\mathcal{M} = (U, m)$ and for every valuation v in $\mathcal{M}$, $\mathcal{M}, v \models x(R\,;(R \cup -R))x$ iff there exists $y \in U$ such that $(v(x), y) \in m(R)$ and $(y, v(x)) \in m(R) \cup -m(R)$ iff there exists $y \in U$ such that $(v(x), y) \in m(R)$ iff φ is satisfied in $\mathcal{M}^\tau$ by v.

Applications of relational definability of first-order formulas are discussed in Sects. 25.6 and 25.7.

It is known that not every formula of the first-order language with identity and with only binary predicates, F_2, is definable in the language of logic $\mathsf{RL}(1, 1')$ of Sect. 2.7. One such formula is presented in [TG87]:

$$\forall x \forall y \forall z \exists u (u \neq x \wedge u \neq y \wedge u \neq z).$$

It is also proved there that the three-variable fragment of the logic F_2 is equivalent to the Tarski calculus of relations (see [Tar41]). However, as it is presented in Chap. 5, first-order logic with function symbols is interpretable in the relational logic based on fork algebras.

25.5 Dual Tableaux

Typically, a dual tableau for a relational logic L consists of the rules which apply to finite sets of relational formulas and of the axiomatic sets of formulas. The axiomatic sets take the place of axioms. The rules reflect properties of operations and constants of the language under consideration. As in the classical logic presented in Chap. 1, there are two groups of rules: *decomposition rules* and *specific rules*. Decomposition rules enable us to decompose formulas into some simpler formulas. Decomposition rules are designed in such a way that they encode semantics of the operations assumed in the language of the given logic. As a result of decomposition we usually obtain finitely many new sets of formulas (although systems with infinitary rules are also in use, see Sect. 19.2). The specific rules enable us to modify formulas which they are applied to. These rules are intended to reflect constraints placed on constants. In other words, the rules reflect the restrictions that we place on the meaning function in the models of the logic.

All the rules mentioned above have the following general form:

$$\text{(rule)} \quad \frac{\Phi(\overline{x})}{\Phi_1(\overline{x}_1, \overline{y}_1, \overline{z}_1) \mid \ldots \mid \Phi_j(\overline{x}_j, \overline{y}_j, \overline{z}_j) \mid \ldots}$$

where $j \in J$ for some set J of indices, $\Phi(\overline{x})$ is a finite (possibly empty) set of formulas whose object variables are among the elements of set$(\overline{x})$; every $\Phi_j(\overline{x}_j, \overline{y}_j, \overline{z}_j)$ is a finite non-empty set of formulas, whose object variables are among the elements of set$(\overline{x}_j) \cup$ set$(\overline{y}_j) \cup$ set$(\overline{z}_j)$; set$(\overline{x}_j) \subseteq$ set$(\overline{x})$; set$(\overline{y}_j)$ consists of the variables that may be instantiated to arbitrary object symbols, usually to the object symbols that appear in the set to which the rule is being applied; set$(\overline{z}_j)$ consists of variables that must be instantiated to new, i.e., not appearing in the formulas of the set to which the rule is being applied, pairwise different variables distinct from all object variables from set$(\overline{y}_j)$. If the cardinality of the set J is $n \geq 2$, then (rule) is the n-fold branching rule. The set of formulas above the line is referred to as the *premise* of (rule) and the set(s) below the line is (are) its *conclusion(s)*. As usual, we write premises and conclusions of the rules as sequences of formulas rather than sets.

A rule is said to be *applicable* to a finite set of formulas, X, whenever $\Phi(\overline{x}) \subseteq X$. The result of an application of a rule of the form (rule) to the set X are the sets of the form $(X \setminus \Phi(\overline{x})) \cup \Phi_j(\overline{x}_j, \overline{y}_j, \overline{z}_j)$, for $j \in J$. Note that $\Phi_j(\overline{x}_j, \overline{y}_j, \overline{z}_j)$ may include the set $\Phi(\overline{x})$. In cut-like rules which are discussed in Sect. 25.9 the set $\Phi(\overline{x})$ is empty.

Let L be a relational logic. We extend the notion of validity to finite sets of formulas in the following way. A finite set X of L-formulas is an L-*set* whenever for every L-model $\mathcal{M}$ and for every valuation v in $\mathcal{M}$, there is a formula in X satisfied by v in $\mathcal{M}$. This is equivalent to the F-validity of the universally quantified first-order disjunction of the translations of the formulas of X into first-order formulas obtained as described in the previous section. If $\mathcal{K}$ is a class of L-structures (not necessarily L-models), then a finite set X of L-formulas is said to be a $\mathcal{K}$-*set* whenever for every L-structure $\mathcal{M}$ of $\mathcal{K}$ and for every valuation v in $\mathcal{M}$, there exists a formula in X satisfied by v in $\mathcal{M}$.

A rule of the form (rule) is L-*correct* whenever for every finite set X of L-formulas, $X \cup \Phi(\overline{x})$ is an L-set iff for each $j \in J$, $X \cup \Phi_j(\overline{x}_j, \overline{y}_j, \overline{z}_j)$ is an L-set. If $\mathcal{K}$ is a class of L-structures, then we define $\mathcal{K}$-correctness in a similar way, that is a rule of the form (rule) is $\mathcal{K}$-*correct* whenever for every finite set X of L-formulas, $X \cup \Phi(\overline{x})$ is a $\mathcal{K}$-set iff for each $j \in J$, $X \cup \Phi_j(\overline{x}_j, \overline{y}_j, \overline{z}_j)$ is a $\mathcal{K}$-set. Every relational proof system for logic L includes the two decomposition rules denoted by (#) and (−#), respectively, for every relational operation symbol # of the language of L. These rules reflect semantics of formulas built with #.

If φ is an L-formula whose validity is in question, we construct a *proof tree* for φ. We place φ at the root and generate a tree by applying deduction rules each of which yields a set of formulas or branches to yield several sets of formulas. A branch of a proof tree *closes* whenever it contains a node with an axiomatic set. A tree is *closed* whenever all of its branches are closed. A *complete branch* of a proof tree is a branch b which is either closed or satisfies the following completion condition, (Cpl rule), for every rule of the form (rule) of the proof system in question. As usual, if b is a branch of a proof tree, the notation $\varphi \in b$ means that the formula φ appears in a node of the branch b.

(Cpl rule) If every formula from $\Phi(\overline{x})$ appears in some node of b, then there exists $k \in J$ such that for every formula φ in Φ_k, the following conditions are satisfied:

- If $\bigcup\{\text{set}(\overline{y}_j) : j \in J\} \neq \emptyset$ and $\bigcup\{\text{set}(\overline{z}_j) : j \in J\} = \emptyset$, then for every $y \in \bigcup\{\text{set}(\overline{y}_j) : j \in J\}$ and for every $t \in \mathbb{OS}$, if y appears in φ, then $\varphi(y/t) \in b$;
- If $\bigcup\{\text{set}(\overline{z}_j) : j \in J\} \neq \emptyset$ and $\bigcup\{\text{set}(\overline{y}_j) : j \in J\} = \emptyset$, then for every $z \in \bigcup\{\text{set}(\overline{z}_j) : j \in J\}$ there is $t \in \mathbb{OV}$ such that if z appears in φ, then $\varphi(z/t) \in b$;
- If $\bigcup\{\text{set}(\overline{y}_j) : j \in J\} \neq \emptyset$ and $\bigcup\{\text{set}(\overline{z}_j) : j \in J\} \neq \emptyset$, then for all $y \in \bigcup\{\text{set}(\overline{y}_j) : j \in J\}$, $z \in \bigcup\{\text{set}(\overline{z}_j) : j \in J\}$, and for every $t_1 \in \mathbb{OS}$ there exists $t_2 \in \mathbb{OV}$ such that if y or z appear in φ, then $\varphi(y/t_1, z/t_2) \in b$;
- If $\bigcup\{\text{set}(\overline{y}_j) : j \in J\} = \bigcup\{\text{set}(\overline{z}_j) : j \in J\} = \emptyset$, then $\varphi \in b$.

In the first item of the above condition $\varphi(y/t)$ denotes the formula obtained from φ by substituting t for every occurrence of y, and similarly in the remaining items.

A branch of a proof tree is *open* whenever it is complete and does not close. Intuitively, a branch is open whenever all of the rules that can be applied to formulas on the branch have been applied and the branch does not close. A *complete proof tree* is a proof tree such that all of its branches are complete. The completion conditions must guarantee that for every formula there exists a complete proof tree for it.

We say that a formula is L-*provable* whenever there is a closed proof tree for it. As usual, soundness of a dual tableau is a property saying that if a formula is provable, then it is valid. To prove soundness it is sufficient to show that each of the rules is L-correct and all the axiomatic sets are L-valid. Completeness of a dual tableau is a property saying that if a formula is valid, then it is provable. To prove completeness it is sufficient to show that every complete L-proof tree of an L-valid formula can be closed. A discussion of the details of the completeness proofs of dual tableaux is presented in Sect. 25.8.

25.6 Constraint–Rule Correspondence

In this section we present correspondences which show how to define a rule, given a constraint. Our approach will be to look at first-order formulas or relational formulas expressing constraints and develop corresponding correct rules.

If $\overline{x} = x_1, \ldots, x_n$ is a sequence of object variables, then by $\forall\overline{x}$ we mean $\forall x_1 \ldots \forall x_n$, and similarly for the quantifier $\exists$.

Constraints Defined in a First-Order Language

The condition (c1)

The condition (c1) is of the form:

(c1) $\forall \overline{x}(A \rightarrow B)$,

where $A = A_1 \wedge \ldots \wedge A_s$, $s \geq 1$, $A_1, \ldots, A_s$ are literals, $B = B^1 \vee \ldots \vee B^i$, $i \geq 1$, and for every $j \in \{1, \ldots, i\}$, $B^j = B_1^j \wedge \ldots \wedge B_{k_j}^j$, $k_j \geq 1$, $B_1^j, \ldots, B_{k_j}^j$ are literals, $\mathbb{OV}(A) = \mathrm{set}(\overline{x}) \neq \emptyset$, $\emptyset \neq \mathbb{OV}(B) \subseteq \mathbb{OV}(A)$.

Then the corresponding rule is an $s + i$-fold branching rule:

$$(\mathrm{rc1}) \quad \frac{}{A_1 \mid \ldots \mid A_s \mid \neg B_1^1, \ldots, \neg B_{k_1}^1 \mid \ldots \mid \neg B_1^i, \ldots, \neg B_{k_i}^i}$$

where the variables that appear in A or B may be instantiated to any object symbol.

Examples of rules corresponding to conditions of type (c1) are:

- Rules (rM5), ..., (rM8), and (rM12) for conditions (M5), ..., (M8), and (M12), respectively, in relevant logics in Sect. 9.3 and in Sect. 9.4;
- Rule (dis R, S) for condition $R \cap S = \emptyset$ in logic $\mathsf{RL}_{\mathsf{CI}}$ in Sect. 11.5;
- Rule (rReC0) of $\mathsf{F}_{\mathsf{BAC}}$-dual tableau presented in Sect. 18.3;
- Rule (rReδ1) of $\mathsf{F}_{\mathsf{BA}\delta}$-dual tableau presented in Sect. 18.5.

The condition (c2)

If a constraint has the form:

(c2) $\forall \overline{x} B$

where B is as in the constraint (c1) and $\mathbb{OV}(B) = \mathrm{set}(\overline{x})$, then the relational counterpart of (c2) is provided by a rule of the form:

$$(\mathrm{rc2}) \quad \frac{}{\neg B_1^1, \ldots, \neg B_{k_1}^1 \mid \ldots \mid \neg B_1^i, \ldots, \neg B_{k_i}^i}$$

where the variables that appear in B may be instantiated to any object symbol.

Examples of rules corresponding to conditions of type (c2):

- Rule (rS2) for condition $R_\emptyset = U \times U$ in logic $\mathsf{RL}_{\mathsf{LFS}}$ in Sect. 12.3;
- Rule (rW2) for condition $R_\emptyset = \emptyset$ in logic $\mathsf{RL}_{\mathsf{LFW}}$ in Sect. 12.3;
- Rule (0) for the empty relation condition in logic $\mathsf{RL}_{\mathsf{MTL}}$ in Sect. 14.4;
- Rule (irref $<$) for irreflexivity of $<$ in logic $\mathsf{RL}_{\mathsf{OMR}}$ in Sect. 15.3.

The condition (c3)

The condition (c3) is of the form:

(c3) $\forall \overline{x} \exists \overline{z} B$

where B is as in the constraint (c1) and $\mathbb{OV}(B) = \mathrm{set}(\overline{x}) \cup \mathrm{set}(\overline{z})$, $\mathrm{set}(\overline{x}) \neq \emptyset$, $\mathrm{set}(\overline{z}) \neq \emptyset$, and $\mathrm{set}(\overline{x}) \cap \mathrm{set}(\overline{z}) = \emptyset$.

The corresponding rule is an i-fold branching rule:

$$\text{(rc3)} \quad \frac{}{\neg B_1^1, \ldots, \neg B_{k_1}^1 \mid \ldots \mid \neg B_1^i, \ldots, \neg B_{k_i}^i}$$

where each of the variables from set($\overline{x}$) may be instantiated to any object symbol and all variables from set($\overline{z}$) must be instantiated to new pairwise different variables that are distinct from the variables of set($\overline{x}$).

Examples of rules corresponding to conditions of type (c3) are:

- Rule (rMTL4) for condition (MTL4) in logic $\mathsf{RL}_{\mathsf{MTL}}$ in Sect. 14.4;
- Rules (rReBA0), (rReBA2#), and (rReBA10) of $\mathsf{F}_{\mathsf{BAC}}$-dual tableaux for spatial theories based on a Boolean algebra presented in Sect. 18.3.

The condition (c4)

The condition (c4) is of the form:

(c4) $\forall\overline{x}(A \rightarrow \exists\overline{z}B)$

where A and B are as in the constraint (c1), $\overline{x}$ and $\overline{z}$ are finite sequences of object variables, not necessarily of the same length, $\mathbb{OV}(A) = \text{set}(\overline{x}) \neq \emptyset$, $\mathbb{OV}(B) \neq \emptyset$, $(\mathbb{OV}(B) \setminus \text{set}(\overline{z})) \subseteq \mathbb{OV}(A)$, and $\text{set}(\overline{x}) \cap \text{set}(\overline{z}) = \emptyset$.

The corresponding rule is an $s + i$-fold branching rule:

$$\text{(rc4)} \quad \frac{}{A_1 \mid \ldots \mid A_s \mid \neg B_1^1, \ldots, \neg B_{k_1}^1 \mid \ldots \mid \neg B_1^i, \ldots, \neg B_{k_i}^i}$$

where each of the variables from set($\overline{z}$) occurring in any B_m^l must be instantiated to a new variable distinct from the variables in set($\overline{x}$), for $l \in \{1, \ldots, i\}$ and $m \in \{1, \ldots, k_i\}$.

Examples of rules corresponding to conditions of type (c4) are:

- The rule (rMTL3) for condition (MTL3) in logic $\mathsf{RL}_{\mathsf{MTL}}$ in Sect. 14.4;
- The rule (ext C) for extensionality axiom in spatial theories based on a plain contact relation in Sect. 18.2;
- The rules (rReC8), (rReC10), and (rReC11) of $\mathsf{F}_{\mathsf{BAC}}$-dual tableaux for spatial theories based on a Boolean algebra presented in Sect. 18.3;
- The rules (r1ReRCC5($\rightarrow$)), (r1ReRCC5($\leftarrow$)), (r2ReRCC5($\rightarrow$)), and the rules (r2ReRCC5($\leftarrow$)), (rReRCC7($\rightarrow$)), and (rReRCC7($\leftarrow$)) of $\mathsf{F}_{\mathsf{RCC}}$-dual tableaux for spatial theories based on a Boolean algebra presented in Sect. 18.4.

In many dual tableaux presented in the book, we admit another form of the rules corresponding to conditions (c3) and (c4). Assume that formula $\exists\overline{z}B$ appearing in the conditions (c3) and (c4) is definable in a relational language with a formula $T_{\exists\overline{z}B}(\overline{x})$. Then the rule corresponding to the condition (c4) is:

$$\text{(rc4)'} \quad \frac{}{A_1 \mid \ldots \mid A_s \mid -T_{\exists\overline{z}B}(\overline{x})}$$

where each of the variables from set($\overline{x}$) may be instantiated to any object symbol.

Example. Let L_{TL} be a basic temporal logic presented in Sect. 16.2 such that the relation R is dense in all L_{TL}-models, i.e.:

$$(\text{den } R) \quad \forall x \forall y (xRy \to \exists z (xRz \wedge zRy)).$$

The relational representation of $\exists z(xRz \wedge zRy)$ is:

$$T_{\exists z(xRz \wedge zRy)}(x, y) = x(R\,;R)y.$$

Then the rule corresponding to condition (den R) is:

$$(\text{den } R) \quad \frac{}{xRy \mid x{-}(R\,;R)y} \quad \text{for all object symbols } x \text{ and } y$$

In much the same way, we can obtain the rules corresponding to conditions of type (c3) and (c4). Examples of the rules corresponding to these conditions are:

- Rule (rL_2) for condition (L_2) in logic $\mathsf{RL}_{\mathsf{L}_2}$ in Sect. 8.4;
- Rules (un R), (ser R), and (wdir R) for the condition of unboundness, seriality, and directness of R, respectively, in logic $\mathsf{RL}_{\mathsf{L}_{\mathsf{TL}}}$ in Sect. 16.3.

Constraints Defined in a Relational Language

The condition (c5)

The condition (c5) is of the form:

(c5) $\forall \overline{x}(A \to T)$

where $\overline{x}$ is a finite sequence of object variables, A is a conjunction of relational literals, $\mathbb{OV}(A) = \text{set}(\overline{x}) \neq \emptyset$, and T is a relational formula such that all of its object variables are in $\text{set}(\overline{x})$. The rule corresponding to condition (c5) has the form:

$$(\text{rc5}) \quad \frac{}{A_1 \mid \ldots \mid A_s \mid -T}$$

where each of the variables from $\text{set}(\overline{x})$ may be instantiated to any object symbol.

After the translation of T into a first-order formula, some instances of condition (c5) are neither of the form (c1) nor (c4) as the following example shows.

Example. In Sect. 16.3 the relational logic $\mathsf{RL}_{\mathsf{L}_{\mathsf{TL}}}$ for the basic temporal logic with discrete models is considered. Discreteness is expressed as follows:

(dis R) $(x, y) \in R$ implies (1) there exists z such that $(x, z) \in R$ and for all t if $(x, t) \in R$, then $(z, t) \in R$, and (2) there exists z such that $(z, y) \in R$ and for all t if $(t, y) \in R$, then $(t, z) \in R$.

In the relational logic the above condition is equivalent to:

$(x, y) \in R$ implies:

(1) $(x, y) \in (-(R; -R^{-1}); R^{-1})$, and
(2) $(y, y) \in (-(R^{-1}; -R); R)$.

Hence, the corresponding rules are:

$$(\text{dis}_1\ R) \quad \frac{}{xRy \mid x{-}(-(R; -R^{-1}); R^{-1})x}$$

x, y are any object variables

$$(\text{dis}_2\ R) \quad \frac{}{xRy \mid y{-}(-(R^{-1}; -R); R)y}$$

x, y are any object variables

In much the same way, the rule (fun R) for the condition of functionality of R in logic $\mathsf{RL}_{\mathsf{L_{TL}}}$ in Sect. 16.3 is obtained.

Theorem 25.6.1. *Let* L *be a relational logic and let* $\mathcal{K}$ *be a class of* L*-structures. Then, the following hold:*

1. *The rule (rci) is* $\mathcal{K}$*-correct iff condition (ci) is satisfied in every* $\mathcal{K}$*-structure, for every* $i \in \{1, 2\}$*;*
2. *If the subformula* $\exists\overline{z}B$ *appearing in the constraint (ci) has a relational representation in the logic* L*, then the rule (rci) is* $\mathcal{K}$*-correct iff condition (ci) is satisfied in every* $\mathcal{K}$*-structure, for every* $i \in \{3, 4\}$*;*
3. *If the subformula* B *appearing in the constraint (c5) has a relational representation in the logic* L*, then the rule (rc5) is* $\mathcal{K}$*-correct iff condition (c5) is satisfied in every* $\mathcal{K}$*-structure.*

Proof. Let L be a relational logic and let $\mathcal{K}$ be a class of L-structures. By way of example, we prove 2.($\rightarrow$) for (rc3) and 2.($\leftarrow$) for (rc4).

2.($\rightarrow$) Assume that the rule (rc3) is $\mathcal{K}$-correct. Since $\text{set}(\overline{z}) \cap \text{set}(\overline{x}) = \emptyset$ and all the variables from $\text{set}(\overline{z})$ are new, $\mathcal{K}$-correctness of (rc3) implies that for every finite set X of L-formulas, X is a $\mathcal{K}$-set iff for every $j \in \{1, \ldots, i\}$ the formula $\forall\overline{x}\forall\overline{z}(\delta_X \vee \neg B_1^j(\overline{y}, \overline{z}) \vee \ldots \vee \neg B_{k_j}^j(\overline{y}, \overline{z}))$ is valid, where $\text{set}(\overline{y}) \subseteq \text{set}(\overline{x})$. We recall that δ_X is the disjunction of formulas from the set X. Let $T_{\exists\overline{z}B}(\overline{x})$ be a relational formula defining $\exists\overline{z}B$ such that $\mathbb{O}\mathbb{V}(T_{\exists\overline{z}B}(\overline{x})) = \text{set}(\overline{x})$. Then $\{T_{\exists\overline{z}B}(\overline{x}), \neg B_1^j, \ldots, \neg B_{k_j}^j\}$ is a $\mathcal{K}$-set for every $j \in \{1, \ldots, i\}$, because $T_{\exists\overline{z}B}(\overline{x})$ implies $\exists\overline{z}(B_1^j(\overline{y}, \overline{z}) \wedge \ldots \wedge B_{k_j}^j(\overline{y}, \overline{z}))$. Thus, by the assumption, $\{T_{\exists\overline{z}B}(\overline{x})\}$ is a $\mathcal{K}$-set, which means that the condition (c3) is satisfied in every $\mathcal{K}$-structure.

2.($\leftarrow$) Assume (c4) holds in every $\mathcal{K}$-structure. Let X be a finite set of L-formulas. We assume that variables from $\text{set}(\overline{z})$ do not occur in X and $\text{set}(\overline{z}) \cap \text{set}(\overline{x}) = \emptyset$. Clearly, if X is a $\mathcal{K}$-set, then each of the sets below the line is valid. Now, assume that $X \cup A_j$ and $X \cup \{\neg B_1^l, \ldots, \neg B_{k_l}^l\}$ are $\mathcal{K}$-sets for all $j \in \{1, \ldots, s\}$, $l \in \{1, \ldots, i\}$. Then since $\text{set}(\overline{z}) \cap \text{set}(\overline{x}) = \emptyset$ and all the variables from $\text{set}(\overline{z})$ are new, for every $l \in \{1, \ldots, i\}$ the formula $\forall\overline{x}\forall\overline{z}(\delta_X \vee \neg B_1^l(\overline{y}, \overline{z}) \vee \ldots \vee \neg B_{k_l}^l(\overline{y}, \overline{z}))$ is valid, where $\text{set}(\overline{y}) \subseteq \text{set}(\overline{x})$. Suppose

X is not a $\mathcal{K}$-set. Then, by the assumption, there exists a $\mathcal{K}$-structure $\mathcal{M}$ such that $\mathcal{M} \models A_j$ for every $j \in \{1, \ldots, s\}$ and $\mathcal{M} \not\models \exists\overline{z}B$, a contradiction with the condition (c4). □

The assumption made in the condition 2. of the above theorem is not very restrictive from the point of view of applicability of the relational proof systems. Most often we pose the constraints on the relations in the models of a relational logic in terms of formulas of that logic.

The completion conditions for the rules corresponding to the constraints considered in this section are:

Cpl(rc1) For every object symbol t and for every $y \in \text{set}(\overline{x})$ either there exists $j \in \{1, \ldots, s\}$ such that $A_j(y/t) \in b$ or there exists $l \in \{1, \ldots, i\}$ such that all $B_1^l(y/t), \ldots, B_{k_l}^l(y/t)$ are in b;

Cpl(rc2) For every object symbol t and for every $y \in \text{set}(\overline{x})$ there exists $l \in \{1, \ldots, i\}$ such that all $B_1^l(y/t), \ldots, B_{k_l}^l(y/t)$ are in b;

Cpl(rc3) For all $x \in \text{set}(\overline{x})$, $z \in \text{set}(\overline{z})$, and for every object symbol t there exists an object variable u such that for some $l \in \{1, \ldots, i\}$ all $\neg B_1^l(x/t, z/u), \ldots, \neg B_{k_l}^l(x/t, z/u)$ are in b;

Cpl(rc4) For all $x \in \text{set}(\overline{x})$, $z \in \text{set}(\overline{z})$, and for every object symbol t, either there exists $j \in \{1, \ldots, s\}$ such that $A_j(x/t) \in b$ or there exists an object variable u such that for some $l \in \{1, \ldots, i\}$ all $\neg B_1^l(x/t, z/u), \ldots, \neg B_{k_l}^l(x/t, z/u)$ are in b;

Cpl(rc5) For every object symbol t and for every $y \in \text{set}(\overline{x})$, either there exists $j \in \{1, \ldots, s\}$ such that $A_j(y/t) \in b$ or $-T(y/t) \in b$.

25.7 Definition–Rule Correspondence

In this section we show how to define a rule given a definition of a relational operation or a relational constant.

First-Order Definition of a Relational Operation

Let L be a logic based on a relational language. Let a k-ary relational operation # be defined by a first-order formula:

For every sequence $\overline{x}$ of object symbols,

$$\text{(def\#)}\ (\#(R_1, \ldots, R_k))(\overline{x}) \text{ iff } Q\overline{z}\varphi(R_1, \ldots, R_k)(\overline{x}, \overline{z}),$$

where $R_i, i = 1, \ldots, k$, are atomic relational terms, Q is a string of quantifiers such that all of them are either universal or existential, φ is an open first-order formula built with predicates $R_1, \ldots, R_k$, the elements of $\text{set}(\overline{x}) \cap \mathbb{OV}_{\mathsf{L}}$ are the free variables in $Q\overline{z}\varphi$, the elements of $\text{set}(\overline{z})$ are the bound variables in $Q\overline{z}\varphi$, and $\text{set}(\overline{x}) \cap \text{set}(\overline{z}) = \emptyset$.

Consider the relational logic L' which is a signature extension of L such that the language of L' is obtained from the language of L by endowing it with the

k-ary operation #. An L'-structure is an L-model $\mathcal{M} = (U, m)$ in which $m(\#(R_1, \ldots, R_k)) \subseteq U^k$. Models of L' are L'-structures such that the meaning of the terms built with operation # is determined by (def #):

$$\text{(m\#)} \quad m(\#(R_1, \ldots, R_k)) = \{\overline{a} : Q\overline{z}\varphi(m(R_1), \ldots, m(R_k))(\overline{a}, \overline{z})\}$$

The proof system for L' is obtained from a proof system for L by adding two decomposition rules (#) and (−#) associated to the operation #:

$$(\#) \quad \frac{(\#(R_1, \ldots, R_k))(\overline{x})}{H_1, (\#(R_1, \ldots, R_k))(\overline{x}) \mid \ldots \mid H_p, (\#(R_1, \ldots, R_k))(\overline{x})}$$

$$(-\#) \quad \frac{(-\#(R_1, \ldots, R_k))(\overline{x})}{K_1, (-\#(R_1, \ldots, R_k))(\overline{x}) \mid \ldots \mid K_q, (-\#(R_1, \ldots, R_k))(\overline{x})}$$

where $R_1, \ldots, R_k$ are relational terms and sets $H_1, \ldots, H_p$ (respectively $K_1, \ldots, K_q$) are obtained in the following way:

(Step 1) We successively apply the decomposition rules of F-dual tableau presented in Sect. 1.3 to formula φ (respectively $\neg\varphi$), with the restriction that if rule ($\forall$) or ($\neg\exists$) is applied, then the repetition of the formula $(\#(R_1, \ldots, R_k))(\overline{x})$ (resp. $(-\#(R_1, \ldots, R_k))(\overline{x})$) is not introduced. The tree obtained in this way is finite;

(Step 2) For $H_1, \ldots, H_p$ (resp. $K_1, \ldots, K_q$) we take the leaves of the decomposition tree obtained in (Step 1).

Observe that the restricted application of ($\forall$) and ($\neg\exists$) in (Step 1) does not make the process incorrect, because the formula $(\#(R_1, \ldots, R_k))(\overline{x})$ (resp. $(-\#(R_1, \ldots, R_k))(\overline{x})$) occurring in the set under the line of the rule (#) (resp. (−#)) contains implicitly all the repetitions required by those rules.

Then the completion conditions, Cpl(#) and Cpl(−#), are obtained according to the definition of Cpl(rule) in Sect. 25.5.

Theorem 25.7.1 (Correspondence). *Let $\mathcal{K}$ be a class of L'-structures. If $Q\overline{z}\varphi(R_1, \ldots, R_k)(\overline{x}, \overline{z})$ is L-definable, then the following conditions are equivalent:*

1. *The rules (#) and (−#) are $\mathcal{K}$-correct;*
2. *$\mathcal{K}$ is a class of L'-models.*

The proof follows from correctness of the rules of F-dual tableau and the condition (def#). More exactly, the rules (#) and (−#) reflect the definition of #, namely, the rule (#) is correct whenever definiens i.e., the defining formula, implies definiendum i.e., the concept being defined, and the rule (−#) is correct whenever definiendum implies definiens. If $Q\overline{z}\varphi(R_1, \ldots, R_k)(\overline{x}, \overline{z})$ is not L-definable, then we have the weaker theorem:

Theorem 25.7.2. *The rules (#) and (−#) are* L'*-correct.*

The rules for the following relational operations are constructed in this way:

- Operations ^{c} and : in Peirce logic in Sect. 4.4;
- Operation ∇ in the fork logic in Sect. 5.4;
- Projection Π and division $\div$ in the logic of typed relations in Sect. 6.4;
- Operations $\rightarrow$ and $\odot$ in relevant logics in Sect. 9.4 and in the logic $\mathsf{RL_{MTL}}$ in Sect. 14.4;
- Iteration * in the propositional dynamic logic in Sect. 19.2;
- Demonic union || in the logic of demonic nondeterministic programs in Sect. 19.4.

Example. Consider logic RL presented in Sect. 2.5 and define relational operations $\Rightarrow$ and $\Leftarrow$ in the following way:

$$x(R \Rightarrow P)y \text{ iff } \forall z(R(z,x) \rightarrow P(z,y));$$

$$x(R \Leftarrow P)y \text{ iff } \forall z(R(y,z) \rightarrow P(x,z));$$

where the symbol $\rightarrow$ on the right side of these definitions denotes the ordinary first-order implication. These operations are needed in a relational logic for resituated semigroups.

Let RL' be a signature extension of RL such that its language is obtained from the language of RL by endowing it with the binary relational operations $\Rightarrow$ and $\Leftarrow$. RL'-structures are RL-models $\mathcal{M} = (U, m)$ such that $m(R \Rightarrow P)$ and $m(R \Leftarrow P)$ are binary relations on U, and RL'-models are RL'-structures such that the meanings of the operations $\Rightarrow$ and $\Leftarrow$ are defined according to the above definitions, respectively.

The respective decomposition rules are:

$$(\Rightarrow) \quad \frac{x(R \Rightarrow P)y}{z{-}Rx, zPy} \quad z \text{ is a new object variable}$$

$$(-\Rightarrow) \quad \frac{x{-}(R \Rightarrow P)y}{zRx, x{-}(R \Rightarrow P)y \mid z{-}Py, x{-}(R \Rightarrow P)y}$$

z is any object symbol

$$(\Leftarrow) \quad \frac{x(R \Leftarrow P)y}{y{-}Rz, xPz} \quad z \text{ is a new object variable}$$

$$(-\Leftarrow) \quad \frac{x{-}(R \Leftarrow P)y}{yRz, x{-}(R \Leftarrow P)y \mid x{-}Pz, x{-}(R \Leftarrow P)y}$$

z is any object symbol

It is easy to see that these rules are RL'-correct, that is they preserve and reflect the meanings of the operations $\Rightarrow$ and $\Leftarrow$ in RL'-models. Moreover, since both operations are definable in the language of the logic RL, by Theorem 25.7.1, if $\mathcal{K}$ is a class of RL'-structures, then $\mathcal{K}$-correctness of these rules implies the intended meanings of the operations in the structures, that is $\mathcal{K}$-correctness implies that $\mathcal{K}$ is a class of RL'-models. By way of example, we show it for $\Rightarrow$.

Let $X_1 \stackrel{\text{df}}{=} \{x(R^{-1};-P)y\}$. Then $X_1 \cup \{z-Rx, zPy\}$ is a $\mathcal{K}$-set. Thus, by $\mathcal{K}$-correctness of the rule $(\Rightarrow)$, $X_1 \cup \{x(R \Rightarrow P)y\}$ is also a $\mathcal{K}$-set. Therefore, $(R \Rightarrow P) \supseteq -(R^{-1};-P)$. Now, let $X_2 = \{z-Rx, zPy\}$. Then, $X_2 \cup \{zRx, x-(R \Rightarrow P)y\}$ and $X_2 \cup \{z-Py, x-(R \Rightarrow P)y\}$ are $\mathcal{K}$-sets. Thus, by $\mathcal{K}$-correctness of the rule $(- \Rightarrow)$, $X_2 \cup x-(R \Rightarrow P)y$ is also a $\mathcal{K}$-set. Thus, for every RL'-structure $\mathcal{M}$ and for every valuation v in $\mathcal{M}$, if $(v(z), v(x)) \in m(R)$ and $(v(z), v(y)) \notin m(P)$, then $(v(x), v(y)) \notin m(R \Rightarrow P)$. Hence, $(R \Rightarrow P) \subseteq -(R^{-1};-P)$, which completes the proof.

The completion conditions for the above rules can be defined according to the method presented in Sect. 25.5.

Definition of a Relational Operation by a Relational Term

Let L be a relational logic. Let a k-ary relational operation # be defined by:

For every sequence $\overline{x}$ of object symbols,

$$\text{(def'\#)} \quad (\#(R_1,\ldots,R_k))(\overline{x}) \text{ iff } T(R_1,\ldots,R_k)(\overline{x})$$

where $R_i, i = 1,\ldots,m$, are relational terms and $T(R_1,\ldots,R_k)$ is a relational term of L built with $R_1, \ldots, R_k$. As in the case of the first-order definition of #, let L' be the relational logic which is a signature extension of L such that the language of L' is obtained from the language of L by endowing it with the operation #. An L'-structure is an L-model $\mathcal{M} = (U, m)$ in which $m(\#(R_1,\ldots,R_k)) \subseteq U^k$. Models of L' are L'-structures such that the meanings of the terms built with operation # are determined by (def' #). The dual tableau for L' is obtained from a dual tableau for L by adding two decomposition rules (#) and (−#) associated to the relational definition of the operation #:

$$(\#)\quad \frac{(\#(R_1,\ldots,R_k))(\overline{x})}{T(R_1,\ldots,R_k)(\overline{x})} \qquad\qquad (-\#)\quad \frac{(-\#(R_1,\ldots,R_k))(\overline{x})}{-T(R_1,\ldots,R_k)(\overline{x})}$$

The completion conditions, Cpl(#) and Cpl(−#), are obtained according to the definition of Cpl(rule) in Sect. 25.5.

The theorems analogous to Theorems 25.7.1 and 25.7.2 hold.

In some dual tableaux presented in the book, we admit another form of the rules (#) and (−#) associated to the relational definition of the operation #. Namely, we decompose term $T(R_1,\ldots,R_k)(\overline{x})$ (resp. $-T(R_1,\ldots,R_k)(\overline{x})$) applying the rules of dual tableau for L. Then, as a conclusion(s) of the rule (#) (resp. (−#)) we take the leaf (the leaves) of the decomposition tree for $T(R_1,\ldots,R_k)(\overline{x})$ (resp.

$-T(R_1, \ldots, R_k)(\overline{x})$). However, there are some restrictions in this process, in particular we stop expanding the decomposition tree whenever it contains a variable which must be instantiated to a new object variable and the only rule that can be applied introduces a variable which is to be instantiated to an arbitrary object symbol.

The rules for the following operations are constructed in this way:

- The operations *Since*, *Until*, and *Next* of the dual tableau for the temporal logic $\mathsf{TL_{SU}}$ in Sect. 16.4;
- The operation of demonic composition ;; and the operation of demonic iteration $^{d(*)}$ of the dual tableau for the logic of demonic nondeterministic programs in Sect. 19.4.

Below we present one more example of that kind.

Example. Consider logic RL presented in Sect. 2.5 and define a ternary relational operation # in logic RL as:

$$\#(R_1, R_2, R_3) \stackrel{\text{df}}{=} -(R_1; -(R_2; R_3)).$$

Let $\mathsf{RL}(\#)$ be a signature extension of RL such that its language is the language of RL endowed with the relational operation #. $\mathsf{RL}(\#)$-structures are RL-models $\mathcal{M} = (U, m)$ such that $m(\#(R_1, R_2, R_3))$ is a binary relation on U, and $\mathsf{RL}(\#)$-models are $\mathsf{RL}(\#)$-structures $\mathcal{M} = (U, m)$ such that the meaning of the operation # is defined as $m(\#(R_1, R_2, R_3)) = -(R_1; -(R_2; R_3))$.

The rules for the terms built with the operation # have the following forms:

$$(\#) \quad \frac{x\#(R_1, R_2, R_3)y}{x{-}R_1 z, z(R_2; R_3)y}$$

z is a new object variable

$$(-\#) \quad \frac{x{-}\#(R_1, R_2, R_3)y}{xR_1 z, x{-}\#(R_1, R_2, R_3)y \mid z{-}R_2 w, w{-}R_3 y, x{-}\#(R_1, R_2, R_3)y}$$

z is any object symbol, w is a new object variable such that $z \neq w$

By Theorem 25.7.1, for every class $\mathcal{K}$ of $\mathsf{RL}(\#)$-structures the following holds: the rules (#) and (−#) are $\mathcal{K}$-correct iff $\mathcal{K}$ is a class of $\mathsf{RL}(\#)$-models.

The completion conditions corresponding to the above rules are:

Cpl(#) If $x\#(R_1, R_2, R_m)y \in b$, then for some object variable z both $x{-}R_1 z \in b$ and $z(R_2; R_3)y \in b$, obtained by an application of the rule (#);

Cpl(−#) If $x{-}\#(R_1, R_2, R_m)y \in b$, then for every object symbol z either $xR_1 z \in b$ or there exists an object variable w such that both $x{-}R_2 w \in b$ and $w{-}R_3 y \in b$, obtained by an application of the rule (−#).

Definition of a Relational Constant

In a similar way we construct the rules corresponding to definitions of relational constants. Namely, if a definition of a relational constant R has the form:

For every sequence $\overline{x}$ of object symbols,

$$\text{(def } R\text{)} \;\; R(\overline{x}) \text{ iff } Q\overline{z}\varphi(R_1, \ldots, R_k)(\overline{x}, \overline{z}),$$

with the assumptions on φ as in (def #), then the corresponding specific rules (R) and $(-R)$ have in the upper sets the formulas $R(\overline{x})$ and $-R(\overline{x})$, respectively, and their lower sets are obtained from φ and $\neg\varphi$, respectively, in the way described in (Step 1) and (Step 2) above. The theorems analogous to Theorems 25.7.1 and 25.7.2 hold. The rules for the following relational constants are constructed in this way:

- C in the relational logics with point relations introduced with definitions in Sect. 3.3;
- R_1 and R_2 in relevant logics in Sect. 9.4;
- $R_{P \cup Q}$ in logics of relative frames in Sect. 12.3;
- $\sqsubset$, C^-, and C^+ in the logic of order of magnitude reasoning in Sect. 15.3;
- B and E in the relational logic for Halpern–Shoham logic in Sect. 17.5; D, M, P, and O in the relational logic for interval temporal logics in Sect. 17.6;
- Mereological relations in the logic $\mathsf{RL}_{\mathsf{Mer}}$ in Sect. 18.2;
- run in the relational logic for event structure logics in Sect. 19.5.

25.8 Branch Model and Completeness Proof

Let L be a relational logic. In order to prove completeness of an L-dual tableau, we suppose that an L-valid formula does not have an L-proof. It follows that there exists a complete L-proof tree for this formula with an open branch, say b. We construct a branch structure $\mathcal{M}^b = (U^b, m^b)$ as follows:

- U^b is the set of object symbols of the language of L;
- $m^b(c) = c$, for $c \in \mathbb{OC}_{\mathsf{L}}$;
- $m^b(R) = \{\overline{x} \in (U^b)^n : n \text{ is the arity of } R \text{ and } R(\overline{x}) \notin b\}$, for any relational variable $R \in \mathbb{RV}_{\mathsf{L}}$;
- For any relational constant $C \in \mathbb{RC}_{\mathsf{L}}$, if C is definable in L, that is if there exists a term T_C such that C does not appear in T_C and in every L-model $m(C) = m(T_C)$, then we set $m^b(C) \stackrel{\text{df}}{=} m^b(T)$ and otherwise $m^b(C) \stackrel{\text{df}}{=} \{\overline{x} \in (U^b)^n : n \text{ is the arity of } C \text{ and } C(\overline{x}) \notin b\}$;
- For every compound relational term T, $m^b(T)$ is defined as in L-models.

Example. In Sect. 2.7 the language of logic $\mathsf{RL}(1, 1')$ includes the relational constant $1'$ which is assumed to be an equivalence relation in all $\mathsf{RL}(1, 1')$-models. In the branch structure we define:

$$m^b(1') \stackrel{\text{df}}{=} \{(x, y) \in U^b \times U^b : x1'y \notin b\}.$$

In Sect. 18.2 dual tableaux for spatial theories based on a plain contact relation are presented, including a dual tableau for the relational logic $\mathsf{RL_{Mer}}$ for reasoning about the mereological relations. The language of $\mathsf{RL_{Mer}}$ contains relational constants P, PP, O, PO, EC, TPP, $NTPP$, DC, and DR whose definitions in $\mathsf{RL_{Mer}}$-models are given. The meaning of constant P is defined as $m(P) \stackrel{\text{df}}{=} -m(C;-C)$, where C is a relational constant. Then in the branch structure $\mathcal{M}^b = (U^b, m^b)$ we postulate $m^b(P) \stackrel{\text{df}}{=} -m^b(C;-C)$.

One of the necessary prerequisites for getting completeness is that the rules of the L-dual tableau guarantee the following condition:

Closed Branch Property
For every branch of an L-proof tree, if $R(\overline{x})$ and $-R(\overline{x})$ for an atomic term R belong to the branch, then the branch can be closed, i.e., $R(\overline{x})$ and $-R(\overline{x})$ will appear in a node of the branch.

Usually, the closed branch property follows from an essential property of relational dual tableaux that any application of the rules, in particular an application of the specific rules, preserves the formulas built with atomic terms or their complements. However, in some dual tableaux there are the rules which do not satisfy such a preservation property, for example, the rule $(\text{sym})^2$ presented in Sect. 2.8 and the rules $(-R_1)$ and $(-R_2)$ presented in Sect. 9.4. In these cases there is no general method of proving the closed branch property, it must be proved locally, depending on a rule which violates the preservation property. In case of rule $(\text{sym})^2$ the closed branch property can be proved as in Proposition 2.8.1. In case of rules $(-R_1)$ and $(-R_2)$ their completion conditions are sufficient for proving the satisfaction in branch model property (see Proposition 9.4.3).

Given a branch structure $\mathcal{M}^b$, we need to show that $\mathcal{M}^b$ is an L-model. This property is referred to as *branch model property*.

Branch Model Property
Let b be an open branch of a complete L-proof tree. Then a branch structure $\mathcal{M}^b$ is an L-model.

For that, we need to show that all the constraints assumed in the L-models on relational variables and the relational constants are satisfied in the branch structure. This must be guaranteed by the rules of L-dual tableau and their corresponding completion conditions. A branch structure which is an L-model is referred to as the *branch model*.

Then, we define the valuation v^b in the model $\mathcal{M}^b$ as the identity function, i.e., $v^b(x) = x$ for every object symbol x, and we prove the so called *satisfaction in branch model property*:

Satisfaction in Branch Model Property
Let b be an open branch of a complete L-proof tree. Then for every L-formula φ, the branch model $\mathcal{M}^b$ and valuation v^b in $\mathcal{M}^b$ satisfy: if $\mathcal{M}^b, v^b \models \varphi$, then $\varphi \notin b$.

The proof of this property is by induction on the complexity of relational terms. The idea of the proof is as follows. First, we prove that the proposition holds for formulas built with atomic relational terms and for formulas built with complements of atomic relational terms.

If the interpretation of an atomic relational term R in the branch model $\mathcal{M}^b = (U^b, m^b)$ is given by $m^b(R) \stackrel{\text{df}}{=} \{\overline{x} \in (U^b)^n : R(\overline{x}) \notin b\}$, then for $\varphi = R(\overline{x})$ the proposition holds by the definition of $m^b(R)$. Assume $\mathcal{M}^b, v^b \models -R(\overline{x})$, that is $R(\overline{x}) \in b$ and, by the closed branch property, $-R(\overline{x}) \notin b$. Therefore, the proposition holds for $\varphi = -R(\overline{x})$. The proof of that kind for the relational variables in logic RL is presented in Sect. 2.5 (see Proposition 2.5.5).

If the interpretation of an atomic relational term R in the branch model $\mathcal{M}^b = (U^b, m^b)$ is defined as in L-models, then L-dual tableau contains the rules (R) and $(-R)$ that reflect the required semantic condition. Then, due to the completion conditions Cpl(R) and Cpl($-R$), the proposition holds for R and $-R$. Then, applying the induction hypothesis, we can show the proposition for any compound relational term T and its complement, by using the completion conditions associated with the rules relevant for the relational operations appearing in T. For example, consider the constant P from the language of relational logic $\mathsf{RL}_{\mathsf{Mer}}$ for reasoning about the mereological relations (see Sect. 18.2). Recall that $m^b(P) \stackrel{\text{df}}{=} -m^b(C;-C)$. The completion conditions determined by the rules for the constant P and its complement are:

Cpl(P) If $xPy \in b$, then for some object variable z both $x{-}Cz \in b$ and $zCy \in b$, obtained by an application of the rule (P);

Cpl($-P$) If $x{-}Py \in b$, then for every object symbol z, either $xCz \in b$ or $z{-}Cy \in b$, obtained by an application of the rule $(-P)$.

Assume $\mathcal{M}^b, v^b \models xPy$. Thus, $(v^b(x), v^b(y)) \in m^b(P)$. Suppose $xPy \in b$. Then, by the completion condition Cpl(P), for some $z \in U^b$, both $x{-}Cz \in b$ and $zCy \in b$. Thus, by the induction hypothesis, for some $z \in U^b$, both $(v(x), v(z)) \in m^b(C)$ and $(v(z), v(y)) \notin m^b(C)$. Hence, $(v(x), v(y)) \in m^b(C;-C)$, that is $(v(x), v(y)) \notin m^b(P)$, a contradiction.

Now, assume $\mathcal{M}^b, v^b \models x{-}Py$. Thus, $(v^b(x), v^b(y)) \notin m^b(P)$. Suppose $x{-}Py \in b$. Then, by the completion condition Cpl($-P$), for every $z \in U^b$, either $xCz \in b$ or $z{-}Cy \in b$. Thus, by the induction hypothesis, for every $z \in U^b$, either $(v(x), v(z)) \notin m^b(C)$ or $(v(z), v(y)) \in m^b(C)$. Hence, $(v(x), v(y)) \notin m^b(C;-C)$, that is $(v(x), v(y)) \in m^b(P)$, a contradiction.

Now, the branch model property and the satisfaction in branch model property enable us to prove:

Completeness
If φ is an L-valid formula, then φ is L-provable.

The idea of the proof is as follows. Let φ be an L-valid formula and suppose it is not L-provable. Then there is no any closed L-proof tree for φ. Let b be an open branch of a complete L-proof tree for φ. By satisfaction in branch model property, φ is not true in the branch model $\mathcal{M}^b$. However, since by the branch model property $\mathcal{M}^b$ is an L-model, φ is not L-valid, a contradiction.

25.9 Alternative Forms of Rules

In this section we discuss alternative forms of the rules presented in Sect. 25.6.

'More Analytic' Rules

Sometimes relational proof systems require cut-like rules in order to get completeness. Cut rules in relational proof systems have the form:

$$\frac{}{\varphi_1 \mid \ldots \mid \varphi_k}$$

where $\{\varphi_1, \ldots, \varphi_k\}$ is an unsatisfiable set of literals.

A cut rule of the form:

$$\frac{}{R(\overline{x}) \mid -R(\overline{x})}$$

where R is an atomic relational term and $\overline{x}$ is a sequence of object symbols, is referred to as a *standard cut rule*. In general, we use analytic cut rules; this means that their application is restricted to R's and x's appearing in the formulas of a set to which the rule is applied. Any such a rule is correct in relational logics. Often, restricted cut rules are used in which literals have a special form, for example they may be built with some relational constants. These rules are referred to as *specialized cut rules*. If a cut rule is present in a dual tableau, then we require that:

(Cpl cut) Every complete branch of a proof tree contains either of φ_i, $i = 1, \ldots, k$, where the variables appearing in φ_i may be instantiated to any object symbol.

Some of the rules presented in Sect. 25.6 are specialized cut rules. In some cases we may replace these rules with analytic rules or axiomatic sets together with standard cut rules for some literals without loosing completeness.

One of such specialized cut rules is the rule (rc1) discussed in Sect. 25.6:

$$(\mathrm{rc1})\quad \frac{}{A_1 \mid \ldots \mid A_s \mid \neg B_1^1, \ldots, \neg B_{k_1}^1 \mid \ldots \mid \neg B_1^i, \ldots, \neg B_{k_i}^i}$$

where the variables that appear in A or B may be instantiated to any object symbol. This rule corresponds to the condition (c1):

(c1) $\forall x(A \to B)$.

where $A = A_1 \wedge \ldots \wedge A_s$, $s \geq 1$, $A_1, \ldots, A_s$ are literals, $B = B^1 \vee \ldots \vee B^i$, $i \geq 1$, and for every $j \in \{1, \ldots, i\}$, $B^j = B_1^j \wedge \ldots \wedge B_{k_j}^j$, $k_j \geq 1$, $B_1^j, \ldots, B_{k_j}^j$ are literals, $\mathbb{OV}(A) = \mathrm{set}(\overline{x}) \neq \emptyset$, $\emptyset \neq \mathbb{OV}(B) \subseteq \mathbb{OV}(A)$.

Let B be a disjunction of literals, i.e., $B = B_1 \vee \ldots \vee B_k$, $k \geq 1$, Then, instead of the rule (rc1) we may admit two rules of the form:

$$(\mathrm{r_1c1})\quad \frac{B_1, \ldots, B_k}{A_1, B_1, \ldots, B_k \mid \ldots \mid A_s, B_1, \ldots, B_k}$$

where the variables that appear in A_i, $i \in \{1, \ldots, s\}$, and do not appear in B may be instantiated to any object symbol

$$(\mathrm{r_2c1})\quad \frac{}{B_i \mid \neg B_i} \qquad \text{for every } i \in \{1, \ldots, k\}$$

where the variables in B_i, $i \in \{1, \ldots, k\}$ may be instantiated to any object symbol.

The completion conditions corresponding to these rules are:

Cpl($\mathrm{r_1c1}$) If $B_1 \in b, \ldots, B_k \in b$, then there exists $i \in \{1, \ldots, s\}$ such that $A_i \in b$, obtained by an application of the rule ($\mathrm{r_1c1}$);
Cpl($\mathrm{r_2c2}$) For every $i \in \{1, \ldots, k\}$, either $B_i \in b$ or $\neg B_i \in b$.

If L is a logic whose models satisfy the condition (c1), then the rules obtained in this way are L-correct. Moreover, the following can be proved:

Theorem 25.9.1. *Let* L *be a relational logic. Then, the rule* $\frac{}{A_1 \mid \ldots \mid A_s \mid \neg B_1 \mid \ldots \mid \neg B_k}$ *is* L*-correct iff the rule* $\frac{B_1, \ldots, B_k}{A_1, B_1, \ldots, B_k \mid \ldots \mid A_s, B_1, \ldots, B_k}$ *is* L*-correct.*

Proof. Let L be a relational logic. By way of example, we show the implication from left to right. Assume that the rule $\frac{}{A_1 \mid \ldots \mid A_s \mid \neg B_1 \mid \ldots \mid \neg B_k}$ is L-correct, that is for every finite set X of formulas, X is an L-set iff $X \cup \{A_i\}$ and $X \cup \{\neg B_j\}$ are L-sets, for every $i \in \{1, \ldots, s\}$ and for every $j \in \{1, \ldots, k\}$. Let Y be any finite set of formulas and let $X \stackrel{\mathrm{df}}{=} Y \cup \{B_1, \ldots, B_k\}$. By the assumption, $Y \cup \{B_1, \ldots, B_k\}$ is an L-set iff $Y \cup \{A_i, B_1, \ldots, B_k\}$ and $Y \cup \{\neg B_j, B_1, \ldots, B_k\}$ are L-sets, for every $i \in \{1, \ldots, s\}$ and for every $j \in \{1, \ldots, k\}$. Observe that $Y \cup \{\neg B_j, B_1, \ldots, B_k\}$ is an L-set for every $j \in \{1, \ldots, k\}$. Thus, for every finite set Y of formulas, $Y \cup \{B_1, \ldots, B_k\}$

is an L-set iff $Y \cup \{A_i, B_1, \ldots, B_k\}$ are L-sets, for every $i \in \{1, \ldots, s\}$. Hence, the rule $\frac{B_1,\ldots,B_k}{A_1,B_1,\ldots,B_k \mid \ldots \mid A_s,B_1,\ldots,B_k}$ is L-correct. □

The specialized cut rule (r_2c1) is needed in the proof of completeness whenever for some $i \in \{1, \ldots, k\}$, B_i is a negative literal or B_i is definable in L. Thus, if a dual tableau contains the rule (r_1c1), then it must also contain the rule (r_2c1). However, if B is a disjunction of positive literals which are not definable in L, then the specialized cut rule (r_2c1) is not needed.

Examples of rules of the form (r_1c1) and their corresponding conditions of the form (c1) with a disjunction of positive literals are:

- Rules (sym P) and (tran P) for symmetry and transitivity condition, respectively, in logic $\mathsf{RL}_{\mathsf{EQ}}$ in Sect. 6.6;
- Rule (rher') for heredity condition and rule (ideal) for right ideal relations in logic $\mathsf{RL}_{\mathsf{INT}}$ in Sect. 8.2;
- Rule (rQ1) and (rQ2) for conditions (Q1) and (Q2), respectively, in logic RL_{J} in Sect. 8.3;
- Rule (rL_1) for condition (L_1) in logic $\mathsf{RL}_{\mathsf{L}_1}$ in Sect. 8.4;
- Rule (SR) for condition $S \subseteq R$ in logic $\mathsf{RL}_{\mathsf{PLL}}$ in Sect. 8.5;
- Rule (ideal) for right ideal relations and rules (rM2i), (rM2'ii), and (rM3) for conditions (M2)(i), (M2')(ii), and (M3), respectively, in logic $\mathsf{RL}_{\mathsf{RLV}}$ in Sects. 9.3 and 9.4; rules (rM10) and (rM11) for conditions (M10) and (M11), respectively, in relevant logics in Sect. 9.5;
- Rules (rI1 $\subseteq$) and (rI1 $\supseteq$) for condition (I1) and rule (rI6) for condition (I6) in logic $\mathsf{RL}_{\mathsf{NIL}}$ in Sects. 11.3 and 11.4; rules (rI7), (rI8'), (rI9'), (rI10), (rI11'), (rI12) for conditions (I7), (I8'), (I9'), (I10), (I11'), (I12), respectively, in logic $\mathsf{RL}_{\mathsf{IL}}$ in Sects. 11.3 and 11.4; rule (3-tran R) for 3-transitivity condition in logic $\mathsf{RL}_{\mathsf{CI}}$ in Sect. 11.5;
- Rules (S) and (R) for the condition $S = R^{-1}$ in logic $\mathsf{RL}_{\mathsf{FCL}}$ in Sect. 13.3;
- Rule (ideal) for right ideal relations, rule (tran $\leq$) for transitivity condition, and rule (rher') for heredity condition in logic $\mathsf{RL}_{\mathsf{MTL}}$ in Sect. 14.4; rules (rMTL1), (rMTL5), and (rMTL6) for conditions (MTL1), (MTL5), and (MTL6), respectively, in logic $\mathsf{RL}_{\mathsf{MTL}}$ in Sects. 14.3 and 14.4;
- Rules (r(i$\prec$)), (r(ii$\prec$)), (r(iii$\prec$)), and (r(iv$\prec$)) for the conditions (i$\prec$), (ii$\prec$), (iii$\prec$), and (iv$\prec$), respectively, in logic $\mathsf{RL}_{\mathsf{OMR}}$ in Sect. 15.3;
- Rules (Euc R) and (pfun R) for the condition of Euclidean and partial functionality of R, respectively, in logic $\mathsf{RL}_{\mathsf{L}_{\mathsf{TL}}}$ in Sect. 16.3;
- Rules (rReC1), ..., (rReC6), and (rReC9) for conditions (ReC1), ..., (ReC6), and (ReC9), respectively, in logic $\mathsf{F}_{\mathsf{BAC}}$ in Sect. 18.3.

Example. Consider logic L_5 which is an axiomatic extension of logic RLV discussed in Sect. 9.5. A relational dual tableau for the logic L_5 includes the rule of the form:

$$\text{(rM5)} \quad \frac{}{R_1(x, y, z, t) \mid -R_2(x, y, z, t)}$$

This rule reflects the condition (M5) which is of type (c1). According to the method described above, instead of the rule (rM5), the following rules can be admitted:

For any object terms x, y, z, and t,

$$(\mathrm{r_1M5}) \quad \frac{R_2(x,y,z,t)}{R_1(x,y,z,t), R_2(x,y,z,t)}$$

$$(\mathrm{r_2M5}) \quad \frac{}{R_2(x,y,z,t) \mid -R_2(x,y,z,t)}$$

Figure 9.4 in Sect. 9.5 depicts $\mathsf{RL}_{\mathsf{L}_5}$-proof of the relevant formula $(p \odot q \to t) \to (p \to (q \to t))$. In Fig. 25.1 we present a relational proof of this formula in a dual tableau with the rules ($\mathrm{r_1M5}$) and ($\mathrm{r_2M5}$) instead of the rule (rM5).

Now, consider a specialized cut rule (rc4):

$$(\mathrm{rc4}) \quad \frac{}{A_1 \mid \ldots \mid A_s \mid \neg B_1^1, \ldots, \neg B_{k_1}^1 \mid \ldots \mid \neg B_1^i, \ldots, \neg B_{k_i}^i}$$

where each of the variables from $\mathrm{set}(\overline{z})$ occurring in any B_m^l must be instantiated to a new variable distinct from variables in $\mathrm{set}(\overline{x})$, for $l \in \{1, \ldots, i\}$ and $m \in \{1, \ldots, k_i\}$.

This rule corresponds to the condition (c4):

(c4) $\forall \overline{x}(A \to \exists \overline{z} B)$

where A and B are as in the constraint (c1), $\overline{x}$ and $\overline{z}$ are finite sequences of object variables, not necessarily of the same length, $\mathbb{OV}(A) = \mathrm{set}(\overline{x}) \neq \emptyset$, $\mathbb{OV}(B) \neq \emptyset$, $(\mathbb{OV}(B) \setminus \mathrm{set}(\overline{z})) \subseteq \mathbb{OV}(A)$, and $\mathrm{set}(\overline{x}) \cap \mathrm{set}(\overline{z}) = \emptyset$.

The rule (rc4) may be replaced with the two rules of the form:

$$(\mathrm{r_1c4}) \quad \frac{\neg A_1, \ldots, \neg A_s}{\neg B^1, \neg A_1, \ldots, \neg A_s \mid \ldots \mid \neg B^i, \neg A_1, \ldots, \neg A_s}$$

where for every $j \in \{1, \ldots, i\}$, $\neg B^j = \neg B_1^j, \ldots, \neg B_{k_j}^j$, and the variables that appear in B^j, $j \in \{1, \ldots, i\}$, and do not appear in A must be instantiated to new pairwise distinct object variables

$$(\mathrm{r_2c4}) \quad \frac{}{A_i \mid \neg A_i} \qquad \text{for every } i \in \{1, \ldots, s\}$$

where the variables in A_i, $i \in \{1, \ldots, s\}$, may be instantiated to any object symbol.

The completion conditions corresponding to these rules are:

Cpl($\mathrm{r_1c4}$) If $\neg A_1 \in b$, ..., $\neg A_s \in b$, then there exists $j \in \{1, \ldots, i\}$ such that for every $i \in \{1, \ldots, k_j\}$, $\neg B_i^j \in b$, where the variables that appear in B_i^j and do not appear in A must be instantiated to new pairwise distinct object variables, obtained by an application of the rule ($\mathrm{r_1c4}$);

Cpl($\mathrm{r_2c4}$) For every $i \in \{1, \ldots, s\}$, either $A_i \in b$ or $\neg A_i \in b$.

If L is a logic whose models satisfy the condition (c4), then the rules obtained in this way are L-correct. Moreover, the following can be proved:

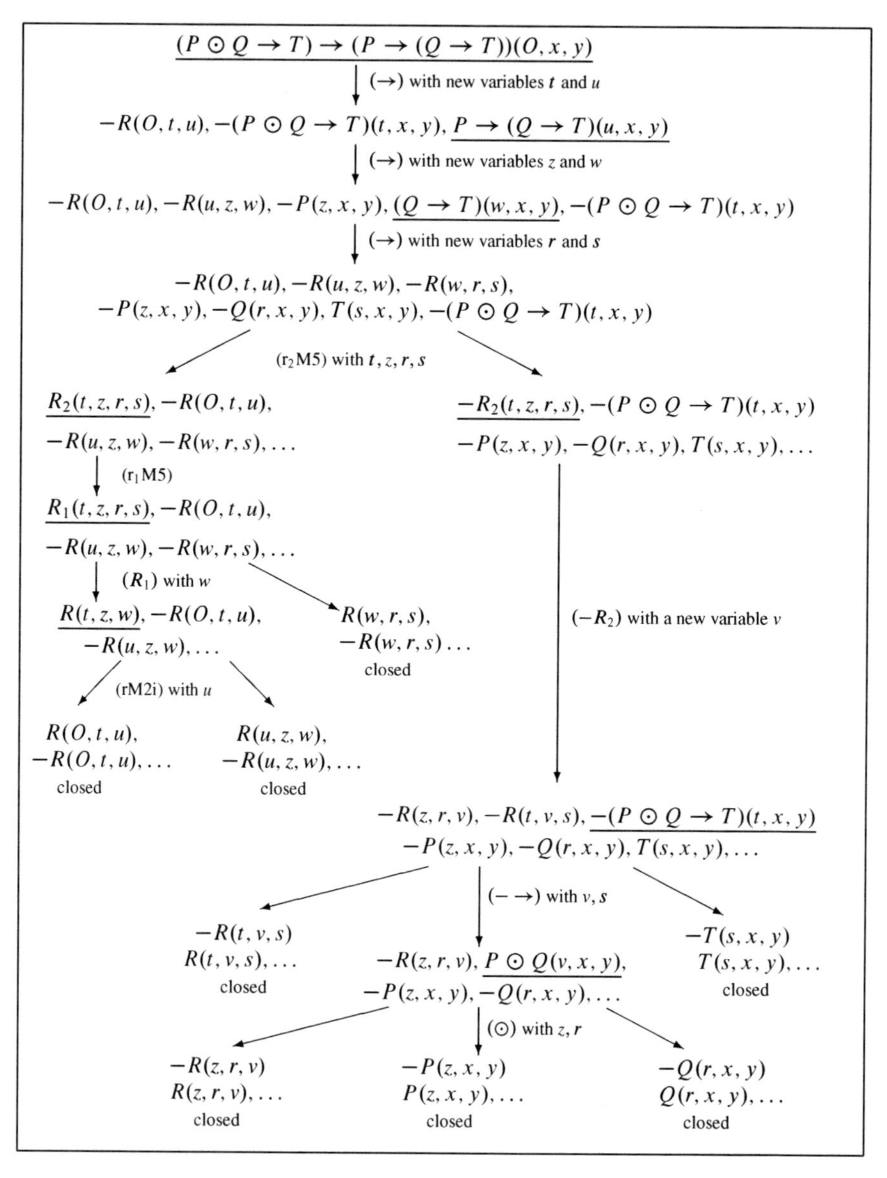

Fig. 25.1 An alternative relational proof of the formula $(p \odot q \to t) \to (p \to (q \to t))$

Theorem 25.9.2. *Let* L *be a relational logic. Then, the rule (rc4) is* L*-correct iff the rule* (r_1c4) *is* L*-correct.*

The above theorem can be proved in a similar way as Theorem 25.9.1.

The specialized cut rule (r_2c4) is needed in the proof of completeness whenever for some $i \in \{1, \ldots, s\}$, A_i is a positive literal or A_i is definable in L. Thus, if a dual tableau contains the rule (r_1c4), then it must also contain the rule (r_2c4). However, if A is a conjunction of negative literals which are not definable in L, then the specialized cut rule (r_2c4) is not needed.

Example. Consider the intermediate logic $\mathsf{INT_{L_2}}$ presented in Sect. 8.4. Models of the logic $\mathsf{INT_{L_2}}$ satisfy the following condition:

$$(\mathsf{L_2})\quad \forall x \forall y \forall z[((x, y) \in R \wedge (x, z) \in R) \rightarrow \exists t((y, t) \in R \wedge (z, t) \in R)].$$

The specific rule reflecting the condition ($\mathsf{L_2}$) has the following form:

For all object variables x, y, and z,

$$(\mathrm{r}\mathsf{L_2})\quad \frac{}{xRy \mid xRz \mid y{-}(R\,;R^{-1})z}$$

According to the method described above, in $\mathsf{RL_{INT_{L_2}}}$-dual tableau we may admit the following two rules instead of the rule ($\mathrm{r}\mathsf{L_2}$):

For all object variables x, y, and z,

$$(\mathrm{r_1}\mathsf{L_2})\quad \frac{x{-}Ry, x{-}Rz}{y{-}Rt, z{-}Rt, x{-}Ry, x{-}Rz} \quad \text{for a new object variable } t \neq x, y, z$$

$$(\mathrm{r_2}\mathsf{L_2})\quad \frac{}{xRy \mid x{-}Ry}$$

Figure 8.4 in Sect. 8.4 depicts an $\mathsf{RL_{INT_{L_2}}}$-proof of the formula $\neg p \vee \neg\neg p$. In Fig. 25.2 we present a relational proof of this formula in a dual tableau with the rules ($\mathrm{r_1}\mathsf{L_2}$) and ($\mathrm{r_2}\mathsf{L_2}$) instead of the rule ($\mathrm{r}\mathsf{L_2}$).

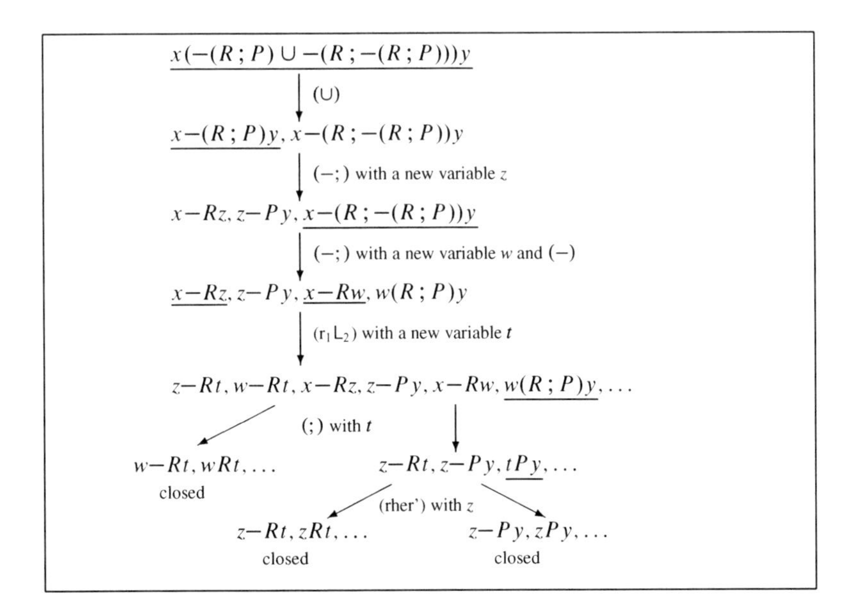

Fig. 25.2 An alternative relational proof of the formula $\neg p \vee \neg\neg p$.

Rules for Conjunctive Constraints

Now, consider a constraint which is the conjunction of some constraints discussed in Sect. 25.6. Then, in general, to get a dual tableau for a relational logic whose models satisfy the condition of that kind, we construct the rules corresponding to each constraint in a conjunction. However, in many cases we may admit one rule instead of multiple rules.

By way of example, let L be an extension of the logic $\mathsf{RL}(1,1')$ (see Sect. 2.7) such that its language is the $\mathsf{RL}(1,1')$-language endowed with the relational constant R interpreted in $\mathsf{RL}(1,1')$-models as a function. Thus, an L-model is an $\mathsf{RL}(1,1')$-model $\mathcal{M} = (U, m)$ that satisfies the following conditions:

(fun$_1$) For every $x \in U$ there exists $y \in U$ such that $(x, y) \in m(R)$;

(fun$_2$) For all x, y, and z in U, if $(x, y) \in m(R)$ and $(x, z) \in m(R)$, then $(y, z) \in m(1')$.

Note that the equivalent form of the condition (fun$_1$) is:

(fun$_1$)' For all $x \in U$, $(x, x) \in m(R\,;1)$.

The rules reflecting the conditions (fun$_1$)' and (fun$_2$) are:

(fun$_1$ R)' $\dfrac{}{x{-}(R\,;1)x}$ for every object variable x

(fun$_2$ R) $\dfrac{}{xRy \mid xRz \mid y{-}1'z}$ for all object variables x, y, and z

According to the method described in this section, the rule (fun$_2$ R) may be replaced with a more analytic rule:

(fun$_2$ R)' $\dfrac{y1'z}{xRy,\, y1'z \mid xRz,\, y1'z}$ for every object variable x

It can be proved that the dual tableau with the rules (fun$_1$ R)' and (fun$_2$ R)' is sound and complete.

The rules that reflect functionality of the relations R_- and $R_\#$, for $\# \in \{\vee, \wedge\}$, admitted in dual tableaux for spatial theories based on a contact relation on a Boolean algebra described in Sect. 18.3 are constructed in such a way.

We can admit yet another form of the rule reflecting functionality of R and preserving completeness of a dual tableau for a relational logic where R is assumed to be functional. The rules (fun$_1$ R)' and (fun$_2$ R)' may be combined to one rule of the form:

(fun R) $\dfrac{}{xRy,\, x{-}(R\,;1)x \mid xRz,\, x{-}(R\,;1)x \mid y{-}1'z,\, x{-}(R\,;1)x}$

for all object variables x, y, and z

Such a rule is admitted in dual tableaux for semantic extensions of basic temporal logic presented in Sect. 16.3.

Rules Versus Axiomatic Sets

Consider the specialized cut rule (rc2) discussed in Sect. 25.6:

$$\text{(rc2)} \quad \frac{}{\neg B_1^1, \ldots, \neg B_{k_1}^1 \mid \ldots \mid \neg B_1^i, \ldots, \neg B_{k_i}^i}$$

where B is as in the constraint (c1) and the variables that appear in B may be instantiated to any object symbol.

This rule corresponds to the condition (c2):

(c2) $\forall \overline{x} B$

Let B be a disjunction of literals, i.e., $B = B_1 \vee \ldots \vee B_k$, $k \geq 1$, Then, instead of the rule (rc1) we may admit the axiomatic set and the rule of the form:

(Ax'c2) $\{B_1, \ldots, B_k\}$

(r'c2) $\dfrac{}{B_i \mid \neg B_i}$ for every $i \in \{1, \ldots, k\}$

where the variables that appear in B_i, $i \in \{1, \ldots, k\}$, may be instantiated to any object symbol.

The completion condition corresponding to the rule (r'c2) is:

Cpl(r'c2) For every $i \in \{1, \ldots, k\}$, either $B_i \in b$ or $\neg B_i \in b$.

If L is a logic whose models satisfy the condition(c2), then the axiomatic set (Ax'c2) is an L-set and the rule (r'c2) is L-correct. Moreover, the following can be proved:

Theorem 25.9.3. *Let* L *be a relational logic. Then, the rule* $\frac{}{\neg B_1 \mid \ldots \mid \neg B_k}$ *is* L*-correct iff every superset of* $\{B_1, \ldots, B_k\}$ *is an* L*-set.*

If a dual tableau contains an axiomatic set of the form (Ax'c2) and for some $i \in \{1, \ldots, k\}$, B_i is a negative literal or is definable in L, then the dual tableau must also contain the specialized cut rule (r'c2) which is needed in the proof of completeness.

If B is a disjunction of positive literals that are nor definable in L, and in addition, all the rules preserve literals B_i, $i \in \{1, \ldots, k\}$, then the relational counterpart of (c2) is provided by an axiomatic set and the specialized cut rule (r'c2) is not needed in the proof of completeness.

Examples of axiomatic sets corresponding to the conditions of the form (c2) with a disjunction of positive literals are:

- $\{x1'x\}$ for reflexivity of $1'$ in logic $\mathsf{RL}(1')$ in Sect. 2.7;
- $\{R(O, x, x)\}$, $\{R(O, x^{**}, x)\}$, $\{R(O, x, x^{**})\}$ for conditions (M1) and (M4) in logic $\mathsf{RL}_{\mathsf{RLV}}$ in Sect. 9.4;

- $\{xRy, xSy\}$ for condition $R \cup S = U$ in logic $\mathsf{RL_{CI}}$ in Sect. 11.5;
- $\{x \leq x\}$ for reflexivity condition in logic $\mathsf{RL_{MTL}}$ in Sect. 14.4;
- $\{x < y, y < x, x1'y\}$ for linearity of $<$ in logic $\mathsf{RL_{OMR}}$ in Sect. 15.3;
- $\{R_{\vee}(x, 0, x)\}$ and $\{R_{\wedge}(x, 1, x)\}$ for conditions (ReBA8) and (ReBA9), respectively, in logic $\mathsf{F_{BAC}}$ in Sect. 18.3.

Example. Let L be a relational logic, let R be a relational constant, and let an L-model $\mathcal{M} = (U, m)$ be such that $m(R)$ is irreflexive, i.e., $(x, x) \notin m^b(R)$, for every $x \in U$; this condition is an instance of (c2).

According to the method described in Sect. 25.6, the dual tableau for L includes the rule of the form:

$$(\text{irref } R) \quad \frac{}{xRx} \quad \text{for every object symbol } x$$

We can replace the rule (irref R) with an axiomatic set (Ax'c2) of the form $\{x{-}Rx\}$ and the rule (r'c2) of the form:

$$\frac{}{xRx \mid x{-}Rx}$$

To show the branch model property, we have to prove that in the branch model $\mathcal{M}^b = (U^b, m^b)$, $m^b(R)$ is irreflexive. For suppose otherwise, then there is $x \in U^b$ such that $(x, x) \in m^b(R)$. Thus, by the definition of $m^b(R)$, $xRx \notin b$. Now, by the completion condition Cpl(r'c2), $x{-}Rx \in b$. Since every superset of $\{x{-}Rx\}$ is an axiomatic set, b is closed, a contradiction. Note, that the completion condition Cpl(r'c2) is crucial in this proof.

Rules for Relational Equations

Now, we discuss alternative forms of the rules that correspond to equational constraints or to definitions of relational constants.

Consider the logic $\mathsf{RL_{FCL}}$ presented in Sect. 13.3. Its language contains two relational constants R and S whose interpretations in $\mathsf{RL_{FCL}}$-models $\mathcal{M} = (U, R, S, m)$ satisfy the equation $R = S^{-1}$. The dual tableau for $\mathsf{RL_{FCL}}$ presented in Sect. 13.3 contains two rules (S) and (R):

$$(S) \ \frac{xSy}{yRx, xSy} \qquad (R) \ \frac{xRy}{ySx, xRy}$$

They reflect the conditions $R \subseteq S^{-1}$ and $S^{-1} \subseteq R$, respectively, that is for every class $\mathcal{K}$ of $\mathsf{RL_{FCL}}$-structures $\mathcal{M} = (U, R, S, m)$ the following hold (see Theorem 13.3.2):

- The rule (S) is $\mathcal{K}$-correct iff $R \subseteq S^{-1}$ holds in every structure of $\mathcal{K}$;
- The rule (R) is $\mathcal{K}$-correct iff $S^{-1} \subseteq R$ holds in every structure of $\mathcal{K}$.

Note that in both of the rules, the formula from the premise of the rule is repeated in the conclusion. However, we may eliminate the repetition in the rules without loosing completeness. Namely, instead of the rules (S) and (R), we admit the rules of the following forms:

$$(S)' \ \frac{xSy}{yRx} \qquad (R)' \ \frac{xRy}{ySx}$$

The completion conditions corresponding to the rules $(S)'$ and $(R)'$ are exactly the same as those corresponding to the rules (S) and (R), respectively:

Cpl$(S)'$ If $xSy \in b$, then $yRx \in b$;
Cpl$(R)'$ If $xRy \in b$, then $ySx \in b$.

It follows that the completeness proof for the dual tableau with the rules $(S)'$ and $(R)'$ is the same as for the dual tableau with the rules (S) and (R). The difference is in the correspondence theorem. The rules $(S)'$ and $(R)'$ are correct with respect to the class of structures satisfying the equation $R = S^{-1}$, not only one of the inclusions. Namely, for every class $\mathcal{K}$ of $\mathsf{RL}_{\mathsf{FCL}}$-structures $\mathcal{M} = (U, R, S, m)$ the following hold:

- The rule $(S)'$ is $\mathcal{K}$-correct iff $R = S^{-1}$ holds in every structure of $\mathcal{K}$;
- The rule $(R)'$ is $\mathcal{K}$-correct iff $R = S^{-1}$ holds in every structure of $\mathcal{K}$.

Each of the rules $(S)'$ and $(R)'$ reflects the equation $R = S^{-1}$, thus in order to have a rule directly corresponding to the equation, it suffices to have one of these rules. However, to get completeness both of these rules must be present, since both completion conditions are needed to prove the branch model property.

In a similar way we can construct alternative forms of the rules for the relations F and P of temporal logic TL in Sect. 16.2 and for the relational constants defined in terms of Boolean operations. In particular, we may delete repetitions of the formula from the premise in the conclusions of the rules corresponding to mereological relations *PP*, *PO*, *EC*, *NTPP*, *DC*, and *DR* (see Table 18.1) presented in Sect. 18.2.

Positive Versus Negative Forms of the Rules

In Sects. 1.8 and 2.7 we discussed negative forms of the rules for the identity in first-order logic and in the relational logic $\mathsf{RL}(1, 1')$, respectively. Recall that the standard rules for the identity in $\mathsf{RL}(1, 1')$-dual tableau have the following forms:

$$(1'1) \quad \frac{xRy}{xRz, xRy \mid y1'z, xRy} \quad z \text{ is any object symbol}$$

$$(1'2) \quad \frac{xRy}{x1'z, xRy \mid zRy, xRy} \quad z \text{ is any object symbol}$$

The negative rules for identity are:

$$(1'1)^2 \quad \frac{x{-}1'y,\, y{-}Rz}{x{-}Rz,\, x{-}1'y,\, y{-}Rz} \qquad (1'2)^2 \quad \frac{x{-}1'y,\, z{-}Rx}{z{-}Ry,\, x{-}1'y,\, z{-}Rx}$$

$$(\mathrm{sym})^2 \quad \frac{x{-}1'y}{y{-}1'x} \qquad (\mathrm{ref})^2 \quad \frac{}{x{-}1'x}$$

where R is any relational variable or relational constant, and x, y, z are any object symbols.

The above negative specific rules are dual to the standard rules. As shown in Sect. 2.7, the dual tableau with negative rules is complete. Observe that contrary to the standard rules, negative rules do not branch a proof tree and do not involve introduction of a variable which makes them more suitable for implementation.

In a similar way, we can modify specific rules for other relational constants. Let L be an extension of the logic RL (see Sect. 2.5) such that its language is the RL-language endowed with the relational constant R interpreted in RL-models as a transitive relation. Recall that an L-structure is a system $\mathcal{M} = (U, R, m)$ such that (U, m) is an RL-model and $R \subseteq U \times U$, that is R is not necessarily a transitive relation.

The standard rule that reflects transitivity of R is:

$$(\mathrm{tran}\ R) \quad \frac{xRy}{xRz,\, xRy \mid zRy,\, xRy} \quad z \text{ is any object symbol}$$

The negative form of the rule is:

$$(\mathrm{tran}\ R)' \quad \frac{x{-}Ry,\, y{-}Rz}{x{-}Rz,\, x{-}Ry,\, y{-}Rz}$$

The completion condition corresponding to the rule (tran R)' is:

Cpl(tran R)' If $x{-}Ry \in b$ and $y{-}Rz \in b$, then $x{-}Rz \in b$.

The following can be proved:

Proposition 25.9.1. *For every class $\mathcal{K}$ of* L*-structures the following conditions are equivalent:*

1. *The rule (tran R) is $\mathcal{K}$-correct;*
2. *The rule (tran R)' is $\mathcal{K}$-correct.*

Proof. First, we show that the rule (tran R)' is $\mathcal{K}$-correct iff R is transitive in every $\mathcal{K}$-structure. Assume the rule (tran R)' is $\mathcal{K}$-correct. Let $X \stackrel{\mathrm{df}}{=} \{xRz\}$. Then, $X \cup \{x{-}Rz, x{-}Ry, y{-}Rz\}$ is a $\mathcal{K}$-set. Thus, by the assumption, $X \cup \{x{-}Ry, y{-}Rz\}$ is also a $\mathcal{K}$-set. So, for every $\mathcal{K}$-structure $\mathcal{M}$ and for every valuation v in $\mathcal{M}$, either $\mathcal{M}, v \models xRz$ or $\mathcal{M}, v \models x{-}Ry$ or $\mathcal{M}, v \models y{-}Rz$, which means that R is transitive.

Now, assume that R is transitive in every $\mathcal{K}$-structure. Let X be any finite set of L-formulas. If $X \cup \{x{-}Ry, y{-}Rz\}$ is a $\mathcal{K}$-set, then so is $X \cup \{x{-}Rz, x{-}Ry, y{-}Rz\}$. Let $X \cup \{x{-}Rz, x{-}Ry, y{-}Rz\}$ be a $\mathcal{K}$-set and suppose that $X \cup \{x{-}Ry, y{-}Rz\}$ is not a $\mathcal{K}$-set. Then, there exist a $\mathcal{K}$-structure $\mathcal{M}$ and a valuation v in $\mathcal{M}$ such that $(v(x), v(y)) \in R$ and $(v(y), v(z)) \in R$ but $(v(x), v(z)) \notin R$, which contradicts the assumption. Due to Proposition 6.6.1, we can prove that the rule (tran R) is $\mathcal{K}$-correct iff R is transitive in every $\mathcal{K}$-structure, which completes the proof. □

In order to prove completeness of a dual tableau with rule (tran R)', we must change the definition of the branch structure presented in Sect. 25.8. Given an open branch b of an L-proof tree, we define a branch structure $\mathcal{M}^b = (U^b, m^b)$ as in the previous section except for the clause for the meaning of R:

$$m^b(R) = \{(x, y) \in U^b \times U^b : x{-}Ry \in b\}.$$

To prove completeness it suffices to show that the branch model property and the satisfaction in branch model property hold. Let us prove that the branch structure defined above is an L-model, that is $m^b(R)$ is transitive. Let $x, y, z \in U^b$ and suppose that $(x, y) \in m^b(R)$ and $(y, z) \in m^b(R)$, but $(x, z) \notin m^b(R)$. Then, by the definition of $m^b(R)$, $x{-}Ry \in b$, $y{-}Rz \in b$, and $x{-}Rz \notin b$. By the completion condition Cpl(tran R)', $x{-}Rz \in b$, a contradiction. Hence, the branch model property holds. The satisfaction in branch model property can be proved in a similar way as in RL-dual tableau.

The method described above can be applied to most of the rules presented in the book, with the restriction that if we intend to construct a rule in a negative form reflecting some property of a relational constant, then all the specific rules reflecting the properties of that constant must be in the negative forms. For otherwise, the meaning of this constant in the branch structure would not be appropriately defined.

25.10 Implementations

A Tool for Translation of a Theory into Relational Formalism

A Prolog-based implementation of a tool, named *transIt*, which uniformly carries out translations from various modal logics to the relational formalism is described in [FOO06]. This tool offers a high degree of uniformity: *transIt* is able to treat many modal logics, all by the very same machinery.

Source languages accepted by the translator are the languages of the logics which employ binary accessibility relations in their Kripke-style models. The translator does not deal with the languages of relevant logics or the logics with binary modalities – requiring ternary relations in their models. The main target language which the translator supports is the algebra of binary relations. Given a formula φ of a source language, the system produces a relational term $\tau(\varphi)$ belonging to an algebraic language. The translation τ preserves validity. The translator takes a formula of a specific source language as an input. The first transformation yields an internal representation of the formula. Then, a sequence of rewritings and simplifications is performed. Finally, the last step generates the final rendering of the translation. More specifically, below we list the salient phases which usually lead to the translation, although some of them may be skipped in specific cases:

Lexical and syntactical analysis
This phase accepts a formula only if it is syntactically correct and its constructs belong to the specific language in question. The syntax-directed translation implemented through this stage is described by an attributed definite clause grammar. Hence, any extension to other logics can be achieved by simply adding a suitable set of grammar rules which characterize the new well-formed formulas. The outcome of this stage is an intermediate representation of the abstract syntax tree, AST, of the input formula.

Generation of an internal representation
By means of rewriting process which acts in a bottom-up recursive fashion, the outcome of the preceding phase is turned into an internal representation of the AST in the form of a Prolog term, independent of the source language.

Abstract propositional evaluation
The internal representation of the given formula is analyzed in order to extract its propositional schema. If possible, this schema is then simplified through replacements of some of its subformulas by equivalent ones.

Reduction to primitive constructs
In this phase the formula is rewritten in terms of a small repertoire of propositional operations, to be regarded as being 'primitive'. For instance, biimplication $\leftrightarrow$ is rewritten as a conjunction of two implications. The aim of this transformation is to make the next phase easier.

Propositional simplifications
Through this phase the internal representation of the formula is simplified by applying a number of propositional simplifications to it, mainly aimed at reducing the size of the formula by elimination of tautological subformulas and applications of the double negation law.

Relational translation
This is the main step of the translation process: the internal representation of the given formula is translated into a term of the logic $\mathsf{RL}(1, 1')$. The rewriting rules employed depend on the source language of the input formula. The outcome of this phase is a relational term.

Relational simplifications
The overall translation process ends with a series of relational simplifications applied to the relational term produced in the preceding step. The simplest among

these rewritings take care of the idempotence, absorption or involution properties of the relational constructs. The process can easily be extended to perform more complex simplifications.

Implementations of Dual Tableaux for Relational Logics

In [FNA06] a Prolog implementation of the dual tableau for the relational logic $\mathsf{RL}(1, 1')$ (see Chap. 2) is described. Actually, it is a part of a prototypical tool supporting assisted and automated relational reasoning that can be used to verify validity of modal formulas as well. The proof of a modal formula in this implementation follows the following phases:

- The given formula is translated into a relational term;
- The relational formula obtained in this way is processed to generate a relational decision graph;
- The relational decision graph is normalized using a term rewriting system in order to impose an order in the nodes of the graph;
- The normalized relational decision graph is compiled into a Prolog program;
- The execution of such a Prolog program performs a search for the proof using a bounded depth-first iterative deepening strategy.

The architecture of the whole system is presented in Fig. 25.3. In particular, it is composed of: a user-friendly mouse-oriented interface; a translator producing an optimized relational rendering of the given formula; the dual tableau for relational logic $\mathsf{RL}(1, 1')$. At each stage of the development of a proof, the user can choose between exploiting the assistance of the tool and developing a proof by her own, or leaving the system to proceed in an autonomous way.

The adoption of an approach based on declarative programming allows developing the system in an incremental way and ensures high modularity and extensibility of the application.

Another SWI Prolog implementation of relational dual tableaux, ReVAT, is described in [LLMS02]. In the report [OS04] a Haskell implementation for relational logics is presented based on the multilevel relational reference language as proposed in [Sch03].

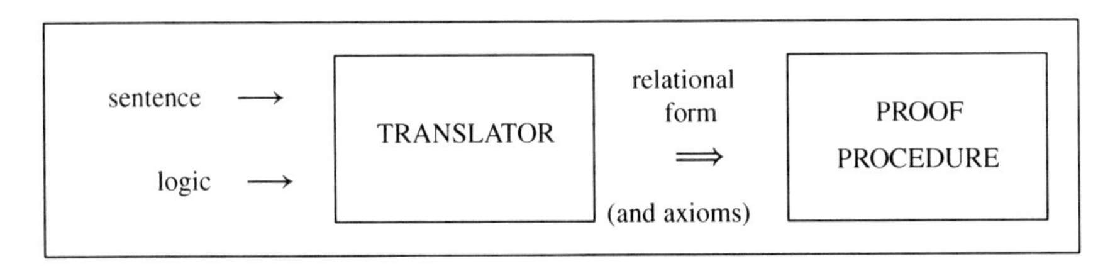

Fig. 25.3 Basic architecture of the implementation of dual tableaux for relational logics

An Automated Theorem Prover of Dual Tableaux for Logics for Order of Magnitude Reasoning

In [BMOO09] the authors describe a Prolog implementation of a dual tableau for the relational logic $\mathsf{RL_{OMR}}$ presented in Sect. 15.3. The main steps of the implementation are as follows.

First, the relations are encoded as predicates. An OMR-formula is represented using the Prolog fact: *formula*OMR(*formula*). The argument is an OMR-formula. A relational formula xTy in logic $\mathsf{RL_{OMR}}$, where x, y are object variables and T is a relational term, is represented as the Prolog fact $rel(address, T, x, y)$. The first argument contains a list of integers which defines the position of the node in the proof tree generated during the proof process. The Prolog predicate `omToreom` reads OMR-formulas and renders the set of $\mathsf{RL_{OMR}}$-formulas.

Once the system receives as an input a relational formula to be checked, it generates a proof tree, whose leaves contain sets of relational formulas whose disjunctions are to be proved. The input formula gets proved when Prolog closes all the leaves in the proof tree.

The addresses of the open leaves are stored in a list, which is handled by the predicate `open_leaves`. The predicate `open_leaves`$([n])$ states that it is necessary to prove validity of the set of formulas stored in node $[n]$.

When Prolog detects a relation representing an axiomatic set, the corresponding leaf is deleted and the user is informed by means of the `remove_leaf` predicate. For example, if $x1'x$ (i.e., the expression *rel*(`Leaf`, *equal*, `X`, `X`)) occurs in the set of relations of the leaf `Leaf`, it is deleted because of the occurrence of an axiomatic set.

The implementation of a rule can be roughly stated as follows: first, the premise of the rule is checked, in order to verify if the rule is applicable; if so, and provided that the rule has not been previously applied with the same arguments, the rule is displayed on the screen and stored as used; finally, the leaf is branched and new labels are attached to each new leaf. Rules are encoded in Prolog and implemented using some Prolog codes.

As an example, consider the rule (r(ii$\prec$)) from Sect. 15.3:

$$(\mathrm{r(ii}\prec)) \qquad \frac{x \prec z}{x \prec y, x \prec z \mid y < z, x \prec z} \qquad \text{for any object variable } y$$

The code for the application of the rule (r(ii$\prec$)) to a set of $\mathsf{RL_{OMR}}$-formulas works as follows. It checks whether $x \prec z$ is in the given set of formulas, the relations introduced by the rule are new, and the rule has not been previously applied. If all these three conditions are satisfied, then the rule is to be applied. The variable y in the conclusions has to be any of the variables or object constants occurring in the branch. For that the predicate `any_variable` chooses a constant or a variable occurring in the branch. Then the predicate `branch(Leaf, 2)` branches the current leaf into two new leaves, and copies all the formulas of the current leaf to the two new leaves. Finally, the predicate `update_leaf` appends $x \prec y$ to the first leaf and $y < z$ to the second leaf.

The main predicate in the implementation of the proof procedure is `run_engine`. It examines the first leaf of the tree and tries to apply the rules to the formulas contained in this leaf. Leaves that have not been closed are stored by the predicate `open_leaves`. If it is a non-empty list, then the corresponding nodes must be visited and the predicate `apply_rules` is called recursively as long as the tree has open leaves.

The rules are organized into several categories which are ordered as follows: first, the rules that do not branch a leaf, then the rules that branch a leaf into several leaves and, finally, explosive rules that introduce variables. Whenever a non-closed leaf does not admit any of the rules in the list, then the system asks the user about considering some cut-like rule. The unrestricted use of cut-like rules might generate excessively big trees, thus its use is strictly controlled.

After an application of the procedure, and provided that a closed tree has been obtained, the system provides a list of the rules used in the proof.

There are several rules of dual tableau for logic $\mathsf{RL}_{\mathsf{OMR}}$ which exhibit the same behavior that rule (r(ii)$\prec$) regarding an introduction of variables. In particular, the rule (r(ii$\prec$)) produces two new leaves with formulas containing 'any variable' y. In principle, there are as many instantiations of the rule as the values that can be chosen for y. Thus, if the application of the rule is without any restriction, the proof tree might grow in an uncontrolled manner. Instead of the naive approach that allows the user to introduce a particular variable when a rule is being applied, in the implementation of $\mathsf{RL}_{\mathsf{OMR}}$-dual tableau a much smarter solution is used. A non-instantiated variable, so-called *phantom variable*, is introduced and the actual instantiation of a variable is delayed until we get a guarantee that, by a unification process, an axiomatic set will be generated. Thus, a phantom variable is a special variable whose possible instantiations are constrained to belong to the set of object variables or object constants occurring in the leaf. In this way, the growth of the tree is controlled.

The implementation introduces also a predicate which allows to switch between a fully automated handling of phantom variables and an interactive mode in which the procedure stops in order to get feedback from the user as to an appropriate choice of an object symbol to be used in the instantiation. When an application of a rule requires an introduction of a variable already occurring on the current branch, and the program is in the interactive mode, the system provides a list of formulas occurring in that branch and the user can choose an adequate variable for the instantiation. The user may refuse to provide feedback and let the system to introduce a phantom variable. For the rules with a behavior similar to that of rule (r(ii$\prec$)), the implementation should choose a variable for the instantiation. If eventually the formula could not be proved, the system would have to return to the previous leaf and choose another variable. The use of phantom variables is crucial for an efficient performance of the implementation.

Implementations of various manipulations with relations can be found in [BS94, HBS94, Sin00, FOP02, CFOZ03, BN05, FOO05, GPBV08].

25.11 Towards Decision Procedures

Apart from the decidable subclasses of formulas of logic RL presented in Sect. 2.12, some decidable classes of formulas of logic $\mathsf{RL}(1, 1')$ can be obtained in view of the following general principle.

Let L_1 be a logic having a decidable validity problem and let L_2 be an undecidable logic. Assume that there is a validity preserving translation τ from formulas of L_1 into formulas of L_2, i.e., for every L_1-formula φ, φ is L_1-valid if and only if $\tau(\varphi)$ is L_2-valid. Consider a subclass of the L_2-formulas defined as $\tau(\mathsf{L}_1) \stackrel{\text{df}}{=} \{\tau(\varphi) : \varphi \text{ is } \mathsf{L}_1\text{-formula}\}$. Then, $\tau(\mathsf{L}_1)$ has a decidable validity problem provided that there is an effective procedure such that for every L_2-formula ψ, it decides whether $\psi \in \tau(\mathsf{L}_1)$ or $\psi \notin \tau(\mathsf{L}_1)$.

Let L_1 be any of the logics considered in Chaps. 7, 8, 11–13, 15, 16, and 19. Most of these logics are known to be decidable. Let L_2 be the corresponding relational logic $\mathsf{RL}_{\mathsf{L}_1}$. The set $\tau(\mathsf{L}_1)$ of those $\mathsf{RL}_{\mathsf{L}_1}$-formulas that are the images under the corresponding translation τ of L_1-formulas into $\mathsf{RL}_{\mathsf{L}_1}$-formulas consists of formulas xTy, where T is a member of the set of $\tau(\mathsf{L}_1)$-terms defined as:

- $P;1$ is a $\tau(\mathsf{L}_1)$-term for every relational variable P of the $\mathsf{RL}_{\mathsf{L}_1}$-language;
- If T is a $\tau(\mathsf{L}_1)$-term, then so is $-T$;
- If T_1 and T_2 are $\tau(\mathsf{L}_1)$-terms, then so are $T_1 \cup T_2$ and $T_1 \cap T_2$;
- If T is a $\tau(\mathsf{L}_1)$-term, then for every relational constant R representing an accessibility relation from the L_1-models, $R;T$ is a $\tau(\mathsf{L}_1)$-term.

Hence, if L_1 has a decidable validity problem, then since the membership problem for $\tau(\mathsf{L}_1)$ is decidable, the set $\tau(\mathsf{L}_1)$ has a decidable validity problem.

This opens the way for construction of uniform relational decision procedures for a large class of non-classical logics.

25.12 Conclusion

The relational logics provide a general framework for specification and reasoning in a number of theories. Often it is more sensible to implement one generic logic within which many other logics can be expressed than to continually reimplement logics from scratch. The relational logics can play the role of such a generic logic, which is one of the main motivation for following the relational approach. The techniques of relational representation have been applied successfully to a wide variety of theories with a wide application domain. Relational logics may be of interest both to theoreticians, who seek to find underlying connections among various theories in order to establish model theoretic or proof theoretic results and also to practitioners, who wish to use a specific logic to determine validity, satisfiability or entailment, in such diverse areas as spatial reasoning, model checking and verification, reasoning under uncertainty, and databases. We may summarize the paradigm in which we have been working as follows. Relational logic:

- Is a general framework for developing formal methods of reasoning;
- Can be seen as an application of algebras of relations to the formalization of a variety of theories;
- Comes equipped with a dual tableau-style method of deduction;
- Has the advantages of uniformity, modularity, and naturality;
- Is appropriate for the development of a general purpose theorem prover.

Given a logic L of relations, we may extend L to a logic L' obtained from L by:

(1) Extending the language of L by adding some relational operations and/or relational constants and/or object constants;
(2) Adding some constraints on the relations and/or object constants in the models of L.

Both kinds of extensions will necessitate a change in the set of models. Also in both cases we must augment the deduction system of L in order to get the system for L'. In case (1) we add decomposition rules corresponding to the new operations and either specific rules or axiomatic sets characterizing the new constants. In case (2) we add specific rules and/or axiomatic sets characterizing the additional constraints.

To ensure modularity, we require that the following conditions must be satisfied by the extended dual tableau:

- All the added rules are correct in L' and the axiomatic sets are valid in L';
- A rule added must provide a sufficient condition to prove that its corresponding constraint holds in the branch model constructed in the completeness proof;
- The extension is conservative, that is, if φ is an L-formula provable in L'-dual tableau, then it is provable in L-dual tableau;
- The set of formulas provable in L' is a superset of the set of formulas provable in L.

The case studies presented in the book show that dual tableaux are semantics-based systems. Therefore, the developments of discrete duality, that is a duality between classes of algebras and classes of relational systems which is established without using a topology, may be helpful in constructing dual tableaux. Given a theory presented as a class of algebras, having a discrete duality for that class we may present the theory as a logic with the Kripke-style semantics determined by the class of relational systems which are dual to those algebras. Then, the methods of the correspondence theory discussed in Sect. 25.6 will enable us to construct the dual tableau rules reflecting that semantics. The principles of discrete duality are presented in [OR07]. Some recent developments of discrete duality can be found in [DO08, OR08, OR09a, OR09b, OR10, DOR10], among others.

References

[AB75] A. R. Anderson and N. D. Belnap. *Entailment: The Logic of Relevance and Necessity. Volume I*. Princeton University Press, Princeton, 1975.

[ABD92] A. R. Anderson, N. D. Belnap, and J. M. Dunn. *Entailment: The Logic of Relevance and Necessity. Volume II*. Princeton University Press, Princeton, 1992.

[Ack54] W. Ackermann. *Solvable Cases of the Decision Problem*. North-Holland, Amsterdam, 1954.

[AHK02] R. Alur, T. A. Henzinger, and O. Kupferman. Alternating-time temporal logic. *Journal of the ACM*, 49(5):672–713, 2002.

[AK01] A. Avron and B. Konikowska. Decomposition proof systems for Gödel-Dummett logics. *Studia Logica*, 69(2):197–219, 2001.

[Ake78] S. B. Akers. Binary decision diagrams. *IEEE Transactions on Computers*, 27(6):509–516, 1978.

[All83] J. F. Allen. Maintaining knowledge about temporal intervals. *Communications of the ACM*, 26(11):832–843, 1983.

[AMN91] H. Andréka, J. D. Monk, and I. Németi, editors, *Algebraic Logic, Proc. Conf. Budapest (1988)*, volume 54 of *Colloq. Math. Soc. J. Bolyai*. North-Holland, Amsterdam, 1991.

[Bal02] Ph. Balbiani. Emptiness relations in property systems. In de Swart [dS02a], pages 15–34.

[BBS94] C. Brink, K. Britz, and R. A. Schmidt. Peirce algebras. *Formal Aspects of Computing*, 6(3):339–358, 1994.

[BD03] D. S. Bridges and L. Dediu. Apartness spaces as a framework for constructive topology. *Annals of Pure and Applied Logic*, 119(1-3): 61–83, 2003.

[BdRV01] P. Blackburn, M. de Rijke, and Y. Venema. *Modal Logic*. Cambridge University Press, Cambridge, 2001.

[Bec05] B. Beckert, editor, *Automated Reasoning with Analytic Tableaux and Related Methods, International Conference, TABLEAUX 2005, Koblenz, Germany, September 14–17, 2005, Proceedings*, volume 3702 of *Lecture Notes in Computer Science*, Springer, Heidelberg, 2005.

E. Orłowska and J. Golińska-Pilarek, *Dual Tableaux: Foundations, Methodology, Case Studies*, Trends in Logic 36, DOI 10.1007/978-94-007-0005-5,

[Ber91] G. Bergman. Actions of Boolean rings on sets. *Algebra Universalis*, 28:153–187, 1991.

[Bet59] E. W. Beth. *The Foundations of Mathematics. A Study in the Philosophy of Sciences*. Studies in Logic. North-Holland, Amsterdam, 1959.

[BFHL96] G. Baum, M. F. Frias, A. M. Haeberer, and P. E. Martínez López. From specifications to programs: A fork-algebraic approach to bridge the gap. In W. Penczek and A. Szałas, editors, *Mathematical Foundations of Computer Science 1996, 21st International Symposium, MFCS'96, Cracow, Poland, September 2–6, 1996, Proceedings*, volume 1113 of *Lecture Notes in Computer Science*, pages 180–191, Springer, Heidelberg, 1996.

[BG85] J. P. Burgess and Y. Gurevich. The decision problem for linear temporal logic. *Notre Dame Journal of Formal Logic*, 26:115–128, 1985.

[BGPO06] D. Bresolin, J. Golińska-Pilarek, and E. Orłowska. Relational dual tableaux for interval temporal logics. *Journal of Applied Non-Classical Logics*, 16(3-4):251–277, 2006.

[BK99] M. Białasik and B. Konikowska. A logic for non-deterministic specifications. In Orłowska [Orł99], pages 286–311.

[Bla90] P. Blackburn. *Nominal Tense Logic and Other Sorted Intensional Frameworks*. PhD thesis, Department of Computer Science, University of Edinburgh, 1990.

[BM05] D. Bresolin and A. Montanari. A tableau-based decision procedure for right propositional neighborhood logic. In Beckert [Bec05], pages 63–77.

[BMOO09] A. Burrieza, A. Mora, M. Ojeda-Aciego, and E. Orłowska. An implementation of a dual tableaux system for order-of-magnitude qualitative reasoning. *International Journal of Computer Mathematics*, 86 (10–11):1852–1866, 2009.

[BMS04] R. Berghammer, B. Möller, and G. Struth, editors, *Relational and Kleene-Algebraic Methods in Computer Science: 7th International Seminar on Relational Methods in Computer Science and 2nd International Workshop on Applications of Kleene Algebra, Bad Malente, Germany, May 12–17, 2003, Revised Selected Papers*, volume 3051 of *Lecture Notes in Computer Science*, Springer, Heidelberg, 2004.

[BMS07] D. Bresolin, A. Montanari, and G. Sciavicco. An optimal decision procedure for right propositional neighborhood logic. *Journal of Automated Reasoning*, 38(1–3):173–199, 2007.

[BMVOA06] A. Burrieza, E. Muñoz-Velasco, and M. Ojeda-Aciego. Order of magnitude qualitative reasoning with bidirectional negligibility. In R. Marín, E. Onaindia, A. Bugarín, and J. Santos, editors, *Current Topics in Artificial Intelligence, 11th Conference of the Spanish Association for Artificial Intelligence, CAEPIA 2005, Santiago de Compostela, Spain, November 16–18, 2005, Revised Selected Papers*, volume 4177 of *Lecture Notes in Computer Science*, pages 370–378, Springer, Heidelberg, 2006.

[BN05] R. Berghammer and F. Neumann. RELVIEW – an OBDD-based computer algebra system for relations. In V. G. Ganzha, E. W. Mayr, and E. V. Vorozhtsov, editors, *CASC 2005*, volume 3718 of *Lecture Notes in Computer Science*, pages 40–51, Springer, Heidelberg, 2005.

[BO97] W. Buszkowski and E. Orłowska. Indiscernibility-based formalization of dependencies in information systems. In Orłowska [Orł97a], pages 293–315.

[BOA05] A. Burrieza and M. Ojeda-Aciego. A multimodal logic approach to order of magnitude qualitative reasoning with comparability and negligibility relations. *Fundamenta Informaticae*, 68(1–2):21–46, 2005.

[Boo47] G. Boole. *The Mathematical Analysis of Logic. Being an Essay Towards a Calculus of Deductive Reasoning*. MacMillan, Barclay and London, 1847.

[Boo79] G. Boolos. *The Unprovability of Consistency*. Cambridge University Press, Cambridge, 1979.

[Boo93] G. Boolos. *The Logic of Provability*. Cambridge University Press, Cambridge, 1993.

[BOO04] A. Burrieza, M. Ojeda-Aciego, and E. Orłowska. Relational approach to order of magnitude reasoning. *Lecture Notes in Artificial Intelligence*, 3040:431–440, 2004.

[Böt92a] M. Böttner. State transition semantics. *Theoretical Linguistics*, 18:239–286, 1992.

[Böt92b] M. Böttner. Variable-free semantics for anaphora. *Journal of Philosophical Logic*, 21:375–390, 1992.

[BQA03] V. Beiu, J. M. Quintana, and M. J. Avedillo. VLSI implementation of threshold logic – a comprehensive survey. *IEEE Transactions on Neural Networks*, 14:1217–1243, 2003.

[Bra90] R. Brady. The Gentzenization and decidability of RW. *Journal of Philosophical Logic*, 19:35–73, 1990.

[Bra03a] R. Brady, editor, *Relevant Logics and their Rivals. Volume II*. Aldershot, Ashgate, 2003.

[Bra03b] R. Brady. Semantic decision procedure for some relevant logics. *Australiasian Journal of Logic*, 1:4–27, 2003.

[Bri81] C. Brink. Boolean modules. *Journal of Algebra*, 71(2):291–313, 1981.

[Bri88] K. Britz. Relations and programs. Master's thesis, University of Stellenbosch, South Africa, 1988.

[Bro07] L. E. J. Brouwer. On the foundations of mathematics. Thesis, Amsterdam, 1907. English translation in A. Heyting, editor, *L. E. J. Brouwer: Collected Works 1*. Philosophy and Foundations of Mathematics, Elsevier, Amsterdam, 1975.

[BS72] S. L. Bloom and R. Suszko. Investigation into the sentential calculus with identity. *Notre Dame Journal of Formal Logic*, 13(3):289–308, 1972.

[BS73] D. J. Brown and R. Suszko. Abstract logics. *Dissertationes Mathematicae CII*, Warsaw, 1973.

[BS85] R. J. Brachman and J. G. Schmolze. An overview of the KL−ONE knowledge representation system. *Cognitive Science*, 9(2):171–216, 1985.

[BS94] R. Berghammer and G. Schmidt. RELVIEW: A computer system for the manipulation of relations. In Nivat et al. [NRRS94], pages 403–404.

[BT83] S. L. Bloom and R. Tindell. Varieties of "if-then-else". *SIAM Journal of Computing*, 12(4):677–707, 1983.

[Bub77] J. Bubenko. The temporal dimension in information modelling. In G. Nijssen, editor, *Architecture and Models in Data Base Management Systems*. North-Holland, Amsterdam, 1977.

[Bul70] R. Bull. An approach to tense logic. *Theoria*, 36:282–300, 1970.

[Bur79] J. P. Burgess. Logic and time. *Journal of Symbolic Logic*, 44(4):566–582, 1979.

[Bur82] J. P. Burgess. Axioms for tense logics II: Time periods. *Notre Dame Journal of Formal Logic*, 23:375–383, 1982.

[BvBW06] P. Blackburn, J. van Benthem, and F. Wolter, editors, *Handbook of Modal Logic*. Elsevier, Amsterdam, 2006.

[BZ86] R. Berghammer and H. Zierer. Relational algebraic semantics of deterministic and nondeterministic programs. *Theoretical Computer Science*, 43:123–147, 1986.

[CC08] L. M. Cabrer and S. A. Celani. Kripke semantics for monoidal t-norm based logics MTL and IMTL. In *Centre for Logic, Epistemology and the History of Science*, volume 8(6), pages 1–3, 2008.

[CFOZ03] D. Cantone, A. Formisano, E. G. Omodeo, and C. G. Zarba. Compiling dyadic first-order specifications into map algebra. *Theoretical Computer Science*, 293(2):447–475, 2003.

[CH73] N. Cat-Ho. Generalized Post algebras and their applications to some infinitary many-valued logics. *Dissertationes Mathemticae*, 57:1–76, 1973.

[CH97] Z. Chaochen and M. R. Hansen. An adequate first order interval logic. In W. P. de Roever, H. Langmaack, and A. Pnueli, editors, *Compositionality: The Significant Difference, International Symposium, COMPOS'97, Bad Malente, Germany, September 8–12, 1997. Revised Lectures*, volume 1536 of *Lecture Notes in Computer Science*, pages 584–608, Springer, Heidelberg, 1997.

[Che80] B. Chellas. *Modal Logic: An Introduction*. Cambridge University Press, Cambridge, 1980.

[CHR89] N. Cat-Ho and H. Rasiowa. Plain semi-Post algebras as a poset-based generalization of Post algebras and their representability. *Studia Logica*, 48:509–530, 1989.

[Cie80] K. Ciesielski. Generalized threshold logic. *Bulletin of the Polish Academy of Sciences, Mathematics*, 28(5-6):219–228, 1980.

[Cla81] B. L. Clarke. A calculus of individuals based on 'connection'. *Notre Dame Journal of Formal Logic*, 22:204–218, 1981.

[Cod70] E. F. Codd. A relational model of data for large shared data banks. *Communications of the ACM*, 13(6):377–387, 1970.

[CZ97] A. Chagrov and M. Zakharyaschev. *Modal Logic*. Oxford University Press, Oxford, 1997.

[Dem00] S. Demri. The nondeterministic information logic NIL is PSPACE-complete. *Fundamenta Informaticae*, 42(3–4):211–234, 2000.

[Der65] M. Dertouzos. *Threshold Logic: A Synthesis Approach*. MIT Press, Cambridge, 1965.

[DG00a] S. Demri and D. M. Gabbay. On modal logics characterized by models with relative accessibility relations: Part I. *Studia Logica*, 65(3):323–353, 2000.

[DG00b] S. Demri and D. M. Gabbay. On modal logics characterized by models with relative accessibility relations: Part II. *Studia Logica*, 66:349–384, 2000.

[Did87] E. Diday. Introduction a l'approche symbolique en analyse des donnees. In Actes des journees symboliques numeriques pour l'apprentissage de connaissances a partir des donnes, 1987.

[Did88] E. Diday. Generating rules by symbolic data analysis and application to soil feature recognition. in Actes des 8emes Journees Internationales: Les systemes experts et leurs applications, 1988.

[Dij68] J. W. Dijkstra. GO TO statement considered harmful. *Communications of the Association for Computing Machinery*, 11(3):147–148, 1968.

[dL22] T. de Laguna. Point, line and surface as sets of solids. *Journal of Philosophy*, 19:449–461, 1922.

[DM71] M. Dunn and R. K. Meyer. Algebraic completeness results for Dummett's LC and its extensions. *Zeitschrift für Mathematische Logik und Grundlagen der Mathematik*, 17:225–230, 1971.

[DM01] I. Düntsch and S. Mikulás. Cylindric structures and dependencies in relational databases. *Theoretical Computer Science*, 269(1–2):451–468, 2001.

[DO96] S. Demri and E. Orłowska. Logical analysis of demonic nondeterministic programs. *Theoretical Compututer Science*, 166(1–2):173–202, 1996.

[DO00a] I. Düntsch and E. Orłowska. Logics of complementarity in information systems. *Mathematical Logic Quarterly*, 46(2):267–288, 2000.

[DO00b] I. Düntsch and E. Orłowska. A proof system for contact relation algebras. *Journal of Philosophical Logic*, 29:241–262, 2000.

[DO01] I. Düntsch and E. Orłowska. Beyond modalities: sufficiency and mixed algebras. In Orłowska and Szałas [OS01], pages 263–285.

[DO02] S. Demri and E. Orłowska. *Incomplete Information: Structure, Inference, Complexity*. Monographs in Theoretical Computer Science. An EATCS Series. Springer, Heidelberg, 2002.

[DO04] I. Düntsch and E. Orłowska. Boolean algebras arising from information systems. *Annals of Pure and Applied Logic*, 127(1–3):77–98, 2004.

[DO07] S. Demri and E. Orłowska. Relative nondeterministic information logic is EXPTIME-complete. *Fundamenta Informaticae*, 75 (1–4):163–178, 2007.

[DO08] I. Düntsch and E. Orłowska. A discrete duality between apartness algebras and apartness frames. *Journal of Applied Non-Classical Logics*, 18(2–3):209–223, 2008.

[Dob96a] W. B. Dobrowolska. Decidable classes in relational logic generated by decidable classes in first order logic. Technical Report, Universita degli Studi di Milano, 1996.

[Dob96b] W. B. Dobrowolska. Some decidable classes in relational logic. *Bulletin of the Polish Academy of Sciences, Mathematics*, 44:87–102, 1996.

[DOR10] I. Düntsch, E. Orłowska, and I. Rewitzky. Structures with multirelations, their discrete dualities and applications. *Fundamenta Informaticae*, 100(1–4):77–98, 2010.

[DOW01] I. Düntsch, E. Orłowska, and Hui Wang. Algebras of approximating regions. *Fundamenta Informaticae*, 46(1–2):71–82, 2001.

[dR95] M. de Rijke. The logic of Peirce algebras. *Journal of Logic, Language and Information*, 4:227–250, 1995.

[dR99] M. de Rijke. A modal characterization of Peirce algebras. In Orłowska [Orł99], pages 109–123.

[dS02a] H. C. M. de Swart, editor, *Relational Methods in Computer Science, 6th International Conference, RelMICS 2001, and 1st Workshop of COST Action 274 TARSKI Oisterwijk, The Netherlands, October 16–21, 2001, Revised Papers*, volume 2561 of *Lecture Notes in Computer Science*, Springer, Heidelberg, 2002.

[DS02b] S. Demri and U. Sattler. Automata-theoretic decision procedures for information logics. *Fundamenta Informaticae*, 53(1):1–22, 2002.

[DSW01] I. Düntsch, G. Schmidt, and M. Winter. A necessary relation algebra for mereotopology. *Studia Logica*, 69(3):381–409, 2001.

[Dum59] M. Dummett. A propositional calculus with denumerable matrix. *Journal of Symbolic Logic*, 24(2):97–106, 1959.

[Dun01a] M. Dunn. A representation of relation algebras using Routley-Meyer frames. In C. A. Anderson and M. Zeleny, editors, *Logic, Meaning and Computation*, volume 305 of *Synthese Library*, pages 77–108, Kluwer, Dordrecht, 2001.

[Dün01b] I. Düntsch. Contact relation algebras. In Orłowska and Szałas [OS01], pages 113–133.

[Dün05] I. Düntsch. Relation algebras and their application in temporal and spatial reasoning. *Artificial Intelligence Review*, 23(4):315–357, 2005.

[DV07] I. Düntsch and D. Vakarelov. Region-based theory of discrete spaces: A proximity approach. *Annals of Mathematics and Artificial Intelligence*, 49(1–4):5–14, 2007.

[DW08] I. Düntsch and M. Winter. A representation theorem for boolean contact algebras. Technical Report CS-03-08, Brock University, 2008.

[DWM99] I. Düntsch, H. Wang, and S. McCloskey. Relation algebras in qualitative spatial reasoning. *Fundamenta Informaticae*, 39(3):229–248, 1999.

[DWM01] I. Düntsch, H. Wang, and S. McCloskey. A relation – algebraic approach to the region connection calculus. *Theoretical Computer Science*, 255(1–2):63–83, 2001.

[Efr52] V. Efremovic. The geometry of proximity. *Matematiceskij Sbornik (New Series)*, 31:189–200, 1952. In Russian.

[EFT94] H. D. Ebbinghaus, J. Flum, and W. Thomas. *Mathematical Logic*. Springer, Heidelberg, 1994.

[EG01] F. Esteva and L. Godo. Monoidal t-norm based logic: towards a logic for left-continuous t-norms. *Fuzzy Sets and Systems*, 124(3):271–288, 2001.

[Eng67] E. Engeler. Algorithmic properties of structures. *Mathematical Systems Theory*, 1(3):183–195, 1967.

[EO67] A. Ehrenfeucht and E. Orłowska. Mechanical proof procedure for propositional calculus. *Bulletin of the Polish Academy of Sciences, Mathematics*, 15(1):25–30, 1967.

[Eps60] G. Epstein. The lattice theory of Post algebras. *Transactions of the American Mathematical Society*, 95(2):300–317, 1960.

[ER90] G. Epstein and H. Rasiowa. Theory and uses of Post algebras of order $\omega + \omega*$. Part I. In *Proceedings of the 20th International Symposium on Multiple-Valued Logic*, pages 42–47, Charlotte, NC, USA, 1990.

[ER91] G. Epstein and H. Rasiowa. Theory and uses of Post algebras of order $\omega + \omega*$. Part II. In *Proceedings of the 21th International Symposium on Multiple-Valued Logic*, pages 248–254, Victoria, Canada, 1991.

[FBH98] M. F. Frias, G. Baum, and A. M. Haeberer. Representability and program construction within fork algebras. *Logic Journal of the IGPL*, 6(2):227–257, 1998.

[FBH01] M. F. Frias, G. A. Baum, and A. M. Haeberer. A calculus for program construction based on fork algebras, design strategies and generic algorithms. In Orłowska and Szałas [OS01], pages 37–58.

[FBHV93] M. F. Frias, G. A. Baum, A. M. Haeberer, and P. A. S. Veloso. A representation theorem for fork algebras. Technical Report, Pontificia Universidade Catòlico do Rio de Janeiro, 1993.

[FBHV95] M. F. Frias, G. A. Baum, A. M. Haeberer, and P. A. S. Veloso. Fork algebras are representable. *Bulletin of the Section of Logic*, 24(2): 64–75, 1995.

[FBM02] M. F. Frias, G. A. Baum, and T. S. E. Maibaum. Interpretability of first-order dynamic logic in a relational calculus. In de Swart [dS02a], pages 66–80.

[Fey65] R. Feys. *Modal Logics*. Louvain E. Nauwelaerts, Paris, 1965.

[FGSB06] M. F. Frias, R. Gamarra, G. Steren, and L. Bourg. Monotonicity analysis can speed up verification. In R. A. Schmidt, editor, *Relations and Kleene Algebra in Computer Science, 9th International Conference on Relational Methods in Computer Science and 4th International Workshop on Applications of Kleene Algebra, RelMiCS/AKA 2006, Manchester, UK, August 29–September 2, 2006, Proceedings*, volume 4136 of *Lecture Notes in Computer Science*, pages 177–191, Springer, Heidelberg, 2006.

[FHV97] M. F. Frias, A. M. Haeberer, and P. A. S. Veloso. A finite axiomatization for fork algebras. *Logic Journal of the IGPL*, 5(3):1–10, 1997.

[Fit90] M. Fitting. *First-order Logic and Automated Theorem Proving*. Springer, Heidelberg, 1990.

[FL79] M. J. Fischer and R. E. Ladner. Propositional dynamic logic of regular programs. *Journal of Computer and System Sciences*, 18(2):194–211, 1979.

[FM95] M. Fairtlough and M. Mendler. An intuitionistic modal logic with applications to the formal verification of hardware. In Pacholski and Tiuryn [PT95], pages 354–368.

[FNA06] A. Formisano and M. Nicolosi-Asmundo. An efficient relational deductive system for propositional non-classical logics. *Journal of Applied Non-Classical Logics*, 16(3–4):367–408, 2006.

[FO95] M. F. Frias and E. Orłowska. A proof system for fork algebras and its applications to reasoning in logics based on intuitionism. *Logique et Analyse*, 150–152:239–284, 1995.

[FO98] M. F. Frias and E. Orłowska. Equational reasoning in non-classical logics. *Journal of Applied Non-Classical Logics*, 8(1–2):27–66, 1998.

[FO03] M. Fitting and E. Orłowska, editors, *Beyond Two: Theory and Applications of Multiple-Valued Logic*. Springer, Heidelberg, 2003.

[FOO05] A. Formisano, E. G. Omodeo, and E. Orłowska. A PROLOG tool for relational translation of modal logics: A front-end for relational proof systems. In Beckert [Bec05], pages 1–10.

[FOO06] A. Formisano, E. G. Omodeo, and E. Orłowska. An environment for specifying properties of dyadic relations and reasoning about them II: Relational presentation of non-classical logics. In H. C. M. de Swart, E. Orłowska, G. Schmidt, and M. Roubens, editors, *Theory and Applications of Relational Structures as Knowledge Instruments II, International Workshops of COST Action 274, TARSKI, 2002–2005, Selected Revised Papers*, volume 4342 of *Lecture Notes in Computer Science*, pages 89–104, Springer, Heidelberg, 2006.

[FOP02] A. Formisano, E. G. Omodeo, and A. Policriti. Automated validation of three-variable formulations of set pairing. In J. G. F. Belinfante, editor, *2002-AMS and MAA Spring Southeastern Section Meeting*, Atlanta, 2002.

[FP06] M. F. Frias and C. López Pombo. Interpretability of first-order linear temporal logics in fork algebras. *Journal of Algebraic and Logic Programming*, 66(2):161–184, 2006.

[Fri02] M. F. Frias. *Fork Algebras in Algebra, Logic and Computer Science*, volume 2 of *Advances in Logic*. World Scientific, Singapore, 2002.

[Gab76] D. Gabbay. *Investigations in Modal and Tense Logics*. Reidel, Dordrecht, 1976.

[Gal75] D. Gallin. *Intensional and Higher Order Modal Logic*. North-Holland, Amsterdam, 1975.

[Gen34] G. Gentzen. Untersuchungen über das logische schliessen. *Mathematische Zeitschrift*, 39:405–431, 1934.

[Gia85] S. Giambrone. TW^+ and RW^+ are decidable. *Journal of Philosophical Logic*, 14:235–254, 1985.

[Gir87] J.-Y. Girard. Linear logic. *Theoretical Computer Science*, 50:1–102, 1987.

[Gli29] V. Glivenko. Sur quelques points de la logique de M. Brouwer. *Bulletins de la classe des sciences*, 5(15):183–188, 1929.

[GM87] I. Guessarian and J. Meseguer. On the axiomatization of "if-then-else". *SIAM Journal of Computing*, 16(2):332–357, 1987.

[GMS03a] V. Goranko, A. Montanari, and G. Sciavicco. A general tableau method for propositional interval temporal logics. In M. Cialdea Mayer and F. Pirri, editors, *Automated Reasoning with Analytic Tableaux and Related Methods, International Conference, TABLEAUX 2003, Rome, Italy, September 9–12, 2003. Proceedings*, volume 2796 of *Lecture Notes in Computer Science*, pages 102–116, Springer, Heidelberg, 2003.

[GMS03b] V. Goranko, A. Montanari, and G. Sciavicco. Propositional interval neighborhood temporal logics. *Journal of Universal Computer Science*, 9(9):1137–1167, 2003.

[GMS04] V. Goranko, A. Montanari, and G. Sciavicco. A road map of interval temporal logics and duration calculi. *Journal of Applied Non-Classical Logics*, 14(1–2):9–54, 2004.

[GMSS06] V. Goranko, A. Montanari, P. Sala, and G. Sciavicco. A general tableau method for propositional interval temporal logics: Theory and implementation. *Journal of Applied Logic*, 4(3):305–330, 2006.

[Göd33] K. Gödel. Zum intuitionistischen aussagenkalkül. *Ergeb. Math. Koll*, 4:40, 1933.

[Gol87] R. Goldblatt. *Logics of Time and Computation*, volume 7 of *CSLI Lecture Notes*. Center for the Study of Language and Information, Stanford, 1987.

[Gol93] R. Goldblatt. *Mathematics of Modality*, volume 43 of *CSLI Lecture Notes*. Center for the Study of Language and Information, Stanford, 1993.

[Gor95] L. Gordeev. Cut free formalization of logic with finitely many variables. Part I. In Pacholski and Tiuryn [PT95], pages 136–150.

[Gor97] R. Goré. Cut-free display calculi for relation algebras. In *CSL '96: Selected Papers from the10th International Workshop on Computer Science Logic*, pages 198–210, Springer, Heidelberg, 1997.

[Gor01] L. Gordeev. Proof systems in relation algebra. In Orłowska and Szałas [OS01], pages 219–237.

[Got00] S. Gottwald. *A Treatise on Many-Valued Logics*, volume 9 of *Studies in Logic and Computation*. Research Studies Press, Baldock, 2000.

[GP07] J. Golińska-Pilarek. Rasiowa-Sikorski proof system for the non-fregean sentential logic SCI. *Journal of Applied Non-Classical Logics*, 17(4):511–519, 2007.

[GPBV08] J. Golińska-Pilarek, A. Mora Bonilla, and E. Munoz Velasco. An ATP of a relational proof system for order of magnitude reasoning with negligibility, non-closeness and distance. In T. B. Ho and Z. H. Zhou, editors, *PRICAI 2008*, volume 5351 of *Lecture Notes in Artificial Intelligence*, pages 128–139, Springer, Heidelberg, 2008.

[GPH05] J. Golińska-Pilarek and T. Huuskonen. Number of extensions of non-fregean logics. *Journal of Philosophical Logic*, 34(2):193–206, 2005.

[GPO07a] J. Golińska-Pilarek and E. Orłowska. Relational reasoning in formal concept analysis. In *FUZZ-IEEE 2007, IEEE International Conference on Fuzzy Systems, Imperial College, London, UK, 23–26 July, 2007, Proceedings*, pages 1–6, London, 2007. IEEE.

[GPO07b] J. Golińska-Pilarek and E. Orłowska. Tableaux and dual tableaux: Transformation of proofs. *Studia Logica*, 85(3):283–302, 2007.

[GPO10] J. Golińska-Pilarek and E. Orłowska. Dual tableau for monoidal triangular norm logic MTL, Fuzzy Sets and Systems (2010), doi: 10.1016/j.fss.2010.09.007.

[Grz60] A. Grzegorczyk. Axiomatization of geometry without points. *Synthese*, 12:228–235, 1960.

[GW99] B. Ganter and R. Wille. *Formal Concept Analysis: Mathematical Foundation*. Springer, Heidelberg, 1999.

[Gyu95] V. Gyuris. A short proof of representability of fork algebras. *Logic Journal of the IGPL*, 3(5):791–796, 1995.

[Häh01] R. Hähnle. Advanced many-valued logics. In D. M. Gabbay and F. Guenthner, editors, *Handbook of Philosophical Logic*, volume 2, pages 297–395. Kluwer, Dordrecht, 2001.

[Häh03] R. Hähnle. Complexity of many-valued logics. In Fitting and Orłowska [FO03], pages 211–233.

[Háj98] P. Hájek. *Metamathematics of Fuzzy Logic*. Kluwer, Dordrecht, 1998.

[HBS93] A. M. Haeberer, G. Baum, and G. Schmidt. On the smooth calculation of relational recursive expressions out of first-order non-constructive specifications involving quantifiers. In D. Bjorner, M. Broy, and I. V. Pottosin, editors, *Formal Methods in Programming and Their Applications, International Conference, Akademgorodok, Novosibirsk, Russia, June 28–July 2, 1993, Proceedings*, volume 735 of *Lecture Notes in Computer Science*, pages 281–298, Springer, Heidelberg, 1993.

[HBS94] C. Hattensperger, R. Berghammer, and G. Schmidt. RALF: A relation-algebraic formula manipulation system and proof checker. In Nivat et al. [NRRS94], pages 405–406.

[HC68] G. E. Hughes and M. J. Cresswell. *An Introduction to Modal Logic*. Methuen, London, 1968.

[HC84] G. E. Hughes and M. J. Cresswell. *A Companion to Modal Logic*. Methuen, London, 1984.

[Hen80] M. Hennessy. A proof system for the first-order relational calculus. *Journal of Computer and System Sciences*, 20(1):96–110, 1980.

[Hey30] A. Heyting. *Die Formalen Regeln der Intuitionistischer Logik*, pages 42–71, 158–169. Sitzungsber. Preuss. Acad. Wiss., Berlin, 1930. English translation in P. Mancosu, editor, *From Brouwer to Hilbert: The Debate on the Foundations of Mathematics in 1920's*, pages 311–327. Oxford University Press, Oxford, 1998.

[HH02] R. Hirsch and I. Hodkinson. *Relation Algebras by Games*. Elsevier, Amsterdam, 2002.

[Hir07] R. Hirsch. Peirce algebras and Boolean modules. *Journal of Logic and Computation*, 17(2):255–283, 2007.

[HKT00] D. Harel, D. Kozen, and J. Tiuryn. *Dynamic Logic*. MIT Press, Cambridge, 2000.

[Hoa69] C. A. R. Hoare. An axiomatic basis for computer programming. *Communications of the ACM*, 12(10):576–580, 1969.

[HR06] J. Hodkinson and M. Reynolds. Temporal logic. In Blackburn et al. [BvBW06], pages 655–720.

[HS91] J. Y. Halpern and Y. Shoham. A propositional modal logic of time intervals. *Journal of the ACM*, 38(4):935–962, 1991.

[HS99] U. Hustadt and R. A. Schmidt. On the relation of resolution and tableaux proof systems for description logics. In T. Dean, editor, *Proceedings of the Sixteenth International Joint Conference on Artificial Intelligence, IJCAI 99, Stockholm, Sweden, July 31–August 6, 1999. 2 volumes, 1450 pages*, pages 110–117, Stockholm, Morgan Kaufmann, 1999.

[Hum79] I. L. Humberstone. Interval semantics for tense logic: some remarks. *Journal of Philosophical Logic*, 8:171–196, 1979.

[HV91a] A. M. Haeberer and P. A. S. Veloso. A finitary relational algebra for classical first order logic. *Bulletin of the Section of Logic*, 20(2):52–62, 1991.

[HV91b] A. M. Haeberer and P. A. S. Veloso. Partial relations for program derivation: adequacy, inevitability, and expressiveness. In *Constructing Programs from Specifications, Proceedings of the IFIP TC2 Working Conference*, pages 319–371, Amsterdam, North-Holland, 1991.

[IL84] T. Imieliński and W. Lipski. The relational model of data and cylindric algebras. *Journal of Computer and System Sciences*, 28(1):80–102, 1984.

[IO06] L. Iturrioz and E. Orłowska. A Kripke-style and relational semantics for logics based on Łukasiewicz algebras. *Multiple-Valued Logic and Soft Computing*, 12(1–2):131–147, 2006.

[Itu82] L. Iturrioz. Modal operators on symmetrical Heyting algebras. In T. Traczyk, editor, *Universal Algerbra and Applications*, volume 9 of *Banach Center Publications*, pages 289–303. Polish Scientific Publishers, Warsaw, 1982.

[Itu83] L. Iturrioz. Symmetrical Heyting algebras with operators. *Zeitschrift für Mathematische Logik und Grundlagen der Mathematik*, 29:33–70, 1983.

[JM02] S. Jenei and F. Montagna. A proof of standard completeness for Esteva and Godo's logic MTL. *Studia Logica*, 70(2):183–192, 2002.

[Joh36] I. Johansson. Der minimalkalkül, ein reduzierter intuitionistischer formalismus. *Compositio Mathematica*, 4:119–136, 1936.

[JT52] B. Jónsson and A. Tarski. Boolean algebras with operators, Part II. *American Journal of Mathematics*, 74:127–162, 1952.

[Kam68] J. A. Kamp. *Tense logic and the theory of linear order.* PhD thesis, University of California, Los Angeles, 1968.

[Kan57] S. Kanger. *Provability in Logic*, volume 1 of *Stockholm Studies in Philosophy*. Acta Universitatis Stockholmiensis, Stockholm, 1957.

[KM00] E. P. Klement and R. Mesiar. *Triangular Norms*. Kluwer, Dordrecht, 2000.

[KMO98] B. Konikowska, C. Morgan, and E. Orłowska. A relational formalisation of arbitrary finite-valued logics. *Logic Journal of the IGPL*, 6(5):755–774, 1998.

[KO01] B. Konikowska and E. Orłowska. A relational formalisation of a generic many-valued modal logic. In Orłowska and Szałas [OS01], pages 183–202.

[Kon87] B. Konikowska. A formal language for reasoning about indiscernibility. *Bulletin of the Polish Academy of Sciences, Mathematics*, 35:239–249, 1987.

[Kon97] B. Konikowska. A logic for reasoning about relative similarity. *Studia Logica*, 58:185–226, 1997.

[Kon02] B. Konikowska. Rasiowa-Sikorski deduction systems in computer science applications. *Theoretical Computer Science*, 286(2):323–366, 2002.

[KP81] D. Kozen and R. Parikh. An elementary proof of the completness of PDL. *Theoretical Computer Science*, 14:113–118, 1981.

[Kri63] S. A. Kripke. Semantical considerations on modal logic. *Acta Philosophica Fennica*, 16:83–94, 1963.

[Kri65] S. A. Kripke. Semantical analysis of intutionistic logic. In J. Crossley and M. A. E. Dummett, editors, *Formal Systems and Recursive Functions*, pages 92–130. North-Holland, Amsterdam, 1965.

[Lee59] C. Y. Lee. Representation of switching circuits by binary-decision programs. *Bell System Technical Journal*, 38:985–999, 1959.

[Leś16] S. Leśniewski. Podstawy ogólnej teorii mnogości. Prace Polskiego Koła Naukowego w Moskwie, Sekcja Matematyczno-Przyrodnicza 2, 1916.

[Leś29] S. Leśniewski. Grundzüge eines neuen Systems der Grundlagen der Mathematik. *Fundamenta Mathematicae*, 14:1–81, 1929. English translation in S. J. Surma, J. T. Srzednicki, D. I. Bernett, and V. F. Rickey, editors, *Stanisław Leśniewski: Collected works*, 1992.

[Leś31] S. Leśniewski. O podstawach matematyki. *Przegląd Filozoficzny*, 30–34, 1927–1931.

[Lew20] C. I. Lewis. Strict implication – an emendation. *Journal of Philosophy*, 17:300–302, 1920.

[Lip76] W. Lipski. Informational systems with incomplete information. In *Proceedings of the 3rd International Symposium on Automata, Languages and Programming, Edinburgh, Scotland 1976*, pages 120–130, 1976.

[Lip79] W. Lipski. On semantic issues connected with incomplete information databases. *ACM Transactions on Database Systems*, 4(3):262–296, 1979.

[LL59] C. I. Lewis and C. H. Langford. *Symbolic Logic*. Dover Publications, New York, 1959. First published in 1932.

[LLMS02] G. Lee, R. Little, W. MacCaull, and B. Spencer. ReVAT – relational validation via analytic Tableaux. Technical Report, St. Francis Xavier University, 2002.

[LM87] P. B. Ladkin and R. Maddux. The algbera of convex time intervals. Technical Report, Kestrel Institute, 1987.

[LM94] P. B. Ladkin and R. D. Maddux. On binary constraint problems. *Journal of the ACM*, 41(3):435–469, 1994.

[Lod00] K. Lodaya. Sharpening the undecidability of interval temporal logic. In J. He and M. Sato, editors, *Advances in Computing Science - ASIAN 2000, 6th Asian Computing Science Conference, Penang, Malaysia, November 25-27, 2000, Proceedings*, volume 1961 of *Lecture Notes in Computer Science*, pages 290–298, Springer, Heidelberg, 2000.

[LT87] K. Lodaya and P. S. Thiagarajan. A modal logic for a subclass of event structures. In T. Ottmann, editor, *Automata, Languages and Programming, 14th International Colloquium, ICALP87, Karlsruhe, Germany, July 13–17, 1987, Proceedings*, volume 267 of *Lecture Notes in Computer Science*, pages 290–303, Springer, Heidelberg, 1987.

[Łuk20] J. Łukasiewicz. O logice trójwartościowej. *Ruch Filozoficzny*, 5:170–171, 1920. English translation in L. Borkowski, editor, *Selected Works of Jan Łukasiewicz*, North-Holland, Amsterdam and Polish Scientific Publishers, Warsaw.

[Lyn50] R. C. Lyndon. The representation of relational algebras. *Annals of Mathematics (Series 2)*, 51:707–729, 1950.

[Mac97] W. MacCaull. Relational proof system for linear and other substructural logics. *Logic Journal of the IGPL*, 5(5), 1997.

[Mac98] W. MacCaull. Relational semantics and a relational proof system for full Lambek calculus. *Journal of Symbolic Logic*, 63(2):623–637, 1998.

[Mac99] W. MacCaull. Relational tableaux for tree models, language models and information networks. In Orłowska [Orł99].

[Mac00] W. MacCaull. A proof system for dependencies for information relations. *Fundamenta Informaticae*, 42(1):1–27, 2000.

[Mac01] W. MacCaull. A tableaux procedure for the implication problem for association rules. In Orłowska and Szałas [OS01], pages 73–91.

[Mad83] R. D. Maddux. A sequent calculus for relation algebras. *Annals of Pure and Applied Logic*, 25:73–101, 1983.

[Mad91] R. D. Maddux. Introductory course on relation algebras, finite dimensional cylindric algebras, and their interconnections. In Andréka et al. [AMN91], pages 361–392.

[Mad06] R. D. Maddux. *Relation Algebras*. Elsevier, Amsterdam, 2006.

[Mak73] L. L. Maksimowa. Structures with implication. *Algebra and Logic*, 12:445–467, 1973.

[Mal93] G. Malinowski. *Many-Valued Logics*. Oxford University Press, Oxford, 1993.

[Man93] E. G. Manes. Adas and the equational theory of if-then-else. *Algebra Universalis*, 30:373–394, 1993.

[McC63] J. A. McCarthy. A basis for a mathematical theory of computation. In P. Braffort and D. Hirschberg, editors, *Computer, Programming and Formal Systems*, pages 33–70. North-Holland, Amsterdam, 1963.

[Men90] M. Mendler. Constrained proofs: A logic for dealing with behavioural constraints in formal hardware verification. In G. Jones and M. Sheeran, editors, *Designing Correct Circuits*, pages 1–28, Springer, Heidelberg, 1990.

[Men93] M. Mendler. *A modal logic for handling behavioural constraints in formal hardware verification*. PhD thesis, Department of Computer Science, Edinburgh University, 1993.

[MO02a] W. MacCaull and E. Orłowska. Correspondence results for relational proof systems with application to the Lambek calculus. *Studia Logica*, 71(3):389–414, 2002.

[MO02b] F. Montagna and H. Ono. Kripke semantics, undecidability and standard completeness for Esteva and Godo's logic $\mathsf{MTL}\forall$. *Studia Logica*, 71(2):227–245, 2002.

[MO04] W. MacCaull and E. Orłowska. A calculus of typed relations. In Berghammer et al. [BMS04], pages 187–199.

[MO06] W. MacCaull and E. Orłowska. A logic of type relations and its applications to relational databases. *Journal of Logic and Computation*, 16(6):789–815, 2006.

[MOG09] G. Metcalfe, N. Olivetti, and D. Gabbay. *Proof Theory for Fuzzy Logics*, volume 36 of *Applied Logic Series*. Springer, Heidelberg, 2009.

[Mon64] D. Monk. On representable relation algebras. *Michigan Mathematical Journal*, 11:207–210, 1964.

[Mor64] A. De Morgan. On the syllogism: IV, and on the logic of relation. *Transactions of the Cambridge Philosophical Society*, 10:331–358, 1864.

[Mos83] B. Moszkowski. Reasoning about digital circuits. Technical Report, Department of Computer Science, Stanford University, Stanford, CA, 1983.

[MS87] G. Mirkowska and A. Salwicki. *Algorithmic Logic*. Reidel, Dordrecht, 1987.

[Mur71] S. Muroga. *Threshold Logic and its Applications*. Wiley-Interscience, New York, 1971.

[Ngu91] T. Nguyen. A relational model of demonic nondeterministic programs. *International Journal of Foundations of Computer Science*, 2:101–131, 1991.

[Nou99] A. Nour. Sémantique algébrique d'un systèmes logique basé sur un ensemble ordonné fini. *Mathematical Logic Quarterly*, 45:457–466, 1999.

[NPW81] M. Nielsen, G. D. Plotkin, and G. Winskel. Petri nets, event structures and domains, Part I. *Theoretical Computer Science*, 13:85–108, 1981.

[NRRS94] M. Nivat, C. Rattray, T. Rus, and G. Scollo, editors, *Algebraic Methodology and Software Technology: Proceedings of the Third International Conference AMAST'93*. Springer, Heidelberg, 1994.

[NW70] S. A. Naimpally and B. D. Warrack. *Proximity Spaces*. Cambridge University Press, Cambridge, 1970.

[OdS93] W. M. J. Ophelders and H. C. M. de Swart. Tableaux versus resolution a comparison. *Fundamenta Informaticae*, 18:109–127, 1993.

[Ono] H. Ono. The finite embeddability property for MTL. Unpublished note.

[Ono85] H. Ono. Semantical analysis of predicate logics without the contraction rule. *Studia Logica*, 44(2):187–196, 1985.

[Ono93] H. Ono. Semantics for substructural logics. In K. Došen and P. Schroeder-Heister, editors, *Substructural Logics*, pages 259–291. Oxford University Press, Oxford, 1993.

[OOP04] E. Omodeo, E. Orłowska, and A. Policriti. Simulation and semantic analysis of modal logics by means of an elementary set theory treated à la Rasiowa-Sikorski. In Berghammer et al. [BMS04], pages 215–226.

[OP84] E. Orłowska and Z. Pawlak. Representation of nondeterministic information. *Theoretical Computer Science*, 29:27–39, 1984.

[OR07] E. Orłowska and I. Rewitzky. Discrete duality and its applications to reasoning with incomplete information. In M. Kryszkiewicz, J. F. Peters, H. Rybiński, and A. Skowron, editors, *Rough Sets and Intelligent Systems Paradigms, International Conference, RSEISP 2007, Warsaw, Poland, June 28-30, 2007, Proceedings*, volume 4585 of *Lecture Notes in Computer Science*, pages 51–56, Springer, Heidelberg, 2007.

[OR08] E. Orłowska and I. Rewitzky. Context algebras, context frames, and their discrete duality. In J. F. Peters et al., editor, *Transactions on Rough Sets IX*, volume 5390 of *Lecture Notes in Computer Science*, pages 212–229, Springer, Heidelberg, 2008.

[OR09a] E. Orłowska and A. Radzikowska. Discrete duality for some axiomatic extensions of MTL algebras. In P. Cintula, Z. Haniková, and V. Švejdar, editors, *Witnessed Years – Essays in Honour of Petr Hájek*, number 10 in College Publications, pages 329–344. King's College London, London, 2009.

[OR09b] E. Orłowska and I. Rewitzky. Discrete duality for relation algebras and cylindric algebras. In R. Berghammer, A. Jaoua, and B. Möller, editors, *Relations and Kleene Algebra in Computer Science, 11th International Conference on Relational Methods in Computer Science, RelMiCS 2009, and 6th International Conference on Applications of Kleene Algebra, AKA 2009, Doha, Qatar, November 1–5, 2009. Proceedings*, volume 5827 of *Lecture Notes in Computer Science*, pages 291–305, Springer, Heidelberg, 2009.

[OR10] E. Orłowska and I. Rewitzky. Algebras for Galois-style connections and their discrete duality. *Fuzzy Sets and Systems*, 161(9):1325–1342, 2010.

[Orł69] E. Orłowska. Mechanical theorem proving in a certain class of formulae of the predicate calculus. *Studia Logica*, 25:17–29, 1969.

[Orł74] E. Orłowska. Threshold logic. *Studia Logica*, 33:1–9, 1974.

[Orł76] E. Orłowska. Threshold logic II. *Studia Logica*, 35:243–247, 1976.

[Orł82] E. Orłowska. Dynamic information systems. *Fundamenta Informaticae*, 5:101–118, 1982.

[Orł83] E. Orłowska. Semantics of vague concepts. In G. Dorn and P. Weingartner, editors, *Foundations of Logic and Linguistics. Problems and Solutions, Selected contributions to the 7th International Congress of Logic, Methodology and Philosophy of Science, Salzburg (1983)*, pages 465–482. Plenum Press, New York, 1983.

[Orł84] E. Orłowska. Logic of indiscernibility relations. 208:177–186, 1984.

[Orł85a] E. Orłowska. Logic of nondeterministic information. *Studia Logica*, 44:93–102, 1985.

[Orł85b] E. Orłowska. Mechanical proof methods for Post logics. *Logique et Analyse*, 110–111:173–192, 1985.

[Orł87] E. Orłowska. Algebraic approach to database constraints. *Fundamenta Informaticae*, 10:57–66, 1987.

[Orł88] E. Orłowska. Kripke models with relative accessibility and their application to inferences from incomplete information. In G. Mirkowska and H. Rasiowa, editors, *Mathematical Problems in Computation Theory*, volume 21 of *Banach Center Publications*, pages 329–339. Polish Scientific Publishers, Warsaw, 1988.

[Orł91] E. Orłowska. Relational interpretation of modal logics. In Andréka et al. [AMN91], pages 443–471.

[Orł92] E. Orłowska. Relational proof system for relevant logics. *Journal of Symbolic Logic*, 57(4):1425–1440, 1992.

[Orł93] E. Orłowska. Dynamic logic with program specifications and its relational proof system. *Journal of Applied Non-Classical Logics*, 3(2):147–171, 1993.

[Orł95] E. Orłowska. Temporal logics in a relational framework. In L. Bolc and A. Szałas, editors, *Time and Logic – a Computational Approach*, pages 249–277. University College London Press, London, 1995.

[Orł97a] E. Orłowska, editor, *Incomplete Information: Rough Set Analysis*. Springer, Heidelberg, 1997.

[Orł97b] E. Orłowska. Relational formalisation of non-classical logics. In C. Brink, W. Kahl, and G. Schmidt, editors, *Relational Methods in Computer Science*, pages 90–105. Springer, Vienna, 1997.

[Orł99] E. Orłowska, editor, *Logic at Work. Essays dedicated to the memory of Helena Rasiowa*, volume 24 of *Studies in Fuzziness and Soft Computing*. Springer, Heidelberg, 1999.

[ORR10] E. Orłowska, A. Radzikowska, and I. Rewitzky. Discrete representability and discrete duality. A draft of a book, 2010.

[OS01] E. Orłowska and A. Szałas, editors, *Relational Methods for Computer Science Applications*, volume 65 of *Studies in Fuzziness and Soft Computing*. Springer, Heidelberg, 2001.

[OS04] E. Orłowska and G. Schmidt. Rasiowa-Sikorski proof systems in relation algebra. Technical Report, Universität der Bundeswehr München, 2004.

[Pas84] S. Passy. *Combinatory dynamic logic*. PhD thesis, University of Sofia, 1984.

[Paw82] Z. Pawlak. Rough sets. *International Journal of Information and Computer Sciences*, 11:341–356, 1982. Also available as Technical Report 435, Institute of Computer Science, Polish Academy of Sciences, Warsaw, 1981.

[Paw91] Z. Pawlak, editor, *Rough Sets*. Kluwer, Dordrecht, 1991.

[Pea99] D. Pearce. Stable inference as intuitionistic validity. *Journal of Logic Programming*, 38(1):79–91, 1999.

[Pei83] C. S. Peirce. Note B: the logic of relatives. In C. S. Peirce, editor, *Studies in Logic by Members of the Johns Hopkins University*, pages 187–203. Little Brown, Boston, 1883.

[Pen88] W. Penczek. A temporal logic for event structures. *Fundamenta Informaticae*, 11:297–326, 1988.

[PM60] M. C. Paull and E. J. McCluskey. Boolean functions realizable with single threshold devices. *Proceedings IRE*, 48:1335–1337, 1960.

[Pnu77] A. Pnueli. The temporal logic of programs. In *18th Annual Symposium on Foundations of Computer Science*, pages 46–57, IEEE, Providence, 1977.

[Pos20] E. L. Post. Determination of all closed systems of truth tables. *Bulletin of the American Mathematical Society*, 26:437, 1920.

[Pos21] E. L. Post. Introduction to a general theory of elementary propositions. *American Journal of Mathematics*, 43:163–185, 1921.

[Pra76] V. R. Pratt. Semantical considerations on Floyd-Hoare logic. In *17th Annual Symposium on Foundations of Computer Science*, pages 109–121, IEEE, Houston, 1976.

[Pra78] V. R. Pratt. A practical decision method for propositional dynamic logic: Preliminary Report. In *Conference Record of the Tenth Annual ACM Symposium on Theory of Computing*, pages 326–337, ACM, San Diego, 1978.

[Pra79] V. R. Pratt. Models of program logics. In *20th Annual Symposium on Foundations of Computer Science*, pages 115–122, IEEE, San Juan, 1979.

[Pre97] S. Prediger. Symbolic objects in formal concept analysis. Technical Report, Technische Hohschule Darmstadt, 1997.

[Pri57] A. N. Prior, editor, *Time and Modality*. Clarendon Press, Oxford, 1957.

[Pri67] A. N. Prior, editor, *Past, Present and Future*. Oxford University Press, Oxford, 1967.

[PS95] J. Possega and P. H. Schmitt. Automated deduction with Shannon graphs. *Journal of Logic and Computation*, 5(6):697–729, 1995.

[PT85] S. Passy and T. Tinchev. PDL with data constants. *Information Processing Letters*, 20(1):35–41, 1985.

[PT95] L. Pacholski and J. Tiuryn, editors, *Computer Science Logic, 8th International Workshop, CSL '94, Kazimierz, Poland, September 25–30, 1994, Selected Papers*, volume 933 of *Lecture Notes in Computer Science*, Springer, Heidelberg, 1995.

[Ras73] H. Rasiowa. On generalized Post algebras of order ω^+ and ω^+-valued predicate calculi. *Bulletin of the Polish Academy of Sciences, Mathematics*, 21:209–219, 1973.

[Ras85] H. Rasiowa. Topological representations of Post algebras of order ω^+ and open theories based on ω^+-valued Post logic. *Studia Logica*, 44(4):353–368, 1985.

[Ras91] H. Rasiowa. On approximation logics: a survey. *Jahrbuch, 1990, Kurt Gödel Gesssellschaft*, pages 63–87, 1991.

[Ras94] H. Rasiowa. Axiomatization and completeness of uncountably valued approximation logic. *Studia Logica*, 53(1):137–160, 1994.

[RCC92] D. A. Randell, Z. Cui, and A. G. Cohn. A spatial logic based on regions and connection. In *Proceedings of the 3rd International Conference on Principles of Knowledge Representation and Reasoning*, pages 165–176, Morgan Kaufmann, Cambridge, 1992.

[Res96] G. Restall. Information flow and relevant logics. In J. Seligman and D. Westerståhl, editors, *Logic, Language and Computation: The 1994 Moraga Proceedings*, Volume 1 of *CSLI Lecture Notes*, pages 463–478. Center for the Study of Language and Information, Stanford, 1996.

[RM73] R. Routley and R. K. Meyer. The semantics of entailment. In H. Leblanc, editor, *Truth, Syntax and Modality*, pages 199–243. North-Holland, Amsterdam, 1973.

[RMPB83] R. Routley, R. K. Meyer, V. Plumwood, and R. Brady, editors, *Relevant Logics and their Rivals I*. Ridgeview, Atascadero, 1983.

[Röp80] P. Röper. Intervals and tenses. *Journal of Philosophical Logic*, 9:451–469, 1980.

[Ros42] P. C. Rosenbloom. Post algebras I. Postulates and general theory. *American Journal of Mathematics*, 64(1):167–188, 1942.

[Rou69] G. Rousseau. Logical systems with finitely truth values. *Bulletin of the Polish Academy of Sciences, Mathematics*, 17:189–194, 1969.

[Rou70] G. Rousseau. Post algebras and pseudo-Post algebras. *Fundamenta Mathematicae*, 67:133–145, 1970.

[RS60] H. Rasiowa and R. Sikorski. On Gentzen theorem. *Fundamenta Mathematicae*, 48:57–69, 1960.

[RS63] H. Rasiowa and R. Sikorski, *Mathematics of Metamathematics*. Polish Scientific Publishers, Warsaw, 1963.

[RT52] J. B. Rosser and A. R. Turquette. *Many-Valued Logics*. North-Holland, Amsterdam, 1952.

[Sal70] A. Salwicki. Formalized algorithmic languages. *Bulletin of the Polish Academy of Sciences, Mathematics*, 18:272–232, 1970.

[Sal72] Z. Saloni. Gentzen rules for the m-valued logic. *Bulletin of the Polish Academy of Sciences, Mathematics*, 20:819–826, 1972.

[Sch91] E. Schröder. *Vorlesungen über die Algebra der Logik (Exakte Logik)*. Algebra und Logik der Relative III, Part 1. Teubner, Stuttgart, 1891. Reprinted by Chelsea, New York, 1966.

[Sch82] W. Schönfeld. Upper bounds for a proof-search in a sequent calculus for relational equations. *Zeitschrift für Mathematische Logik und Grundlagen der Mathematik*, 28:239–246, 1982.

[Sch91] R. A. Schmidt. Algebraic terminological representation. Master's thesis, Department of Mathematics, University of Cape Town, Cape Town, South Africa, 1991. Available as Thesis-Reprint TR 011. Also as Technical Report MPI-I-91-216, Max-Planck-Institut für Informatik, Saarbrücken, Germany.

[Sch03] G. Schmidt. Relational language. Technical Report, Universität der Bundeswehr München, 2003.

[Sch06] R. A. Schmidt. Developing modal tableaux and resolution methods via first-order resolution. In G. Governatori, I. M. Hodkinson, and Y. Venema, editors, *Advances in Modal Logic 6, papers from the sixth conference on "Advances in Modal Logic", held in Noosa, Queensland, Australia, on 25–28 September 2006*, pages 1–26, College Publications, London, 2006.

[Seg71] K. Segerberg. *An Essay in Classical Modal Logic*. University of Uppsala, Uppsala, 1971.

[Seg77] K. Segerberg. A completeness theorem in the modal logic of programs. *Notices of the AMS*, 24(6):A–552, 1977.

[Sha38] C. E. Shannon. A symbolic analysis of relay and switching circuits. *Transactions of the AIEE*, 57:713–723, 1938.

[Sha48] C. E. Shannon. A mathematical theory of communication. *Bell System Technical Journal*, 27:379–423, 623–656, 1948.

[Sim87] P. Simons. *Parts. A Study in Ontology*. Clarendon Press, Oxford, 1987.

[Sin00] C. Sinz. System description: ARA – an automatic theorem prover for relation algebras. In D. A. McAllester, editor, *Proceedings of the 17th International Conference on Automated Deduction*, volume 1831 of *Lecture Notes in Computer Science*, pages 177–182, Springer, Heidelberg, 2000.

[Smu68] R. M. Smullyan. *First Order Logic*. Springer, Heidelberg, 1968.

[SOH04] R. A. Schmidt, E. Orłowska, and U. Hustadt. Two proof systems for Peirce algebras. In Berghammer et al. [BMS04], pages 238–251.

[SR92] A. Skowron and C. Rauszer. The discernibility matrices and functions in information systems. In R. Słowiński, editor, *Intelligent Decision Support: Handbook of Applications and Advances of Rough Set Theory*, volume 11 of *System Theory, Knowledge Engineering and Problem Solving*, pages 331–362. Kluwer, Dordrecht, 1992.

[SS63] B. Schweizer and A. Sklar. Associative functions and abstract semigroups. *Publ. Math. Debrecen*, 10:69–81, 1963.

[SS83] B. Schweizer and A. Sklar. *Probabilistic Mertic Space*. North-Holland, Amsterdam, 1983.

[SS93] G. Schmidt and T. Ströhlein. *Relations and Graphs – Discrete Mathematics for Computer Scientists*. Monographs in Theoretical Computer Science. An EATCS Series. Springer, Heidelberg, 1993.

[Ste00] J. G. Stell. Boolean connection algebras: A new approach to the Region-Connection Calculus. *Artificial Intelligence*, 122(1–2):111–136, 2000.

[Sto98] T. Stokes. Radical classes of algebras with B-action. *Algebra Universalis*, 40:73–85, 1998.

[Sus68] R. Suszko. Non-fregean logic and theories. *Analele Universitatii Bucuresti, Acta Logica*, 9:105–125, 1968.

[Sus71a] R. Suszko. Identity connective and modality. *Studia Logica*, 27:7–39, 1971.

[Sus71b] R. Suszko. Quasi-completeness in non-fregean logic. *Studia Logica*, 29:7–14, 1971.

[Sus71c] R. Suszko. Semantics for the sentential calculus with identity. *Studia Logica*, 28:77–81, 1971.

[Sus72] R. Suszko. Abolition of the Fregean axiom. In R. Parikh, editor, *Logic Colloquium: Symposium on Logic held at Boston, 197273*, volume 453 of *Lecture Notes in Mathematics*, pages 169–239, Springer, Heidelberg, 1972.

[Sus73] R. Suszko. Adequate models for the non-fregean sentential calculus SCI. In *Logic, Language and Probability. A selection of papers of the 4th International Congress for Logic, Methodology and Philosophy*, pages 49–54, Reidel, Dordrecht, 1973.

[Tar41] A. Tarski. On the calculus of relations. *Journal of Symbolic Logic*, 6(3):73–89, 1941.

[TG87] A. Tarski and S. R. Givant. *Formalization of Set Theory without Variables*, volume 41 of *Colloquium Publications*. American Mathematical Society, Providence, 1987.

[TMR68] A. Tarski, A. Mostowski, and A. Robinson. *Undecidable theories*. Studies in Logic and the Foundations of Mathematics. North-Holland, Amsterdam, 1968.

[Urq72] A. Urquhart. Semantics for relevant logics. *Journal of Symbolic Logic*, 37(1):159–169, 1972.

[Urq84] A. Urquhart. The undecidability of entailment and relevant implication. *Journal of Symbolic Logic*, 49:1059–1073, 1984.

[Urq96] A. Urquhart. Duality for algebras of relevant logics. *Studia Logica*, 56:263–276, 1996.

[Vak87] D. Vakarelov. Abstract characterization of some knowledge representation systems and the logic NIL of nondeterministic information. In P. Jorrand and V. Sgurev, editors, *Artificial Intelligence: Methodology, Systems, Applications*, pages 255–260, North-Holland, Amsterdam, 1987.

[Vak89] D. Vakarelov. Modal logics for knowledge representation systems. In A. R. Meyer and M. A. Taitslin, editors, *Logic at Botik '89, Symposium on Logical Foundations of Computer Science, Pereslav-Zalessky, USSR, July 3-8, 1989, Proceedings*, volume 363 of *Lecture Notes in Computer Science*, pages 257–277, Springer, Heidelberg, 1989.

[vB83] J. van Benthem. *The Logic of Time*, volume 156 of *Synthese Library*. Reidel, Dordrecht, 1983.

[vB85] J. van Benthem. *A Manual of Intensional Logic*, volume 1 of *CSLI Lecture Notes*. Center for the Study of Language and Information, Stanford, CA, 1985.

[vB95] J. van Benthem. Temporal logic. In D. Gabbay, C. J. Hogger, and J. A. Robinson, editors, *Handbook of Logic in Artificial Intelligence and Logic Programming*, volume 4, pages 241–350. Clarendon Press, Oxford, 1995.

[vB96] J. van Benthem. *Exploring Logical Dynamics*. CSLI Lecture Notes. Center for the Study of Language and Information & Cambridge University Press, Stanford, Cambridge, 1996.

[VDB01] D. Vakarelov, I. Düntsch, and B. Bennett. A note on proximity spaces and connection based mereology. In C. Welty and B. Smith, editors, *Proceedings of the 2nd International Conference on Formal Ontology in Information Systems (FOIS01)*, pages 139–150, ACM, New York, 2001.

[VDDB02] D. Vakarelov, G. Dimov, I. Düntsch, and B. Bennett. A proximity approach to some region-based theories of space. *Journal of Applied Non-Classical Logics*, 12(3–4):527–559, 2002.

[Ven90] Y. Venema. Expressiveness and completeness of an interval tense logic. *Notre Dame Journal of Formal Logic*, 31(4):529–547, 1990.

[Ven91] Y. Venema. A modal logic for chopping intervals. *Journal of Logic and Computation*, 1(4):453–476, 1991.

[VHF95] P. A. S. Veloso, A. M. Haeberer, and M. F. Frias. Fork algebras as algebras of logic. *Bulletin of Symbolic Logic*, 1(2):265–266, 1995.

[Vis82] A. Visser. On the completeness principle: A study of provability in Heyting's arithmetic. *Annals of Mathematical Logic*, 22:263–295, 1982.

[vW65] G. H. von Wright. And Next. *Acta Philosophica Fennica, Fast.*, 18:293–304, 1965.

[Wad75] W. W. Wadge. A complete natural deduction system for the relational calculus. Technical Report, Coventry, 1975.

[WDB98] H. Wang, I. Düntsch, and D. Bell. Data reduction based on hyper relations. In R. Agrawal, P. Stolorz, and G. Piatetsky-Shapiro, editors, *Proceedings of the 4th International Conference on Knowledge Discovery and Data Mining*, pages 349–353, 1998.

[WDG00] H. Wang, I. Düntsch, and G. Gediga. Classificatory filtering in decision systems. *International Journal of Approximate Reasoning*, 23:111–136, 2000.

[Whi29] A. N. Whitehead. *Process and Reality*. MacMillan, Barclay and London, 1929.

[Wil82] R. Wille. Restructuring lattice theory: an approach based on hierarchies of concepts. In I. Rival, editor, *Ordered Sets*, pages 445–470. Reidel, Dordrecht, 1982.

[Win80] G. Winskel. *Events in computation*. PhD thesis, Department of Computer Science, University of Edinburgh, 1980.

[Win86] G. Winskel. Event structure semantics for CCS and related languages. In volume 224 of *Lecture Notes in Computer Science*, pages 510–584, Springer, Heidelberg, 1986.

[Wol94] A. Wolf. Optimization and translation of tableau-proofs into resolution. *Elektronische Informationsverarbeitung und Kybernetik*, 30(5–6):311–325, 1994.

[WS92] W. A. Woods and J. G. Schmolze. The KL–ONE family. *Computers and Mathematics with Applications*, 23(2–5):133–177, 1992.

[Yan59] J. Yanov. On equivalence of operator schemes. *Problems of Cybernetic*, 1:1–100, 1959.

[Zem73] J. Zeman. *Modal Logic, the Lewis-Modal Systems*. Oxford University Press, Oxford, 1973.

Index

Breinigsville, PA USA
05 December 2010

250560BV00007B/26/P

9 789400 700048